OUR ENVIRONMENT

THIRD EDITION

OUR ENVIRONMENT
A Canadian Perspective

DIANNE DRAPER
University of Calgary

MAUREEN G. REED
University of Saskatchewan

THOMSON

NELSON

Australia   Canada   Mexico   Singapore   Spain   United Kingdom   United States

THOMSON

NELSON

**Our Environment: A Canadian Perspective,
Third Edition**

by Dianne Draper and Maureen G. Reed

Associate Vice-President, Editorial Director:
Evelyn Veitch

Publisher, Social Sciences and Humanities:
Chris Carson

Marketing Manager:
Rodney Burke

Senior Developmental Editor:
Edward Ikeda

Permissions Coordinator:
Indu Ghuman

Photo Researcher:
Indu Ghuman

Senior Production Editor:
Bob Kohlmeier

Copy-Editor:
Lisa Berland

Proofreader:
Wayne Herrington

Indexer:
Dennis A. Mills

Production Coordinator:
Helen Locsin

Creative Director:
Angela Cluer

Interior-Design Modifications:
Peter Papayanakis

Cover Design:
Liz Harasymczuk

Cover Image:
Daryl Benson/Masterfile

Compositor:
Tammy Gay

Printer:
Transcontinental Printing Inc.

National Library of Canada Cataloguing in Publication

Draper, Dianne Louise, [date]
Our environment : a Canadian perspective / Dianne Draper and Maureen G. Reed. — 3rd ed.

Includes bibliographical references and index.
ISBN 0-17-641581-5

1. Canada—Environmental conditions. 2. Environmental policy—Canada. 3. Environmental degradation—Canada. I. Reed, Maureen Gail, [date] II. Title.

GE160.C3D73 2004   363.7'00971
C2004-902752-2

BRIEF CONTENTS

CONTENTS

<table>
<tr><td>PART 3</td><td>RESOURCES FOR CANADA'S FUTURE</td></tr>
</table>

Chapter 5
Our Changing Atmosphere   122

Chapter 6
Agroecosystems and Land Resources   178

Chapter 7
Fresh Water **224**

Chapter 12
Wild Species and Natural Spaces 436

PART 4 GETTING TO TOMORROW

Chapter 13
Sustainability and the City 480

Chapter 14
Meeting Environmental Challenges   516

PREFACE

Environmental education is a lifelong process, and no one textbook or course can cover in depth every element relevant to a full understanding of environmental issues. This book is intended as a learning tool so that students, today and tomorrow, may be better informed about—and able to make scientifically grounded and socioeconomically balanced decisions on—environmental issues. Canadian students generally care a great deal about the world they live in. Many want to be challenged to think critically about, and to act responsibly with respect to, the environment. We sincerely hope that the third edition of this book helps make environmental science meaningful and relevant to Canadian (and other) students so that they may meet the challenge of sustaining a healthy and productive Earth environment.

OBJECTIVES

The purpose of this third edition of *Our Environment: A Canadian Perspective* is to provide a contemporary introduction to scientific concepts that are important to the study and understanding of the ecological functioning of our global environment. The book also aims to present current information on environmental issues we face in Canada. If our efforts to resolve these issues are to succeed, we need to think critically and in an integrated fashion about them and about the relationships between people and Earth's ecosystems. Thus, information presented here integrates physical and human dimensions. It reflects a broad, interdisciplinary approach to the study of global and local environments. This is why discussion of issues such as water appears in chapters dealing not only with water but in chapters on agriculture, forests, and energy resources.

Canadians' experiences with environmental issues vary from east to west and north to south. This book is intended to help students understand the evolution of environmental concerns in Canada as well as the variability of environmental problems in different resource sectors and regions of the nation. In addition, the range of examples selected from across the country and internationally is intended to help students appreciate the range of opinion on, and approaches that have been taken to resolve, environmental concerns in Canada and beyond.

Canadians have achieved some notable successes in certain environmental matters, but we cannot become complacent about what remains to be done. In this regard, a number of key themes thread through the text, including the importance of observation and critical reflection on environmental matters, the place of stewardship and cooperative problem solving in environmental action, and a focus on sustainability and on the future.

ORGANIZATION

The book contains fourteen chapters. These are grouped into four parts. In Part I, students are introduced to the broad field of environmental science through discussion of our constantly changing global environment and the major causes of environmental problems. The role of science in understanding our environment and in moving toward a sustainable future is considered, as are the roles of worldviews, environmental values, and ethics.

In Part II, the focus is on fundamental features of, and interconnections in, the ecosphere, as well as on human population issues and the effects that human numbers have on the environment, both nationally and internationally.

Part III focuses on Canada's natural resource base. Chapters on the atmosphere, agroecosystems and land resources, fresh water, oceans and fisheries, forests, mining, energy, and wild species and natural spaces each deal with human activities and impacts on these resource sectors and environments. Each chapter also identifies a range of Canadian responses to the challenges encountered through resource development and use of environments. These responses identify international and national actions, and discuss the activities, partnerships, and local actions of individuals, environmental nongovernmental organizations, industry, and government.

Part IV considers issues associated with the effects of our lifestyle choices and some of the ways we can move toward sustainable environments in the future.

FEATURES

Our Environment, Third Edition, contains several special features to help students in their environmental education. Each chapter begins with a list of *Chapter Objectives* so that students are aware of the concepts they will come to understand as they read and study the chapter. *Key terms* appear in bold throughout the text; definitions for these terms are found in the *Glossary. Italicized terms* are discussed within the text. *Case studies* in each chapter not only provide real illustrations of environmental problems but also demonstrate how people have applied the principles and concepts discussed in the text to the resolution of environmental issues at several geographic scales. These cases should help students understand the economic, social, political, and environmental interconnections in environmental science.

Enviro-Focus boxes identify a range of personal interest/impact issues ranging from increases in skin cancer cases in Canada and the revival of agricultural hemp for use in clothing, to the competition between wild species and humans for habitat that affects grizzly bears in Banff National Park. *Future Challenges* sections, in the resource sector chapters, are intended to stimulate and extend our thinking about local environmental problems to the future of the world we live in and share with others.

Each chapter ends with a list of *Chapter Questions*; some questions review material presented in the chapter, others engage students in development of observational and critical thinking skills. A list of *References* containing research sources also appears at the end of each chapter.

WHAT'S NEW IN THIS EDITION

Many changes in our understanding of environmental issues have occurred since the publication of the second edition. To reflect these changes and to emphasize how dynamic a process it is to deal with environmental issues, new topics and updated statistics have been added to make the text as current as possible. We have also placed increased emphasis on social issues that attend changes in environmental conditions and policy.

Chapter 1 offers a Canadian-centred focus on our role in over-consumption of resources and natural systems. An expanded discussion of ecological footprints has been added to increase student awareness of consumption and the principles of sustainability.

Chapter 2 provides an explanation of the relationship among various disciplines in environmental studies and sustainable development. Updated discussions of several topics, including Canada's Environment Week, have been added.

Chapter 3 focuses on fundamental features and interconnections of the Earth's life-support systems. This chapter helps students understand why learning to work with nature is essential in achieving a sustainable future.

Chapter 4 includes an expanded discussion on global population dynamics including the role of consumption and technology as pressures on the Earth's resources and ecosystems. Updated information is provided on the impact of AIDS/HIV and other disease epidemics. A discussion of whether Africa is over- or under-populated is used to generate interest in the many dimensions of population issues.

Chapter 5 provides updated information and data on atmospheric science and its relation to climate change and air quality, as well as information on modelling climate change.

Chapter 6 considers the complex nature and importance of the agriculture industry to Canadians. Several new topics have been introduced, including the bovine spongiform encephalopathy (BSE) and chronic wasting disease issues in Canada, integrated pest management, fair trade, and the World Trade Organization and agricultural reform.

Chapter 7 contains updated information and an expanded discussion of water quality issues in urban areas including withdrawal and instream uses, impacts of urban form on hydrological processes, and water and sewage treatment. Discussion of Walkerton and North Battleford tainted-water tragedies has been updated as impacts became evident and final results of the inquiries became available.

Chapter 8 provides new and expanded information on oceans and fisheries, including aquacultural operations, offshore drilling, and marine protected areas.

Chapter 9 expands the discussion of social and cultural issues that attend forest use and harvesting practices, and includes new information about Aboriginal involvement in forestry, eco-certification, and community forestry experiments across Canada. Impacts of harvest changes on workers, uses of forests beyond timber, and impacts of tourism as an alternative economic generator in forested ecosystems are discussed.

Chapter 10 illustrates the economic importance of mining in Canada, including in northern economies. This chapter includes new content on the expansion of Canada's diamond mining industry, the use of extended producer responsibility legislation, and impact-benefit agreements.

Chapter 11 has updated material on energy use, production, and sustainability; the Hibernia project; wind and solar power; and nuclear fuel (and waste). There is also new content on the Sydney Tar Ponds, the Light Up the World project, and oil sands development.

Chapter 12 introduces biosphere reserves and information related to new parks in Canada. Initiatives to protect biodiversity, including stewardship on private lands, habitat alteration, and biodiversity protection initiatives in urban environments are new in this edition.

Chapter 13 provides updated coverage on air quality, water use and wastewater treatment, sustainable housing, and water supply and water quality. Discussion about ecological footprints is expanded, as is consideration of social elements of sustainability, such as cooperative housing, car co-ops, and sustainability planning at universities.

Chapter 14 includes the responses to a questionnaire designed specifically for this edition of *Our Environment* and sent to every federal, provincial, and territorial Minister of the Environment and to several ENGOs. Environmental protection regulations and initiatives and ENGO actions to safeguard the environment are presented in this and preceding chapters. This chapter reinforces positive changes that are being made to protect the environment and illustrates the importance of environmental protection initiatives undertaken by both individuals and collectives. Efforts to measure change (such as those undertaken by GPI Atlantic) are highlighted as positive ways to evaluate our relative success in achieving sustainability.

INSTRUCTOR'S RESOURCES TO FACILITATE TEACHING AND LEARNING

A rich variety of instructional resources supplement the book, giving instructors the tools to create a dynamic, exciting, and effective course.

Instructor's CD-ROM. The Instructor's CD-ROM contains the Instructor's Manual/Test Bank, Computerized Test Bank, and PowerPoints. See below for more details about each supplement.

Instructor's Manual/Test Bank. The instructor's manual consists of two parts. Part 1 includes a section on using the text's support package and integrating items such as the video and website into the classroom. The bulk of Part 1 provides teaching suggestions, chapter lecture outlines, activities, and exercises for students on a chapter-by-chapter basis. Part 2 contains the Test Bank. Multiple-choice, true/false, short-answer, and fill-in-the-blank questions are included for every chapter.

Computerized Test Bank. Available only on the Instructor's CD-ROM, all questions from the Test Bank are available in Examview for Windows and Mac format. Examview allows instructors to create, deliver, and customize texts (both print and online) in minutes with this easy-to-use assessment and tutorial system.

PowerPoint® Slides. PowerPoint slideshows recap the major topics discussed in each chapter and incorporate figures from the text to enhance classroom presentation. These files are available on the Instructor's CD-ROM and can also be downloaded from the password-protected Instructor's Resource area of our website.

Environmental Science on the World Wide Web. The dynamic website that accompanies this text can be found at www.ourenvironment3e.nelson.com. The site gives students access to helpful learning resources such as practice quizzes, chapter-specific Web links, study tips, information on degrees and careers in environmental science, and much more.

Infotrac® College Edition. Gain access to Infotrac, our online database of scholarly and popular journals available for downloading from the Web. Simply enter your password (included on the card in this text) for immediate access to full-text articles from *Maclean's, The Ecologist, Environmental Action Magazine, Geographical Journal,* and other environmental and geographical magazines, reviews, and journals.

ACKNOWLEDGMENTS

The authors have benefited greatly from the assistance of many people in preparing this edition of *Our Environment.* We would like to extend special thanks to Sandy McAndrews who again provided such effective research assistance to this project. We would like to thank her for her knowledge, dedication, enthusiasm, and useful insights. In addition, research assistance was provided by Jana Berman, Karen Butler, and Suzanne Mills.

Many people responded to requests for information, including the federal minister and most of the provincial ministers of Environment, industry personnel, and representatives of ENGOs; thank you for your contributions to this edition. Lorn Fitch, Dave Irvine-Halliday, and Lawrence Nkembirim kindly took the time to write specific materials for this edition; thank you for your willingness to contribute your expertise.

We are very grateful for the assistance of the talented members of the editorial and production team at Thomson Nelson. We would like to express our sincere gratitude to Evelyn Veitch, Chris Carson, Edward Ikeda, and Bob Kohlmeier for their advice and professional assistance. Others at Nelson helped in other ways, and we'd like to thank them, too: Susan Calverley, Marnie Benedict, Tammy Gay, Katherine Strain, Peter Papayanakis, Susan Calvert, and Helen Locsin.

We would like to thank the reviewers who pointed out errors and made important suggestions for improvements in this book. Although remaining errors and deficiencies are ours, we extend our thanks to Darren Bardati, Bishop's University; Peter Brown, McGill University; Ronald Chopowick, Seneca College; Alan Diduck, University of Winnipeg; John Middleton, Brock University; Anthony Price, University of Toronto; Hélène Savard, Sir Sandford Fleming College; Joseph Shorthouse, Laurentian University; Stephen Swales, Ryerson University; and Susan Vajoczki, McMaster University. Thank you for your help.

Finally, we thank our families for the commitment and support they have given us during this project. Maureen would especially like to thank Michael, Louis, and Bruce. We would like to dedicate this edition to all who have nurtured our love for this Earth and our place within it.

Dianne Draper
Calgary, Alberta

Maureen G. Reed
Saskatoon, Saskatchewan

PART 1

Our Environment

Our Environment: Problems and Challenges

Chapter Contents

"... although it is only a little planet it is hugely beautiful and surely the finest place in the world to be."

Lawrence Collins (in Brower, 1975, p. 127)

Chapter Objectives

After studying this chapter you should be able to

- identify a range of local, regional, and international environmental issues of relevance to Canadians and all of Earth's citizens
- appreciate the ways humans are linked with Earth's ecosystems
- describe the root causes of environmental problems
- discuss the concepts of sustainable development and sustainability
- identify and summarize the guiding principles of sustainability

INTRODUCTION

From the feel of earth under our feet, to the light of sun and sky in our eyes; from the flaming colours that greet us in autumn woods, to the shimmering oceans, lakes, and streams that flow around us—Canadians know they inhabit some special places on a beautiful planet. The challenge is that many of our activities in this country and on this planet are rapidly altering its beauty and its ecological functioning. Exponential growth in resource consumption and in global population has caused forests, wetlands, and grasslands to disappear, topsoil to be blown or washed away, oceans and water bodies to be poisoned, and wildlife species to be driven to extinction. The good news is that we are learning to think and act differently to sustain, rather than degrade, our planet.

This book is about environmental problems and challenges that face us as Canadians and global citizens. It also discusses efforts that individuals, groups, industries, and government agencies are making to solve existing problems and to improve the ways we interact with the ecosystems of the planet. Although many problems are significant ones, and some people feel powerless to change them, it is important to remember that individual and combined actions do make a difference. Each one of us can improve our ecological knowledge and understanding of how the world works so that our individual and collective decisions and actions in the future will be less harmful to the environment. Industries and governments, too, can be challenged to develop new approaches to conserving the planet's basic life-support systems and to ensure the long-term sustainability of species and resources.

In addition to providing an overview of the nature of environmental challenges facing people in Canada and elsewhere (Figure 1–1), in this chapter we introduce some of the linkages humans have with Earth's interconnected ecosystems and identify some of the root causes of environmental problems. We also consider concepts associated with the term sustainability and some of the principles that help us work toward it.

Earthweek: A Diary of the Planet

By Steve Newman

City Birds Song

German researchers have determined that birds living in large cities have learned to sing louder so they can be heard by other birds over the din of traffic and other manmade noises. A study of nightingales in metropolitan Berlin showed that the male birds were forced to raise their voices even in the middle of the night so females could hear them. Results also showed that males at noisier locations sang with higher sound levels than birds in territories less affected by background sounds, according to a research team headed by Dr. Henrik Brumm of the Institute for Biology at Berlin's Freie University. Brumm's team also found the birds sang at a lower level on weekends, when traffic noise was reduced.

Earthquakes

Emergency officials in western China's Qinghai province rushed tents and other supplies to an area around the city of Delingha after a magnitude 5.5 temblor damaged more than 3,100 houses. Early reports indicate there were no casualties due to the quake.
- Two Taiwanese died in a landslide triggered by a magnitude 5.8 temblor that also caused buildings to sway in the capital Taipei. The quake was centered on the island's eastern coast.
- Earth movements were also felt in Borneo, the Andaman Islands, Sicily, western areas of Guatemala and El Salvador, coastal Chile and California's central coast region.

Tropical Cyclones

An area of disturbed weather spinning off the coast of southwestern India formed into Cyclone 01A as it moved on a slow north-northwesterly course. It was predicted to strengthen to hurricane force and pass well offshore from Bombay before entering the Arabian Sea.
- Weak cyclone 23S developed near the Indian Ocean outpost of Diego Garcia, threatening shipping lanes with storm-force winds and high seas.

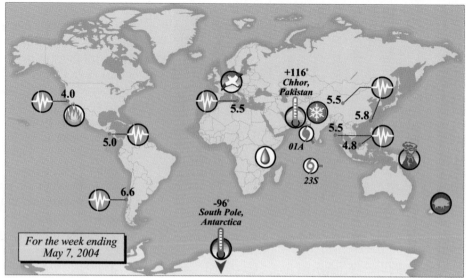

+116° Chhor, Pakistan

4.0

5.5

5.5

5.8

5.5

4.8

5.0

01A

23S

6.6

-96° South Pole, Antarctica

For the week ending May 7, 2004

East African Downpours

Heavy seasonal rainfall in parts of East Africa unleashed flash flooding that drowned 15 people in various parts of Kenya. The *Daily Nation* reported that floods forced thousands of people in the slums of Nairobi to move to higher ground, while flood deaths occurred in several areas from the interior Rift Valley to the coast near Mombasa. April and May are the wettest months of the year for Kenya, and are critical for a successful growing season and the survival of wildlife in the country's famed game reserves.

Eruption

Authorities in Papua New Guinea placed residents in central Bougainville on alert as Mount Bagana threw up fountains of lava over its slopes. Pilots reported that the stream appeared to be flowing toward village of Torokina, six miles away. Bagana is one of the world's most active volcanoes with eruptions almost every year.

Southland Wildfires

A long stretch of record-breaking temperatures across Southern California sparked wildfires that destroyed at least 14 houses and forced thousands from their homes. Cooler weather late in the week, and lighter winds, allowed firefighters to gain the upper hand on a trio of blazes that burned more than 22,000 acres, mainly to the southeast of Los Angeles. Fire officials said that due to the unusually dry winter, the blazes were burning with an intensity not usually seen until late in the fire season.

Himalayan Spring

A freak storm across northern India produced the latest snowfall for Kashmir and neighboring Himachal Pradesh state in 20 years, trapping 100 tourists on a Himalayan mountain pass. The country's meteorological department said the late-season storm dumped more than 2 feet of snow over Rohtang Pass, and triggered hailstorms that damaged some

crops in the lower valleys. The 13,050-foot pass had been opened almost two months earlier than usual on April 10 because of lighter snowfalls over the past winter.

Pork Wellington

An invasion of wild pigs around New Zealand's capital of Wellington is frightening residents and damaging gardens in a suburb only 10 minutes from the city center. Animal control officer Ken Wright told reporters that about 10 swine, including a boar and a sow with piglets, had moved into the suburb of Karori. He warned that the boar and sow could become aggressive to residents if they were accidentally cornered. The feral pigs are descendants of animals released by Captain James Cook during his voyages to New Zealand in the late 18th century, Wright said.

Distributed by: Universal Press Syndicate
E-mail: feedback@earthweek.com

©2004 Earth Environment Service

Figure 1–1

A week on Earth

SOURCE: Adapted from *Earthweek: A Diary of the Planet,* S. Newman, May 2004, http://www.earthweek.com

THE CHANGING GLOBAL ENVIRONMENT

Television and video programs have brought each of us face to face with the tragic images of environmental refugees fleeing famine, disease, and death as ecological deterioration overtakes their homelands. Through news photography we have seen images of the aftermath of nuclear explosions at Chernobyl and Three Mile Island, cyanide and heavy-metals mining spills on the Danube River system, burning tropical rain forests in Central and South America, oil escaping from the Exxon *Valdez,* and elephants, rhinos, and other African wildlife slaughtered by poachers. Canadian news media have carried stories about the shrinking Arctic ice pack; severe smog events in major cities; illness and death from contaminated water supplies; the collapse of cod and salmon fisheries; the spread of "mad cow" disease and the "foot and mouth" virus; and the use of genetically altered animals such as goats and sheep to produce human proteins for medicinal purposes. These images and stories carry the same message—in a very short time, humans have greatly accelerated environmental change.

While it is true that Earth's history reveals periods of major environmental changes (such as when glaciation transformed landscapes, and catastrophic volcanic eruptions and floods destroyed species and ecosystems), these events generally occurred over long time frames and provided for relative stability in ecological processes. In contrast, human activities have increased the pace of environmental change and have had dramatic impacts on the quality and productivity of the planet's **ecosystems**. Canadians are becoming increasingly aware of how human demands for, and increased consumption of, resources have resulted in events such as the collapse of the Atlantic cod fishery as well as the reduction of old-growth temperate rain forests by over 50 percent, prairie grasslands by about 80 percent, and Ontario's Carolinian forests to less than 15 percent of their former expanses. These numbers are estimates because there is considerable debate about definitions and measurement techniques that are used to derive these figures.

As people have pursued their economic development goals, concerns about environmental quality have increased. For instance, acidic atmospheric pollutants from Canadian and U.S. industrial sources have destroyed fish in many lakes in eastern Canada; Arctic **country foods** (local meat and fish) have been contaminated by polychlorinated biphenyls (PCBs); and Labrador has been the recipient of long-range transport of radioactive particles from Chernobyl. Polluted soils and water, Arctic haze, large die-offs of neotropical migrant birds, and increased levels of greenhouse gases such as carbon dioxide in the atmosphere have resulted from discharges of industrial and community wastes as well as from agricultural, forestry, and mining practices.

Such results of our activities have alerted people to the fact that we are not separate from our **environment** but are an integral part of Earth's interconnected ecosystems.

Photo 1–1a

Photo 1–1b

Photo 1–1c

Photo 1–1d

Dimensions of our environment: landforms, wildlife, vegetation, atmosphere, and water bodies.

Photo 1–2a

Photo 1–2b

The productive capabilities of different environments are affected by natural limitations (deserts) and human activities (deforestation).

Perhaps we can understand the profound impacts that population growth and economic development have on the global ecosystem (the **ecosphere**) when we realize that "the world's population has multiplied almost fivefold since the early 1900s [and that during that same time period] the world's economy has grown by 20 times, the consumption of fossil fuels by 30 times, and industrial production by 50 times" (Government of Canada, 1996).

Ecosystems have finite productive capacities and assimilative abilities, and they are affected by human activities. As human populations grow, for example, they increase demands on energy supplies for industrial development, transportation, and housing. If the productivity and quality of our environment are to be retained, it is important to acknowledge our complex relationships with Earth's ecosystems and to act in ways that will not put

intolerable pressure on natural resources and life-support systems of the ecosphere. Impacts of human activities often transcend political boundaries and can lead to serious social, economic, and environmental problems. This is why Vancouverites should care about what happens in the Arctic, why New Brunswickers should be concerned about pollution in Eastern Europe, and why Inuit in the Northwest Territories should be interested in the Brazilian rain forest.

Many people in the Western world are troubled by the current impacts of their own activities on the environment, and worried about how future environmental changes might affect their health and socioeconomic well-being. Among these concerns are effects of nuclear waste disposal, toxic emissions, acid precipitation, genetically modified plant crops, deforestation, habitat loss and fragmentation, greenhouse gases, and global climate change in general.

At the individual level, many Canadians are concerned about whether their well water is contaminated with pesticides or agricultural runoff; whether enough trees are being replanted to replace all those that have been cut; whether sufficient high-quality habitat is being protected for wildlife; the degree to which pollution from other countries is affecting Canada's air and water quality; and whether exotic species introduced accidentally (or deliberately) can be controlled. Increasingly, people around the world are asking how they can balance the need for economic and social development with the need to sustain those resources on which such development rests. To put it another way, people want to know what actions to take individually and collectively, as well as how to achieve a balance among the social, economic, and environmental dimensions of activities such as mining, fishing, agriculture, forestry, and tourism.

Photo 1–3

One example of the impact that human activity can have on wildlife.

The 1993 "World Scientists' Warning to Humanity" identified critical stresses facing the Earth's environment and noted what we must do to avoid irretrievably mutilating our planet (see Enviro-Focus 1). This warning was part of a long-term campaign by the Union of Concerned Scientists to increase awareness of the threat that global environmental degradation poses to humanity's life-support systems. By 1993, more than 1670 scientists from 71 countries, including 104 Nobel laureates, had signed

the warning. Among Canadian scientists who signed the warning were Paul-Yves Denis, Gerhard Herzberg, Digby McLaren, Brenda Milner, Lawrence Mysak, John Polanyi, and Betty Roots. In addition to these scientists, many individuals are aware that achieving the future we desire for human society and the plant and animal kingdoms means we must improve our knowledge of the ecosphere. Better understanding will enable all of us to cooperatively plan and manage environmental resources.

ENVIRO-FOCUS 1

World Scientists' Warning to Humanity

Introduction Human beings and the natural world are on a collision course. Human activities inflict harsh and often irreversible damage on the environment and on critical resources. If not checked, many of our current practices put at serious risk the future that we wish for human society and the plant and animal kingdoms, and may so alter the living world that it will be unable to sustain life in the manner that we know. Fundamental changes are urgent if we are to avoid the collision our present course will bring about.

The Environment The environment is suffering critical stress:

The Atmosphere. Stratospheric ozone depletion threatens us with enhanced ultraviolet radiation at the earth's surface, which can be damaging or lethal to many life forms. Air pollution near ground level and acid precipitation are already causing widespread injury to humans, forests, and crops.

Water Resources. Heedless exploitation of depletable groundwater supplies endangers food production and other essential human systems. Heavy demands on the world's surface waters have resulted in serious shortages in some 80 countries, containing 40 percent of the world's population. Pollution of rivers, lakes, and groundwater further limits the supply.

Oceans. Destructive pressure on the oceans is severe, particularly in the coastal regions, which produce most of the world's food fish. The total marine catch is now at

or above the estimated maximum sustainable yield. Some fisheries have already shown signs of collapse. Rivers carrying heavy burdens of eroded soil into the seas also carry industrial, municipal, agricultural, and livestock waste—some of it toxic.

Soil. Loss of soil productivity, which is causing extensive land abandonment, is a widespread byproduct of current practices in agriculture and animal husbandry. Since 1945, 11 percent of the earth's vegetated surface has been degraded—an area larger than India and China combined—and per capita food production in many parts of the world is decreasing.

Forests. Tropical rain forests, as well as tropical and temperate dry forests, are being destroyed rapidly. At present rates, some critical forest types will be gone in a few years, and most of the tropical rain forest will be gone before the end of the next century. With them will go large numbers of plant and animal species.

Living Species. The irreversible loss of species, which by 2100 may reach one-third of all species now living, is especially serious. We are losing the potential they hold for providing medicinal and other benefits, and the contribution that genetic diversity of life forms gives to the robustness of the world's biological systems and to the astonishing beauty of the earth itself.

Much of this damage is irreversible on a scale of centuries, or permanent. Other processes appear to pose additional threats. Increasing levels of gases in the atmosphere from human activities, including carbon dioxide released from fossil fuel burning and from deforestation, may alter climate on a global scale. Predictions of global warming are still uncertain—with projected effects ranging from tolerable to very severe—but the potential risks are very great.

(continued)

Our massive tampering with the world's inter-dependent web of life—coupled with the environmental damage inflicted by deforestation, species loss, and climate change—could trigger widespread adverse effects, including unpredictable collapses of critical biological systems whose interactions and dynamics we only imperfectly understand.

Uncertainty over the extent of these effects cannot excuse complacency or delay in facing the threats.

Population The earth is finite. Its ability to absorb wastes and destructive effluent is finite. Its ability to provide food and energy is finite. Its ability to provide for growing numbers of people is finite. And we are fast approaching many of the earth's limits. Current economic practices which damage the environment, in both developed and underdeveloped nations, cannot be continued without the risk that vital global systems will be damaged beyond repair.

Pressures resulting from unrestrained population growth put demands on the natural world that can overwhelm any efforts to achieve a sustainable future. If we are to halt the destruction of our environment, we must accept limits to that growth. A World Bank estimate indicates that world population will not stabilize at less than 12.4 billion, while the United Nations concludes that the eventual total could reach 14 billion, a near tripling of today's 5.4 billion. But, even at this moment, one person in five lives in absolute poverty without enough to eat, and one in ten suffers serious malnutrition.

No more than one or a few decades remain before the chance to avert the threats we now confront will be lost and the prospects for humanity immeasurably diminished.

Warning We the undersigned, senior members of the world's scientific community, hereby warn all humanity of what lies ahead. A great change in our stewardship of the earth and the life on it is required, if vast human misery is to be avoided and our global home on this planet is not to be irretrievably mutilated.

What We Must Do Five inextricably linked areas must be addressed simultaneously:

1. **We must bring environmentally damaging activities under control to restore and protect the integrity of the earth's systems we depend on.**
   We must, for example, move away from fossil fuels to more benign, inexhaustible energy sources to cut greenhouse gas emissions and the pollution of our air and water. Priority must be given to the development of energy sources matched to Third World needs—small-scale and relatively easy to implement.

   We must halt deforestation, injury to and loss of agricultural land, and the loss of terrestrial and marine plant and animal species.

2. **We must manage resources crucial to human welfare more effectively.** We must give high priority to efficient use of energy, water, and other materials, including expansion of conservation and recycling.

3. **We must stabilize population. This will be possible only if all nations recognize that it requires improved social and economic conditions, and the adoption of effective, voluntary family planning.**

4. **We must reduce and eventually eliminate poverty.**

5. **We must ensure sexual equality, and guarantee women control over their own reproductive decisions.**

The developed nations are the largest polluters in the world today. They must greatly reduce their over-consumption, if we are to reduce pressures on resources and the global environment. The developed nations have the obligation to provide aid and support to developing nations, because only the developed nations have the financial resources and the technical skills for these tasks.

Acting on this recognition is not altruism, but enlightened self-interest: whether industrialized or not, we all have but one lifeboat. No nation can escape from injury when global biological systems are damaged. No nation can escape from conflicts over increasingly scarce resources. In addition, environmental and economic instabilities will cause mass migrations with incalculable consequences for developed and underdeveloped nations alike.

Developing nations must realize that environmental damage is one of the gravest threats they face, and that attempts to blunt it will be overwhelmed if their populations go unchecked. The greatest peril is to become trapped in spirals of environmental decline, poverty, and unrest, leading to social, economic, and environmental collapse.

Success in this global endeavour will require a great reduction in violence and war. Resources now devoted to the preparation and conduct of war—amounting to over $1 trillion annually—will be badly needed in the new tasks and should be diverted to the new challenges.

A new ethic is required—a new attitude toward discharging our responsibility for caring for ourselves and for the Earth. We must recognize the Earth's limited capacity to provide for us. We must recognize its fragility. We must no longer allow it to be ravaged. This ethic must motivate a great movement, convincing reluctant leaders and reluctant governments and reluctant peoples themselves to effect the needed changes.

The scientists issuing this warning hope that our message will reach and affect people everywhere. We need the help of many.

We require the help of the world community of scientists—natural, social, economic, political;

We require the help of the world's business and industrial leaders;

We require the help of the world's religious leaders; and

We require the help of the world's peoples.

We call on all to join us in this task.

SOURCE: "World Scientists' Warning to Humanity," The Union of Concerned Scientists, 1993, Cambridge, MA: Author. Reprinted with permission.

LINKAGES: HUMANS AS PART OF ECOSYSTEMS

Although our intellectual characteristics distinguish us from other species, humans play essentially the same role as any other species within ecosystems. That is, humans rely—as do other species—on clean air, water, soil, and a continuing supply of plant and animal products, some of which they consume and convert to meet their own physiological needs. The wastes humans discharge become part of the Earth's cycles of decay and renewal (see Chapter 3).

Humans, however, are unlike other species in that we have the ability to cause drastic changes in the ecosystems on which we depend. Our technologies enable us not only to extract and use resources and ecosystem products faster than the **biosphere** can renew them, but also to discharge wastes from our production processes faster than the ecosphere can assimilate them. Such actions cause a decline in the natural productivity of ecosystems that may be permanent and irreversible. If this happens, we are said to be living off the capital of our environment rather than the interest. This is a significant reason for learning to sustain environmental resources so that they will be available in the future.

Unlike other species, humans have been successful in minimizing or overcoming factors such as extreme climates, other predators, and diseases that formerly limited our numbers. These successes have contributed to an expanding world population that continues to put increasing pressure on Earth's natural resources, life-support, and socioeconomic systems. It is important to recognize that humans are an integral and interdependent part of Earth's ecosystems and are subject to the same constraints as other living creatures. The way we lead our lives has a significant impact on the environment we share with them.

Consider how quickly our lives would end without the plants and animals that supply the products we consume every day—the oxygen we breathe, the food we eat, the timber for our houses, our cotton and linen clothing, the fibres in the pages of this book, our shoe leather, and the wool of our sweaters and socks, to name a few (Figure 1–2). People have always used the environment's resources to provide food, clothing, shelter, and other commodities. Often, we measured our progress by our success in exploiting these resources—fish, animals, land, water, and trees—but without realizing that all of these things were interconnected. The use of one environmental resource always affects the status of another resource or ecosystem, either immediately or in the longer term.

Figure 1–2

Human–environment connections:
An anthropocentric view of planet
Earth's provision for human life
and activities

Humans can be viewed as being
part of three integrated systems
or worlds:
• natural systems
• biological systems
• sociocultural systems.

Human production, consumption,
and waste processes have a
variety of direct and indirect
effects on Earth's systems that
induce environmental change.

As our understanding of how the Earth works as a system continues to improve, people are realizing how they are a part of the ecosystem and are beginning to appreciate the importance of caring for the entirety of their ecosystem, their Earth home. Some people are also realizing how much knowledge is needed to understand the effects of our ongoing activities so that we can determine which activities are sustainable (that is, which ones will help maintain environmental resources so they continue to provide benefits for people and other living things).

Today, even though many people remain ignorant or unconcerned about environmental matters, many others have accepted the environment as one of humanity's major social and political issues. With the help of technology, "ordinary" citizens, scientists, businesspersons, and government agencies are trying to find solutions to the problems our consumerist activities have created. Part of the task of determining appropriate solutions is to understand the major causes of the environmental problems we face.

Major Causes of Environmental Problems

There is general agreement among environmental scientists that the complex and interrelated root causes of envi-

ronmental problems include human population growth, over-consumption of resources and natural systems, and pollution. Each of these major causes and several related themes are discussed in the following sections.

HUMAN POPULATION GROWTH

Many scientists believe that exponential doubling of the Earth's human population is the fundamental issue of the environment—environmental damage occurs simply because of the very large number of people now on Earth. For most of the world's history, human numbers have been low: within the past 150 years, however, a dramatic increase in population has occurred (Figure 1–3). It was not until 1800 that the human population reached its first billion, and it took another 130 years to double to 2 billion. Between 1930 and 1975, in only 45 years, the population doubled again to 4 billion. The six billionth person was added to Earth's population in 1999 and, by about 2050, almost 9 billion people could inhabit the planet (see Chapter 4).

In the financial world, exponential rates of increase are considered to be a good thing. For example, if you were promised compensation for a 30-day job, starting at one cent for the first day and doubling each succeeding day, on the final day you would be paid over $5 million! As far as population is concerned, however, exponential rates of increase intensify the competition among people as they try to gain their share of the Earth's water, land, food, and other resources. United Nations statistics indicate

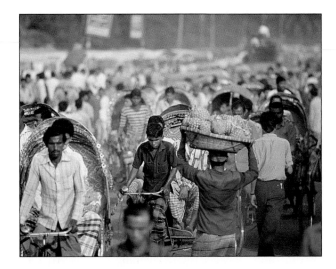

Photo 1–4
Developing countries such as Bangladesh struggle with the problems associated with overcrowding.

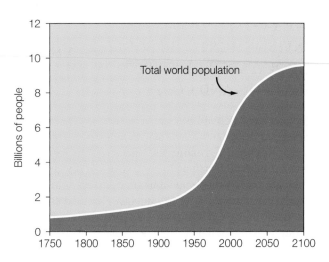

Figure 1–3
Growth in total world population since 1750 and projected from the late 1990s to 2100

that about four-fifths of the more than six billion people now living on Earth do not have adequate food, housing, and safe drinking water. Each day, almost 110 000 people die from starvation or related illnesses. While most of the severe stress occurs in developing countries, even in industrialized nations such as Canada there are people who do not have enough food to eat and who cannot afford a warm, dry place in which to live.

When human populations exceed their environmental resources, particularly the capacity of land to produce food resources, malnutrition and starvation result. Following the Sahel drought (1973–74) in Africa, for example, about 500 000 people died of starvation and several million more were affected permanently by malnutrition. In parts of Africa in the 1980s, human population growth created such demand for food from agricultural lands that their future productivity appeared to be threatened.

This situation illustrates a classic conflict in values: Which is more important, the survival of people alive today, or the conservation of the environment on which future food production and human life depend? Part of the answer requires technical and scientific knowledge to determine whether agricultural production can continue to increase without destruction of the land on which it is based. Such knowledge provides an important basis for a decision that depends on our values. Ultimately, however, unless we can limit the total number of people on Earth to a number the environment can sustain, we cannot expect to solve other environmental problems. Given that the issue of the rapidly increasing human population underlies debates about almost all environmental problems, we revisit population topics in more detail in Chapter 4.

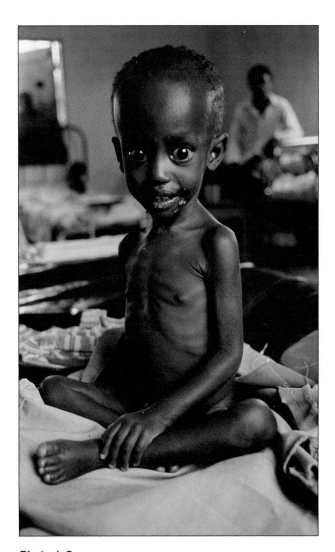

Photo 1–5
Children are among the hardest-hit victims of food and other resource shortages.

OVER-CONSUMPTION OF ENVIRONMENTAL RESOURCES AND INEQUITABLE DISTRIBUTION OF BENEFITS AND COSTS

Humans have defined resources as anything useful that serves our needs and is available at a price we are willing to pay. **Renewable resources** (also known as renewable **natural capital**) such as forests, solar energy, and livestock are replaced by environmental processes in a time frame that is meaningful to humans. Provided they are not used more quickly than they are restored, these resources will continue to supply goods and services that meet human needs into the future. **Nonrenewable resources** are finite in supply or are replaced so slowly (in human terms) by the environment that, practically speaking, their supply is finite. Another important characteristic of nonrenewable natural capital, such as coal, oil, and other fossil fuels, is that their supply is depleted with use. Yet, our practices often have resulted in unsustainable levels of use of renewable resources. In some cases, these practices have so decimated local populations (e.g., Atlantic cod), that their ability to renew themselves has been seriously undermined.

Frequently, people have used living environmental resources such as trees, fish, and wildlife faster than they were replenished by natural processes. Humans also have extracted nonliving resources such as oil, minerals, and groundwater without thinking about their limits or the need for recycling. Not surprisingly, the combination of resource demands and human numbers has resulted in humans exerting an environmental impact greater than any other species.

Technology has allowed humans to extract and use resources so efficiently that some people have been able to satisfy their wants in addition to their needs. This is particularly true of the affluent population that lives in developed nations (Western Europe, the former USSR, the United States, Australia, Japan, and Canada); comprising about 20 percent of the world's population, people in these nations consume nearly 80 percent of the world's resources each year. The remaining 80 percent of the world's population survives on about 20 percent of the world's ecological output. Over-consumption of natural capital by rich North Americans, Europeans, and others affects not only the quality of life of poorer people but also threatens the Earth's productive capacity (Wackernagel & Rees, 1996).

Canadians are among the world's most formidable consumers. Although there are significant differences among people of different socioeconomic status, energy and material consumption in Canada is typically three to five times the world average (see Table 1–1). In the next two to three decades, as developing nations strive to attain standards of living closer to those of Canada and other developed nations, resource use and associated impacts are expected to rise sharply, significantly increasing the risk to Earth's ecological assets and biophysical systems.

POLLUTION

Media coverage of dramatic events such as the 1984 release of methyl isocyanate from Union Carbide's pesticide production facility in Bhopal, India, the 1986 Chernobyl nuclear plant disaster in the Soviet Union, the 1989 Exxon *Valdez* oil spill in Prince William Sound in Alaska, the 1991 oil slick in the Persian Gulf, the 1995 tailings dam break at the Omai gold mine in Guyana, the 2000 *Esmeralda* cyanide mining spill into the Danube River, the 2001 oil spill from the tanker *Jessica* off the Galápagos Islands, and the sinking of the *Prestige* off the coast of Spain in 2002 has promoted international awareness of pollution problems.

Far more insidious, perhaps, are the less dramatic but longer-term releases of toxic chemicals, sewage, carcino-

TABLE 1–1
AVERAGE CONSUMPTION OF SELECTED RESOURCES IN CANADA, THE UNITED STATES, INDIA, AND THE WORLD

Consumption per Person in 1991	Canada	USA	India	World
Fresh water withdrawal (m^3/yr)	1 688	1 868	612	644
Fossil energy use (gigajoules/yr)	234	287	5	56
Paper consumption (kilograms/yr)	247	317	2	44
Vehicles per 100 persons	46	57	0.2	10

SOURCE: Adapted from *Our Ecological Footprint: Reducing Human Impact on the Earth*, M. Wackernagel & W. Rees, 1996, Gabriola Island, BC: New Society Publishers, p. 85. Copyright © M. Wackernagel and W. Rees. www.newsociety.com

gens, hormone residues, pesticides, nuclear contaminants, and other harmful substances into our atmosphere, rivers, oceans, and soils. These sorts of industrial and municipal pollution problems have resulted in dead fish on stream banks, vanishing species, recreational beach closures, restrictions on shellfish consumption, dying lakes and forests, birth defects, and other debilitating or fatal conditions such as organ and nerve damage, and lung and bone marrow cancers.

Pollutants—substances that adversely affect the physical, chemical, or biological quality of the Earth's environment or that accumulate in the cells or tissues of living organisms in amounts that threaten the health or survival of these organisms—may originate from natural or human (anthropogenic) sources. Volcanoes, fires ignited by lightning, decomposition of swamp materials, and other natural processes contribute to pollution of Earth's air, water, and soil. But as human populations have grown, technological capabilities advanced, and consumption of resources and consumer goods increased, large numbers and volumes of anthropogenic pollutants—including substances not previously found in nature—have been released into the environment. It has become increasingly difficult for the environment to absorb and process these substances.

Many pollution problems are global in scope: increases in greenhouse gases, a decline in stratospheric ozone, and acidification of soils, forests, and lakes are among the most prominent of current global pollution issues. International cooperative action, including overcoming political inertia and socioeconomic barriers that hamper progress, is required to solve these global pollution problems. Other pollution issues occur on regional scales (such as industrial pollution in Eastern Europe) and local scales (such as air pollution in Mexico City) and also require cooperative efforts to resolve.

RELATED THEMES

In identifying major causes of environmental problems, we must not forget that life on planet Earth involves complex interrelationships among living things—the land, ocean waters and fresh waters, and the atmosphere. The effects of human activity on Earth are now so extensive that global (not just local) environmental change is under way, and no one is entirely sure how these interrelationships will evolve. We know that the actions of many people in different locations have contributed to changes in concentrations of stratospheric ozone and greenhouse gases; we do not know exactly how these global changes will alter the Earth's climate. What is certain, however, is that because human actions are changing the environment at the global level, we need to ensure that our ways of thinking about and understanding the Earth and its systems have a global perspective.

We need to remember, too, that most people in the developed world, including most Canadians, live in urban areas, in cities and towns that continue to expand over the land, spreading over farmland and natural areas, altering drainage patterns, endangering wildlife, and so on. Urban areas have not been studied intensely from an environmental quality perspective, but they do experience air pollution, waste-disposal problems, social unrest, and other environmental stresses. Increasingly, the livability and

Photo 1–6a

Photo 1–6b

Historically, humans have discharged their wastes without due regard for the long-term and cumulative impacts on the environment. Our challenge for the future is to change our attitudes and work toward sustainability as a primary goal.

Photo 1–7
Even in Canada, where population pressures are less intense, urban growth and development continue to encroach on limited agricultural land and natural habitats.

quality of the urban world, and the balance we strike between economic development and urbanization, will be important focuses in future sustainable environments.

To solve contemporary environmental problems, and to move toward future sustainable environments, we need more than facts and scientific understanding regarding a particular issue. While knowledge is an integral part of environmental decision making, so too are our value systems and ethical concerns about social justice and moral commitments to fellow human beings as well as to other living things, both for today and for future generations. Sustainability is a function of how many people depend on an environment, the ways they use that environment, and the values these people hold. Several kinds of questions emerge for consideration: for example, What kind of life do we want for ourselves and our descendants? What kind of environment do we want for them? For people in developing nations? For our neighbours living in inner cities? If we know what our values are and which potential solutions are socially just, we should be able to apply our scientific knowledge and determine an acceptable, sustainable solution for each specific issue.

ENVIRONMENTAL SUSTAINABILITY CHALLENGES

Environmental sustainability is a major challenge related to the future of our lives and of the planet. As a concept, environmental **sustainability** refers to the ability of an ecosystem to maintain ecological processes and functions, **biodiversity**, and productivity over time (Kaufmann et al., 1994). In practice, the meaning of sustainability has varied, but there is agreement that people must learn how to sustain environmental resources so that they continue

to provide benefits to us, to other living things, and to the larger environment of which we are a part.

Considerable importance is attached to the concept of environmental sustainability because it offers a holistic approach to decisions where equal care and respect are given to people and to the enveloping ecosystem of which everyone is a part (Hodge et al., 1995). If Canadians as well as people in other countries and other species that inhabit the Earth are to enjoy an acceptable quality of life in the future, our actions need to reflect the message of sustainability.

The concept of **sustainable development** entered the public domain with the 1980 publication of the World Conservation Strategy (International Union for the Conservation of Nature and Natural Resources [IUCN], United Nations Environment Programme [UNEP], & World Wildlife Fund [WWF], 1980). The World Conservation Strategy (WCS) aimed to achieve sustainable development through the conservation of living resources. Living resource conservation had three main objectives: (1) maintaining essential ecological processes and life-support systems, (2) preserving genetic diversity, and (3) ensuring the sustainable utilization of species and ecosystems.

The WCS promoted integration of conservation and development to ensure that "modifications to the planet do indeed secure the survival and wellbeing of all people" and "to meet the needs of today without foreclosing the achievement of tomorrow's [needs]" (IUCN et al., 1980, p. 1). To be sustainable, the WCS declared, development must be sensitive to short- and long-term alternatives; take social, economic, and ecological factors into account; and include living and nonliving resources.

The publication of *Our Common Future* (World Commission on Environment and Development, 1987), also known as the Brundtland Report, propelled the concept of sustainable development into widespread usage as a global objective. Generally, most definitions of this concept mention the need for people to live equitably within the means of the ecosphere. Unfortunately, in popular usage the term *sustainable development* sometimes is misused or misunderstood because the two words contained in the term mean different things to different people. Some people tune in to the *sustainable* part of the term and understand that sustainable development calls for ecological and social transformation to a world of environmental stability and social justice. Other people identify more with the *development* part of the term and interpret it to mean more growth, or more sensitive growth, or a reformed version of the status quo.

The trouble with the term *sustainable development* is its failure to distinguish clearly between mere growth and true development. Growth generally means increasing in size, while development means improving quality or getting better. If we apply this understanding to the term, sustainable development means progressive social betterment without growing beyond the ecological

Photo 1–8a

Photo 1–8b

The 1992 Earth Summit illustrates that despite the appearance of cooperation among international leaders, political inertia can often be overcome only by the actions of individuals and nongovernmental organizations.

carrying capacity. Put another way, sustainable development means improving the quality of life while remaining within the carrying capacity of supporting ecosystems (World Conservation Union et al., 1991).

To reduce misunderstanding, some people use the broad term *sustainability*. Sustainability echoes the need to balance environmental and developmental concerns, to take an ecological approach to decisions, and to stay within carrying capacity. Sustainability also suggests that excessive material and energy consumption by people in wealthy societies compromises the opportunities for consumption by others. Not everyone accepts the implications of such a concept, namely that levels of consumption in wealthy nations must decrease so that people in poorer nations may consume more in order to improve their quality of life. And, if the world is already at the limit of its carrying capacity, providing adequately and fairly for everyone will be a considerable challenge.

If we are to appropriate sustainability as a means of ensuring a better future, we must realize that sustainability may well affect the economic choices we make. Even though our economies are rooted in the ecosphere, and our lives depend on the maintenance of ecological life support, as citizens of wealthy nations we may find it difficult to acknowledge that if we over-consume today we will have less natural capital and lower natural income tomorrow. We also may find it difficult to acknowledge that "every decision resulting in the appropriation of more resources by those who already consume more than their fair share is a conscious choice against ecological, social and economic sustainability" (Wackernagel & Rees, 1996, p. 156).

As people work toward a sustainable future, one important mechanism to help them achieve their goals is increased cooperation and collaborative effort among individuals, organizations, and nations. A good example of cooperative global action toward sustainability was the development and subsequent adoption of treaties dealing with climate change and biological diversity, as well as Agenda 21 and the Earth Charter, at the 1992 Rio de Janeiro Earth Summit. National representatives attending the Earth Summit (officially, the second United Nations Conference on Environment and Development [UNCED]) discussed and debated a number of international environmental problems, including deterioration of the Earth's atmosphere and oceans from pollution, forest destruction, and loss of biodiversity.

Negotiations on these topics, treaties, and agreements began years before the Earth Summit was held. Agenda 21, for example, is a complex document outlining actions and programs to support sustainable development for the 21st century. Although developed nations did not commit nearly enough financial assistance to help developing countries industrialize without harming the environment, Agenda 21 itself was developed through the collective action and shared responsibility of both developed and developing nations. For a brief overview of Agenda 21, see the summary description in Box 1–1.

Attended by more than 100 heads of state, including Canada's prime minister, the Earth Summit was the largest international gathering ever to concentrate on serious environmental issues. Because the Earth Summit and the Global Forum (a parallel citizens' conference held in conjunction with the United Nations meeting) received so much international attention, they not only increased worldwide awareness of global issues, but also became the most visible expression of international awareness of sustainability ever seen. Progress (or lack of it) in achieving the objectives established at the 1992 Earth Summit was considered at the Earth Summit +5 sessions held in New York in 1997 and in the World Summit on Sustainable Development in Johannesburg, South Africa in 2002 (see Box 5–3).

Canada has been involved in international efforts to define and refine the concept of sustainability and in

Agenda 21 is a plan of action for the world's governments and citizens. It sets forth strategies and measures aimed at halting and reversing the effects of environmental degradation and promoting environmentally sound and sustainable development throughout the world. The Agenda contains some 40 chapters and totals more than 800 pages. In the words of UNCED's Secretary General Maurice Strong, "It is the product of intensive negotiations among Governments on the basis of proposals prepared by the UNCED Secretariat, drawing on extensive inputs from relevant United Nations agencies and organizations, expert consultations, intergovernmental and nongovernmental organizations, regional conferences and national reports, and the direction provided through four sessions of the Preparatory Committee of the Conference." Agenda 21 is "based on the premise that sustainable development is not just an option but an imperative, in both environmental and economic terms and that while the transition towards sustainable development will be difficult, it is entirely feasible."

UNCED has grouped Agenda 21's priority actions under seven social themes:

1. The Prospering World (revitalizing growth with sustainability)
2. The Just World (sustainable living)
3. The Habitable World (human settlements)
4. The Fertile World (global and regional resources)
5. The Shared World (global and regional resources)
6. The Clean World (managing chemicals and waste)
7. The People's World (people participation and responsibility)

For an overview of Agenda 21 and summaries of these themes, refer to UNCED's publication, *The Global Partnership for Environment and Development: A Guide to Agenda 21.* Canada's International Development Research Centre has also published *IDRC, An Agenda 21 Organization: A Backgrounder on Current Activities.* In implementing Agenda 21, IDRC focuses on capacity building, that is, helping local communities in developing nations to enhance their environmental knowledge, decisions, and policies.

SOURCE: *Agenda 21: Green Paths to the Future,* International Research Development Centre, 1993, Ottawa: IRDC. Reprinted with permission.

domestic efforts to promote the concept. Table 1–2 identifies some of the major milestones in international growth of the sustainability concept as well as selected Canadian efforts to support sustainability.

GUIDING PRINCIPLES OF SUSTAINABILITY

Development activities that increase the capacity of the environment to meet human needs, that improve the quality of human life, and that protect and maintain life-support systems are appropriate points from which to progress toward sustainability in Canada. Progressive improvement in human and environmental affairs should lead to (1) people who are healthy, well nourished, adequately clothed and housed, employed in productive work, and able to enjoy leisure and recreational pursuits; and (2) biologically diverse environments that are fully functional from an ecological perspective and that can be monitored for sustainability. These and other desired benefits, however, must be maintained indefinitely to be considered sustainable.

Efforts to achieve sustainability likely will entail a complex mixture of activities incorporating various social, economic, and environmental objectives; the exploitation of Earth's natural capital assets; and the intellectual resources to help people reach their full potential and enjoy a reasonable standard of living. Table 1–3 illustrates

a variety of elements associated with five sustainability objectives identified by the Canadian federal government.

One of the major principles of sustainable action, found in the initial WCS and reiterated in *Our Common Future* and Agenda 21 of the Earth Summit, involves ethical issues such as equity and respect for the rights and welfare of other people (Box 1–2). Other major principles are those of ecological sustainability, social sustainability, and economic sustainability (Table 1–4), as well as the precautionary principle. These principles are discussed briefly below.

Ecological Sustainability

Ecosystems are the source of all life's vital requirements—water to drink, air to breathe, sun for warmth, soils for plant growth—and are the structures within which these life-supporting processes occur (see Chapter 3). If an ecosystem is damaged (for instance, if soils cannot regenerate, or if carbon, oxygen, and other elements cannot circulate), its ability to sustain the life of people, plants, and animals may be reduced or destroyed. Since people depend on Earth's ecosystems, we need to learn how ecosystems support us and how environmental change affects ecosystems. Furthermore, because "the human economy is a fully dependent sub-system of the ecosphere" (Wackernagel & Rees, 1996, p. 4), it is vital that our developments incorporate the concept of ecological sustainability. To that end, we briefly consider the concept of carrying capacity, examine

TABLE 1-2

MAJOR MILESTONES: SUSTAINABLE DEVELOPMENT AND SUSTAINABILITY

Year(s)	Document or Process	Agency
1980	World Conservation Strategy	International Union for the Conservation of Nature and Natural Resources, United Nations Environment Programme, and World Wildlife Fund
1987	*Our Common Future* (the Bruntland Report) • Sustainable development gained international prominence as a global objective. • In 1985, the Brundtland Commission held hearings in Canada.	World Commission on Environment and Development
1989	Preparatory discussions began for the United Nations Conference on Environment and Development (UNCED).	United Nations
1989, 1990	Roundtables on the environment and the economy • National, provincial, and territorial roundtables established in response to Bruntland Report; cross-sectoral issues discussed.	National, provincial, and territorial governments in Canada
1991	*Caring for the Earth* • A more broadly based version of the first WCS. • In 1986, Canada supported the IUCN Conference on Conservation and Development that evaluated the WCS and led to publication of *Caring for the Earth*.	World Conservation Union (IUCN)
1992	Earth Summit (UNCED) in Rio de Janeiro. Treaties and agreements signed included *Climate Change*: to curb CO_2 emissions and reduce greenhouse effects; *Biological Diversity*: to decrease the rate of extinction of species; *Agenda 21*: an actionplan for sustainable development; and *Earth Charter* (the Rio Declaration): a statement of philosophy about environment and development.	United Nations
1994	A new Earth Charter Initiative was launched in The Hague to complete the Earth Charter (unfinished business,1992 Rio Earth Summit).	Maurice Strong, Secretary General of the Earth Summit and Chairman of the Earth Council, and Mikhail Gorbachev, President of Green Cross International, with support from the Dutch government
1997	Earth Charter Commission established to oversee completion of the Earth Charter.	Earth Charter Secretariat was established at the Earth Council, Costa Rica
2000	Final version of the Earth Charter released, providing fundamental principles for building a just, sustainable, and peaceful global society. • Individuals and organizations from all regions of the world helped draft this document. • The Earth Charter circulated globally, promoting awareness and commitment to sustainable ways of life.	Earth Charter Commission
2002	World Summit on Sustainable Development established three key documents, including the Johannesburg Declaration on Sustainable Development affirming the importance of this concept to the world community, an implementation plan outlining actions and priorities, and a statement of partnerships for sustainable development that provided commitments to coalitions among private, civil, and public-sector organizations.	United Nations

SOURCE: Adapted from *The State of Canada's Environment—1996,* Government of Canada, 1996, Ottawa: Supply and Services Canada; "Higher Education and the Earth Charter Initiative," Association of University Leaders for a Sustainable Future, 2000, *The Declaration, 4*(1), pp. 1, 16–20; and *Canada at the World Summit on Sustainable Development,* Government of Canada, 2002, www.canada2002earth-summit.gc.ca

TABLE 1-3
SUSTAINABILITY OBJECTIVES: CANADIAN FEDERAL GOVERNMENT

Sustainability Objectives and Associated Elements

1. Sustain our natural capital (resources)
   - sustainable renewable resources development
   - efficient use of nonrenewable resources
2. Protect the health of Canadians and ecosystems
   - eliminate toxic substances
   - adopt a pollution prevention approach
   - protect representative areas
3. Meet our international obligations
   - protect the ozone layer
   - reduce greenhouse gas emissions
   - conserve biodiversity
   - reduce acidic precipitation and long-range transport of hazardous materials
   - improve management of high-seas fisheries and forestry industries
4. Promote equity
   - ensure fair distribution of costs and benefits between generations
   - ensure fair distribution of current costs and benefits of sustainability
5. Improve quality of life and well-being
   - improve productivity through environmental efficiency (waste minimization, energy and water efficiency)
   - support innovation toward sustainability (long-term time horizon, flexibility)
   - include nonmonetary dimensions of progress

SOURCE: Adapted from *A Guide to Green Government*, Government of Canada, 1995, Ottawa, pp. 4–17.

the idea of ecological footprints, and note the need for an ecosystem approach in making decisions about sustainability.

Carrying Capacity In simplistic terms, **carrying capacity** is the number of organisms that an area can support indefinitely. An expanded definition, which reflects the importance of all ecosystem components and ecological processes, indicates that carrying capacity is the capability of an ecosystem to support healthy organisms while maintaining its productivity, adaptability, and capability for renewal (World Conservation Union et al., 1991).

The concept of carrying capacity is sometimes used to suggest the number of people who can be supported by the environment over time, or to define what levels of human impact might be sustainable. However, the idea of carrying capacity was developed by population biologists in their study of nonhuman populations, so the concept applies to human populations only by analogy. As a result, and because there is no universally agreed-on equation to calculate the carrying capacity for human populations, we do not know the human carrying capacity of the Earth at this time. Carrying capacity calculations increase in complexity when wealthier places import locally scarce resources (such as water, fertilizers, and fossil fuels) to supplement their natural carrying capacity. The carrying capacity of ecosystems in poorer places is more likely to be defined by their natural limitations.

Despite these difficulties, the carrying capacity concept is useful in debating whether ecosystems define natural limits to growth and whether humans already have exceeded the Earth's carrying capacity. Some observers believe that growth in per capita consumption is the most significant reason that Earth's carrying capacity is shrinking. Mathis Wackernagel and William Rees coined the phrase "ecological footprint" to describe a measurement of the land area required to sustain a population. They analyzed water and energy use; uses of land for infrastructure, agriculture, energy, and other materials; as well as waste assimilation, and converted these uses into an estimate of "land-equivalent."

An ecological footprint analysis (see discussion below) by Wackernagel & Rees (1996) showed two conflicting trends: one in which the available ecologically productive land (and water) has decreased and the second in which our demand on the Earth's resources has grown. Calculations undertaken in 2002 based on 1999 data suggested that existing biological capacity of the Earth, the area of land and water that is biologically productive, is 1.9 hectares per person. This figure does not consider the need to protect land and resources for other species. By contrast, on a global basis, our consumption of biological resources is 2.3 hectares per person.

As illustrated in Table 1–5, both biologically productive areas and rates of use vary tremendously from country to country. While growth in population is a significant issue in some countries, these data suggest that rates of consumption in high-income countries need to be controlled if sustainability objectives such as greater equity among the world's people and other species are to be achieved. Why do you think Canada, one of the largest per capita consumers of energy, water, and other resources does not register an ecological deficit? Calculate your own footprint using the guide at the Redefining Progress website (www.redefiningprogress.org).

In contrast to the viewpoint that consumption is the major contributor to the loss of carrying capacity, many people believe that technology and resource substitution provide endless possibilities for procuring the raw materials for continued economic growth. This perspective equates sustainable development with expansion and rapid economic growth in all nations. The assumption is that eco-

The environment is something in which all people share a common interest and by which all may be affected. Not surprisingly, ethical concerns lie at the heart of many environmental issues.

Ethics refers to the principles that define a person's duty to other people and, indeed, to other living things with which we share the planet. Such principles are based on respect for the rights and welfare of other people, including those whom we may never meet and those who are as yet unborn. People look to ethics to decide if an action is right or wrong.

Socioeconomic development that affects the environment and even entire ecosystems also has ethical dimensions. Decisions about development are usually based on estimates of the present and future costs and benefits to society. Such assessments take into account the likely environmental, economic, and social effects of the development. To assess options, every effort is made to define effects in measurable terms and to identify who may gain and who may lose. Such a process is an indispensable means of assembling relevant information and can provide an invaluable basis for identifying the ethical dimensions involved.

However, many ethical issues cannot be quantified. It is difficult to fit ethical factors into the conventional framework of environmental decision making in the same way as more tangible factors such as economic and biophysical data. What is necessary, then, is for people to consider the broad implications of any development—its present and future impacts on other people, on other species, and on ecosystems. Access to full and accurate information about the development and about experiences with similar activities will guide people in deciding about the options that are available. Views on ethics are an essential supplement to the hard, measurable data in the decision-making process.

SOURCE: *The State of Canada's Environment—1996*, Government of Canada, 1996, Ottawa.

---

nomic growth will lead to diversification and better distribution of income. In combination, these factors will assist developing countries to mitigate environmental stresses that derive from economic growth. Unfortunately, according to the United Nations, income disparities between countries and within countries widened throughout the 20th century. In countries such as Canada and the United States, these disparities have also increased, particularly since 1961.

Regardless of which perspective people support, there can be little debate about the impacts of human, resource-related, technological, and economic development: widespread environmental degradation, serious depletion of some resources and resource stocks, and loss of many species. Many of these environmental changes are permanent and irreversible, and, since the economy is dependent on the ecosphere (the bottom line of the economy, so to speak), it follows that we may have overshot local and

T A B L E   1 – 4
PRINCIPLES OF SUSTAINABILITY

Ecological Sustainability	Social Sustainability	Economic Sustainability
• Conserve life-support systems • Conserve biological diversity (genes, species, and ecosystems) • Anticipate and prevent adverse environmental impacts • Practise full-cost accounting (environmental and social costs) • Accept responsibility to protect the global environment • Respect the intrinsic value of nature	• Aim for equitable distribution of benefits and costs of resource use and decisions (among nations and generations) • Ensure future generations have environmental assets equal to or greater than previous generation • Promote a good quality of life (through opportunities for livelihood, education, training, social, cultural, and recreational services, and a quality environment) • Involve the public in land use and related resource and environmental decisions	• Increase capacity to meet human needs and improve quality of life, while using a constant level of physical resources • Encourage diversified, efficient resource use • Ensure renewable resources are used sustainably and nonrenewable resources continue to provide for future generations • Assimilate wastes from economic activity within ecosystem capacity • Promote environmentally sound economic activity (through education, political, and legal instruments)

SOURCE: Adapted from *Off Course: Restoring Balance between Canadian Society and the Environment*, D.M. Taylor, 1994, Ottawa: International Development Research Centre. Reprinted with permission.

global carrying capacity and that current economic growth patterns may not be sustainable. If we believe that social and economic benefits attained through economic growth (such as increased average life spans, reduced physical work effort, and the multitude of consumer products available) are outweighed by the environmental costs and the danger of living at the ecological edge, then we must conclude that humans are living unsustainably, especially those in material-rich Western societies.

Ecological Footprint Analysis As most Canadians live in cities and towns, we tend to forget that human life is very tightly entwined with nature. Urban living tends to break our connections with the ecosystems that provide us with a host of basic materials we need for life—from the air we breathe to the wood we use to build houses and make paper products. Ecosystems also absorb our wastes, protect us from ultraviolet radiation, and provide a wide range of other functions (see Figure 1–4).

A dramatic metaphor for humanity's continuing dependence on nature, an ecological footprint allows people to visualize the impacts of their consumption patterns and activities on ecosystems. The ecological footprint for a particular human population or economy is an estimate of the total area of land and water (ecosystems) needed to produce all the resources consumed and to assimilate all the wastes discharged by that population or economy. Box 1–3 briefly describes the ecological footprint concept using the footprint of the Lower Fraser Valley in British Columbia as an example.

Calculating an ecological footprint is a complex process, given that world trade in foodstuffs, forest resources, and minerals, as well as the products manufactured from those raw materials, facilitates international interaction between people and ecosystems. The ecological footprint model discussed in Box 1–3 and in Chapters 4 and 13 interprets this complexity by examining the amount of resources an individual, city, or nation consumes and by determining how much land area is required to provide those resources on a sustainable basis. How much land does the average Canadian need in order to provide the materials, food, and energy resources he or she consumes? An ecological footprint analysis suggests that our lifestyle is not sustainable. If every individual on the Earth consumed as many resources as does the average Canadian, we would need 4.6 more planets to

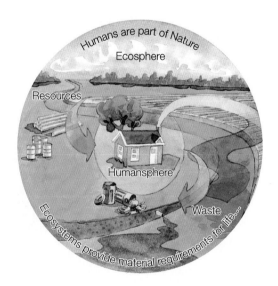

Figure 1–4

Humans and the Ecosphere

BOX 1-3

THE ECOLOGICAL FOOTPRINT CONCEPT

Most of us have looked at our footprints in the sand on a beach, in the dirt of a field, or in new-fallen snow, but how many of us have considered looking at the footprints our communities leave on their surroundings? That's one of the ideas William Rees and Mathis Wackernagel have explored since 1990 through the Task Force on Planning Healthy and Sustainable Communities at the University of British Columbia.

In considering a city's impact, or footprint, on the environment, it is important to consider our reliance on ecosystems as well as on the more obvious buildings, roads, industrial parks, and housing areas that are spread over the landscape. Ecological footprints measure the load that any given community or population imposes on the ecosystems that support it. Footprints are expressed in terms of the land area necessary to sustain current levels of resource consumption and waste discharge by that population. According to Wackernagel and Rees (1996, p. 9), "ecological footprint analysis is an accounting tool that enables us to estimate the resource consumption and waste assimilation requirements of a defined human population or economy in terms of a corresponding productive land area."

Much of the work involved in determining the footprint is aimed at estimating appropriated carrying capacity—the amount of land it would take to produce all the goods and services that are used by people living in a city. It is impossible to take every factor into account, and the calculations are complex, but a rough idea of a city's—or your individual—ecological footprint can be revealing.

Wackernagel and Rees estimated the ecological footprint for residents of British Columbia's densely populated Lower Fraser Valley (Vancouver to Hope) and determined that this region depends on a land area 19 times larger than that contained within its boundaries to satisfy current consumption levels of

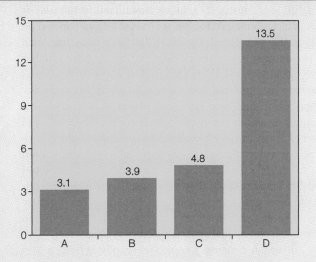

Box Figure 1–2

Comparison of ecological footprints by type of household

A Single parent with child—annual household expenditure $16 000

B Student living alone—annual household expenditure $10 000

C Average Canadian family, 2.72 people—annual household expenditure $37 000

D Professional couple, no children—annual household expenditure $79 000

SOURCE: *Response to the Environmental Petition No. 55 by the Sierra Legal Defence Fund under the Auditor General Act and Review of the Government of Canada's Policies, Laws and Regulations Concerning Air Pollution and Air Quality*, Government of Canada, 2003.

food, forest products, energy, and carbon dioxide assimilation. By their calculations, each individual Canadian requires 4.27 hectares of land to support her or his current lifestyle. The footprint for an average American was estimated at 5.1 hectares. "If everyone lived like today's North Americans, it would take at least two additional planet Earths to produce the resources, absorb the wastes, and otherwise maintain life-support. Unfortunately, good planets are hard to find" (Wackernagel & Rees, 1996, p. 15).

SOURCES: *The State of Canada's Environment—1996*, Government of Canada, 1996a, Ottawa: Supply and Services Canada; *1996 Report of Canada to the United Nations Commission on Sustainable Development*, Government of Canada, 1996b, http://www.ec.gc.ca/agenda21/96/part1.html; *Our Ecological Footprint: Reducing Human Impact on the Earth*, M. Wackernagel & W. Rees, 1996, Gabriola Island, BC: New Society.

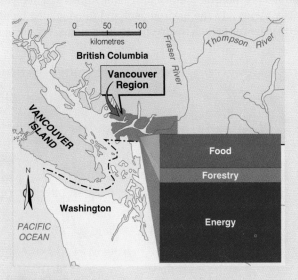

Box Figure 1–1

Lower Mainland/Greater Vancouver area

SOURCE: *How Big Is Our Ecological Footprint? Using the Concept of Appropriated Carrying Capacity for Measuring Sustainability*, M. Wackernagel, 1993 (pamphlet), Vancouver: Task Force on Planning Healthy and Sustainable Communities.

provide for everyone (see Figure 1–5). Each year, human society consumes 30 percent more of the Earth's natural capital than is regenerated.

The concept of an ecological footprint is important in guiding our efforts to achieve sustainability because it helps us understand that our individual actions can and do affect the global environment. While it is true that small communities in the developing world may have smaller ecological footprints (because they rely less on trade and technology) and may live at a level close to the carrying capacity of their local environmental resources, the same cannot be said of small communities in the industrialized world. Even in the Canadian North, for instance, where just a few decades ago people in small communities lived in harmony with their local environment and extracted quality country food, they now purchase imported food-stuffs and goods from southern Canada. This is partly because of concern about bioaccumulation of toxins in some fish and animal species eaten (see Chapter 3).

In urban centres, people's everyday activities and actions contribute to disruptions in ecological processes. Our individual actions and those activities associated with our local and regional industrial processes have affected the global environment—emissions of carbon dioxide and chlorofluorocarbons (CFCs), destruction of stratospheric

ozone, and global climate change are among the results. Whether it is disposal of community waste or discharge of industrial pollutants, experience has shown us that environmental carelessness at any one location may have impacts that extend over considerable distances and time periods. Realizing that our individual actions contribute to these environmental problems is an important step to understanding that all humans are connected to the one Earth that supports us. In that one Earth we share a common natural heritage and have a responsibility to care for that heritage. The ecological footprint approach helps us see the ecological reality of our actions and challenges and directs us toward more sustainable lifestyles (see Chapter 13).

An Ecosystem Approach   If we think of ecological footprint analysis as a practical tool to help us assess the sustainability of our current activities, then we can conceive of an ecosystem approach to help us apply our understanding of ecological constraints to make more effective decisions about achieving sustainability. An ecosystem approach requires a fundamental shift in thinking; it reflects a whole-Earth ethic in which our decisions and actions take into account the dependencies that exist within and among ecosystems as well as the connections between the biological and physical components of the ecosphere. This, too, is a practical element in working toward a sustainable future. For example, we cannot improve the management of our fish stocks unless we understand both the aquatic environment of which the fish are a part and the human activities that affect that aquatic environment.

To achieve our desired sustainable future, we need to promote a holistic approach to human interactions with ecosystems and the environment and to be guided by an ecosystem approach that respects ecosystems, organisms,

Figure 1–5
Wanted: 4.6 planet Earths

SOURCE: Adapted from *Our Ecological Footprint: Reducing Human Impact on the Earth,* by M. Wackernagel & W. Rees, 1996, Gabriola Island, BC: New Society Publishers, p. 15; "Ecological Footprint of Nations, November 2002 Update: How Much Nature Do They Use? How Much Nature Do They Have?" in *Redefining Progress* by Mathis Wackernagel, Chad Monfreda, and Diana Deumling, 2002, available at www.redefiningprogress.org

Photo 1–9
The ecological footprint for residents of the Lower Fraser Valley has been estimated at 19 times the land area they occupy.

Photo 1–10
Urban infrastructure such as transportation systems contributes to a city's ecological footprint.

and people. By adopting an ecosystem approach (see Chapter 12) and adhering to ethical principles (see Chapter 2), we should see new attitudes and practices combining to safeguard the future of the Earth.

As we have seen in our discussion of ecological footprints and carrying capacity, however, it is difficult to measure all the dimensions of humanity's load on the Earth, and there is no easy way to determine the balance between environmental costs and benefits. However, ecological sustainability is a fundamental principle in the drive to contribute to improved decisions that will enable the world to move toward sustainable futures. Social sustainability, the subject of the next section, is also a guiding principle of sustainability.

Social Sustainability

Social constraints on development are just as difficult to measure as ecological limits, and just as important. Reflecting the values attached to human health and well-being, social constraints (or social norms) are based on the traditions, religions, and customs of people and their communities. Sometimes these powerful social values are written in law. Often, they are invisible and not put in writing. For example, beliefs about what is considered right or wrong in a community or in relation to the environment may be strongly held, yet they are not given formal legal status. Other social norms may be made more concrete through the passing of policies, programs, and funding arrangements to provide social welfare, language, and education.

Like many other nations, Canada is undergoing rapid social and economic changes. Today's pervasive pressures for change, as well as the multicultural nature of Canada's varied communities, have compounded the difficulties in defining, measuring, and evaluating the changes occurring in social norms and in the limits they entail regarding development. Development activities may be considered socially sustainable if they conform with a community's social norms or do not stretch them beyond the community's tolerance for change. Conversely, socially unsustainable development may be indicated when antisocial behaviour occurs, such as property damage, violence, and other community disruptions.

While most norms will change over time, some social norms are extremely persistent; no matter what development activity is proposed, people will oppose or resist it. Even though they are not measured easily (if at all), these are the norms or constraints that must be respected in assessing whether social sustainability will be achieved. For instance, in communities where traditional ways of life depend on the natural resource base (minerals, fish, land, and forests), people often are extremely resistant to change. Their strong social norms insist, for example, that the values they place on fish and fishing for a livelihood must be respected, even though outsiders (often developers) do not appreciate these same values. Unless these values are addressed in development decision making, people in such communities will continue to resist change.

Attention to social sustainability means that actions taken to protect environmental resources must consider the effects on local human populations. People who live in rural places who are involved in the extraction and processing of environmental resources often share similar social histories and cultural norms. For example, efforts to curb logging will affect the livelihood and the well-being of loggers and all who have helped to build forestry communities—e.g., families, community service providers, shopkeepers in forestry towns, professional foresters, and public servants. Environmental decisions affect not only the pocketbooks of these people, but also their identities as rural people. Often these people share common ideas about rural work and life that include living close to nature, doing hard physical work, and being self-reliant and independent. For example, decisions to reduce access to timber for harvesting and to impose job losses in forestry directly challenge the ability of people to live their lives based on these ideas and thereby threaten the very fabric of collective identity. The strength of social norms and their potential to place constraints on development activity means that they must be taken into account in planning for sustainability. The people in affected communities must be part of the process of defining social limits to development and to sustainability (refer to Table 1–4). As many of the examples discussed in later chapters demonstrate, the importance of cooperation and collaboration with groups or communities concerned cannot be underestimated if we are to move toward a sustainable future.

Economic Sustainability

Even though it is as difficult to predict and is affected by as many variables, economic sustainability is more easily measured than ecological or social sustainability. This is because economic sustainability requires economic benefits (defined usually in monetary terms) to exceed or at least balance costs.

Economic sustainability reflects the interplay of supply and demand factors. In general, supply-side factors include the availability, cost, and transport of raw materials, as well as the energy, labour, and machinery costs relating to their extraction or processing. Since ecological factors ultimately limit sustainability, economic development must use resources in ways that do not permanently damage the environment and must not impair the replenishment capacity of renewable resources (Government of Canada, 1996). This statement suggests that economic sustainability will be attained only if we use nonrenewable resources sparingly, if we reduce the content of nonrenewable resources in the goods we produce, and if we reduce the energy consumed in their production.

On the demand side, sustainability is threatened because of the increasing rate of human consumption of resources. Nowadays, whether prices are high, because of the heightened value of a scarce resource, or low, because of efficient harvesting technology, stocks continue to be overharvested. If prices stay high, harvesters may maintain pressure on the resource in order to generate maximum profit, and if prices are low while demand is high, overharvesting may reduce stocks to below-recovery levels. Neither high nor low market prices for resources are good indicators of sustainability. Instead of considering only price, it is possible to consider a set of more provocative indicators (e.g., proximity of residents to public transportation) that could be used to track consumption levels more closely and thereby demonstrate how to put values associated with sustainability into action. Some Canadians have been involved in efforts to identify sustainability indicators appropriate for their communities. We consider some of these efforts in more detail in Chapter 13.

Balancing costs and benefits to achieve sustainability is a complex task. In the pollution control field, for example, regulations may require a plant owner to install costly new equipment or to adjust production procedures to meet new emissions standards. Balancing costs and benefits is difficult when, as is frequently the case, unexpected cost savings derive from installation of such new equipment or procedures.

Economic, ecological, and social sustainability are equally important principles. Economic sustainability, however, may receive disproportionate emphasis (in the media, at least) because it is the basis of all forms of income distribution and returns on investment within a society (Government of Canada, 1996). Given this, we may say that development activity is unsustainable unless it supports socially acceptable income levels. Similarly, development that emphasizes jobs and income to the exclusion of ecological benefits is not sustainable either.

The Precautionary Principle

Canada and many other nations lack adequate data and records regarding environmental change. Environmental information collected in the past is not always relevant today because it was not directed toward the problems and issues that concern us now, nor was it collected with sustainability objectives in mind. Appreciating that lack of data is currently (and, for some time into the future, likely to continue to be) a serious hindrance to environmental protection and achievement of sustainability, countries attending the 1992 Earth Summit, including Canada, adopted the precautionary principle.

The precautionary principle states that when there are threats of serious or irreversible damage to the environment, the "lack of full scientific certainty shall not be used as a reason for postponing cost-effective measures to prevent environmental deterioration" (Government of Canada, 1996). This principle means that, as long as the weight of evidence suggests action is appropriate, a country should take such action to protect its environment. For instance, if action were not already being taken in regard to global climate change, the precautionary principle would provide impetus for such action to commence even though disagreement existed about the specific details regarding climate change.

Environmental Stewardship

Environmental stewardship is demonstrated by the many local groups that have provided conservation, rehabilitation, and other care for threatened and special places in their communities. As well, these groups have pressured decision makers to act in environmentally responsible ways (Lerner, 1994). Through active participation in caring for and maintaining the well-being of a place, many people have learned that they are part of nature. They have learned also how vital it is to maintain ecosystems not only for the benefits they bring to people and other organisms, but also for the sake of ecosystems themselves.

Stewardship is an important principle in achieving sustainability. The perception that governments and corporations in Canada have not been able to protect the environment effectively has persisted over the past decade or so. That perception, as well as fears that damage to the environment and human health could be irreversible, has prompted individuals and groups of people to demand comprehensive, proactive planning for environmental quality and sustainability in their region and the country (Lerner, 1994).

Photo 1–11

Toronto residents demonstrate stewardship in a local effort to clean up the Don River.

Stewardship permits people to take leadership roles and act responsibly when a threat to a locally valued place or environment occurs. Local people who feel a sense of stewardship or responsibility may be found cleaning up their natural areas, challenging polluters to do things differently, and demanding action and accountability from their governments. Stewardship also promotes working in partnership with other individuals and environmental nongovernmental groups, as well as with government or private-sector programs. Stewardship helps everyone understand the importance of accountability in regard to ecosystem sustainability. In short, stewardship is "active earthkeeping" (Lerner, 1993) that helps promote and attain the public good.

MONITORING FOR SUSTAINABILITY

As the preceding descriptions of ecological, social, and economic sustainability principles have noted, there is a great deal of uncertainty associated with attaining sustainable futures. However, every action we take to move toward sustainability can teach us about what works and what does not work. If we consistently record, monitor, evaluate, and report on these efforts, we may learn how to improve our progress. Using an ecosystem approach to monitor development activities and the policies that guided them and assessing the long-term and cumulative impacts of development activity will contribute to improved decision making in the future. Monitoring and related actions will also provide the background for comparison of environmental, social, and economic parameters of change.

Another important outcome of monitoring efforts could be the establishment of a set of comprehensive questions about the expected environmental, social, and economic impacts of proposed developments. If all decision makers routinely applied such a tool, sustainability would be enhanced and our thinking and planning for the future could be improved. Monitoring could also provide one means to enhance global understanding and cooperation and, thereby, to move toward sustainability.

TOWARD SUSTAINABILITY

Attaining sustainability is an important Canadian goal that requires political will as well as individual and collective action. New ways of thinking—about the ecosphere and about our local environment, about ecosystems and the interactions and interdependencies that characterize them, and about how sustainability can be achieved now and for the future—are bringing about new means of achieving desired, sustainable futures. These new ways of thinking involve cooperative stewardship efforts and, as we will see throughout the chapters regarding Canada's resources, have been delivering some exciting results.

The notion of sustainable development as an achievable goal provides the only possible basis for a viable future for Canadians. More profits, jobs, and goods will be to no avail if they are gained at the cost of a compromised life-support system. Sustainability is a concept whose time has come. We must seize the challenge. The ideal may elude us, but as long as we continue to search, consult, and face up to hard choices, we will leave behind a better world for our children and our grandchildren (Government of Canada, 1996, pp. 1–14).

Chapter Questions

1. Briefly compare and contrast the concepts of sustainable development and sustainability.

2. Discuss the ways in which the five major recommendations found in the World Scientists' Warning to Humanity could contribute to restoring balance between society and the environment.

3. Is the world facing a crisis of numbers of people (population growth) or a crisis of consumption by people (affluence)? Give reasons for your answer.

4. From the environmental issues cited in the chapter, list at least three that affect Canada at each of the local, regional, and international levels. Compile two or three additional examples of different scales of environmental problems (from your hometown, your campus, or an issue you read about in a local or national newspaper).

5. Which one of the environmental problems you listed in Question 4 do you think is most serious? Discuss the reasons why you identified this as the most serious problem. What is the root cause of this problem?

6. Discuss the guiding principles of sustainability, commenting specifically on how each principle contributes in the effort to move toward sustainability.

references

Brower, D. (Ed.). (1975). *Only a little planet.* New York: Friends of the Earth/Ballantine Books.

Government of Canada. (1996). *The state of Canada's environment—1996.* Ottawa: Supply and Services Canada.

Hodge, T. S., Holtz, S., Smith, C., & Hawke Baxter, K. (1995). *Pathways to sustainability: Assessing our progress.* Ottawa: National Round Table on Environment and the Economy.

International Union for the Conservation of Nature and Natural Resources (IUCN), United Nations Environment Programme (UNEP), & World Wildlife Fund (WWF). (1980). *World Conservation Strategy: Living resource conservation for sustainable development.* Gland, Switzerland: International Union for the Conservation of Nature and Natural Resources.

Kaufmann, M. R., Graham, R. T., Boyce, Jr., D. A., Moir, W. H., Perry, L., Reynolds, R. T., Bassett, R. L., Mehlhop, P., Edminster, C. B., Block, W. M., & Corn, P. S. (1994). *An ecological basis for ecosystem management.* Fort Collins, CO: U.S. Department of Agriculture Forest Service, Rocky Mountain Forest and Range Experiment Station and Southwestern Region. USDA Forest Service General Technical Report RM-246.

Lerner, S. (Ed.). (1993). *Environmental stewardship: Studies in active earthkeeping.* Waterloo, ON: University of Waterloo Department of Geography Publication Series No. 39.

Lerner, S. (1994). Local stewardship: Training ground for an environmental vanguard. *Alternatives, 20*(2), 14–19.

Wackernagel, M., & Rees, W. (1996). *Our ecological footprint: Reducing human impact on the earth.* Gabriola Island, BC: New Society Publishers.

World Commission on Environment and Development. (1987). *Our common future.* Oxford: Oxford University Press.

World Conservation Union (IUCN), United Nations Environment Programme (UNEP), & World Wildlife Fund (WWF). (1991). *Caring for the earth: A strategy for sustainable living.* Gland, Switzerland: International Union for the Conservation of Nature and Natural Resources.

"While science should not be unreasonably limited in its endeavors—pure science has produced many spin-off benefits—we need a newly expansive framework within which science can pursue its new mission, to safeguard the planet; and society within it."

Norman Myers (1990, p. 174)

CHAPTER 2

Environmental Studies: Science, Worldviews, & Ethics

Chapter Contents

INTRODUCTION

In Chapter 1 we noted some fundamental features of life at the beginning of the 21st century: how people have changed the global environment; how ecosystems are under increasing threat from human population growth and resource consumption, pollution, and other environmental abuses; how uncertainty about global changes highlights the importance of sustainability as a counterbalance to these changes; and how both knowledge and values are vital in solving contemporary environmental problems.

In this chapter we discuss the historical and current roles and influences of science, worldviews, and ethics in environmental studies. The significance of interdisciplinary science in generating new environmental knowledge and understanding is stressed, as is the importance of the social sciences in understanding human attitudes, values, and beliefs in order to address root causes of environmental problems. Attention also is given to the growth in support for environmental action, particularly the "environmental revolution."

SCIENCE AND THE ENVIRONMENT

Early inhabitants of Canada depended on their knowledge of the environment for their survival; careful observations of wildlife habits and migration routes, weather patterns, and other natural events were crucial to their continued well-being. To some extent this reliance on environmental knowledge continues among occupations that depend on the natural environment (such as fishing, hunting, and trapping), but for most Canadians, especially those living in urban settings, the environment is not something we think about scientifically each day. However, the environment continues to play a vital role in our everyday lives.

As unaccustomed as we might be to thinking scientifically about the environment, there are several reasons why science, as a process for refining our understanding of how the natural world works, is important. The systematic observation and analysis underlying both science and social science has provided us with a great deal of understanding about ecosystems, their functioning, and human interactions with our environment. These are the kinds of knowledge that can be applied to prevent negative effects or better manage human impacts on the environment. With improved knowledge and understanding, we can learn how to avoid inappropriate choices and

Photo 2–1
Before European settlement, indigenous peoples depended directly on the land for life's necessities.

identify courses of action that will be positive and sustainable in the long term.

Environmental decision making reflects the interactions of society, politics, culture, economics, and values, as well as scientific information. To demonstrate why thinking critically (scientifically) about the environment is important to making decisions about environmental sustainability, the next sections discuss what science is (and is not), the importance scientists place on measurement, the methods of science, and the role of science in decision making. Since science alone is not sufficient to resolve environmental problems, some social science aspects of environmental management are considered later in the chapter. Different worldviews (ways of perceiving reality) are explored, as are the role of the 1960s environmental revolution and the subsequent (1985–onward) wave of environmentalism in stimulating a reassessment of human relationships with the natural world.

WHAT IS SCIENCE?

Science is a process used to investigate the world around us—a systematic attempt to understand the universe. By repeatedly asking questions about how the natural environment works, conducting orderly observations (including experiments), and analyzing their findings, scientists attempt to reduce the complexity of our world to

general principles that provide new insights or solve problems. Thinking about environmental issues involves thinking scientifically.

While science provides one way of looking at the world, it is not the only way of viewing the world. Religious, moral, aesthetic, cultural, and personal values, for example, provide different and valuable ways of perceiving and making sense of the world around us. In general, scientists focus on rigorous observation, experimentation, and logic in developing an integrated and objective understanding of our world. However, science is not value-free: that is, scientists have their own values, interests, and cultural backgrounds that may influence their interpretation of data. This means that we need to be able to evaluate the statements they make about the environment: Are these statements based on observations and data? Are they a systematic, or "expert," interpretation of data, or are they subjective, personal opinion or assertion? In order to determine an answer for ourselves, we need to understand more about what science is and is not.

Modern science does not deal with metaphysical questions (What is the purpose of life?) or with questions that are answered by morals or values (What is beauty?), but rather with things that are observable and testable. In terms of the natural world, scientists must be able to make observations from which they develop a **hypothesis** (explanation) that can be accepted until it is disproved. Such hypotheses are based on certain assumptions that scientists make about the world.

Photo 2–2
Scientific views, such as Darwin's theory of evolution, can conflict with or complement generally accepted social beliefs.

Assumptions in Science

Scientists make five basic assumptions about the natural world they study: (1) people can understand the patterns of events in the natural world through careful observation and analysis; (2) the same rules or patterns of behaviour that describe events in the natural world apply throughout the universe; (3) science is based on inductive reasoning that begins with specific observations of, and extends to generalizations about, the natural world; (4) these generalizations can be subjected to tests that try to disprove them—if no such test can be devised, then a generalization still cannot be treated as a scientific statement (because it is true only until new, contrary evidence is found); and (5) existing scientific theories can be disproved by new evidence, but science can never provide absolute proof of the truth of its theories (Botkin & Keller, 1995).

Thinking Scientifically

Scientists use two kinds of reasoning: inductive and deductive. When scientists draw conclusions about their observations of the natural world by means of logical reasoning, they are engaging in **deductive reasoning** (thinking). In this process, a specific conclusion flows logically from the definitions and assumptions (the premises) set initially by the scientists. If a conclusion follows logically from the premises, it is said to be proved. Deductive proof does not require that the premises be true, only that the reasoning be logical. This means that logically valid but untrue statements can result from false premises, as the following example illustrates:

Humans are the only tool-using organisms.

The Egyptian vulture uses tools.

Therefore, the Egyptian vulture is a human being.

The final statement must be true if the two preceding statements are true, but we know that this conclusion is untrue. If the second statement is true (and it is: some African populations of Egyptian vultures break ostrich eggs by dropping rocks on them), then the first statement cannot be true and the conclusion must be false. Because the rules of deductive logic deal only with the process of moving from premises to conclusions, in this example the conclusion that the Egyptian vulture is human follows logically from the series of statements (even though it is nonsensical).

This problem of false conclusions is why science requires not only logical reasoning but also correct premises. If these three statements were expressed conditionally, they would be scientifically correct:

If humans are the only tool-using organisms,

and the Egyptian vulture uses tools,

then the Egyptian vulture is a human being.

When a scientist (or anyone else) draws a general conclusion based on a limited set of observations, that person is engaging in **inductive reasoning**. Let us say, for example, that we are observing fruits with particular characteristics, such as tomatoes, and we observe that such fruits are always red. We may make the inductive statement (or generalization) that "all tomatoes are red." This really means that all the tomatoes we have ever seen are red. Since it is highly unlikely we will observe all the world's tomatoes, we do not know if the next observation we make may turn up a fruit that is like a tomato in all respects except that it is yellow. In inductive reasoning, then, people (scientists included) can only state what is usually true—for example, that there is a very high degree of probability that all swans are white—but we cannot say so with absolute certainty. Inductive reasoning produces new knowledge, but is error prone.

Probability is one way scientists express how certain (or uncertain) they are about the quality of their observations and how confident they are of their predictions. If they are highly confident in their conclusions, scientists may state the degree of certainty as "There is a 99.9 percent probability that ..." This is about as close as scientists can get to stating proof of their theories—but it is still not proof.

The concept of correlation provides another way in which scientists can express their uncertainty. Correlation describes the degree to which one variable is related to another variable; for example, we know that climate and crops are closely correlated.

Proving something using inductive reasoning is different from demonstrating proof using deductive reasoning, and science needs both types to help analyze whether the conclusions reached are valid. Both ways of thinking are complementary and important in improving understanding of our environment.

Scientific Measurement

Perhaps more than anyone else, scientists appreciate that every measurement they make is an approximation. Depending on the instruments used and the people who use the instruments, all measurements contain some limitations. These limitations or uncertainties can be reduced, but they can never be eliminated completely.

To make their measurements meaningful, particularly to decision makers, scientists provide an estimate of the uncertainty associated with their measurements. Consider the case of a wildlife biologist who is asked to determine the impact on a caribou herd if the flow rate of a river they cross during their migration were to double due to a proposed hydroelectric development. If the wildlife biologist calculates that doubling the flow in the river will reduce the caribou population by 100 animals, decision makers still do not have enough information to determine whether construction of the hydro plant should proceed. In addition to information about the average population size of the particular herd affected (what proportion of the herd does 100 animals represent?), decision makers need to know the uncertainty associated with the loss of 100 animals. In this case, the statistical definition of significance likely would be used in interpreting results. *Statistical significance* describes the probability of whether or not the event or relationship has a greater potential to occur with more frequency than simple random chance. If the uncertainty were 1 percent, their decision might be different than if it were 10 percent.

In addition to uncertainty, scientists may encounter systematic and random measurement errors during their research. Random errors occur by chance, but if a scientist's instrument had been calibrated incorrectly and consistently provided inaccurate readings, a systematic error would result. *Standard error* calculations show us the degree to which results vary from sample to sample and are used to help determine whether errors are random or systematic. (Numerically, standard error is the square root of variation in the sample population divided by the individual sample size.) Obviously it is important to avoid such errors and to make and report measurements as accurately and precisely as possible.

If a scientist's measurement is accurate, it will correspond to the value scientists have already accepted for that feature; if that scientist's measurement is precise, the feature will have been measured with a high degree of exactness. Note that it is possible to make very precise measurements that are not accurate. For instance, the accepted value for the boiling point of water at sea level is 100.000°C (or 212.000°F); if you measure the boiling point of water as 99.885°C, your measurement is as precise as the accepted one (because both are recorded to the nearest 0.001°), but your measurement is still (slightly) inaccurate. It is important, too, not to mislead others by reporting measurements with more precision than is warranted.

Careful choice of research and statistical procedures, adoption of standard measurement procedures, and improvements in instrumentation will all help to reduce measurement errors and uncertainties. Given that errors will continue to occur in research, however, it is important to be informed about and understand the nature of measurement uncertainties so that we may read reports of scientific research critically and evaluate their validity.

The increased use of aerial photography and satellite remote sensing during the past 20 years has contributed greatly to our ability to map, inventory, and monitor the environment (thus reducing error and uncertainty). Similarly, the data storage, display, and analysis capabilities of geographic information systems (GISs) have been vital in enabling researchers to contribute to environmental management decisions. As satellite transmitters became small enough to slip over the shoulders of peregrine falcons, researchers were able to track some of these birds as they migrated more than 14 000 kilometres from their breeding grounds in Alaska and northern Canada to their wintering grounds in Central and South America. The data gathered using this technology not only fill large gaps in knowledge based on previous leg-banding programs, but also enable production of flight-path maps to determine whether peregrines are travelling to areas where potentially dangerous pesticides are used. In turn, this information has helped in decision making about removing the bird from the endangered species list (Yoon, 1996).

Photo 2–3
The drowning of 9600 caribou in northern Quebec in October 1984 is thought to have resulted from increased flow in the Caniapiscau River, due in part to the operation of dam systems.

THE METHODS OF SCIENCE

As noted above, **observations**—made through any of our five senses or instruments that extend those senses—are the fundamental basis of science. When scientists check for accuracy in science, they compare their observations with those made by many other scientists, and when all (or almost all) of them agree that an observation is correct, they call it a **fact**.

Observations and facts provide a basis for **inferences**—conclusions derived either by logical reasoning from premises and/or evidence, or by insight or analogy based on evidence. Before inferences are accepted as facts, they must be tested (accepting untested inferences is sloppy thinking) and they must be repeatable (when the same methodology is followed). When scientists test an inference, they convert it into a hypothesis, a statement that they can try to disprove. Often, the hypothesis is stated in negative terms, called the null hypothesis, because in science it is assumed that there are no absolute truths (this is why assumptions 4 and 5 regarding generalizations and scientific theories can only be disproved, not proved). By attempting systematically to demonstrate that certain statements are not valid—that is, are not consistent with what has been learned from observation—scientists learn which general principles governing the operation of the natural world are true. Invalid statements are rejected, while statements that have not been proved to be invalid are retained until such time as they are found to be incorrect.

Hypotheses are stated typically in the form of "If ... then" statements—for example, "If I apply more fertilizer, then my pumpkin plants will produce larger pumpkins." This statement relates two conditions, namely the amount of fertilizer applied and the size of pumpkins produced. Because each of these conditions can vary, they are called variables. The size of pumpkins is called the *dependent* (or *responding* or *y*-axis) variable because it is assumed to depend on the amount of fertilizer applied, which is the *independent* (or *manipulated* or *x*-axis) variable.

Many variables may exist in growing pumpkin plants. Some can be assumed to be irrelevant, such as the position of the planet Mars, while others, such as length of the growing season, average temperatures during the growing period, and daylight hours, are potentially relevant. To test the stated hypothesis (above), a scientist would want to run a **controlled experiment**. A controlled experiment is designed to test the effects of independent variables on a dependent variable by changing only one independent variable at a time. For each variable tested, there are two setups—an experiment and a control—that are identical except for the independent variable being tested. If there are differences in the outcome between the experiment and the control (relating to the dependent variable), then these differences are attributed to

Photo 2–4
Data are an integral part of the scientific method. Here a researcher tests for dissolved oxygen.

the effects of the independent variable tested. It can be a challenge to properly design control tests and to isolate a single variable from all other variables.

It is also important to ensure that variables are defined in ways that enable their exact meaning to be understood by all scientists. The vagueness of the variable "size of pumpkins," for example, might cause one scientist to interpret it as weight, another as diameter. Both independent and dependent variables must be defined operationally before an experiment is carried out. **Operational definitions** tell scientists what to look for or what to do in order to carry out the measurement, construction, or manipulation of variables. In that way, scientists know how to duplicate an experiment and how to check on the results reported: for example, they would know whether "size of pumpkins" was to be measured in kilograms or in centimetres.

Keeping accurate records of independent and dependent variables during experiments is an important element in science. These values, or data, are referred to as either *quantitative data* (numerical) or *qualitative data* (non-numerical). In the pumpkin example, above, qualitative data would record the size of pumpkins as small, medium, or large, while quantitative data would record each pumpkin's weight in grams or diameter in centimetres. There is a long-standing bias in science toward quantitative data, but many fields with relevance to environmental issues collect data in qualitative forms (for instance, sociology, psychology, animal behaviour, human geography, and environmental policy analysis).

Scientific research continues to contribute to the growing body of knowledge pertaining to the natural environment. As knowledge accumulates in both quantitative and qualitative dimensions, scientists develop explanations or models to illustrate how the natural environment works. Different types of models exist, from actual working models to mental, computer, mathematical, and laboratory models. All models may be revised or replaced as knowledge increases and currently accepted hypotheses are modified in light of new understanding. Models that offer broadly conceived, logically coherent, and well-supported concepts are labelled *theories*. Einstein's theory of relativity and Newton's theory of gravity are examples of strongly supported theories that are unlikely to be rejected in the future. However, science does not guarantee that future evidence will not cause even these theories to be revised.

It is worth noting that while theories usually grow out of research, theories also may guide research. In fact, scientists make their observations in the context of existing theories. On occasion, scientific revolutions occur when a growing discrepancy between observations and accepted theories forces replacement of old theories by new or revised ones. Perhaps the shift in thinking about Lovelock's Gaia concept—its shift from hypothesis to theory—may be viewed in this light.

Photo 2–5

Dolly the sheep, the first mammal to be cloned from an adult cell (genetic engineering), was euthanized in February 2003 at the age of six and one-half years. She had developed arthritis and a progressive lung disease, a problem common in 10- to 12-year-old sheep. DNA evidence in some cloned mammals, including Dolly, indicates they age much faster than their non-cloned counterparts.

MISUNDERSTANDINGS ABOUT SCIENCE

Use of Language

As we have seen, researchers use a variety of scientific methods, as well as their imagination and insight, to increase knowledge and understanding about the natural world. We also appreciate that while scientists may disprove things, they cannot establish absolute truth or proof. So, when you encounter statements that something has been "proved scientifically," it is important to recognize this inaccurate use of scientific language. The claim being made falsely implies that science yields absolute proof or certainty. This situation may occur when people do not understand the nature and limitations of science, or when language is misused deliberately.

Also, be alert to use of the term **theory**. In everyday language, people use the term to indicate a guess or a lack of knowledge ("it's just a theory") and do not accord theory the prestige it is given in science. Development of a theory is among the greatest achievements in science. It is only after considerable debate, speculation, and sometimes controversy that scientists develop theories based on their consensus about explanations of phenomena. The significance of scientific consensus is often undervalued by the public and the media as well; this consensus gives even more credence to the message of the "World Scientists' Warning to Humanity" (see Chapter 1).

Value-Free Science

Earlier in this chapter we noted that science is not value-free, that scientists are influenced by their social environment. This does not mean that objectivity is not a goal of scientists, but it means we need to recognize that scientists have biases that must be identified explicitly.

Philanthropy, Biotechnology, and Edible Vaccines

The Boyce Thompson Institute (BTI) for Plant Research was founded more than 75 years ago by philanthropist William Boyce Thompson. BTI's broad goal is to improve the understanding of plants for the benefit of the environment and the quality of human life.

In the early 1990s, BTI and other scientists proposed to use biotechnology to engineer tomatoes or bananas to produce human vaccines that could be delivered to people through their food, thereby improving human health. Initially thought to be a convenient, economical way to distribute vaccines in countries where health care systems were not fully developed, the practicalities of using edible material as medicines have caused scientists to rethink this strategy. Initiated in 1999, edible vaccine trials, using a potato-based vaccine for hepatitis B (a virus that causes fatal liver disease), revealed two major challenges:

- difficulty in obtaining a consistent dose of vaccine—because different plants and different fields yield variable amounts of vaccine; and

- concern that vaccine recipients might develop antigen tolerance, which would prevent their immune system from developing the appropriate immune response when exposed to the hepatitis B virus.

Scientists continue to work on plant-based vaccines, including the idea of overcoming the dose consistency problem by freeze-drying transgenic plant products and then making powders or pellets that contain well-defined, consistent dosages of the vaccine. In developing countries, part of the appeal of raising and freeze-drying edible vaccine crops is associated with potential employment and economic benefits. However, clinical groups and pharmaceutical companies have expressed little interest in edible vaccines. Lack of interest means they do not contribute funding, raising another challenge for BTI.

SOURCES: "Edible Vaccines: Not Quite Ready for Prime Time," L. Bonetta, 2002, *Nature Medicine, 8*(2), p. 94; "BTI's Vision and Mission," Boyce Thompson Institute home page, 2003: http://bti.cornell.edu/bti2/bti2_page.taf?page=missionstatement

Perhaps we need to estimate the effects of these biases, too. Controversial issues, from endangered species preservation to genetic engineering (see Enviro-Focus 2) and vehicle emission standards, give rise to conflicts among science, scientists, technology, and society. While it is appropriate for different scientists and other individuals to express their values, science does not permit sloppy or fuzzy thinking. It is necessary to think critically and logically about science and social issues, for without such thinking it is more likely that pseudoscientific (false) ideas of the Earth and how it works will be believed. Pseudoscience constitutes a weak basis for making important environmental decisions with long-term, serious consequences.

Given that science is an open process of continual investigation and advances in knowledge, it is sometimes difficult to determine which scientific ideas will become accepted and which will not. Evidence for ideas and models at the frontiers of science is more ambiguous than for those ideas accepted by the scientific community, but some of these frontier-type ideas will be picked up before they have been verified fully (or discarded). In particular, media reports on scientific issues frequently deal with new discoveries, frontier science, and science beyond the fringe. As potential consumers of such information, we need to be able to analyze both media and scientific reports and decide if they are based on objective interpretation of observations and data or on subjective opinion (see Box 2–1). While expert opinion is valuable, accepting a statement as fact simply because it was made by a scientist is contrary to the nature of science (particularly if that person has not studied the topic as a scientist).

BOX 2-1
THINKING CRITICALLY

With all the competing views and claims about environmental issues, how can we know which ones to believe? How can we determine what piece of evidence or interpretation is valid and what we should do about an issue? Critical thinking skills help us avoid jumping to conclusions by developing a rational basis for systematically recognizing and evaluating the messages that are conveyed. Critical thinking skills include the ability to recognize assumptions, hidden ideas, and meanings; to separate facts and values; and to assess the reasons and conclusions presented in arguments. These are useful tools in our everyday lives, not just in terms of the environment.

Each one of us has used critical thinking skills at some point. A common example is questioning the information conveyed on television ads. What does a particular beverage company mean when it labels its product as "good tasting"? According to whose taste buds? What does the product's "new and improved taste" involve—more salt and sugar? In addition to your "well-being," what motivations are behind the ads? If you have ever asked questions like these, you have used critical thinking skills.

Critical thinking involves questioning and synthesizing what has been learned; it is a deliberate effort to think and plan how to think about or analyze a problem rationally. To do this well, we need to be willing to question authority (because even experts are wrong sometimes); to be open-minded and flexible enough to consider different points of view and explanations; to seek full information about an issue; to focus on the main point(s); to be sensitive to the knowledge, feelings, and positions of people involved in the discussion of an issue; and to take a stand on an issue when the evidence warrants it, remembering that we, too, could be wrong and have to rethink our assessment later.

There are a number of steps we need to practise, and many questions to ask ourselves, if our critical thinking is to be effective:

1. Examine the claims made: On what premise or basis are they made? Is there evidence to support the claims? What conclusions are drawn from the evidence? Are the conclusions true?

2. Identify and clarify use of language: Are the terms used clear and unequivocal or do they have more than one meaning? Is everyone using the same meaning? Are there any ambiguities in the language used? Could those ambiguities be deliberate? Are all the claims true simultaneously?

3. Separate facts and values or opinions: If the claims that are made can be tested, then they are factual statements and should be verifiable by evidence. If claims are made about the worth or lack of worth of something, they are value statements or opinions and may not be verified objectively.

4. Identify the assumptions and determine potential reasons for the assumptions, evidence, or conclusions people present: Does anyone have a personal agenda or "an axe to grind" on a particular issue? Are there gender, racial, economic, or other issues clouding the discussion?

5. Establish the reliability or credibility of an information source: What special knowledge or information do the experts bring? What makes the experts qualified in this specific issue? How can the accuracy, truth, or plausibility of the information they offer be determined? Is information being withheld?

6. Recognize the basic beliefs, attitudes, and values that each person, group, or agency holds: In what ways do these beliefs and values influence the way these people view themselves and the world around them? Are any of these beliefs and values contradictory?

If we use the preceding questions to examine the logic of the arguments people offer about an environmental issue, we should be able to determine what to believe even when facts seem confused and experts disagree. This is not to say that critical thinking is easy, but the skills come with practice and will improve with time. Look for opportunities to think for yourself and use those critical thinking skills.

SOURCE: Adapted from unpublished work by Karen J. Warren, Philosophy Department, MacAlester College, St. Paul, MN. Used by permission.

The Scientific Method

Usually students are informed that the series of steps scientists take to carry out their research is called the **scientific method**, and that it consists of steps similar to those in Figure 2–1. However, it is important to realize that not all research fits into such a neatly defined, step-by-step process. While discovering general principles about how the world works is quite often a result of inductive reasoning, accidental discoveries (serendipity), creativity, and flashes of insight also play roles in advancing knowledge and understanding.

The myth of a single scientific method dies hard, even though the disciplines of science undertake research differently. For instance, chemists or physicists use a different research logic to guide their research than biologists or geographers. Even within single disciplines, evolutionists conduct their research differently than ecologists, and hydrologists work differently than geomorphologists. In reality, there are many methods of science rather than one scientific method.

Complexity, Values, and Worldviews

In science, the more complex a system or problem being studied, the less certain are the hypotheses, models, and theories used to explain it. When we combine this fact

a. Common Steps in the Scientific Method

1. Observe and develop a question about your observations.
2. Develop a hypothesis—a tentative answer to the question.
3. Design a controlled experiment or model defining independent and dependent variables to test your hypothesis.
4. Collect data and record it in an organized manner (such as a table or graph).
5. Interpret the data.
6. Draw a conclusion from the data.
7. Compare your conclusion with your hypothesis to determine whether your results support or disprove your hypothesis.
8. If you accept your hypothesis, conduct further tests to support it.
9. If you reject your hypothesis, make additional observations and construct a new hypothesis.

b. Feedback Processes in Scientific Investigation

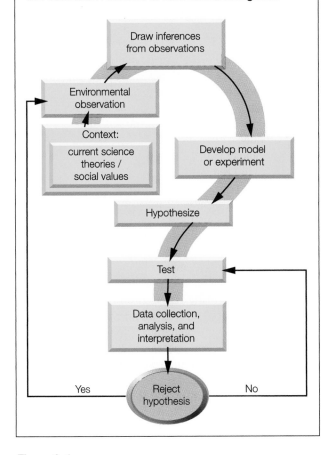

Figure 2–1
The scientific method

with the knowledge that most environmental problems involve complex mixtures of data (or lack of data), hypotheses, and theories in the physical and social sciences, we realize that we really do not have sufficient information to understand them well or fully. It is in this context that advocates of any particular action (or inaction) regarding an environmental issue or problem can use incomplete information to support their beliefs, claiming scientific support for their perspective. Alternatively, an insistence that we fully understand a problem before taking action may lead to "paralysis by analysis." If we think this way, the inherent limitations of science and the complexities of environmental problems force us into an irresolvable situation.

Since environmental problems are not about to go away, there comes a point when people have to evaluate available information and make a political or economic decision about what to do (or not do). Often, these decisions are based on intuition, values, or common sense (claimed by all sides, of course!). This explains why we find different values and worldviews at the heart of most environmental controversies—people with different worldviews and values can take the same information, examine it in a logically consistent fashion, and come to completely different conclusions. (Worldviews, values, and ethics are examined later in this chapter.)

SCIENCE AND ENVIRONMENTAL DECISION MAKING

Just as we can identify a series of steps in the scientific method, we can portray the process of making decisions about the environment as a series of steps (see Table 2–1). While this procedure can help guide rational decision making, it is rather simplistic. In real-world environmental issues, for example, both social and scientific data frequently are incomplete and their interpretation sometimes is controversial. Consequences of particular courses

TABLE 2–1
A SIMPLIFIED DECISION-MAKING PROCESS

Steps in Environmental Decision Making

1. State the issue as clearly and concisely as possible.
2. Research the issue, gathering pertinent social and scientific data and information.
3. Identify all possible courses of action (including alternatives developed through public participation processes).
4. Predict the outcome of each course of action with reference to the positive and negative consequences of each.
5. Predict the probability of occurrence of each course of action.
6. Evaluate the alternatives and choose the most sustainable one (followed by implementation).

Photo 2–6
Environmental decision making should incorporate all possible worldviews and values. Matthew Coon-Come spoke for the Cree during the James Bay hydroelectric development hearings.

of action are very difficult to anticipate, and unintended consequences are even more difficult to envision, particularly if decisions are based on a personal perspective or only on the economic bottom line. The trend to incorporate public participation processes in environmental decision making may appear to accentuate the conflicting, often emotionally charged, interests of different groups. Often, however, highly effective decisions result after stakeholders have been given an opportunity to build a consensus about appropriate courses of action (assuming implementation strategies are operationalized).

While there are no easy answers here, it is essential to develop sound, critically evaluated approaches to environmental decisions that include sustainability principles and that reflect a commitment to environmental democracy (Mason, 1999). In practice, environmental democracy recognizes the basic right of humans to a safe and healthy environment and incorporates "an attitude of respect towards other living things, nonviolence and global forms of democratic self-understanding" (Mason, 1999, p. 51). Environmental and social sciences have key roles in positively promoting societal awareness and understanding of the importance of making environmentally sound decisions. For instance, scientific research has identified a range of health effects of human exposure to toxic substances. Governments have used these data to make difficult decisions about regulating human exposure to hazardous substances. This is a good example of how

addressing environmental issues from scientific and societal perspectives can be a powerful means of achieving mutual gain as well as educating citizens about their common interests in accepting responsibility for the social and environmental costs of their consumption.

In Canada, governments play important roles in science and environmental decision making through involvement in international research agreements, and through domestic support of various research agencies, research councils, Crown agencies, and private-sector interests. One example is the Technology Partnerships Canada (TPC) program supported by the Ministry of Industry. A technology investment fund that contributes financial support to environmental technologies, TPC helps to develop sustainable alternatives, prevent pollution via clean process technologies, and devise pollution remediation technologies and methodologies.

Since its inception, TPC has invested almost $300 million in 61 different environmental technology initiatives. It is interesting to note, however, that environmental technologies represent the smallest proportion—just over 14 percent—of the $2 billion invested to date (Technology Partnerships Canada, 2003a). One partnership, with Vancouver-based Westport Innovations, Inc., supports the research and development of high-performance, low-emission engine fuel systems (see Chapter 11), designed to displace diesel fuel with natural gas or hydrogen in order to reduce carbon dioxide emissions by up to 25 percent (Technology Partnerships Canada, 2003b). TPC also has invested in Premier Tech of Rivière-du-Loup, Quebec, to advance its peat-based biofiltration wastewater treatment systems. International interest in the potential of such technologies illustrates how scientific research can provide information that contributes directly to environmentally sound decision making.

Even though government downsizing has reduced monetary support for environmental research, government agendas still drive many research programs. Since most environmental policy decisions are made through the political process, decisions generally are made by political leaders and citizens, and only rarely by the scientists who possess some of the best understanding of environmental problems. This gap emphasizes the need for environmental and scientific education of all those in government and business, particularly policymakers, as well as all citizens.

WORLDVIEWS AND VALUES

Worldviews are "sets of commonly shared values, ideas, and images concerning the nature of reality and the role of humanity within it" (Taylor, 1992, pp. 31–32). Each society's worldview is reflected in and transmitted through its culture. Beliefs, ideas, values, and assumptions about knowledge that each culture transmits help to shape attitudes toward nature and human–environment relationships. These attitudes, in turn, lead to lifestyles and behaviours that may or may not be compatible with natural systems and that may or may not cause environmental problems.

Groups of many political persuasions—from ecofeminists and deep ecologists to advocates of maximum resource development—have adopted the term *sustainable development* as a guiding force in their activities. However, each of these groups operates with a different, sometimes conflicting, worldview. Different worldviews lead to different interpretations of sustainability and, in turn, to different decisions about use of the environment to achieve various goals. The two major competing worldviews that characterize Western society—expansionist and ecological—are described briefly below.

EXPANSIONIST AND ECOLOGICAL WORLDVIEWS

Two approaches to conservation in the early part of the 20th century have evolved into two major competing worldviews that exist today. The first of these, the expansionist or Western worldview, is based on the values of the 18th-century Enlightenment tradition. The newer and still-evolving ecological worldview is based on values of the Counter-Enlightenment and Romantic traditions. A little of the historical nature of both worldviews follows in order that we may understand their contemporary forms and their links to sustainability.

The Expansionist Worldview

The Enlightenment was a period of profound economic, political, and social changes in society. Capitalism, an economic system based on accumulation of personal wealth,

Photo 2–7
Crowded and unsanitary urban conditions in 19th-century Europe were a direct result of the Industrial Revolution.

Photo 2–8
Early European immigrants to North America extensively exploited surrounding natural resources.

gained wide acceptance. Democracy, established through political revolutions in North America and France, asserted the rights of individuals to determine their own destinies through law making, the ownership of property, and the development of resources on private property.

The Industrial Revolution (beginning in England at the end of the 18th century) brought urbanization, accelerated use of resources, and pollution. As workers clustered in industrial areas and became separated physically from direct, daily contact with the land, knowledge of nature was no longer transmitted to succeeding generations. The quality of life in urban areas declined as coal-burning industries polluted the air, as sewage and other wastes were poured into rivers and streams, and as contagious diseases spread quickly in the crowded, unsanitary conditions.

Many Europeans who migrated to North America during this era took their expansionist worldview with them. The roots of this worldview emphasized the following: faith in science and technology to control nature for human ends; belief in the inherent rights of individuals; accumulation of wealth so that material wants could be satisfied and progress could occur; and exploitation of nature and resources to achieve these ends. Arriving in a new land that seemed to have unlimited natural resources, and having the technological means to make maximum use of this resource base, these settlers aggressively exploited their surroundings. With their frontier mentality, European settlers spread across the continent, trying to tame the wilderness. Many of the Native North Americans whose land was taken over and whose cultures were fragmented or destroyed as settlement spread had lived lifestyles based on a very different worldview—a deep respect for the land, its animals, and other resources.

The Ecological Worldview

People who espouse the contemporary ecological worldview have built their opposition to the fundamental assumptions of the expansionist worldview on historical and philosophical traditions from both Western and non-Western sources. To varying degrees, people have accepted concepts from India and China that stress the unity of human life with nature; beliefs from Aboriginal people about the importance of kinship and the relatedness of all life forms; and ideas from the early animistic and mystic traditions of Celtic, Nordic, and Germanic societies (Taylor, 1992).

As well, the Counter-Enlightenment and Romantic thought of the late 18th and early 19th centuries influ-enced the ecological worldview. In particular, the ecological viewpoint protested the Enlightenment assumption that the universe was a great machine that rationalized and mechanized humans and nature and separated them from their intrinsic spiritual value. According to the expansionist worldview at that time, quantities (measurability) mattered, not qualities. Values, emotions, instincts, and all nonmeasurable aspects of the environment were of secondary importance compared with science and reason. Body and mind, and spirit and nature, were separate entities. On the contrary, Romanticism (part of the Counter-Enlightenment position) emphasized the importance of emotions, instincts, and the irrational. Romantics reacted against urbanization and technology and celebrated the world of nature and all that was not artificial. They tried to

BOX 2-2

SIR CLIFFORD SIFTON (1861–1921): FATHER OF CONSERVATION IN CANADA

Clifford Victor Sifton was born in 1861 on a farm about two kilometres east of the tiny village of Arva, just north of London, Ontario. Although little is known of his early childhood, he had an outstanding record at school and won the gold medal when he graduated from Victoria College in Cobourg, Ontario, in 1880.

Ambitious and well educated, Sifton articled with a Winnipeg law firm and was called to the Manitoba bar in 1882. His legal practice in Brandon soon flourished, and he became the city solicitor. Like his father before him, Clifford Sifton entered politics, running as a Liberal candidate in the North Brandon riding. Successful in his bid for the seat, Sifton entered the Manitoba legislature in 1888, rising to become attorney general in the government of Thomas Greenway in 1891.

In January 1896, Sifton masterminded Greenway's victory in the provincial election, and in June 1896, the Liberals under Wilfrid Laurier won the federal election. In late 1896, Laurier appointed Clifford Sifton minister of the interior and superintendent general of Indian affairs in the federal Cabinet.

As minister of the interior, Sifton earned his place in Canadian history for his aggressive promotion of immigration to settle the west. But he also was one of the few public officials who knew about the conservation movement in other countries and realized Canada should do something—even though its resources seemed inexhaustible, they were not unlimited. Almost as soon as he became minister of the interior, Sifton placed forests under federal control and in 1902 created a separate forestry branch of the Department of the Interior. Sifton had also organized the Canadian Forestry Association in 1900.

By this time, conservation had gained national importance in Canada, as it had in the United States. In May 1909, the Canadian government established a Commission of Conservation and appointed Sifton as chairman. Although it had no formal power, the Commission set up a wide range of studies on topics including fisheries, forestry, lands, minerals, game and fur-bearing animals, water and water power, and public health.

In Sifton's day, conservation focused more on efficient management and the best way to exploit resources, rather than on preservation. The forest industry, for example, was governed by only a few, poorly enforced government regulations. The Commission of Conservation was determined to demonstrate that Canada would benefit from sensible conservation regulations in the long term. Sifton believed that forests should be preserved not just for the sake of their beauty but also because, in practical business terms, conservation techniques made economic sense. A successful businessman himself, Sifton argued that private enterprise operated in its own self-interest and that government control was needed to protect the public interest.

Sifton campaigned strongly against free trade and the practice of selling Canada's resources to the United States. In 1910, on behalf of the Commission, Sifton persuaded the Canadian government to veto a U.S. plan to construct a hydroelectric dam on the St. Lawrence River at the Long Sault Rapids above Cornwall. Sifton was convinced the project would lock Canada into permanently supplying power for U.S. industries. Following a similar dispute over another power project, Sifton resigned from the Commission in 1918. Without his leadership, the Commission carried little influence and was abolished by the government in May 1921.

Sifton's public contributions to Canadian life were recognized by King George V, who knighted him on January 1, 1915. Following the First World War, Sifton was no longer in government, but he served from 1924 to 1928 on the Canadian National Advisory Committee on the development of the St. Lawrence Seaway for shipping and hydroelectric purposes. Once again Sifton argued that the international section of the river should not be developed for hydropower until Canadian demand was sufficient to warrant development.

Sir Clifford Sifton died in a New York hospital from the effects of abdominal cancer on April 17, 1929. Two days later, the father of conservation in Canada was buried in Toronto's Mount Pleasant Cemetery.

SOURCES: *Clifford Sifton*, D. J. Hall, 1976, Don Mills, ON: Fitzhenry & Whiteside; "Sir Clifford Sifton," D. J. Hall, 1988, *Canadian Encyclopedia* (2nd ed.), pp. 1999–2000.

unify those elements expansionists had separated—body and mind, and the supernatural and natural (Taylor, 1992).

During this Counter-Enlightenment period, the roles of human emotions, independence, and freedom of expression were elevated, sometimes above the claims of reason and science. Poets such as William Wordsworth extolled the values of beauty and tranquillity in nature and denounced artificial and urban realms. Other writers of the 1800s, such as Thoreau and Emerson, promoted individuals' rights to access universal truths through personal communion with nature and criticized the existing political structures for standing in the way of these truths. Writing in the 1860s, when deforestation was the major environmental preoccupation, George Perkins Marsh, a physical geographer, warned of the destructive effects of dominant cultural beliefs and practices on the environment.

Conservation in the Early 20th Century

The conservation movement was a reaction to "the excesses and wastefulness of an expanding industrial society," but by the early 20th century conservationists were viewing the problems from two competing world-views (Taylor, 1992, p. 30). "Wise-management" conservationists such as Clifford Sifton in Canada (see Box 2–2) and Gifford Pinchot in the United States were allied with the expansionist worldview and pitted against the "righteous-management" conservationists such as Americans John Muir and, later, Aldo Leopold and Rachel Carson (see Figure 2–2). Both Leopold and Carson spoke of human responsibility for the Earth, echoing the older Christian concept of **stewardship**. Some of the key per-

spectives of each approach to conservation are presented in Table 2–2. (For a definitive history of the conservation movement, see the book by Hays [1959] listed in the References section of this chapter.)

In addition to the features noted in Table 2–2, members of the wise-management school of preservation made it known that they were not preservationists but promoted sustainable exploitation, in which forests, soils, water, and wildlife could be harvested as if they were renewable crops. In contrast, righteous-management conservationists rejected the expansionist emphasis on viewing the world principally in economic and utilitarian terms. Preservationists believed that, in nature, humans could realize their inner spiritual, aesthetic, and moral sensibilities. Believing that physical nature, particularly wilderness, was a benchmark against which to judge the state of human society, the preservationists suggested that large areas of the natural world should be preserved and protected against human interference (Taylor, 1992).

During the period 1900 to 1960, the values of the Enlightenment tradition remained dominant in conservation theory in both Canada and the United States. Nature was seen as a storehouse of resources used to satisfy the continually increasing material needs of an ever-growing population. The expansionist worldview equated material growth with development, which, in turn, was seen as a prerequisite for human happiness and prosperity. Proponents of the expansionist worldview continue to believe that scientific and technological advances will ensure increased global standards of living, employ renewable and other environmentally friendly sources of energy, increase food production, solve the problems created by

---

TABLE 2-2
A COMPARISON OF EARLY-20TH-CENTURY APPROACHES TO CONSERVATION

Expansionist Worldview	Ecological Worldview
"Wise management" is based on the values of the Enlightenment tradition: • Nature is a resource to be used, not preserved. • Conservation must work together with the dominant values of the surrounding society, not against them. • The primary value of natural areas lies in their value to modern society. • Conservation should work against the wastefulness and environmentally disruptive excesses of a developing society. • Conservation is equated with sustainable exploitation.	"Preservation" or "righteous management" is based on the values of the Counter-Enlightenment tradition: • The universe is nondualistic, a totality with all of its parts interrelated and interlocked. • The biotic community and its processes must be protected. • Nature is intrinsically valuable—animals, trees, rock, etc., have value in themselves. • Human activities must work within the limitations of the planet's ecosystems. • Preservation works against the dominant societal values. • Nature provides a forum to judge the state of human society.

SOURCE: Adapted from "Disagreeing on the Basics: Environmental Debates Reflect Competing World Views," D. M. Taylor, 1992, *Alternatives, 18*(3), p. 29. Reprinted courtesy of Alternatives Journal: Environmental Thought, Policy and Action.

Figure 2–2

Key early figures in
20th-century conservation

John Muir
(1838–1914)

• explorer, naturalist, writer
• crusaded for establishment of
  parks and preservation of forests
• instrumental in establishment of
  Yosemite and Sequoia National Parks

Gifford Pinchot
(1865–1946)

• forester and politician
• introduced principles of scientific
  forest management
• worked with F. D. Roosevelt in
  establishing many National Forests

Clifford Sifton
(1861–1929)

• lawyer and politician
• Minister of Interior—created separate
  forestry branch (1902)
• considered father of conservation
  in Canada

founded
Sierra Club

The Yosemite

*The Fight for
Conservation*

estab. Cdn. Forestry Assoc.

| *1890* | 1892 | | 1898 | 1900 | | 1909 | 1910 | 1912 | | 1918 | *1920* | | *1930* |

Chief of U.S. Forest Service

Chair of Commission of Conservation

previous technologies, create substitutes for depleted resources, and replace damaged environments (Taylor, 1992). Although the expansionist worldview originated with capitalism, today both capitalist and socialist countries apply the basic tenets of the expansionist position.

ENVIRONMENTALISM

In the 1960s and 1970s, a reassertion of Counter-Enlightenment and Romantic values occurred within the conservation movement. In fact, during this time of environmental revolution, both Canadian and U.S. governments began to recognize the value of qualitative and eco-centric approaches to conservation. The late 1960s and early 1970s were years of idealism and optimism—environmental awareness (including interest in worldviews held by other cultures) was growing among all segments of the population, major issues such as pollution and nuclear power were receiving media attention (see Table 2–3), and various pieces of legislation respecting environmental protection were promoted. The first Earth Day and environmental teach-in, bringing together scientists, environmentalists, politicians, students, government officials, and citizen groups, was held in the United States on April 20, 1970. In 1971 the Canadian government established the Department of the Environment. Since then, Environment Week in Canada

has been an important way to promote environmental awareness and the benefits of environmental protection among Canadians (see Box 2–3).

Photo 2–9

A windmill and inflatable globe serve as visual props in encouraging ecology at an Earth Day celebration on the lawn of the Capitol Building, Washington, D.C.

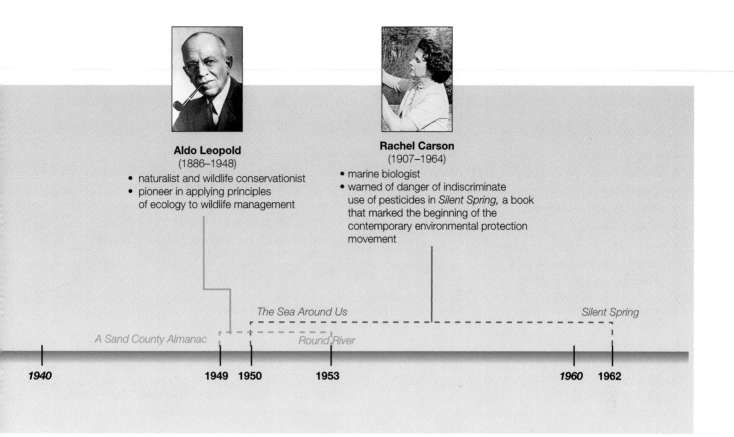

Aldo Leopold
(1886–1948)
- naturalist and wildlife conservationist
- pioneer in applying principles of ecology to wildlife management

Rachel Carson
(1907–1964)
- marine biologist
- warned of danger of indiscriminate use of pesticides in *Silent Spring,* a book that marked the beginning of the contemporary environmental protection movement

The Sea Around Us

Silent Spring

A Sand County Almanac

Round River

1940   1949 1950   1953   1960 1962

BOX 2-3
ENVIRONMENT WEEK IN CANADA

Canadian Environment Week is held in early June each year to coincide with World Environment Day (proclaimed by the United Nations in 1972, and celebrated on June 5). Canada's Environment Week was established through a private member's bill introduced by British Columbia Member of Parliament Tom Goode. Given royal assent in March 1971, Environment Week in Canada provides an opportunity for concerned citizens to focus on environmental issues and to help conserve and protect our country's natural heritage. A key purpose of Environment Week is to urge all Canadians to make activities that preserve the Earth a part of their daily lives.

The theme of Environment Week varies from year to year. For example, the 2004 theme was conserving Canada's natural legacy. This theme includes activities that help ensure the existence of healthy ecosystems with productive and protected habitats, and clean air and water.

Over the more than 30 years of its history, two themes have emerged and are now annual events during Canada's Environment Week: the Commuter Challenge and Clean Air Day. Provincial and municipal jurisdictions, workplaces, and a variety of other organizations issue challenges to citizens, employees, and members to walk, cycle, roller blade, skateboard, use public transportation, and car pool during Environment Week. In 2004, the Commuter Challenge was linked closely to the federal government's One Tonne Challenge. The One Tonne Challenge is an initiative that encourages every Canadian voluntarily to reduce their current GHG emissions by one tonne each. Since half of Canada's GHG emissions come from vehicles, we personally can reduce GHGs significantly by driving less, driving fuel-efficient vehicles, and using ethanol-enhanced fuel. Clean Air Day aims both to increase our awareness of air quality and climate change and to celebrate and promote actions, such as tree planting and recycling, that improve air quality and address climate change (see http://www.climatechange.gc.ca/plan_for_canada/challenge).

Be sure to watch for Environment Week activities in your community. Better yet, become involved personally. To find out more about Environment Week activities, go to your local and provincial government environment websites or to the Canadian Environment Week website: http://www.ec.gc.ca/e-week/index_e.htm.

TABLE 2-3

ENVIRONMENTAL ISSUES AND CHARACTERISTICS OF THE FIRST AND SECOND WAVES OF ENVIRONMENTALISM

Environmental Issues	Characteristics and Emphases
First Wave (1968–76)	**First Wave (1968–76)**
• pollution	• tendency for individuals and groups to alienate themselves, to detach from social, political, and economic order
• energy crisis	
• offshore oil drilling, tanker spills	• anti-technological character
• nuclear power	• tendency to millennialism (escapism)
• population	• regulatory, "end-of-pipe" solutions favoured by decision makers (standards for emissions)
• resource depletion, especially of oil	
• urban neighbourhood preservation	• building awareness of problems
Second Wave (1985 onward)	**Second Wave (1985 onward)**
• global warming	• reemergence of preservationist issues
• ozone depletion	• globalized concerns
• new wilderness and habitat concerns: old-growth forests, tropical rain forests, animal rights	• acceptability of some environmental ideas within economic and political elites
• waste reduction, recycling	• professional character of major environmental organizations
• hazardous wastes, carcinogens, pollution	• split between those inclined to compromise and those opposed
• resource depletion, especially of forests, fisheries, and biodiversity	• multiple tools approach
• oil tanker spills	
• urban planning, automobiles, land use	
• indoor air quality	

SOURCE: Adapted from "Eco-history: Two Waves in the Evolution of Environmentalism," R. Paehlke, 1992, *Alternatives, 19*(1), p. 22. Reprinted courtesy of Alternatives Journal: Environmental Thought, Policy and Action.

However, long-lasting change in understanding nature and the place of humans in the environment did not occur during this first wave of environmentalism (1968–76). When inflation and an economic downturn began to drive the monetary costs of a clean environment upward, many people (especially in economically disadvantaged regions) began to argue that jobs were more important than environmental controls (Paehlke, 1992).

By the 1980s, the idealism of the 1960s and 1970s had given way to an emphasis on individual well-being, especially economic well-being. In both Canada and the United States, eco-centric values in conservation were increasingly difficult to maintain as the idea of limits to growth was rejected, eco-centric values were pitted against development values, and economic goals took precedence in environmental and resource management decisions (Taylor, 1992). For example, in 1987, when *Our Common Future* (the Brundtland Report) was published and nations were challenged to develop sustainable development strategies, the Canadian government replied "within the framework of the Expansionist" worldview (Taylor, 1992, p. 28). Although neoconservative or expansionist values were dominant among federal politicians, poll after poll showed North Americans wanted to continue efforts to ensure a clean, safe environment for their children and themselves (see Dunlap, 1987).

This latent interest in environmental matters and concern for the state of the environment generated a second wave of environmentalism from 1985 onward. Not only were Counter-Enlightenment and Romantic values back again, but this time deep ecology and sustainable development entered the debate (Paehlke, 1992; Taylor, 1992). As illustrated in Table 2–3, issues of concern in the second wave of environmentalism often were global in perspective (particularly global warming and ozone depletion), and frequently dealt with nature, wilderness, and biodiversity. Earth Day 2000, the 30th anniversary event, was celebrated worldwide as a global expression of support for environmental action. The theme of Earth Day 2003 was "who says you can't change the world?" During 2003–2004,

The World Water Council

Established in 1996, and headquartered in Marseilles, France, the World Water Council (WWC) labels itself an international think tank dedicated to "strengthening the world water movement for an improved management of the world's water resources and water services." With more than 300 members, including industry, government agencies, and nonprofit organizations, the WWC's mission is to "promote awareness, build political commitment and trigger action on critical water issues at all levels, including the highest decision-making level to facilitate the efficient conservation, protection, development, planning, management and use of water in all its dimensions on an environmentally sustainable basis for the benefit of all life on earth" (World Water Council, n.d.).

Formed because of the serious challenges facing the world's freshwater resources, including growing water scarcity, deteriorating water quality, and increasing difficulty in accessing clean and reliable sources of water for drinking and food production, the WWC is concerned about the lack of an integrated global institutional framework to deal efficiently and effectively with common world water problems (Shady, 1996). Through its World Water Forum (held every three years) and other activities, the WWC attempts to provide advice and information to "institutions and decision makers on the development and implementation of comprehensive pro-poor policies and strategies for sustainable water resources and water services management...." (World Water Council, n.d.).

In an entry in its *Disinfopedia,* the Centre for Media and Democracy notes there are claims that the WWC "has an industry bias and that it advocates pro-privatization water policies...." By using the term "public-private partnerships" instead of "privatization," the WWC is accused of promoting "the illusion that leaving water supply to transnational corporations is the only way forward. In fact the experience of the last decade has shown that these profit-seeking corporations are incapable of delivering water to the world's poorest" (Centre for Media and Democracy, 2003). CTV's recent documentary, which included discussion of the challenges associated with a transnational corporation's "privatization" of the water supply system in Saint John, NB, suggests this concern may be valid. If nothing else, such opposing characterizations of the WWC identify the importance we need to place on challenging and carefully examining the arguments and evidence pertaining to each of these perspectives.

World Water Day

Established in 1992 by the United Nations General Assembly, World Water Day provides an opportunity to raise awareness of global water challenges. In recognition of World Water Day, March 22, 2003, seven people from different UN agencies, including UN University, wrote an editorial that began with bleak messages about water shortages, contaminated water, waterborne diseases, and the prediction of "impending calamity." However, they also offered a message of hope that

- the cooperation evidenced in watershed management will continue
- investment will continue in water treatment facilities that have helped formerly "dead" rivers in North America and Europe to support fish spawning and migration
- improvements in sanitation systems in Asia (that provided clean water to 220 million people in 1990) will be extended to the 800 million people still without safe and healthy supplies of drinking water

These authors noted clearly that the world does not need any more declarations on how to achieve sustainability. Rather, what is needed is action on these declarations, such as the Plan of Implementation established at the 2002 World Summit on Sustainable Development. Responsible stewardship to safeguard the world's freshwater resources will require integrated management at all levels, from the individual to the international.

SOURCES: "World Water Council," *Disinfopedia,* Centre for Media and Democracy, 2003, http://www.disinfopedia.org/ wiki.phtml?title=World_Water_Council; "A Glass Half Empty? An editorial by United Nations Agencies and UN University for World Water Day, 22nd March 2003," N. Desai, M. M. Brown, K. Toepfer, K. Matsuura, A. Tibaijuka, C. Bellamy, & H. van Ginkel, 2003, http://www.waterday2003.org/OPED.htm; "The World Water Council Is Now Official," A. Shady, 1996, *Water News, 15*(2), 10; *About WWC,* World Water Council, n.d., http://www.worldwatercouncil.org/about.shtml

Earth Day's campaign highlighted critical water issues such as water access, health, and usage. This campaign works in concert with World Water Day and World Water Council concerns (see Box 2–4) to improve stewardship of local, national, regional, and global water resources.

Deep Ecology, Sustainable Development, and Green Alternatives

Neither *deep ecology* nor *sustainable development* were familiar terms prior to the second wave of environ-

mentalism. Deep ecology states that humans are only one species among many, and that nature and nonhuman species are as valuable in their own right as are humans (Devall & Sessions, 1985). These assertions are labelled, respectively, the *principle of self-realization* (an awareness of one's ultimate inseparability and wholeness with the nonhuman world) and the *principle of biocentric equality* (all organisms and entities in the ecosphere have intrinsic worth and are part of the interrelated web of life). In the deep ecologist's view, at least implicitly, these principles mean that sometimes wild nature must

Photo 2–10

The long-standing protest at Greenham Common, in England, against U.S. nuclear missile storage may be seen as an early expression of ecofeminism. Following decommissioning of the U.S. nuclear facilities, area residents formed Greenham Common Trust and purchased the area in 1997. The Common has been restored and reopened to the public to enjoy again.

be chosen over human habitat and human well-being. In contrast, sustainable development gives priority to global human needs (Paehlke, 1992).

Throughout North America and Europe, other green alternatives have developed that criticize the assumptions underlying modern society. Like deep ecology, ecofeminism is based on the biocentric equality principle but also tries to address the "hierarchical and dominance relationships that it sees as endemic to patriarchy" (Taylor, 1992, p. 30). Ecofeminism argues that the ongoing domination of nature and the ongoing domination of women are systemically related (Salleh, 1984; Hessing, 1993). Similarly, social ecologists argue that all forms of human domination are related directly to the issue of ecology. That is, as long as hierarchy and domination occur in human society, then the domination of nature will continue and lead the planet to ecological extinction (Bookchin, 1980).

Ecological worldview adherents also have found the study of general systems theory to be useful in that its view of the universe as a systemic hierarchy, or as organized complexity, pictures a "myriad of wholes within wholes, all of them interconnected and interacting" (Taylor, 1992, p. 31; Prigogine & Stengers, 1984). This perspective has appeared in James Lovelock's Gaia hypothesis, which explained the Earth and its living organisms in terms of a single, indivisible, self-regulating process (Lovelock, 1988). Lovelock's theory provided impetus to the concern that human expansionist activities at local levels were threatening Gaia's health and all the life contained on the planet.

Environmentalism is, primarily, a social movement; it embodies numerous environmentalist ideologies (as above) and organizational diversity (as you will note throughout this book). The major contribution that environmentalism has made to sustainability has emanated from efforts of numerous people to have nonhuman species and the interests of future generations included on political agendas and within democratic decision-making processes (Mason, 1999). In challenging existing models of democratic politics, environmentalism has created "tension between democratic means and green ends ..." (Mason, 1999, p. 31). Since ecological degradation affects everyone, we all share a common interest in addressing environmental problems. How society prioritizes these environmental interests that constitute the basis of our physical survival and well-being, and how society includes human rights to a healthy and safe environment, are part of the challenge of achieving environmental democracy (see the section "Science and Environmental Decision Making" earlier in this chapter).

Toward the Future

The second wave of environmentalism brought about remarkable changes. Environmental ideas are now widely promoted within the North American and international political elites. Today, blue boxes, composting, and other

Photo 2–11

Increasing awareness and changing attitudes result in a broader acceptance of environmental initiatives, such as the Blue Box recycling campaign.

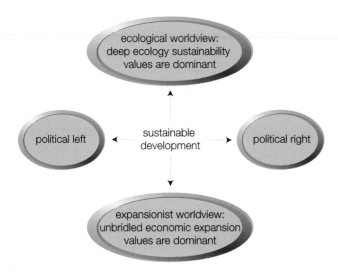

Figure 2–3

Worldviews in the political sphere at the turn of the 21st century

SOURCE: After "Eco-history: Two Waves in the Evolution of Environmentalism," R. Paehlke, 1992, *Alternatives, 19*(1), p. 22. Reprinted courtesy of Alternatives Journal: Environmental Thought, Policy and Action.

recycling programs generate economic returns for municipalities. In the business world, being perceived as environmentally concerned is important to a company's corporate image and perhaps to business success. Unfortunately, in some instances these changes reveal that the concept of sustainable development has been "co-opted by individuals and institutions to perpetuate many of the worst aspects of the expansionist model under the masquerade of something new" (Taylor, 1992, p. 32). However, sustainability does carry the hope that society will be able to transform its political, economic, and social institutions in keeping with what is socially and environmentally sustainable.

The expression of worldviews in the political sphere of the 1990s is illustrated in Figure 2–3. The horizontal axis shows that political opinions held by environmentalists or nonenvironmentalists can be either left or right. The vertical axis places ecological (sustainability) and expansionist (economic) worldviews at opposite ends, with sustainable development somewhere in the middle. There is tension between these ecological–sustainable development–expansionist values and ways of looking at the world; the challenge for the future is to identify and reach win-win conditions that will allow simultaneous improvement in equity, economy, and ecology.

The term "sustainable development" provides an opportunity for a dialogue between enthusiasts of economic growth, environmentalists, and advocates of greater equity among peoples and within nations.

However, the real challenge lies in integrating new ideas for ecological protection and sustainability with a workable set of transformations of society and economy (Paehlke, 1992; Mason, 1999). For many people, current environmental problems reflect a cultural or worldview crisis, a concern for current national and international policies and values as they affect the long-term viability of social and natural systems. Thus, understanding the struggle between the entrenched expansionist worldview and the emerging ecological worldview is important in helping us move toward both environmental and societal sustainability in the future.

Making decisions about our environment and planning effectively to meet the needs of the future require us to place a value on all aspects of our environment. As we attempt to deal with the exploding human population, meet the needs of an urban world, sustain resources for future generations, and preserve our environment on a global level, we need to know how we are changing the environment, how we can rectify the problems we have caused, and what factors are most important to us. To these ends, the following section provides a brief discussion of environmental values and environmental ethics.

Photo 2–12

Family fishing endeavours in Canada's coastal communities have suffered as a result of declining fish stocks.

ENVIRONMENTAL VALUES AND ETHICS

ENVIRONMENTAL VALUES

Placing a value on some aspect of our environment—for example, healthy fish populations in Canada's rivers and oceans, clean air in industrialized regions of the country, scenic beauty, or the preservation of natural landscapes—may be based on utilitarian, ecological, aesthetic, or moral categories. The utilitarian and ecological categories deal with practical reasons such as economic benefit or our own survival. A **utilitarian justification** for the conservation of nature states that the environment, ecosystem, habitat, or species provides individuals with direct economic benefits or is directly necessary to their survival. Fishers, for example, derive their livelihood from the oceans and need a supply of fish so that they may continue to earn their living.

An **ecological justification** for conserving nature is based on the knowledge that a species, ecological community, ecosystem, or the Earth's biosphere provides specific functions necessary to the persistence of our life. The ability of trees in forests to remove carbon dioxide produced through burning fossil fuels is a public benefit and an important element in the argument to maintain large areas of forest. Enlightened self-interest also would suggest that dealing with the problems of polluted air in parts of Eastern Europe (caused by burning lignite and poor-quality coal) would be beneficial ecologically. Similarly, even if individuals did not benefit immediately, there is value in action designed to counter the production of greenhouse gases that may lead to climate change affecting the entire Earth.

Aesthetic arguments for the conservation and protection of nature are made on the basis that nature is beautiful and that beauty is of profound importance and value to people. In an effort to add beauty to their surroundings, many people spend hours gardening, and city workers plant trees, shrubs, and flowers in parks and on boulevards. On a larger scale, many people find the Canadian wilderness beautiful and would rather live in a world with wilderness than without it. Psychological, medical, and social benefits accrue from the aesthetic values of the environment. For example, research has shown that patients recover more quickly if their hospital room has a view of trees and natural landscapes (Krakauer, 1990). The value of natural sounds and areas as restorative environments (Hartig, Mang, & Evans, 1990) and of wilderness as sacred space (Graber, 1976) are examples of how important the environment's aesthetic values are.

A **moral justification** for conserving nature is that aspects or elements of the environment have a right to exist, independent of human desires, and that it is our moral obligation to allow them to continue or to help them persist. An example of moral justification is the assertion that the Fraser River in British Columbia and the Red Deer River in Alberta (examples of the few remaining "wild" rivers in populated areas of Canada) have the right to exist as free-running waters. Similar moral arguments have been extended to many nonhuman organisms such as trees and wildlife. For instance, in 1982 the United Nations General Assembly World Charter for Nature stated that species have a right to exist. From a human perspective, a strong moral claim for working toward ecological sustainability exists because humans have rights to a healthy environment as well as rights to participate in environmental decision-making processes.

A new discipline, environmental ethics, analyzes these issues and the concerns about our moral obligations to future generations with respect to the environment. Environmental ethics is introduced briefly in the following section.

ENVIRONMENTAL ETHICS

During the 1970s, philosophers began to develop **environmental ethics**, a field that studies the value of the physical and biological environment. Because it is a large and complex academic subject, only a few aspects of the field are considered here.

The need for new environmental ethics has arisen because of the diverse changes that human activities and technology are bringing to the world. For instance, humans are having new effects on nature, are developing new knowledge of the environment, and are creating an expanded set of moral concerns. Environmental ethics indicates that an examination of the utilitarian, ecological, aesthetic, and moral consequences of these developments is necessary.

With regard to our new knowledge of nature, science and social science have been able to show us how we have changed our surroundings in ways that were not understood previously. For example, scientists have demonstrated that the burning of fossil fuels and the large-scale clearing of forests have changed the amount of carbon dioxide in the atmosphere, which may change the global climate. This global perspective provides an impetus to examine new moral issues. Also, the extension of moral and legal rights to animals, trees, and objects such as rocks is seen as a natural expansion of civilization as it begins to incorporate the environment in ethics and politics.

Concern with environmental ethics involves discussion of the rights of animals and plants, of nonliving things, and of large systems that are important to our life

Photo 2–13
The use of living animals in scientific experimentation generates controversy. Is their use justifiable?

support. One important statement of environmental ethics is Aldo Leopold's land ethic. In *A Sand County Almanac*, Leopold (1949) affirmed that all resources (plants, animals, and earth materials) have a right to exist, to continue to exist, and to continue to exist in a natural state in at least some locations. This land ethic indicates that humans are no longer conquerors of the land but are citizens and protectors of the environment. As citizens and protectors, humans should show love and reverence in their relationships with the land; land is not merely an economic commodity to be used up and discarded. Leopold's land ethic assumes that we are ethically responsible to other individuals and society, as well as to the larger environment of which we are a part (including plants, animals, soils, the atmosphere, and water). Note that such responsibility places some limits on the freedom of individuals and societies in their struggle for existence.

Leopold's land ethic suggests that each of us is a steward of our environment; our role as stewards includes the moral responsibility to sustain nature for ourselves and for future generations. This philosophy has certain implications; for example, although the land ethic assigns rights to animals to survive as a species, it does not necessarily assign those same rights of survival to an indi-

vidual member of a species (a deer or chicken, for example). This means we must distinguish between an ideal and a realistic land ethic. Another implication is that, because the wilderness has intrinsic value, we, as morally responsible stewards, must maintain it for itself and because our own survival depends on it. Whether or not we agree with the land ethic, we need to consider whether ethical values should be extended to nonhuman biological communities. Our position will depend on our values and our understanding of natural systems and other environmental factors.

Given that human effects on today's environment have consequences for the future, any discussion of environmental ethics also involves the rights of future generations and what we owe them. This issue has become increasingly important because the impacts of technology have the potential to affect the environment for hundreds or thousands of years to come. Radioactive waste from nuclear plants, long-term climate changes resulting from land use changes and technology, extinction of large numbers of species as a result of human activities, and the direct effects of human population increases are among the issues of concern for the future. Depending on what we know about these issues, we will make value judgments about them, about the rights of future citizens, and about the idea of stewardship of the Earth. If we think it is important to consider the future in our decision making, then we are more likely to consider ourselves as merely the latest in a long line of humans who are the stewards of the Earth.

In response to the potentially major changes that these issues imply, some people have reacted by "hiding their heads in the sand," seeking a simpler life, and rejecting all science, technology, and progress. A more useful, longer-term response would be to use science and technology to the best of our abilities, keeping in mind our environmental ethics. Ernest Partridge (1981), a philosopher concerned with environmental ethics, has commented that scientific knowledge and discipline need to be augmented by a critical moral sense and passionate moral purpose if we are to ensure a better future.

Ultimately, there are many ways in which each of us can become involved in environmental ethics. Environmentalism encompasses a wide range of approaches, from the conservative to the radical, to making a difference in the world around us. In simple ways, every one of us can address environmental issues personally by adjusting our lifestyles and attitudes toward consumerism. By using the best available scientific instruments and methods, and being aware of the ethical implications of our actions, we can better understand and contribute to the appropriate management of our environment and its resources.

Chapter Questions

1. What is the scientific method? Why is it appropriate to think of it as a general guide to scientific thinking?

2. Scientific knowledge may be acquired through inductive and deductive reasoning; identify the ways in which these processes are similar and different.

3. Discuss the value of critical thinking in the environmental science field.

4. Select a current environmental controversy and identify the social, economic, aesthetic, and ethical issues it raises.

5. What are the most important environmental benefits and harmful conditions passed on to you by previous generations? What obligations, if any, do you have to future generations? For how many generations do your responsibilities extend?

6. Find an article from a newspaper about a controversial topic and make a list of any ambiguous or loaded words (words that convey an emotional reaction or value judgment) used in the article. In your opinion, is the article a systematic one or a personal opinion piece? Why or why not?

references

Bookchin, M. (1980). *Toward an ecological society.* Montreal: Black Rose Books.

Botkin, D. B., & Keller, E. A. (1995). *Environmental science: Earth as a living planet.* New York: John Wiley & Sons.

Devall, B., & Sessions, G. (1985). *Deep ecology: Living as if nature mattered.* Salt Lake City, UT: Peregrine Books.

Dunlap, R. E. (1987, July/August). Public opinion on the environment in the Reagan era. *Environment, 29,* 7–11, 32–37.

Graber, L. H. (1976). *Wilderness as sacred space.* Washington, DC: Association of American Geographers.

Hartig, T., Mang, M., & Evans, G. W. (1990). Perspectives on wilderness: Testing the theory of restorative environments. In A. T. Easley, J. F. Passineau, & B. L. Driver (Compilers), *The use of wilderness for personal growth, therapy and education* (pp. 86–95). Fort Collins, CO: U.S. Department of Agriculture, Rocky Mountain Forest and Range Experiment Station, General Technical Report RM-193.

Hays, S. (1959). *Conservation and the gospel of efficiency.* Cambridge, MA: Harvard University Press.

Hessing, M. (1993). Women and sustainability: Ecofeminist perspectives. *Alternatives, 19*(4), 14–21.

Krakauer, J. (1990). Trees aren't mere niceties—they're necessities. *Smithsonian, 21,* 160–171.

Leopold, A. (1949). *A Sand County almanac.* New York: Oxford University Press.

Lovelock, J. (1988). *The ages of Gaia: A biography of our living Earth.* New York: Norton.

Mason, M. (1999). *Environmental democracy.* London: Earthscan.

Myers, N. (1990). *The Gaia atlas of future worlds: Challenge and opportunity in an age of change.* Toronto: Doubleday.

Paehlke, R. (1992). Eco-history: Two waves in the evolution of environmentalism. *Alternatives, 19*(1), 18–23.

Partridge, E. (1981). *Responsibilities to future generations: Environmental ethics.* Buffalo, NY: Prometheus Books.

Prigogine, I., & Stengers, I. (1984). *Order out of chaos: Man's new dialogue with nature.* Toronto: Bantam Books.

Salleh, A. K. (1984). Deeper than deep ecology: The eco-feminist connection. *Environmental Ethics, 6*(4), 65–77.

Taylor, D. M. (1992). Disagreeing on the basics: Environmental debates reflect competing world views. *Alternatives, 18*(3), 26–33.

Technology Partnerships Canada. (2003a). http://strategis.ic.gc.ca/SSG/tp00170e.html

Technology Partnerships Canada. (2003b). http://strategis.ic.gc.ca/SSG/tp00331e.html

Yoon, C. K. (1996, August 28). Peregrine migration secrets unfold. *The Globe and Mail,* p. A6.

PART 2

The Ecosphere We Live In

"All life exists in a thin layer wrapped around the globe, caught between the molten heat of the earth's interior and the cold immensities of space. The biosphere, the only part of the entire universe known to support life ... proportionately is no thicker than the shine on a billiard ball."

Lean & Hinrichsen (1992, p. 11)

Chapter Contents

Chapter Objectives

After studying this chapter you should be able
to

- identify and describe the Earth's major
  components

- outline the components and structure of
  ecosystems

- discuss ecosystem functions and their interconnections

- explain how ecosystem population
  dynamics work

- identify the major forces of change and
  adaptation affecting the Earth

- identify the key features of living systems
  that help humans learn to live sustainably

INTRODUCTION

The environmental problems and challenges we face have no simple solutions. However, improving our understanding of how the world works, and applying that knowledge, may help us make decisions that are directed toward sustainable environments and futures for ourselves, our communities, and the enveloping ecosystem of which everyone is a part. With that broad goal in mind, this chapter focuses on some fundamental features and vital interconnections of Earth's life-support systems.

MATTER AND ENERGY: BASIC BUILDING BLOCKS OF NATURE

MATTER

Matter is the material of which things are made, the stuff of life. Everything on Earth is composed of matter— everything that is solid, liquid, or gaseous, including our bodies, the air we breathe, oceans or lakes we fish or swim in, animals we see grazing, vegetables grown in our gardens or farmers' fields, and minerals extracted from the Earth. Matter is anything that has **mass** (weight) and takes up space. Essentially, the Earth is a **closed system** for matter. With the possible exception of meteors and meteorites that enter the Earth's atmosphere and add matter to the biosphere, most of the matter that will be incorporated into objects in future generations already is present and has been present since the planet came into being.

Scientists note that Earth's matter has two chemical forms: **elements** (the simplest building blocks of matter that make up all materials) and **compounds** (two or more different elements held together in fixed proportions by the attraction in the chemical bonds between their atoms). All matter is built from the 109 known chemical elements (92 naturally occurring and 17 synthetic; see Appendix A). While each element has its own unique atomic structure, elements can combine to form a seemingly boundless number of compounds. So far, chemists have identified more than 10 million compounds.

Scientists also tell us that all elements (and all matter) are composed of three types of building blocks: atoms, ions, and molecules. **Atoms** are the smallest particles that exhibit the unique characteristics of a particular element. In turn, atoms consist of subatomic, electrically charged particles known as **ions**. These ions are of three types: **protons** (which are positively charged), **neutrons** (uncharged or electrically neutral), and **electrons** (negatively charged) (see Figure 3–1a). A set number of protons and neutrons, which have approximately the same mass, cluster in the centre of the atom and make up its nucleus. Electrons, which have little mass compared with protons and neutrons, continually and rapidly orbit the nucleus; they are held in orbit by attraction to the positive charge of the nucleus.

An atom of any given element, say hydrogen, is distinguished from that of other elements such as carbon or oxygen by the number of protons in that atom's nucleus (called its **atomic number**). Hydrogen, which is the simplest element, has only one proton in its nucleus and has an atomic number of 1. Carbon, in contrast, has six protons and an atomic number of 6. An even larger atom, uranium, has 92 protons and an atomic number of 92.

Scientists describe the mass of an atom in terms of its **mass number**, which is the number of neutrons (n) plus the number of protons (p) in its nucleus. All atoms of a particular element have the same number of electrons and protons but they may have different numbers of neutrons, which change the mass or weight of the atom (Figure 3–1b). These different forms of the same atom are called isotopes of that element, and they may exhibit dif-

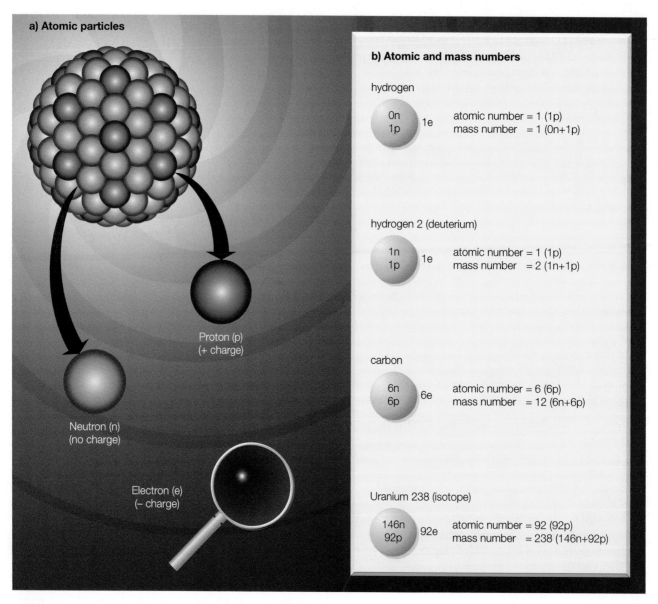

Figure 3–1
Atomic structure

ferent characteristics (such as radioactivity in the case of the isotopes of hydrogen and uranium).

Molecules are formed when two or more atoms of the same or different elements combine. In nature, some elements are found as molecules, including oxygen (O_2), nitrogen (N_2), and hydrogen (H_2). In chemical notations such as these, O alone would represent one atom of oxygen; 2O would represent two atoms of oxygen that had joined to form a molecule of oxygen. Molecules composed of two or more different elements are known as compounds. When we see the **chemical formula** H_2O, we know that this molecular compound, water, is formed of two hydrogen atoms chemically bonded to one oxygen atom. Similarly, glucose sugar, $C_6H_{12}O_6$, is a compound formed of 6 carbon atoms, 12 hydrogen atoms, and 6 oxygen atoms. An ionic compound such as sodium chloride (table salt) consists of oppositely charged ions (Na^+ and Cl^-) held together by the forces of attraction between opposite electrical charges.

As noted previously, most matter on Earth exists as compounds, both organic and inorganic. **Organic compounds** contain atoms of the element carbon and usually are combined with each other and with atoms of one or more other elements such as hydrogen, oxygen, nitrogen, sulphur, phosphorus, chlorine, and fluorine. Among the compounds that are important in our lifestyles are vitamins and aspirins, plastics and detergents, oil and natural gas, table sugar, and penicillin. Since carbon is organic, both CO and CO_2 are organic compounds.

Hydrocarbons, chlorinated hydrocarbons, chlorofluorocarbons (CFCs), and simple sugars are examples of different types of organic compounds. Hydrocarbons are compounds of carbon and hydrogen atoms. Methane (CH_4), for example, is the major component of natural gas. Chlorinated hydrocarbons are compounds of carbon, hydrogen, and chlorine. They are used in materials such as the insecticide DDT ($C_{14}H_9C_{15}$) and PCBs such as $C_{12}H_5C_{15}$, which have been used as insulation in electrical transformers. Freon-12 ($CC_{12}F_2$), a coolant used in older refrigerators and air conditioners, is an example of a chlorofluorocarbon, a compound of carbon, chlorine, and fluorine atoms. Simple sugars such as glucose ($C_6H_{12}O_6$) are also organic compounds that most plants and animals break down in their cells to obtain energy.

All other compounds, including water (H_2O) and sodium chloride (NaCl), are **inorganic compounds**. Examples of inorganic compounds you will encounter in this book include ammonia (NH_3), carbon monoxide (CO), carbon dioxide (CO_2), nitric oxide (NO), nitrogen dioxide (NO_2), sulphur dioxide (SO_2), and sulphuric acid (H_2SO_4).

Matter Quality

The term "matter quality" reflects the relative value that humans place on the resources they use; this term does not reflect the value of the contribution that a resource makes to an ecosystem. Based on its availability and concentration, people classify matter as being of high or low quality. **High-quality matter** (such as coal and salt deposits) is usually found near the Earth's surface in an organized or concentrated form, so that its potential for use as a resource is great. **Low-quality matter** usually has little potential for use as a resource because it is dispersed or diluted (in the oceans or atmosphere) or hard to reach (deep underground). An aluminum can is a more concentrated, higher-quality form of aluminum than the aluminum ore from which it was derived. That is why it is less costly in terms of energy, water, and money to recycle an aluminum can than to produce a new one from aluminum ore.

ENERGY

Energy, the ability or capacity to do work, is what enables us to move matter (such as our arms or legs or a basketball) from one place to another or to change matter from one form to another (such as to boil water to produce steam, or to cook food on a barbecue using natural gas). Forms of energy include light, heat, electricity, chemical energy in coal and sugar, moving water and air masses, and nuclear energy from isotopic nuclei. Among other things, we use energy to build, heat, and cool our homes and businesses, to process and transport food, and to keep our body's cells active and functioning properly.

Energy is either kinetic or potential. Matter has **kinetic energy** because it moves and has mass. Wind (a moving air mass), for example, has kinetic energy, as do flowing streams, moving cars, heat, and electricity. Forms of electromagnetic radiation such as visible light, microwaves, ultraviolet radiation, and cosmic rays also are types of kinetic energy. **Potential energy**, stored and potentially available for use, includes the chemical energy stored in gasoline molecules and food molecules, and in water stored behind a dam. Burning gasoline in a car engine changes the chemical bonds of its molecules into heat, light, and kinetic (or mechanical) energy that moves the cars.

Energy Quality

Energy quality is a measure of energy's ability to perform useful work. **High-quality energy** derives from natural sources such as high-velocity wind and coal, or is generated by using other forms of high-quality energy. For example, energy is produced from concentrated sunlight, natural gas, gasoline, or nuclear fission (uranium). Such concentrated energy sources have great utility in industrial processes and in running electronic devices, lights, and electric motors. In contrast, **low-quality**

Photo 3–1a

Photo 3–1b

Generating power from water or wind sources incurs different environmental effects. Knowledge of such effects is important in making informed choices among competing alternatives.

energy is dispersed and has little ability to do useful work. For example, even though it contains more stored heat than all of Saudi Arabia's high-quality oil deposits, the Atlantic Ocean's energy is too widely dispersed to accomplish tasks such as moving vehicles or heating things to high temperatures (Miller, 1994). However, we can use low-quality energy from dispersed geothermal sources, for instance, to heat our homes and other buildings (to temperatures of 100°C or less). If we match the quality of energy used to the specific task we need to perform, we will avoid wasting energy unnecessarily and save money, too.

PHYSICAL AND CHEMICAL CHANGES IN MATTER

When we melt snow to boil water for a cup of tea or hot chocolate during a winter camping trip, we do not alter the chemical composition of the H_2O molecules, but we change water from a solid to a liquid state. This **physical change** causes the water molecules to organize themselves differently in space. In lighting our camp stoves and burning the fuel, however, we initiate a **chemical change** or reaction between the carbon contained in the fuel and oxygen from the atmosphere. This chemical reaction produces carbon dioxide gas and energy ($C + O_2$ yields CO_2 + energy). In addition to showing that camp stove fuel is a high-quality, useful energy resource, this example demonstrates how burning carbon-containing compounds such as wood, coal, or natural gas adds carbon dioxide, a greenhouse gas, to the atmosphere.

The Law of Conservation of Matter

Under ordinary circumstances, matter is neither created nor destroyed, but is recycled repeatedly. Matter is transformed and combined in different ways, but it does not disappear—everything goes somewhere. That is why it is inaccurate to talk about consuming or using up resources, because we are using, discarding, reusing, or recycling the same atoms. We can physically rearrange the atoms into different spatial patterns or chemically combine them into different combinations, but we are not creating or destroying them. This is the **law of conservation of matter**, and, in affluent societies such as Canada's, it means that every disposable consumer good thrown away remains with us in one form or another. This is why we hear about the need to emphasize waste reduction and pollution prevention.

First and Second Laws of Energy

The **first law of thermodynamics** (or the **first law of energy**) states that during a physical or chemical change energy is neither created nor destroyed. However, it may change form and it may be moved from place to place. When one form of energy is converted to another form in any physical or chemical change, energy input always equals energy output—we cannot get something for nothing in terms of energy quantity.

The **second law of thermodynamics** (or the **second law of energy**) indicates that with each change in form, some energy is degraded to a less useful form and given off to the surroundings, usually as low-quality heat. That is, in the process of doing work, high-quality energy is converted to more dispersed, disorganized, and lower-quality energy. If we used all of our camp stove fuel in making our cup of tea or hot chocolate, we would have lost energy quality (the amount of useful energy available for the future).

To consider how we lose energy quality, consider the incandescent light bulb. When electrical energy flows through the filament wires, it changes into about 5 percent useful light and 95 percent low-quality heat. This heat enters the environment and is dispersed by the random motion of air molecules (thus the suggestion that light bulbs really should be called heat bulbs). With each transfer of energy in these processes, heat is given off to the immediate surroundings and dissipates to the external environment and, eventually, through Earth's atmosphere to space. Effectively, the second law of energy means that we can never recycle or reuse high-quality energy to perform useful work.

Energy is constantly flowing from high-quality, concentrated, useful forms to low-quality, dispersed, and less useful forms. This tendency toward dispersal or disorganization is called **entropy**. Entropy is a measure of disorder: high-quality energy has a low entropy in contrast to low-quality energy, which has a high entropy. Entropy is increasing continuously in the universe (and at some point, billions of years from now, all energy will be uniformly distributed as low-quality heat). The second law of thermodynamics indicates that entropy (disorder) in a system tends to increase over time; the more energy we use (and waste) the more entropy (disorder) we create in the environment.

The apparent ability of living things to maintain a high degree of organization as they grow and develop conceals the fact that living organisms maintain that degree of order over time only with the constant input of energy. This is why animals must eat and plants must photosynthesize. Inevitably, plants and animals die and the system tends toward entropy.

EARTH'S LIFE-SUPPORT SYSTEMS

EARTH'S MAJOR COMPONENTS

The Earth's environment consists of four interconnected environmental spheres or layers. Surrounding the inner (solid) and outer (molten) cores and the mantle of the Earth are the **lithosphere**, **hydrosphere**, **atmosphere**, and **biosphere** or **ecosphere** (see Figure 3–2).

Although knowledge of the interior of the Earth is incomplete and imperfect, the **lithosphere** generally is said to consist of the upper zone of the Earth's mantle (to a depth of about 40 to 50 kilometres beneath the crust) as well as the inorganic mixture of rocks and mineral matter contained in the Earth's crust.

On the crust of the Earth lies the **hydrosphere**, the Earth's supply of moisture in all its forms—liquid (both fresh and saltwater), frozen, and gaseous. The hydrosphere includes the surface waters in oceans, lakes, rivers, and swamps; underground water wherever it is located; frozen water in the form of ice, snow, and high cloud crystals; water vapour in the atmosphere; and the moisture that is stored temporarily in the tissues and organs of all living organisms. The hydrosphere impinges on and overlaps significantly with the other spheres.

The **atmosphere** completely surrounds the solid and liquid Earth. Relative to Earth's radius, the atmosphere is a very thin layer of gases consisting mostly of nitrogen (78 percent) and oxygen (21 percent) plus small quantities of water vapour and argon, and minute amounts of other gases such as carbon dioxide and ozone (see Table 3–1).

Due to the forces of gravity and compressibility of gases, the two lowest layers of the atmosphere—the troposphere and stratosphere—together make up about 99 percent of the atmosphere's mass. The **troposphere** (containing about 80 percent of the atmospheric mass) is the lowest layer of the atmosphere and the zone in which most weather events occur. Its height above sea level varies from an average of about 6 kilometres at the poles to about 18 kilometres over the equator, and it also varies seasonally, being higher in summer than in winter. The next layer, the **stratosphere**, contains about 19 percent of the atmospheric mass and extends to about 50 kilometres above the Earth's surface.

The stratosphere and troposphere have very similar compositions except that in the stratosphere the volume of water vapour is about 1000 times lower, and ozone is nearly 1000 times higher than in the troposphere. Stratospheric ozone protects life on Earth's surface by absorbing most incoming solar ultraviolet radiation. If the **ozone layer** were a band of pure gas surrounding the globe at sea-level pressure and temperature, it would be no more than three millimetres thick, or about the thickness of three Canadian dimes (Government of Canada, 1991).

The biosphere (ecosphere) consists of the incredibly diverse plant and animal organisms that inhabit the Earth and their interactions with each other and with the atmosphere, hydrosphere, and lithosphere. While most living things inhabit the interface between atmosphere and lithosphere (a zone about five kilometres thick), some live largely or entirely within the hydrosphere or atmosphere and many others move freely from one sphere to another. The ecosphere extends vertically about 32 kilometres from the ocean floors to above the tops of the highest mountains. Yet, if the Earth were an apple, the web of life within which we live would be no thicker than the apple's skin (Miller, 1994). The field of ecology tries to determine how this thin skin of air, water, soil, and organisms interacts.

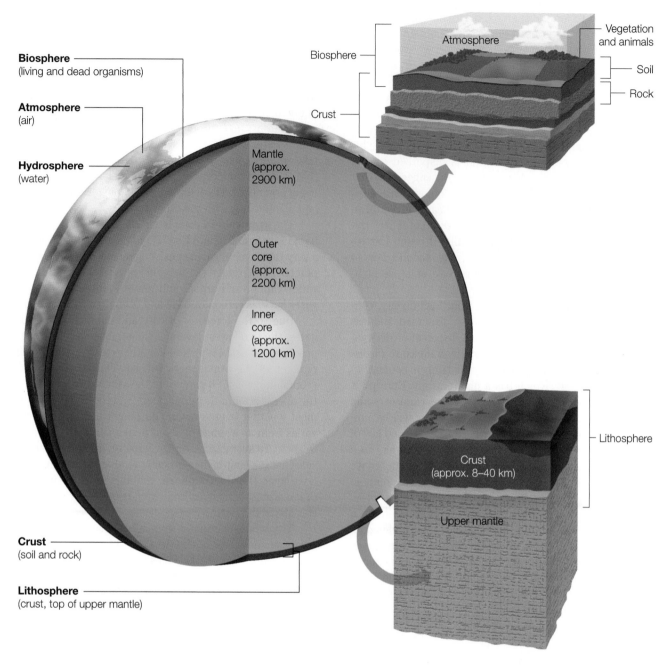

Biosphere
(living and dead organisms)

Atmosphere
(air)

Hydrosphere
(water)

Mantle
(approx.
2900 km)

Outer
core
(approx.
2200 km)

Inner
core
(approx.
1200 km)

Biosphere

Atmosphere

Crust

Vegetation
and animals

Soil

Rock

Crust
(soil and rock)

Lithosphere
(crust, top of upper mantle)

Crust
(approx. 8–40 km)

Upper mantle

Lithosphere

Figure 3–2
The general structure of the Earth

Connections on Earth

Life on Earth depends on three pervasive and interconnected factors: energy flow, matter cycling, and gravity. In particular, it is the one-way flow of high-quality energy from the sun through the materials and living things of the ecosphere, and then into the environment as low-quality energy (and eventually back out into space) that is the ultimate source of energy in most ecosystems. In addition, living organisms require the cycling of critical elements such as carbon, phosphorus, nitrogen, water,

and oxygen through the ecosphere. Gravity is important in that it keeps the planet's atmospheric gases from escaping into space and it draws chemicals downward in the matter cycles.

Although energy reaches the Earth continuously as sunlight, less than 0.023 percent of the total energy reaching the atmosphere each day actually is captured by living things through photosynthesis (Kaufman & Franz, 1993). The remainder of the energy is reflected by cloud cover and does not reach the surface, or is radiated by the Earth's surface back into space as heat (see Figure 3–3).

TABLE 3-1
PRINCIPAL GASES OF EARTH'S ATMOSPHERE

Component	Symbol or Formula	Percent of Volume of Dry Air	Concentration (in parts per million of air)
Uniform Gases[1]			
molecular nitrogen	N_2	78.08	
molecular oxygen	O_2	20.94	
argon	Ar	0.934	
neon	Ne	0.00182	18.2
helium	He	0.00052	5.2
methane	CH_4	0.00015	1.5
krypton	Kr	0.00011	1.1
molecular hydrogen	H_2	0.00005	0.5
Important Variable Gases[1]			
water vapour	H_2O	0–4	
carbon dioxide	CO_2	0.03	353
carbon monoxide	CO		100
ozone	O_3		2
sulphur dioxide	SO_2		1
nitrogen dioxide	NO_2		0.2

[1] The chemical composition of pure, dry air at lower elevations is unvarying through time, and its components are identified in the uniform gases section of this table. A number of minor gases, however, vary markedly according to location and time, as does the amount of moisture in the air. These are identified as variable gases in this table.

SOURCES: *Environmental Science: A Global Concern* (3rd ed.), W. P. Cunningham & B. W. Saigo, 1995, Dubuque, IA: Brown, p. 353; *Physical Geography: A Landscape Appreciation* (3rd ed.), T. McKnight, 1990, Englewood Cliffs, NJ: Prentice Hall, p. 53.

Photo 3–2
Because they occupy aquatic and terrestrial habitats, amphibians are sensitive indicators of environmental health.

The Earth is an open system for energy, continuously receiving and using energy from the sun and radiating waste heat into space.

ECOLOGY

Basic ecological knowledge is an important foundation of environmental awareness and is a basis for using and managing Earth's resources in an environmentally sound and sustainable manner. Part of such ecological knowledge is knowing how the biosphere works and how natural systems function and respond to change.

Ecology, from the Greek words *oikos* (house, or place to live) and *logos* (study of), is defined as the study of the interactions of living (biotic) organisms with one another and with their nonliving (abiotic) environment of matter and energy. As part of determining how Earth's living systems maintain the integrity of the ecosphere,

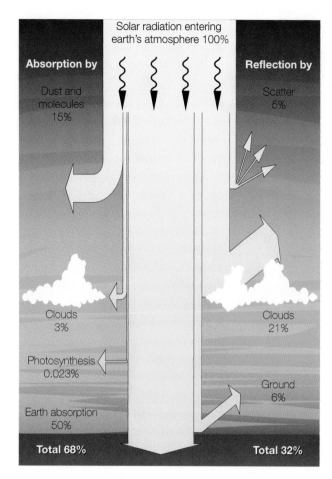

Figure 3–3

Schematic diagram of energy balance on the Earth

SOURCE: Adapted from *Human Geography: Landscapes of Human Activities,* J. Fellman, A. Getis, & J. Getis, 1995, Dubuque, IA: Brown, p. 454.

ecologists may study individual species as well as the structure and function of natural systems at the population, community, and ecosystem levels. Other scientists and social scientists such as geologists, earth scientists, and geographers focus on understanding the patterns and distributions of living and nonliving elements of the environment and on the interrelationships of people, other organisms, and their environments.

One of the characteristics of life on Earth is its high degree of organization (see Figure 3–4). Starting at the simplest level, atoms are organized into molecules, which in turn are organized into cells. In multicelled organisms, cells are organized into tissues, tissues into organs (such as the brain or liver), organs into organ or body systems (such as the nervous system or digestive system), and organ systems into individual multicellular **organisms** (such as bears, whales, humans, orchids, and cacti).

An individual organism is a single member of a **species**, defined as a group of organisms that resemble one another in appearance, behaviour, chemical makeup and processes, and genetic structure and that produce fertile offspring under natural conditions. While estimates as to the number of species on Earth vary between 5 and 100 million (mostly insects, microscopic organisms, and small sea creatures), most of the world's species remain unknown. Only about 1.4 million species have been discovered and described (Huyghe, 1993), and for the great majority of these species little is known about their roles and interactions.

Canola in a prairie field, brook trout in a particular stream, or people in Canada—a group of individuals of the same species living and interacting in the same geographic area at the same time—is called a **population**. Although all members of the same population share common structural, functional, and behavioural traits, individuals in a population vary slightly in their genetic makeup and thus exhibit slightly different behaviours and appearances. This is known as **genetic diversity**. The place where the organism or population lives—whether oceans, forests, streams, or soils—is its **habitat**. Populations of different species interact, making up a biological **community**, such as an alpine meadow community or a prairie community.

A community and its members interact with each other and with their nonliving environment of matter and energy, making up an **ecosystem**. Wetlands, estuaries, and the Great Lakes are examples of aquatic (water) ecosystems, while grasslands, high mountain deserts, and Carolinian forests are examples of terrestrial (land) ecosystems. A broad, regional type of ecosystem characterized by distinctive climate and soil conditions and distinctive communities of plants, animals, and microorganisms adapted to those conditions is referred to as a **biome**. The geography of the biosphere—that is, of global terrestrial ecosystems or world biomes—is illustrated in Figure 3–5. Canada has five terrestrial biomes (tundra, boreal forest, temperate deciduous and rain

Photo 3–3

These zebras are visually similar, but each individual is genetically different.

Figure 3–4
Levels of biological organization

forest, grassland, mountain complexes) and both aquatic biomes (freshwater and marine). In Canada, these major biomes have been subdivided into more specific "eco-zones" (15 terrestrial and 5 marine) that reflect Canadian ecological conditions including dominant landforms, soils, climate, natural ecosystems and species (see Figure 3–6). Finally, the highest level of organization is the biosphere or ecosphere, which consists of all communities of living things on Earth, of all Earth's ecosystems together.

Generally the biosphere is not studied as a single large system; instead, the smaller but still globally inter-related ecological systems frequently are the focus of investigation. Ecosystems vary in size and location, but all are dynamic entities, always changing as a result of changes in their external environments. These changes may be natural, as in grassland fires or changes in precip-itation levels in a forest, or they may be human induced, such as by spraying a field or forest with an insecticide. Unless the change is one that exceeds the threshold (tol-erance) limits of the individual ecosystem or the bios-phere, ecosystems can usually compensate for the stresses of external changes without incurring permanent change. The buildup of greenhouse gases, stratospheric ozone depletion, and the rapid loss of biological diversity,

Photo 3–4
Mixed-grass prairie in Grasslands National Park, in southern Saskatchewan.

however, are examples of potentially serious and perma-nent changes to the ecosphere.

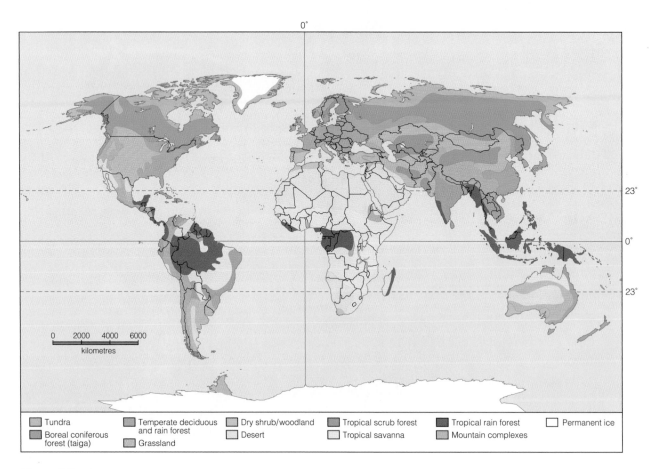

Figure 3–5
Earth's major biomes

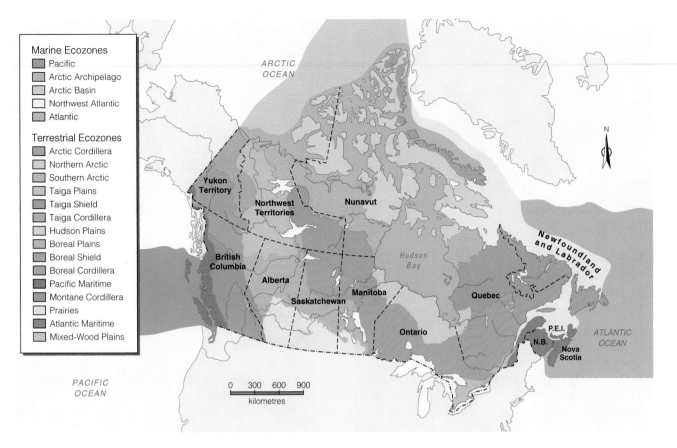

Figure 3–6

Ecozones of Canada

SOURCE: Adapted from *The State of Canada's Environment—1996,* Environment Canada, Ottawa: Supply and Services Canada, Figure II.2. Reprinted with permission of the Minister of Public Works and Government Services Canada, 2004.

BIODIVERSITY

Biological diversity, or biodiversity, involves three different concepts: genetic diversity, species diversity, and ecological diversity. Genetic diversity, as noted previously, is the variation in genetic makeup among individuals within a single species. **Species diversity** refers to the number of different species (that is, species richness) and their relative abundance in different habitats on Earth. **Ecological diversity**, sometimes called habitat diversity, is the variety of biological communities—oceans, lakes, streams, wetlands, forests, grasslands, deserts—that interact with one another and with their physical and chemical (nonliving) environments. Every currently living species not only represents a form of life best suited to survive present conditions but also contains stored genetic information relating to its adaptation to Earth's changing environmental conditions through time (plus the raw material for future adaptations). Given these characteristics, biodiversity provides nature's insurance against ecological disasters.

In addition to generating billions of dollars annually for the global economy, Earth's abundant variety of genes, species, and ecosystems provides humans with food and medicinal products, energy, fibres, raw materials, and industrial chemicals. We depend on this largely unknown biocapital for life forms and ecosystems, as well as for recycling, purification, and natural pest-control services.

Human cultural diversity is sometimes included as part of Earth's biodiversity. In the same manner as genetic material contained within other living species provides for future adaptability, so might the variety of human cultures represent our adaptability and survival options in the face of changing conditions.

TYPES OF ORGANISMS

For hundreds of years, biologists categorized living things into two broad categories: plants and animals. When microscopes revealed that many organisms did not fit well into either category at the cellular level, a five-kingdom classification system was devised. This system, while not perfect, consists of Prokaryotae, Protista, Fungi, Plantae, and Animalia. Bacteria are neither plants nor animals; they are **prokaryotic**, meaning they lack a nuclear envelope and other internal cell membranes. They have their own kingdom, Prokaryotae (see Figure 3–7). The four

Photo 3–5a

Photo 3–5b

Photo 3–5c

Photo 3–5d

Diversity within genus and species is illustrated in these photos. All four bears belong to the genus *Ursus*: black bear (*Ursus americanus*) (3–5a), polar bear (*Ursus maritimus*) (3–5b), grizzly bear (*Ursus arctos horribilis*) (3–5c), and Kodiak brown bear (*Ursus arctos middendorffi*) (3–5d). The latter two are subspecies of the brown bear, whose coastal and island populations (the Kodiak browns) are genetically distinct from the interior and arctic populations (the grizzlies).

remaining kingdoms are composed of organisms with a **eukaryotic** cell structure. Eukaryotic cells have a high degree of internal organization: a nucleus (genetic material surrounded by a membrane) and several other internal parts enclosed by membranes.

Most of the major groups (phyla) of eukaryotic organisms are single-celled or relatively simple multicellular organisms. They are classified as members of the kingdom Protista and include algae, protozoa, slime moulds, and water moulds. In addition to the Protista, three specialized groups of multicellular organisms form the Fungi, Plantae, and Animalia kingdoms.

These organisms differ from one another in several ways, including their nutrition. Members of the Fungi kingdom (such as mushrooms and yeasts) secrete digestive enzymes into their food and then absorb the predigested nutrients, while members of the Plantae kingdom (such as ferns, conifers, and flowering plants) use radiant energy to manufacture food molecules by photosynthesis. Members of the Animalia kingdom ingest their food and digest it inside their bodies. Most members of the Animalia kingdom are **invertebrates** (they have no backbone, such as jellyfish, worms, insects, and spiders).

Animals with backbones, the **vertebrates**, include fish (shark, tuna), amphibians (frogs, salamanders), reptiles (turtles, alligators), birds (eagles, robins, puffins), and mammals (elephants, whales, bats, and humans).

COMPONENTS AND STRUCTURE OF ECOSYSTEMS

The ecosphere and its ecosystems can be divided into two parts: the living or **biotic** components, such as plants and animals, and the nonliving or **abiotic** components, such as water, air, solar energy, and nutrients necessary to support life. Living organisms in ecosystems are usually classified as producers, consumers, or decomposers, depending on their nutritional needs and feeding type (Figure 3–8).

Sunlight is the source of energy that powers almost all life processes on Earth, and **producers** or **autotrophs** (self-feeders) are self-nourishing organisms that perform photosynthesis. Using solar energy, autotrophs convert relatively simple inorganic substances such as water, carbon dioxide, and nutrients into complex chemicals such as carbohydrates (sugars and starches), lipids (oils,

Kingdom

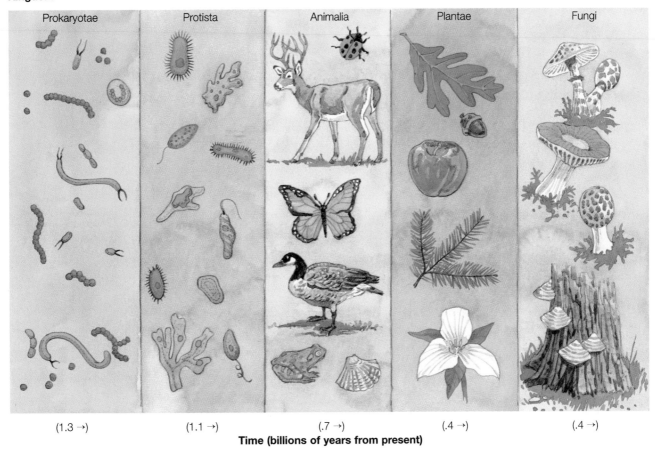

| Prokaryotae | Protista | Animalia | Plantae | Fungi |

(1.3 →)   (1.1 →)   (.7 →)   (.4 →)   (.4 →)

Time (billions of years from present)

Figure 3–7
The five-kingdom system of classification. Prokaryotes were the first organisms to appear; fungi are the most recent.

waxes), and proteins. Hundreds of chemical changes take place sequentially during photosynthesis; in most temperate-zone plants, photosynthesis can be summarized in the following way: $6H_2O + 6CO_2$ + solar energy yields $C_6H_{12}O_6$ (sugar) + $6O_2$. By incorporating the chemicals they produce into their own bodies, producers become potential food resources for other organisms. On land, green plants are the most significant producers; in aquatic ecosystems, algae and certain types of bacteria are important producers. Producers are the foundation of ecological productivity and ecosystem function.

A special group of bacteria (**chemotrophs**) converts the energy found in inorganic chemical compounds in aquatic and other environments into energy without sunlight. In the pitch-black thermal vent areas of deep oceanic trenches, for instance, nonphotosynthetic bacteria use the heat energy (generated by decay of radioactive elements deep in the Earth's core) from the vents to convert dissolved hydrogen sulphide (H_2S) and carbon dioxide (CO_2) into more complex nutrient molecules. Through **chemosynthesis**, these bacteria produce food energy for their nutritional needs and for other consumers in this special ecosystem (Lutz, 2000).

All other organisms in ecosystems are **consumers**, or **heterotrophs**, eating the cells, tissues, or waste products of other organisms. Heterotrophic organisms obtain the food energy and bodybuilding materials they need either directly or indirectly from autotrophs and thus indirectly from the sun. Unable to manufacture their own food, heterotrophs live at the expense of other plants and animals. They are categorized broadly as macroconsumers or microconsumers.

Macroconsumers, who feed by ingesting or engulfing particles, parts, or entire bodies of other organisms (living or dead), include herbivores, carnivores, omnivores, scavengers, and detrivores. **Herbivores** (plant-eaters), or **primary consumers**, such as deer, eat green plants directly. **Carnivores** (meat eaters), or **secondary consumers**, such as bobcats and certain snakes, feed indirectly on plants by eating herbivores. Most carnivores are animals, but the Venus flytrap is an example of a plant that traps and consumes insects. **Omnivores**, or consumers that eat both plants and animals, include black bears, pigs, and humans. **Tertiary consumers** are carnivores such as hawks that eat secondary (other carnivorous) consumers.

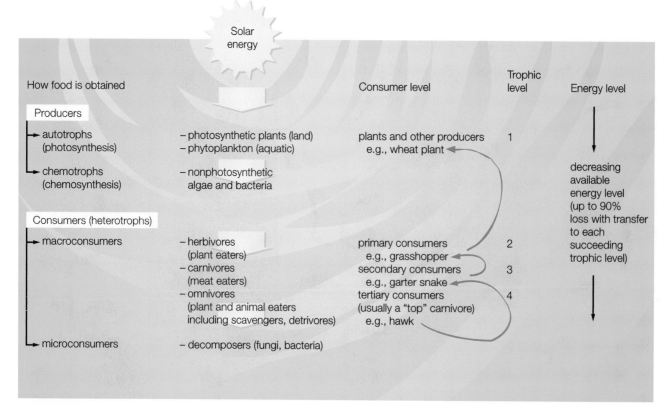

Solar energy

How food is obtained		Consumer level	Trophic level	Energy level

Producers

autotrophs (photosynthesis) — photosynthetic plants (land) — phytoplankton (aquatic) | plants and other producers e.g., wheat plant | 1

chemotrophs (chemosynthesis) — nonphotosynthetic algae and bacteria

decreasing available energy level (up to 90% loss with transfer to each succeeding trophic level)

Consumers (heterotrophs)

macroconsumers — herbivores (plant eaters) | primary consumers e.g., grasshopper | 2

— carnivores (meat eaters) | secondary consumers e.g., garter snake | 3

— omnivores (plant and animal eaters including scavengers, detrivores) | tertiary consumers (usually a "top" carnivore) e.g., hawk | 4

microconsumers — decomposers (fungi, bacteria)

Figure 3–8
Classification of organisms and trophic levels in ecosystems

Many heterotrophs consume dead organic material. Those that consume the entire dead organism, such as vultures and hyenas, are known as **scavengers**. Consumers that ingest fragments of dead or decaying tissues or organic wastes are called **detrivores**, or **detritus feeders**. Examples are earthworms, shrimp, dung beetles, and maggots. **Microconsumers** or **decomposer** organisms such as fungi and bacteria live on or within their food source, completing the final breakdown and recycling of the complex molecules in detritus into simpler compounds (which we call rot or decay). Decomposers play the major role in returning nutrients to the physical environment, providing an important source of food for worms and insects in the soil and water.

Awareness of the crucial importance of decomposers to the continuation of life in ecosystems has developed relatively recently. As understanding of their performance and function has improved, so has appreciation of the vital link that decomposers play in the cycle that returns chemical nutrients to the physical environment in a form that can be used by producers. Given that organisms constantly remove necessary chemicals from the environment, it is not difficult to imagine how quickly nutrients in the soil would be depleted if decomposers did not recycle nutrients continually after the death of producers and consumers. The Earth would quickly be covered in plant litter, dead animal bodies, animal wastes, and garbage if decomposers did not act on them.

Both producers and consumers use chemical energy stored in glucose and other nutrients to drive their life processes. This energy is released by **aerobic respiration**, which uses oxygen to convert nutrients such as glucose back into carbon dioxide and water. A complex process, the net chemical change for aerobic respiration

Photo 3–6
Decomposers, including these fungi, play an important role in the process of returning nutrients to ecosystems.

($C_6H_{12}O_6 + 6O_2$ yields $6H_2O + 6CO_2$ + released energy) is the reverse of photosynthesis.

Any individual organism depends on the flow of matter and energy through its body, while the community of organisms in an ecosystem survives by a combination of matter recycling and one-way energy flow. Energy, chemicals, and organisms are the main structural components of an ecosystem and are linked by energy flow and matter recycling (Figure 3–9).

Tolerance Ranges of Species

Every population in an ecosystem exhibits a range of tolerance to variations in its physical and chemical environment. While a speckled trout population may do best at water temperatures between 14°C and 19°C, for example, a few individual trout can survive temperatures as high as 24°C or 25°C for short periods of time because of small differences in their genetic makeup, health, and age (Kaufman & Franz, 1993). Beyond the range of tolerance, however, no trout will survive. The **law of tolerance** notes that the presence, number, and distribution of a species in an ecosystem are determined by whether the levels of one or more physical or chemical factors fall within the range tolerated by the species.

Some organisms have wide ranges of tolerance to some factors and narrow ranges of tolerance to other factors. Generally, the least tolerance is exhibited during the juvenile or reproductive stages of an organism's life cycle. Highly tolerant species can live in a range of habitats with different conditions, and other species can adjust their tolerance to physical factors such as temperature, if change is gradual. This adjustment to slowly changing conditions, **acclimation**, is a protective device, but it has limits. Each adjustment brings a species closer to its absolute limit until, without warning, the next small change triggers a **threshold effect**. This is a harmful or even fatal reaction to exceeding the tolerance limit (similar to adding the straw that broke the camel's back). This threshold effect explains why many environmental problems seem to arise so suddenly. Maple trees in Quebec and Ontario may suddenly seem to have begun dying in droves, but part of the cause may be exposure to numerous air pollutants, including acid precipitation, for

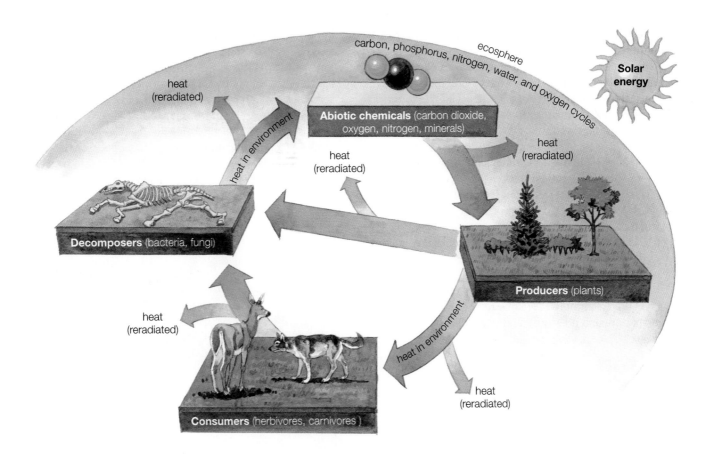

Figure 3–9
Energy flows and matter recycling connect energy, chemicals, and organisms in an ecosystem.

Photo 3–7

In contrast to other trees in the region, these Ontario sugar maples have not been damaged noticeably by air pollution.

decades. The concern about exceeding thresholds also explains why efforts must be made to prevent pollution and to ensure that biodiversity is maintained.

Limiting Factors in Ecosystems

An ecological principle related to the law of tolerance is the **limiting factor principle**. This principle asserts that too much or too little of any abiotic factor can limit or prevent growth of a population even if all other factors are at or near the optimum range of tolerance. Limiting factors in land ecosystems include precipitation, humidity, wind, temperature, light and shade, fire, salinity, available space, and soil nutrients. If you were to plant corn in soil that was lacking in phosphorus, even though all other factors were at optimum levels, the corn would not grow once it had used up the available phosphorus. Just as phosphorus levels determine how much corn will grow in a field, growth can be limited by an excess of an abiotic factor such as too much water or too much fertilizer. A limiting factor in aquatic ecosystems is *salinity* (the amounts of various salts dissolved in a given volume of water). So, too, is the **dissolved oxygen content** (the amount of oxygen gas dissolved in a given volume of water at a particular temperature and pressure), and availability of nutrients.

ROLES OF SPECIES IN ECOSYSTEMS

TYPES OF SPECIES IN ECOSYSTEMS

From a human perspective, the species in an ecosystem may be categorized into four types: native (or endemic),

immigrant (or exotic), indicator (or bellwether), and keystone species. In a particular ecosystem, a given species may be more than one of these types. **Endemic species** are those that normally live and thrive in a particular ecosystem, while **immigrant** or **exotic species** are those that migrate into or are introduced into an ecosystem, deliberately or accidentally, by humans. Despite the benefits bestowed on humans by intentionally introduced species, many of them have negative effects on native species or native ecosystems because they lack natural predators in their new habitat. **Indicator species** such as neotropical migratory songbirds and amphibians may provide early warnings of environmental damage to communities or ecosystems. The current decline in numbers of neotropical songbirds in North America indicates that both their summer habitats here and their winter habitats in Latin American and Caribbean tropical forests are disappearing. In a similar way, many of the world's amphibians (frogs, toads, and salamanders) are declining, possibly due to increased ultraviolet radiation, chemical contamination, and other causes (see Chapter 12).

While all species play important roles in maintaining the structure and function of their ecosystems, some scientists consider the roles of certain species to be more important than their abundance or biomass would suggest. Bees, bats, and hummingbirds have been labelled **keystone species** because they play crucial roles in tropical forests by pollinating flowering plants, dispersing seeds, or both. Top predators, such as alligators, wolves, and giant anteaters, are designated as keystone species because they exert a stabilizing influence on their ecosystems by feeding on and regulating the populations of certain other species. Sea otters of the Pacific Ocean feed on sea urchins and other shellfish, helping to reduce the sea urchins' destruction of kelp beds, thereby providing a

Photo 3–8

Introduction of the exotic zebra mussel to some North American aquatic environments has had devastating effects on crayfish endemic to the Great Lakes.

Figure 3-10
Pacific Ocean sea otters: a keystone role

larger habitat for many other species and indirectly increasing species diversity (see Figure 3-10). Given that the long-term resilience of an ecosystem is linked to the activities of this species, if the sea otter is removed or its keystone role within the ecosystem changes, the basic nature of the community changes (with more sea urchins there is less kelp and fewer species). On the Canadian prairies, the ground squirrel also is considered a keystone species (Box 3-1). Population crashes and extinctions of other species that depend on the keystone species can send ripple effects through the entire ecosystem.

Ecological Niche

Each species meets the challenge of survival in its own unique fashion. The way an organism interacts with other living things and with its physical environment defines that organism's **ecological niche**, or role within the structure and functions of an ecosystem. An ecological niche includes all the environmental (physical, chemical, and biological) conditions an organism or species needs to live, interact, reproduce, and adapt in an ecosystem. **Specialist species** have narrow niches, meaning a species may be able to live only in one type of habitat, eat a few types of food only, or tolerate a narrow range of climatic or other environmental conditions (see Enviro-Focus 3).

In tropical and Canadian temperate rain forests (see Chapter 9), diverse plant and animal species occupy specialized ecological niches within the distinct layers of the forest. These specialized niches enable species to mini-mize or avoid competition for resources with other species, thus preserving species diversity. Canadian examples include salamanders, which depend on habitat found in old-growth forests, and the insectivorous pitcher and sundew plants found only in sphagnum moss/peat bogs. **Generalist species** have broad niches and are able to live in many different places while tolerating a wide range of environmental conditions. Humans are considered a generalist species, as are flies, mice, raccoons, and white-tail deer.

Interactions between Species

Different species in an ecosystem often interact and develop close associations with one another. The major types of species interactions are interspecific competition, predation, parasitism, mutualism, and commensalism.

If commonly used resources are abundant, different species are able to share them and to come closer to occupying their fundamental niches. A **fundamental niche** is the full range of physical, chemical, and biological factors each species could use if there were no competition from other species. In most ecosystems, however, each species faces competition from other species for one or more of the same limited resources of food, sunlight, water, soil nutrients, or space. This is **interspecific competition**, in which parts of the fundamental niches of different species overlap significantly. Since no two species can occupy the same niche in the same community indefinitely, one species may occupy more of its fundamental niche than the other species as a result of competition between them. This **competitive exclusion principle** means that one of the competing species must migrate to another area if possible, shift its feeding habits or behaviour, suffer a sharp decline in population numbers, or become extinct.

The degree of fundamental niche overlap may be reduced by **resource partitioning**. Dividing up of scarce resources occurs in order that species with similar requirements can use the resources in different ways, in different places, and at different times. Resource partitioning occurs between owls and hawks that feed on similar prey; owls hunt at night while hawks hunt during the day. Similarly, some species of warblers hunt for insects in different parts of the same coniferous tree (Figure 3-11). Sharing the wealth among competing species results in each species occupying a **realized niche** (that portion of the fundamental niche actually occupied by a species).

Predation, in which members of a **predator** species feed on parts or all of an organism of a **prey** species, is the most obvious form of species interaction. A turtle eating a fish in a freshwater pond ecosystem, a fox feeding on a rabbit in a field ecosystem, and a killer whale culling a sick seal in a marine ecosystem are all examples of

BOX 3–1

IN DEFENCE OF "GOPHERS" (RICHARDSON'S GROUND SQUIRRELS)

The golden rodent with the impressive name—*Spermophilus richardsonii*—is a keystone species in the Canadian prairies. "Eliminating gophers would wipe out a whole suite of wildlife on the prairies," says biologist Cliff Wallis (Barnett, 1996, p. A4). A long list of species depends on the Richardson's ground squirrel (sometimes called gopher) for survival, from wild tomatoes that grow on ground squirrel mounds to salamanders that use ground squirrel burrows as shady shelters on their treks between wetlands. Without these ground squirrels, many birds of prey and mammals such as swift foxes and badgers would disappear for lack of food. According to Alberta provincial biologist Steve Brechtel, rare ferruginous hawks depend on Richardson's ground squirrels for 90 percent of their diet and eat up to 480 annually when raising their young. He suggests that prairie falcons and endangered burrowing owls (that use ground squirrel burrows for their homes) are in decline in part because the ground squirrel population is in decline.

Richardson's ground squirrels have long been targets of eradication efforts by prairie farmers who claim the rodents cause millions of dollars in damage to their fields, grain, and other crops. Highly toxic strychnine is the poison of choice: it takes a mere 0.7 milligrams to kill a ground squirrel, the equivalent of a couple of grains of sand. Once the ground squirrel dies, the poison remains active and kills animals that eat the ground squirrel carcass. Since 1994, the federal government has restricted the amount of strychnine used to kill ground squirrels because of fears of contaminating groundwater and poisoning other animals.

Cliff Wallis points out that poor land management practices are a major contributor to farmers' problems with ground squirrels. People allow overgrazing to occur, clear all the trees that provide nesting sites for birds of prey, and cultivate right to the edges of fields so there is no room for coyote dens (ground squirrels are a staple of coyote diets). Clearly, many of the problems with ground squirrels have their origin in human actions.

University of Lethbridge biologist Gail Michener notes that Richardson's ground squirrels have survived for tens of thousands of years, far longer than humans have been commercially cultivating the prairies. They are "definitely part of the prairies" and will not be driven easily from their homeland (Dempster, 1996, p. A4). Nor should ground squirrels be eradicated because, in addition to providing food to predators, ground squirrels "balance the grazing pressure by eating species of plants that cows are less interested in" (Holyrood, cited in Barnett, 1996, p. A4). Given the principle of connectedness and the unknown or unanticipated effects people encounter when they simplify ecosystems, Richardson's ground squirrels deserve greater respect for their role as a keystone prairie species.

In part, education about Richardson's ground squirrels will help to improve understanding and contribute to that greater respect. A graduate student in geography at the University of Lethbridge, Adela Kincaid, conducted survey research on rural and urban area residents' knowledge and attitudes toward ground squirrels. She found that urban people had a more positive attitude than did rural people. Rural people, who had greater knowledge about Richardson's ground squirrels, perceived problems caused by the rodents to be more serious than their urban neighbours did and supported more lethal management practices than urbanites. Urban residents "more often supported alternative management practices such as capture and relocation or the introduction of predators" (Kincaid, 2003, vi). Results from such research may help decision makers develop suitable public educational programs.

Photo 3–9
Richardson's ground squirrel: *Spermophilus richardsonii*.

SOURCES: "Biologist Rises to Defence of Gopher," V. Barnett, April 27, 1996, *Calgary Herald,* p. A4; "Gopher Guru Says Critters Here to Stay," L. Dempster, April 17, 1996, *Calgary Herald*, p. A4; *A Study of Attitudes Pertaining to the Richardson's Ground Squirrel,* A. Kincaid, 2003, unpublished master's thesis, University of Lethbridge; "Richardson's Ground Squirrels," G. Michener, (n.d.), University of Lethbridge website: http://people.uleth.ca/~michener/

predator–prey relationships. Another type of predator–prey interaction is **parasitism**, a symbiotic relationship in which the parasite benefits by obtaining nourishment from the host and the host is weakened or perhaps killed by the parasite preying on it. Parasites such as ticks, mosquitoes, and mistletoe plants live outside the host's body, while other parasites such as tapeworms and disease-causing organisms (pathogens) live within the host.

Symbiosis is any intimate relationship between individuals of two or more different species. Sometimes symbiosis can take an extreme form. In the case of three-toed sloths, their fur is often occupied by green algae and pyralid moths that feed on the algae, as well as by house mites, a number of beetle species, and several other kinds of arthropods. A single sloth can be home to over 900 beetles (Perry, 1986).

Mutualism is another symbiotic relationship in which interacting species, such as honeybees and certain flowers, both benefit. In the process of feeding on a flower's nectar, honeybees also pick up pollen and pollinate the female flowers. Sometimes, mutualistic partners can be completely dependent on one another. In the case of the yucca

Pacific Yew: Trash Tree Now Coveted as Cancer Treatment

For decades, the Pacific yew (a small shrub-like tree) was considered worthless in comparison to the highly profitable Douglas fir. The two species co-exist in forested areas in western North America. To accommodate clear-cutting of Douglas fir, however, massive cutting and burning of the Pacific yew took place. Impacts on Pacific yew populations as a whole were particularly severe during the 1970s and 1980s when Douglas fir timber production was high. It was not until the mid-1980s and the discovery of taxol that the Pacific yew became a product valued by society.

Taxol is a successful cancer-fighting agent and is particularly useful in treating ovarian cancer. Since the Pacific yew is the only known source of taxol, harvesting of the species is now widespread. This harvesting pressure may decline as new synthetic sources of taxol are developed, but until then, careful management and harvesting of the species is essential. Unfortunately, the Pacific yew is a very slow-growing tree; research has shown that even in undisturbed populations, these trees show little change in size and structure over periods of several decades. The United States Department of Agriculture Forest Service (Busing & Spies, 1995) esti-

mated that disturbed populations may require centuries to recover the population size and structure characteristic of old-growth forest stands.

In 1992, the United States introduced management guidelines for Pacific yew conservation (Busing & Spies, 1995). These guidelines require the establishment of genetic reserves for yew, replanting after harvest, and retention of some live yew trees in harvesting areas. Although the Pacific yew is but one of the millions of species sharing planet Earth, its transition from trash tree to valued resource underscores the need to preserve and protect even the most seemingly insignificant species.

Photo 3–10
Pacific yew bark: one person's treatment requires the bark from 6 to 100 trees.

SOURCE: *Modeling the Population Dynamics of the Pacific Yew,* R. T. Busing & T. A. Spies, 1995, United States Department of Agriculture, Forest Service. Research Note PNW-RN-515.

plant and the yucca moth, for instance, the moth transfers pollen between plants and the plants provide both food and a safe habitat for the moth larvae, which hatch from eggs laid inside the flower. Without the specialist species of yucca moth, pollination (successful reproduction) would not occur in the yucca, and without the yucca plant, the moth would be unable to reproduce successfully because it lays its eggs only inside yucca flowers.

Another type of symbiotic species interaction, **commensalism**, occurs when one species benefits while the other is neither helped nor harmed. On land, a good example of commensalism is the relationship between a

tropical tree and its **epiphytes** (air plants) that live attached to the bark of the tree's branches. Epiphytes do not obtain nutrients or water directly from the tree to which they are anchored, but their position on the tree allows them to receive adequate light, water (by rainfall dripping down the branches), and minerals (washed out of the tree's leaves by rainfall). The epiphytes benefit from the association while the tree remains generally unaffected.

Recent research in old-growth forest canopies of the Pacific Northwest has revealed that epiphytes (lichens, mosses, and liverworts) contribute to the creation of treetop soil (from decaying remains of leaves, epiphytes,

Figure 3–11
Resource partitioning
Resource partitioning and niches among *Dendroica* species
(wood warblers). Each species spends most of its feeding time in
a distinct portion of the trees it frequents. The shaded regions
identify where each species spends at least half its foraging time.

and needles). Since soils in temperate rain forests tend to
be nutrient poor because of leaching by heavy rains, tree
roots actually tap the soil up in their own canopies. When
they are wet, nitrogen-fixing lichens are particularly
important in canopies because they release excess
nitrogen that may be absorbed by other epiphytes or the
tree itself. The mutualism that yields this critical source of
forest nutrition is declining as logging of old-growth
forests (colonized most heavily by nitrogen-fixing lichens)
is causing these lichens to decline (Moffett, 1997).

In a marine environment, commensalism occurs
between various species of clownfish and sea anemones.
The stinging tentacles of the anemones paralyze most fish
that touch them, but clownfish gain protection by living
unharmed among the tentacles and feeding on the
detritus left from the meals of their host anemones. The
sea anemones seem neither to be harmed by this relation-

ship nor to benefit from it. A similar relationship occurs
between manta rays and remora fish.

ENERGY FLOW IN ECOSYSTEMS

FOOD CHAINS AND FOOD WEBS

One way in which individuals in a community interact is
by feeding on one another. Through feeding, energy,
chemical elements, and some compounds are transferred
from organism to organism along **food chains** (the
sequence of who feeds on or decomposes whom in an
ecosystem). Ecologists have assigned every organism in
an ecosystem to a feeding or **trophic level**, depending on
whether it is a producer or a consumer and what it eats
or decomposes (refer to Figure 3–8).

The first trophic level, and the basic provider of food for
the rest of the food web, consists of producers, the plants
that start the food chain by capturing the sun's energy
through photosynthesis. Primary consumers (and omnivores) eat the plants and make up the second trophic level;
secondary consumers (and omnivores) eat the herbivores
and form the third trophic level; and tertiary consumers (and
omnivores) belong to the fourth trophic level. Detrivores, or
decomposers, process detritus from all trophic levels.

Photo 3–11
The meeting of two links in the food chain.

Simple food chains such as described above occur rarely in nature because few organisms eat just one other kind of organism. Typically, the flow of energy and materials through terrestrial, aquatic, and oceanic ecosystems occurs on the basis of a range of food choices on the part of each organism involved. These organisms form a complex network of feeding relationships called a **food web**. In a simplified Great Lakes example, the food web shows relationships between some of the better known species including lake trout, salmon, herring gulls, bald eagles, and humans (see Figure 3–12).

An important feature of energy flow in ecosystems is that it is one-way. Energy can move along the food chain until it is used; at that point, it is unavailable for use by any other organism in the ecosystem. Depending on the types of species and ecosystem involved, there can be up to a 90 percent energy loss with each transfer from one trophic level to the next (refer to Figure 3–8). The more trophic levels or steps that exist in a food chain or web, the greater the cumulative loss of usable energy. (Reductions in energy are the result of the second law of thermodynamics.)

The large energy loss in moving to each successive trophic level explains why most food chains and webs rarely have more than four consecutive links or energy transfers. Energy flows also demonstrate why the eating habits of people can influence how many humans planet Earth can support. If people eat at lower trophic levels by directly consuming grains such as rice, rather than eating grain eaters such as beef cattle, they will be located on a shorter food chain and obtain more energy than meat eaters do from the same amount of plant material.

PRODUCTIVITY OF PRODUCERS

The rate at which producers in an ecosystem capture and store chemical energy as **biomass** (the amount of living or organic matter contained in living organisms) is that

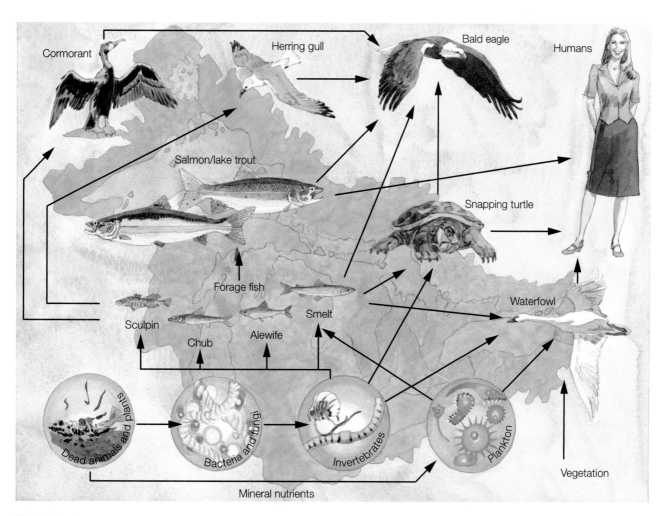

Figure 3–12

A simplified Great Lakes food web

SOURCE: *The State of Canada's Environment—1991,* Environment Canada, Ottawa: Supply and Services Canada, p. 18-8. Reprinted with permission of the Minister of Public Works and Government Services Canada, 2004.

ecosystem's **gross primary productivity**. Since plants must respire to provide energy for their life processes, and respiration uses up some biomass, energy that remains after respiration occurs is known as net primary productivity. Essentially, **net primary productivity** represents the rate at which organic matter is incorporated into plant bodies so as to produce growth. Expressed simply, net primary productivity (or plant growth) equals gross primary productivity (total photosynthesis) minus plant respiration. Net primary productivity usually is reported in terms of kilocalories (of energy fixed by photosynthesis) per square metre per year, or in terms of dry weight (grams of carbon incorporated into tissue) per square metre per year.

In different parts of the Earth, net primary productivity varies greatly. Terrestrial communities are generally more productive than aquatic communities, partly because of the greater availability of light for photosynthesis and partly because of higher concentrations of available mineral nutrients. On land, estuaries, swamps and marshes, and tropical rain forests are most productive, while Arctic and alpine tundra and desert scrub ecosystems are least productive. Water surrounding aquatic ecosystems moderates temperatures, but the productivity of these systems is limited by scarce mineral nutrients and low light intensity.

It has been estimated that humans have destroyed outright about 12 percent of the terrestrial net primary productivity and currently use an additional 27 percent. Humans have appropriated nearly 40 percent of the terrestrial food supply, leaving only 60 percent for the millions of Earth's other land-based plants and animals (Postel, 1996). Keeping in mind that humans have taken the 40 percent that was most easily acquired, and that human populations continue to grow, concern arises about the planet's carrying capacity and the sustainability of human actions with regard to productivity in ecosystems.

MATTER CYCLING IN ECOSYSTEMS

NUTRIENT CYCLES

If an organism such as a plant is to live, grow, and reproduce, it must take in variable amounts of different **nutrients**. Nine **macronutrients**—carbon, oxygen, hydrogen, nitrogen, phosphorus, potassium, calcium, magnesium, and sulphur—are the major constituents of the complex organic compounds found in all living organisms. Macronutrients are required in large amounts by most forms of life to make proteins, fats, and carbohydrates. Many organisms also require trace amounts of numerous **micronutrients** such as boron, copper, iron, zinc, molybdenum, chlorine, and manganese.

Multidirectional **nutrient cycles**, or **biogeochemical cycles**, are the means by which these nutrient elements and their compounds cycle continually through the Earth's atmosphere, hydrosphere, lithosphere, and biosphere, connecting past, present, and future life. Unfortunately, toxins can also circulate through these same cycles (see Box 3–2). The five main cycles, driven directly or indirectly by incoming solar energy and gravity, are the carbon, oxygen, nitrogen, phosphorus, and hydrologic (water) cycles. While much research remains to be done before we fully understand major biogeochemical cycles, aspects of the major cycles, including the rock cycle, are discussed briefly below.

Carbon and Oxygen Cycles

Proteins, carbohydrates, and other molecules essential to life contain carbon; living organisms must have carbon available to them. Carbon dioxide gas, which makes up about 0.03 percent of the volume of the troposphere and also is dissolved in water, is the basis of the carbon cycle. Producers remove CO_2 from the atmosphere or water and, using photosynthesis, convert it into complex chemical compounds such as glucose ($C_6H_{12}O_6$). In doing so, they produce oxygen (O_2) and release it to the environment (see Figure 3–13). Simultaneously, the compounds are used as fuel for aerobic cell respiration in producers, consumers, and decomposers. Respiration breaks down the glucose and other nutrient compounds and converts the carbon back to CO_2 in the atmosphere or water, where producers can reuse it.

Sometimes the carbon in biological molecules is not recycled back to the abiotic environment for a long time—the carbon stored (sequestered) in the wood of trees, for instance, may remain there for several hundred years. Both forests and oceans are major carbon sinks (carbon storage sites) and are important elements in the Kyoto Protocol (see Chapter 5). Generally, carbon circulates in the ecosphere as a result of the linkage between photosynthesis in producers and aerobic respiration in producers and consumers. Oxygen and hydrogen, which together with carbon form the compounds of life, cycle almost in step with carbon.

Oscillations in the balance of O_2 and CO_2 in the atmosphere have occurred throughout Earth's history. However, human activities that have disturbed the balance of the carbon cycle, in ways that add more carbon dioxide to the atmosphere than oceans and plants can remove, have increased markedly since the 1950s. People have cleared extensive areas of forests and brush, leaving less vegetation available to absorb CO_2, and people also may have exceeded threshold limits by burning increasing amounts of fossil fuels and woods, releasing CO_2 into the

BOX 3-2
TOXINS AND THE GREAT LAKES ECOSYSTEM

In 1971, a biologist at Scotch Bonnet Island in Lake Ontario found only 12 herring gull chicks where there should have been 100. That disturbing discovery of the reproductive problems of the ubiquitous herring gull became a symbol of the problems afflicting wildlife in the Great Lakes ecosystem.

In the early 1970s, a program designed to monitor persistent toxic chemicals in the eggs of herring gulls soon showed that water birds in the Great Lakes were among the most heavily contaminated in the world. High levels of chlorinated organic contaminants in the gulls' eggs coincided with high embryonic mortality and behavioural changes in adults (such as inattentiveness) that resulted in lower hatching success and physiological abnormalities in embryos and chicks.

Contaminants in herring gull eggs declined through the 1970s as a result of regulations implemented to control use and production of chlorinated organic compounds. However, in 1981 and 1982, levels increased briefly again, indicating that persistent contaminants continued to cycle through the ecosystem. The less easily controlled sources of these contaminants included leaching from landfill sites, disturbance of lake sediments, and deposition from the atmosphere. As herring gulls are sensitive to the presence of biologically significant concentrations of chemicals in the Great Lakes, monitoring of contaminant levels in herring gull eggs continues and is now the longest-running annual wildlife contaminants monitoring program in the world. Recently, a newly discovered family of chemicals, polybrominated diphenyl ethers (PBDEs), has increased dramatically in Great Lakes gulls and has been added to the approximately 100 compounds analyzed in the monitoring program.

An indicator species, bald eagles are extremely sensitive monitors of ecosystem quality. The fact that nesting pairs reintroduced to both the north and south shores of Lake Erie continue to survive confirms that ecosystem quality in those parts of the Great Lakes has improved. The fact that many of their eggs are fertile also is evidence of improvements in environmental quality. In 1991, however, 8 of 12 hatchlings died of wasting by the age of four weeks—wasting is a syndrome linked with persistent toxic substances. In 1993, bald eagles were hatched with twisted beaks and deformed talons, a problem caused by persistent

Photo 3–12
Eaglet hatched with twisted beak.

toxic substances. These events indicate that efforts made since the late 1970s to reduce some contaminant levels have made substantial progress, but not sufficient progress to restore the viability of bald eagle chicks in the populations of bald eagles nesting near the shoreline of the Great Lakes. By implication, there is potential danger to the dense human population around the Great Lakes.

Since it is not yet possible to retrieve or remove completely a persistent toxic substance once it has entered the environment, the focus must be on preventing the generation of such substances in the first place, rather than on trying to control their use, release, and disposal after they are produced. For those substances that persist and bioaccumulate in the environment, there is no safe level. The challenge is to implement the goal of zero discharge for persistent toxic substances—that is, to stop their generation, use, and release into the environment.

SOURCES: *The State of Canada's Environment—1991*, Government of Canada, 1991, Ottawa: Supply and Services Canada, p. 18-16; Sixth Biennial Report under the Great Lakes Water Quality Agreement of 1978 to the Governments of the United States and Canada and the State and Provincial Governments of the Great Lakes Basin, International Joint Commission, 1992, Ottawa: Author; "Keeping the Zero in Zero Discharge," P. Muldoon & J. Jackson, 1994, *Alternatives, 20*(4), 14–20; "Thirty Years of Monitoring Great Lakes Herring Gulls," Enviornment Canada, 2004, *EnviroZine, 43*.

---

atmosphere at a rate greater than the natural carbon cycle can handle. (Discussion of the impacts of rising CO_2 in the atmosphere is found in Chapter 5.)

Nitrogen Cycle

One of the most significant macronutrients, nitrogen is crucial for all living things because it is an essential part of amino acids, which make up proteins (important struc-

tural components of cells) and nitric acids (which store genetic information). Although Earth's atmosphere is about 78 percent nitrogen gas (N_2), nitrogen gas is comparatively unreactive, so most plants and animals cannot use nitrogen gas directly from the atmosphere. The nitrogen must be "fixed" (that is, combined with oxygen or hydrogen) to provide compounds that plant roots are able to use. The five steps in the nitrogen cycle, whereby bacteria convert nitrogen gas into water-soluble compounds containing nitrogen, are illustrated in Figure 3–14.

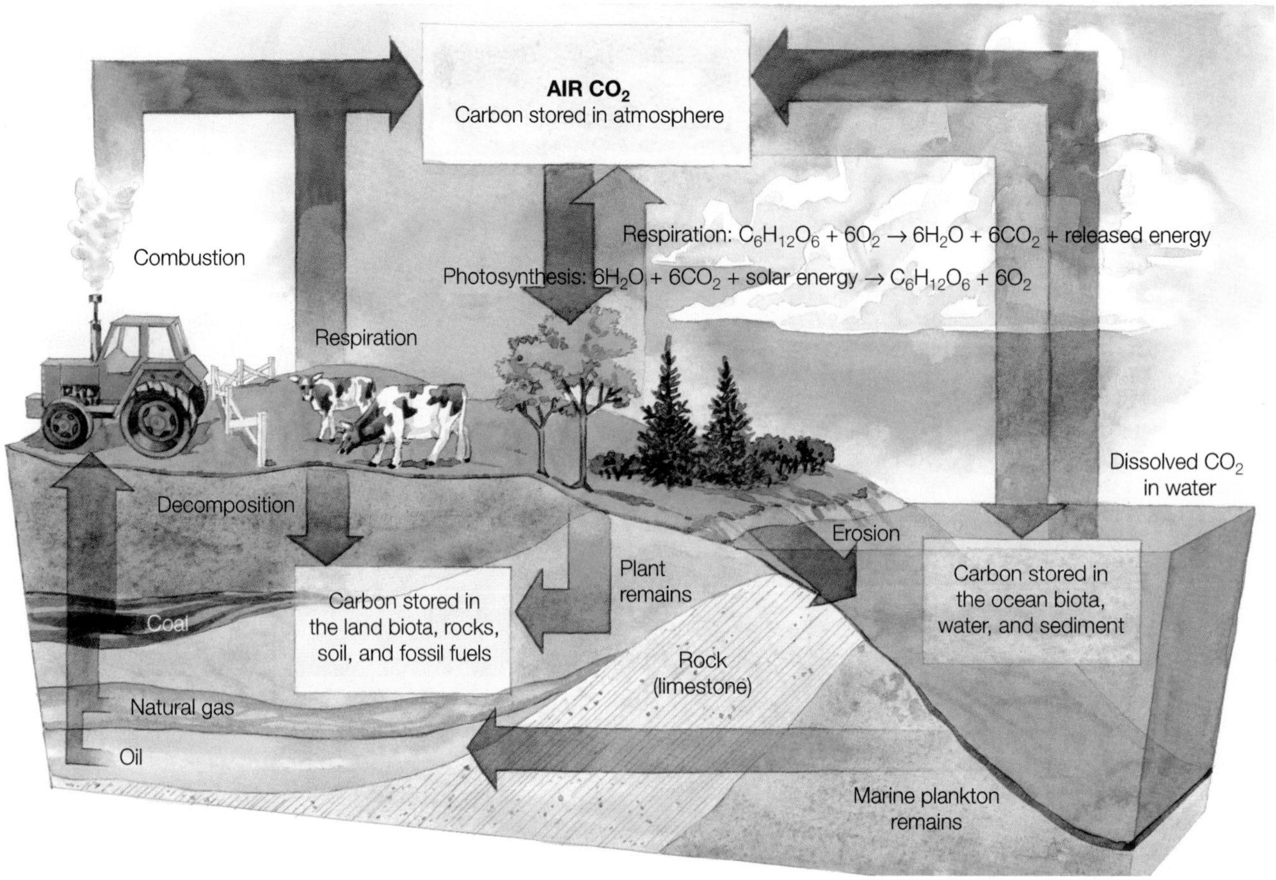
AIR CO_2
Carbon stored in atmosphere
Respiration: $C_6H_{12}O_6 + 6O_2 \rightarrow 6H_2O + 6CO_2$ + released energy
Photosynthesis: $6H_2O + 6CO_2$ + solar energy $\rightarrow C_6H_{12}O_6 + 6O_2$
Combustion
Respiration
Dissolved CO_2 in water
Decomposition
Erosion
Plant remains
Carbon stored in the land biota, rocks, soil, and fossil fuels
Carbon stored in the ocean biota, water, and sediment
Coal
Rock (limestone)
Natural gas
Oil
Marine plankton remains

Figure 3–13
A simplified diagram of the carbon and oxygen cycles

The complex nitrogen cycle starts with **nitrogen fixation**, a process that converts atmospheric nitrogen (N_2) into ammonia (NH_3). This conversion is done mostly by nitrogen-fixing bacteria such as cyanobacteria (a type of photosynthetic bacterium) in soil and aquatic environments, and by the important *Rhizobium* bacteria, which live in special nodules or swellings on the roots of legumes such as beans, peas, and alfalfa. Ammonia also is found in organic nitrogen contained in plant and animal tissue. The most important pathway for plants to acquire nitrogen is through the cycling of organic nitrogen from dead plant matter into mineralized nitrogen. Thus, ammonia must be converted from organic to mineralized nitrogen.

In the second phase of the nitrogen cycle, called ammonification, ammonia is converted to ammonium ions (NH_4^+). In the third step of the cycle, nitrification, ammonium ions are oxidized into nitrate and nitrite ions. Ammonium, nitrite, and nitrate ions are deposited in the soil solution and bound to soil particles from which they easily are taken up to plants in the fourth step, assimilation. Increasingly, as a result of human activities, atmospheric ammonium and nitrate ions are deposited in soils through precipitation. Finally, unused mineralized nitrogen compounds are released as gases in the fifth stage of the cycle, denitrification.

Humans have intervened in the nitrogen cycle in several ways. One important intervention occurs when burning fuels release large quantities of nitric oxide (NO) into the atmosphere. When nitric oxide combines with oxygen to form nitrogen dioxide (NO_2) gas, it can then react with water vapour to form nitric acid (HNO_3), a component of acid deposition (acid rain) that can damage trees and aquatic systems. Water pollution is another significant impact of human intervention in the nitrogen cycle, particularly as it relates to the use of nitrogen fertilizers. Overuse of commercial fertilizers on land can lead to excess nitrogen compounds in agricultural runoff and in the discharge of municipal sewage. Nitrogen-based fertilizers stimulate the growth of algae and aquatic plants that, when they subsequently decompose, deplete the water of dissolved oxygen and cause other aquatic organisms, including fish, to die of suffocation. Nitrates from fertilizers also can leach or filter down through the soil and contaminate groundwater, causing concern about the quality of drinking water.

Phosphorus Cycle

Phosphorus, one of the essential elements for life and a major constituent of agricultural fertilizers, does not exist in a gaseous state and does not circulate in the

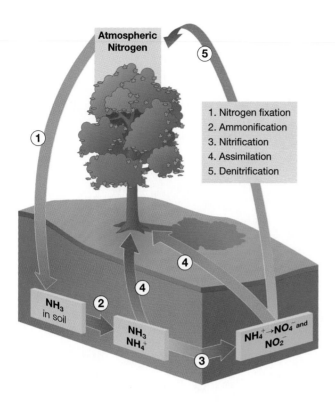

Figure 3–14

A simplified diagram of the nitrogen cycle

SOURCES: *Terrestrial Ecofunctions* (2nd ed.), J. Aber & J. Melillo, 2001, San Diego: Harcourt, pp. 256–260; *The Nature and Property of Soils* (12th ed.), N. C. Brady & R. R. Weil, 1999, Upper Saddle River, N.J.: Prentice Hall, pp. 495–496.

atmosphere. Instead, via the long-term geologic processes of the rock cycle, phosphorus slowly cycles from phosphate deposits on land to shallow sediments in the oceans to living organisms and then back to the land and oceans (see Figure 3–15). Although the Earth's crust contains a large amount of phosphorus, only a small fraction is available through conventional mining techniques. As a result, phosphorus is often a limiting nutrient for plant growth.

As water runs over rocks containing phosphorus, inorganic phosphate molecules are released into the soil where they are taken up by plant roots. Phosphorus moves through the food chain in the same way as carbon and nitrogen do when one organism consumes another. Eventually, decomposers release phosphorus, which becomes part of the inorganic phosphate in the soil that can be reused by plants. A similar process occurs in aquatic communities.

Human influences on the phosphorus cycle derive from activities that accelerate long-term losses of phosphorus from the land, including agricultural crop production that absorbs phosphates from the soil, removal of

vegetation and clear-cutting of timber that leads to soil erosion, and mining phosphate rock to produce detergents and commercial fertilizers (to compensate for the steady loss of phosphate from farmers' fields). Other human activities that affect the phosphorus cycle are the addition of excess phosphorus to aquatic ecosystems through runoff of animal wastes from feedlots, from runoff of commercial phosphate fertilizers from cropland, and from discharge of municipal sewage.

Most freshwater lakes and streams contain little phosphorus. When excess phosphorus compounds (phosphates) are added to such aquatic systems, the increased phosphorus concentration stimulates plant growth and algal blooms: **eutrophication**. Eventually when these plants and algae die, they sink to the bottom and decompose, using up much of the dissolved oxygen (anoxic conditions), and limiting the growth of many aquatic species. Massive fish kills can occur in extreme cases of eutrophication.

Hydrologic Cycle

In the hydrologic (or water) cycle, Earth's fixed supply of water circulates continuously from the oceans to the atmosphere to the land and back to the oceans, providing a renewable supply of purified water on land (see Figure 3–16). The water cycle, driven by solar energy, involves seven main recycling and purifying processes: evaporation, transpiration, condensation, precipitation, infiltration, percolation, and runoff. Clouds form in the atmosphere when water evaporates from the oceans, soil, streams, rivers, and lakes, and transpires from land plants. After water vapour has condensed into liquid water droplets, water moves from the atmosphere to the land and oceans in the form of precipitation (rain, snow, hail).

Once on land, water can infiltrate the soil and percolate down through the soil and permeable rock formations to groundwater storage areas, or it can run off in rivers and streams to coastal estuaries and back into the sea to begin the cycle again. Regardless of its physical form or its length of storage time in oceans, streams, reservoirs, or glaciers, every water molecule eventually moves through the hydrologic cycle. Great quantities of water—on the order of 400 000 cubic kilometres—are cycled annually between the Earth and its atmosphere.

Humans affect the hydrologic cycle in two main ways: by withdrawing large quantities of fresh water from surface and underground sources and by clearing vegetation from land. Withdrawals in heavily populated areas have led to groundwater depletion or saltwater intrusion into underground water supplies. Clearing vegetation for agriculture, mining, roads, residences, and other developments not only reduces seepage (which recharges groundwater supplies) but also increases flooding risks and increases speed of surface runoff (producing more soil erosion and landslides).

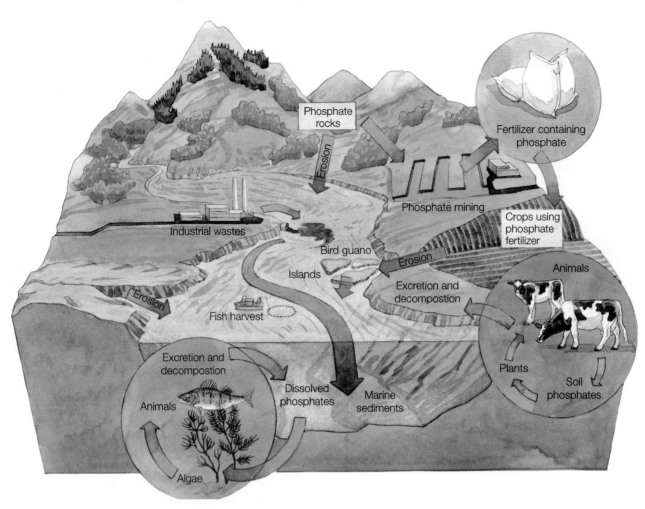

Figure 3–15

A simplified diagram of the phosphorus cycle in terrestrial and aquatic environments

Rock Cycle

The slowest of Earth's cyclic processes, the rock cycle consists of several processes that produce rocks and soil and redistribute chemical elements within and at the surface of the Earth. Based on the way they form, rocks have been classified into three classes: igneous, sedimentary, and metamorphic. **Igneous rock** is produced from molten materials crystallizing at the Earth's surface (such as lava from volcanoes) or beneath the surface (such as granite). **Sedimentary rock** forms when small bits and pieces of matter and sediments are carried by wind or rain and then deposited, compacted, and cemented to form rock. As the deposited layers grow larger, increasing pressure causes their particles to bond together to form sedimentary rocks such as shale, limestone, and bituminous coal.

Metamorphic rock is formed when preexisting rocks lying deep below the Earth's surface are subjected to high temperatures, high pressures, chemically active fluids, or a combination of those agents, causing the rocks' crystal structure to change. For example, marble is formed from limestone that has been heated and recrystallized.

Rocks located at or near the surface are subject to physical and chemical processes of weathering that, over time, can change them. For instance, freeze–thaw actions can break rock apart by repeated expansion and contraction, and rocks may be dissolved by weak acids that form in the presence of carbon dioxide, organic material, and water; such actions produce sediments that may be transported by wind, water, or ice. The interaction of all processes that change rocks from one type to another is the rock cycle; it is responsible for concentrating the mineral resources on which humans depend.

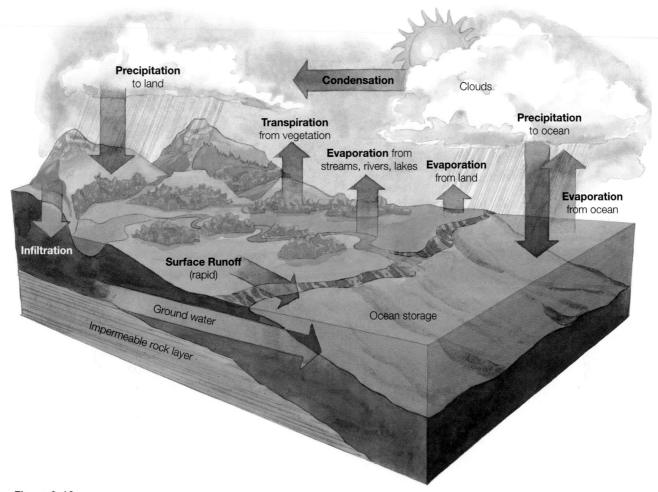

Figure 3–16
A simplified diagram of the hydrologic cycle

TERRESTRIAL AND AQUATIC ECOSYSTEMS

THE GEOGRAPHY OF LIFE

For centuries, people have been fascinated by geographic variations in the kinds and numbers of species found in various parts of the world. Unless they are in zoos, polar bears do not inhabit southwestern British Columbia, nor do arbutus trees grow in the Northwest Territories. Geographic variations occur, in part, because the distributions of plant and animal species are governed by the ability of each species to tolerate the environmental conditions of its surroundings. On a global scale, distribution patterns of species have been recognized for centuries; in each major kind of climate a distinctive type of vegetation develops, and certain animals and other kinds of organ-

isms are associated with each major type of vegetation. Similarly, certain aquatic organisms assemble in each of the Earth's major aquatic ecosystems.

LIFE ON LAND: MAJOR TERRESTRIAL BIOMES

A biome is a kind of ecosystem—a large, relatively distinct region such as a desert, tropical rain forest, tundra, or grassland characterized by certain climatic conditions, soil characteristics, and plant and animal inhabitants regardless of where on Earth it occurs. More than any other factor, the climate—particularly differences in average temperature and average precipitation—influences the boundaries of Earth's major biomes (refer to Figure 3–5) and Canada's ecozones (refer to Figure 3–6).

Precipitation generally determines whether a land area is forest, grassland, or desert. For instance, we find generally that forest areas receive more than 100 centimetres of precipitation per year, grasslands receive 25 to 75 centimetres, and desert areas receive less than 25 centimetres annually. When temperature and precipitation factors are combined, we find hot areas that receive 100 or more centimetres of rainfall per year sustain tropical savannas, whereas temperate areas with the same amount of precipitation support deciduous forests. If temperate forest soil types are added as a limiting factor, we find that maples and beeches are more successful on high-nutrient soils while oaks and hickories are more successful on low-nutrient soils. Acting together, then, average annual precipitation and temperature along with soil type lead to deserts, grasslands, and forests (and associated vegetative and animal species) in tropical, temperate, and polar areas.

Although maps show the divisions between Earth's major biomes as sharp boundaries (refer to Figure 3–5), in reality, biomes blend into each other (see Figure 3–17). While various experts identify different numbers of classes of biomes, typically, terrestrial biomes are named for their dominant plant species (such as coniferous forest or grasslands), for the dominant shapes and forms (physiognomy) of the dominant plants (forest versus shrub land), or for the dominant climatic conditions (cold desert versus warm desert). Aquatic biomes usually are identified on the basis of their dominant animals.

Although the relationship between climate and vegetation is complex, over the long term, differences in average precipitation and temperature establish the type and amount of natural vegetation (biomes) that would grow in any particular area if human activities were not present.

Tundra, the northernmost biome, occurs under harsh polar climates characterized by low rainfall and low average temperature. The dominant vegetation of both *arctic tundra* (occurs at high latitudes) and *alpine tundra* (occurs at high elevations) includes grasses,

Photo 3–13
Arctic tundra in the Yukon.

sedges, mosses, lichens, flowering dwarf shrubs, and mat-forming plants. All are adapted to extreme cold and a very short growing season. The dominant animal species differ, however, with the large land areas of arctic tundra supporting important large mammals such as caribou as well as small mammals, birds, and insects. Small rodents and insects dominate the alpine tundra, which occurs in small, isolated areas. Parts of the tundra are characterized by **permafrost** (a permanently frozen layer of subsoil). Ecologically fragile, these areas may be changed permanently by human developments such as mines, roads, and recreational trails.

The *boreal forest,* or *taiga,* is found south of the tundra and is dominated by dense stands of relatively small (under 30 metres high) coniferous trees. Spruce, fir, larch, and certain pines that are adapted to the cold, dry,

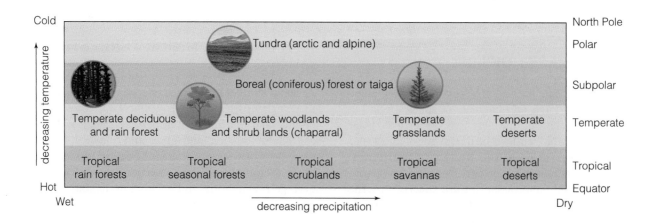

Figure 3–17
Simplified climatic effects on natural vegetation

subpolar winters and short but warm growing seasons of high latitudes and high altitudes are dominant, but aspen, poplar, and birch are important flowering trees within the boreal forest. Although they cover a very large area in North America, boreal forests contain only about 20 major tree species that are the source of much lumber and pulp and paper (see Chapter 9). Dominant animals of boreal forests include some large mammals such as moose, deer, wolves, and bears; small rodents such as squirrels and rabbits; many insects; and migratory birds. Fires, windstorms, and insect infestations are common disturbances in boreal forests.

Temperate deciduous and rain forests occur where precipitation is relatively high and temperatures are slightly warmer than those of the boreal forests. Dominant tree species in the deciduous forests—important economically for hardwood production and products such as furniture—include tall maples, beeches, oaks, hickories, and chestnuts. Although deer are present, in temperate deciduous forests the low density of large mammals results partly from the deep shade of the forest interior (less food for ground-dwelling mammals). Small mammals such as squirrels and mice, as well as birds and insects, are abundant.

Rare *temperate rain forests*, such as those found on the northwest coast of North America, produce giant (over 70 metres tall), long-lived (over 400 years) conifer trees including redwood, Douglas fir, and western cedar. These trees grow in moderate temperature regimes where the rainfall exceeds 250 centimetres per year. The species diversity of plants and animals in temperate rain forests is low, partly because the climatic conditions favour specialized species and partly because the abundant growth of the dominant vegetation produces a very deep shade in which few other plants can grow (limited food for herbivores). This biome is important economically as the large tree species continue to be a major timber source.

Temperate woodlands and shrub lands are located in areas with slightly drier climates than those found in temperate deciduous forest areas. Small trees such as piñon junipers, evergreen oaks, and pines characterize temperate woodlands. Fire is a common disturbance in these open, dry woodlands that frequently are used for recreational activities.

Temperate shrub lands, such as the dense chaparrals that occur along the coast of California, in Chile, South Africa, and the Mediterranean, are located in dry Mediterranean climates with low rainfall concentrated in the cool season. Important for water sheds and erosion control, these rapidly regenerating thickets of trees and small-leafed shrubs are adapted to fires. Much of the vegetation, including sage, is aromatic. Although few large mammals exist here, reptiles and small mammals are common. Humans find this climatic regime attractive, and housing developments are expanding rapidly into chaparral lands.

In areas of moderate temperature and 25 to 60 centimetres of seasonal precipitation (too dry for forests and too moist for deserts), *temperate grasslands* thrive on deep, mineral-rich soil. Located in the prairies of North America, the pampas of South America, the steppes of Eurasia, and the plains of eastern and southern Africa, the dominant species are grasses and other flowering plants (many are perennials with extensive root systems). Grasslands support the greatest diversity and highest abundance of large mammals, ranging from the wild horses and antelopes of Eurasia, the huge herds of bison that used to roam the North American prairies, to the kangaroos of Australia and the antelopes of Africa.

Tropical biomes occur when the average annual temperature exceeds 18°C. *Tropical rain forests* grow in the wettest climates and in areas where copious, almost daily rainfall occurs. Even though soils are mineral-poor (nutrients are held in the living vegetation), tree species diversity is high—hundreds of species may be found within a few square kilometres in lightly disturbed rain forests of Central and South America, Indonesia, the Philippines, Borneo, parts of Malaysia, northeastern Australia, and Hawaii. Generally, tropical rain forests support at least three storeys of forest foliage, including many epiphytes. Animal species also are abundant; mammals frequently live in the trees, although some are ground dwellers. Insects and other invertebrates show great diversity. Many undiscovered species live in these relatively unknown tropical rain forests.

Tropical seasonal forests and *tropical savannahs* occur where rainfall is high (up to 120 centimetres) and a pronounced dry season occurs, such as in India, Southeast Asia, Africa, and South and Central America. *Tropical savannas* (grasslands with scattered trees) in Africa exhibit the greatest abundance of large mammals remaining in the world, and plant species diversity is high also. Fire and grazing are common sources of disturbance.

Deserts occur in both temperate and tropical areas where there is little precipitation (less than 50 centimetres per year) and high evaporation. North Africa's Sahara Desert and deserts of the southwestern United States, Mexico, and Australia occur at low elevations. Cold deserts occur in areas such as the high country in Nevada and Utah and in parts of western Asia. Communities and organisms in deserts, including invertebrate and vertebrate animals, possess specialized water-conserving adaptations. Nonmammalian vertebrates such as snakes and reptiles are the dominant animal species; desert mammals are usually small, such as the kangaroo mouse in North American deserts. Desert soils frequently lack organic matter but have abundant nutrients, needing only water to be highly productive. Common disturbances come in the form of occasional fires, occasional cold weather, and infrequent but sudden, intense rains that cause flooding.

Climate and vegetation vary with latitude (distance from the equator) and altitude (height above sea level). If we were to travel from the equator to the North or South poles or from low to high altitudes, we would see parallel changes in vegetation (see Figure 3–18). That is, changes in the type and distribution of vegetation we would see while hiking up a mountain (gaining altitude) would be similar to the changes in vegetation seen while travelling toward the North Pole (increasing latitude). Hiking some of the trails in the Canadian Rockies, for example, we would be able to see the temperate deciduous forest at the base of the mountains give way to subalpine coniferous species that, with increasing elevation gain, would give way to alpine tundra below the permanent ice and snow of the peaks.

LIFE ON EARTH: MAJOR AQUATIC BIOMES

Oceans

Oceans cover about 71 percent of the Earth's surface, serving as a huge reservoir for carbon dioxide, helping to regulate the temperature of the troposphere, and pro-viding habitats for about 250 000 species of marine animals and plants (many of which are eaten by humans and other organisms). Oceans contain many valuable resources, including sand and gravel, oil and natural gas, iron, phosphates, and magnesium. As long as their tolerance limits are not exceeded, their size and currents enable oceans to mix and dilute many human-produced wastes that flow or are dumped into them, rendering the wastes less harmful or even harmless.

The coastal zone and the open sea are the two major life zones of the oceans. Constituting less than 10 percent of the ocean's area, the **coastal zone** contains about 90 percent of all marine species and is where most of the large commercial marine fisheries are located. The coastal zone is the nutrient-rich, relatively warm shallow water that extends from the high-tide mark on land to the gently sloping, relatively shallow edge of the continental shelf at a depth of about 200 metres.

Coastal zones are among the most densely populated and most intensely used and polluted ecosystems on Earth. Coral reefs, estuaries, coastal wetlands, and barrier islands are among the most highly productive ecosystems within the coastal zone. Their high net primary productivity per unit of area is due to the ample sunlight and the nutrients deposited from land and stirred up by wind and ocean currents.

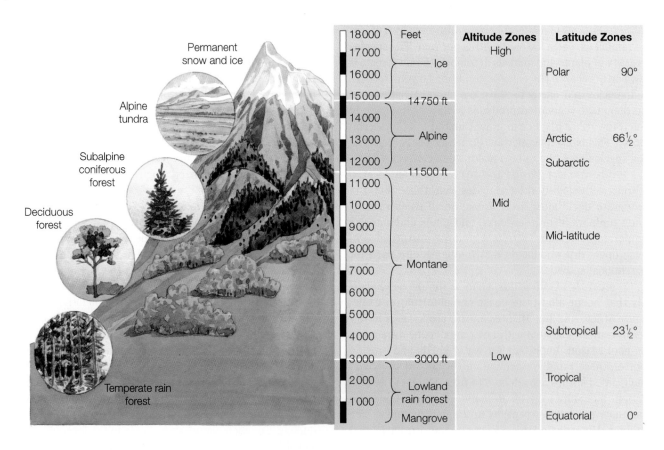

Figure 3–18

Generalized effects of altitude and latitude on climate and biomes

Photo 3–14
Rivers carry nutrients to the ocean, where the nutrients support many coastal and marine species. Fertile deltas also support human agriculture.

Found in warm tropical and subtropical oceans, slow-growing **coral reefs** are as rich in species as tropical rain forests; a single reef may contain more than 3000 species of corals, fish, and shellfish. Not only do almost one-third of all the world's fish live on coral reefs, providing critical fishing grounds for many countries, but coral reefs also reduce the energy of incoming waves, thus protecting about 15 percent of the world's shorelines from storms. In spite of their importance, coral reef formations are being degraded and destroyed by pollution, siltation from clear-cutting operations inland, land-reclamation efforts, tourism, and mining of coral formations for building materials.

As bodies of coastal water partly surrounded by land, **estuaries** have access to the open sea and a large supply of fresh water from rivers. A combination of several factors—nutrients from the land transported into the estuary by rivers and streams, ocean current action that rapidly circulates the nutrients and filters out waste products, and the presence of many plants whose roots and stems mechanically trap food material—provides important nursery conditions for the larval stages of most commercially important shellfish and fin fish species.

Coastal wetlands and **mangrove swamps** also are breeding grounds and habitats for marine organisms (oysters, crabs), waterfowl, shore birds, and other wildlife. Here, too, human ignorance has caused the loss of estuarine environments as people have used coastal wetlands and tidal marshes as dumps or have filled them with dredged material to form artificial land for residential and industrial developments. Mangrove swamps often are destroyed to provide firewood and agricultural land.

Barrier islands, the long, low, narrow offshore islands of sediment that run parallel to much of North America's Atlantic and Gulf coasts, help protect coastal wetlands, lagoons, estuaries, and the mainland from storm damage by dispersing wave energy. These islands also come under stress from human recreational and developmental activities.

The second major life zone of the oceans, the open sea, has two main divisions: the **benthic environment** (ocean floor) and the **pelagic environment** (ocean water). The pelagic environment has three vertical zones—the euphotic, bathyl, and abyssal zones—that are based chiefly on the penetration of sunlight (see Figure 3–19). The euphotic region (the upper 100 to 200 metres of ocean water beyond the continental shelf) allows enough sunlight to penetrate for photosynthesis to take place. Here, large populations of producers (microscopic cyanobacteria and protists) known as phytoplankton are eaten by slightly larger primary consumers (zooplankton) that are then eaten by other consumers such as baleen whales, herring, sardines, and other small fish. In turn, larger predators such as tuna, mackerel, seals, and orcas eat these small fish. Dead and decaying organisms fall to the ocean floor to feed microscopic decomposers and scavengers such as crabs and sea urchins.

The bathyl zone (from a depth of 200 to about 2000 metres) is lit dimly, much like twilight. Many species of large animals are found here, but at low density. Beyond about 2000 metres, into the abyssal zone, very little light penetrates; the few organisms living in this zone are adapted to darkness and scarcity of food. For example, chemosynthetic bacteria surrounding hydrothermal vents (that release hydrogen sulphide) provide support for giant clams, worms, and other unusual life forms. Except at upwellings, where the upward flow of water brings nutrients to the surface (particularly in the Arctic and Antarctic—see Chapter 8), the productivity of the open sea is quite low, principally due to the lack of sunlight and the fairly low levels of nutrients such as nitrogen and phosphorus.

Freshwater Ecosystems

Freshwater ecosystems include the standing water in lakes and ponds, the flowing water in rivers and streams, and freshwater wetlands (marshes and swamps). Lakes are large bodies of standing water formed when precipitation, overland runoff, or groundwater flowing from springs fills depressions on the earth's surface. Typically, the water in large lakes would have three basic life zones—littoral, limnetic, and profundal—while small lakes would lack a profundal zone. Just as in marine environments, each of these zones provides habitats and niches for different species.

The **littoral zone** is the shallow-water and vegetated area along the shore of a lake or pond (see Figure 3–20). Because nutrient availability and photosynthesis are greatest here, this is the most productive zone of the lake. Frogs and tadpoles, turtles, worms, crustaceans, insect

CHAPTER 3: EARTH'S LIFE-SUPPORT SYSTEMS

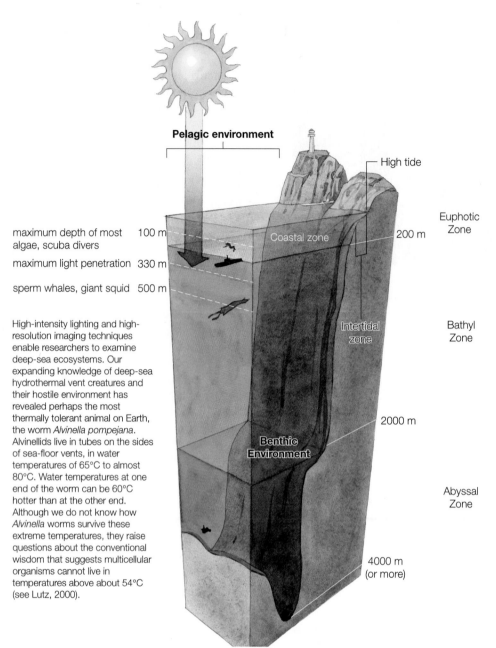

High-intensity lighting and high-resolution imaging techniques enable researchers to examine deep-sea ecosystems. Our expanding knowledge of deep-sea hydrothermal vent creatures and their hostile environment has revealed perhaps the most thermally tolerant animal on Earth, the worm *Alvinella pompejana*. Alvinellids live in tubes on the sides of sea-floor vents, in water temperatures of 65°C to almost 80°C. Water temperatures at one end of the worm can be 60°C hotter than at the other end. Although we do not know how *Alvinella* worms survive these extreme temperatures, they raise questions about the conventional wisdom that suggests multicellular organisms cannot live in temperatures above about 54°C (see Lutz, 2000).

Figure 3–19

Zonation in the marine environment

larvae, and many fish as well as insects are found in this zone. Extending downward as far as sunlight penetrates, the **limnetic zone** is the open-water area away from the shore. Here the main organisms are phytoplankton (photosynthetic cyanobacteria and algae) and zooplankton (nonphotosynthetic organisms including protozoa—animal-like protists—and small animals including the larval stages of many animals that are large as adults). Larger fish spend most of their time in the limnetic zone, although they may feed and breed in the littoral zone. There is less vegetation in this zone due to its depth.

The deepest zone of a lake is its **profundal zone**, where lack of light means that no producers live there.

Considerable food drifts into the profundal zone from the other two zones and decay bacteria decompose the dead plants and animals, freeing the minerals from their bodies. With no producers to absorb and incorporate these minerals into the food chain, however, the profundal zone habitat becomes rich in minerals and anaerobic, and few organisms live there.

Normal lakes that have minimal levels of nutrients are unenriched or **oligotrophic**, whereas a lake that is enriched (with nitrates and phosphates) in excess of what producers need is termed **eutrophic**. Many lakes fall somewhere between the two extremes of nutrient enrichment and are called **mesotrophic** lakes.

Photo 3–15
Aquatic vegetation provides cover and nursery conditions for fish and shellfish.

Freshwater Rivers and Streams

Movement of water from the mountains to the seas gives rise to different environmental conditions throughout a river system and from stream to stream. Headwater streams (the sources of a river) usually are shallow, swiftly flowing, highly oxygenated, and cold. Organisms here frequently exhibit adaptations such as suckers to avoid being swept away by the current, or flattened bodies to slip under or between rocks. Farther from its headwaters, the river usually widens and becomes deeper, slower-flowing, less oxygenated, and not as cold. Although their shapes are streamlined to reduce resistance when moving through water, organisms in these larger, slow-moving water bodies do not need the same adaptations as those in faster waters. In fact, where the current is slow, plants and animals characteristic of lakes and ponds replace those of the headwaters.

Flowing-water ecosystems are different from freshwater ecosystems not only because of water currents but also in their dependence on the land for much of their energy. Up to 99 percent of the energy input in headwater streams comes from leaves and other detritus (dead organic matter) carried by wind or surface runoff into the stream. Farther downstream, rivers have more producers and are less dependent on detritus as an energy source.

Water moving downhill is associated with landform creation: over the years, the friction from sediment-laden waters can level mountains and cut deep canyons into the landscape. Rocks and sediments removed by the water are deposited in low-lying areas, contributing to (salt) marshes and building up deltas and other landforms.

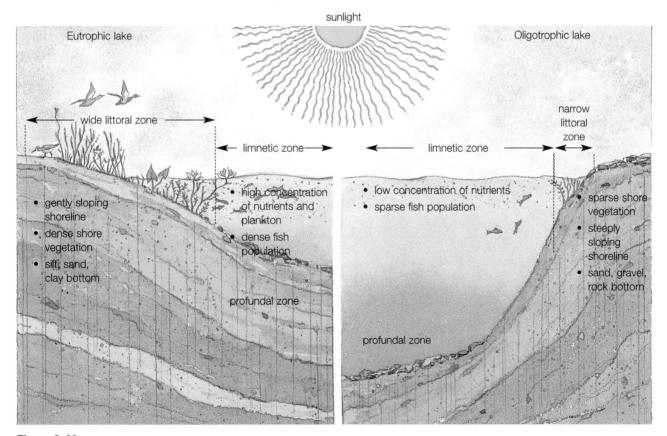

Figure 3–20
Basic life zones in eutrophic and oligotrophic lakes

CHAPTER 3: EARTH'S LIFE-SUPPORT SYSTEMS

Photo 3–16
A flowing-water ecosystem in the Queen Charlotte Islands.

Inland Wetlands

Transitional between aquatic and terrestrial ecosystems, **wetlands** are usually covered with fresh water at least part of the year and have characteristic soils and water-tolerant vegetation. Inland wetlands include marshes, dominated by grass-like plants; swamps, dominated by woody plants; bogs, including peat moss bogs; prairie pot-holes; mud flats; floodplains; fens; wet meadows; and the wet Arctic tundra in summer. While some wetlands are covered with water year-round, others are seasonal, including prairie potholes (small shallow ponds formed when glacial ice melted at the end of the last ice age) and floodplain wetlands.

Inland wetlands provide habitat for game fish, migratory waterfowl, beaver, otters, muskrats, and other wildlife, and they improve water quality by acting as a sink to filter, dilute, and degrade sediments and pollutants as

Photo 3–17
Cranberry fields are one example of the agricultural use of wetlands.

water flows through them. Wetlands help control flooding by storing excess water during periods of heavy rainfall, or when rivers flood their banks, by slowly releasing the water back into the rivers. Not only do wetlands help to provide a steady flow of water throughout the year, they also reduce riverbank erosion and flood damage. Through infiltration, wetlands serve an important function as groundwater recharge areas. Freshwater wetlands produce many commercially important products such as wild rice, cranberries, and peat moss. Many people also enjoy fishing, hunting, boating, photography, and nature study in wetland areas.

In spite of the ecological importance of permanent and seasonal wetlands, they are threatened increasingly by their conversion to cropland and by pollution, dredging and mining, engineering (dams and highways), and urban development.

RESPONSES TO ENVIRONMENTAL STRESS

THE CONSTANCY OF CHANGE

The "balance of nature" is a misnomer—the key thing happening in nature is constant change, caused by both natural and human-related forces and adjustments to environmental stresses. When we talk about sustainability as a concept in the environmental sciences, we include the idea that populations, communities, ecosystems, and human systems all have the capacity to adapt to new and changing conditions. In the following sections, we consider the kinds of changes that have been going on and continue to go on in the world around us.

Changes in Population Size

Changes in human population sizes are governed by four variables: births, deaths, immigration, and emigration. Births and immigration contribute to gains in a population, while deaths and emigration lead to declines in population numbers. A simple formula describes this relationship:

Population change = (births + immigration) – (deaths + emigration).

Clearly, the influence of these variables is affected by changes in resource availability or other environmental changes.

Populations have a variable capacity for growth. Under ideal conditions—that is, when there are no limits on its growth—the **biotic potential** of a population is the maximum rate (r_{max}) at which it can increase. Under ideal conditions, populations of all organisms will increase exponentially. If any organism's biotic potential is plotted against time, a characteristic J curve will emerge (see Figure 3–21). In animal species, however, different biotic potentials exist because of variations in their reproductive span (when reproduction starts and stops), frequency of reproduction, litter size, and the number of offspring that survive to reproductive age. Generally, larger organisms such as elephants and whales have lower biotic potentials, whereas microorganisms have the greatest biotic potentials. The limiting factor principle asserts that in nature there are always limits to growth, and no population can continue to grow indefinitely at its biotic potential. Instead, a population will reach a size limit imposed by a shortage of one or more of the limiting factors of light, water, space, and nutrients. Essentially, the number of organisms in a population is controlled by the ability of the environment to support that population.

The limits set by the environment that prevent organisms from reproducing indefinitely at an exponential rate are known collectively as **environmental resistance**. **Carrying capacity** represents the highest population that can be maintained for an indefinite period of time by a particular environment as characterized by its particular limiting factors. Many factors determine carrying capacity, including predation, competition among species, migration, and climate. As well, seasonal or abnormal changes in the weather or food supplies, water, nesting or calving sites, and other crucial environmental resources can alter the carrying capacity for a population. If tracked over long periods of time, the rate of population growth for most organisms decreases to about zero; this levelling out occurs at or near the limit of the environment's ability to support a population. The curve on a graph of popula-

tion numbers plotted against time has a characteristic S shape that also shows the population's initial exponential increase (note the J shape at the start), followed by a levelling out as the carrying capacity of the environment is approached (see Figure 3–21).

Sometimes, because of a reproductive time lag (the time required for the birth rate to fall and the death rate to rise in response to environmental resource limits), a population temporarily will overshoot the carrying capacity (see Figure 3–22a). Unless large numbers of individuals can avoid local environmental degradation by moving to an area with more favourable conditions, the population will crash. Often such a population will fall back to a lower level that fluctuates around the area's carrying capacity. Also during the overshoot period, if degradation and destruction of local environmental resources occur, the area's carrying capacity may be lowered.

In nature, three general types of population change curves may be observed: relatively stable, irruptive, and cyclic (Figure 3–22b). If we were to examine many of the species found in undisturbed tropical rain forest habitats, we would find their population sizes fluctuate slightly above and below their carrying capacities. If we assume no significant changes in those capacities, these species would be exhibiting relatively stable population sizes. If we considered a species such as the raccoon, which normally has a fairly stable population, we would observe an occasional explosion or irruption of the population to a high peak, followed by a crash to a relatively stable lower level. The population explodes in response to factors (such as better weather, more food, or fewer predators) that temporarily increase carrying capacity for the population. In some other instances, species undergo sharp increases and periodic crashes in their numbers; the actual causes of these boom and bust cycles are not well understood, although predators sometimes are blamed.

Species reproduction strategies vary widely. Some, such as algae, bacteria, rodents, many fish, most insects, and annual plants, produce a great many offspring early in their life cycle. The offspring usually are small, mature rapidly, and have short life spans, many dying before they can reproduce. Such species are opportunistic and reproduce rapidly when favourable conditions exist or when a new habitat or a new niche becomes available (a newly plowed field, a recently cleared forest). If environmental conditions are not favourable, however, such populations will tend to go through boom and bust cycles.

At the other end of the reproductive strategy spectrum are species such as humans, elephants, whales, sharks, birds of prey, and long-lived plants whose offspring are few in number, fairly large, and often nurtured for lengthy periods to ensure that most reach reproductive age. Populations of these species exhibit an S-shaped curve as their numbers are maintained near the carrying capacity of their fairly stable habitats.

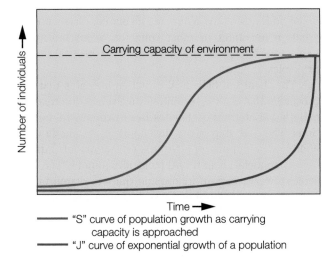

Figure 3–21
S and J curves of population growth

Figure 3–22

Selected population change dynamics

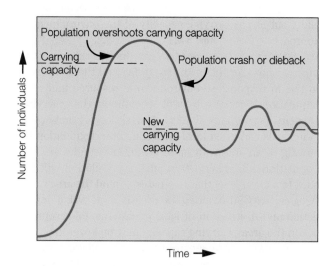

a) A reproductive time lag and population crash

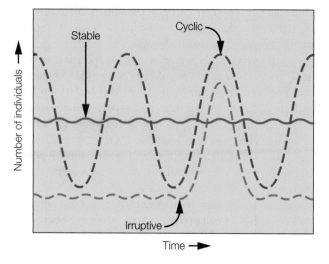

b) Three idealized types of population change curves

Human populations, too, are affected by population biology dynamics. If we look back to events such as the 1845 destruction of the Irish potato crop (by a fungus infection), we can see a human population crash in which about one million people died and three million people emigrated. Today, changes in technology, society, and culture appear to have extended Earth's carrying capacity for the human species—we seem to have controlled many diseases, increased food production, and used energy and matter resources at rapid rates in order to make habitable and usable many formerly uninhabitable and unusable parts of the Earth. On a planet that has finite resources and space, a critical question is how long we will be able to keep extending this carrying capacity.

Biosphere II was a US$30 million experiment that attempted to develop a self-sustaining ecological system (effectively extending carrying capacity). Occupying more than three acres near Tucson, Arizona, the area was completely enclosed in glass and steel and its occupants were sealed off from the Earth. Powered by solar energy, and theoretically containing everything required to sustain life, Biosphere II housed more than 3800 species of macroscopic plants and animals, including eight human researchers, who spent two years (1991–93) living inside this artificial environment. The simulated Earth environments contained within Biosphere II were intended to provide food and the means to recycle water and wastes and to purify air and water.

Scientists and other observers criticized the project for having a secretly installed carbon dioxide recovery system (to augment Biosphere II's air recycling system), for letting outside air in, and for having a large supply of food hidden inside. While the creators of Biosphere II denied the charges, the experience showed how difficult it is to maintain natural cycles and species diversity even in such a simplified biosphere. This lesson should cause us to pause and think about how well complex natural ecosystems work and how appropriate our efforts to manage them better really are.

Biological Evolution, Adaptation, and Natural Selection

Understanding biological diversity is important if we are to meet the challenges for our sustainable future. In this context, it is important to appreciate that **biological evolution** (change in inherited characteristics of a population from generation to generation) is touted as the driving force of adaptation to environmental change. According to the theory of evolution, it is the evolution of populations from earlier forms that explains the diversity of life on Earth today. While this theory conflicts with the creation accounts of most religions, most biologists explain the changes in life on Earth based on evolutionary principles.

The theory of biological evolution indicates that the processes that lead to evolution involve the interplay of genetic variation and changes in environmental conditions. The first step is the development of genetic variability in a population. **Genes** (segments of various deoxyribonucleic acid [DNA] molecules found in chromosomes) impart certain inheritable traits to organisms. A population's **gene pool** is the sum of all genes possessed by the individuals of a population, but individuals within the population of a particular species do not share exactly the same genes. **Mutations**, the random and unpredictable changes in DNA molecules that can be transmitted to offspring, bring about this variability. Every time a cell divides, the DNA is reproduced so that each new cell gets a copy, but sometimes the copying is inaccurate, resulting in a change in the DNA and a subsequent change in inherited characteristics. External environmental agents such as radiation (X-rays, ultraviolet light) or certain toxic organic chemicals also may come in contact with the DNA and alter the molecule. The result of millions of random changes in the DNA molecules of individuals in a population is genetic variability.

Offspring with a mutation frequently fail to survive, but in other cases mutations result in new genetic traits that give individuals and their offspring improved chances for survival and reproduction. Any genetically controlled characteristic—structural, physiological, or behavioural—that enhances the chance for members of a population to survive and reproduce in its environment is called an **adaptation**.

Structural adaptations include coloration (which enables individuals to hide from predators or to sneak up on prey), mimicry (which allows individuals to look like a dangerous or poisonous species), protective cover (shell or thick skin, bark, thorns), and gripping mechanisms (hands with opposable thumbs). Physiological adaptations include the ability to poison predators, to give off chemicals that repel prey, and to hibernate during cold weather. The ability to fly to a warmer climate during winter is an example of behavioural adaptation, as are resource partitioning and species interactions such as parasitism, mutualism, and commensalism.

Individuals with one or more adaptations that enable them to survive under changed environmental conditions are more likely to produce more offspring with the same favourable adaptations than are individuals without such adaptations; this is known as **differential reproduction**. **Natural selection** is the process by which organisms whose biological characteristics (beneficial genes) better fit them to the environment are better represented by descendants in future generations than are those organisms whose characteristics are less fit for the environment. Natural selection has four major characteristics that over time demonstrate the combination of factors leading to evolution: genetic variability (inheritance of traits in succeeding generations, with some variation in those traits); environmental variability; differential reproduction that varies with the environment; and the influence of the environment on survival and reproduction.

Limits to adaptation exist in nature. Human lungs and livers, for example, do not quickly become more capable of dealing with air pollutants and other toxins to which we are exposed. Human skin cannot evolve sufficiently rapidly to be more resistant to the harmful effects of ultraviolet radiation. The reasons for our inability to adapt quickly relate chiefly to our reproductive capacity. Unlike mosquitoes, rats, and species of weeds, humans and other large species such as sharks, elephants, and tigers cannot produce large numbers of offspring rapidly,

so our ability to adapt quickly to particular environmental changes through natural selection is very limited. In addition, since a population can adapt only for traits that are present in its gene pool, even if a favourable gene were present, most members of the population would have to die or become sterile so that individuals with the trait could become dominant and pass on the trait. This is hardly a desirable solution! We must remember, also, that species adaptations develop in the context of their ecological situation (relationships with other organisms and the environment) and that our knowledge of this complexity is incomplete.

Speciation and Extinction

Two other processes, speciation and extinction, are believed to have affected the millions of species on Earth. **Speciation** is the formation of two or more species from one as the result of divergent natural selection in response to changes in environmental conditions.

One mechanism that generates speciation is geographic isolation, a situation in which populations of species become separated for a long time in areas with different environmental conditions. This may occur naturally (as when part of a group migrates for food) or through human-related activities. Highway construction, for example, may quickly separate a small group of individuals from a larger population. If these groups remain geographically separated for a long time and do not interbreed, they may begin to diverge in their genetic makeup because of different selection pressures. Continued reproductive isolation may mean that members of the separated populations become so different that they cannot interbreed and produce fertile offspring. If this occurs, one species becomes two, as perhaps occurred with arctic and grey foxes. It is thought that these two species were once one but separated into northern and southern populations that adapted to different environmental conditions (the arctic fox with its white, heavy fur and short ears, legs, and nose adapted to cold conditions, and the grey fox with its lightweight fur and long ears, legs, and nose adapted to heat).

Extinction is the second process affecting Earth's species. A species is eliminated from existence when it cannot adapt genetically and reproduce successfully under new environmental conditions. The continuous, low-level extinction of species that has occurred throughout much of the history of life is termed **background extinction**. **Mass extinction**, in contrast, is the disappearance of numerous species over a relatively short period of geological time. Fossil and other geological evidence points to catastrophic global events that eliminated major groups of species simultaneously. While extinction is a crisis for the affected species, extinction provides an opportunity for other species to evolve to fill new or vacant ecological niches in changed environments (a process called adaptive radiation). Species biodiversity, one of the Earth's most important resources, may be expressed in an equation:

Species biodiversity = number of species + speciation – extinction.

Ecological Succession

A community of organisms does not spring into full-blown existence but develops gradually through a series of stages until it reaches maturity. The process of community development over time, during which the composition and function of communities and ecosystems change, is called **succession**. Ecological succession is a normal process in nature, and is in constant flux as a community or ecosystem responds to natural or human disturbance. A landscape that includes several successional stages represents maximum habitat for the greatest number of species.

Succession is usually described in terms of the species composition of the vegetation of an area, although each successional stage also has its own characteristic animal life. Ecologists recognize primary and secondary types of ecological succession, depending on conditions at a particular site at the outset of the process. **Primary succession** involves the development of biotic communities in a previously uninhabited and barren habitat with no or little soil. A bare rock surface recently formed by volcanic lava or scraped clean by glacial action, or a new sandbar deposited by shifting ocean currents, or a surface-mined area from which all overburden (soil) has been removed, are examples of conditions under which primary succession begins. Usually the first signs of life are the hardy pioneer species such as microbes, lichens, and mosses that are quick to establish large populations in a new area.

Secondary succession begins in an area where the natural vegetation has been removed or destroyed but where soil is present. Burned or cut forests and abandoned farm fields are common sites where secondary succession occurs. The presence of soil permits new vegetation to spring up quickly, often in a matter of weeks, and subsequent successional stages to develop over the next 100 or more years.

Ecological succession is not necessarily an orderly process, with each successional stage leading inevitably to the next more stable stage until the area is occupied by a mature or **climax community**. The exact sequence of species and community types that appear during the course of succession in a given area can be highly variable. Even the climax community is not permanent; fires, floods, landslides, and extreme weather events all present opportunities for new species to colonize altered space, once again illustrating the constancy of change in nature.

HUMAN IMPACTS ON ECOSYSTEMS

Over the past century, human activities have simplified natural ecosystems. Subsequent to plowing native prairie grasses, clear-cutting forests, and filling in wetlands, we replaced the complexity of thousands of plant and animal interrelationships in these ecosystems with monocultures of wheat, canola, or commercially valuable trees, or with highways, parking lots, and buildings.

Humans spend a lot of money defending such monocultures from invasion by pioneer species because weeds (plants), pests (insects or other animals), and pathogens (fungi, viruses, or disease-causing bacteria) can destroy an entire monoculture crop unless it is protected by pesticides or some form of biological control. Given that fast-breeding insect species undergo natural selection and develop genetic resistance to pesticides, people end up using stronger doses or switching to new pesticides. Ultimately, natural selection in the pests increases to the point that these chemicals are ineffective. This danger is seen in the comeback of malaria, principally because of mutant forms of the mosquito-borne microbe that are becoming increasingly resistant to drugs (Nichols, 1997).

These processes highlight the important point that when humans intrude into nature, we can never affect just one thing; instead, we cause multiple effects, many of which are unpredictable in their outcomes. The **principle of connectedness** specifies that everything is connected to and intermingled with everything else—we are all in this together. The multiple effects of human activities result from our limited understanding of how nature works and our inability to identify which connections are strongest and most important in the sustainability and adaptability of ecosystems.

Cultivating monocultures is not the only way people simplify ecosystems. Many Canadian ranchers continue to poison ground squirrels that compete with livestock for grass and damage crops and fields, inadvertently eradicating species such as the burrowing owl and swift fox (refer to Box 3–1). Ranchers frequently shoot wolves, coyotes, bears, and other predators that occasionally kill livestock. During the winter of 1995, some Saskatchewan sheep ranchers put out insecticide-laced meat to try to kill coyotes; at least 15 bald eagles were killed instead. One rancher was fined only $430, less than the cost of one guard dog (Dambrofsky, 1996).

Nomadic herders and ranchers around the world have permitted livestock to overgraze grasslands to the point where erosion turns these ecosystems into less productive deserts or wastelands. Furthermore, the cutting of large tracts of tropical and temperate rain forests has destroyed part of the Earth's biodiversity, and people have fished and hunted some species to extinction or near extinction; all of these activities simplify ecosystems. Even burning fossil fuels in our vehicles, homes, and industrial plants simplifies forest and aquatic ecosystems by creating air pollutants that not only degrade the atmosphere but also kill or weaken trees and fish.

Maintaining a balance between simplified human ecosystems and the more complex natural ecosystems that surround us is a challenge, particularly given the rate at which we have been altering nature for human purposes.

WORKING WITH NATURE

Living systems have six key features—interdependence, diversity, resilience, adaptability, unpredictability, and limits—that suggest humans could learn to live sustainably if they understood and mimicked how nature perpetuates itself. Learning to live sustainably begins with recognizing the following: humans are a part of, and not separate from, the dynamic web of life on Earth; human economies, lifestyles, and ultimate survival depend totally on the sun and the Earth; and everything is connected to everything else, although some connections are stronger and more important than others.

The law of conservation of matter and the second law of energy tell us that as we use resources, we add some waste heat and matter to the environment. In recent decades, the rapid economic growth and rising standard of living in Canada and other industrialized countries have been supported by accelerated resource consumption and increased use of Earth's materials and energy. Our attempts to maximize short-term economic gain have generated a legacy of degraded water, air, soil, forests, and biological diversity. "As the world becomes ecologically over-loaded, conventional economic development actually becomes self-destructive and impoverishing. Many scholars believe that continuing on this historical path might even put our very survival at risk" (Wackernagel & Rees, 1996, pp. 2–3).

One of the first things we must do is acknowledge and accept the reality of ecological limits and the resultant changes that such acceptance will bring to our socioeconomic systems. Even if we were to recycle more materials so that economic growth could continue without depleting resources and without producing large volumes of pollutants, we need to remember that recycling matter resources requires high-quality energy, which cannot be recycled. If we were to become a recycling society as an interim stage in moving toward achieving a sustainable society, we would need not only an inexhaustible supply of affordable high-quality energy, but also an environment with an infinite capacity to absorb, disperse, dilute, and degrade heat and waste. Since such efforts can be temporary solutions at best, the scientific principles and ecological functioning identified in this chapter help show us that human society, including the human economy, is a subsystem of the ecosphere. If we are to achieve sustainability, we must shift our emphasis "from 'managing resources' to managing ourselves" and "learn to live as part of nature." Such a shift in emphasis would mean that "[e]conomics ... becomes human ecology" (Wackernagel & Rees, 1996, p. 4).

Such a shift would require us to undertake a range of actions to achieve sustainability, including reduction in use of matter and energy resources through more efficient and appropriate use of energy, a shift to renewable energy sources, less waste of renewable and nonrenewable resources of all types, and an emphasis on pollution prevention and waste reduction. Achieving a sustainable future will require that all Earth's citizens learn to work with nature.

Chapter Questions

1. Using the second law of energy, explain why some may choose to become vegetarians.

2. Distinguish between an ecosystem and a biological community.

3. Why is a realized niche usually narrower (more restricted) than a fundamental niche?

4. Explain the difference between a habitat and a niche.

5. Using the second law of energy, explain why there is such a sharp decrease in usable energy as energy flows through a food chain or web. Does the energy loss at each step violate the first law of energy?

6. Why is the cycling of matter essential to the continuance of life on Earth?

7. Why is an understanding of biogeochemical cycles important in environmental science? Use one or two examples to justify your answer.

8. Compare and contrast the geochemical cycles for phosphorus and nitrogen.

9. What climate and soil factors produce each of the major terrestrial biomes?

10. Why are coral reefs and coastal and inland wetlands such important ecosystems? Why have human activities destroyed so many of these vital ecosystems?

11. Is the human species a keystone species? Explain.

12. Why are natural ecosystems less vulnerable to harm from insects, plant diseases, and fungi than human-modified ecosystems?

references

Dambrofsky, G. (1996, April 8). Ranchers bothered by killer coyotes. *Calgary Herald*, p. A3.

Fellman, J., Getis, A., & Getis, J. (1995). *Human geography: Landscapes of human activities.* Dubuque, IA: Brown.

Government of Canada. (1991). *The state of Canada's environment—1991.* Ottawa: Supply and Services Canada.

Huyghe, P. (1993). New species fever. *Audubon, 95*(2), 88–92, 94–96

Kaufman, D. G., & Franz, C. M. (1993). *Biosphere 2000: Protecting our global environment.* New York: HarperCollins.

Lean, G., & Hinrichsen, D. (1992). *Atlas of the environment.* Oxford: Helicon.

Lutz, R. A. (2000). Deep sea vents. *National Geographic, 198*(4), 116–127.

Miller, G. T. (1994). *Sustaining the Earth: An integrated approach.* Belmont, CA: Wadsworth.

Moffett, M. W. (1997). Climbing an ecological frontier: Tree giants of North America. *National Geographic, 191*(1), 44–61.

Nichols, M. (1997, May 19). Malaria's comeback. *Maclean's*, 57.

Perry, D. (1986). *Life above the jungle floor.* New York: Simon & Schuster.

Postel, S. (1996). Carrying capacity: Earth's bottom line. In J. L. Allen (Ed.), *Environment 96/97* (pp. 28–36). Guilford, CT: Dushkin Publishing Group/Brown and Benchmark.

Wackernagel, M., & Rees, W. (1996). *Our ecological footprint: Reducing human impact on the Earth.* Gabriola Island, BC: New Society.

CHAPTER 4

Human Population Issues and the Environment

"Since 1940 world numbers have grown by 3 billion. By 1960 environmental stresses were already showing: severe soil erosion, spreading deserts, and shrinking forests. By 1970 there were added the problems of gross-scale pollution and, by 1980, acid rain and mass extinction of species, plus suspicion of the greenhouse effect and ozone-layer depletion. Today it is plain there is biospheric breakdown of multiple sorts. If present trends continue until 2025, will we be looking out on a biosphere needing centuries or millennia to repair?"

N. Myers (1990, p. 39)

"The contentious issue of population growth and 'overpopulation' looms large in our collective imagination. Since Malthus, and more insistently in the last four decades, we have been barraged with strident claims that proclaim 'over population' is at the root of all our problems... 'They' are the problem. 'We' are absolved of all responsibility."

Jael Silliman (1999, p. vii)

Chapter Contents

Chapter Objectives

After studying this chapter you should be able to

- explain why some people consider the human population issue to be the most important environmental issue

- understand how population size is affected by rates of birth, death, fertility, and migration

- discuss the importance of exponential growth rates, doubling times, and other measures of population dynamics

- describe the four major phases in the demographic transition

- construct an age structure diagram and use it to predict future population growth rates

- discuss the approaches taken in response to world population growth

INTRODUCTION

Many have observed that human population growth and environmental damage have coincided. This link has led to claims by some scientists that the human population issue is *the* environmental issue, while others, like Jael Silliman quoted above, contend that such a link is simplistic and leads to erroneous conclusions. The tremendous increase in human numbers is testament to the biological success of humans as a species. But when we combine our biological success with our great technological power to cause environmental change, we realize that humans are using up, depleting, and degrading Earth's limited resources, threatening not only our own sustainability but that of other plant and animal species that share this planet. There is no doubt that we live on a human-dominated planet (Vitousek et al., 1997). Pollution of natural environments, depleted energy reserves, reduced biodiversity, and wildlife extinctions are serious problems. Do these arise from the growth of human populations or do they arise because of the ways human populations use their environments? This is a subtle, yet significant, distinction.

This chapter examines fundamental concepts about human population, the state of the current human population, and how the human population has changed over time. We also consider the consequences of human domination of the planet and approaches that address human population changes in different parts of the world.

BASIC POPULATION CONCEPTS

We face a paradox. If you are the typical age of most university or college students, you probably were born in the mid- to late 1980s. In 1980, the world contained close to 4.5 billion people. By 1999, the world's population reached 6 billion people, and it is possible—if the annual growth rate remains at about 1.3 percent—that during your lifetime human numbers will grow to about 8.9 billion. At that rate, about 78 million people are added to the Earth's population each year, more than 2.5 times the entire population in Canada! About 90 percent of these 78 million people are born in the developing nations of Africa, Asia, and Latin America (see Box 4–1). From 1995 to 2000, the average annual population growth rate in these developing regions was about 1.6 percent, while in developed nations such as Canada, the United States, Japan, and those of Western Europe, the growth rate was less (sometimes much less) than 1 percent (Kent & Crews, 1990). (See Appendix B for selected Canadian

BOX 4-1

DISTINGUISHING BETWEEN DEVELOPED AND DEVELOPING COUNTRIES

Countries can be classified into two groups—developed and developing—based on a number of factors.

Developed countries (also known as post-industrial, highly developed countries [HDCs] or, in some cases, overdeveloped countries) have low rates of population growth and the longest doubling times (see page 102), are highly industrialized, and have high per capita incomes relative to the rest of the world. Countries such as Canada, the United States, Japan, Sweden, Australia, and Germany are HDCs. They are also the largest consumers of the world's resources, particularly fossil fuels. Developed countries have the lowest birth rates in the world, and in some, such as Germany, birth rates are just below those needed to sustain their populations and thus their populations are declining slightly. HDCs also have very low infant mortality rates.

Developing countries can be categorized as either moderately developed countries (MDCs) or less developed countries (LDCs). Moderately developed countries include Mexico, Thailand, and most South American nations. In these countries, birth rates and infant mortality rates are higher than those of highly developed countries, their level of industrialization is moderate, and their average per capita incomes are lower than those of highly developed nations.

Less developed countries include Bangladesh, Ethiopia, Laos, and Niger; these countries have the highest birth rates, the shortest doubling times, the highest infant mortality rates, the shortest life expectancies, and the lowest average per capita incomes in the world.

population statistics.) Yet, in terms of consumption, these figures are reversed. The countries of Canada, the United States, Japan, and Western Europe are by far the greatest consumers of the world's resources. For example, they consume about two-thirds of the world's energy resources. Africa, for example, with about 13 percent of the world's population, uses only 3 percent of the world's energy resources. Some have estimated that "the annual commute by car into New York City alone uses more oil than the whole of Africa (excluding South Africa) in one year" (Elliott 1999, p. 40). In 1987, the World Commission on Environment and Development estimated that to achieve sustainable development we had to increase the standard of living of developing countries. To do so, there would have to be a four-fold increase in output of product and services! Yet calculations of our ecological footprint suggest that if all people consumed as much as "average Canadians," we would require several more worlds to meet our demands (see Chapters 1 and 13). How can we achieve stable human population and stable consumption in aid of sustainable development?

POPULATION AND TECHNOLOGY

The threats that humans pose to the sustainability of the environment depend on the total number of people on Earth and the impact each person has on the environment. In the past, when there were fewer people on Earth and when their technology was limited, the human impact on the environment was small in scale and confined to local areas. If a local resource was overused, the effects were not large. Long-term impacts were almost nonexistent. Now there are so many people with such powerful technologies that human impacts have the potential for important, large-scale, and long-term environmental changes.

Paul Ehrlich (1971), a well-known American expert in population biology and ecology, suggested that the total environmental effect of the human population on the Earth's environment may be expressed as the product of the impact per individual times the total number of individuals. He is known for developing the IPAT formula, where

Impact = Population + Affluence + Technology

According to this formula, either an increase in the total number of people or an increase in the individual impact each of us has on the environment results in an increase in the total human impact. Nevertheless, several scientists have emphasized population as the most important element in the equation. This approach to understanding population and environmental relationships has been criticized for oversimplifying the relationships among the variables. Other approaches highlight the social, cultural, institutional, and political contexts in which population and environment relationships occur. Today, scientists still struggle to measure and explain many of the basic relationships among population, development, and the environment (Livernash & Rodenburg, 1998).

Advances in technology have greatly increased the impacts people have on the environment. Whereas earlier peoples lived by hunting and gathering and used simple wooden or stone tools for crop production, today millions of car owners drive vehicles that use fossil fuels. Vehicle use generates significantly increased demands for oil and steel, and results in the release of air pollutants. Only many years after the invention of chlorofluorocarbons (CFCs) for use as a coolant in air conditioners and refrigerators and as a propellant in spray cans did we realize CFCs caused depletion of the ozone layer in the upper atmosphere. Immediately after World War II, dichlorodiphenyltrichloroethane (DDT) was used widely for agricultural and commercial application to combat

Photo 4–1a

Photo 4–1b

Photo 4–1c
Population concentration and congestion are visible in the development of Canadian cities. Ste. Catherine Street, Montreal, is shown in 1901 (Photo 4–1a), 1952 (4–1b), and 2000 (4–1c).

insect-borne diseases such as malaria. It was used heavily in North America until it was banned in Canada (1969) and in the United States (1972) because of its deadly and long-lasting effects on other plant and animals species, including humans. Clearly, the combination of rapid increases in both population and technology has increased human effects on the environment.

The addition of each new individual in the population of a developed or industrialized country results in a greater effect on the environment than it does in a developing nation. Canadians often express their concerns about the environmental impacts of developing nations whose populations continue to grow so rapidly compared with the Canadian population. However, we need to be aware that Canadians and residents of other smaller industrialized nations have greater per capita effects on the environment because of our higher standards of living and greater use of technology. Countries such as the United States, Japan, and those of the European Union, which have large populations and high levels of technology use, cause even greater environmental effects. For example, the military is an intensive user of contemporary technology. In the United States, 20 000 military sites are considered the most polluted hazardous waste sites in the country. The Pentagon generates a tonne of toxic waste per minute, more toxic waste than the five largest U.S. chemical companies together, making it the largest polluter in the United States. This figure does not include the Department of Environment's nuclear weapons plants and the Pentagon's civilian contractors (Hynes, 1999).

For Canadians and others who place a great value on international aid to developing countries, this aspect of the population–technology issue poses a troubling dilemma. Various aid agencies commonly strive to achieve two fundamental objectives: an improvement in the standard of living and a decline in overall human population growth. However, in supporting efforts to improve standards of living, we could be contributing to increases in the total environmental impact, a result that is counterproductive to the environmental benefits of a reduction in population growth. This does not mean we should not support international aid, but it does indicate a need to better understand the dynamics of human population growth so that our aid efforts are effective.

HUMAN DEMOGRAPHY

To improve our understanding of human population impacts on the biosphere and its resources, we need to study some basic human demography. **Demography** is the scientific study of the characteristics and changes in the size and structure of human populations (as well as nonhuman populations such as wildlife, fish, and trees). Demographers use a variety of tools, terms, and concepts to understand important population dynamics. One such

A census is an official count of the number of people in a country, including information about their age, gender, and livelihood. In Western civilization, the first estimates of population were conducted in the Roman era (particularly for taxation purposes), and there were occasional efforts to estimate numbers of people during the Middle Ages and the Renaissance. The first modern census was taken in 1655 in the Canadian colonies by the French and English.

Sweden began to undertake the first series of regular censuses in 1750, and the United States began in 1790 to undertake a census every ten years. However, many countries do not take censuses or conduct them only irregularly. For example, in much of the developing world, census coverage has been sporadic or inaccurate. Reasons for this include insufficient funds; lack of trained census personnel; high rates of illiteracy; isolated populations; poor transportation; and widespread suspicion of government, which limits the kinds of questions census takers are able to ask.

Even in countries like Canada, census information may contain gaps and biases because of differences in collecting and analyzing data or because of changes in definitions over time. It is wise to remember that the limitations noted above cause various types of errors, omissions, or limitations when using census data. When relying on census data, one should acknowledge such potential for error or identify its limitations.

---

tool is the national census (see Box 4–2). Some of the terms and concepts we consider here include birth and death rates, age structure, the demographic transition, and total fertility.

HUMAN POPULATION GROWTH

Historical growth of the human population is often described in terms of four major periods or stages (see Table 4–1). The first stage represents the early period of hunters and gatherers when total world population, population density, and average rate of growth were very low. Stage 2, beginning about the time of agricultural settlement, was characterized by the first major increase in the total world population and a much greater density of people. Stage 3 began about the time of the Industrial Revolution and saw a rapid increase in the human population resulting from improvements in health care and food supplies. Stage 4 represents today's increasingly urbanized world, in which the rate of population growth has declined in wealthy, industrialized nations but has continued to rise rapidly (until recently) in poorer, developing countries.

Although the total human population has increased with each succeeding stage, the modern era is unprecedented in terms of population growth (see Table 4–2). The dramatic increase in the total human population since 1950 and the contribution of developing regions to that growth are evident in Figure 4–1. Although there was little or no change in the maximum length of a human lifetime during the second and third stages in the history of human population growth, there were changes in birth, death, and population growth rates, in age structure, and in average life expectancy. To ensure we understand these changes in population sizes and the causes of these changes, let us consider population dynamics.

POPULATION DYNAMICS

As discussed in Chapter 3, changes in population sizes occur through births, deaths, immigration (arrivals from elsewhere), and emigration (individuals leaving to go elsewhere). How rapidly a population grows depends on the difference between the **crude birth rate (CBR)** and the **crude death rate (CDR)**. This difference is known as the **crude growth rate (CGR)**, and may be expressed as

$$CGR = CBR - CDR$$

In this simple equation, the crude birth rate (or, more simply, the birth rate) is the annual number of live births per 1000 population. It is "crude" because it relates births to total population without regard to the age or sex composition of that population. This means we cannot make accurate predictions about the future dynamics of the population based on such data.

A measure of the rate of population increase is

$$\frac{\Delta N}{N} \times 100$$

where ΔN is population change in one year.

This is the annual growth rate.

The change in the population over the total number of people in one year, times 100 percent, will yield the change in population over one year. Using the formula above, a country with a population of 4 million and with 80 000 births per year would grow at a rate of 2 percent per year.

TABLE 4–1

A BRIEF HISTORICAL OVERVIEW OF HUMAN POPULATION GROWTH STAGES

Stage	Population Time Period	Total Human Density	Population	Average Rate of Growth
1. Hunters and gatherers	• From first human on Earth to beginning of agriculture	• About 1 person per 130–260 km^2 in the most habitable areas	• As low as 250 000 to less than a few million	• At this time, the average annual rate of increase over the entire history of human population was less than 0.00011% per year.
2. Early, preindustrial agriculture	• Beginning between 9000 and 6000 B.C. and lasting until about the 16th century A.D.	• Domestication of plants and animals and the rise of settled villages increased human population density to 1 or 2 people per km^2.	• About 100 million by A.D. 1 and 500 million by A.D. 1600	• The growth rate was about 0.03%, large enough to increase the human population from 5 million in 10 000 B.C. to about 100 million in A.D. 1 (the Roman Empire accounted for about 54 million). • From A.D. 1 to A.D. 1000, the population increased to between 200 and 300 million.
3. The machine age (Industrial Revolution)	• Beginning about 1600 with the Renaissance in Europe and continuing until the 1950s	• The transition from agricultural to literate societies took place; better medical care and sanitation reduced the death rate; urban densities continued to increase.	• About 900 million in 1800, almost doubling in the next century, and doubling again to about 3 billion by 1960	• By 1600, the growth rate was about 0.1% per year, increasing about 0.01% every 50 years until 1950. • Growth resulted from discovery of causes of diseases, invention of vaccines, sanitation improvements, and medical and health advances; improvements in agriculture led to a rapid increase in production of food, clothing, and shelter.
4. The modern era	• From the 1950s onward	• Increasing urbanization of population	• Estimated 6.2 billion in 2002	• Growth rate reached 2.1% between 1965 and 1970, declining slightly to between 1.7 and 1.8% in the 1980s and 1990s. • HIV/AIDS pandemic threatens health in all regions.

SOURCE: Adapted from *Environmental Science: Earth as a Living Planet*, D. B. Botkin & E. A. Keller, 1995, New York: John Wiley & Sons, p. 87. Copyright © 1995. Reprinted by permission of John Wiley & Sons, Inc.

TABLE 4-2
WORLD POPULATION MILESTONES AND VITAL EVENTS

World Population Milestones Reached and Predicted World Population

1 billion in	1804	
2 billion in	1927	(123 years later)
3 billion in	1960	(33 years later)
4 billion in	1974	(14 years later)
5 billion in	1987	(13 years later)
6 billion in	1999	(12 years later)
7 billion in	2013	(14 years later)
8 billion in	2028	(15 years later)
9 billion in	2054	(26 years later)

World Vital Population Events

Time Unit	Births	Deaths	Natural Increase
Year	131 144 457	53 930 540	79 213 917
Month	11 095 371	4 494 212	6 601 160
Day	364 779	147 755	217 024
Hour	15 199	6 156	9 043
Minute	253	103	151
Second	4.2	1.7	2.5

SOURCES: *2002 World Population Datasheet,* Population Reference Bureau, 2003, p. 3.
http://www.prb.org/pdf/WorldPopulationDS02_Eng.pdf

The crude death rate, also called the mortality or death rate, is calculated the same way as the crude birth rate: the annual number of deaths per 1000 population. The crude growth rate of a population is the net change, or simply the difference between the crude birth rate and the crude death rate.

Death rates can be calculated for specific age groups. For instance, the **infant mortality rate** is the ratio of deaths of infants under 12 months of age per 1000 live births. The 1995 to 2000 infant mortality rate in the developing world was estimated at 57 per 1000 live births; in contrast, in the developed countries, the estimate was 9 per 1000 (United Nations, 1999). Even though infant mortality worldwide has fallen significantly since 1970, in 1998, 7.7 million infants died before reaching their first birthday. This represents 14 percent of all deaths in that year (U.S. Census Bureau, 1999). These overall figures also hide enormous regional variations. Canada's infant mortality rate in 2000 was 6 per 1000, while that of Sierra Leone was 180 per 1000. In addition, in those world areas most affected by the HIV/AIDS epidemic (sub-Saharan Africa, India, and Southeast Asia), child mortality rates may double by 2010, reversing the hard-won improvements in child survival achieved through immunization and public health campaigns (Gelbard, Haub, & Kent, 1999).

National birth and death rates vary widely. In 2002, for instance, Niger recorded 55 births per 1000 population, while Italy and Canada recorded 9 and 11 births per 1000, respectively (Population Reference Bureau, 2003). In 1999, infant mortality rates for all of Africa reached 86 per 1000, but the rate was much higher in Sierra Leone (Population Reference Bureau, 2003). Even within nations, there can be variations in these rates. Infant mortality rates in Canada, the United States, and Western European nations ranged from 5 to 10. Note that these low rates can obscure locally higher rates of infant deaths due, for example, to prenatal exposure to mercury, lead, or dioxins from industrial pollution, such as may have occurred in the Peace River health region of Alberta (Pederson, 1997). In the mid-1990s, the infant mortality rates for Aboriginal people were twice as high as the rest of the Canadian population.

EXPONENTIAL GROWTH

Exponential growth in numbers of any species occurs when growth takes place at a constant rate per time period (see Chapter 3 discussion and Figure 3–21). During the first half of the 20th century, the modern stage of human population history, population actually increased at a rate faster than an exponential rate (Figure 4–1). This occurred when the population growth rate peaked at 2.04 percent between 1965 and 1970 (United Nations, 1998). This increased rate reflected gains in health care and sanitation practices, and improvements in food production. Since then, the human growth rate declined globally to 1.33 percent between 1995 and 2000 (United Nations, 1999). Table 4–3 illustrates selected average annual growth rates for the world from 1950 to 2050.

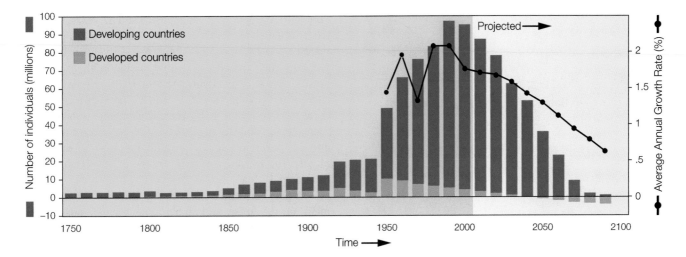

Figure 4–1

World population growth, 1750–2000, with average annual growth rates, 1950–2000, with projections

NOTE: For each decade it is possible to identify how many million people were added each year (for example: for the 1980–90 decade, about 82 million people were added each year for a total increase of 820 million people in that time). Note also that from about 2040, it is predicted that developed regions will experience a negative growth rate and their absolute numbers will decline.

PROJECTING FUTURE POPULATION GROWTH

Because human population growth is such an important environmental issue, we need to know what will happen to our population in the future. Projections of future population growth are "what if" exercises to forecast how quickly the world's population might increase, and whether or when it might stop increasing. Given certain assumptions about future tendencies in birth and death rates as well as migration, we can calculate a nation's population for a given number of years into the future. While such calculations indicate any upcoming population changes, they must be interpreted carefully because they vary according to the assumptions made. One of the simplest approaches to making population projections is to calculate doubling time.

DOUBLING TIME

Doubling time is the length of time required for a population to double in size. Table 4–4 provides doubling times for different growth rates. Figure 4–2 shows these figures graphically, illustrating that even with vigorous policies to reduce population growth rates, populations will continue to grow for some time.

In populations that are growing exponentially, doubling times change very quickly as the growth rate changes. The following examples highlight the effect that small changes in population growth rates have on doubling

TABLE 4–3
SELECTED AVERAGE ANNUAL GROWTH RATES FOR THE WORLD (1950–2050)

Year	Population	Average Annual Growth Rate (%)
1950	2 555 898 461	1.44
1956	2 832 536 024	1.95
1960	3 038 930 391	1.33
1965	3 344 855 925	2.07
1970	3 705 987 692	2.07
1975	4 087 382 478	1.76
1980	4 457 593 483	1.71
1985	4 854 659 097	1.68
1990	5 281 672 973	1.58
1995	5 691 012 889	1.41
2000	6 090 912 914	1.28
2010	6 862 796 548	1.10
2020	7 601 785 909	0.92
2030	8 276 375 737	0.77
2040	8 877 430 755	0.61
2050	9 368 223 050	

NOTE: Given the difficulties that may be encountered in conducting a national census (refer to Box 4–2), the accuracy implied by these precise statistics may be inappropriate. Consider these data as illustrative rather than definitive.

SOURCE: *Total Midyear Population for the World: 1950–2050,* U.S. Census Bureau, International Data Base, 1996, http://www.census.gov/ipc/www/worldpop.html

TABLE 4–4
DOUBLING TIMES FOR DIFFERENT GROWTH RATES

Annual Growth Rate (%)	Doubling Time (years)
0.5	139
1.0	69
1.5	46
2.0	35
2.5	28
3.0	23
4.0	17

times. In 1999, the natural increase of the Canadian population rose by approximately 0.4 percent. If this rate represented total population growth, Canada's doubling time would be about 162 years (assuming a constant rate of natural increase). In Western European nations, annual rates of population growth by natural increase are about 0.1 percent. At this rate, nations such as Austria, France, and Belgium would have doubling times of about 577, 693, and 2310 years, respectively. The two most populous nations on earth, China and India (which surpassed the 1 billion population mark in 1999), were growing at rates of about 0.9 and 1.9 percent in 1999. They would be expected to double their populations in about 78 and 37 years, respectively. Populations in Jamaica, Haiti, and Grenada grew at rates of 1.7, 2.1, and 2.3 percent, respectively, in 1999; at that rate their populations would double in about 40, 33, and 30 years, respectively. Also in 1999, populations in Nigeria and Ghana grew at 3.0 and 2.9 percent rates, doubling in about 23 and 24 years, while the populations in Gaza, Oman, and the West Bank grew at rates of 4.4, 3.9, and 3.3 percent, leading to doubling times of approximately 16, 18, and 21 years, respectively (Population Reference Bureau, 2000).

In part, the effect of doubling times on the environment, on resources, and on sustainability depends on the absolute size of the initial population. For instance, the doubling of a population of 1000 will have less impact than the doubling of a population of 10 000 or 100 000. Even though growth rates do not remain constant, and doubling times do not predict accurately the real growth of a population, the doubling time is a valuable measure because it allows us to consider possible future social and environmental conditions if present growth rates were to continue.

In countries such as Canada, where there is low population growth (because both birth and death rates are low), the effects of immigration and emigration can play a role in changing the national growth rate and doubling time. The nature of immigration policies, such as those intended to attract wealthier immigrants who can promote economic growth and development, also have important implications for future population growth rates and consumption trends.

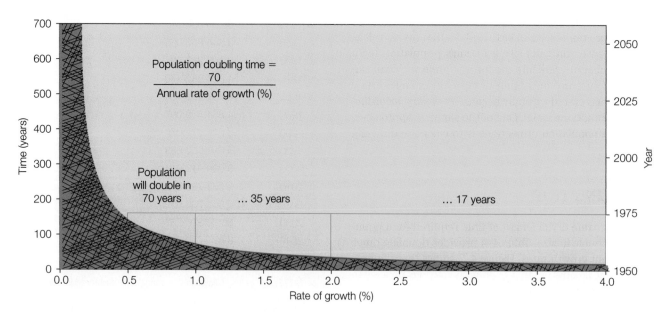

Figure 4–2

Doubling times and population growth rates

SOURCE: Adapted from *Biosphere 2000: Protecting Our Global Environment,* D. G. Kaufman & C. M. Franz, 1993, New York: HarperCollins College Publishers, p.134. Reprinted by permission of HarperCollins Publishers.

THE LOGISTIC GROWTH CURVE

As was noted in Chapter 3, no population can continue to grow indefinitely at an exponential rate and be able to sustain itself. Eventually it will run out of food and space. Unchecked human population growth also would stress the country's social, economic, and political systems. If exponential rates of growth do not occur, what will the changes in a population be like as time passes?

One long-standing idea is that over longer periods of time the human population would follow a smooth S-shaped curve known as the **logistic growth curve** (see Figure 4–3). In this case, the population increases exponentially at the outset (note the J shape at the start) and then levels out as the carrying capacity of the environment is reached. Once a population reaches the limits of its environment, the population tends to remain at or near the carrying capacity with a rate of population growth around zero. Although the logistic growth or S curve simplifies actual population changes over time, it does seem to describe population growth patterns observed in animal populations (in nature and the laboratory).

If death rates continue to decline owing to continuing improvements in food supplies, health, and medicine, it is unlikely that the logistic growth curve will be useful in projecting the maximum future human population size. Instead, if a human population is to achieve zero population growth (a stabilized or nongrowing population where birth and death rates are equal), it must pass through the demographic transition. Demographic transition is the subject of the following section.

DEMOGRAPHIC TRANSITION

THE FOUR-STAGE MODEL

Based on their observations of what happened to the birth and death rates of the European population as it urbanized and industrialized during the 19th century, demographers developed a four-stage model of population change (see Figure 4–4). Since all of the highly developed nations with advanced economies have gone through this **demographic transition**, demographers suggest this model is a key to understanding how human populations stabilize.

In the preindustrial stage, harsh living conditions gave rise to a high birth rate (to compensate for high infant mortality) and a high death rate; there was little population growth. Although no countries are in the first stage today, Finland in the late 1700s would have been in the first demographic stage. As a result of more reliable food and water supplies as well as improved health care, the second or transitional stage is characterized by a decline in the death rate. However, because the birth rate

Photo 4–2
Many developing nations face problems providing adequate housing and sanitation for their growing populations.

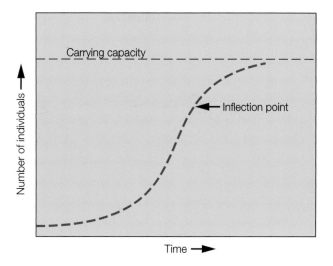

Figure 4–3
Logistic growth curve

This S-shaped (logistic growth) curve shows the population increasing exponentially at the outset (note the J shape) and then levelling out as the carrying capacity of the environment is reached. The inflection point is where the S curve changes slope and curves toward the horizontal.

is still high, the population grows rapidly (by about 2.5 to 3 percent per year). By the mid-1800s, Finland was in the second stage of the demographic transition, as are much of Latin America, Asia, and Africa today. Much of Europe "solved" the problem of rapid population growth by encouraging emigration to other parts of the world.

Some developing countries seem to be caught in a **demographic trap**, unable to break out of the second stage. Partly, this is due to the very rapid changes in medical practices. In Europe, it took 150 years to move through the demographic transition, whereas many African countries only began this process in the last 60 years. Furthermore, different challenges such as AIDS have meant that the transition is not a smooth one; its trajectory may have unexpected peaks and valleys. Whether some of these nations will be able to achieve a lower birth rate before reaching disastrously high population levels is still unknown.

The third or industrial demographic stage is characterized by a declining birth rate that eventually approaches the newer, lower death rate. As it declines, population growth may fluctuate depending on economic conditions. The relatively low death rate, combined with a lower birth rate, slows population growth in the third stage. Reasons for the decline in births include better access to birth control, declines in infant mortality, improved educational and job opportunities for women, and the high cost of raising children. These reasons represent a cultural shift in social norms and values that take time to change. By the early 1900s, Finland had reached this stage.

In the fourth or postindustrial phase of the demographic transition, birth rates decline further to equal (or fall below) the low death rates. Zero population growth (or even population decrease) is achieved partly because parents understand the benefits of small family size. It is at this point that emphasis can shift from unsustainable to sustainable forms of economic development. Ironically, countries having passed through these stages have the highest rates of consumption of the world's resources. As birth and death rates drop and become synchronized at a lower level, typically living standards and economic expectations rise. The paradox is that pressure on local environments may be displaced by countries that import resources to meet demands for exotic foods, consumer goods, and fossil fuels.

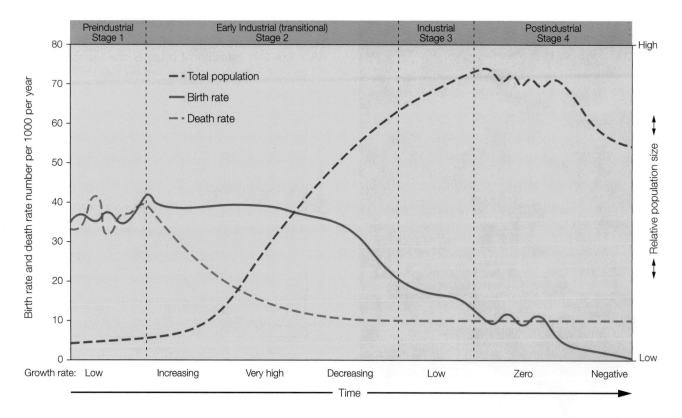

Figure 4–4

Demographic transition

This generalized model identifies the four stages of the demographic transition and how variations in birth and death rates over time result in changes in the population growth rate.

DISEASES AND DEATH IN INDUSTRIAL SOCIETY

One of the key reasons why economically developed societies have been able to complete the demographic transition is because modern medicine has been able to greatly reduce the number of deaths due to epidemic diseases. Typically, when an epidemic disease occurs in a population, a large percentage of people are affected by it—consider the number of people who contract cholera, measles, mumps, or influenza during an outbreak. In part, the spread of resistant forms of diseases such as tuberculosis, malaria, and cholera reflects the high density of populations of poor people.

Acquired Immune Deficiency Syndrome (AIDS) is also a factor in the increasing incidence of deaths due to epidemics (because individuals with AIDS lack resistance to diseases). Worldwide, by 2001, about 23 million people had died from HIV/AIDS, and an additional 37.1 million people were infected with the virus. Three-quarters of the most affected countries are located in sub-Saharan Africa; in Botswana, the country with the highest HIV prevalence, more than one out of every three adults is HIV-positive. The United Nations estimates that 90 percent of children now infected with HIV were born in Africa. In Zimbabwe, where one in three adults is also infected, life expectancy at birth was estimated at 41 years in 1995–2000. This figure is 25 years lower than it might have been without AIDS and it is expected to decline again to 33 years in 2000–2005. It is estimated that by 2050, Zimbabwe's population will be 61 percent lower than its without-AIDS projection (United Nations Population Division, 2003).

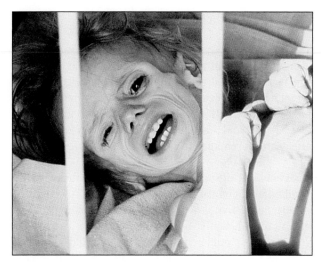

Photo 4–4
A young AIDS victim in Romania.

Animal populations also are susceptible to disease epidemics. In early 1997, at least 50 000 pigs in Taiwan were slaughtered in an attempt to control foot and mouth disease. Because humans can carry this disease (but do not die from it), Taiwan's largest zoo and all national parks were closed to protect animals there from potential human carriers (Kohlenberg, 1997). Bovine spongiform encephalopathy (BSE), or mad cow disease, also led to the killing of thousands of British, Swiss, and German cattle. Even where the disease is not widespread, it can have devastating economic impacts. In May 2003, in Canada, one case of mad cow disease discovered in a cow sent to a processing plant in Alberta led to the closure of the Canada–U.S. border so that ranchers could not export their beef (see Box 6–1). Indeed, Canada could not export its beef to any country To test for the disease, 1400 cows were slaughtered. By August 2003, the Canada–U.S. border was reopened to some beef, but closed to live cattle. A cow that normally would have sold for $1300 could be purchased for $15. While the border was to open in December 2003, on December 23, the United States announced its first apparent case of BSE in a cow in Washington state. It was estimated that the full closure of the border cost producers $11 million per day. Linked to the human brain-wasting disorder Creutzfeldt-Jakob disease, BSE is spread by feed containing ground parts of cows, sheep, and other mammals. In August 1997, the Canadian government amended the Health of Animals regulations and banned domestic cattle food that contains mammalian material. While the link has not been established definitively, microbiologists warned that BSE could cross the species barrier from cows to humans (Drohan, 1997). Recent reports of a 20 to 30 percent increase in deaths from the human form of BSE may reflect this concern ("Human Deaths," 2000).

Photo 4–3
In postindustrial nations the aging of the population is evident.

ZERO POPULATION GROWTH

The logistic growth curve has been used to project the eventual maximum population size in specific nations as well as the world. Despite the fact that the logistic curve permits simple calculations of a future carrying capacity, its use is unrealistic—it ignores potential changes in the environment and possible technological changes that may affect human population growth in the future, including zero population growth.

A number of social and economic consequences have resulted from population stability in Europe. In a situation of zero population growth, when births plus immigration exactly equal deaths plus emigration, there are fewer young people, an increasing proportion of older citizens, and a rise in the median age of the population. If there is actual population decline, as is now common in Europe, reduced demand for facilities such as schools and univer-

sities results in their permanent reduction. Instead of schools, governments must provide pensions and various social services for the roughly 25 percent of their citizens who are aged 60 and over, through the taxation of a diminishing workforce. In 1991, for instance, there were 4 pensioners in Germany for every 10 workers; by 2030, the numbers are expected to be equal. Canada's political leaders and social planners have noted the potential for a similar zero population growth situation to arise in this country, and are aware of the continuing need for future planning, including pension reform.

Carrying Capacity

When we talk about the environment's carrying capacity for humans, we are talking about the maximum number of the human species that a particular habitat (or the planet) can support sustainably in the long term. If a pop-

---

BOX 4-3
THOMAS MALTHUS ON THE HUMAN POPULATION PROBLEM

Long before most scholars were concerned about overpopulation, Thomas Malthus was writing his warning about it. Born in Surrey, England, Malthus studied theology at Cambridge and became an ordained minister. While still a minister, he began writing *An Essay on the Principle of Population,* which was published in 1798 and read by Darwin and Marx, among many others. Gradually, writing and lecturing became his major interest. In 1805 he was appointed a professor of modern history and political economy at Laileybury College; he held this position until his death.

In his essay, Malthus clearly noted the potential of

Photo 4–5

Thomas Robert Malthus
(1766–1834)

geometric growth in the human population to outstrip the arithmetical increase in food supply. To prevent this, Malthus felt population control was necessary. If people failed to control their own numbers, Malthus believed that "natural controls" of warfare, famine, and disease would solve the problem.

The premises on which Malthus based his arguments are included in the following excerpts from his essay:

I think I may fairly make two postulata.

First, that food is necessary to the existence of man.

Secondly, that the passion between the sexes is necessary, and will remain nearly in its present state.... Assuming, then, my postulata as granted, I say, that the power of population is indefinitely greater than the power in the earth to produce subsistence for man.

Population, when unchecked, increases in a geometrical ratio. Subsistence only increases in an arithmetical ratio. A slight acquaintance with numbers will show the immensity of the first power in comparison of the second.

By that law of our nature which makes food necessary to the life of Man, the effects of these two unequal powers must be kept equal.

This implies a strong and constantly operating check on population …

[The power of population growth is so great] that premature death must in some shape or other visit the human race. The vices of mankind are active and able ministers of depopulation, but should they fail, sickly seasons, epidemics, pestilence and plague, advance in terrific array, and sweep off their thousands and ten thousands. [Should these fail,] gigantic famine stalks in the rear, and with one mighty blow, levels the population with the food of the world.

For more on Malthus, see *Malthus Past and Present,* J. Dupaquier & A. Fauve-Chamoux (Eds.), 1983, New York: Academic Press.

SOURCE: *The Human Mosaic: A Thematic Introduction to Cultural Geography,* T. G. Jordan & L. Rowntree, 1990, New York: Harper & Row, p. 43.

---

ulation exceeds that carrying capacity, it changes the environment in ways that decrease the future population size. If we want to avoid human and environmental degradation, it is important to be able to estimate carrying capacity.

What is the carrying capacity of a nation or the planet for humans? No one knows for sure how many humans can be supported by the Earth, and calculations vary widely depending on the assumptions that have been made. Almost 200 years ago, the English clergyman, historian, and political economist Thomas Malthus made the earliest—and perhaps some of the most famous—statements about the threat of uncontrolled human population growth (see Box 4–3). In *An Essay on the Principle of Population* (1798), Malthus stated that the human ability to multiply far exceeded our ability to increase food production. His pessimistic conclusion was that unless people found a more humane solution to stabilizing their populations, famine, disease, and warfare would be inevitable because they would act to curb population growth.

Critics of Malthus point out that his predictions have not come true, that technology has enabled humans to live at increasing densities, and that technology will continue to provide a way out of a Malthusian fate. Even though Malthus did not anticipate the capability of technological advances and that the Earth may be able support many more people than it did then, in the long run there are several environmental signals that suggest an upper limit to the number of people the Earth can support. We are faced with trying to find ways to limit population increases and distribute environmental resources in ways that combine our knowledge of environmental science with arguments about rights, values, and ethics, and to do so within a global perspective. For example, to calculate a carrying capacity we would need to determine and agree on minimum standards of living and quality of life, while being sensitive to differences in how these might be accomplished. At least to some extent, these are questions about values and how we define sustainable futures.

Limiting Factors

The factors that eventually will limit human populations can be categorized as short term (1 year), intermediate term (greater than 1 year and less than 10 years), and long term (10 or more years). Some factors may have a combination of effects, for example, both short- and intermediate-term effects.

Short-term factors that limit human populations include those events that disrupt a country's food supply and distribution. Factors such as a local crop failure, a shortage of energy to transport foodstuffs, weather changes such as drought or heavy rains, or political events such as civil wars could lead to food disruptions. On the global level, major disasters such as the outbreak of a new disease, a new strain of a disease that was previously under control (see Enviro-Focus 4), or the spread of a toxic chemical are other examples of short-term effects on food availability and, ultimately, on the human population.

Intermediate-term factors that could limit human populations include declines in the supply of firewood or other cooking and heating fuels; some climate changes; a system-wide dispersion of toxic pollutants into freshwater bodies, oceans, and fisheries; desertification; widespread disease or war; and energy shortages that affect food production and distribution. Long-term factors include soil erosion; declines in the quantity and quality of groundwater supplies; global warming; and widespread pollution such as acid precipitation. In addition, social and cultural factors such as the availability of public education to boys and girls, increased income, and attitudes toward marriage, family, and women's roles also affect population over the long term.

Many people believe that the quality of human life will continue to improve as our technology advances. However, if one of the limiting factors is the amount of biological resources available per person, some data indicate we may already have passed the peak of resource availability (see Table 4–5) and have approached the limits of productive capacity with our existing technology. This suggests we have exceeded the long-term carrying capacity of the Earth for people.

AGE STRUCTURE

Since it is difficult to forecast the future of population growth through use of doubling times, exponential growth, and logistic curves, we can consider the **population age structure**, which identifies the distribution of the population by age. Analysis of the population age structure provides information about current and future birth, death, and growth rates; about current and future social and economic status; and about our likely impact on the environment.

For any population, the number of males and the number of females at each age, from birth to death, can be represented in a generalized age structure (or population pyramid) diagram (see Figure 4–5). The diagram is divided vertically in half; typically the right side represents the females in a population, and the left side represents the males. The bottom one-third of the diagram represents prereproductive humans (from birth to 14 years of age); the middle one-third represents humans in their reproductive years (15 to 44), and the top one-third represents postreproductive humans (45 years and older). The width of each of these segments is proportional to the population size, so that the greater the width, the larger the population.

"The Cell from Hell" and Other Epidemics

"The Cell from Hell"

Toxic dinoflagellates (*Pfiesteria piscicida*) have been keeping humans company from ancient times. Unicellular parasites belonging to both animal and plant kingdoms, and able to live in both salt and fresh water—the more polluted the better—dinoflagellates have been responsible for red tides (algal blooms) since biblical times. Since the early 1990s, dinoflagellates have killed billions of fish on the eastern seaboard of the United States, and afflicted hundreds of divers, fishers, sailors, and swimmers with festering skin sores, faulty memories, immune failure, and even personality changes. Able to take on 24 different shapes (from hibernating cyst to two-tailed flagellates that stun prey with neurotoxins), the airborne vapours from dinoflagellates have caused scientists to lose consciousness. Dinoflagellates deserve their biohazard level-three status!

As much like the latest *X Files* threat as dinoflagellates seem to be, their emergence off the coast of North Carolina appears to be in response to nutrient-rich human and animal wastes that have been spilled into coastal rivers and flowed to the sea. North Carolina's booming hog industry (15 million pigs concentrated on clay-based soils) produces more waste than the people of New York City. When nitrogen-rich pig manure escapes into water courses (through leaching, or following collapse of a holding pond wall, for example), it becomes food for plankton, which in turn attract the voracious fish-flesh-eating dinoflagellates.

In spite of warnings from scientists about the unpleasant microbial consequences of, and need to clean up the waste created by, corporate pig-farming practices, North Carolina's health department denied there were any water quality problems and dismissed the toxic hangovers of fishers as hearsay. In what appeared to be an effort to protect the billion-dollar pig industry and other polluters, scientific findings were denigrated regularly. Human arrogance, poor government, and irresponsible economics are the chief reasons why dinoflagellates are flourishing in the ocean off North Carolina's coast (Barker, 1997).

Malaria

Endemic in sub-Saharan Africa, where often more than 50 percent of the population in rural areas is infected, malaria puts about 40 percent of the world's population at risk. Of the estimated one million people who die annually from malaria, about 800 000 are children under five years of age. In spite of massive efforts to eradicate it, malaria is making a comeback. There seems to be an upward trend in the number of cases in the Americas and some Asian countries. The increase is due partly to the growing resistance of malaria-carrying mosquitoes to insecticides and of the *Plasmodium* parasites to anti-malarial drugs (World Resources Institute, 1992).

A growing number of Canadian travellers are bringing malaria home with them; in 1995, 637 cases of malaria were reported in Canada (Nichols, 1997). Infectious-disease specialists are alarmed that too many travellers are going into malaria-infected regions of Africa, Asia, and Latin America without accurate information or proper antimalarial medication. Chloroquine is the least costly and most widely used drug against malaria. Very effective when used in conjunction with a good primary health-care system, chloroquine administered outside of health-care systems has actually contributed to the resurgence of malaria. People who were unknowingly affected by malaria and who stop taking their antimalarial drugs too soon have low doses of the drugs in their system. This provides an ideal breeding

Photo 4–6
Family members attend to a malaria victim.

ground for the development of mutations. Mefloquine, one of the main replacements for chloroquine, is more costly and may have side effects.

Cholera

In other parts of the world, the threats to human health posed by environmental deterioration were evident in early 1991, when for the first time in the 20th century a cholera epidemic struck six Latin American countries. More than 300 000 cases of cholera and 3200 deaths were reported by September 1991, mostly in Peru, but also in Ecuador, Colombia, Mexico, Guatemala, and Brazil.

Cholera is an acute intestinal infection caused by the *Vibrio cholerae* bacterium and transmitted princi-

pally through contaminated water and food, particularly raw vegetables and seafood. Children are highly susceptible to this disease, which spreads rapidly in overpopulated communities with poor sanitation and unsafe drinking water. The outbreak in Peru appeared almost simultaneously in communities along a 1200 kilometre stretch of coastline. Cholera was felt to be a side effect of the rapid urbanization of that country. The growth of crowded slums and the lack of safe water and sanitation facilities have been prime factors in these outbreaks. While cholera is treatable with rehydration salts, the ultimate solution requires improvements in water, sanitation, health and education, and food safety (World Resources Institute, 1992).

SOURCES: *And the Waters Turned to Blood: The Ultimate Biological Threat*, R. Barker, 1997, New York: Simon & Schuster; "Malaria's Comeback," M. Nichols, May 19, 1997, *Maclean's*, p. 57; *The Fourth Horseman: A Short History of Plagues, Emerging Viruses and Other Scourges*, A. Nikiforuk, 1991, Toronto: Viking; *World Resources 1992–93*, World Resources Institute, 1992, New York: Oxford University Press.

TABLE 4-5
PER CAPITA AVAILABILITY OF RESOURCES

Resource	Peak Per Capita Production	Year of Peak Production
Wool	0.86 kg/person (1.9 lb/person)	1960
Wood	0.67 m³/person (0.88 yd³ person)	1967
Fish	5.5 kg/person (12.1 lb/person)	1970
Mutton	1.92 kg/person (4.21 lb/person)	1972
Beef	11.81 kg/person (26.0 lb/person)	1977
Cereal crops	342 kg/person (754.1 lb/person)	1977

SOURCE: Adapted from *World Development Report*, World Bank, 1985, New York: Oxford University Press.

We can use population age structure diagrams to predict population. The overall shape of the diagram reveals whether the population is expanding, stable, or declining. The age structure diagram of a country with a very high growth rate, such as Kenya or Nigeria, is shaped like a pyramid. In contrast, the age structure diagrams of coun-

tries with stable or declining populations, such as Italy and Germany, respectively, have narrower bases (see Figure 4–6).

In countries that have rapidly expanding populations, the largest proportion of the population is in the prereproductive age group. It is estimated that about one-third of the population worldwide is under 15 years of age. This means that the probability of future population growth is great; when all these children mature, they will become the parents of the next generation. Even if the birth rate does not increase and these parents have only two children, the population growth rate will continue to increase simply because there are more people reproducing.

In contrast, the narrower bases of the age structure diagrams of countries with stable or declining populations indicate that a smaller proportion of children will become the parents of the next generation. The age structure diagrams of countries with stable populations (neither growing nor shrinking) illustrate that the numbers of people at prereproductive and reproductive ages are approximately the same. In stable populations, also, a larger percentage of the population is in the postreproductive age group than in countries with rapidly increasing populations. Many nations in Europe have stable populations. Other European nations, such as Germany and Hungary, have populations that are

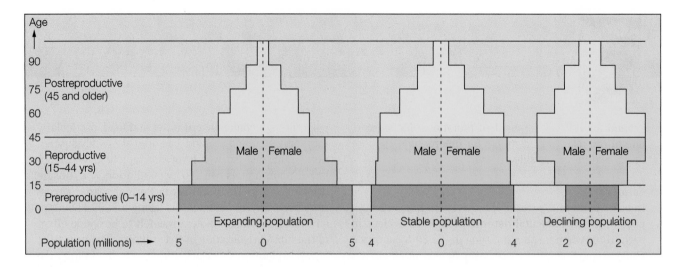

Figure 4–5

Generalized age structure diagrams for expanding, stable, and declining populations

NOTE: These generalized age structure diagrams represent an expanding population, a stable population, and a population that is decreasing in size. In each diagram, the left half represents the males in the population while the right half represents the females. The horizontal divisions in the diagram separate the prereproductive, reproductive, and postreproductive age groups, while the width of each segment represents the population size of each group.

shrinking in size. In these countries, the prereproductive age group is smaller than either the reproductive or postreproductive age groups.

As is evident in Figure 4–1, most of the worldwide population increase since 1950 has occurred in the devel-

oping countries, reflecting the younger age structure as well as the higher-than-replacement-level fertility rates of their populations. (Fertility rates are considered in a later section.) In 1950, about 66.8 percent of the world's population was found in developing countries in Africa, Asia

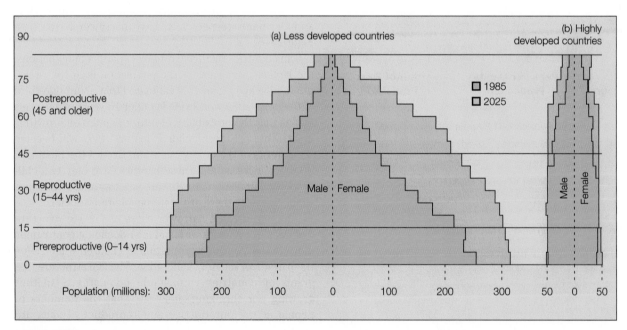

Figure 4–6

Age structure diagrams for less developed and highly developed countries

NOTE: The green region represents the actual age distribution in 1985. The orange region represents the projected increase in population and age distribution in 2025. Clearly there is a much higher percentage of young people in less developed countries than in highly developed countries. A much greater population growth is projected for less developed countries when their large numbers of young people enter their reproductive years.

(excluding Japan), and Latin America; the remaining 33.2 percent was in developed nations in Europe, the former Soviet Union, Japan, Australia, and North America. Since 1950, the world's population more than doubled, and the number of people in developing countries rose to 80 percent of the worldwide population in 1998 (United Nations, 1998). By 2002, world population was growing at a rate of 1.2 percent annually, implying a net addition of 77 million people per year. Six countries account for half of that increase: India (21 percent), China (12 percent), Pakistan (5 percent), Bangladesh, Nigeria, and the United States (4 percent each). In its *World Population Prospects: The 2002 Revision*, the United Nations estimated for the first time that by 2050, three out of every four countries in less developed regions will be experiencing below-replacement (2.1) fertility. The impact of the HIV/AIDS epidemic in the most affected countries is anticipated to be more serious and prolonged than previously estimated (United Nations Population Division, 2003).

The proportion of a country's population in each age group strongly influences the demand for certain types of goods and services within that national economy. For instance, a country with a high proportion of young people has a high demand for educational facilities and certain types of health delivery services, whereas a population with a high percentage of elderly people requires medical goods and services specific to that age group.

The Dependency Ratio

The **dependency ratio** is another useful measure (based on age structure) to forecast the condition of the future human population. The dependency ratio is a simple measure of the number of dependants, young and old, that each 100 people in their economically productive years (usually 15 to 64) must support. The ratio is calculated as

$$\frac{\text{Sum of the number of people under 15 and over 65 years of age}}{\text{Number of people between ages 15 and 65}}$$

Statistics Canada (2004) indicated that in 2003, an estimated 5 781 600 Canadians were under 15 years of age, and 4 060 200 were aged 65 years and over, for a total of 9 841 800 dependent-age people. With 21 787 800 working-age people, the dependency ratio was 0.45 (9 841 800 divided by 21 787 800) or 45 dependent-age people for every 100 working-age people.

The dependency ratio affects the economic conditions of a population: the larger the ratio, the lower the present and near-future living standards. In preindustrial societies, where average lifetimes are short, younger persons care for the elderly within the same family. In modern technological societies, however, family size is

reduced and the elderly are supported through the collection and distribution of taxes (that is, those who work provide funds to care for those who cannot work). Provision for the elderly is a necessary factor in achieving zero population growth because, without it, parents would rely on a large number of children to ensure their future well-being. But the shift from an age structure like that of Kenya (where about 50 percent of the population was under 15 years of age in 1991) to that of Austria (which is experiencing a decline in population growth) results in a decrease in the tax base for care of the elderly because of the smaller percentage of the population that is working.

These situations pose problems for national governments, particularly with regard to promoting sustainability. Governments are often overwhelmed by the need to provide hospitals, clinics, health care, schools, housing, roads, and employment opportunities for their rapidly growing populations. One way to increase tax income to support these ventures is to increase the percentage of young people in the population; in turn, this promotes rapid population growth. Short-term economic pressures facing national governments can lead to political policies supporting rapid population growth that are not in the long-term interests of the nation. Yet if governments are unable to raise the finances to support their population needs, the cycle of poverty, illiteracy, and unemployment continues and strengthens, and the potential to achieve sustainability declines.

A large proportion of seniors can strain a nation's social services network. Health care, nursing homes, transportation infrastructure, and social security systems are all stressed by an increasing proportion of older citizens. In the future, if fertility and growth rates remain low, care of the aging baby boomers (people born between 1946 and 1964) will fall to a declining number of younger workers. This situation, too, may generate economic and political policies that may not be appropriate in the long term.

FERTILITY RATES AND TIME-LAG EFFECTS

Another set of measures useful in understanding changes in human population growth are fertility rates and time-lag effects. In general, measures of fertility, or the actual bearing of offspring, are reasonably accurate indicators of the potential for future population growth. The **general fertility rate** is the number of live births per 1000 women of childbearing age per year. Ages 15 to 44 are commonly cited as the childbearing years. This indicator, however, is not commonly employed. A more helpful indicator, from the perspective of predicting potential future population growth, is the **age-specific fertility rate**, the number of live births per 1000 women of a specific age group per year.

Changes in the growth of human populations can be delayed by time-lag effects. One of the most important lag effects is the **total fertility rate (TFR)**, which is the average number of children expected to be born to a woman during her lifetime (TFR is based on the current age-specific fertility rate and assumes that current birth rates remain constant throughout the woman's lifetime). A TFR of 2.1 births per woman enables each woman to replace herself and her mate and allows for the death of some female children before they reach their reproductive years. However, from the day that a population achieves replacement fertility, that population will continue to grow for several generations—a phenomenon known as **population momentum** or **population lag effect**.

Population momentum occurs because a population that has had a very high fertility in the years before reaching replacement level will have a much younger age structure than a population that has had lower fertility before reaching the replacement threshold. In a population that is approaching replacement-level fertility, the proportion of young people is important because the size of the largest recently born generation as well as the size of the parent generation determine the ultimate size of the total population when births and deaths finally are balanced. Even after reaching replacement-level fertility, as long as the generation that is producing children is disproportionately larger than the older generation (where most deaths occur), births will continue to outstrip

deaths, and the population will continue to grow. Clearly, this time lag is very important to human populations as it has tremendous implications for resource use, environmental impacts, and sustainability.

In many less developed countries, the average replacement fertility is 2.5 or more, reflecting the greater risk of death faced by children in those countries. Figure 4–7 illustrates TFRs for a number of developing and developed nations. In 2004, for example, the total fertility rate in Afghanistan was 6.8; this means that a woman in her childbearing years would be expected to have about 7 children by the time she reached 49 years of age. If Afghanistan were to maintain a TFR of 6.8, its population would be expected to grow very rapidly, whereas if Canada maintained its TFR of 1.5 over a long time, the population would decline. Populations in Europe are expected to decline in the 21st century.

A comparison of developed and developing nations suggests that the TFR declines as income increases. In countries such as Bangladesh, Nigeria, and Nepal, where the average income per person is a few hundred dollars per year, total fertility rates are high. In developed nations such as Canada, Japan, and Denmark, where per capita annual incomes are above US$10 000, total fertility rates are at or below replacement level. Since about 1950 there has been a downward trend in the world's TFR; this has been reflected in declining rates of population growth worldwide (although the world's population is still growing). Figure 4–8 illustrates the current and projected

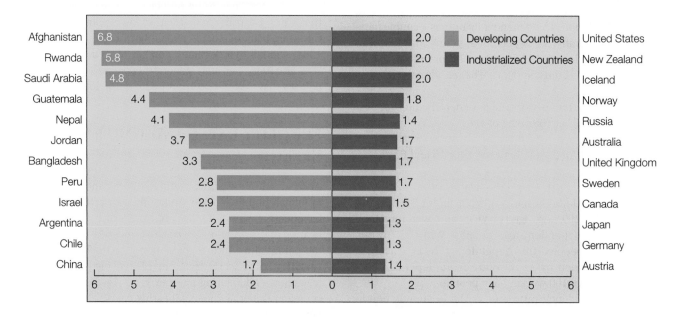

Figure 4–7
Total fertility rates (TFR) for selected countries, 2004

SOURCE: *2004 World Population Datasheet,* Population Reference Bureau, 2004. Reprinted with permission of the Population Reference Bureau. http://www.prb.org/pdf04/04WorldDataSheet_Eng.pdf

future decline in the world's total fertility rate from 1950 to 2020 as the world's population increases. As shown in Figure 4–8, by the year 2020, a woman would have about two children during her lifetime, a major decline from an average of greater than three in the 1990s (Kent & Crews, 1990).

Future Population Trends

The human population has reached a turning point; although our numbers continue to increase, the global rate of population growth has declined during the past few decades and is predicted to continue to decline. Even if replacement fertility is reached rapidly on a global basis, it will take many years for the world population to stabilize (because of the momentum of the world's present age structure). Demographic experts at the United Nations have noted that with the continuation of fertility decline and increase in life expectancy, the population of the world will age much faster in the next 50 years. Three trends characterize this aging:

1. The median age of the world's population is increasing (from 23.5 years in 1950, to 26.1 years in 1998, to a projected 37.8 years by 2050).

2. The proportion of children under 15 years old is declining (from 34 percent in 1950, to 30 percent in 1998, to 15 percent in 2050).

3. The proportion of older persons (60 or over) has increased from 8 to 10 percent between 1950 and 1998.

Europe will remain the world area most affected by aging of its population. In 1998, for instance, Italy was the oldest country in the world, with 1.6 persons aged 60 or above for each person below 15 years of age. In Africa, the youngest population region of the world in 1998, children made up 43 percent of the population while older persons constituted 5 percent. By 2050, Africa will still have twice as many children as elderly persons, although the effects of AIDS on life expectancy may alter this projection. Even in countries such as Canada, elder care is frequently considered a private, domestic responsibility entrusted to extended family members.

Cultural Factors

In the developed world, a decrease in family size is recognized as an important step toward reduction in the global population growth rate. Many cultures of the world, however, place a positive value on large families. Their reasoning is pragmatic: a large family provides such benefits as protection from enemies, a greater chance of leaving descendants, and a type of insurance for retirement. In societies without life and disability insurance and retirement programs, family members provide care for the elderly.

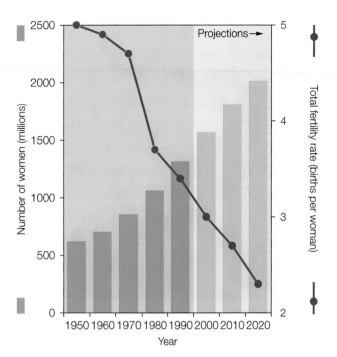

Figure 4–8
World decline in total fertility rate, 1950–2020

SOURCE: Adapted from *World Population: Fundamentals of Growth,* M. M. Kent, & K. A. Crews, 1990, Washington, DC: Population Reference Bureau. Reprinted with permission of the Population Reference Bureau.

Other cultural arguments are offered for large families. In Africa, for example, cultural beliefs link men's virility with the number of offspring they produce. Additional reasons for large families in Africa include the formerly high rates of infant and childhood mortality, low education levels (particularly for women), young age at marriage, and lack of access to family planning services and safe, reliable means of birth control. Religious and moral arguments are also part of the discussion of why large families are considered valuable. The influence of these arguments is noted briefly in the following section on birth control in developing nations.

ADDRESSING HUMAN POPULATION ON A GLOBAL BASIS

It is often difficult to separate the rate of growth in numbers of the human species inhabiting Earth from issues relating to food supply, land and soil resources, water resources, and the net primary production of the world's lands and oceans (such as forests and plankton). If population growth is slowed, and resource consumption per

Photo 4–7
Traditionally, large families have played an important social, cultural, and economic role in developing nations.

oping nations in Asia, illustrates this relationship. In Sri Lanka, the average marriage age is 25 and the average number of children born per woman is 2.5. In Bangladesh, in contrast, the average age at marriage is 16, and the average number of children born per woman is 4.9. In 1950, China first set laws for minimum marriage ages; men could marry at 20 and women at 18. In 1980 these were revised upward so that the minimum marriage age was 22 for men and 20 for women. As a result of this and other population control programs that promoted a national goal of reaching zero population growth by the year 2000, China's birth rate dropped from 32 to 16 per thousand people, and the fertility rate dropped from 5.7 to 1.8 children during the period 1972–98. Nevertheless, few countries would accept such regimes.

Increasing Education Levels

In developing countries there is a strong correlation between the fertility rate and the amount of education a woman receives. Typically, women with more education tend to marry later and have fewer children. In Egypt, for example, 56 percent of women with no schooling become mothers in their teens, compared with just 5 percent of women who remained in school past the primary level (Brown, Renner, & Halweil, 2000). In part, education provides women with greater knowledge of ways to control their fertility and the means to improve the health of their families, thereby decreasing infant mortality. Education also opens doors to new careers and ways of attaining status for women besides motherhood. As family incomes of educated people have increased (because of their ability to earn more money), their standards of living also have increased and smaller family sizes have resulted.

Birth Control in Developing Countries

In the developing world, considerable emphasis is placed on family planning centres and their programs to provide birth control services. Both traditional methods (such as abstinence, breast-feeding, and induction of sterility with natural agents) and modern methods (including oral contraceptives, injectables and implants, surgical sterilization, mechanical devices, and medically provided abortions) have been used to decrease birth rates. Interestingly, the World Bank noted that traditional practices such as breast-feeding (because they delay the resumption of ovulation) sometimes have provided more protection against conception in developing countries than have family planning programs (Guz & Hobcraft, 1991). This is because of cultural taboos against intercourse while women are breast-feeding and because breast-feeding is affordable to all segments of the population and, in some cases, it is more culturally acceptable and therefore more likely to be adopted than other means that rely on medical technology. Abortion is

person is decreased, the world will be in a better position to tackle many of its most serious environmental problems. However, our experience suggests that where population growth has slowed, our consumption has increased. Some scholars have placed emphasis on slowing the rate of population growth to provide more time to address environmental problems. Others have placed emphasis on other factors such as improving access to land and resources for those in developing countries, opposing militarism and war, and decreasing consumption as ways to address the population/resource consumption imbalance (see Box 4-4). Below, measures to reduce population growth are described, while recognizing that these are not the only ways to address global environmental degradation and depletion of environmental resources.

SLOWING POPULATION GROWTH

Increasing the Marriage Age

One way to slow population growth is to delay the age of marriage and of first childbearing by women. This pattern tends to occur as more women enter the paid workforce and as education levels and standards of living increase. Sometimes, social pressures that lead to deferred marriage and childbearing are also effective in the medium to long term.

The average age at which women marry varies widely among all countries, but there is always a correlation between marriage age and a nation's fertility rate. The difference between Sri Lanka and Bangladesh, two devel-

BOX 4 – 4

IS AFRICA UNDER- OR OVERPOPULATED? A SCHOLARLY DEBATE

"Africa is overpopulated now ... its soils and forests are rapidly being depleted ..." (Ehrlich & Ehrlich, 1990, p. 39). The Ehrlichs promote a popular viewpoint. Africa is a continent in trouble. There is a simple logic in the claim made above. High rates of human population growth lead to environmental and social problems such as poverty and soil degradation. This is an assertion; however, it is not a fact. By contrast, the link between human population numbers and environmental degradation is a point of scholarly and practical debate.

There are at least two opposing views on population and local environmental change. One approach, which has long been advanced by Paul and Ann Ehrlich (1971, 1990) and Garret Hardin (1968, 1993), sees population growth leading to environmental decline. In this viewpoint, environmental degradation occurs because more people lead to increased use of land for farming and livestock rearing. As a result, increased pressures are placed on the land base, particularly as land is subjected to shorter fallow periods. In addition, people may add more cattle; overstocking of rangeland results. In both cases, soil degradation begins, nutrients are exhausted or leach out, and soils become subject to erosion and ultimately, to desertification. Incomes and food production fall, and poverty increases.

An alternative viewpoint is that population growth can have a positive influence on the productivity of environments. This viewpoint is associated with Ester Boserup (1965, 1981) who based this argument on observations from many countries and time periods. She agrees that as populations grow, demand for food and marketable crops increases and fallow periods shorten. But she suggests that there are many examples where these triggers spur the application of more labour and *complementary* inputs per unit of land. In developing countries, new technologies that are labour-intensive are applied to soils. These inputs are made to maintain and enhance yields, and the quality of land is improved. Thus, she suggests that population growth can lead to adaptation where there are resources, incentives, and opportunities to invest in the natural resource base. These incentives and practices may be most disrupted by poverty, war, and overexploitation arising from international economic markets or governmental policy.

These differences of opinion have encouraged scholars to determine what other factors aside from human population growth might be important in explaining environmental degradation and resource depletion. In fact, there are many variations on these themes. Geographers have pointed out that we need to identify where population growth rates are the greatest. They also suggest that we need to distinguish between changes that are naturally occurring from those that are human-induced, although given the dominance of human influence on the planet, this distinction is increasingly difficult to make (Vitousek et al., 1997).

There is no doubt that the current rates of population growth in Africa are among the highest in the world. However, these growth rates are very unevenly dispersed, concentrated on coasts, in cities, and in highland areas. For example, the mean national population density of Zambia is 13 people per square kilometre, with significant differences within the country as a result of the history of Portuguese colonialism and other human factors (Stock, 2004). As Africa is a large continent, it is important to look for specific evidence in specific locations.

There is evidence that global economic and political circumstances may exacerbate human use of the environment. For example, during the 1950s and 1960s in Nigeria and Senegal, international demand and favourable climatic conditions (wet years) led to increases in the cultivation and export of ground nuts. The wet decades were followed by a natural cycle of dry years. Soil that was broken during the wet decades was susceptible to erosion during drought periods that followed. While these cycles occur naturally, human intervention in these cycles resulted in heightened soil erosion and devastating economic impacts for residents of those countries. In the 21st century, farmers on the Canadian prairies may also be facing the vagaries of similar cycles. Settlements formed on the Prairies in the wet years of the early 20th century will be susceptible as the climate warms, although so far, some farmers have been somewhat shielded from these effects by technological inputs such as fertilizers and irrigation systems.

Today, many countries put more land into agricultural production to meet international demands and to provide economic exchange rather than to meet local food requirements. When international prices for commodities decline, there are greater pressures to put more land into production to gain some form of income. For example, coffee prices in the early part of the 21st century are at an all-time low. As a result, Ethiopia, Cameroon, Burundi, Ivory Coast, and Tanzania, among others, earn less income from growing coffee than they receive in development aid or debt relief (Stuart & McAlpine, 2003). Falling prices place pressures on producers to cultivate more land in order to eke out a reasonable standard of living.

In the 1980s and 1990s, the World Bank and the International Monetary Fund introduced structural adjustment programs to developing countries with high debt loads. These programs were designed to help these countries restructure their economies, but they also required many countries to intensify production and to undertake cutbacks in health and education. These reductions had a dual effect, placing increased pressure on the land base and reducing the number of people who would have access to education and basic health care. Thus, public education programs that might lead to improving infant and child mortality rates and reducing fertility rates in the short term as well as literacy and educational programs that might reduce population growth rates over the longer term were made less accessible.

In the 20th century, the cost of preparing for and waging war became steadily more expensive, although obtaining accurate and current data is difficult. The value of arms production between 1950 and 1980, for example, increased by 3.3 percent per capita while population increased 1.7 percent. The financial burden per capita for arms doubled between 1950 and 1980. According to Patricia Hynes (1999, p. 49), estimates suggest that "military energy use can jump from 2 to 3 percent to 15 to 20 percent in wartime, estimates that do not include energy demand for weapons manufacture. Altogether the world's military may use as many petroleum products as Japan, the world's

(continued)

BOX 4-4
(CONTINUED)

second-largest economy." About 20 percent of all global environmental degradation is due to military and related activities (Hynes, 1999). Monies spent on arms purchase are not available for health care, education, and other social programs that might address population issues. Furthermore, military use of land during wartime and peacetime also degrades land quality and introduces toxic chemicals. Thus, both the availability and the quality of land and soil resources for food production are significantly diminished.

A different kind of example suggests that sometimes we do not "read" the local landscape accurately. Europeans often view rainforests in Africa as "virgin" or "primary" forests that are key for maintaining biodiversity. However, much of Africa's tropical rainforests are secondary forests that have been cultivated for

generations. Research in Guinea, using aerial photographs and anthropological field work, established that as human settlements became more concentrated, forest islands were planted by local people, who managed them as part of complex cultivation systems. As settlement in the country spread, more, rather than fewer, forest islands were created. Thus, human population and cultivation improved rather than destroyed soils and biodiversity (Fairhead & Leach, 1996).

As populations increase, will ecosystems collapse? Is Africa over- or underpopulated? Are environmental degradation and resource depletion due to increased numbers of people or other factors? You decide. But to make a convincing argument either way, you must make good use of evidence (see also Box 2-1).

SOURCES: *The Conditions of Agricultural Growth: The Economics of Agrarian Change under Population Pressure,* by E. Boserup, 1965, London: Allen & Unwin; *Population and Technological Change: A Study of Long-term Trends,* by E. Boserup, 1981, Chicago: University of Chicago Press; *Society and Ecology in a Forest Savanna Mosaic,* by J. Fairhead & M. Leach, with research collaboration of D. Millimouno and M. Kamano, 1996, Cambridge and New York: Cambridge University Press; "Taking Population out of the Equation: Reformulating I=PAT," by P. Hynes, 1999, in J. Silliman & Y. King (Eds.), *Dangerous Intersections: Feminist Perspectives on Population, Environment and Development* (pp. 39–73), 1999, Cambridge MA: South End Press; *Africa South of the Sahara: A Geographical Interpretation,* 2nd ed., by R. F. Stock, 2004, New York: Guilford Press; "Wake Up and Smell the Coffee," by R. Stuart & S. McAlpine, June 12, 2003, *The Globe and Mail,* p. A17; "Human Domination of Earth's Ecosystems," by P. M. Vitousek, H. A. Mooney, J. Lubchenco, & J. M. Melillo, 1997, *Science,* 277(5325), 494–499.

one of the most controversial methods of birth control from a moral perspective, even though it is widely used: between 30 and 50 million abortions are performed annually worldwide.

The use of contraceptives and contraceptive devices is widespread in many parts of the world. More than 60 percent of married women use these and additional family planning methods in Brazil, Mexico, Thailand, and many other developing nations. However, less than 10 percent of women in Mali, and less than 20 percent in Pakistan, use family planning (Gelbard et al., 1999). This low rate of use results from factors such as cultural barriers to family planning, fear of adverse health effects from specific methods, and the low social status and educational levels of women, including the inability to afford contraceptives.

In many developing African countries, women and children grow the subsistence crops, graze animals, gather wood and water, use most of the household's energy in cooking, and care for the immediate environment of the household. Women are the primary resource and environmental managers in the household, and yet their health and education are often neglected compared with men. Women's education helps in improving the health of children and also enables women to plan their families and increase birth spacing (thus reducing maternal mortality risk).

Aside from the difficulty of achieving widespread social acceptance of the idea of fewer children and smaller families, many women need assurance that the children they do have will survive. The fact that Canadian infants are 30 times more likely to survive than those born in Sierra Leone is a telling example. Improved access to affordable family planning services is important in this context. In addition, in many male-dominated African societies, family planning centres provide information primarily to women. These centres might be more effective if both women and men were counselled on birth control alternatives. Similarly, if unmarried adolescents could gain access to family planning services, and delay their first birth until they are in their twenties, the interval between generations would increase and help lower average fertility. By delaying marriage and childbearing, family planning education also helps reduce high-risk births to teenaged mothers.

A common problem in many male-dominated societies is that even if women do not want more children, they often do not use contraceptives because their husbands want more children, particularly sons. Many men in developing nations disapprove of family planning practices. Condom use is avoided (except with prostitutes, out of concern for AIDS) as it runs counter to cultural expressions of manhood and the desire to build up their ethnic group (Stock, 1995).

Photo 4–8

Family planning education helps women to control family size and the interval between births.

In various parts of the world, religion has an important influence on forms of birth control used. Many Roman Catholics in developing countries consider rhythm the only acceptable form of birth control. This has been publicly promoted by the Pope, the spiritual leader of the Catholic Church. But this does not hold true in more developed countries where sterilization, the pill, and even abortion are among the most popular methods. Political influences also are important; in 1984, for instance, when developing nations appeared ready to benefit from family planning, the U.S. administration under Ronald Reagan denied funding to the International Planned Parenthood Federation, one of the largest nongovernmental organizations providing family planning assistance to LDCs. The U.S. prohibited funding for organizations involved in abortion-related activities or for countries where family planning activities were considered to be coercive (such as payment for undergoing sterilization).

National Birth Rate Reduction Programs

Because the choice of population control methods involves social, moral, and religious beliefs, which vary widely among nations and individuals, it is difficult to generalize about a world approach to reduce birth rates. In 1974, however, representatives to the World Population Conference approved a plan that recognized the right of individuals to decide freely the number and spacing of their children, and their right of access to information to help them achieve their goals. The 1994 Program of Action of the United Nations International Conference on Population and Development (ICPD) reaffirmed these basic rights. The ICPD also directed attention to the need to increase people's—especially women's—access to information, education, skill development, employment, and high-quality health services. Empowerment of women, elimination of inequality

between men and women, and education are three important factors in attaining sustainable futures throughout the world.

The historic agreements reached at the ICPD were reaffirmed at subsequent UN conferences in the 1990s. These included the World Summit for Social Development in 1995, the Fourth World Conference on Women also in 1995, and the World Food Summit in 1996. As a result of the ICPD, governments in some LDCs have changed their policies and institutions to reflect a broader emphasis on women's status and health (Gelbard et al., 1999).

In 1952, India became the first nation to establish a government-sponsored family planning program. In part because of the diversity of cultures, languages, religions, and customs in different regions of the country, India did not experience immediate results from population control efforts. However, India's action was a stimulus to the introduction of family planning programs in almost every other nation worldwide. In 1976, the Indian government moved more aggressively to introduce new incentives to control population growth and to force compulsory sterilization on any man with three or more living children. Compulsory sterilization failed to affect the birth rate significantly and was very unpopular.

Following the ICPD, more recent efforts to integrate development and family planning projects have been undertaken; for instance, adult literacy and population education programs have been combined. Such programs now emphasize reproductive health rather than demographic targets such as limiting family size. Voluntary birth control has been promoted through multimedia advertisements and educational channels. As well, emphasis on lowering the infant mortality rate, on improving the status of women, and on increasing the spacing between births has been successful. India's total fertility rate has declined from about 5.3 in 1980 to 3.4 in 1999.

In 1978, recognizing that its rate of population growth had to decline or the quality of life for everyone in China would decrease, the Chinese government implemented a plan to push China into the third demographic stage (characterized by a decline in the birth rate and a relatively low death rate). Incentives to promote later marriages and one-child families were prominent in this aggressive plan, as were penalties (such as fines, surrender of privileges) for the second and any succeeding child. Again, such drastic measures raise serious ethical questions, particularly as measures may have greatest impacts on women.

While these measures compromised individual freedom of choice, they did bring about a drastic reduction in China's fertility from about 5.8 births per woman in 1970 to about 1.8 births per woman in 1999. In thousands of suspected cases, however, parents killed newborn baby girls in preference for male babies (keeping alive the tradition that sons provide old-age security for their parents). Even

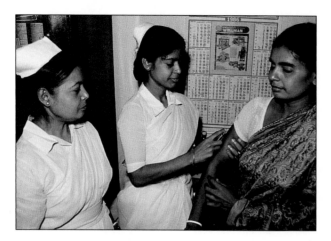

Photo 4–9
Access to good public health care for women may improve their control over their own fertility.

if these actions were not reported, evidence of the practice of female infanticide can be seen in demographic data. For example, during the 1960s and 1970s, the sex ratio at birth remained about 106 males per 100 females. In 1982, the ratio was 108.5; in 1987 it was 110.9, and in 1990 the ratio was 111.3. By 2000, sex ratio at birth was reported at 116.9 males per 100 females. In Haninan and Guandong provinces in South China, these ratios were 135.6 and 130.3 respectively. Although sex-selective abortion is strictly forbidden, it is still practised, particularly in rural areas. The result is millions of missing girls and women, a prospect that may result in future unrest.

From simple provision of information, to promotion and provision of some of the means of birth control, to offering rewards and implementing penalties, a wide variety of approaches has been used to reduce the rate of population increase in the developing world. Each approach has logistical, cultural, and ethical challenges that requires careful consideration before it is introduced on a mass scale.

OTHER FACTORS AFFECTING GLOBAL POPULATION GROWTH

Migration

In the past, migration was an important way by which people could escape degradation of their local environment. In Western Europe, emigration in the 19th century was an important way to reduce the pressures of rapidly growing populations on newly industrializing nations. For example, the potato famine in Ireland brought migrants to many places around the world, including Canada. Between 1846 and 1932, approximately 50 million Europeans left for the Americas and Australasia (Hall, 1995). Between 1881 and 1910, about 20 percent of

Europe's population increase emigrated to the "New World." Today, few countries are as welcoming of new residents. Nevertheless, migrants will play an increasingly important role in population change in the next century (Gelbard et al., 1999). Migrants are increasing in number and in diversity, moving from and to more countries and for more varied reasons than ever before. For instance, in some countries, family reunification has surpassed employment as a leading reason for immigration.

Immigrants challenge receiving nations in diverse ways, including the need to cope with expressions of anti-immigrant sentiment that may develop among residents, and demands to provide economic support for and extend political rights to immigrants. Refugees and involuntary migrants, cut off from traditional networks that provide socioeconomic support, are vulnerable to persecution and exploitation. Agencies such as the UN High Commission for Refugees will need to address the issues faced by refugees and other displaced people around the world. Even in Canada, where immigration targets set by government have never been met, individual incidents related to illegal migration (e.g., "boat people") often bring out public sentiments that Canada is being overrun by newcomers.

POPULATION AND ENVIRONMENTAL SUSTAINABILITY

Canadians need to be concerned about the impact of human population on the Earth's ecosystems. Many environmental problems, including air and water pollution

Photo 4–10
China's one-child policy is an example of an official government response to the need for population control.

and global climate change, transcend national boundaries—as population grows or becomes wealthier, demands for resources increase, resulting in pollution and waste being added to the biosphere. Increased resource use also has an effect on population: environmental pollutants that mimic human hormones may be one cause of the declining ratio of boys to girls born in Canada since 1970 ("Pollutants May Be Linked," 1997).

More population and increased consumption mean increased energy use and the resultant escalation of problems such as global warming, acid precipitation, oil spills, and nuclear waste. More population means more land is required for agriculture, housing, industrial sites, and transportation links, leading to deforestation, soil erosion, and less land for agriculture and fewer habitats for other species (contributing to their extinction, perhaps). As lands are cultivated, they become less productive than naturally vegetated areas. More population means a greater demand on water resources for drinking, irriga-

tion, waste disposal, and industrial processes. Often there is a parallel decline in quantity or quality of the body of water used for these purposes.

As we shall see in Part 3 of this book, human demands on resources are numerous and multifaceted. If Canadians and other inhabitants of this planet are to experience a reasonable quality of life, we need to take steps to protect human and environmental health, to prevent resource abuse through conservation, and to preserve living systems. While it may seem an impossible task, many agree that it is a worthwhile goal to provide all people, whatever the size of the global population, with the opportunity for a life of quality and dignity.

The following chapters outline what is happening to natural resources in Canada, how Canadians can influence resource use and environmental change, what is being done (or has been done) about the various issues, and what needs to be done to ensure sustainability of Canada's part of the global environment.

Chapter Questions

1. How do the crude birth rate and the fertility rate differ? Which measure is a more accurate statement of the amount of reproduction occurring in a population?

2. The world population in 1995 was 5.6 billion and the growth rate was 1.4 percent. Calculate the doubling time of this population. If the doubling time of the human population is decreasing, can the human population be growing strictly according to an exponential curve? Explain why or why not.

3. Why is it important to consider the age structure of a human population?

4. Why is demographic momentum a matter of interest in population projections?

5. What is meant by the demographic transition? When would one expect replacement fertility to be achieved—before, during, or after the demographic transition?

6. Debate the following statements: "Canada must increase immigration in order to avoid a collapse in our population." "Based on consumption patterns, the United States is overpopulated."

7. What forms of development assistance to LDCs might help in reducing population pressures on the environment?

8. How does war affect human population rates and human consumption of environmental resources?

9. Many LDCs are very reluctant to introduce measures to reduce population growth in their countries. Why is this the case?

10. What population size do you believe would allow the world's people to have a good quality of life? Justify your response.

Brown, L. R., Renner, M., & Halweil, B. (2000). *Vital signs 2000: The environmental trends that are shaping our future.* New York: Norton.

Drohan, M. (1997, June 27). British beef returns to McDonald's menu. *The Globe and Mail,* p. A11.

Ehrlich, P. (1971). *The population bomb* (Rev. ed.). New York: Ballantine Books.

Ehrlich, P., & Ehrlich, A. (1990). *The population explosion.* London: Hutchinson.

Elliott, J. (1999). *An introduction to sustainable development* (2nd ed.). London and New York: Routledge.

Gelbard, A., Haub, C., and Kent, M. M. (1999). World population beyond six billion. *Population Bulletin, 54.* http://www.prb.org/pubs/bulletin/bu54-1.htm

Guz, D., & Hobcraft, J. (1991). Breastfeeding and fertility: A comparative analysis. *Population Studies, 45,* 91–108.

Hall, R. (1995). Stabilizing population growth: The European experience. In P. Sarre & J. Blunden (Eds.), *An overcrowded world? Population resources and the environment* (pp. 109–160). Oxford: Oxford University Press in association with The Open University.

Human deaths from mad cow disease increasing. (2000, July 19). *Calgary Herald,* p. A21.

Hynes, P. (1999). Taking population out of the equation: Reformulating I=PAT. In J. Silliman & Y. King (Eds.), *Dangerous intersections: Feminist perspectives on population, environment and development* (pp. 39–73). Cambridge, MA: South End Press..

Kent, M. M., & Crews, K. A. (1990). World population: Fundamentals of growth. Washington, DC: Population Reference Bureau.

Kohlenberg, L. (1997, April 7). And this little piggy went … *Time,* p. 12.

Livernash, R., & Rodenburg, E. (1998). Population change, resources, and the environment. *Population Bulletin, 13,* 5–6.

Myers, N. (1990). *The Gaia atlas of future worlds: Challenge and opportunity in an age of change.* London: Gaia Books.

Pedersen, R. (1997, July 14). High birth defects probed. *Calgary Herald,* p. A1.

Pollutants may be linked to declining male births. (1997, January 8). *Calgary Herald,* p. A9.

Population Reference Bureau. (2000). *1999 world population data sheet.* Washington, DC: Author.

Population Reference Bureau. (2003). *2002 world population datasheet.* http://www.prb.org/pdf/WorldPopulationDS02_Eng.pdf

Silliman, J. (1999). Introduction. In J. Silliman & Y. King, *Dangerous intersections: Feminist perspectives on population, environment and development* (pp. viii–xxiv). Cambridge, MA: South End Press.

Statistics Canada. (1999). *Population by sex and age.* http://www.statcan.ca/english/Pgdb/People/Population/demo10a.htm

Statistics Canada. (2004). *Population by sex and age group 2003.* http://www.statcan.ca/english/Pgdb/People/Population/demo10a.htm

Stock, R. (1995). *Africa south of the Sahara.* New York: Guilford Press.

Stock, R. (2004). *Africa south of the Sahara* (2nd ed.). New York: Guilford Press.

Stuart, R., & McAlpine, S. (2003, June 12). Wake up and smell the coffee. *The Globe and Mail,* p. A17.

UNAIDS & World Health Organization. (1998). *Report on the global HIV/AIDS epidemic.* Geneva: United Nations.

United Nations. (1998). *1998 revision of the world population estimates and projections.* New York: Author. http://www.popin.org/pop1998/

United Nations. (1999). *Demographic indicators 1995–2000.* http://www.undp.org/popin/wdtrends/p98/tp98iwld.htm

United Nations Population Division. (2003). *World population prospects: The 2002 revision.* http://www.un.org/esa/population/publications/wpp2002.

U.S. Census Bureau. (1999). *World population profile: 1998 highlights.* http://www.census.gov/ipc/www/wp98001.html

Vitousek, P. M., Mooney, H. A., Lubchenco, J., & Melillo, J. M. (1997). Human domination of Earth's ecosystems. *Science, 277*(5325), 494–499.

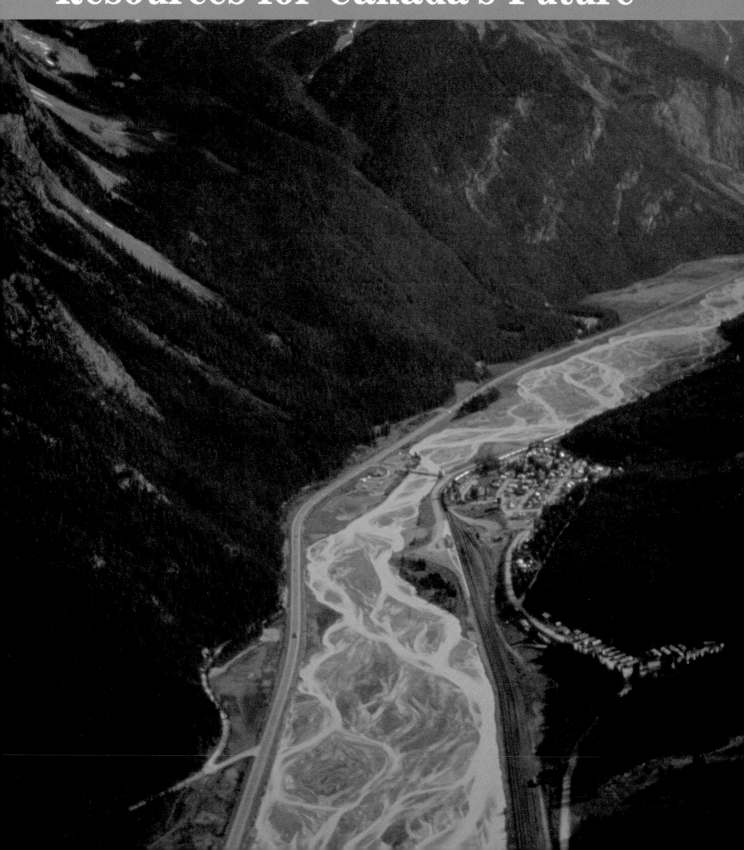

Resources for Canada's Future

Our Changing Atmosphere

Chapter Contents

"We, the members of the National Forum on Climate Change, believe that climate change will touch the life of every Canadian. Decisions taken today on this complex and controversial issue will have implications for our communities, our children, and future generations ... Every Canadian has a role to play in reducing greenhouse gas emissions. The time for action is now."

National Round Table on the Environment and the Economy (1998)

Chapter Objectives

After studying this chapter you should be able to

- understand the main issues and concerns relating to Canada's atmosphere

- identify a range of human uses of the atmosphere

- describe the impacts of human activities on the atmosphere around us

- appreciate the complexity and interrelatedness of issues relating to the atmosphere

- outline Canadian and international responses to the need to protect the atmosphere

- discuss challenges to a sustainable future for the atmosphere

INTRODUCTION

As far as we know, Earth's atmosphere is unique in its ability to provide all inhabitants of the ecosphere with the range of gases crucial to sustaining life's biological processes. In addition to providing oxygen for respiration, carbon dioxide for plant photosynthesis, nitrogen for nitrogen-fixing bacteria and ammonia-manufacturing plants, and other trace gases (noted in Chapter 3), the atmosphere protects all of Earth's inhabitants from the harmful effects of cosmic rays from outer space and ultraviolet (UV) radiation from the sun. As a fundamental part of the hydrological cycle, the atmosphere transports water from the oceans to land. Another service atmospheric gases perform is regulating the Earth's surface temperatures, in part so that liquid water is available for living organisms. Each of these long-standing services of the atmosphere is crucial to the Earth's climate, and to life on the planet.

Human-induced changes began to occur in the composition of the thin film of gases that make up our atmosphere when energy from fossil fuels helped to ignite the Industrial Revolution beginning about 1750. Subsequent growth in human productivity and technological creativity, as well as an unprecedented increase in the human population, resulted in the increasing use of the atmosphere as a dumping ground for many pollutants that, in turn, augmented the natural geological and biological forces of atmospheric change. The complexity of issues related to atmospheric pollution is staggering. Once addressed as separate disciplines, researchers now understand that a multiple range of effects can be attributed to one pollutant. The consequences of such an interrelationship mean, for example, that a scientist cannot study climate change without accounting for photochemical smog, ozone depletion, and acidic deposition. Although atmospheric issues are treated separately in this text, the reader is cautioned that, in practice, impacts associated with atmospheric pollution cannot be isolated into exclusive categories.

Examples of human activities that have induced changes in the composition of the Earth's atmosphere are summarized in Table 5–1. Each of these activities has contributed to major atmospheric effects, including increased acidity in the atmosphere, production of pollutant oxidants (such as photochemical smog) in localized areas of the lower troposphere, heightened levels of infrared-absorbing greenhouse gases, and threats to the ultraviolet-filtering ozone layer in the stratosphere

TABLE 5-1
EXAMPLES OF HUMAN ACTIVITIES ALTERING EARTH'S ATMOSPHERE

Activity	Atmospheric Pollutants Contributed	Potential and Confirmed Effects
Industrial activities	• Includes all criteria air contaminants[1] (CACs) and all greenhouse gases[2] (GHGs), chlorofluorocarbons (CFCs), and toxic heavy metals • Largest sectoral contributor to Canada's SO_x[1] (at 77 percent), largely concentrated in aluminum, cement and concrete, and nonferrous mining and smelting industries • Upstream oil and gas industries discharge well over half of the NO_x[1] and VOCs[1] generated by this sector	• Photochemical smog • Ground-level ozone pollution • Ozone depletion • Predicted climate change effects • Acid deposition • Acute toxicity in plants, animals, and humans
Stationary combustion sources (burning large quantities of fossil fuel)	• Includes electricity and heat generation facilities that (in 2001) produced half of Canada's CO_2 equivalents; coal fired plants are a major source of CO_2 and a major contributor of CO • Contributes polycyclic aromatic hydrocarbons (PAHs), most CACs and some GHGs; the largest sectoral contributor of GHGs (48 percent)	• Predicted climate change • Localized carbon monoxide toxicity • Photochemical smog • Ground-level ozone pollution • Ozone depletion
Transportation practices	• Introduce PAHs, some GHGs, and all CACs in significant proportions • Transportation practices were responsible for 25% of Canada's total CO_2 equivalents in 2001 (down from 30 percent in 1990) and more than half the NO_x emissions • CO emissions associated with transportation practices are almost 7.5 times higher than any other sector	• Predicted climate change • Photochemical smog • Ground-level ozone pollution • Ozone depletion • Carbon monoxide toxicity
Alteration of land surfaces	• Creates more CO_2 equivalent sinks than emissions, resulting in a GHG impact of −0.3 percent • Changes in woody biomass stock and abandonment of managed lands result in the storage of CO_2 • Forest and grassland fires and deforestation contribute substantially to CO and CO_2 total emissions • Two-thirds of all particulate matter (CACs) released into Canada's atmosphere are from construction sites and road dust • PAHs released from land alteration are a concern	• Photochemical smog • Increase in condensation nuclei • Carbon storage
Agricultural practices	• Generate four CACs and four GHGs, primarily particulate matter, CO, and CH_4 • Agricultural soils are important carbon dioxide sinks—in 2001 they stored 300Kt of CO_2	• Increased condensation nuclei • Localized carbon monoxide toxicity • Predicted climate change • Carbon storage

[1] Criteria Air Contaminants (CACs) are a group of seven common air pollutants released into the air by various sectors. They are total particulate matter (TPM), particulate matter ≤ 10 microns (PM10), particulate matter ≤ 2.5 microns (PM2.5), sulphur and nitrogen oxides (SO_x and NO_x), volatile organic compounds (VOCs), and carbon monoxide (CO). Facilities emitting any of these contaminants beyond a specified threshold are required by Environment Canada to report these emissions to the National Pollutant Release Inventory (Environment Canada, 2002).

TABLE 5-1
(CONTINUED)

[2] Greenhouse gases (GHGs) "are a gas that absorbs infrared radiation and in turn emits it in the atmosphere. The net effect is a local trapped energy and a tendency to warm the earth's surface" (Environment Canada, Greenhouse Gas Division, 2003). Environment Canada conducts annual inventories on the six primary greenhouse gases in the Earth's atmosphere: carbon dioxide (CO_2), methane (CH_4), nitrous oxide (N_2O), hydrofluorocarbons (HFCs), perfluorocarbons (PFCs), and sulfahexafluoride (SF_6) (Environment Canada, 2003a).

SOURCES: The National Pollutant Release Inventory, Criteria Air Contaminant Handout, Environment Canada, 2002, http://www.ec.gc.ca/pdb/npri/documents/html/CAC_Handout_e.cfm; *Information on Greenhouses Gas Sources and Sinks,* by Environment Canada, Greenhouse Gas Division, 2003, http://www.ec.gc.ca/pdb/ghg/glossary_e.cfm#g; *1990–2001 National and Provincial GHG Emissions,* Environment Canada, 2003a, http://www.ec.gc.ca/pdb/ghg/_e.cfm; *Criteria Air Contaminants Emission Summaries,* Environment Canada, 2003b, http://www.ec.gc.ca/pdb/ape/cape_home_e.cfm; *Environmental Chemistry* (6th ed.), by S. E. Managhan, 1994, Boca Raton, FL: CRC Press.; *National Action Program on Climate Change,* Environment Canada, 1996, http://www.ec.gc.ca/climate/resource/cnapcc/c1part04.html

Since climate is the major factor influencing the planet's biodiversity as well as the locations where humans can live, grow food, and have adequate water, climate change is a key threat to continued effective functioning of the biosphere. In the sections that follow, we explore some of the forces of change associated with Earth's climate, including increases in concentrations of greenhouse gases and thinning of the ozone layer. As well, we briefly discuss selected issues of ambient air pollution and long-distance transportation of pollutants in order to understand the ways in which human activities have caused atmospheric change and the implications of such changes. We also consider international and Canadian actions undertaken to reduce emissions and some challenges that occur in ensuring atmospheric quality.

RECENT CAUSES OF CLIMATE CHANGE

During the past century, several processes external to Earth's climate system are believed to have influenced trends in global climate. Five major processes appear to be involved: changes in solar intensity, changes in concentrations of stratospheric aerosols, increases in concentrations of greenhouse gases, increases in concentrations of tropospheric aerosols, and thinning of the ozone layer. Table 5–2 briefly summarizes the nature and effects of each of these causes of climate change (called *climate forcings*).

Developing policy regarding climate change is difficult as each of the forces of change (listed above) evolves differently and has a unique effect, in time and space, on Earth's climate system. As you can see in Table 5–2, some activities warm the Earth, while others cool it. Such varia-

tion in effect and scale poses significant challenges to modelling and predicting climate change. At first glance, modellers' predictions appear contradictory: while temperatures will continue to increase globally, some local regions may encounter warmer climates, still others may become cooler. These apparent contradictions are explored in this chapter. Although we still have a lot to learn about the various influences on climate and climate change, it appears that all of these forces are similar in magnitude to the combined influence of greenhouse gases and the direct effect of sulphate aerosols (Hengeveld, 2000).

EARTH'S NATURAL CLIMATE SYSTEM

Life forms on Earth are intimately connected with Earth's climate, and both work in conjunction to maintain environmental conditions suitable for life on the planet. A basic understanding of the Earth's natural climate system and the natural greenhouse effect is fundamental to appreciating human-induced changes in the atmosphere and the enhanced greenhouse effect. Our understanding of changes in the Earth's natural climate system is increased through such activities as evaluating the geological record, monitoring, and mathematical modelling.

THE GLOBAL CLIMATE SYSTEM

The global climate system is a complex matrix of interactions among a dynamic atmosphere, a circulating ocean, and the changing surface of the Earth (Ackerman, 2000; McBean & Hengeveld, 1998). However, individual elements of the system are simple to understand. For

TABLE 5–2
RECENT CLIMATIC INFLUENCES

Source of Change	Nature of Change
Changes in solar intensity	• Solar intensity fluctuates slightly from decade to decade and may have important effects on climate. • Researchers suggest that an increase in solar intensity over the past 300 years has increased the amount of energy flowing into the lower atmosphere by about 0.2 watts/m^2 over the past century.
Changes in concentrations of stratospheric aerosols	• Large explosive volcanic eruptions (such as Mt. Pinatubo, 1991) inject sufficient highly reflective sulphur-based aerosols directly into the stratosphere (10 to 50 km above Earth's surface) to cause temporary net surface cooling (until aerosols settle out of the atmosphere in 3 to 5 years). • Lower than average volcanic activity between 1920 and 1960 allowed more sunlight to reach the Earth's surface, perhaps contributing to slight surface warming.
Increase in concentrations of greenhouse gases	• An increase in radiative forcings[1] (believed to be the result of a global increase in the concentrations of long-lived, well-mixed greenhouse and other trace gases) has amplified heat energy globally by about 2.5 watts/m^2 during the past century. • Additional increases in heat energy, of between 2 and 8 watts/m^2, are predicted if GHG concentrations continue to increase (see statistics cited in "Changes in Atmospheric Composition").
Increases in tropospheric aerosols	• During the past 50 years, concentrations of anthropogenic sulphate-based aerosols and other particulate matter found in the troposphere (0 to 10 km above Earth's surface) have increased substantially, particularly in the largely industrialized northern hemisphere. • Tropospheric aerosols are relatively short-lived and reflect solar energy away from the Earth's surface; they also induce changes in cloud properties that have additional effects on climate. Estimates indicate these aerosols have reduced solar energy at the Earth's surface by 0 to 1.5 watts/m^2; this effect likely will be compounded by increasing emissions from developing countries.
Thinning of the ozone layer	• Thinning of the stratospheric ozone layer, exacerbated by anthropogenic sources of persistent chlorine-based ozone-destroying substances (ODSs), has resulted in two notable, seemingly contradictory, effects: (1) an increase in UV rays reaching Earth's surface, boosting solar radiation that reaches Earth's surface (see "Thinning of the Ozone Layer"); and (2) a reduction, estimated at 0.2 watts/m^2, in the net amount of heat energy retained by the climate system. • Research suggests this trend may reverse with the elimination of ODSs under the Montreal Protocol (see Box 5–7).

[1] See Box 5–1 and Box 5–6
SOURCES: *Meteorology Today* (5th ed.), D. C. Ahrens, 1994, St. Paul, MN: West; *Projections for Canada's Climate Future: A Discussion of Recent Simulations with the Canadian Global Climate Model,* H. G. Hengeveld, 2000, Ottawa: Minister of Public Works and Government Services; *Climate Change 2001: The Scientific Basis,* J. T. Houghton, T. Ding, D. J. Griggs, M. Noguer, P. J. van der Linden, X. Dai, K. Maskell, & C.A. Johnson, eds., 2001, Cambridge: Cambridge University Press.

instance, when the sun's energy enters the Earth's atmosphere, that energy encounters ozone, clouds, and aerosols that reflect some of the sun's radiation back to space and scatter and absorb some of the energy within the atmosphere. About half of the solar energy (short-wave radiation) penetrates through the atmosphere (refer to Figure 3–3), heating up both the atmosphere and the Earth's surface—more at the equator than at the poles. This differential heating causes the atmosphere and oceans to circulate and sets the weather systems in motion. To avoid overheating, the Earth releases energy back into space as heat (long-wave infrared radiation).

Natural Greenhouse Effect

The atmosphere insulates the Earth's surface from the loss of heat to space through a process popularly known as the *greenhouse effect*. The role played by trace or greenhouse gases—the most important being water vapour, carbon dioxide (CO_2), methane (CH_4), ozone (O_3), and nitrous oxide (N_2O)—in this energy exchange is a critical one. Greenhouse gases retard the loss of heat from Earth to space, raising the temperature of the Earth's surface and surrounding air. By keeping the planet's mean surface temperature at about 15°C (about 33°C warmer than it would be otherwise), this natural greenhouse effect is crucial to life as we know it on this planet.

Greenhouse gases enter the atmosphere through several natural processes. Water vapour, for example, enters the atmosphere via evaporation and transpiration processes. Ozone is produced through various chemical reactions within the atmosphere, and carbon dioxide comes mainly from plant and animal respiration, combustion, and the decay of organic matter in soils. Most of the naturally occurring methane is produced from the decay of organic matter in wetlands, while most naturally produced nitrous oxide enters the atmosphere from chemical reactions in soil.

Eventually, natural processes also remove greenhouse gases from the atmosphere. When carbon dioxide is absorbed into the oceans, or when forests and agricultural crops remove carbon dioxide as they grow, or when water vapour returns to the Earth as precipitation, these gases are said to have reached their destinations, or "sinks." When sources and sinks are in balance, atmospheric concentrations of greenhouse gases remain stable, but if the balance is upset, concentrations will change until a new balance is reached.

Assuming all other factors remain constant, when greenhouse gas concentrations change, temperatures at the Earth's surface also change. Studies of the Earth's past climate—through proxy indicators of climate (including tree rings, corals, and fossilized air bubbles trapped in ancient ice)—show that concentrations of greenhouse gases were lower during glacial periods and higher during interglacial periods such as today. This finding reinforces the concept that concentrations of greenhouse gases are linked clearly to the type of climate we can expect on Earth (Intergovernmental Panel on Climate Change [IPCC], 1995). Data from polar ice cores indicate that during the past 10 000 years (the current interglacial period), concentrations of carbon dioxide and methane have been remarkably stable ... or they were, until about 200 years ago (see Box 5–2). By that time, following the start of the Industrial Revolution, humans unwittingly had begun an "experiment" with Earth's life support systems, known now by such terms as the enhanced greenhouse effect, global warming, and climate change.

Excluding water vapour and particulates, the atmosphere still consists of 99.9 percent nitrogen, oxygen, and argon, but the remaining 0.1 percent of atmospheric trace gases that are being altered by human activities are of great concern. Since the concentrations of these gases in the atmosphere are so low, it is possible for anthropogenic emissions to have a significant effect on them. Serious consequences for the stability of ecosystems and the well-being of human societies can arise from even small changes in the atmospheric concentrations of greenhouse gases. And since climate is a result of the exchanges of energy and moisture within the Earth–ocean–atmosphere system, anything that alters the distribution of energy within the system or the amount of energy entering or leaving the Earth's atmosphere inevitably changes the planet's climate. The reality is that atmospheric changes have enormous potential consequences for life and life-support systems on Earth (Hare, 1995; Hengeveld, 2000). This is why climate-change issues demand our immediate attention and action.

CHANGES IN ATMOSPHERIC COMPOSITION

In this part of the chapter we examine two major changes in atmospheric composition, specifically the enhanced greenhouse effect and thinning of the ozone layer. Changes in both greenhouse gases and the ozone layer are noted, as are selected effects of enhanced greenhouse gases and ozone depletion. The language of climate change is quite specific. Refer to Box 5–1 for clarification of terms.

ENHANCED GREENHOUSE EFFECT

Since about 1800, atmospheric concentrations of greenhouse gases have increased substantially in all areas of the world. For example, concentrations of carbon dioxide, methane, and nitrous oxide have increased by 31 ± 4 percent, 151 ± 25 percent, and 17 ± 5 percent, respectively; they now exceed any past levels detectable in the fossilized air bubbles of ice cores over at least the past 420 000 years. Also, the *rate* of increase is unprecedented in the last 20 000 years (Albritton, Filho et al., 2001; Intergovernmental Panel on Climate Change, 2001). Taking into account the uncertainties associated with the magnitude of climatic feedback from the terrestrial biosphere, climate projections suggest that, within the next 100 years, CO_2 concentrations almost certainly will double and may triple or quadruple those of preindustrial levels (from 0.03 percent today to between 0.05 and 0.12 percent). These values translate to increases in CO_2 levels of between 75 and 350 percent above 1750 values (IPCC, 2001; Hengeveld, 2000; McBean & Hengeveld, 1998). This is known as the *enhanced greenhouse effect*.

To scientists studying them, the sources of these increases clearly are human. In 2001, the Intergovernmental Panel on Climate Change (refer to Box 5–1) noted that most

BOX 5-1
THE LANGUAGE OF CLIMATE CHANGE

A significant increase in specific gases associated with the beginning of the Industrial Revolution is believed to have influenced global climate trends, a process now referred to as climate change. Understanding climate change requires an understanding of its language. The information following provides a description of the major institutions, terms, and concepts related to the discussion of climate change in this text.

Category	Description
Terms	
Greenhouse effect	• Insulation of the Earth's surface by atmospheric gases that selectively absorb and reradiate much of the Earth's outgoing infrared radiation. Naturally occurring greenhouse gases (GHGs) include water vapour, carbon dioxide, methane, and nitrous oxide.[1]
Direct GHGs	• Anthropogenically generated gases that reach Earth's atmosphere and create the greenhouse effect. They include carbon dioxide, methane, nitrous oxide, hydrofluorocarbons, perfluorocarbons, and sulphur hexfluoride.[1]
Indirect GHGs	• Anthropogenically generated atmospheric pollutants that indirectly affect terrestrial radiation by influencing the destruction and formation of tropospheric and stratospheric ozone. More specifically, indirect GHGs include sulphur oxides, nitrogen oxides, carbon monoxide, nonmethane volatile organic compounds, and aerosols. Many of these pollutants are associated with ozone depletion, acidic deposition, and photochemical smog.[2]
Measurements	
CO_2 equivalent	• The amount of CO_2 that would cause the same effect as a given amount or mixture of other greenhouse gases. Carbon dioxide is assigned a value of one. CO_2 equivalent is used to compare the influence of various GHGs and is expressed as a function of mass.[3]
Global warming potential (GWP)	• A comparison of the relative radiative heating of various greenhouse gases indexed against 1 kg CO_2. The numeric value takes into account the atmospheric lifetime of each gas and its radiative forcing.[4]
Radiative forcings	• Changes in radiant energy (solar + long wave), after allowing for stratospheric temperature readjustment to radiative equilibrium, over a defined area in the tropopause (boundary between troposphere and stratosphere). Positive radiative forcing may induce surface warming while negative forcing may induce cooling of the earth's surface.[1, 5]
Isotopic markers	• Isotopes of the same atom have the same number of protons and electrons, but a different number of neutrons. They have similar *chemical properties* but slightly different *physical properties*. $\delta^{13}C/C^{12}$ (sigma carbon 13/sigma carbon 12) identifies carbons associated with the burning of fossil fuels while $\Delta^{14}C$ (delta carbon 14) is carbon from non-fossil fuel sources. Isotopes play a critical role in distinguishing between GHGs from natural and anthropogenic sources.[6, 7, 8]
Proxy indicators	• Because we do not have written records of Earth's climate history, researchers use proxy indicators such as tree rings, corals, ice cores, fossilized air bubbles and ground temperatures to determine past GHG concentrations and climate conditions.[9]
Concepts	
Certainty	• A scale of confidence levels that represents the collective judgment of IPCC scientists on the validity of a conclusion based on observational evidence, modelling results, and theory that they have examined. The terms in this scale range from virtually certain to exceptionally unlikely. They are used as qualifiers for predictive statements about climate change.[9]

BOX 5-1

CONTINUED

Inertia

- Relates to the relative slowness of bureaucratic institutions to respond to environmental and social challenges in general and, more specifically, to climate change. In a social context, inertia refers to the delay (perhaps even years to decades) between perceiving a need to respond to a major challenge, and the research and planning that go into developing and implementing a solution to that challenge.[9]

Institutions

United Nations Framework on Climate Change Convention (UNFCCC)

- The predominant international organization charged with the ratification, negotiation, and implementation of the Kyoto Protocol. The organization houses the Conference of the Parties (CoP), the primary decision-making body of the United Nations Framework on Climate Change Convention. Representatives from all signatories of the Kyoto Protocol have met annually to negotiate implementation measures, chart progress, review new technologies, and provide support in reaching Kyoto commitments.

Intergovernmental Panel on Climate Change (IPCC)

- A group of leading scientists from over 30 countries, organized jointly by the World Meteorological Organization and the United Nations Environment Programme in 1988, to study global climate change. Comprised of government-appointed scientists, the IPCC's role lies in assessing peer reviewed and published scientific and technical literature. Through consultation, negotiation, and evidence testing, the IPCC scientists provide consensus-based policy advice and scientific review. The IPCC is considered to be the authoritative voice for climate change information at the international level.[10, 11]

Conventions

Kyoto Protocol

- An international agreement in which signatories agreed to reduce their GHG emissions; if met, these emission reductions in developed countries will be 5.2 percent below 1990 levels by the period 2008–2012. Canada ratified the Kyoto Protocol in 2002 and is legally bound to a 6 percent reduction of 1990 emissions.

[1] Ahrens, C. D. (1994). *Meteorology today*, 5th ed. St. Paul-Minneapolis: West.

[2] United States Department of State. (2002). *U.S. climate action report 2002*. Washington, D.C.

[3] Environment Canada. (n.d.). *Information on greenhouse gas sources and sinks. Glossary.* http://www.ec.gc.ca/pdb/ghg/glossary_e.cfm

[4] United States Environmental Protection Agency. (2002). Greenhouse and global warming potential values. Excerpt from the *Inventory of U.S. greenhouse gas emissions and sinks: 1990–2000.*

[5] Houghton, J. T., Ding, T., Griggs, D. J., Noguer, M., van der Linden, P. J., Dai, X., et al., eds. (2001). *Climate change 2001: The scientific basis.* Cambridge: Cambridge University Press.

[6] Daub, G. W., & Seese, W. S. (1996). *Basic chemistry,* 7th ed. Upper Saddle River, NJ: Prentice-Hall.

[7] Levin, I., & Kromer, B. (1997). Records from Schauinsland. In *Trends: A compendium of data on global change.* Oak Ridge, Tennessee: Carbon Dioxide Information Analysis Center, Oak Ridge National Laboratory, U.S. Department of Energy.

[8] Blasing, T. J., Broniak, C. T., and Marland, G. (2003). Estimates of monthly emissions and associated $^{13}C/^{12}C$ values from fossil-fuel consumption in the USA. In *Trends: A compendium of data on global change.* Oak Ridge, Tennessee: Carbon Dioxide Information Analysis Center, Oak Ridge National Laboratory, U.S. Department of Energy.

[9] Barnola, J.-M., Raymond, D., Lorius, C., and Barkov, N. I. (2003). Historical CO2 record from the Vostok ice core. In *Trends: A compendium of data on global change.* Oak Ridge, Tennessee: Carbon Dioxide Information Analysis Center, Oak Ridge National Laboratory, U.S. Department of Energy.

[10] Intergovernmental Panel on Climate Change. (2001). *Climate change 2001: Synthesis report.* http://www.ipcc.ch/pub/un/syreng/spm.pdf

[11] Intergovernmental Panel on Climate Change. (n.d). *About IPCC.* http://www.ipcc.ch/about/about.htm

CHAPTER 5: OUR CHANGING ATMOSPHERE

of the observed warming over the last 50 years likely (66–90 percent chance) was due to the increase in greenhouse gas concentrations (IPCC, 2001). In addition, the IPCC indicated it was very likely (90–99 percent chance) that warming during the 20th century (that caused widespread loss of land ice and thermal expansion of sea water) contributed to the observed rise in sea level (Houghton et al., 2001). Primary sources of carbon dioxide include the combustion of fossil fuels for energy and the clearing of forests for agriculture and other uses. Sources of methane and nitrous oxide include biological processes associated with agriculture and various industrial sources (refer to Table 5–1). In addition, particularly since the 1970s, small doses of very potent, long-lasting synthetic greenhouse gases such as **halocarbons** have been added to the atmosphere for the first time. We do not yet understand all the interactions among these gases, and, when we add to the greenhouse experiment the interactions of other human experiments (such as those causing ozone depletion, acid precipitation, and smog), additional complications arise. Most alarming, perhaps, is that we know so little about how our experiments will turn out and how we will be affected (McBean & Hengeveld, 1998). Uncertainty, however, is not a reason to delay personal and political action on global climate change (Dotto, 2000).

Before we consider climate change projections for the future, let us consider some characteristics of changes in greenhouse gases so that we can appreciate the effects human activities have on them and, in turn, how greenhouse gases affect climate change (see Box 5–2).

Changes in Carbon Dioxide

In 2001, it was estimated that Canadians contributed 566 megatonnes (Mt) of carbon dioxide (CO_2) to the atmosphere. This estimate represented a *decrease* of 577 Mt over 2000 levels (Environment Canada, 2003a). Total human emissions of carbon dioxide into the Earth's atmosphere are estimated to be about 28 billion tonnes (Gt) annually (Hengeveld, 1998). This represents about 5 percent of the average natural flow of CO_2 into the atmosphere through plant and soil respiration and venting from surface waters of the oceans (see Figure 3–13). Natural emissions of CO_2 are offset by their absorption by plants for photosynthesis and by direct uptake by the oceans. Just like a bank account, the "balance" of the global carbon budget reflects changes in the amount of CO_2 in the atmosphere that are determined by the size of the average imbalance between inflow (sources) and outflow (sinks), not by the magnitude of the flows themselves.

Carbon dioxide releases constitute the largest of all greenhouse gas emissions resulting from human activities. Between 1989 and 1998, annual global emissions from fossil fuel and cement production averaged 6.3 Gt, an increase of 14 percent over the annual average (5.5 Gt) of the previous decade (1980–1989) (IPCC, 2000). Deforestation and mismanagement of agricultural lands also are major sources of atmospheric CO_2, adding between 2 and 9 Gt of carbon annually (Hengeveld, Bush, & Edwards, 2002). Although these amounts may seem small when compared with the 150 Gt of carbon released to and removed from the atmosphere each year by the carbon cycle, they have been sufficient to disrupt the natural balance between carbon sources and carbon sinks.

Data from 40 monitoring stations around the world indicated that by 1997, average atmospheric concentrations of CO_2 had reached 364 parts per million by volume (ppmv), about 1.3 and 3.1 ppmv higher than in 1996 and 1995, respectively. By the end of 1999, atmospheric concentrations of CO_2 reached 368 ppmv (Hengeveld, 2001). From 1980 onward, the average annual rate of increase in concentration of CO_2 has been 1.5 ppmv. Ice core records provide clear evidence that for at least 1000 years prior to the Industrial Revolution, the atmospheric concentration of CO_2 varied by only a few percentage points from an average value of 280 parts per million. This means that, on average, the natural carbon budget was well balanced (inflow equalled outflow) during that time period. Together with other sources of evidence, these data indicate that cumulative effects of humans' introduction of small but increasing imbalances into the carbon budget are the principal causes for the 30 percent increase in CO_2 concentrations noted over the past several centuries (Hengeveld, 1998; Hengeveld & Edwards, 2000).

Part of the evidence supporting the contention that human activities have caused the increase in atmospheric concentrations of CO_2 is the timing of the increases— about 60 percent of the increase in CO_2 has occurred since 1958, when fossil fuels powered a rapid growth in the postwar global economy (see Figure 5–1). In addition to ice core data, other evidence comes from carbon isotopes in the atmosphere. Changes in their relative mix point to an increase in carbon isotopes that originate from burning fossil fuels and forests (IPCC, 1995). Since burning 1 kilogram (kg) of pure carbon releases 3.6 kg of CO_2, it is not difficult to appreciate why CO_2 levels have increased rapidly. All these changes in greenhouse gas concentrations also show that humans, in effect, have moved some gases from the uniform category to the variable category (see Table 3–1).

Changes in Methane

Research based on high-resolution ice core data (see Box 5–2) has indicated that, between 1000 and 1800 A.D., methane concentrations in the Earth's atmosphere exhibited a mean value of 695 ppbv (± 40 ppbv). During the subsequent industrialized period, methane (CH_4) concentrations increased rapidly. By 2000, global CH_4 con-

For decades, climatologists have expressed concern that the rate of change in climatic conditions is not within historic global norms and that this rate of change is a result of human-induced increases in greenhouse gases, particularly CO_2. What proof is there of rising atmospheric CO_2 concentrations?

Researchers use proxy indicators of climate, including tree rings, corals, ice cores and ground temperatures, to show that concentrations of GHGs were lower during glacial periods and higher during interglacial periods such as today. Instrument-collected temperature data from the past 120 years coincides very well with proxy data that, when extended back, indicates the 1900s were the warmest century, and that 1998 and 2001 were the two single warmest years, during the last millennium. The close alignment between proxy data and recorded temperature data during the past 120 years provides the basis for climate change models (see Box 5–6). Of particular importance is the fossil record held in polar ice sheets. Fossilized air bubbles provide evidence of historic atmospheric chemical composition.

In 1998, a team of climate change scientists from Russia, the United States, and France withdrew an ice core from the Vostok, Antarctica, ice sheet (the core was drilled to a depth of 3623 metres, the deepest ever recorded). The air bubbles in this ice core contained evidence of atmospheric conditions through four climate cycles dating back more than 400 000 years before present (BP). Analysis of the air bubbles found within the ice core indicated that the concentration of atmospheric CO_2 had been very stable from 250 to 10 000 BP but had increased by 37 percent since 1750 (when the Industrial Revolution began).

Scientists know the source of CO_2 increase is anthropogenic through evidence from isotopic markers (see Box 5–1). Isotopes of the same atom have the same number of protons and electrons, but a different number of neutrons. They have similar *chemical properties* but slightly different *physical properties*. Carbon in methane and carbon dioxide from fossil fuel combustion are examples of a $\delta^{13/12}$ carbon; CO_2 from plant and animal respiration is an example of a $\Delta^{14}C$. Scientists compare isotopic CO_2 from present-day natural and human-induced sources to each other and with the fossil records to determine relative contributions from each source. This type of comparison enables climatologists to conclude that while concentrations from natural sources of CO_2 have remained fairly constant over time (present-day natural-source emissions are similar to those held in the ice record), those contributed by human actions have increased substantially. Climate change scientists conclude that human generated GHGs are the source of the increase in atmospheric GHGs we currently are experiencing.

SOURCES: "Historical CO_2 Record from the Vostok Ice Core," by J.-M. Barnola, D. Raymond, C. Lorius, & N. I. Barkov, 2003, in *Trends: A Compendium of Data on Global Change*, Oak Ridge, TN: Carbon Dioxide Information Analysis Center, Oak Ridge National Laboratory, U.S. Department of Energy; "Estimates of Monthly Emissions and Associated $\delta^{13}C$ Values from Fossil-Fuel Consumption in the USA," by T. J. Blasing, C. T. Broniak, & G. Marland, 2003, in *Trends: A Compendium of Data on Global Change*; *Basic Chemistry* (7th ed.), by G. W. Daub & W. S. Seese, 1996, Upper Saddle River, NJ: Prentice-Hall; *Frequently Asked Questions about Climate Change*, by H. G. Hengeveld, E. Bush, & P. Edwards, 2002, Ottawa: Environment Canada; "$\Delta^{14}C$ Records from Schauinsland," by I. Levin & B. Kromer, 1997, in *Trends: A Compendium of Data on Global Change*.

centrations were about 1750 ppbv and continuing to rise (but at about one-third the rate reached during the 1980s). Researchers suggest that if human emissions could be reduced by 5 percent, the global methane budget might approach a new equilibrium at approximately 1800 ppbv (Houghton et al., 2001; Dlugokencky et al., 1998).

Through a combination of measurement techniques, including aircraft and tethered balloon data, natural sources of CH_4 emissions (such as wetlands, termites, and oceans) are estimated to discharge 160 Mt/yr. Human-related CH_4 emissions are estimated to be between 300 and 450 Mt annually. Among the largest anthropogenic sources of methane are cultivated rice paddies (CH_4 is released by anaerobic activity in flooded rice lands), livestock-related enteric fermentation (which occurs in the digestive tracts of cattle and sheep), cooking fires, and **biomass burning**. Canadian scientists recently identified that the tens of thousands of artificial lakes created for power generation, water supply, irrigation, recreation, flood control, and other purposes are contributing large amounts of CO_2 and CH_4 to the atmosphere. Because the surface areas of these reservoirs have increased over time, they should be accounted for in calculations of greenhouse gas emissions (St. Louis et al., 2000).

Other important sources of CH_4 emissions include fossil fuel production (including direct leakage of natural gas, byproduct emissions from petroleum recovery, and coal mining), shallow hydroelectricity reservoirs in carbon-rich peatlands, landfill sites (where bacteria degrade organic matter in municipal refuse), and animal waste management systems in developed countries (Hengeveld & Edwards, 2000; IPCC, 1995; Rhode, 1990). In Canada, for instance, recent estimates suggest that manure slurries annually release almost 1 million tonnes of CH_4, a significantly higher volume than had been estimated previously (Kaharabata, Schuepp, & Desjardins, 1998). This finding highlights a difficulty in methane research, namely gaining an understanding of the complex microbial processes that generate methane.

Atmospheric reaction with hydroxyl ions ($OH-$) is the primary sink for methane, and consumption in soils is an important secondary sink.

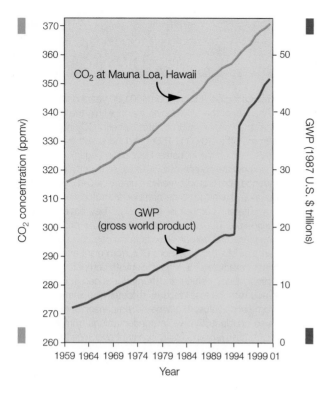

Figure 5–1

Atmospheric concentration of carbon dioxide since 1959, and gross world product since 1960

NOTE: Measurements of CO_2 are taken at Mauna Loa Observatory, HI, far from urban areas and from the direct effects of human and other biological activity (where CO_2 levels would be elevated because of factories, power plants, and motor vehicles).

SOURCES: *World Trade as Percentage of Gross World Product, 1970–2001*, Global Policy Forum, 2001, http://www. globalpolicy.org/globaliz/charts/tradepertable.htm; *Atmospheric CO2 Concentrations (ppmv) Derived from in Situ Air Samples Collected at Mauna Loa Observatory, HI*, by C. D. Keeling & T. P. Whorf, 2003, La Jolla, CA: Scripps Institute of Oceanography, University of California.

Changes in Nitrous Oxide

As is the case with other greenhouse gases, atmospheric concentrations of nitrous oxide are increasing. In 1999, nitrous oxide (N_2O) reached 315 ppbv, an increase of about 12 percent over the past century (Hengeveld, 2001). Concentrations appear to be increasing at a sustained average rate of about 0.7 ppbv per year (Hengeveld & Edwards, 2000). Including the largest source, soils, global emissions from all N_2O sources are estimated to be about 18 Mt/yr (Hengeveld, 2001). About 30 percent of these emissions are related to human food production activities and vary depending on such agricultural factors as soil conditions, temperature, field management techniques, timing of fertilizer applications, and fertilizer type. For example, plowing of grasslands dramatically increases N_2O emissions from soils for several years after the disturbance. Rice paddy flooding methods also affect N_2O emissions.

While more research remains to be done regarding nitrous oxide, studies have shown that forest landscapes, soils and oceans are natural sources of N_2O. When cleared by burning, forests generate a large, sustained pulse of emissions, which continue at sustained high levels for extended periods after the burn. Through the nitrogen cycle, soils contribute about 60 percent of natural source N_2O. Oceans also are a large natural source of N_2O that respond quite rapidly to climate influences. Through isotopic studies (refer to Box 5–1) we are learning that several previously unknown processes may be involved in explaining these responses (Albritton, Filho et al., 2001; Hengeveld & Edwards, 2000).

Changes in Chlorofluorocarbons and Halons

Human-made chlorofluorocarbons (CFCs), also known by their trademark name, Freons, are compounds that con-

Photo 5–1a

Photo 5–1b

Rice paddies and livestock are major human-related sources of methane gas emissions. Canada is the source of about 1 percent of the world's methane contributions to global warming.

tain chlorine (Cl) and fluorine (F) bonded to carbon (C). The most widely manufactured chlorofluorocarbons include CCl_3F (CFC-11) and CCl_2F_2 (CFC-12). Bromine-containing compounds, called **halons**, are related to CFCs and used mostly in fire extinguishers.

Given that they are synthetic and their production is well documented, estimates of chlorofluorocarbons (CFCs) and halon emissions are reasonably precise. While these chemicals will continue to enter the atmosphere for some time after their production has stopped, some CFC emissions are peaking or declining because of production phase-outs established through the 1987 Montreal Protocol on Substances that Deplete the Ozone Layer (and later amendments). Concentrations of CFC replacements, while currently low, are increasing rapidly.

Some fluorine compounds, known as perfluorocarbons (PFCs), are byproducts of aluminum and magnesium smelting. These gases are present in the atmosphere at concentrations in the low parts per trillion, but because they absorb radiation in a highly efficient manner, and have molecular lifetimes of thousands and even tens of thousands of years, they contribute to the additional warming created by increased concentrations of greenhouse gases. (Additional information regarding chlorofluorocarbons and halons is found later in this chapter in the discussion of the thinning ozone layer.)

Changes in Aerosols

Atmospheric aerosols are solid or liquid particles smaller than 100 micrometres (μm) in diameter that are suspended in the air. They consist of a variety of materials of natural origin such as sea salt nuclei, wind-blown soil dust, and fog, as well as anthropogenic materials such as cement dust, pulverized coal, and sulphuric acid mist. As noted in Table 5–2, aerosols affect both the stratosphere and the troposphere. Anthropogenic aerosols add to the complexity of atmospheric sciences; they contribute to photochemical smog, acidic deposition, and climate change, and they also have negative effects on human health and the environment. Given that they exhibit negative radiative forcings, anthropogenic aerosols offset the positive radiative forces of well-mixed GHGs associated with global warming. Concentrations of anthropogenic aerosols are expected to decrease over time, and although it is anticipated that their negative effects on human health and the environment will decrease, their ability to mitigate global warming will decrease as well. The lack of certainty regarding the effects of atmospheric aerosols on future climate trends is a major reason for the uncertainty associated with long-term climate prediction models (see Box 5–1 and Tables 5–2 and 5–3) (Ajavon et al., 2002).

Following Mt. Pinatubo's eruption in 1991, concentrations of sulphate aerosols in the stratosphere peaked over the high northern hemisphere in spring 1992. These aerosols significantly influenced the regional formation of polar stratospheric clouds (see the section "Antarctic Ozone Depletion" later in this chapter). Analysis of polar ice cores suggests that other large volcanic eruptions, such as Krakatoa in 1883, have had similar or greater effects on climate in the past. Human sources such as direct aircraft emissions also have been implicated in the slow, long-term trend toward higher stratospheric aerosol loading.

Sulphates in the mid-troposphere, formed when sulphur dioxide and other sulphurous gases combine with oxygen, affect global warming. Sulphur emissions from human sources (mostly from the burning of fossil fuels) have increased since the 1860s from less than 3 Mt annually to approximately 80 Mt in 1980. Between 1980 and 1995, sulphur dioxide emissions have decreased to just under 47 Mt (World Resources Institute, 1999). Generated principally in the industrialized regions of North America and Europe, and more recently in China, sulphur emissions also are of concern because of their contribution to acid precipitation (see the discussion of acidic deposition later in this chapter). Presently, about 35 percent of aerosol emissions, principally in the northern hemisphere, are sulphates.

Photo 5–2

Aerosols from aircraft emissions may be becoming more important in both the troposphere and stratosphere.

Sulphate aerosols can mask the regional effects of increased greenhouse gas concentrations (Hengeveld, 1999). Particularly in conjunction with fine aerosols from the burning of forests, one of the direct effects of sulphates is to produce a cooling effect that may moderate the warming from greenhouse gases (Taylor & Penner, 1994). Sulphate aerosols may promote cooler surface temperatures by providing condensation surfaces to aid cloud formation, thereby increasing reflection of solar radiation back to space, but this process is complex and not understood fully (see also the section "Effects of Ozone Depletion" later in this chapter). Some regions that are within a few thousand kilometres downwind of industrialized areas actually show cooling as a result of the substantial effects sulphate aerosols have on incoming solar energy (Kerr, 1995).

Using human-produced aerosols to counteract increased greenhouse gas concentrations is not a practical solution, however, because the complexity of the effects and influences of these aerosols are not well understood. Aerosols do not simply offset the much more uniform effects of greenhouse gases (Hengeveld, 1997). Also, aerosols are not a solution to greenhouse warming because of air quality impacts from burning fossil fuels and the costly effects that acid deposition incurs.

In China, the world's leading producer and consumer of coal, acid precipitation (resulting from sulphur dioxide emissions from industrial and domestic use of coal) destroys hundreds of millions of dollars worth of crops and forest annually (Schoof, 1996). Even in Taiyuan, a city of three million people in the northern province of Shaanxi, where natural gas has replaced coal for most home heating, smog continues to be a human health hazard. By 1996, particulate levels there were down to 540 micrograms (pg) from 1200 pg per cubic metre reached about 15 years ago, but particulates remained at a level at least six times the maximum recommended by the World Health Organization (Schoof, 1996). While industrialized North American and European emitters continue to reduce their sulphur emissions, it is possible their reductions will be offset by increases in sulphur emissions from developing countries such as China, which relies on its large reserves of coal to meet its increasing energy demands.

Smoke from biomass burning is a dominant, seasonal source of black carbon aerosols in tropical and southern hemisphere regions. Satellite monitoring in 1992 revealed fine smoke and sooty aerosol concentrations over remote tropical South Atlantic regions that were 10 times greater than those in the northern hemisphere. This increase was linked to the transport of emissions from biomass burning in South America and Africa and was estimated to have caused significant regional cooling, particularly over savanna regions. While total human emissions of all atmospheric aerosols still comprise only a small fraction

Photo 5–3
As the world's top producer and consumer of coal, China experiences high levels of acid precipitation, smog, and particulate pollution.

of those from natural sources, it is expected that they will grow rapidly during the next 50 years and become comparable in quantity with natural source emissions (Hengeveld, 1999).

Satellite systems are becoming important contributors to our improved understanding of the complex and highly variable global distribution of aerosols and their effects on climate. When satellite and ice core data are compared, for instance, measurements suggest that ice core data underestimate past emissions. Such information can help in building more accurate climate projection models (discussed briefly later in this chapter).

Radiative and Residence Characteristics of Greenhouse Gases

The ultimate effect on climate of higher concentrations of greenhouse gases depends on the radiative characteristics of these gases (that is, the types and amounts of energy they absorb, or their absorptive capacity) and their residence time in the atmosphere. Carbon dioxide, the least efficient absorber of infrared radiation, has had the greatest climatic impact of all the human-related greenhouse gases, not only because emissions of CO_2 are much greater than those of other gases, but also because additional quantities of CO_2 may take from 50 to 200 years to return to sinks in the oceans and forests.

In contrast, methane absorbs 15 to 27 times as much infrared radiation over a 100-year time frame as does CO_2, but methane's direct impact on global warming is much smaller. This is because methane emissions are about 20 times smaller than CO_2 emissions and methane remains in the atmosphere for between 9 and 15 years only. Nitrous oxide emissions are about 1000 times lower than carbon dioxide emissions, but nitrous oxide absorbs 310 times as much infrared radiation as an equal mass of CO_2 over a 100-year time frame and remains in the atmosphere for 120 years (Government of Canada, 1996). CFCs and PFCs are the most powerful greenhouse gases: over a 100-year time frame, a tonne of CFC-12 (with an estimated atmospheric lifetime of 102 years) would absorb 8500 times as much infrared radiation as a tonne of CO_2.

In order to evaluate the potential climate effects of equal emissions of each of the greenhouse gases, the concept of relative **global warming potential** (GWP) has been developed to take into account the differing times that gases remain in the atmosphere. This index defines the warming effect due to an instantaneous release of one kilogram of a given greenhouse gas in today's atmosphere, relative to that of carbon dioxide (Intergovernmental Panel on Climate Change, 1990).

Table 5–3 illustrates the effect over 100 years of emissions of greenhouse gases relative to carbon dioxide. Although CO_2 is the least effective greenhouse gas per kilogram emitted, its contribution to global warming, which depends on the product of the GWP and the amount of gas emitted, is largest. Given a number of difficulties in devising and calculating the values of GWPs, including inadequate inclusion of feedbacks such as changing atmospheric composition, these values should be considered estimates and subject to change. The GWPs, however, help us understand more about the significance of each greenhouse gas and where to direct remedial actions.

When we take absorptive capacity, atmospheric lifetime, and other factors into consideration, carbon dioxide has been estimated to account for 64 percent of the additional greenhouse warming that has occurred since preindustrial times, methane for 19 percent, nitrous oxide for 6 percent, and CFCs and halons for 11 percent (Shine et al., 1995). We may discover, as we attempt to reduce CO_2 emissions and find substitutes for other greenhouse gases, that rates of increase of emissions other than CO_2

TABLE 5–3
EFFECTS OF HUMAN-INDUCED GREENHOUSE GAS EMISSIONS (2001)

Greenhouse Gas	Atmospheric Lifetime (years)	100-Year Global Warming Potential (GWP)[1]	Radiative Forcing, 2000 (W/m²)	Emissions, 2001 (Kt)[2, 4]	Relative Contribution over 100 Years[3] (%)
Carbon dioxide	50–200	1	1.50	566 000	64.0
Methane	12 ± 3	21	0.49	93 000 [d]	19.0
Nitrous oxide	120	310	0.15	51 000 [d]	6.0
Hydrofluorocarbons e.g., HCFC-22	1.50–264	140–11 700 1500	0.002	900 [d]	11.0 0.5
Perfluorocarbons	3200–50 000	6500–9200	0.002	6000 [d]	n.a.
Sulphur hexafluoride	3200	23 900	0.003	2000 [d]	n.a.

[1] See Box 5–1.

[2] Kt : kilotonne = 1000 tonnes.

[3] 2001 data do not total 100% due to rounding.

[4] CO_2 equivalent is a measure used to compare emissions from various GHGs based on their global warming potential compared to CO_2 (European Environmental Agency, 2003). See also Box 5–1.

SOURCES: *2001 Greenhouse Gas Emissions Summary for Canada,* Environment Canada, 2003, http://www.ec.gc.ca/pdb/ghg/query/index_e.cfm; *European Environmental Agency Multilingual Environmental Glossary,* European Environmental Agency, 2003, http://glossary.eea.eu.int/EEAGlossary/searchGlossary; *Climate Change 2001: The Scientific Basis,* by J. T. Houghton et al., eds., 2001, Cambridge: Cambridge University Press; "Greenhouse and Global Warming Potential Values," excerpt from the *Inventory of U.S. Greenhouse Gas Emissions and Sinks: 1990–2000,* United States Environmental Protection Agency, 2002, http://yosemite.epa.gov/OAR/globalwarming.nsf/UniqueKeyLookup/SHSU5BUM9T/$File/ghg_gwp.pdf

may change too, requiring different responses and adaptations.

Clearly, future levels of greenhouse gas emissions will reflect the influence and interplay of factors such as population growth, economic growth, and deforestation rates on the one hand, and higher energy prices, improvements in energy efficiency, the availability of practical alternatives to fossil fuels, the development of policies and controls for regulating greenhouse gas emissions, and preserving reservoirs of carbon on the other hand (Government of Canada, 1996). Even if it were possible to hold carbon dioxide emissions at 1990 levels, given current trends, atmospheric concentrations would continue to rise to a level about 60 percent higher than preindustrial levels (450 ppmv) by 2050, and about 85 percent higher (520 ppmv) by 2100 (IPCC, 1995).

These trends suggest that without significant progress in controlling emissions from fossil fuel use and burning or destruction of forests, a doubled carbon dioxide atmosphere appears inevitable by the end of the 21st century. With about 90 million people being added to the world population every year, and potential growth in the economies of India and China (fuelled largely by coal), the world community must be vigilant in controlling the upward pressure on greenhouse gas emissions. At the same time, full and careful consideration of the social, economic, and environmental dimensions of this issue is required. As well, uncertainties exist regarding the future size and behaviour of some of the natural sinks that remove greenhouse gases from the atmosphere. There are concerns about the ability of oceans and terrestrial ecosystems to continue to absorb nearly half the carbon dioxide emitted by human activities. Climate change could be affected by unexpected feedbacks as well. For instance, polar regions are estimated to have very large quantities of methane locked away in frozen hydrates. If large areas of permafrost were to thaw, some of this gas would be released and could result in a marked intensification of the greenhouse effect (Bubier, Moore, & Bellisario, 1995; Government of Canada, 1996).

EFFECTS OF ENHANCED GREENHOUSE GASES

Scientists anticipate that an enhanced greenhouse effect will bring about changes in the Earth's climate that will affect many different areas ranging from air temperature to human health and sea levels. Figure 5–2 presents a summary of the range of possible impacts of an enhanced greenhouse effect in Canada

Most atmospheric scientists are confident that human activities are increasing greenhouse gas concentrations and that this will affect our future climate. As we have noted previously, their evidence of climate change

and the role of greenhouse gases in that change comes from several sources, including global weather and ocean temperature records from the past century. When temperature records from land areas and oceans are analyzed, they suggest that the average air temperatures over land areas have increased by slightly more than 0.6 ± 0.2°C and average sea surface temperatures have increased by about 0.4°C (Intergovernmental Panel on Climate Change, 2001).

Although questions remain, the certainty associated with predictions emanating from climate change models has improved during the last five years or so. Researchers from the IPCC developed a scale of confidence levels that represented their collective judgment about the validity of their conclusions (see Table 5–4).

A recent analysis of freeze and thaw records for 26 lakes and rivers in Canada, the United States, Finland, Switzerland, Siberia, and Japan indicates that freeze-up dates now occur about 8.7 days later and ice breakup dates now occur about 9.8 days earlier than they did 150 years ago. These changes in freeze and thaw dates correspond to an air temperature warming of about 1.8°C over 150 years (Magnuson et al., 2000).

As the climate models predicted, warming has been stronger in the middle and high latitudes than in the tropics, and strongest in the continental interiors of the northern hemisphere. In Canada, temperature records show the northwestern interior has warmed by as much as 1.8°C, while the eastern Arctic has cooled over the past 50 years (Government of Canada, 1996). The effects of these predicted temperature changes, as well as other climate change effects, are illustrated in Figure 5–2.

TABLE 5 – 4
CERTAINTY SCALE DEVELOPED BY INTERGOVERNMENTAL PANEL ON CLIMATE CHANGE

Judgment	Confidence level (% chance that a result is true)
Virtually certain	>99
Very likely	90–99
Likely	66–90
Medium likelihood	33–66
Unlikely	10–33
Very unlikely	1–10
Exceptionally unlikely	<1

SOURCE: *Climate Change 2001: Synthesis Report,* Intergovernmental Panel on Climate Change, 2001, http://www.ipcc.ch/pub/un/syreng/spm.pdf

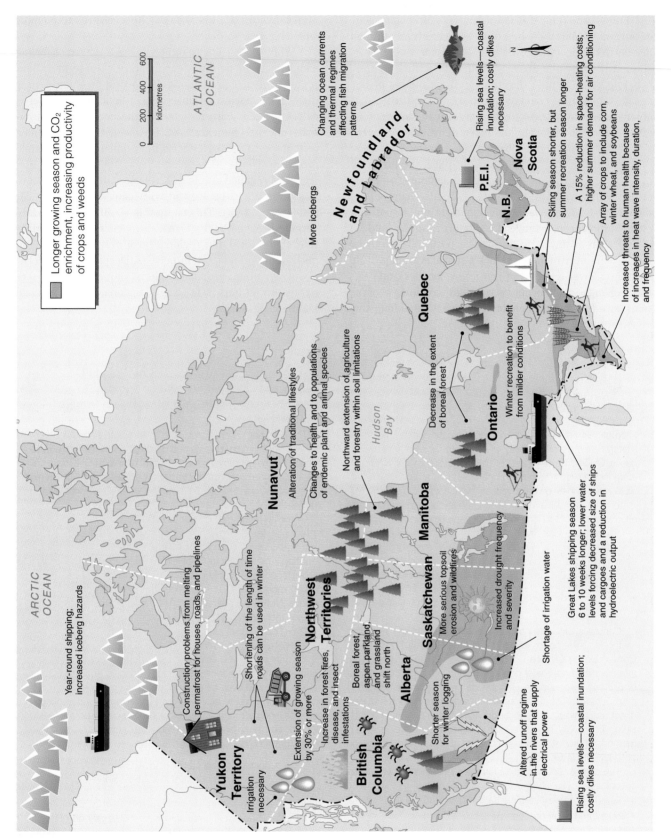

Figure 5–2

Potential effects of climate change

SOURCES: *Global Climate Change Fact Sheet,* Environment Canada, Atmospheric Environment Services, 1995, Ottawa; *Climate Change, Regional Impacts,* Environment Canada, 2002, http://climatechange.gc.ca/english/issues/how_will/regional.shtml

Long-term temperature data are scarce for deep ocean waters, but a cooling of over 1°C in the Labrador Sea between the early 1970s and 1990 has been observed (Lazier, 1996). While this may seem to be a small change, it is important to note that the top three metres of the ocean store as much heat as the entire atmosphere. Small changes in heat storage in the oceans can significantly affect both global climate and ecosystems. (See the section "Other Atmospheric Changes" later in this chapter.)

Although many Canadians might relish the prospect of a warmer climate in this high-latitude country—especially warmer winters—the kinds of climate changes predicted by global circulation models might not be positive ones. For instance, average global temperatures would be higher than at any time during the past 100 000 years (most computer models predict a global warming of at least 3.0°C over a period of just a few decades), and climatic changes could occur very quickly, providing ecosystems with little time to adapt to the new conditions and

humans with little time to develop measures to counter such changes.

While global climate projections are viewed as having a high degree of certainty, modellers are less certain about regional impact projections. Reporting mechanisms for quantifying climate variables are continually being altered as researchers learn more about the constituents of climate change (see Box 5–3). However, these alterations contribute to inconsistencies in regional climate change forecasts. The IPCC suggests that results of climate scenarios at regional and local levels contain key uncertainties and should be viewed with caution (IPCC, 2001; Hengeveld, 2000). This caution is significant and problematic because decision makers need reliable information at the local and regional level in order to develop mitigation and adaptation policies. Despite the concerns about certainty, possible regional climate change scenarios are an important part of regional climate change planning and policy.

B O X   5 – 3
REPORTING ON CANADA'S GREENHOUSE GAS EMISSIONS

International agreements to address climate change require all signatories to report annually on direct and indirect greenhouse gases (GHGs). These reports are necessary, for example, to generate indicators to compare signatories' performance, establish base-line comparison data, assess the accuracy of predictive climate change models, track emissions levels, and evaluate the effectiveness of GHG emissions reduction programs. In Canada, the Greenhouse Gas Division of Environment Canada is charged with the monitoring and reporting of our GHG emissions. To date, Environment Canada has released ten emissions inventories. The most current (2001) inventory reported a 1.3 percent decline in total GHG emissions compared to 2000 levels. What are these inventories based on? Are they accurate? How are they verified? Are they useful?

Canada's GHG inventory is based on international reporting methods developed by the United Nations Framework Convention on Climate Change (UNFCCC). *The Revised 1996 Intergovernmental Panel on Climate Change (IPCC) Guidelines of National Greenhouse Gas Inventories* details the reporting methodology. The IPCC seeks to report all human-induced direct and indirect GHG emissions (by source) and removals (by sink). The IPCC holds regular meetings to refine reporting methods in order to increase both the accuracy and verification of GHG data (IPCC, 1997).

The annual GHG reporting format is based on six emissions sources:

• Energy
• Solvent and other product use
• Industrial processes
• Agriculture
• Land-use change and forestry
• Waste

The reporting mechanisms used to complete Canada's inventory requirements are diverse and complex and vary from source to source. The example provided here is based on GHG inventories in Canada's energy sector; energy-related activities emit more than 80 percent of Canada's GHGs (Environment Canada, 2002a).

Of the six direct GHGs inventoried in Canada, three are associated with the energy sector, namely: carbon dioxide (CO_2), methane (CH_4), and nitrous oxide (N_2O). To avoid double counting, the four indirect GHGs associated with the energy sector (specifically, suphur dioxide (SO_2), nitrogen oxides (NO_x), carbon monoxide (CO), and nonmethane volatile organic compounds [NMVOCs]) are reported as Criteria Air Contaminants and are not included in the energy sector (Environment Canada, 2002a, chap. 2, app. a).

In energy sector accounting, emissions and removal estimates may be derived using the following methods: direct measurement, mass balance, technology-specific emission factor calculations, and average emission factor calculations. The general equation used to calculate emissions from fuel combustion activities (energy, manufacturing, construction, transportation and all work and heat generating combustion activities) is specified below (Environment Canada, 2002b).

$$\text{Quantity of fuel combusted} \times \text{Emission factor per physical unit of fuel} = \text{Emissions}$$

In this equation,

• the *quantity of fuel combusted* is based largely on quarterly reports from Statistics Canada. These reports estimate our energy supply and demand by balancing data about fuel production with data concerning fuel use (such as import/export, residential, and industrial energy use); and

- the *emission factor* is the relationship between the amount of pollution produced and the quantity of fuel processed or burned. Emission factors have been developed from studies conducted by national and international agencies including Environment Canada and the U.S. Environmental Protection Agency. The emission factor for CO_2, for example, is based on the amount of fuel consumed, average carbon content of the fuel, and the portion of the fuel that is oxidized (Environment Canada, 2002a, chap. 2, app. a).

With GHG reporting in its infancy, published emissions totals are neither quantitatively exact nor completely accurate. In its guidelines document, the IPCC predicted there would be initial problems with accuracy in data reporting (because of the newness of the methodology). The IPCC also predicted that accuracy would improve over time as methods were adjusted. Initial problems have been acknowledged in Canada's inventory reports; for instance, very few of the values reported are from direct measurement (most are estimates or calculated values). Although (under stipulations of the National Pollution Release Inventory) Canada has legally binding and direct reporting requirements for the four indirect GHGs emitted by the energy sector, no reporting is required for any of the direct GHGs. The emission factors for many sub-sectors that are critical for emissions calculations are lacking or are based on incomplete data. Some categories, though required by the ICPP, are not reported because the data do not exist. For example, waste gas and flaring emissions from petroleum refining are not reported because data are lacking.

Currently, verification of emissions estimates occurs internally. The IPCC reporting guidelines suggest methods should be well documented, reproducible, and measured against other estimation methods. The IPCC guidelines also suggest that estimates derived from various methodologies be compared to measured data and other peer-reviewed estimates. The Greenhouse Gas Division of Environment Canada has a verification centre that is taking steps to develop consistent GHG standards, protocols, and methodologies for calculating, measuring, and verifying GHG emissions. International verification standards for accuracy do not exist at this time (Environment Canada, 2002c).

Despite their shortcomings, the value of GHG inventories should not be discounted. One of the most valuable outcomes relates to what is learned from conducting the inventories themselves. Data collection using consistent methodologies allows those collecting and analyzing data to assess the effectiveness of current methodologies and to make adjustments accordingly. In turn, this provides decision makers with more precise annual emissions and removal information. Data application is apparent already. Ten years of tracking emissions data has provided evidence of definitive trends that enabled the Greenhouse Gas Division of Environment Canada to determine that reaching Canada's Kyoto commitments required additional steps and to identify where those additional steps need to be taken. For example, GHG inventories provided clear evidence that emissions reductions in the industrial sector were insufficient to offset large increases in the energy sector (the energy sector accounted for 96.6 percent of GHG emission increases between 1990 and 2000). GHG inventories also provided information on which sub-sectors to target for reduction efforts. Vehicle emissions, combined with electrical and steam generation emissions, were responsible for more than half the increase in GHGs during that same time period (Environment Canada, 2002a).

If Canada is to meet its Kyoto Protocol commitments, inventory data are vital in determining where to best place our reduction efforts. Since, in 2000, Canadians contributed about 2 percent of total global GHG emissions (726 megatonnes of CO_2) and Canada ranked ninth in the world for CO_2 emissions (in large part because of our energy-intensive economy), GHG inventories are a valuable tool to help meet the challenges in reducing GHGs.

SOURCES: *Canada's Greenhouse Gas Inventory 1990–2000,* Environment Canada, 2002a, wyswig://26/http://www.ec.gc.ca/pdb/ghg/1990_00_report/appal_e.cfm; *The National Pollutant Release Inventory, Criteria Air Contaminant Handout,* Environment Canada, 2002b, http://www.ec.gc.ca/pdb/npri/documents/Final_CAC_Handout_e.pdf; *Information on Greenhouse Gas Sources and Sinks. Welcome to the Verification Center,* Environment Canada, 2002c, http://www.ec.gc.ca/pdb/ghg/ghg_vc_e.cfm; *The Revised 1996 IPCC Guidelines of National Greenhouse Gas Inventories,* Intergovernmental Panel on Climate Change, 1997, http://www.ipcc-nggip.iges.or.jp/public/gl/invs4.htm

The effects that an enhanced greenhouse effect would have on human societies and natural ecosystems depend on how regional climates and affected societies respond. In turn, regional climate responses will depend not only on how local factors such as evaporation and soil moisture change, but also on how the circulation patterns of the oceans and the atmosphere evolve. Changing oceanic and atmospheric circulation patterns would cause some regions to warm dramatically, others to warm only moderately, and still others (possibly the North Atlantic region off the coast of Labrador) to cool. As storm tracks could shift at the same time, some areas would receive more precipitation and others less.

Prediction of extreme weather events is another key uncertainty identified by the IPCC. Extreme weather events occur on smaller scales than climate models can incorporate and these events occur so infrequently that existing long-term weather data cannot provide the quantitative values needed for modelling (IPCC, 2001; Hengeveld, 2000). Just as the preceding suggestions regarding regional climate change accounted for possibilities rather than certainties, the same is true in the following account of extreme weather events. With an enhanced greenhouse effect, changes in the size or frequency of extreme events such as heat waves, droughts, hurricanes, and thunderstorms could occur. Small changes in climate variability can produce large changes in the frequency of extreme events. For example, a general warming would tend to lead to an increase in the number of days with extremely high temperatures during summer and a decrease in the number of days with extremely low temperatures in winter. One study suggests the frequency of intense five-day heat spells will become eight times more common in Toronto (Colombo, Etkin, & Karney, 1999). This has important implications for direct heat stress, space cooling power demands, and air quality.

Some climate models suggest that precipitation will increase in intensity in some areas, leading to the possibility of more extreme precipitation events, including flooding, while other areas could experience more frequent or severe drought (Intergovernmental Panel on Climate Change, 2001). The risk of flooding in Canada would be expected to increase as a result of a warmer climate, mostly through rainstorm floods, and with heavier rainfall from more (and possibly more severe) thunderstorms and from fewer but larger rainstorms associated with large-scale weather systems (Francis & Hengeveld, 1998).

Research has revealed a tendency toward more extreme precipitation in the northern hemisphere; in particular, heavy rainfalls have increased in the United States, Japan, the former Soviet Union, China, and countries around the North Atlantic rim. Canadian records show heavier precipitation, mostly in the north, since 1940 (Francis & Hengeveld, 1998). Drought has become more common since 1970 in parts of Africa as well as along the coasts of Chile and Peru and in northeastern Australia. Most Canadians are aware of the devastating drought on the Prairies in the 1930s, but the worst drought on Canadian record occurred in 1961, the only year on record outside the range of climatic variability expected on the Prairies. Regional drought is part of the normal long-term climate regime of prairie ecosystems and imposes real hardship on prairie farmers. According to Agriculture and Agri-Food Canada (2003), years that qualify as displaying true drought conditions are 1936, 1937, 1961, 1984, 1988, 2001, and 2002.

Severe winter storms have become more frequent in the Pacific Ocean and, since the 1970s, there has been an increase in the number and destructiveness of storms along the eastern coast of North America. In the past half-century, seven of the eight most destructive storms experienced in this region have occurred within the past 25 years (Francis & Hengeveld, 1998). In part, the magnitude of destruction reflects an increase in elements of the built environment.

Other evidence of widespread warming that will have a variety of effects on human activities includes the long-

Photo 5–4a

Photo 5–4b
It is predicted that global warming will cause greater climatic variability, damaging crops through more severe droughts (top) and more intensive precipitation (bottom).

term sea-level records as well as recent satellite measurements that show increases in sea levels; the worldwide retreat (melting) of mountain glaciers during the past 100 years; thinning of the polar ice cap by 40 percent in just three decades; a modest reduction in the percentage of land covered by snow in the northern hemisphere since the 1970s; and increased precipitation in the mid-latitudes of the northern hemisphere and throughout most of the southern hemisphere.

Even though most of these observations are consistent with the changes projected by general circulation model (AOGCM) experiments, they do not prove that an enhanced greenhouse effect is the cause. The reason they fail to do so is that since the end of the last ice age (10 000 years ago), the world's average surface temperature has varied over a range of almost 2°C from purely natural causes. Given this evidence, some people feel that recent temperature changes could be the effect of natural fluctuations in the climate system.

Photos 5–5a   **Photos 5–5b**

Global warming has resulted in the retreat of Angel Glacier in Jasper National Park. Compare the 1935 photo (left) with the 1991 photo and notice how much the "wings" and "trunk" of the glacier have melted.

Absolute proof of climate change does not exist, although scientific consensus is that the correspondence between trends in climate of the past century and their observed and modelled geographical distribution patterns are "likely … due to an increase in greenhouse gas concentrations due to human activities" (IPCC, 2001, p. 10).

If average global temperatures continue to rise a further 0.5 to 1.0°C beyond current values, the greenhouse warming hypothesis likely would be confirmed as it would be the only explanation that could reasonably account for such an increase (Kerr, 1995). Some recognized Canadian scientists continue to question the climate change hypothesis and charge that the scientific process in Canada is prejudiced by Canadian climate change policy. Some nongovernmental climate scientists have requested a public examination of current climate science by a neutral, independent scientific organization (Ball, 2003). In the meantime, the precautionary principle indicates that action should be taken to prevent atmospheric deterioration. Later in the chapter we review some international and national efforts taken to deal with atmospheric issues.

THINNING OF THE OZONE LAYER

In the following sections, we explore both the nature of the ozone layer and the changes occurring within it due to anthropogenic and natural ozone-depleting substances. In turn, the effects that reductions in the ozone layer have on increasing ultraviolet (UV) radiation and UV's effects on humans and the Earth's ecosystems are identified. Concerns about ground-level ozone are considered in Chapter 13.

The Ozone Layer

By absorbing harmful ultraviolet radiation in the stratosphere and serving as a radiation shield for living things on Earth, ozone (O_3) serves an essential protective function. Produced by a photochemical reaction, ozone is found throughout the atmosphere, although about 90 percent of it occurs in the stratosphere at altitudes of 18 to 35 kilo-

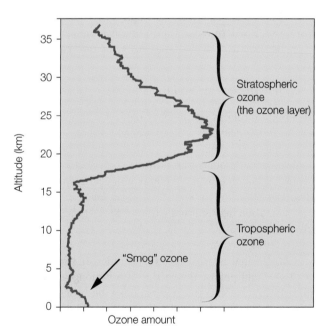

Figure 5–3

Distribution of ozone in the atmosphere

SOURCE: *The State of Canada's Environment—1996,* Environment Canada, Ottawa: Supply and Services Canada. Reprinted with the permission of the Minister of Public Works and Government Services Canada, 2004.

metres (Figure 5–3). The maximum ozone concentration within this band or ozone layer occurs between 20 and 25 kilometres above the Earth's surface (World Meteorological Organization, 1994). Even here, ozone molecules are scattered so thinly (about 300 parts per billion at peak concentrations) that if compressed to ground-level pressure, they would form a band of pure ozone only 3 millimetres thick at sea level. This is equivalent to 300 **Dobson units** (DU), the unit usually used to measure the thickness of the ozone layer in the atmosphere.

Ozone is produced in largest quantities near the equator, where sunlight is most direct and intense. However, varying stratospheric pressures and stratospheric winds cause ozone to move toward the poles, with the result that ozone may be 50 percent thicker at mid- and higher latitudes than in the tropics. Ozone thickness also varies seasonally. Arctic ozone levels are highly variable but generally are highest in late winter or early spring (February or March) and lowest in autumn. Antarctic levels are more predictable and are lowest in the Antarctic spring (October). Most poleward transport of ozone occurs in winter. With little or no ozone-destroying sunlight available at that time of year, stratospheric ozone increases over the winter and peaks near the end of the cold season (February in the northern hemisphere). As sunshine intensifies with the coming of spring, ozone depletion resumes and continues through the summer. In Canada this means lowest ozone thicknesses usually are recorded in the summer and fall, which coincide with lowest annual global levels (Newman, Pyle et al., 2002). In the tropics, where solar energy input is more constant over the seasons, ozone variations are much smaller.

Changes in the Ozone Layer

The human role in ozone layer depletion has been suspected since the late 1960s, when it was argued that water vapour and oxides of nitrogen from proposed subsonic and supersonic aircraft might deplete stratospheric ozone. Although effects of aircraft (and later, space shuttles) were thought to be negligible, the issue of chemical contamination from our industrialized society began to be investigated.

In 1974, two American scientists, Drs. F. S. Rowland and M. Molina, hypothesized that chlorofluorocarbons (CFCs) could persist in the atmosphere long enough to diffuse upward into the stratosphere, be broken up by intense solar radiation, and release active chlorine atoms that, in turn, would destroy ozone. Initially people treated this theory with skepticism. However, growing evidence and discovery of the "ozone hole" over Antarctica in 1985 focused attention on CFCs and other synthetic compounds.

The abundance of ozone is related directly to chemical reactions with **precursor chemicals** and the OH rad-

---

BOX 5–4
CANADIAN ENVIRONMENTAL TECHNOLOGY: THE BREWER OZONE SPECTROPHOTOMETER

Canada is recognized as a world leader among countries involved in research and development of new or improved environmental technologies, including those related to atmospheric science.

The Brewer Ozone Spectrophotometer, a ground-based ozone-monitoring instrument that was patented in 1991 by Environment Canada's Atmospheric Environment Service, is a case in point. Considered to be the world's most accurate ozone-measuring device, the Brewer is produced in Canada and is used here daily to collect and process data from the 12 sites established for the stratospheric ozone and UV monitoring program. About 80 units were in use in more than 38 countries around the world in 2003.

---

ical (the primary sink for ozone). In the lower stratosphere, ozone concentrations over Canada and Europe have declined by about 5 to 10 percent per decade since 1973 (Hengeveld, 1999). It is anticipated that if chlorine concentrations in the stratosphere decline, ozone concentrations in the mid- to low latitudes will recover slowly. But over the next few decades, increased stratospheric cooling is expected to keep stratospheric ozone at low concentrations during polar spring seasons.

Humans appear to have increased tropospheric ozone by as much as 66 percent near Earth's surface, and about 20 percent in the upper troposphere (Hengeveld, 1999). Concentrations of ozone in the troposphere are highly variable, however. For example, O_3 concentrations over the central equatorial Pacific are very low (suggesting a significant regional chemical sink), while over the Andes, tropospheric ozone appears to be increasing by 0.9 to 1.5 percent per year. In the northern hemisphere (outside of the tropics) principal sources of tropospheric ozone are industrial activities and, in the southern hemisphere, biomass burning. In some instances, emission plumes of ozone precursors from biomass burning in Africa have been tracked across the Atlantic and have affected ozone chemistry over the south Atlantic during the following season (Mauzerall et al., 1998). Aircraft emissions of the ozone precursor NO_x (that are expected to increase in future decades) are a significant contributor to ozone increases in the upper troposphere (Hengeveld, 1999; 2000).

Anthropogenic Ozone-Depleting Substances We now know that emissions of chlorofluorocarbons alone account for more than 80 percent of total stratospheric ozone depletion. Together, CFC-11 (used in plastic foam blowing) and CFC-12 (used in vehicle air conditioners and refrigerator coolant) account for about half of the ozone-depleting chlorine entering the stratosphere (Environment Canada, 1999). Other synthetic compounds also contribute to ozone depletion, including carbon tetrachloride, halons, hydrochlorofluorocarbons (HCFCs), methyl bromide, and methyl chloroform. All these chemicals are members of a large class of chlorine- and bromine-containing compounds known as industrial halocarbons.

The most widely used of all the ozone-depleting substances (ODSs), CFCs were researched intensively in the early 1930s as a safe and efficient refrigerant to replace toxic ammonia. The characteristics of CFCs—nontoxic, odourless, nonflammable, noncorrosive, and chemically stable—quickly made them the prime choice as refrigerants. From the late 1950s to the late 1960s, the uses of CFCs multiplied: they were used as blowing agents in plastic foam production (for cushioning, insulation, packaging), as propellants in aerosol spray cans, and as solvents to clean electronic equipment and microchips. Today, CFCs continue to be used widely as coolants in refrigeration and air conditioning, as solvents in degreasers and cleaners, as a blowing agent in foam production, and as an ingredient in sterilant gas mixtures. Canada's 20 million household refrigerators alone contain about 5 million kilograms of CFC-12 (each compressor contains an average charge of 0.25 kg of CFC-12).

Their chemical stability allows CFCs to survive in the atmosphere for several decades to a few centuries (refer to Table 5–3). During this period they diffuse gradually from the troposphere into the stratosphere, where they eventually are broken down by intense ultraviolet (UV) radiation.

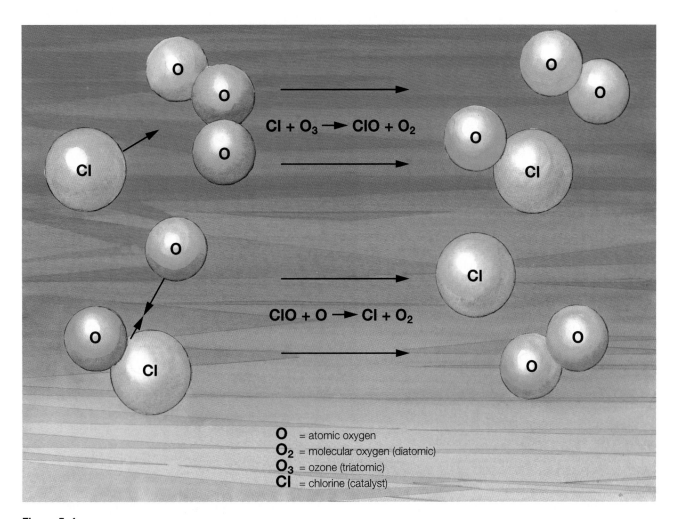

$$Cl + O_3 \rightarrow ClO + O_2$$

$$ClO + O \rightarrow Cl + O_2$$

O = atomic oxygen
O_2 = molecular oxygen (diatomic)
O_3 = ozone (triatomic)
Cl = chlorine (catalyst)

Figure 5–4

How ozone-depleting substances destroy stratospheric ozone

NOTE: Chemical reactions in the stratosphere are considerably more complex than the two equations shown here. However, these equations demonstrate the basic form of a chemical change reaction as occurs in depletion of ozone in the stratosphere.

SOURCE: Adapted from *Understanding Atmospheric Change: A Survey of the Background Science and Implications of Climate Change and Ozone Depletion* (2nd ed.), H. Hengeveld, 1995, SOE Report No. 95–2, Ottawa: Environment Canada.

This process—which is more complex than described here—releases chlorine atoms that react easily with ozone, producing chlorine monoxide and oxygen (Figure 5–4). In turn, the chlorine monoxide breaks down quickly, freeing its chlorine atom to combine again with another molecule of ozone. This property (actually a catalytic chain reaction) enables a single atom of chlorine to destroy approximately 100 000 ozone molecules over a one- to two-year period before it forms a more stable combination with another substance (Rowland, 1989). During the past 100 years, the abundance of chlorine in the stratosphere has increased from a natural background level of about 0.6 parts per billion to about 3.6 parts per billion in 1994.

In 1986, Canadian companies used 19 100 metric tonnes of CFCs, or about 2 percent of the world's total. By 1991, that use had been reduced by 45 percent, to 11 000 metric tonnes. One of only four countries to place an early (1980) ban on major propellant uses of CFCs, Canada had prohibited almost all aerosol uses of CFCs by 1990. About that same time, annual global releases of between 800 000 and 900 000 metric tonnes of CFCs were occurring (Forester, 1991). These CFC emissions were significant as they could trigger a level of ozone depletion 100 times larger than the original emissions.

In 2000, total tropospheric chlorine levels were about 5 percent lower than peak values reached in 1992–94, and *total* chlorine from CFCs is no longer increasing. The abundance of CFC-11 and CFC-13 in the Earth's lower atmosphere peaked in 1994 and is now declining slowly. However, the level of CFC-12 is still increasing, although the rate of increase has slowed. In the stratosphere, however, CFCs are expected to persist for up to 50 years (Ajavon et al., 2002; Environment Canada, 1999).

Other industrial halons contribute to ozone depletion as well. One example, carbon tetrachloride (CCl_4), is used in dry-cleaning, as an industrial solvent, as an agricultural fumigant, and to make CFCs. The strategy to phase out halons in Canada has seen CCl_4 levels decline by about 1 percent per year since 1996. Similarly, methyl chloroform, introduced in the 1950s, is an all-purpose industrial solvent used to clean metal and electronic parts. A substitute for toxic carbon tetrachloride, methyl chloroform is used in large quantities, and much of it is vented directly to the atmosphere during metal cleaning. It was recognized as an important ODS in 1989, and estimated to contribute to about 5 percent of total global ozone depletion. Since the mid-1990s, the atmospheric concentration of methyl chloroform has declined by approximately 20 percent; it is the first restricted halocarbon to show a decrease in atmospheric concentration (Ajavon et al., 2002; Environment Canada, 1999).

Hydrofluorocarbons (HCFCs) contain chlorine, but because they contain hydrogen also, they break down in the lower atmosphere and result in a lower ozone-depletion effect. Used as substitutes for CFCs, HCFCs are known as transitional chemicals because they represent an interim step between strong ODSs and ozone-friendly replacement chemicals. HCFCs account for about 0.5 percent of global ozone depletion, but in Canada, the calculated value of consumption ([production + imports] − exports) of HCFCs was frozen as of January 1, 1996 (Ozone Secretariat, 2000). By 2015, it is expected that HCFC use in Canada will have been reduced by 90 percent. Worldwide, however, HCFCs are still used extensively and emissions continue to increase (Ajavon et al., 2002).

Halons are used primarily as fire suppressants for delicate equipment, computer and electronic equipment facilities, museums, ships, and tanks, and they are in general use in industries, homes, and offices. Halons were produced in large quantities in the 1980s. Levels in the atmosphere have not risen rapidly because most halons have not been vented yet, but remain stored in fire extinguishers. Halons contribute to about 3 percent of global ozone depletion, although their concentrations are increasing (along with concerns for future impacts of these long-lived ODSs).

Methyl bromide has been used as a pesticide since the 1960s but has been recognized as an important ODS only since 1991. Farmers use it to sterilize soil in fields and greenhouses, and to kill pests on fruit, vegetables, and grain before export. Approximately 90 percent of the methyl bromide use in Canada occurs in Ontario and Quebec. Scientists estimate that human sources of methyl bromide are responsible for 5 to 10 percent of global ozone depletion. Although it has a relatively short lifetime, bromine removes ozone very effectively; consequently, methyl bromide is considered a significant contributor to ozone depletion. This short lifetime characteristic suggests that an immediate reduction in consumption of methyl bromide could have a more immediate effect on reducing ozone depletion than would a reduction in consumption of longer-lived substances (but at the same time, this does not imply we should not act on the longer-lived ODSs!). Canada froze the calculated level of consumption of methyl bromide in 1995 and expects to prohibit its consumption by 2005 (Environment Canada, 2002a).

Our use of industrial halocarbons, including CFCs, will continue to have far-reaching effects on the atmosphere. While the chemical interactions and effects of these substances are extremely complex and not yet fully understood, we do know that they portend potentially severe atmospheric changes.

Natural Ozone-Depleting Substances Natural factors relating to changes in ozone levels include the 11-year sunspot cycle (Ram, Stolz, & Koenig, 1997), periodic reversals in wind direction over the equator, and volcanic eruptions. Volcanoes can erupt with sufficient force to inject dust particles and gases into the stratosphere. If that happens, volcanic particles (aerosols) can affect ozone levels because they speed up the chemical reactions that destroy ozone directly. They can also block incoming UV radiation and affect weather patterns that indirectly influence ozone formation and destruction. For the first six months following the eruption of Mt. Pinatubo in the Philippines in June 1991, local stratospheric ozone concentrations were as much as 20 percent below previous levels (Environment Canada, 1997d). The severe Antarctic ozone depletion in 1993 has been attributed partly to the presence of aerosols from this eruption (Manney et al., 1994). However, the impact of volcanic particles is short-lived: in a few years, particulates settle out of the atmosphere, thus posing a reduced threat to the ozone layer.

Spatial Variations in Ozone Depletion

Antarctic Ozone Depletion   The British Antarctic Survey began measuring stratospheric ozone in 1957. In 1985 members of the survey published data showing clearly that ozone concentrations remained at about 300 Dobson units from 1957 to 1970, but after 1970 there was a sharp drop to about 200 DU in 1984 (Farman, Gardiner, & Shanklin, 1985). Since then, ozone concentrations have been quite variable, hitting a high of about 250 DU in 1988 and a low of about 90 DU in 1993. Since 1993, the minimum total ozone has remained at about 100 DU. Chemistry climate models project springtime Antarctic ozone levels will increase by 2010 because halogen levels are expected to have declined by then (Newman, Pyle, et al., 2002; Hamill & Toon, 1991; Stolarski, 1988). Satellite data on ozone concentrations before 1985 confirmed the British Antarctic Survey findings, and the depletion in ozone was dubbed the ozone hole. (Note that there is no actual hole in the ozone shield around the Earth, but there is a decrease in the concentration of ozone that occurs during the Antarctic spring—September to November—each year.)

The most dramatic depletion of the ozone layer occurs over the Antarctic during the southern spring. Two events that occur in the southern polar region during winter are important to the severity of this ozone depletion. One event, known as the **polar vortex**, occurs during the polar winter (night), when the Antarctic air mass is partially isolated from the rest of the atmosphere and circulates around the pole. The second event is the formation of **polar stratospheric clouds** (PSCs); these form in the extremely low temperatures (below –78°C) that develop within the polar vortex as it matures, cools (in the absence of heating by sunlight or by the influx of warmer air from lower latitudes), and descends. Ice crystals in these clouds provide the medium for a complex variety of chemical reactions that lead to rapid depletion of ozone when sunlight returns in the spring (Toon & Turco, 1991).

Antarctic ozone depletion has become more pronounced over time. In 1989, the area of serious depletion (where total ozone thickness was less than 220 DU) covered about 7.5 percent of the southern hemisphere. In October 1993, ozone thicknesses of 91 DU were recorded, and in 1993–94 the area of reduced ozone thickness covered about 10.7 percent of the southern hemisphere (an area about the size of North America). Since then, ozone losses often have started earlier and the affected areas have expanded more rapidly than previously, even extending beyond the tip of South America. Every year, it seems, the ozone hole gets larger. In 1996, the ozone hole reached 26.0 million square kilometres in area; that record was broken in 1998 when the hole encompassed 27.3 million square kilometres, and again in 2000 when the hole was more than 30 million square kilometres (Environment Canada, 1999). In contrast, the 2002 ozone hole was very small (approximately 20 million square kilometres) and broke up earlier than in previous years (Wood & Bodeker, 2003). In 2003, ozone loss again was near record levels, dispelling hopeful predictions that ODS impacts had maximized.

New research highlights the interrelationships among atmospheric issues. Cooling of the Antarctic polar vortex, caused by ozone depletion and warmer temperatures elsewhere, has increased the speed of the polar vortex spin and could exacerbate climate changes. Climatologists in the southern hemisphere are concerned this dynamic could result in a permanent drought in southwestern Australia, where rainfall has been declining for the last 50 years (Byrnes, 2003).

Tropical and Mid-Latitude Ozone Depletion
Because the natural replenishment rate of ozone at low latitudes is high, ozone-depleting chemicals have had the least effect on tropical ozone levels. However, ice particles do occur in the stratosphere over the tropics, and at

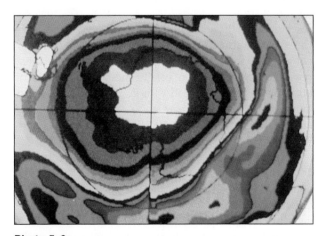

Photo 5–6
A computer-enhanced image of the ozone "hole" over the South Pole region.

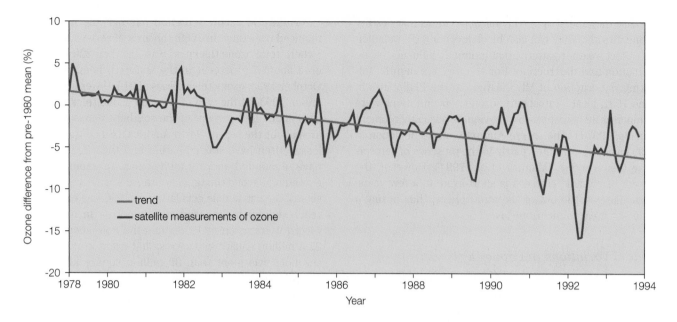

Figure 5–5

Total ozone trend, 30–65°N

SOURCE: *The State of Canada's Environment—1996,* Environment Canada, Ottawa: Supply and Services Canada. Reprinted with permission of the Minister of Public Works and Government Services Canada, 2004.

times there is an abundance of aerosols in the stratosphere from volcanic eruptions. As yet, however, there is no substantial evidence to support the theory that these particles cause ozone depletion in the tropics.

Arctic Ozone Depletion In 2000, the Ozone Secretariat highlighted the pronounced variability in the annual patterns of ozone loss between the North and South poles—ozone depletion in the Antarctic is much more consistent in amount, timing, and pattern of loss than is ozone depletion in the Arctic. (This largely is due to greater variation in Arctic winter temperatures.) In recent winters, Arctic ozone levels increased, generating hope that ozone losses had peaked in 1996–97. However, research has indicated that higher ozone levels were associated with warmer Arctic winter temperatures and were not a reflection of the direct effects of current regulations regarding ODSs. Persistent cold temperatures in the upper atmosphere are associated with greater ozone loss and, during the last several years, the Arctic has experienced comparatively mild winters and a reduction in ozone loss (Ajavon et al., 2002; Newman, Pyle, et al., 2002).

Satellite measurements show that average ozone concentrations over the mid-latitudes of the northern hemisphere have declined by about 7 percent since 1978 (Figure 5–5). Mid-latitude ozone in both hemispheres continues to be depleted, 3 percent below 1980 levels in the northern hemisphere and 6 percent lower in the southern hemisphere. The tropics remain unaffected (Avajon et al., 2002).

However, because the Arctic vortex is much less stable than the southern polar vortex and breaks up sooner, and because the Arctic stratosphere is slightly warmer and less conducive to PSC formation, it is anticipated that ozone depletion on the scale experienced in the Antarctic will not occur in the Arctic. Nevertheless, if an enhanced greenhouse effect were to cool the upper atmosphere as predicted, the conditions for PSC would improve and increase the possibility of more pronounced Arctic ozone depletion. As if to confirm this threat, in March 1997, ozone values over the Canadian Arctic reached their lowest values since monitoring began in the 1960s. We need to appreciate that ozone depletion remains a global concern, from the poles to the tropics.

EFFECTS OF OZONE DEPLETION

When ozone depletion occurs in the stratosphere as a result of the use of CFCs and other industrial halons, that area of the upper atmosphere is cooled. Normally, when ozone absorbs incoming UV radiation, it warms the surrounding atmosphere. But as ozone levels decline as a result of ODSs, the stratosphere cools. In addition, carbon dioxide in the atmosphere (which is on the rise because of burning of fossil fuels) may contribute indirectly to stratospheric cooling and could accelerate the onset of Arctic ozone depletion.

Ozone depletion also may affect global climate indirectly through the loss of phytoplankton (a carbon sink;

see Chapter 3). Threatened by increased levels of UV radiation, the productivity of oceanic phytoplankton—the basis of the ocean's food chain—may be reduced. If their productivity is reduced, their ability to store approximately 80 percent of the CO_2 released into the atmosphere by human activities will decrease. That means atmospheric concentrations of CO_2 may rise, enhancing the greenhouse effect and changing global climate. Phytoplankton also produce dimethyl sulphoxide, a chemical important in the creation of clouds above the oceans. Phytoplankton losses may affect cloud patterns and thereby affect global climate.

As increased levels of UV radiation reach the lower atmosphere, it is anticipated that the reactivity of chemicals such as ground-level ozone, hydrogen peroxide, and acids will increase. Such changes could exacerbate human health problems and could mean more difficulty and expense in achieving current air pollution reduction goals (United Nations Environment Programme, 1998). Ozone depletion will affect plants, including food crops. Ozone depletion also results in an increase in UV radiation reaching the Earth: this is the subject of the following section.

Increased Ultraviolet Radiation

Due to the release of ozone-depleting substances, the ozone layer over Canada is 5 to 10 percent thinner than it was before 1980. This is of concern because when it is sufficiently intense, UV radiation can break stable chemical bonds and damage deoxyribonucleic acid (DNA), the genetic coding material that all living things carry in their cells. Of the three wavelength regions of UV radiation, the longest and least powerful wavelengths are known as UV-A (between 320 and 400 nanometres; 1 nanometre is 1 millionth of a millimetre). These wavelengths pass through the atmosphere almost as easily as visible light and are relatively harmless to most organisms within the normal range of intensity. In contrast, UV-C rays are the shortest (between 200 and 280 nanometres) and most biologically harmful, but they are absorbed almost completely in the upper atmosphere and do not reach the Earth's surface.

A small proportion of the powerful middle wavelengths (UV-B, between 280 and 320 nanometres) reaches the Earth's surface and may cause biological damage to people, plants, and animals. Note that UV-B radiation has always affected people, but, as Canadians increasingly spend more time outdoors (and expose more of their skin while doing so), the effects on human health have become more prevalent. And as more UV-B reaches the Earth as a result of ozone depletion, it may compound the effects of our "sun-worshipping" habits.

How much UV radiation reaches the Earth's surface depends on the angle of the sun, the presence of atmos-

pheric aerosols, the amount and type of cloud, and the thickness of the ozone layer. Ozone, because it absorbs almost all of the most harmful UV-C and most UV-B radiation, is one of the most important determinants of how much UV-B radiation reaches the Earth. It has been calculated that, under clear skies in the mid-latitudes, every 1 percent decrease in the thickness of the stratospheric ozone layer results in an increase in UV-B radiation of about 1.1 to 1.4 percent (McElroy et al., 1994).

Anticipated UV-B radiation increases during the next decade (at least) will continue to affect public health and will impact on terrestrial, freshwater, and marine plants; animals; agricultural crops and livestock; forests; freshwater resources; fisheries; and building materials. While there are widely varying responses from species to species and within different varieties of a single species, many plants will show reduced photosynthesis and growth. For instance, important global food supply crops such as wheat, rice, barley, peas, oats, sweet corn, and soybeans are particularly sensitive to UV-B radiation, as are tomatoes, cucumbers, broccoli, cauliflower, and carrots.

For every 1 percent increase in UV-B radiation reaching the Earth, food production could drop by 1 percent (Environment Canada, 1997e). Vegetable production regions in both British Columbia and Ontario could be affected as UV-B disrupts the way plants use nitrogen. Some livestock species would require protective shelters to avoid reduction in their productivity; other livestock, such as free-range species, would require more land to compensate for the reduced productivity of the plants on which they graze.

While only a few species of Canadian trees have been tested for UV-B sensitivity, increased radiation adversely affected over 45 percent of them, particularly young seedlings. This has important implications for the ability of sensitive replacement species to survive in clear-cut logged areas; if young trees fail to survive, forest sector productivity will decline. Practices such as selective cutting (see Chapter 9) could help reduce potential losses.

Since more than 30 percent of the world's animal protein for human consumption comes from the sea, possible losses caused by ozone depletion would further stress many commercial fish species. Phytoplankton losses, described below, could disrupt fresh- and saltwater food chains and lead to a species shift in Canadian waters. In turn, loss of biodiversity could result in reduced fish yields for sport and commercial fisheries. Even farmed fish raised in shallow ponds with no shade provided could suffer cataracts and lesions (Environment Canada, 1997e).

Organisms such as phytoplankton and zooplankton, living in the surface layers of lakes and oceans, may provide clear evidence of UV radiation damage related to ozone depletion because of their relatively direct exposure to the sun. The blooms of Antarctic phytoplankton, for instance, begin to develop just as ozone thinning is occurring; a 1990 estimate suggested that phytoplankton productivity was 6 to 12 percent lower within the zone of ozone depletion than beyond it (Prézelin, Boucher, & Schofield, 1994). Increased UV-B intensity also may be contributing to global declines in frog and toad populations (see Chapter 12). These examples illustrate the complexity of determining increased UV radiation impacts on natural populations, and of predicting its effects on different ecosystems.

In terms of the public health effects of increased UV radiation, most people realize that too much sun (UV radiation) is dangerous for their skin and health. But how do we know when we should be cautious about being in the sun? Sunburned skin is damaged skin, but how long does it take our skin to burn when exposed? Does it make a difference if we are at the beach or in the mountains?

One tool to help us make decisions about our exposure to UV radiation is the UV index. The UV index was a Canadian first; in 1992, Canadian scientists devised a method to predict the strength of the sun's UV rays, based on daily changes in the ozone layer. That same year, they developed the UV index and Canada became the first nation in the world to issue countrywide daily forecasts of the next day's UV. Environment Canada's UV index is now produced twice daily for at least 48 locations across Canada, as well as holiday destinations, and is available on

Photo 5–7
Excessive exposure to the sun and ultraviolet radiation damages skin and may result in serious problems such as skin cancer.

radio and TV, in the newspaper, through local weather offices, and Environment Canada's Weather Forecast website.

The index measures UV radiation on a scale of 0 to 10, with 10 being a typical midday value for a summer day in the tropics (where the UV is at its highest on Earth). As Table 5–5a indicates, the higher the number on the UV index, the more UV radiation you receive, and the faster your skin burns. Depending on where you live in Canada, your exposure to UV rays will vary. In summer, you will receive roughly three times more UV radiation if you live in southern Canada than if you live in the Northwest Territories (Mills & Jackson, 1995). Table 5–5b indicates some typical summer midday UV index values for places in Canada (and others for comparison).

UV radiation also varies with altitude; at 1000 metres, UV radiation is 7 percent stronger, and at 3000 metres it is 20 percent stronger, than at sea level. Although UV intensity is low during winter, it may be enhanced significantly by reflection: up to 85 percent of UV radiation will be reflected off fresh snow (skiers beware!). Water reflects less than 10 percent of the incoming UV radiation; roughly 40 percent passes through the first 30 centimetres of fresh water and over 80 percent through the same depth in a swimming pool (swimmers take note!). Clouds reduce (but do not eliminate) UV exposure: variable or light cloud cover reduces UV radiation by 10 to 20 percent, whereas heavy, dark, overcast decks of cloud reduce UV by 50 to 80 percent. Also, clear sky radiation reaches a daily maximum at solar noon. During the summer, this usually occurs between 1:00 and 1:45 p.m. local time, depending on location (Mills & Jackson, 1995).

All of this information helps us make informed choices and take actions (or change our behaviour) to reduce our exposure to UV radiation. The UV index helps us plan outdoor activities so we can prevent overexposure to the sun's rays. Not only can excessive sun exposure result in painful sunburn, it can lead to other serious health problems including melanoma, a life-threatening form of skin cancer. Excessive UV exposure can also lead to premature aging of the skin, cataracts, nonmelanoma skin cancers, and immune system suppression (possibly including an increase in some types of infectious diseases and a reduction in the effectiveness of some vaccination programs).

The major exposure to sun in a person's life occurs before age 20; research indicates that the cumulative effects of excessive sun exposure, especially sunburn in young children, can produce skin cancer in later years. Even one or two blistering sunburns during childhood may double the risk of melanoma later in life (United States Environmental Protection Agency, 1995). According to the Canadian Dermatology Association, over 60 000 Canadians develop skin cancer annually (Canadian Safety Council, n.d.). Melanoma is the fastest-rising form of cancer in men and the third fastest in women. Today,

TABLE 5-5
LIVING WITH ULTRAVIOLET

a. The UV Index

UV Index	Category	Sunburn Time
Over 9	Extreme	< 15 minutes
7–9	High	About 20 minutes
4–7	Medium	About 30 minutes
0–4	Low	> 1 hour

b. The UV Index: Typical Summer Midday Values

Location	UV Index	
Tropics	10.0	Extreme
Washington, DC	8.8	High
Toronto	8.0	High
Halifax	7.5	High
Edmonton	7.0	High
Yellowknife	6.0	Moderate
Iqaluit, Nunavut	4.8	Moderate
North Pole	2.3	Low

Strategies to Reduce UV Exposure

- Sunburn times are for light, untanned skin; the times would be somewhat longer for those with darker skin.

Minimize sun exposure

- Plan outdoor activities before 11 a.m. or after 4 p.m., and consult UV Index for daily forecasts of UV intensity.
- Practise sun protection behaviours when outdoors between April and September, between 11 a.m. and 4 p.m. every day.
- In winter, practise sun protection behaviours during periods of extended exposure and/or when you are near fresh/bright snow.
- When visiting warmer climates, remember that UV radiation is more intense there; sun protection is particularly important.
- There is no such thing as a healthy tan; UV radiation from sun and tanning lamps is a major contributor to skin cancer.

Seek shade

- Seek shade, particularly during the 11 a.m. to 4 p.m. period.
- Work toward creating shade in the form of shelters, canopies, and trees.

Cover up

- Wear tightly woven, loose-fitting clothing to cover your arms and legs.
- Wear a hat with a wide brim to shade your face and neck.
- Wear sunglasses that absorb or block 99 to 100 percent of UV radiation.

Use sunscreen

- Use sunscreen in conjunction with shade, clothing, hats, and sunglasses, not instead of them (and reapply every two hours).
- Sunscreens are not intended to increase length of time spent in the sun but to reduce exposure and provide some protection from sunburn when people need to be in the sun.
- Use a sunscreen with sun protection factor (SPF) 15 or higher.

SOURCES: *UV and You,* Environment Canada, n.d., http://www.ns.ec.gc.ca/udo/uv/uvandyou.html; *Workshop Report: Public Education Messages for Reducing Health Risks from UV Radiation,* C. Mills & S. Jackson, 1995, 1995 Saskatchewan On-Line Sun Conference website, http://alep.unibase.com/sunconf/papers/cmills/cmills.html; *Sun Protection for Children,* United States Environmental Protection Agency, 1995, http://www.epa.gov/ozone/uvindex/uvwhat.html

about one in seven Canadians can expect to get some form of skin cancer in his or her lifetime (see Enviro-Focus 5). Because skin cancer takes between 10 and 20 years to develop, we have not yet seen the full impact of post-1980 ozone depletion on Canadian skin cancer rates.

While the incidence of skin cancers in Canada has increased rapidly during the past two decades, perhaps partly as a result of greater awareness and improved diag-

nostics, it is difficult to link human health effects to ozone depletion. For instance, the increase in skin cancers could reflect lifestyle choices, such as the growth in popularity of tropical vacations, rather than an increase in UV-B intensity due to thinning of the ozone layer (Government of Canada, 1996). Nevertheless, a sustained 10 percent thinning of the ozone layer globally is expected to result in nearly two million new cases of cataracts per year and a 26 percent

A Sun Sensitivity Test and "Spot Check"

Your risk of skin cancer is related to your skin type and the amount of time you spend in the sun. How vulnerable are you? Try the Canadian Dermatology Association's Sun Sensitivity Test to determine your chances of getting skin cancer and whether you need to act differently with regard to UV exposure.

Yes	No	
☐	☐	I have red or blond hair.
☐	☐	I have light-coloured eyes—blue, green, or grey.
☐	☐	I always burn before I tan.
☐	☐	I freckle easily.
☐	☐	I had two or more blistering sunburns before I turned 18.
☐	☐	I lived or had long vacations in a tropical climate as a child.
☐	☐	My family has a history of skin cancer.
☐	☐	I work outdoors.
☐	☐	I spend a lot of time in outdoor activities.
☐	☐	I am an indoor worker, but I like to get out in the sun as much as possible when I am able.

- Score yourself 10 points for each "YES."
- Add an additional 10 points if you use tanning devices, tanning booths, or sun lamps.

(80–100) You are in the high-risk zone. Read on to find out how you can protect your skin from the sun.
(40–70) You are at risk. Take all precautions possible.
(10–30) You're still at risk. Carry on being careful.

The Canadian Dermatology Association also produces "Spot Check," a quick reference guide to moles and pigmented spots on the skin (see sidebar). Checking your moles and spots for changes could help catch skin cancer early when it is most easily treated. Most skin cancers are not life threatening, but they can cause extensive disfigurement if left untreated.

For further information, contact the Canadian Dermatology Association.

SOURCE: Canadian Dermatology Association. Reproduced with permission.

spot check

Normal mole: round or oval, even colour. **Many** moles–*increased* risk of melanoma skin cancer.

Atypical mole: mix of browns, smudged border, often bigger than 5 mm. *Increased* risk of melanoma skin cancer.

Melanoma skin cancer: potentially deadly. **Look for *changes* in: Colour:** new colour, black, brown, red, blue, or white.

Shape: irregular, border scalloped but well defined.

Size: enlarges.

Acinic Keratoses: not skin cancer. Indicates excess sun exposure over many years. Red, rough, scaly spots, may itch or sting. *Increased* risk of skin cancer.

Basal cell skin cancer: Can cause disfigurement. Flesh-coloured, red, or black round bump with a pearly border, develops into ulcerating sore.

Squamous cell skin cancer: Can be life threatening. Thickened, red, scaly bump or wart-like growth, develops into a raised crusted sore.

Common skin cancers usually appear on sun-exposed areas.

See your dermatologist if you note any of the above.

increase in the incidence of nonmelanoma skin cancer (Environment Canada, 1997b). The Australian slogan "Slip, Slap, Slop" ("Slip on a T-shirt, slap on a hat, and slop on the sunscreen") is good advice for Canadians, too.

OTHER ATMOSPHERIC CHANGES

Constantly interacting as they circulate, oceans and the atmosphere are closely linked in our efforts to understand long-term climate changes. The cyclical nature of various atmosphere–ocean systems, how they overlap, and the uncertainty of how anthropogenic effects on the atmosphere will influence those systems, add to the complexity of understanding system interactions and predicting how they might be influenced by the variables referred to in this chapter. Slight changes in the temperatures of ocean currents can influence air temperatures and weather patterns worldwide. Perhaps the best-known example of this is El Niño, but oceanographers have identified other, more subtle climatic effects resulting from currents, as well as effects on ocean circulation due to global warming. In the sections that follow, we briefly consider atmospheric changes associated with El Niño and the Southern Oscillation (ENSO), the Pacific Decadal Oscillation (PDO), acidic deposition, and airborne contaminants.

ENSO AND THE PDO

Global climate patterns can be disrupted by the ocean–atmosphere system called El Niño and the Southern Oscillation (ENSO). The El Niño is an invasion of warm surface water from the western equatorial Pacific to the eastern equatorial region and along the coasts of Peru, Ecuador, and northern Chile. Normally, the cold Peruvian current moves northward along these coasts, and southerly winds blowing offshore promote the upwelling of cold, nutrient-rich water that supports large populations of fish, particularly anchovies.

Each year, around Christmas time, a warm current of nutrient-poor tropical water moves south, displacing the cold water. In most years, El Niño is not very warm and lasts perhaps as long as a month. However, about every three to seven years, this phenomenon becomes very strong, persisting for several months. The Japanese Meteorological Agency considers an El Niño to be underway when the tropical Pacific Ocean is a minimum 0.5°C above normal for at least six consecutive months (Nkemdirim & Budikova, 1996). El Niño last occurred in 1997–98, when it was exceptionally strong and disrupted climate around the world.

El Niño changes the sea surface temperature and causes air pressure and wind patterns to change and perhaps reverse. That is, air pressure and wind at opposite ends of the South Pacific oscillate with El Niño; El Niño affects the atmosphere and global temperature by pumping heat energy into the atmosphere. What is strictly a local South American phenomenon is turned into an event with global implications.

Major El Niños cause high mortality in fish and marine plant populations along the Peruvian coast. In the tropics, El Niño events disrupt every aspect of the weather and impact physical and human environments through monsoons and droughts; crops fail, forests burn, and terrestrial and marine habitats are compromised. In western North America and southern Canada, the northward extension of warm tropical waters provides greater than normal water vapour, which is associated with flooding in the west and midwest and unseasonably warm, dry winters in the foothills of the Rockies and western prairies. Floods and droughts both negatively affect agricultural production (Nkemdirim & Budikova, 1996).

Phenology, the study of periodic occurrences in nature and their relationship to climate, may be an accurate and inexpensive way to track biotic responses to climate change. For example, spring phenology surveys of dates for first and full blooms of local Alberta plants (such as purple lilac, aspen poplar, and prairie crocus) have revealed that medium to strong El Niño events correlate with spring arriving about eight days early in Edmonton (Beaubien, 1998). Plantwatch, an expansion of the ongoing Alberta Wildflower Survey database, is a national survey of flowering dates for local plants that involves the public in helping to observe and report plant responses to climate changes.

The best understood of ENSO's complex effects show that in an ENSO occurrence affecting Canada, wet conditions would tend to predominate between December and August in the Pacific northwest area, and the western Arctic area would tend to be warmer and drier than normal. Semi-arid areas in the southwest would tend to be drier than normal in winter and spring. Continued research on El Niño events is important to understanding the potential perturbations that affect global climate.

Satellites and floating sensors are among the arsenal of instruments scientists presently use to help map the circulation of surface and deep currents in the oceans and to improve understanding of ocean–atmospheric interactions and links with climate change. In so doing, researchers have identified major elements of the ocean's circulation. In addition to El Niño, observable shifts in sea surface temperatures affect local weather and may affect global climate patterns (see Table 5–6). The effects of one of these temperature shifts—the Pacific Decadal Oscillation (PDO)—on our society's use of natural resources illustrate the importance of understanding the behaviour of these phenomena.

TABLE 5-6
SELECTED CHARACTERISTICS OF FIVE MAJOR SURFACE OCEAN CURRENTS AFFECTING CLIMATE

Surface Current	Cycle Length (approximate)	Selected Characteristics
Arctic Circumpolar Wave	8 years	Propelled by the Antarctic Circumpolar Current and atmospheric highs and lows, areas of warmer and cooler water circulate around Antarctica from South America to southern Africa.
Pacific Decadal Oscillation (PDO)	20 to 30 years	North Pacific Ocean surface temperature change that seems to have two phases: 1977–1997; the eastern North Pacific was cooler than normal, while the western Pacific was warmer. In the late 1990s this pattern was reversed.
El Niño and the Southern Oscillation	3 to 7 years	Warm surface water from the western equatorial Pacific replaces cooler eastern equatorial water and the cold Peruvian current. It is followed by its cold water counterpart—La Niña.
North Atlantic Oscillation	Indefinite. Current phase has lasted 30 years	Long term but reversible trends in winter atmospheric pressure change surface water temperatures in the Arctic and North Atlantic Oceans. Currently, extended low pressure systems have resulted in warmer than normal winter temperatures in the eastern United States and northwestern Europe.
Tropical Atlantic variability	Indefinite	Reversible patterns of variable sea surface temperatures result in heavier than normal rainfall in some areas, drought in others in South America and Brazil.

SOURCE: "New Eyes on the Oceans," J. Ackerman, 2000, *National Geographic, 198,* 94–96.

Any climatic temperature change has implications for the ways in which people manage their resources. In British Columbia, for example, melting mountain snowpacks provide drinking water and supply hydroelectric power plants. In warm phases of the PDO, when less snow accumulates in the mountains, new sources of fresh water and electric power must be located. Also, the choice of which seedlings to grow in newly reforested areas is determined by temperature; foresters may see growth of seedlings slowed (or the young trees may be killed) by abrupt temperature changes caused by the different phases of PDO.

Salmon stocks, too, are affected by the PDO. During a warm phase, phytoplankton and zooplankton at the base of the food chain decrease, ultimately lowering numbers of top-level predators such as salmon. PDO temperature changes are likely behind historical shifts in salmon runs. For instance, in 1997 (the last time the PDO phase change occurred), salmon runs off British Columbia collapsed while the Alaskan runs increased by over 200 percent.

Tree ring analysis provides clues not only about when the next PDO change might occur (since the PDO changes state every 23 to 26 years, another change should be coming soon), tree ring chronologies also reveal trends in ENSO behaviour over centuries (Codding, 2000).

ACIDIC DEPOSITION

Numerous human activities—including the burning of fossil fuels for transportation, heat, and other energy needs; smelting and refining of metals; pulp and paper processing; and pesticide and fertilizer applications in agricultural operations—introduce both gaseous and particulate contaminants into the air. Whether they are common or more exotic substances, the atmosphere can transport these contaminants long distances from their place of origin. No part of Canada is immune to atmospheric contamination.

In the late 1970s and during the 1980s, "acid rain" became a worrisome environmental issue for a great many Canadians. As early as the 1950s, scientists had detected abnormal acidity in precipitation and in the waters of Nova Scotia lakes. In the 1960s, severe losses among fish populations in acidified lakes southwest of Sudbury, Ontario, were noted. In 1976, prompted by Canadian and international research findings, Environment Canada established a scientific program to study the occurrence and effects of long-range transport of airborne pollutants (LRTAP).

Scientists reported that meteorological conditions in Canada were conducive to long-distance transport of acidic pollutants and that sensitive soils, waters, fish, and forests were susceptible to damage. These meteorological conditions included prevailing winds that moved air masses containing acid-forming gases and other pollutants across the U.S. border from the Ohio Valley and the Cleveland and Detroit areas to Ontario and Quebec. Similarly, air masses containing emissions from central Canada and the United States drifted northeastward to southern New Brunswick and Nova Scotia.

Long-term research by David Schindler and other scientists at the Experimental Lakes Area in northwestern Ontario was instrumental in demonstrating convincingly that acid rain killed trout at acidic levels that U.S. politicians had said were harmless. The immediacy of the issue, and the realization that acid rain could affect everyone, helped ensure that scientists, the media, legislators, environmentalists, and the general public made acid rain a major focus of attention.

What Is "Acid Rain"?

Acid rain occurs when pollutants such as sulphur dioxide (SO_2) and nitrogen oxides (NO_x) are converted chemically to sulphuric acid and nitric acid in the atmosphere, transported, and eventually deposited. Since diluted forms of these acids fall to Earth as rain, hail, drizzle, freezing rain, or snow (wet deposition), or are deposited as acid gas or dust (dry deposition), they are referred to as acidic deposition.

The strength of an acid is described by means of the logarithmic pH scale, where 0 is highly acidic, 7 is neutral, and 14 is basic or alkaline (see Figure 5–6). On this scale, normal rain has a pH value between 5.6 and 5.0, and acid

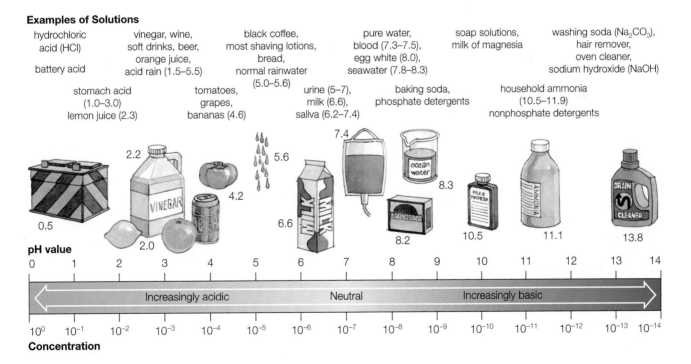

Examples of Solutions

Figure 5–6

The pH scale

NOTE: Values shown are approximate.

rain has a pH below 5.0. Because the pH scale is logarithmic, a pH value of 3 is 10 times more acidic than a pH value of 4, and 100 times more acidic than a pH value of 5. Much of the precipitation that falls over eastern North America and Europe can be 10 to 100 times more acidic than natural rainfall.

Sources of Acidic Pollutants

More than 90 percent of the SO_2 and NO_x emissions occurring in eastern North America are from human activities. In 2000, about 67 percent of the total eastern Canadian SO_2 came from the smelting or refining of sulphur-bearing metal ores and 18 percent from the burning of fossil fuels for energy. In contrast, in 1998 in the United States, about 67 percent of SO_2 emissions came from coal- or oil-fired electrical generating stations. In Canada in 1998, 53 percent of NO_x pollutants were formed during the burning of fossil fuels for on-road transportation, 26 percent in industrial processes, 11 percent in power generation, and the remaining 10 percent from various other sources (Environment Canada, 2001). In 1980, which has served as the reference year for tracking emissions, SO_2 emissions were estimated to be 4.6 million tonnes in Canada and 24 million tonnes in the United States, while emissions of NO_x were 1.8 million tonnes in Canada and 20 million tonnes in the United States. By 1998, those emissions were reduced to 2.7 and 17.7 million tonnes in Canada and the United States, respectively (Environment Canada, 2003b).

Acid rain is not confined to eastern Canada but is continental (and intercontinental) in distribution. A variety of point-source emissions of acidic pollutants associated with local smelters, hydrothermal plants, pulp and paper mills, and oil and gas processing facilities occur in the Prairies, for example, and southwestern British Columbia is exposed to point sources of local and foreign origin. Also, in spite of the lack of any significant emission sources, the Yukon, Northwest Territories, and Nunavut are exposed to acid deposition originating in Eurasian and eastern North American locations through the long-range transport of air pollutants. Southern Scandinavia also receives a significant air pollution load from heavily industrialized areas in Europe.

As the media and public concern about acid rain increased through the 1960s and 1970s—and governments responded with new emission standards—power plants, smelters, and industries began using smokestacks up to 300 metres high (see Box 10–4). These stacks enabled users to reduce local concentrations of air pollutants and to meet government standards without adding expensive air pollution control devices. Once released into the atmosphere, however, acidic pollutants could be carried up to 1000 kilometres by prevailing wind and weather systems, across national and international borders, before being deposited. Effectively, the acid rain and other pollutants became "someone else's problem."

Downwind of these tall stacks, regional pollution levels began to rise. In eastern North America, prevailing winds push pollutants to the northwest and northeast; through computer modelling, it was estimated that more than 50 percent of the acid rain in eastern Canada came from U.S. sources, mainly Ohio, Indiana, Pennsylvania, Illinois, Missouri, West Virginia, and Tennessee. Until the Clean Air Act of 1990 called for a significant reduction in U.S. emissions of SO_2 and a modest reduction of NO_x by the year 2000 (and comparable reductions in Canadian emissions), the large flow of acid deposition from the United States to Canada was the source of political tension between these two countries during the 1980s (McMillan, 1991). Canada continues to negotiate with the United States to reduce the cross-border flow of pollutants.

Throughout this period, environmental groups did a great deal to educate the media and the public about acid rain (and other pollution issues) and to encourage politicians to act against it. While political lobbying about acid rain was unsuccessful during the Reagan administration, it did help achieve significant acid rain amendments to the 1990 American Clean Air Act during the Bush administration.

Effects of Acidic Deposition

In addition to exposure to emissions from thermal generating stations in the United States, areas in the southern Canadian Shield, southern Nova Scotia and New Brunswick, and much of Newfoundland have been exposed to acidic deposition from smelters in Manitoba, Ontario, and Quebec. Because much of the region has little ability to buffer or neutralize acidic pollutants (due to thin, coarsely textured soils and granitic bedrock), many of the more than 700 000 lakes are expected to continue to lose populations of fish and other freshwater species. For example, salmon populations in 31 Nova Scotia rivers have been lost or depleted because acidic precipitation has resulted in the loss of one-third of available Atlantic salmon habitat (Government of Canada, 1996). (See Table 5–7 for an overview of some of the effects of acid deposition.)

Figure 5–7 illustrates the regional capabilities of soils and bedrock to mitigate acidic deposition. Acidic deposition affects some soils by lowering soil pH and impeding plant productivity, while other soils are able to neutralize acidic precipitation, gas, and dust. On the Prairies, chernozem soils contain high amounts of calcium carbonate, an important buffer material that helps reduce adverse effects of acid precipitation. In contrast, studies of northern forests, where soils are already slightly acidic, indicate that anthropogenically acidified soils will take hundreds of years to recover, despite compliance with current emission reductions (Environment Canada, 2003b; Elvingson, 1999).

TABLE 5–7
SOME EFFECTS OF ACIDIC DEPOSITION

Aspect Affected	Nature of Effects
Aquatic ecosystems	• Interactions between living organisms and the chemistry of aquatic habitats are extremely complex; as one species changes in response to acidification, the entire ecosystem is likely affected through predator–prey relationships of the food web. • As water pH approaches 6.0, crustaceans, insects, and some plankton species begin to disappear. • As pH approaches 5.0, major plankton community changes occur; less desirable species of mosses and plankton may invade, and progressive loss of some fish populations is likely (most highly valued species generally are least tolerant of acidity). • Below pH of 5.0, water is mostly devoid of fish; bottom is covered with undecayed material; nearshore areas may be dominated by mosses. • Terrestrial animals dependent on aquatic ecosystems also are affected; for waterfowl dependent on aquatic organisms for nourishment and nutrients, loss of food sources reduces habitat quality and reproductive success declines.
Terrestrial plant life	• Natural vegetation and crops are affected. • Protective waxy surfaces of leaves are affected, lowering resistance to disease, cold, and drought. • Plant germination and reproduction may be inhibited. • Soil weathering and removal of nutrients are accelerated, slowing growth rates and/or killing plant species. • Some toxic elements such as aluminum become more soluble; high aluminum concentrations in soil can prevent plant uptake and use of nutrients.
Animal life	• Effects are hard to assess. • As a result of pollution-induced alteration of habitat or food resources, acid deposition may cause population decline through stress (fewer available resources) and lower reproductive success.
Socioeconomic consequences	• Lower productivity in fisheries, forestry, and agriculture results in fewer jobs for some of Canada's important industries. • Acid deposition causes accelerated corrosion, fracturing, and discoloration of buildings, structures, and monuments.
Human health	• We eat food, drink water, and breathe air that has come in contact with acid deposition. • Acid deposition can increase the levels of toxic metals such as aluminum, copper, and mercury in untreated drinking water supplies. • Premature mortality, chronic bronchitis, respiratory and cardiac hospital admissions, asthma symptoms, and restricted activity days are associated with acidic deposition (see Box 5–5).

SOURCES: *A Primer on Environmental Citizenship,* Environment Canada, 1997, Ottawa; *Acid Rain and ... Water,* Environment Canada, 2003a, http://www.ec.gc.ca/acidrain/acidwater.html; *Acid Rain and ... Forests,* Environment Canada, 2003b, http://www.ec.gc.ca/acidrain/acidforest.html; *Human Health Benefits from Sulfate Reduction,* United States Environmental Protection Agency, 2000, http://www.epa.gov/airmarkt/articles/healtheffects/index.html

About 43 percent of Canada's land area is highly sensitive to acid precipitation (see Figure 5–7). Not only are bodies of fresh water and aquatic life affected by acid precipitation, but other species such as trees can be impacted also. In eastern Canadian forests, for example, 96 percent of the land with high capability for forestry is subject to sulphate deposition in excess of the target of 20 kilograms per hectare annually (although it should be

Chapter 5: Our Changing Atmosphere

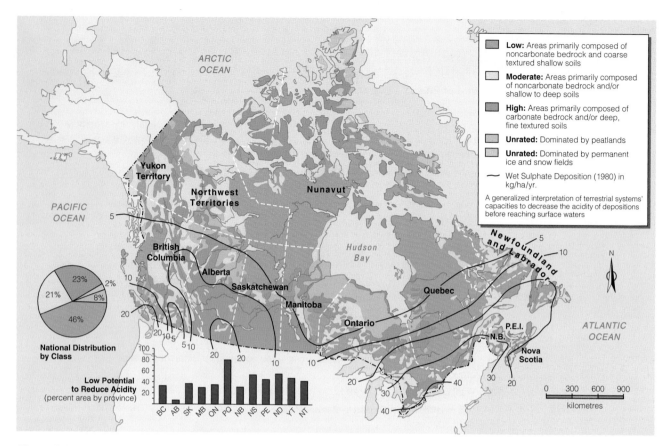

Figure 5–7

Potential of soils and bedrock to reduce acidity of atmospheric deposition in Canada

NOTE: Values shown are approximate.

SOURCE: *Environmental Fact Sheet: Acid Rain: A National Sensitivity Assessment, 1988,* Environment Canada. Reprinted with permission of the Minister of Public Works and Government Services Canada, 2001.

noted that the area receiving 20 kilograms per hectare declined by nearly 59 percent from 1980 to 1993). White birch in southeastern New Brunswick have died or deteriorated as a result of acid fog, as have white birch stands on the shores of Lake Superior (where acid fog occurs frequently) and in the Bay of Fundy region.

Between 1985 and 1987, a survey of more than 2 million hectares of sugar maple stands in Quebec revealed that 3 percent of the area was severely damaged, 47 percent showed moderate damage, and the remaining 50 percent had marginal symptoms of acidic deposition. Canada's Acid Rain National Early Warning System (ARNEWS) has been assessing and monitoring forest health at 150 sites across the country. Scientists at the Laurentian Forestry Centre, one of 32 ARNEWS sites in Quebec, have been able to assess damage to trees caused by insects, diseases, acid deposition, pollutants, and climatic extremes. They determined the effects of drought and soil freezing on the health of maples following their 1981 decline, for instance (Natural Resources Canada, 1997).

In addition to impacting sensitive ecosystems such as forests, acid deposition affects farmlands, wildlife, and freshwater species. For instance, high levels of acidic deposition may cause metals to leach into the water system from soils surrounding acid-sensitive bodies of fresh water. High acidity and elevated levels of metals such as aluminum can seriously impair the ability of water bodies to support aquatic life, resulting in a decline in species diversity. Such effects impact negatively on related water uses such as sport fishing and recreation. In addition, there has been speculation about the role of aluminum in Alzheimer's disease.

Other human health responses to acidic pollutants may lead to aggravation of respiratory ailments such as bronchitis and asthma (Box 5–5). Since more than 80 percent of all Canadians live in areas with high acid deposition–pollution levels, susceptible individuals may experience increased respiratory problems. For example, research has shown a relationship between decreased lung function, increased cardiorespiratory mortality, and long-term exposure to acidic aerosols (Environment Canada, 1996b).

BOX 5–5
AIR QUALITY AND YOUR HEALTH

Poor air quality affects us all …

Knowledge is growing about the wide range of negative effects of atmospheric contamination on personal health and community well-being. However, many people seem unwilling to make changes that may require some personal inconvenience in order to reduce atmospheric pollution—not only for their own benefit but also for the well-being of future generations and the environment. Regardless of the future long-term negative effects of atmospheric contamination, many Canadians appear to be ignoring a very real, present, and personal reason for taking steps to improve air quality: the risk it presents to their personal health. Atmospheric contaminants cause death and chronic illness on a daily basis in Canada. Despite the fact that thousands of premature deaths and annual emergency room visits are attributed to poor air quality, the health industry is not actively involved in promoting reductions in atmospheric pollutants. (See the text for more about the relationship between air quality and your health).

Acid air pollution, smog, and ground-level ozone—alone or in combination—can be major human health hazards in both urban and rural areas. Major health effects of various forms of air pollution are related to impacts on the human respiratory system. For instance, research indicates that ground-level ozone increases the susceptibility of asthmatics to common allergens such as dust mites and moulds that thrive in ordinary buildings. Similarly, people with respiratory problems may suffer more symptoms during periods of high ozone levels. In Ontario, more people are admitted to hospitals for respiratory problems when elevated levels of ozone and/or sulphates occur (however, it is not clear that these pollutants are the only ones responsible for higher hospital admissions; particulates and climate also may play a role).

Research is continuing into the effects of low-level, long-term exposure to ground-level ozone and the decreased ability of people's lungs to ward off disease, particularly if the inhaled ozone has penetrated deeply into the lungs and damaged some of the alveoli (individual air sacs in the lungs where the exchange of oxygen and carbon dioxide takes place). After years of exposure, these small lesions in the lungs of experimental animals have been shown to result in connective tissue damage (scar tissue formation deep in the lungs); the implications of this accelerated aging of lung tissue for humans are being investigated.

While the impacts of acid deposition on the environment have been discussed widely, the effects of acid air pollution on human health have tended to go unrecognized. The suspended acidic particles of compounds such as sulphuric and nitric acids are small enough to penetrate deeply into our lungs when we breathe. There they may cause such effects as coughing, congestion, and constriction of the airways; increased mucus production in the respiratory system; and reduced ability to clear foreign matter from the lungs. Recent studies show that more people are hospitalized with respiratory problems on days when acid air pollution is relatively high.

Health Canada has compared children living in Portage la Prairie, Manitoba, where acid air pollution is low, with children in Tillsonburg, Ontario, where the pollution level is relatively high. On average, the Ontario children had a 2 percent lower lung function and more chest colds, coughs, allergies, and stuffy noses than their prairie counterparts. While not dramatic, this measurable difference was followed up in studies of five Saskatchewan and five Ontario communities, where virtually the same results were found.

Other health problems derive from motor vehicle exhaust and combustion processes. Eye irritation, for example, is a result of two pollutants, peroxyacetyl nitrate and aldehydes. Particulates, originating from diesel exhaust and industrial activities, are of concern for human health for two reasons: they are small enough to be inhaled deeply into the lungs, and they act as a transport medium for compounds such as acids or metals that may adhere to them. Particulates and what is attached to them are known to cause short-term respiratory irritation.

People who exercise in a smoggy environment, such as running along a main thoroughfare during rush hour, may find a decrease in their performance due to carbon monoxide. Emitted from all motor vehicle exhaust, carbon monoxide binds with red blood cells much more readily than oxygen does. In this way, if some red blood cells bind with carbon monoxide, less oxygen may be available to the body's muscles and organs during the exposure to air pollution. Other groups at risk are pregnant women, infants, and people with cardiovascular or respiratory disease, including chronic angina. Smokers may be at particular risk because they have higher levels of carbon monoxide from smoking.

Indoor air quality can be affected as a result of energy conservation measures (more tightly sealed houses), and may present even greater health risks than does exposure to outdoor air pollution in our largest cities. As a result, attention has been given to the design and construction of "healthy houses" that substantially reduce human exposure to a wide range of indoor air pollutants. Healthy-house construction involves planning to achieve energy efficiency and healthy indoor air quality through insulation, specialized windows, and landscaping that captures natural energy. Construction materials are recycled, environmentally benign, and nonallergenic where possible. Although tightly sealed for energy efficiency, a healthy house is well ventilated through efficient air-exchange systems.

In the work environment, a variety of illnesses are attributed to "sick buildings." People report such maladies as minor eye irritation and tearing, nasal congestion and headaches, lethargy, sore throats, and coughs. Often these symptoms may be associated with building renovations, including painting, plastering, and carpeting; when combined with inadequate ventilation, the emissions associated with these activities frequently have an adverse impact on employees and workers.

Little is known about how much pollution we actually are exposed to, and how much really affects our health. Increasingly, researchers have people carry pollution-measuring devices so that "personal exposure monitoring" can be undertaken and relationships between regional air pollution monitoring data and personal exposure levels established. Such data are expected to help in assessing and developing air quality guidelines to protect human health.

Photo 5–8
Acid precipitation damages both natural and human environments, including important heritage structures such as statues and buildings.

Additional reduction of acidic deposition would provide enormous health and financial benefits to Canadians. Acid precipitation causes about $1 billion damage in Canada annually, to lakes, fish habitats, forests, and some of Canada's heritage buildings, including the Parliament Buildings in Ottawa (Environment Canada, 1997f). Sulphur dioxide is an urban smog precursor; when SO_2 combines with water vapour, it forms fine particulate matter, a key component of photochemical smog and another significant health hazard. Elimination of SO_2 emissions by an additional 50 percent would prevent 5000 premature deaths and provide between $0.5 and $5 billion annually in total health benefits (Environment Canada, 2003b; United States Environmental Protection Agency, 2000a).

AIRBORNE CONTAMINANTS

In the Canadian High Arctic, where there are few known local anthropogenic sources of contaminants, PCBs and other organochlorines, polycyclic aromatic hydrocarbons (PAHs), mercury, lead, cadmium, and radionuclides have been found in lakes and rivers. Air currents in the upper troposphere brought these substances from all industrial regions of the northern hemisphere via the long range transport of airborne particles (LRTAP). Studies have shown that atmospheric inputs of mercury (and other contaminants) have increased by a factor of three since 1900 (cited in Government of Canada, 1996). Water

bodies collect the contaminants deposited within their drainage basins and, in turn, aquatic biota bioaccumulate these substances (see Chapter 8). In the Arctic, hexachlorocyclohexane (HCH) is the organochlorine found at highest concentration in fresh waters, while toxaphene, PCBs, and chlordane are the compounds found at highest concentrations in fish. This is one source of the PCBs found in breast milk of Aboriginal women in the Arctic.

Many Canadian cities continue to experience unacceptable air quality, particularly during summer. Ground-level ozone, formed by the action of sunlight on pollutants such as nitrogen oxides, can combine with airborne particles and other types of pollution to form smog. People with respiratory problems, children, the elderly, and people who exercise vigorously outdoors during the summer are at risk, particularly if they live in the lower Fraser Valley of British Columbia, the Windsor–Quebec City corridor (where two-thirds of the Canadian population resides), and southern New Brunswick and southwestern Nova Scotia. These areas are subject to Canadian and American emissions of **volatile organic compounds** and nitrogen oxides, and experience serious episodes of ground-level ozone pollution (Environment Canada, 1996a; Government of Canada, 1996). National parks such as Kejimkujik and Fundy have recorded high levels of pollutants drifting up from the northeastern United States. If climate change causes average air temperatures in these areas to continue to rise, this situation could be aggravated.

In addition to human health concerns, ground-level ozone, sulphur dioxide, and nitrogen dioxide injure plants and damage rubber, paint, plastic, and other materials. Fine airborne particles (with diameters less than one millionth of a metre) are increasingly subject to study as it is believed these particles can adsorb toxic organic compounds and carry them deep into our lungs. Little is known about the effects on human health of trace amounts of the toxic hydrocarbon benzene (found in city air from unburned gasoline in vehicle exhaust). However, recent studies in China and Czechoslovakia found that very low chronic levels of benzene exposure were correlated with a change in the number of white blood cells, chromosomal aberrations, and elevated levels of metabolites in urine (Qu, 2003; Albertinii, 2003). Because benzene is known to cause a specific form of leukemia in humans, the Canadian Council of Ministers of the Environment committed to reducing national levels of benzene emissions by 30 percent between 1995 and 2000. It is estimated that benzene emissions were reduced by 39 percent during the specified time frame. Objectives were met in all but three sectors: petroleum distribution, residential wood combustion, and miscellaneous combustion (Canadian Council of Ministers of the Environment, 2002; Environment Canada, 1996a). (See Chapter 13 for further details on urban air quality issues.)

PREDICTING CLIMATE CHANGE

In their efforts to analyze the climatic effects of increasing concentrations of greenhouse and other gases in our atmosphere, scientists use computer models to predict climate change and future climates. These models are based on the physical laws that govern behaviour of the earth–ocean–atmosphere system. Scientists use mathematical equations describing these laws to conduct experiments on the climate system that would be impossible (or unwise) to carry out in the real world. For instance, much of our understanding of potential climatic impacts of greenhouse warming comes from experiments on (1) greenhouse gases and aerosols (GHG + A) and (2) CO_2 concentrations doubled from 1980 levels at 1 percent per year compounded annually (Hengeveld, 2000).

The most elaborate of these models in the late 1980s were the general circulation models (GCMs), also known as global climate models. In three dimensions and over time, these models simulated the workings and interactions of the sun, atmosphere, oceans, land surfaces, soils, vegetation, and ice. Scientists used these GCMs to explore *equilibrium climate change*—the changes in climate that would be manifested after the climate system had stabilized in response to a given change, such as a doubling in greenhouse gas concentrations. Like any model, these GCMs had their strengths and limitations (see Box 5–6); they could represent some physical processes with precision but others with much less accuracy. For example, these models were unable to predict realistically the effects of global changes in climate on the subcontinental and regional characteristics of future climate and weather (Hengeveld, 1997).

In part due to a lack of computing power and limited

BOX 5-6
MODELLING CLIMATE CHANGE

By Lawrence Nkemdirim, PhD, Intergovernmental Panel on Climate Change delegate

Global average surface air temperature increased by about 0.6°C, give or take 0.2°C, since the last decade of the 19th century. The 1990s were the warmest decade; 1998 the warmest year. Most of the increase occurred in two separate periods; 1910–1945 and 1976–present but with a difference. The 1910–1945 warming was patchy and regional; the present event is global. The largest increases have occurred in the mid- and high latitudes of northern hemisphere continents, where average warming rates have increased from 0.1°C to 2°C per decade.

These figures are in line with model projections of temperature change, due in part to a future doubling of atmospheric carbon dioxide concentration or its equivalent. Based on assembles of climate models, the Intergovernmental Panel on Climate Change (IPCC), the international body especially created to report on climate change, estimates that average surface air temperature will rise between 1.5° and 4.5°C by the end of the 21st century (2070–2100).

What are climate models? How are they used to project future climates? Can they successfully attribute the present warming trend to the anthropogenically driven growth in atmospheric greenhouse gases (GHGs)? What are their shortcomings?

Climate Models
Climate models are physically based mathematical formulations of the various quantities, forces, and processes that determine climate and its variability over time and space. If a model successfully replicates present climate and its regional patterns and tests well against major features of past climates, it bolsters confidence in its use for projecting the climate of the future.

The climate system comprises land-, ocean-, and atmospheric-based quantities. Land quantities include the surface itself, the terrestrial biosphere and the cryosphere (ice sheets, seasonal

snow, glaciers, and permafrost). Because the climate resulting from interactions among land, ocean, and the atmosphere does not impact the sun, the latter is not considered part of the climate system. Yet solar radiation is the most important external drive of the system.

Given the number of the variables involved and their space-time distribution, climate models are highly complex even for modest objectives. Climate models are hierarchical, ranging from the simplest one-dimensional models (latitude and height, for example), to complex three-dimensional ones involving latitudes, longitudes, and height on land and ocean. Modelling can be performed separately for the land-atmosphere system (Atmospheric General Circulation Model—AGCM) or the ocean (OGCM). In the most complex models the AGCM and OGCM are linked to produce a coupled Atmospheric Ocean General Circulation Model (AOGCM). Some AOGCMs treat only the ocean within the thermohaline, the zone where circulation, upwelling, and subsidence occur, while others include the entire depth.

Model Types
Simple models (1-D and 2-D) are useful tools for exploring relationships among the major climatic drivers. They provide insights into the consequences of change in one or more variables upon other variables and help answer questions concerning climate response to perturbation on a *global scale*. Broad latitudinal response may also be indicated. However, they are not powerful enough to simulate the impact of key processes on climate sensitivity (defined below), nor are they suited to the assessment of delays due to oceanic response and slow feedbacks. 3-D models perform those tasks. In addition, they provide transient (time-dependent) data useful for trending change from inception through to *equilibrium* (stability) as well as data suitable for regional analysis. 3-D models are expensive to run. They may

(continued)

BOX 5 – 6

(CONTINUED)

require data not available at many locations, which leads to considerable parameterization (formulations representing possible system-wide impact of variables and processes observed at sub-grid level) and perhaps cumulative errors. Because their resolution is coarse, they do not capture regional patterns well.

Modelling the Impact of GHGs on Climate

General circulation models (GCMs) use data and scenarios (assumptions of patterns of future change in GHGs) to calculate how long it will take for the greenhouse gases combined to reach a level equivalent to double, triple, or quadruple the preindustrial concentration of carbon dioxide. The impact of the new concentration on climate is then assessed using the model. Key calculations include (a) the change in *net radiation* (difference between all incoming and all outgoing radiation fluxes), called *radiative forcing* and (b) *climate sensitivity,* which refers to the impact of long-term change in mean global temperatures following a doubling of atmospheric CO_2 or its equivalent. The term arises because any initial warming will impact several areas of the climate system, resulting in either the amplification of the original warming (positive feedback) or its damping (negative feedback).

Model Runs

Coupled Ocean Mixed layer—AGCM is run with the "present" GHG concentration until its response settles into a stable climate (Run 1). If the present climate is adequately captured, the model is run a second time with a suddenly increased CO_2 concentration (doubling is normal), stopping at the stage when a new equilibrium (statistically constant climate) is reached (Run 2). The difference in climate response between Runs 1 and 2 indicates the change due to CO_2 forcing. A slight modification of this procedure entails the continuation of Run 1 through the same period covered by the Run 2 but without any increase in CO_2 (Run 3). Run 3 is seen as a "control" run capable of capturing any non-GHG related climate perturbation during the time period covered by Run 2. If the coupled mixed-layer AGCM is further coupled with an OGCM representing the full depth of the ocean, a much longer time period will be required to achieve equilibrium because of ocean damping.

Model Validation

The latest generation of AOGCMs generally replicates annual and decadal temperature variability well. A comparison of the results of the Canadian GCM 2 against observed average global temperature data and projections by the Intergovernmental Panel on Climate Change showed that the new results are not in marked departure from earlier ones (GCM1 versus GCM2) especially with respect to trend. This appears to validate the predictive quality of the models. Similar agreement is shown by other models including GFDL (USA), HadCM (UK), and Ham3L (Germany). The small disagreements that occur among model response reflect minor differences in approaches to matters such as flux adjustment, aerosol parameterization, scaling up of small climate system processes, and representation of oceanic behaviour and its time lines. Given the convergence among several models, confidence in temperature prediction for the future at the global scale is high.

Attribution

The IPCC defines attribution as the process of establishing cause and effect with some defined level of confidence, including the assessment of competing hypotheses. In light of the close agreement between model response and observed data, can science attribute the current warming trend and its future to GHGs? To what degree can the current warming trend be attributed to GHG forcing since it may not be exclusively free from natural ones, including external drives such as solar radiation and volcanic aerosols? Scientists have looked for answers in what may be termed the balance of evidence. Such a balance may be sought from data in which surface air temperature is reconstructed from proxy records (coral, ice cores, tree rings, and historical documents) and matched against instrumental data. In a millennial northern hemisphere temperature reconstruction, the IPCC (2001) used instrumental data from 1000 to 1999 A.D. along with proxy records and found there is good agreement between the proxy data and the instrumental ones during the period when both series overlap. Hence the confidence that the temperature series are compatibly drawn. Second, until recently, temperature fluctuation was well within the 95 percent confidence band. This is interpreted as representing a system that is operating within the bounds of natural variability. However, in the last decade of the 20th century, temperature change has broken through that confidence band indicating that it is unlikely to be part of the natural variability of the climate system. Third, the rate of temperature rise over the last decade and a half is unprecedented for a millennium that saw several significant trends over extended periods. Based on physical principles and model simulation, the IPCC concluded that natural forcing alone is unlikely to have produced the recent warming. On the other hand, both quantitative and qualitative consistencies between observed changes in climate and model response suggest that anthropogenic forcing is likely the lead factor in global warming.

SOURCES: The Second-Generation Coupled General Circulation Model (CGCM2), Canadian Meteorological Service, 2003, http://www.cccma.bc.ec.gc.ca/models/cgcm2.shtml; Variations of Cloudiness, Precipitable Water, and Relative Humidity over the United States: 1973–1993, W. P. Elliott and J. K. Angell, 1997, *Geophysical Research Letters, 24,* 41–44; Relationships between Tropospheric Water Vapour and Surface Temperature as Observed by Radiosonde, D. J. Gaffen, 1994, *Geophysical Research Letters, 19,* 1839–1842; *Climate Change 2001—The Scientific Basis,* IPCC, 2001, Oxford University Press; Increases in Middle Atmospheric Water Vapor as Observed by Halogen Occultation Experiment and Ground Based Water Vapor Millimeter-wave Spectrometer from 1991 to 1997, G. E. Nedoluha, Bevilacqua, et al., 1998, *Journal of Geophysical Research, 103,* 3531–3543.

knowledge of ocean processes, these 1980s GCMs could not simulate *transient climate change*—the behaviour of the climate system while it is changing (not after it has changed). Into the 1990s, however, a third generation of climate models known as coupled atmospheric–ocean general circulation models (AOGCMs) or, simply, coupled climate models, was developed. Canada's first coupled climate model, known as CGCMI, was developed by the Canadian Centre for Climate Modelling and Analysis in the mid-1990s. This model has been used internationally and was recognized as one of the leaders in simulation of climate systems (Hengeveld, 2000). Current AOGCM models incorporate all of the previously mentioned variables plus aerosols (AOGCM + A). Remembering that aerosols have negative radiative forcings that offset the warming influences of well-mixed GHGs, we can understand that the incorporation of associated aerosol variables is crucial for accurate modelling projections. Scientific consensus is that these AOGCMs model global climate changes with a confidence level of between 66 and 90 percent, but regional projections are less certain (see Box 5–6 and Table 5–8). Figure 5–8 shows annual temperatures predicted for Canada in the 2080s.

RESPONSES TO ATMOSPHERIC CHANGES

INTERNATIONAL ACTIONS

The transboundary nature of many atmospheric changes highlights the necessity for truly international action to control greenhouse gas emissions, to protect the ozone layer, and to reduce acidic deposition. Although progress continues to be made in each of these areas, as noted below, a number of obstacles remain to be overcome if international actions are to result in adequate protection of the Earth's atmosphere.

Canada has been an international leader in addressing anthropogenic impacts on the atmosphere. Two of the first international conferences that produced significant action to reduce atmospheric pollutants were initiated by Canada. In 1987, Canada hosted the Montreal Protocol on Substances that Deplete the Ozone Layer and, in 1988, in collaboration with UNEP and the WMO, Canada hosted the

TABLE 5–8

PREDICTED GLOBAL CLIMATE CHANGE FROM CANADA'S COUPLED ATMOSPHERIC–OCEAN GENERAL CIRCULATION MODEL

Climate Variable	Predicted Climate Change[1]
Temperature	• Global average increase of 4.5°C; land + 6°C; sea + 3.5°C. • Warmer than any time in the last 125 000 years.
Precipitation	• 4.5 percent increase over oceans; the same or less over land except the higher latitudes of the northern hemisphere, where more rain and less snow are predicted. More precipitation is expected over all.
Ocean circulation	• Slower circulation rate because of decreased salinity (this is due to an increase in the amount of freshwater runoff reaching oceans from increased glacial melt and increased precipitation at higher latitudes). Slower circulation will decrease the amount of warm water flowing north. • Less change to higher southern hemisphere latitudes.
Extreme events	• Tend to be more regional; there is a lower degree of certainty associated with these predictions.

[1] With a 1 percent increase in carbon dioxide compounded annually until double preindustrial levels by 2100 A.D. and a likely to very likely degree of certainty (see Table 5–4).

SOURCE: *Climate Change 2001: The Scientific Basis,* Summary for Policy Makers, J. T. Houghton, Y. Ding, D. J. Griggs, M. Noguer, P. J. van der Linden, X. Dai, K. Maskell, C. A. Johnson, eds., 2001, Cambridge: Cambridge University Press, pp. 5–16. Reprinted by permission of Cambridge University Press.

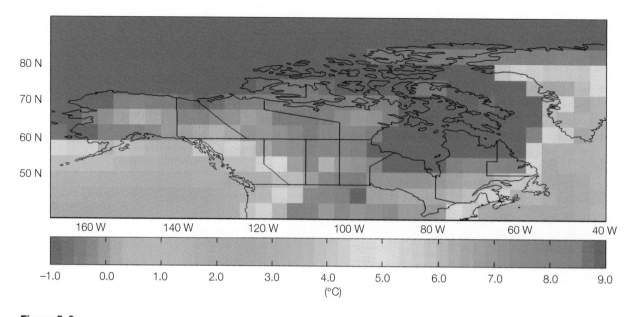

Figure 5–8

Annual temperature projection for Canada during the 2080s, based on Canadian Global Coupled Model 2-A21

SOURCE: *Climate Change Impacts and Adaptation: A Canadian Perspective,* D. S. Lemmen and F.J. Warren, 2004, Natural Resources Canada, http://adaptation.nrcan.gc.ca/perspective/summary-02_e.asp

first international conference to acknowledge the threat of climate change. Ultimately, this conference lead to establishment of the IPCC in May 2002.

Controlling Greenhouse Gas Emissions

The world community largely has recognized that anthropogenic greenhouse gases represent a real risk of climate change. At the 1992 Earth Summit, more than 150 nations (including Canada) signed the United Nations Framework Convention on Climate Change (FCCC). Although it did not set any specific goals for achieving its objectives, the FCCC called for nations to stabilize greenhouse gas concentrations in the atmosphere at a level that would prevent anthropogenic interference with the climate system. To prevent threats to food production and to enable economic development to proceed in a sustainable manner, the FCCC suggested that action should occur within a time frame that would allow ecosystems to adapt naturally to climate change. Given the rapidity with which climate change might impact various ecosystems, the practicality of this goal is dubious.

As a first step toward achieving these broad objectives, most industrialized nations committed to stabilize net greenhouse gas emissions (other than those covered by the Montreal Protocol) at 1990 levels by the year 2000. A variety of measures were considered, but many countries emphasized moderation of energy demand through increased efficiency of energy use. Also, replacement of high-carbon fuels such as coal and gasoline with alterna-

tives such as propane, natural gas, and gasohol (or ethanol- or methanol-blended gasoline) was encouraged. France and Japan planned to increase their reliance on nuclear energy, while Germany and Denmark intended to increase use of renewable, wind-generated, and solar power sources. Some countries have used economic incentives or carbon taxes to influence consumer behaviour, and there have been calls for international emissions trading agreements ("Economists' Statement," 1997). The Earth Summit +5 meetings in 1997 reviewed international progress toward greenhouse gas reduction, and led up to the Kyoto meetings held in late 1997. At the Earth Summit +10, held in Johannesburg, South Africa, in September 2002, Canada's federal government promised it would ratify the Kyoto Protocol later that year and did so in December 2002.

A number of international structures also have been developed over the past few decades to help determine the causes of and appropriate responses to global environmental change, including climate change. Probably the best known of these structures is the Intergovernmental Panel on Climate Change (IPCC), but other international scientific inquiry and decision-making programs include the United Nations Framework Convention on Climate Change (FCCC), the World Meteorological Organization (WMO), the International Geosphere Biosphere Program (IGBP), the International Human Dimensions Program (IHDP), the Scientific Committee on Problems of the Environment (SCOPE), and the International Geographical Union's Commission on Climatology (IGU).

The Kyoto Protocol on Climate Change

In December 1997, the legally binding Kyoto Protocol was established at a meeting in Japan of officials from 160 countries who had signed the FCCC. Under the Kyoto agreement, industrialized countries were required to reduce their collective emissions of greenhouse gases by 5.2 percent by the period 2008–2012 (although there remains no mechanism to deal with lack of compliance). When officials signed the Kyoto Protocol in April 1998, different countries agreed to different reduction targets relative to their 1990 levels; Canada's reduction was set at 6 percent below our 1990 level of greenhouse gases. Since emissions in most countries have increased since 1990, the effort required to get below the 1990s levels means that actual reductions will have to be in the range of 25 to 30 percent or more for Canada and 15 percent for the United States (Dotto, 1999; Environment Canada, 2000e). Furthermore, the IPCC has estimated that stabilizing atmospheric concentrations of greenhouse gases at 1990 levels would require reducing global emissions about 50 to 70 percent. This is well beyond any climate change goal yet contemplated by even the most aggressive, pro-environmental governments (Dotto, 1999).

Developing countries did not adopt specific emission reduction commitments in Kyoto, but an agreement was reached on a Clean Development Mechanism (CDM) that would allow industrialized countries to invest in projects that reduce greenhouse gas emissions in developing countries. Developed countries can then credit those emission reductions against their own Kyoto commitments (because, regardless of the location where they occur, greenhouse gas reductions have the same long-term impact). New methodologies for implementing CDMs were being developed in autumn 2003 (United Nations Framework Convention on Climate Change, 2003).

After the meetings in Kyoto, Canada's provincial and territorial ministers of environment and energy met in Toronto to (1) approve a process to examine the impacts, costs, and benefits of implementing the Kyoto Protocol; (2) establish a credit for early action to reduce greenhouse gas emissions; and (3) strengthen voluntary action. The ministers agreed that no region of the country should be asked to bear an unreasonable burden as Canada acted to reduce greenhouse gas emissions (Environment Canada, 1998c). In February 1998, the prime minister established a Climate Change Secretariat, whose objectives included the responsibility to (1) develop the federal government's domestic policy on climate change, (2) develop a National Implementation Strategy to enable Canada to meet the Kyoto greenhouse gas emission reduction targets, and (3) manage a three-year, $150 million Climate Change Action Fund (CCAF) (Environment Canada, 1998a). Part of the CCAF funding was directed to examining the impacts, costs, and benefits of different ways to address climate change.

On issues as complex as climate change, provincial interests vary. In terms of total Canadian CO_2 production, Alberta is second (27 percent) to Ontario (32 percent) and is the highest per capita emitter of CO_2. The Alberta government views reductions in CO_2 emissions as disproportionately affecting the province's economy, particularly the energy (oil) industry. The province has argued for voluntary industrial initiatives that it believes could achieve real greenhouse gas reductions at a lower cost than legislated actions. To date, however, reliance on voluntary industry initiatives has not achieved the intended targets, and is not likely to achieve the Kyoto objective of reducing emissions to 6 percent less than 1990 levels (Pembina Institute, 2000; Rainham, 1999).

In September 2000, Canada provided the United Nations with national inventory data of emissions of greenhouse gases. These data showed Canada's emissions in 1998 were 13 percent above 1990 levels. In October 2000, Canada's federal, provincial, and territorial energy and environment ministers met to consider Canada's first national business plan on climate change, a plan they all agreed must contain concrete actions to reduce Canada's greenhouse gas emissions, particularly in the electricity and transportation sectors. In 1998, emissions in the electricity sector were 33 percent above 1990 levels (because coal was being used to satisfy the increased demand for electricity). In the transportation sector, emissions were 22 percent above 1990 levels (a result of lack of improvement in fuel efficiency, and an increase in road freight and in the number of sport utility vehicles, vans, and light trucks on the roads) (Environment Canada, 2003a; Environment Canada, 2000a). It was encouraging to note that, by 2001, Canada's CO_2 emissions were down nearly 2 percent over 2000 levels (but still about 20 percent above 1990 levels).

Since then, all Canada's provinces and territories and the federal government have committed to reducing GHG emissions, although not all jurisdictions approved of the federal government's decision to ratify the Kyoto Protocol. Without exception, each province and territory in Canada has developed or is in the process of developing plans to reduce GHG emissions. In the survey conducted for this edition of *Our Environment,* almost every Canadian Minister of the Environment (or equivalent) identified climate change as a major environmental concern in their jurisdiction. In 2002, the federal government presented its plan to address climate change, titled *The Climate Change Plan for Canada* (Government of Canada, 2002).

Critics of Canada's response to the Kyoto Protocol charge that Canada has lagged behind other countries in taking appropriate action to reduce emissions: in 2002, Canada emitted more than twice the amount of GHGs per capita than did the European Union or Japan (Bramley, 2002). Climate scientists have suggested that, in order to stabilize the levels of GHGs in the atmosphere, global

CHAPTER 5: OUR CHANGING ATMOSPHERE

emissions will need to be reduced by more than 50 percent (Bramley, 2002). Possible actions for Canada to take in the fight against climate change and some of the benefits that might accrue from these actions are identified in Table 5–9. Compared with the aggressive actions of other countries, however, Canada's efforts to improve energy efficiency and to introduce renewable energy technologies appear weak. By the end of 1998, for instance, Germany had installed wind energy capacity of 2874 megawatts (MW), while Canada had installed just 317 MW by 2003 (Canadian Wind Energy Association, 2003; Bramley, 2000).

Concerns about the final implementation of the Kyoto Protocol and the world's ability to achieve the necessary GHG targets were raised in early December 2003 when a top Russian bureaucrat threatened that his government would not ratify the Kyoto Protocol. The reason given for this remark was the fear of significant limitations on economic growth that Russia felt the agreement would produce (Jaimet, 2003). Clearly, "[e]mitting greenhouse gases is what makes the world go round, economically speaking, and it's proving hard for us to stop" (Dotto, 1999, p. 178). Unless Canada acts now to expand the economic sectors and infrastructure needed to reduce emissions, it may well risk future competitiveness in a low-GHG global economy (Bramley, 2002).

Protecting the Ozone Layer

Since the early 1980s, Canada has been a world leader in strongly supporting the need for international controls on ozone-depleting substances (ODSs) and in meeting and surpassing its ODS reduction commitments. On June 4, 1986, Canada became the first country to sign the Vienna Convention, a framework for controls on the production and consumption of ODSs. Canada played a leading role in developing the Montreal Protocol on Substances That Deplete the Ozone Layer, signed by 24 nations on September 16, 1987. The Montreal Protocol was the first truly international effort to cooperate on protecting the environment, and was the first international mechanism designed to address a rising global environmental problem (Environment Canada, 1997a). The protocol was the result of unprecedented cooperation between all levels of government, the scientific community, industry, and the Canadian public (Environment Canada, 1996c).

The complex Montreal Protocol came into effect on January 1, 1989, and required each party to the agreement to freeze its production and consumption of CFCs at 1986 levels by July 1, 1989, to reduce them by 20 percent by 1993, and to further reduce them to 50 percent of 1986 levels by 1998. Also, each nation was required to limit its production and consumption of halons to 1986 levels by 1992. Recognizing that developing nations would need more than the specified time to control their emissions of ODSs, the Montreal Protocol permitted these countries a 10-year grace period in which to comply, and established a fund to provide them with financial and technological support. Canada contributes about $5 million per year to this fund and has provided technical assistance to Chile, China, Brazil, India, and Venezuela (Environment Canada, 1997a).

As of December 2003, 184 countries had ratified the (amended) Montreal Protocol (Ozone Secretariat, 2000).

TABLE 5–9
SELECTED EXAMPLES OF MEASURES TO REDUCE GREENHOUSE GAS EMISSIONS AND EXPECTED BENEFITS

Action/Measure	Direct Benefits	Co-Benefits
<ul><li>Fuel efficiency standards for vehicles</li><li>Phased-in increases in fuel taxes (1 cent per litre per year for 10 years)</li><li>Increased public transit</li><li>Fuel switching in electricity generation</li><li>More renewable resources in electricity generation</li><li>Home and commercial building energy efficiency retrofits</li></ul>	These six measures alone would achieve CO_2 reductions of <ul><li>68 million tonnes per year in 2010</li><li>9 percent of the projected national total in 2010 of 748 million tonnes per year</li><li>36 percent of the estimated reduction necessary to meet Canada's Kyoto commitment of 187 million tonnes per year</li></ul>	These measures would <ul><li>Improve energy efficiency of the economy generally</li><li>Reduce human exposure to toxic air contaminants</li><li>Avoid flooding and other land requirements (lower demand for hydroelectricity)</li><li>Avoid community impacts (lower demand for transportation infrastructure)</li></ul>

SOURCE: *Clearing the Air,* The David Suzuki Foundation, (n.d.), www.davidsuzuki.org. Reprinted with permission.

The Montreal Protocol is science-based and relies on assessment panels of the United Nations Environment Programme to guide its revisions. The Montreal Protocol has been amended on four occasions: in London in 1990, in Copenhagen in 1992, in Montreal in 1997, and in Beijing in 1999; the Protocol also has been adjusted once, in Vienna in 1995. Several changes were made to the original agreement at these meetings:

London:

- Tighter control measures for CFCs and halons were agreed on
- Carbon tetrachloride and methyl chloroform were added to the list of controlled substances
- The Interim Multilateral Fund was created to help developing countries (whose annual per capita consumption of controlled substances did not exceed 0.3 kg) to meet the control measures of the Protocol

Copenhagen (ratified[a])

- Phase-out deadlines for several ozone-depleting substances (ODSs) were accelerated
- Resolutions were adopted to encourage recovery, recycling, leakage control, and destruction of ODSs
- Methyl bromide, HCFCs, and HBFCs (hydrobromofluorocarbons) were added to the list of substances subject to control
- A definition of "essential use" was agreed on

Montreal (ratified[a])

- Established and implemented a system for licensing the import and export of ODSs in order to minimize the growing illegal trade in these substances[f]

Beijing (ratified[a])

- Monitor consumption and production of Halon 1011 in order to increase understanding of global use and manufacturing
- Imports and exports of HCFCs are banned to countries that have not ratified the Copenhagen Amendment (which came into effect February 25, 2002)[e]

In 1995, in Vienna, the parties to the Montreal Protocol made some additional adjustments, agreeing to add a phase-out schedule for methyl bromide and a reduction of the cap for the base level of HCFC consumption. In the following table, the updated Montreal Protocol terms are noted, as is Canada's timetable for prohibiting production and importation of ODSs.

This table also reflects the point that, despite progress in controlling ODSs, depletion of the ozone layer is not a fully resolved problem, largely because of the long lifetimes of many ODSs and a lack of compliance by some signatories. It is to be hoped, however, that advancements in the schedule will continue to occur as our understanding of the dynamics of ozone-depleting processes continues to increase.

MONTREAL PROTOCOL AND AMENDMENTS TO 1999[a]		CANADIAN TARGETS AS OF 2003
CFCs	Calculated level of production:[1] 0 by 2010	No longer produced or imported[b]
Halons	Calculated level of production: 0 by 2010	No longer produced or imported[b]
Carbon tetrachloride	Calculated level of consumption:[2] 0 by 1996	Imported only for use as feedstock in chemical production[b]
Methyl chloroform	Calculated level of consumption: 0 by 1996	No longer produced or imported[b]
HCFCs	Calculated level of consumption: 0 by 2030	No longer produced or imported by 2030[c]
HBFCs	Calculated level of consumption: 0 by 1996	Calculated level of consumption: not >0 by 1996[d]
Methyl bromide	Calculated level of production: 0 by 2015	Calculated level of consumption: 0 by 2005[d]

SOURCE: Adapted from "Disagreeing on the Basics: Environmental Debates Reflect Competing World Views," D. M. Taylor, 1992, *Alternatives, 18*(3), p. 29. Reprinted courtesy of Alternatives Journal: Environmental Thought, Policy and Action.

[1] Calculated level of production = Sum of annual production of each controlled substance × ozone-depleting potential of each controlled substance
[2] Calculated level of consumption = Calculated levels of production and imports − Calculated level of exports
SOURCES:
[a] Ozone Secretariat. (2000). The Montreal Protocol on substances that deplete the ozone layer as either adjusted and/or amended. In London, 1990, Copenhagen, 1992, Vienna, 1995, Montreal, 1997, Beijing, 1999. http://www.unep.ch/ozone/ratif.shtml
[b] Canadian Council of Ministers of the Environment. (2001). National Action Plan for the Environmental Control of Ozone Depleting Substances (ODS) and their Halocarbon Alternatives. Winnipeg: Canadian Council of Ministers of the Environment.
[c] Heating, Refrigeration and Air Conditioning Institute of Canada. (n.d). The HCFC Phase-out Schedule. http://www.hrai.ca/hcfcphaseout/
[d] Environment Canada. (2002). Stratospheric Ozone: Substances and Sectors. http://www.ec.gc.ca/ozone/EN/SandS/index.cfm?intCat=186
[e] United Nations Environment Programme. (2001). The OzonAction Programme. http://www.uneptie.org/ozonaction/library/pressrel/beijing.html
[f] Ozone Secretariat. (1999). Licensing Agreement on Trade in Ozone Depleting Substances Enters into Force. http://www.unep.ch/ozone/press-rel/Press-Rel-230899.shtml

Box 5–7 provides details on the phase-out schedule for ODSs included in the Montreal Protocol as well as Canadian targets. In 1996, an important milestone was reached when all developed countries eliminated the production and banned the importation of most new supplies of the most damaging ODSs, such as CFCs. The United Nations has designated September 16 as the International Day for the Preservation of the Ozone Layer (Environment Canada, 1996c).

The tenth anniversary of the Montreal Protocol was celebrated in September 1997, when the annual meeting of the parties was held in Montreal. At that meeting, Environment Canada released an independent economic analysis of the benefits and costs of the protocol. This study was undertaken to answer the question of whether the benefits from the protocol's global efforts outweighed their costs. The answer was an unequivocal "yes." In their efforts to prevent continued deterioration of the ozone layer from 1987 to 2060, nations are expected to avoid harmful impacts on human health, fisheries, agriculture, and building materials. In terms of health, for instance, it is expected that there will be more than 19 million avoided cases of nonmelanoma skin cancer worldwide by 2060, about 1.5 million avoided cases of melanoma skin cancer, and 333 500 avoided skin cancer deaths; about 129 million avoided cases of cataracts; and a significant reduction in illnesses and deaths from infectious diseases. The dollar benefits from reduced ultraviolet radiation damage to fisheries, agriculture, and building materials were estimated at $459 billion. The net benefit of the protocol, shared by all nations, was $224 billion plus health benefits (not quantified) (Environment Canada, 1998b). In addition, technological innovation driven by the protocol is expected to contribute additional economic and environmental benefits.

While there has been progress in fighting ozone depletion, the battle is far from over. Reductions in ODSs take years to be reflected in the stratosphere and, because we still use some ozone-depleting chemicals, ODSs are continuing to build up. Even if all nations meet their international commitments to phase out ODSs, the ozone layer is not expected to return to normal (that is, to pre-1980 levels) until at least the year 2050—but it is expected to recover. Amelioration in the depletion of the ozone layer is expected to begin within 10 years (Ajavon et al., 2002).

The relevance of international agreements to reduce air pollution was mentioned previously. Since the largest external impact on Canadian air quality comes from the United States, Canada and the United States developed specific agreements to address ozone depletion. These agreements include the Ozone Annex of the Canada–United States Air Quality Agreement and the Joint Plan of Action on Transboundary Air Pollution. Negotiating the Ozone Annex to reduce the transboundary flow of ground-level ozone began in February 2000. This agreement recognizes ground-level ozone as an important contributor to smog and air pollution, and a key element in about 5000 premature deaths in Canada's cities each year (Environment Canada, 2000c). The primary goal of the Ozone Annex is to obtain substantial improvements in Canada's air quality as well as the associated public health benefits, such as fewer hospital admissions and doctor and emergency room visits. Attaining these goals requires a reciprocal arrangement between Canada and the United States (because the United States is the source of between 30 and 90 percent of the ground-level ozone problem in eastern Canada, and Canada is responsible for pollution flowing from Ontario and Quebec to the northeastern United States). The Ozone Annex agreement was signed by both countries in December 2000.

Controlling Acidic Deposition

Canada's acid rain reduction program highlights several important themes in environmental sustainability:

- The difficulty in repairing ecosystems: prevention of harm is preferred over remediation of damage, because not all ecosystems are well understood and their response times are slow; additionally, societal response is phased (see "Inertia" in Box 5–1)
- The importance of partnerships: provincially, nationally, and internationally
- The role of adaptive management and the importance of good relationships and flexibility (particularly in light of the preceding points).

Internationally, a number of agreements have been implemented to reduce and prevent long-range transboundary acid deposition. Selected examples of these agreements are discussed below, but given that Canada's major concerns regarding acidic deposition cannot be resolved without significant cooperation from the United States, this discussion deals with international Canadian and U.S. initiatives. Canada and the United States have a history of air quality agreements that date back to 1985. Key agreements in place as of 2003 include the Canada-Wide Acid Rain Strategy for Post-2000, Canada–United States Air Quality Agreement, Eastern Canada Acid Rain Program, and Border Air Quality Agreement.

The reduction of acidic deposition has focused on emissions of the two primary causal agents: sulphur dioxide and nitrogen oxides. Early agreements emphasized a reduction in sulphur dioxide. Canada–U.S. agreements that addressed the required reductions were the 1991 Canada–United States Air Quality Agreement and the Eastern Canada Acid Rain Program, a national agreement signed by all provinces from Manitoba eastward that worked in conjunction with the U.S. Acid Rain Program to cap sulphur dioxide emissions.

Recent evaluations and newly acquired information indicate that the original reduction goals have not achieved the desired results, although the stated goals were successfully achieved. Evidence of acidic deposition destruction still exists throughout eastern Canada. Additional efforts are addressed in two new agreements. The Canada-Wide Acid Rain Strategy for Post-2000 was signed in 1998 and put more emphasis on nitrous oxide reductions as well as increased SO_2 reductions. In April 1997, Canada and the United States developed a Joint Plan of Action on Transboundary Air Pollution.

Much assessment work was carried out during the 1980s by the federal–provincial Research and Monitoring Coordinating Committee. Monitoring of acid levels in more than 200 lakes in Ontario, Quebec, and the Atlantic region (between 1981 and 1994) revealed that 33 percent showed some improvement in acidity, 56 percent showed stable acid levels, and 11 percent (in the Atlantic region) were becoming worse (Environment Canada, 1996b). In areas where substantial reductions in acidic deposition occurred, striking ecological improvement was evident.

Perhaps the best example of this improvement can be seen in lakes in the greater Sudbury area of Ontario, where the majority of monitored lakes showed an improvement in acidity, due mainly to substantial reduction of SO_2 emissions from Sudbury's nickel smelters. Long acknowledged as the largest industrial producer of SO_2 emissions in Canada, Inco Limited's plant at Copper Cliff released SO_2 at a rate of 5500 tonnes per day (2 million tonnes annually) in 1969. Since then, company initiatives and government emission control regulations reduced annual emissions to about 685 000 tonnes in 1992, and efforts were made to reduce them further to 265 000 tonnes by 1994 and even lower in subsequent years. As emissions declined in the Sudbury area, air and water quality improved sufficiently to permit an extensive and successful revegetation program (see Chapter 10 for more information on restoration efforts in Sudbury).

Similarly, the 550 000 tonnes of SO_2 released in 1980 at the Noranda Minerals copper smelter at Rouyn-Noranda, Quebec, were projected to be reduced 90 percent by the year 2000 through efforts such as the use of new smelting technologies and the extraction of sulphuric acid (Elder, 1991).

Acid rain continues to be a problem in eastern Canada. Efforts to mitigate this problem occur through the Canada-Wide Acid Rain Strategy Post-2000 and the Eastern Canada Acid Rain Program. Acidity in some lakes has declined to a modest extent, but other surface water has not responded favourably to emissions reduction efforts because of variation in the ability of specific environments to neutralize acid deposition. Currently, emission caps are now geared toward critical loads, rather than target loads (Environment Canada, 2002b). The need to identify environment-specific critical loads poses

difficulties for decision makers because they cannot implement across-the-board reduction policies (e.g., for SO_x emissions). Critical loads must be environment-specific and, as a result, are more labour intensive and take longer and are more costly to develop and implement. However, if the negative effects of acidic deposition are to be addressed successfully, researchers are convinced that critical load capabilities must be honoured. Environment Canada states that despite the more than 50 percent reductions in SO_x emissions in the United States and Canada, if further reductions do not occur, almost 800 000 km^2 would continue to be damaged by acidic deposition (Environment Canada, 2002b).

Concern also exists regarding previously underestimated risks to forest health and productivity caused by nitrogen deposition and its contribution to acidification. To date, acid rain strategies have not been as successful in reducing nitrogen oxide emissions as they have been in reducing sulphur oxides. Unlike SO_x emissions, which are addressed largely through point source controls, NO_x emissions are from nonpoint sources (vehicles are the largest emitters) and are more difficult to control. Emission reduction success is predicted when both Canada and the United States implement more stringent emission standards for on-road vehicles. If the United States aggressively reduces its NO_x emissions, Canada will benefit through significant air quality improvements in southern Ontario and Quebec, as well as New Brunswick and Nova Scotia (Environment Canada, 2000c).

By 1985, the Eastern Canadian Acid Rain Program had been formalized; the seven provinces east of Saskatchewan agreed to reduce their combined SO_2 emissions to 2.3 Mt per year by 1994. This target was exceeded in 1993, principally because of industrial process changes, installation of scrubbers, and the switch away from high-sulphur fuels.

The Canada–United States Air Quality Agreement was signed by the United States and Canada in 1991. An additional agreement was signed in 1997, to develop a Joint Plan of Action on Transboundary Air Pollution. These two agreements continue to govern transboundary air pollution issues and have led to current negotiations for an Ozone Annex to the 1991 Agreement, specifically to reduce transboundary flows of NO_2 emissions (Environment Canada, 2000b). A primary long-term goal of the strategy is to achieve critical loads or threshold levels (see Chapter 3) for acid deposition across Canada. The Canada–United States Air Quality Agreement obliged Canada to establish a permanent national limit on SO_2 of 3.2 Mt and a 10 percent reduction in projected NO_x emissions from stationary sources, both by the year 2000. Canada met this goal in 1993, and by 1994, national SO_2 emissions were down to approximately 2.7 Mt. Beginning in 1995, the United States committed to reduce its annual SO_2 emissions to 14.4 Mt (9.1 Mt below 1980 levels) and its NO_x emissions by approximately 1.8 Mt by the year 2000. In 2000, SO_x emissions in the United States were still above targeted levels (United States Environmental Protection Agency, 2000b). However, Canada lags behind the United States in meeting its mobile NO_x reductions (International Joint Commission, 1998).

The importance of compliance with international agreements to reduce the effects of highly mobile atmospheric pollutants cannot be overemphasized. Residents and ecosystems in Northern Canada are among the victims of long-range transport of air pollutants: they suffer from the effects of air pollution without having generated the pollutants. In response to the survey conducted for this edition of *Our Environment,* the Northwest Territories Minister of the Environment noted that international atmospheric pollutants were causing significant environmental problems. "International pollutants (i.e., greenhouse gases, persistent organic pollutants, heavy metals) are continuing to impact the northern environment even though they are not produced or released from local sources. It is imperative that international treaties be developed, ratified and implemented if sources of these global pollutants are to be controlled" (Antoine, 2003).

CANADIAN LAW, POLICY, AND PRACTICE

Canada is attempting to develop an integrated strategy for action on clean air. Areas of particular attention include vehicles and the fuels they run on, industrial emissions, negotiations with the United States, and the inclusion of Canadians in the search for solutions. The 1999 Canadian Environmental Protection Act (CEPA) helps this strategy by providing the federal government with additional tools and powers to reduce pollution and eliminate and regulate emissions of toxic substances. Future vehicle emission standards, for instance, can be achieved partly through consultation with those affected (health and environmental groups, petroleum refiners, automotive and engine manufacturers, and alternative fuels representatives) and partly through regulations (such as reductions in sulphur content in gasoline and diesel fuels). Federal investment in a "green infrastructure" to reduce the threat of climate change in urban and rural centres includes public transit and energy services, water and wastewater treatment facilities, and waste management systems (Environment Canada, 2000d). A wide range of strategies and actions is possible, as the following examples of Canadian law, policy, and practice illustrate.

After signing the Montreal Protocol in 1987, Canada worked on putting in place a control program that would meet its international commitments. By 1995, when consultation meetings were held across Canada, the national

Photo 5–9
Planting seedlings for new forests is one way to help remove carbon dioxide from the atmosphere.

Ozone Layer Protection Program was strengthened and target dates for phase-out of ODSs were accelerated. For example, nonrecoverable uses of HCFCs were to be eliminated by 2010, and the consumption of methyl bromide was to be reduced by 25 percent by 1998 and eliminated completely by 2001. Reaching these targets has proved more difficult than originally anticipated. Canada's stated goals are now to eliminate consumption, export, and import of HCFCs by 2030 and to completely eliminate our calculated consumption of methyl bromide by 2005 (see Box 5–7) (Canadian Council of Ministers of the Environment, 2001).

One set of regulations under CEPA, the Ozone-Depleting Substances Regulations, deals with control of ODS production, import, and export. A second set of regulations, the Ozone-Depleting Substances Products Regulations, deals with control of manufactured products (such as aerosols and plastic foam packaging) that contain ODSs. Also under the CEPA, two environmental codes of practice have been developed for use in both public and private sectors. These codes recommend practices for pollution prevention, emission reduction, environmental management, and preventive maintenance regarding halon and fluorocarbon emissions.

The National Action Plan for the Recovery, Recycling and Reclamation of CFCs was endorsed by the Canadian Council of Ministers of the Environment in October 1992. The plan originally targeted the recovery, recycling, and reclamation of CFCs from refrigeration and air conditioning systems, but eventually encompassed all aspects of pollution prevention and all industry sectors that used ODSs. The *Climate Change Plan for Canada, 2002,* is now Canada's premier climate change document. Committed to a made-in-Canada approach, where no region should have to shoulder unfair burdens associated with the implementation of climate change mitigation, the 2002 Plan proposes both short- and longer-term measures to address climate change. Short-term actions include providing Canadians and Canadian businesses with the tools and incentives to make more energy-efficient decisions. Longer-term measures such as investing in more energy-efficient technologies and production methods, and switching to less carbon-intensive forms of energy, also are included in the Plan. Financial incentives are being provided to encourage research and development in areas such as carbon management, biotechnologies, fuel cells, and the hydrogen economy. New regulations and tax measures are other components of the Plan (Environment Canada, 2002c).

The Climate Change Plan for Canada promotes energy efficiency in all sectors and specifically refers to emission reduction targets in industry and a partnership fund that shares the costs of emissions reductions with provinces and territories, and invests in climate change proposals and innovations. Key areas identified in the Action Plan are transportation, industrial emitters, renewable energy and cleaner fossil fuel, low-emissions enterprises and fugitive emissions, agriculture, forestry and landfills, and emission reductions at international, regional, and individual levels.

The Climate Change Plan for Canada was preceded by the National Action Program for Climate Change (NAPCC), which presented a range of detailed, sector-specific, mostly voluntary measures for achieving stabilization of greenhouse gas emissions. Although it did not produce the originally sought-after voluntary GHG emission reductions, the voluntary component of the NAPCC was important in getting some individuals, companies, and organizations to think about and develop adaptation strategies that would limit their net greenhouse gas emissions.

Adaptation strategies offer returns that make them worth doing even if greenhouse gas concentrations continue to increase. For example, in agriculture, reduced vulnerability to climate change can be attained by avoiding monocropping and selecting crops or species that demand less water. Similarly, in forestry, strengthening fire and pest monitoring and intensifying firefighting may be considered. Other sectors such as fisheries, construction, and energy supply could undertake adaptation strategies that would produce environmental and economic benefits in addition to reductions in greenhouse gas emissions.

In addition, the federal government's Efficiency and Alternative Energy Program encompasses dozens of initiatives to reduce greenhouse gas emissions by improving energy efficiency through changes to equipment, appliances, buildings, industrial processes and machinery, and motor vehicles and transportation systems. The use of alternative energy sources such as low-carbon transportation fuels, noncarbon fuels such as hydrogen, and renewable energy sources such as biomass, wind, and solar power also is promoted. Communities and public utilities are being encouraged to place greater emphasis on demand management, co-generation (use of a fuel for both electricity and useful heat), and district heating (see Chapter 11).

Provincial and territorial jurisdictions have developed their own climate change action plans, and some Canadian municipalities have formed the "Twenty Percent Club" in their efforts to share cost-effective strategies to reduce greenhouse gas emissions by 20 percent of 1988 baseline levels by 2005 (see Chapter 13). Canada's 13 environment ministers meet twice annually to find solutions to Canada's environment problems, which include atmospheric pollution (see Box 5–8).

CANADIAN PARTNERSHIPS AND LOCAL ACTIONS

Information transfer and education programs have been part of Canada's efforts to reduce consumption of ODSs. For instance, in partnership with the Knowledge of the Environment for Youth (KEY) Foundation, Environment Canada undertook an education initiative to develop and implement curriculum materials for schools across Canada on protection of the ozone layer. In order to make environmentally responsible decisions, people need to know the purpose and function of the ozone layer and understand how human activities contribute to the ozone-depletion problem. This initiative recognized the important role that education plays in encouraging appropriate actions and discouraging damaging behaviours. The KEY Foundation—itself an educational partnership among people who work in environment, school system, government, and industry sectors—is recognized as a credible source of accurate, balanced, and current education resources on the environment.

As part of the National Action Plan for the Recovery, Recycling and Reclamation of CFCs, more than 95 000 people have been trained in the recovery and recycling of ODSs in servicing refrigerator and air conditioning units. The cooperation of industry has been critical in the success of these ventures, achieving reductions specified under the Montreal Protocol faster than required. Northern Telecom, for instance, formerly the single largest Canadian user of solvents containing ODSs, pledged to eliminate the use of CFC solvents from all its operations by the end of 1991. Working with chemical manufacturers, Northern Telecom spent about $1 million on research and development between 1988 and 1991, and saved about $4 million during the same period on the costs to purchase new and dispose of used solvent. In 1989, Northern Telecom co-founded the International Co-operative for Environmental Leadership, made up of

Made up of environment ministers from the federal, provincial, and territorial governments, the Canadian Council of Ministers of the Environment (CCME) is the major intergovernmental forum in Canada for discussion and joint action on environmental issues of national and international concern. Generally, the 13 environment ministers meet twice a year to discuss national environmental priorities and determine work to be carried out under the auspices of the CCME.

The CCME promotes cooperation on and coordination of interjurisdictional issues. In efforts to achieve a high level of environmental quality across the country, members of the CCME propose nationally consistent environmental standards and objectives. However, the CCME cannot impose its proposals as it has no authority to implement or enforce legislation.

After a review in 2001 the CCME decided that, to remain effective, the council should provide governmental leadership on the environment and provide policies, strategies, and practical solutions relevant to jurisdictional needs. The CCME identified "significant and enduring issues" that will require long-term involvement. These core issues include:

- Issues that cut across traditional environmental media and departmental mandates
- Air, atmosphere, and climate change
- Water and soils
- Waste management

For more information, see the CCME website at http://www.ccme.ca

SOURCE: *The Canadian Council of Ministers of the Environment (CCME) Business Plan 2003/4 to 2005/6,* Canadian Council of Ministers of the Environment, 2003, http://www.ccme.ca/assets/pdf/businessplan.pdf

multinational companies that share information on alternatives to ozone-depleting solvents (Environment Canada, 1997c).

The Great Lakes–St. Lawrence Basin Project is a joint Canada–U.S. research initiative of the Atmospheric Environment Service's Environmental Adaptation Research Group and the National Oceanic and Atmospheric Administration's Great Lakes Environmental Research Laboratory. The project was established to improve understanding of the complex interactions between climate and society so that informed regional adaptation strategies could be developed in response to potential climate change and variability. In May 1997, a binational symposium was held in Toronto to consider issues involved in adapting to climate change and variability in the basin (Great Lakes Commission, 1997).

At the community level, Montreal ran 155 buses on biofuel for one year beginning March 2002 in an effort to determine whether or not biofuel was a viable alternative to conventional diesel fuel. Biofuel is produced by a chemical reaction between methanol and vegetable oil or animal fats; it is renewable, relatively clean, and uses recycled agro-waste. Although the fuel is gaining popularity in Germany and France, it is largely untested for suitability in Canada. The scale of the *Biobus* pilot project was the largest ever conducted in North America. Interest in and support of the project were widespread. The federal government, the province of Quebec, Maple Leaf Food Group, and the Canadian Renewable Fuels Association provided funding. Carbon dioxide emissions were reduced by 2100 tonnes during the *Biobus* pilot project. Other indicators evaluated were the quality of the fuel supply, use and maintenance costs, emissions monitoring, and employee and customer satisfaction. Final conclusions indicated that biofuel is a viable alternative to diesel fuel (Natural Resources Canada, 2002).

By 1996, the city of Calgary had a climate change action plan in place. The city took the position that even a small reduction in greenhouse gases by its municipal services operations would be one step toward the larger goal of benefiting the global community. Today, Calgary's light-rail transit system is powered solely by wind energy (the equivalent of the amount of energy required to operate the light-rail system is supplied to the electrical grid from wind farms in the province).

In a move to reduce air pollution and the health problems that result from smog and other contaminants, British Columbia, in 1995, reintroduced inspections of the exhaust emissions of all cars and light trucks in the province. Reducing vehicle exhaust was expected to help prevent 3600 deaths, 23 000 cases of chronic bronchitis, and 123 000 emergency room visits, and save up to $30 billion in medical bills by the year 2020 (Joyce, 1995). Since older vehicles pollute more than newer ones, the province instituted the Scrap It program, whereby owners of fully operational vehicles made prior to 1982 were offered cash or transit passes to get their cars off the road. Similar efforts have taken place in Ontario and other provinces since then.

And finally, in Winnipeg, the Sierra Club of Canada launched the Zer-O-Zone project to foster public awareness of and support for Manitoba's ozone-protection regulation. Other environmental advocacy groups and

Photo 5–10
Traffic lanes reserved for buses, car pool vehicles, and cyclists are becoming more common in Canadian cities.

consumer groups continue to be active in programs to protect the ozone layer and to control greenhouse emissions on the local level.

FUTURE CHALLENGES

In light of the potential that changes in the composition of Earth's atmosphere have to jeopardize the planet's life-support systems, the level of international cooperation that has developed to control greenhouse gases, ozone-destroying substances, and other pollutants has been a major success. Even though our understanding of atmospheric changes is incomplete, actions to stabilize and reduce greenhouse gases and ozone depletion have become more scientifically based.

However, the social and economic challenges involved in controlling and reducing greenhouse gas and other emissions appear to be formidable. Although the use of CFCs and other ODSs affects only a small part of the global economy, their elimination has been difficult despite the availability of practical alternatives.

Greenhouse gases derive principally from fossil fuels; fossil fuel energy is the basis of our industrial economy. As a result, "reducing greenhouse gas emissions to the level necessary to stabilize their atmospheric concentrations will require a massive reorientation of the world's energy use away from carbon-based fuels and towards more benign alternatives and greater energy efficiency" (Government of Canada, 1996). Such adjustments could have enormous political, social, and economic costs if they are undertaken too quickly, particularly since economically practical fossil fuel alternatives remain scarce. At the same time, without decisive action, greenhouse gas concentrations could rise to levels that pose serious consequences for human societies as well as the natural world.

One approach to shifting our orientation has been proposed by Ralph Torrie (1999), one of Canada's foremost experts in sustainable energy. His vision of a low-carbon future suggests Canadians can cut greenhouse gas emissions by 50 percent of current levels over the next 30 years by using existing energy-efficient technologies and techniques.

Take personal transportation, for example. Canadians travel on average about 20 000 km per year, mostly in personal vehicles. If nothing is done to reduce carbon emissions in this sector, we will have pumped 140 Mt of greenhouse gases into the atmosphere by the year 2030, 90 percent of it coming from personal vehicles and the remainder from airplanes. Five factors influence how much greenhouse gas we emit from our travels: the number of trips, the length of the trips, the mode of travel (walking, cycling, driving an automobile, or taking transit), the fuel efficiency of the vehicle, and the type of fuel used in the vehicle.

In Torrie's low-carbon scenario, changes in each of these factors can lead to reductions in greenhouse gas emissions. Torrie identifies three practical ways to reduce energy consumption: (1) reduce demand, (2) triple fuel efficiency, and (3) expand transit use. Easing demand for fuel can occur when people gain access to what they need and want with fewer and shorter trips. Not only can the Internet help achieve this, so can the design of new neighbourhoods (and the redesign of old ones) to reduce dependency on our vehicles (see Chapter 13). Second, in a low-carbon future, vehicle efficiency and the use of alternative fuels are among the factors that will lead to reduced greenhouse gas emissions from personal transportation. By 2030, fuel efficiency is expected to be two to three times higher than it is today; a new generation of vehicles will be powered by hybrid gasoline/electric engines and engines will run using hydrogen fuel cells. If that hydrogen is made from natural gas, the fuel cycle efficiency would be about 75 percent, several times higher than efficiencies derived from present internal combustion engines. If the hydrogen is derived from hydroelectricity, an almost zero-emission vehicle results.

Torrie's third source of increased fuel efficiency is based on a 10 to 20 percent growth in the number of people taking transit to work. Future transit systems would use highly efficient vehicles, running on alternative fuels, and provide a more diversified and customer-responsive transit system, with door-to-door, on-demand service.

On a personal level, several actions can help reduce the risk of climate change. Individually, for example, we can buy ozone-friendly products and ensure that the technicians who service our refrigerators or air conditioners recover and recycle the CFC coolants. Collectively, we will need to reexamine our attitudes toward the automobile and public transit. If communities can be designed with sustainability in mind, walking and bicycling could reduce reliance on vehicles and fossil fuels. Sustainable agricultural methods (see Chapter 6) as well as energy-efficient housing and transportation alternatives (see Chapter 13) are part of the suite of actions required if greenhouse gas emissions are to be reduced significantly and the ozone layer is to recover fully.

In addition, improved communication between the scientific community and the public is necessary. If research findings are to be applied to reduce the risks of atmospheric change, awareness and understanding must be increased. One way to accomplish this is to restore, maintain, and enhance both national and local state-of-the-environment monitoring and reporting operations. As well, it will be important to disseminate more broadly the findings of these operations. Wider exposure of results from monitoring studies, presented through school, college, and university curricula and in the media, will be important in ensuring that all members of Canadian society are aware of the environmental choices facing them and the substantial amount of work that is needed if the rate of atmospheric change is to be influenced by individual actions.

Knowledge building has been ongoing for decades in the scientific communities associated with atmospheric change, but there is a need for clear and effective translation of scientific information into laypersons' terms. Just as the international community rallied to undertake significant measures to stop the depletion of the ozone layer once they understood the implications of ozone depletion, so too might more members of Canadian society act in more environmentally responsible and sustainable ways if they possessed a better understanding of how their actions affect the environment in which they live.

Photo 5–11
Promotional materials from Toronto's 2001 Bike Week campaign.

Chapter Questions

1. Why is acidic deposition a problem of continuing importance to Canadians? In what ways is your region affected by acidic deposition? What efforts have been made to overcome the problem?

2. Discuss the major causes and effects of ozone depletion (both for the world and for Canada).

3. Discuss the different anthropogenic greenhouse gases in terms of their contributions to global warming.

4. What consumption patterns and other lifestyle choices do you make that directly and indirectly add greenhouse gases to the atmosphere? What actions might you take to reduce your contribution to this problem?

5. What do you think are the most important initiatives to combat ozone depletion and greenhouse gas emissions on an international scale? Within Canada? Justify your choices.

6. In highly technological societies, is 100 percent clean air possible? Is it a feasible air quality standard? Why or why not?

references

Ackerman, J. (2000). New eyes on the oceans. *National Geographic, 198,* 86–115.

Agriculture and Agri-Food Canada. (2003). *Historical Perspective of Droughts on the Canadian Prairies.* http://www.agr.gc.ca/pfra/drought/drhist_e.htm

Ajavon, A. L., Albritton, D., Mégie, G., & Watson, R. (2002). Executive Summary, Final, UNEP/WMO *"Scientific Assessment of Ozone Depletion: 2002."* http://www.wmo.ch/web/arep/reports/execsumm.23%20aug%2002.final.pdf

Albertinii, R. (2003). Biomarkers in Czech workers exposed to 1.3-Butadiene: A transitional epidemiological Study. *HEI Research Report,* 116.

Albritton, D. L., Filho, L. G. M., et al. (2001). *Technical Summary of the Working Group I Report. Climate Change 2001: The Scientific Basis.* Intergovernmental Panel on Climate Change. http://www.grida.no/climate/ipcc_tar/wg1/

Antoine, J. (2003). Ministry of Environment, Northwest Territories, response to questionnaire survey for *Our Environment: A Canadian Perspective.*

Ball, T. (2003, December 18). A chink in the government's climate science armour: Access to Information documents show science driven by pro-Kyoto bias. *Calgary Herald,* p. A.21.

Beaubien, E. (1998). Tracking the biotic response to climate warming. *Encompass, 2*(3), 6–7.

Bramley, M. (2000). Why Canada must act now on climate change. *Encompass, 4*(3), 26–27.

Bramley, M. (2002). *A comparison of current government action on climate change in the U.S. and Canada.* Drayton Valley, AB: Pembina Institute for Appropriate Development.

Bubier, J. L., Moore, T. R., & Bellisario, L. (1995). Ecological controls on methane emissions from a northern peatland complex in the zone of discontinuous permafrost, Manitoba, Canada. *Global Biogeochemical Cycles, 9,* 455–470.

Byrnes, M. (2003, September 23). Scientists see Antarctic Vortex as drought maker. Reuters. http://asia.reuters.com

Canadian Council of Ministers of the Environment. (2001). *National Action Plan for the Environmental Control of Ozone Depleting Substances (ODS) and their Halocarbon Alternatives.* Ottawa: Environment Canada.

Canadian Council of Ministers of the Environment. (2002). *Benzene Canada-wide Standard, Phase 1 National Summary, Annual Progress Report—December 2001.* www.ccme.ca/assets/pdf/bzph1ntnlsmry_15jan02_e.pdf

Canadian Safety Council. (n.d.). National Summer Safety Week (May 1–7). http://www.safety-council.org/SUN.HTM

Canadian Wind Energy Association. (2003). *Canadian wind power—Installed capacity.* http://www.canwea.ca/CanadianProduction.html

Codding, D. (2000, February 18). Tree ring research yields clues to Pacific climate change. *The Ring.* http://web.uvic.ca/ucom/Ring/00feb18/treering.html

Colombo, A. F., Etkin, D., & Karney, B. W. (1999). Climate variability and the frequency of extreme temperature events for nine sites across

Canada: Implications for power usage. *Journal of Climate, 12,* 2490–2502.

Dlugokencky, E. J., Masarie, K. A., Lang, P. M., et al. (1998). Continuing decline in the growth rate of the atmospheric methane burden. *Nature, 393,* 447–450.

Dotto, L. (1999). *Storm warning: Gambling with the climate of our planet.* Toronto: Doubleday.

Dotto, L. (2000). Proof or consequences. *Alternatives Journal, 26*(2), 8–14.

Economists' statement on climate change. (1997). *Delta, 8*(1), 6.

Elder, F. (1991). Acidic deposition: Acid test for the environment. In Government of Canada, *The state of Canada's environment—1991* (pp. 24-1–24-24). Ottawa: Supply and Services Canada.

Elvingson, P. (1999). What hopes for recovery? *Acid News, 2,* 13–14.

Environment Canada. (1996a). Urban air quality. *SOE Bulletin, 96*(1) (Spring). Ottawa.

Environment Canada. (1996b). Acid rain. *SOE Bulletin, 96*(2) (Spring). Ottawa.

Environment Canada. (1996c). *International day for the preservation of the ozone layer: Recognizing human achievement.* http://www.ec.gc.ca/minister/speeches/ozone_s_e.htm

Environment Canada. (1997a). *The Montreal Protocol.* http://www.ec.gc.ca/ozone/protect/sect8e.html

Environment Canada. (1997b). *Countdown to Montreal.* http://www.ec.gc.ca/ozone/Brochure.eng.html

Environment Canada. (1997c). *The business of ozone.* http://www.ec.gc.ca/ozone/protect/sect11e.html

Environment Canada. (1997d). *A primer on ozone depletion.* http://www.ec.gc.ca/ozone/primer/primeroz.htmllife

Environment Canada. (1997e). *Stratospheric ozone.* http://www.ec.gc.ca/ozone/

Environment Canada. (1997f). *A primer on environmental citizenship.* Ottawa.

Environment Canada. (1998a). *Addressing climate change.* http://climatechange.gc.ca/english/html/addressi.html

Environment Canada. (1998b). *Global benefits and costs of the Montreal Protocol: A summary of study results.* http://www.ec.gc.ca/press/protocol_b_e.htm1

Environment Canada. (1998c). *Addressing climate change.* http://climatechange.gc.ca/english/html/addresso.html

Environment Canada. (1999a). Stratospheric ozone depletion. *SOE Bulletin, 99*(2) (Summer). Ottawa.

Environment Canada. (2000a). *Canada's greenhouse emissions for 1990–1998.* http://www.ec.gc.ca/press/000906_m_e.htm

Environment Canada. (2000b). *Acid rain update.* http://www.ec.gc.ca/press/000519d_f_e.htm

Environment Canada. (2000c). *The Ozone Annex to the Canada–United States Air Quality Agreement, 1991.* http://www.ec.gc.ca/press/000519i_f_e.htm

Environment Canada. (2000d). *Government of Canada actions on clean air.* http://www.ec.gc.ca/press/000519h_f_e.htm

Environment Canada. (2000e). *Greenhouse gases—The Kyoto Protocol.* http://www.ec.gc.ca/pdb/ghg/english/ekyoto.html

Environment Canada. (2001). *Acid rain.* http://www.ec.gc.ca/acidrain/index.html

Environment Canada. (2002a). *Stratospheric ozone: Substances and sectors.* http://www.ec.gc.ca/ozone/EN/SandS/index.cfm?intCat=186

Environment Canada. (2002b). *2001 Annual progress report on the Canada-wide acid rain strategy for post-2000.* Ottawa: Author.

Environment Canada. (2002c). *Climate Change Plan for Canada.* http://www.climatechange.gc.ca/plan_for_canada/plan/preface.html

Environment Canada. (2003a). *1990–2001 national and provincial GHG emissions.* http://www.ec.gc.ca/pdb/ghg/_e.cfm

Environment Canada. (2003b). *1990–2001 national and provincial GHG emissions.* Executive Summary. http://www.ec.gc.ca/pdb/ghg/tables_e.cfm

Environment Canada. (2003b). *Acid rain and … water.* http://www.ec.gc.ca/acidrain/acidwater.html.

Farman, J. C., Gardiner, B. G., & Shanklin, J. D. (1985). Large losses of total ozone in Antarctica reveal seasonal ClO_x/NO_x interactions. *Nature, 315,* 207–210.

Forester, A. (1991). Stratospheric ozone: Wearing thin. In Government of Canada, *The state of Canada's environment—1991* (pp. 23-1–23-24). Ottawa: Supply and Services Canada.

Francis, D., & Hengeveld, H. (1998). *Extreme weather and climate change.* Ottawa: Minister of Supply and Services.

Government of Canada. (1996). *The state of Canada's environment—1996.* Ottawa: Supply and Services Canada.

Government of Canada. (2002). *Climate Change Plan for Canada.* http://www.climatechange.gc.ca/plan_for_canada/summary/index.htm

Great Lakes Commission. (1997). *Adapting to climate change and variability in the Great Lakes–St. Lawrence Basin.* http://www.glc.org/announce/97/climate/climate.html

Hamill, P., & Toon, O. (1991). Polar stratospheric clouds and the ozone hole. *Physics Today, 44*(12), 34–42.

Hare, F. K. (1995). Contemporary climatic change. In B. Mitchell (Ed.), *Resource and environmental management in Canada* (pp. 10–28). Toronto: Oxford University Press.

Hengeveld, H. G. (1997). 1994–95 in review: An assessment of new developments relevant to the science of climate change. *CO_2/Climate Report, 97*-1, 52 pp.

Hengeveld, H. G. (1998). Frequently asked questions about the science of climate change. *CO_2/Climate Report, 98*-2, 9 pp.

Hengeveld, H. G. (1999). 1997 in review: An assessment of new research developments relevant to the science of climate change. *CO$_2$/Climate Report*, 99-1, 46 pp.

Hengeveld, H. G. (2000). *Projections for Canada's climate future: A discussion of recent simulations with the Canadian Global Climate Model.* Ottawa: Minister of Public Works and Government Services.

Hengeveld, H. G. (2001). 1999 in review: An assessment of new research developments relevant to the science of climate change. *CO$_2$/Climate Report*, Spring 2001, 26 pp.

Hengeveld, H. G., Bush, E., & Edwards, P. (2002*). Frequently asked questions about climate change science.* Ottawa: Minister of Supply and Services.

Hengeveld, H. G., & Edwards, P. (2000). 1998 in review: An assessment of new research developments relevant to the science of climate change. *CO$_2$/Climate Report*, Spring 2000, 65 pp.

Houghton, J. T., Ding, T., Griggs, D. J., Noguer, M., van der Linden, P. J., Dai, X., Maskell, K., and Johnson, C. A. (Eds.). (2001). *Climate change 2001: The scientific basis.* Cambridge: Cambridge University Press.

Intergovernmental Panel on Climate Change. (1990). *Scientific assessment of climate change.* Geneva: World Meteorological Organization and United Nations Environment Programme.

Intergovernmental Panel on Climate Change. (1995). Climate change 1995. Summary for policymakers: Radiative forcing of climate change. In J. T. Houghton et al. (Eds.), *Climate change 1994—Radiative forcing of climate change and an evaluation of the IPCC 1992 emission scenarios* (pp. 7–34). Cambridge: Cambridge University Press.

Intergovernmental Panel on Climate Change. (1997). *The revised 1996 IPCC guidelines of national greenhouse gas inventories.* http://www.ipcc-nggip.iges.or.jp/public/gl/invs4.htm

Intergovernmental Panel on Climate Change. (2000). *Summary for policymakers: Land use, land-use change, and forestry.* Geneva: World Meteorological Organization and United Nations Environment Programme.

Intergovernmental Panel on Climate Change. (2001). *Summary for policymakers. Climate change 2001: Synthesis report.* Geneva: World Meteorological Organization and United Nations Environment Programme.

International Joint Commission. (1998). *Canada–United States Air Quality Agreement, 1998 Progress Report.* http//www.ijc.org/en/home/Main_acceuil.htm

Jaimet, K. (2003, December 3). Kyoto on verge of collapse. *Calgary Herald*, pp. A1, 5.

Joyce, G. (1995, October 17). B.C. pushes tough standards. *Calgary Herald*, p. A2.

Kaharabata, S. K., Schuepp, P. H., & Desjardins, R. L. (1998). Methane emissions from aboveground open manure slurry tanks. *Global Biogeochemical Cycles, 12*, 545–554.

Kerr, R. A. (1994). Methane increases put on pause. *Science, 263*, 751.

Kerr, R. A. (1995). Studies say—tentatively—that greenhouse warming is here. *Science, 268*, 1567–1568.

Lazier, J. R. N. (1996). The salinity decrease in the Labrador Sea over the past thirty years. In D. G. Martinson et al. (Eds.), *Natural climate variability on decade to century time scales.* Washington, DC: National Research Council, National Academy Press.

Magnuson, J. J., Robertson, D., Benson, B. J., Wynne, R. H., Livingston, D. M., Arai, T., et al. (2000, September 8). Historical trends in lake and river ice cover in the northern hemisphere. *Science, 289,* 1743–1750.

Manney, G. L., Zurek, R. W., Gelman, M. E., Miller, A. J., & Nagatani, R. (1994). The anomalous Arctic lower stratosphere polar vortex of 1992–1993. *Geophysical Research Letters, 21,* 2405–2408.

Mauzerall, D. L., Logan, J. A., Jacob, D. L., et al. (1998). Photochemistry in biomass burning plumes and implications for tropospheric ozone over the tropical South Atlantic. *Journal of Geophysical Research, 103,* 8401–8423.

McBean, G. A., & Hengeveld, H. G. (1998). The science of climate change. *The Climate Network, 3*(3), 4–7.

McElroy, C. T., Kerr, J. B., McArthur, L. J. B., & Wardle, D. I. (1994). Ground-based monitoring of UV-B radiation in Canada. In R. H. Biggs & M. E. B. Joyner (Eds.), *Stratospheric ozone depletion/UV-B radiation in the biosphere* (pp. 271–282). Berlin: Springer-Verlag.

McMillan, T. (1991). *The real story about Canada's acid rain campaign in the United States.* Notes from remarks by the Honourable Tom McMillan, Canadian Consul General to New England, to the National Conference on Government Relations, Ottawa.

Mills, C., & Jackson, S. (1995). Workshop report: Public education messages for reducing health risks from UV radiation. http://alep.unibase.com/sunconf/papers/cmills/cmills.html

National Round Table on the Environment and the Economy. (1998). *Declaration of the National Forum on Climate Change.* Ottawa: Author.

Natural Resources Canada. (1997). Laurentian Forestry Centre. http://www.cfl.forestry.ca/75a.htm

Natural Resources Canada. (2002). *Biobus Project: Biodiesel Demonstration and Impact Assessment with the Société de Transport de Montréal (Stm).* http://www.nrcan-rncan.gc.ca/media/newsreleases2002_e.htm

Newman, P. A., Pyle, J. A., et al. (2002). Polar stratospheric zone: Past and future. In A. L. Ajavon, D. Albritton, G. Mégie, and R. Watson, Executive Summary, Final, UNEP/WMO *"Scientific Assessment of Ozone Depletion: 2002."* Geneva: UNEP/WMO.

Nkemdirim, L., & Budikova, D. (1996). The El Niño–Southern Oscillation has a truly global impact: A preliminary report on the ENSO Project of the Commission on Climatology. *International Geographical Union Bulletin, 46,* 27–37.

Ozone Secretariat, UNEP. (2000). *The Montreal Protocol on Substances That Deplete the Ozone Layer as Either Adjusted and/or Amended in London 1990, Copenhagen 1992, Vienna 1995, Montreal 1997, Beijing 1999.* http://www.unep.ch/ozone/ratif.shtml

Pembina Institute. (2000). *Five years of failure: Federal and provincial government inaction on climate change during a period of rising industrial emissions.* Drayton Valley, AB: Author.

Prézelin, B. B., Boucher, N. P., & Schofield, O. (1994). Evaluation of field studies of UV-B radiation effects on Antarctic marine primary productivity. In R. H. Biggs & M. E. B. Joyner (Eds.), *Stratospheric ozone depletion/UV-B radiation in the biosphere* (pp. 181–194). Berlin: Springer-Verlag.

Qu, Q. (2003). Validation and evaluation of biomarkers in workers exposed to benzene in China. *HEI Research Report, 115.*

Rainham, D. (1999). Global climate change: Is global warming a health warning? *Encompass, 4*(2), 15–17.

Ram, M., Stolz, M., & Koenig, G. (1997). Eleven year cycle of dust concentration variability observed in the dust profile of the GISP2 ice core from Central Greenland: Possible solar cycle connection. *Geophysical Research Letters, 24,* 2359–2362.

Rhode, H. (1990). A comparison of the contributions of various gases to the greenhouse effect. *Science, 263,* 271.

Rowland, F. S. (1989). Chlorofluorocarbons and the depletion of stratospheric ozone. *American Scientist, 77,* 36–45.

St. Louis, V. L., Kelly, C. A., Duchemin, E., Rudd, J. W. M., & Rosenberg, D. M. (2000). Reservoir surfaces as sources of greenhouse gases to the atmosphere: A global estimate. *Bioscience, 50*(9), 766–775.

Schoof, R. (1996, November 4). "A sea of coal": Chinese struggle to wipe out throat-stinging smog. *Victoria Times Colonist*, p. D10.

Shine, K. P., Fouquart, Y., Ramaswamy, V., Solomon, S., & Srinivasan, J. (1995). Radiative forcing. In J. T. Houghton et al. (Eds.), *Climate change 1994—Radiative forcing of climate change and an evaluation of the IPCC 1992 emission scenarios* (pp. 163–203). Cambridge: Cambridge University Press.

Stolarski, R. S. (1988). The Antarctic ozone hole. *Scientific American, 258*(1), 30–36.

Taylor, K. E., & Penner, J. E. (1994). Response of the climate system to atmospheric aerosols and greenhouse gases. *Nature, 369,* 734–737.

Toon, O. B., & Turco, R. P. (1991). Polar stratospheric clouds and ozone depletion. *Scientific American, 264*(6), 68–74.

Torrie, R. (1999). *Powershift: Cool solutions to global warming.* Vancouver: David Suzuki Foundation.

United Nations Environment Programme. (1998). *Environmental effects of ozone depletion: 1998 assessment.* Secretariat for the Vienna Convention for the Protection of the Ozone Layer and the Montreal Protocol on Substances That Deplete the Ozone Layer. Nairobi.

United Nations Framework Convention on Climate Change. (2003). *Proposed new methodologies submitted to the Executive Board.* http://cdm.unfccc.int/EB/Panels/meth/PNM_Recommendations/index.html)

United States Environmental Protection Agency. (1995). *Sun protection for children.* http://www.epa.gov/ozone/uvindex/uvwhat.html

United States Environmental Protection Agency. (2000a*). Human health benefits from sulfate reduction.* http://www.epa.gov/airmarkt/articles/healtheffects/index.html

United States Environmental Protection Agency. (2000b). *National air pollutant emission trends: 1990–1998.* EPA 454/R-00-002.

Wood, S. & Bodeker, G. (2003, September 22). Antarctic ozone hole-near record levels. New Zealand Institute of Water and Atmospheric news release. http://www.niwa.cri.nz

World Meteorological Organization. (1994). *Scientific assessment of ozone depletion 1994.* WMO Global Ozone Research and Monitoring Project, Report No. 37. Geneva.

World Resources Institute. (1999). *Environmental change and human health, 1998–1999.* http://www.wri.org/wri/wr-98-99/#description

Agroecosystems and Land Resources

"In the last decade there has been much discussion on creating a more stable, more sustainable agricultural industry ... that would be less polluting, would maintain (even enhance) our healthy and attractive landscape, and would be less stressful on farm operators ... even a brief scan should convince anyone that 'sustainable agriculture' ... is quite different from the present pattern, and that even moving in the direction of achieving [sustainability] objectives requires major changes in much of our present society."

I. McQuarrie (1997, pp. 54–55)

Chapter Objectives

After studying this chapter you should be able to

- understand the main issues and concerns relating to Canada's land resources and their agricultural use

- identify a range of agricultural uses of our land resources

- describe the effects of agricultural activities on land and affected water resources

- appreciate the complexity and interrelatedness of issues relating to land and agroecosystems

- outline some Canadian and international responses to the need for sustainable agriculture and agroecosystems

- discuss challenges to a sustainable future for land resources, agroecosystems, and agriculture

INTRODUCTION

Historically, Canada has depended on its land resources (agriculture, forestry, mining, energy) for its economic well-being. Since the 1960s, however, there has been a shift away from resource-based to service-based industries in Canada. From 1961 to 1996, agricultural, forest, and mineral products industries dropped from 22 percent to 13 percent of Canada's GDP (gross domestic product), while personal and business service industries increased their share of GDP from 13 to 21 percent (Statistics Canada, 2002a). In the agricultural sector, some evidence of this shift is found in an almost 11 percent drop in the number of farms between 1996 and 2001, the largest decline since 1971 (Statistics Canada, 2003a).

Despite the drop in number of farms, in 2001, the value of Canada's agri-food exports was $26.6 billion; in 2002 that export value dropped to $25.9 billion, and dropped again to $24.3 billion in 2003 due to effects of drought and lower grain prices. During 2001, wheat exports accounted for $3.9 billion ($3.0 billion in 2002), oilseed exports, including canola, amounted to $2.2 billion ($1.96 billion in 2002), and exports of live animals and animal products totalled $8.3 billion ($8.9 billion in 2002) (Statistics Canada, 2002a, 2003a). Beyond the significance of Canada's agricultural lands to our economy, their diversity and quality are linked closely to our sense of national identity.

In this chapter, the main focus is on agricultural uses of land resources and on the impacts that food production activities have on sustainability of land and affected water resources. When agriculture is carried out in a sustainable manner, the natural resource base (land) is protected; degradation of soil, water, and air quality are prevented; the economic and social well-being of Canadians is enhanced; a safe and high-quality supply of agricultural products is assured; and the livelihoods and well-being of agricultural workers and their families are safeguarded (Environment Bureau, n.d.).

As the environmental archaeology of many ancient societies demonstrates, all civilizations depend on the ecological viability of their agricultural base (Taylor, 1994). In the early 1990s in Canada, there was concern that "economic sustainability [was being] jeopardized by the neglect of the physical and biological resource on which agriculture depends" (Science Council of Canada, 1992, p. 13). We examine the nature of these concerns and consider actions undertaken and efforts required to move toward environmental as well as economic sustainability of Canadian agricultural lands.

CANADA'S AGRICULTURAL LAND BASE

With a total surface area (land plus fresh water) of 9 970 610 km^2, Canada occupies 7 percent of the world's land mass and supports about 0.5 percent of the world's people. A wide range of landscapes and ecological zones across Canada (see Chapter 3) results in great diversity of climate, landform, vegetation, mineral, and hydrocarbon resources, and supports a variety of economic activities. Environmental concerns associated with agricultural activities include soil fertility, water quality, loss of wildlife habitat and wetlands, pesticide and herbicide use, and the effects of biotechnology on biodiversity. Other effects of industrialization of agriculture and agro-ecosystems relate to globalization of the food system and food security. (Some of these concerns also are discussed in other chapters.)

It is important to have reliable information about the quality and suitability of land resources for particular uses if we are to use and manage Canada's lands wisely and sustainably for agricultural production as well as other activities. One of the largest land inventories undertaken in the world, the Canada Land Inventory (CLI) focused on information about our most productive lands, mostly in the southern, heavily populated portions of the country. Completed in the 1970s, the CLI provides information regarding the long-term capabilities of about 2.6 million km^2 of land to support agriculture, outdoor recreation, forestry, waterfowl, and ungulates.

The CLI categorized Canada's soil capability for agriculture into seven classes. The supply of soil capability classes 1 to 3, which provide Canada's dependable agricultural land base for crop production, is 454 630 km^2 or about 5 percent of our total land area. Saskatchewan and

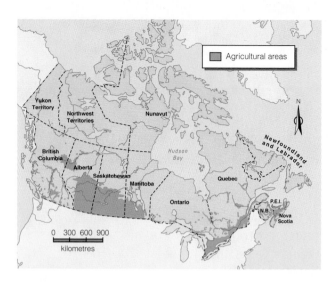

Figure 6–1

Agricultural areas within Canada

SOURCE: *The State of Canada's Environment—1996,* Government of Canada, 1996, Figure 11.1. Reprinted with permission of the Minister of Public Works and Government Services Canada, 2004.

Alberta contain the largest areas of dependable land, with 162 988 km^2 and 107 289 km^2, respectively. Ontario follows, with 72 833 km^2. Far more limited than most Canadians appreciate, class 1 agricultural capability lands occupy less than 0.05 percent (41 461 km^2) of Canada's total land area. Fifty-two percent of this class 1 land (21 568 km^2) is located in southern Ontario. Saskatchewan (9997 km^2) and Alberta (7865 km^2) also contain significant prime land areas (Statistics Canada, 2000). Another 2 percent of land (although not as suitable as classes 1, 2, or 3 land) is used for agriculture in Canada (Hoffman, 2001). Figure 6–1 illustrates the location of agricultural areas within Canada, while the

Photo 6–1a

Photo 6-1b

Canada depends on a small percentage of good-quality land to support agriculture. Balancing competing demands for land is a necessity if both a healthy environment and a prosperous economy are to exist.

amount of agricultural land (in CLI classes 1 to 3) in each province is identified in Table 6–1.

It is worth noting that Canada does not have any vast agricultural reserves; virtually all the land that is amenable to agricultural production and that has not been built on or paved over is in agricultural use. As Canadian cities and towns expanded between 1971 and 1996, the amount of dependable agricultural land paved over or built on increased substantially. Of the almost 15 000 km^2 of land occupied by urban uses, such as suburbs and shopping centres, almost half (5900 km^2) is located on dependable agricultural land. By 1996, 19 percent of Ontario's class 1 agricultural land had been displaced by urban uses, while Alberta had lost 6 percent and Saskatchewan had lost less than 1 percent (Statistics Canada, 2000). In practical terms, this land is permanently lost to agriculture.

In spite of the preceding statements, there are limited data about how much land suitable for agriculture is actually being farmed, and there is no current national database identifying how much land really is available for agricultural use. This is an important gap in knowledge, given that expanding human needs and economic activities have put increasing pressures on land, and that achieving sustainable agricultural uses of land depends, in part, on reliable data.

Competing and conflicting land uses have heightened many Canadians' awareness of the necessity of balancing competing demands on our limited land if we are to maintain both a healthy environment and a prosperous economy. The resulting challenge for sustainable agriculture is to maintain and enhance the quality of Canada's finite agricultural soils as well as the air, water, and biodiversity resources that are part of our agroecosystems.

Agroecosystems are communities of living organisms, together with the physical resources that sustain them, such as biotic and abiotic elements of the underlying soils and drainage networks, that are managed for the purpose of producing food, fibre, and other agricultural products. These are complex and dynamic ecological processes and systems, with many interrelationships; any action in one component of an agroecosystem affects other components and ecosystems. For example, a farmer's decision to use management practices such as fertilizers or pesticides is influenced by his or her access to technology as well as by the economics of the marketplace and government policy. Such management practices affect the health of agroecosystems and, ultimately, the productivity and sustainability of agriculture in Canada.

THE CHANGING NATURE OF CANADIAN AGRICULTURE

Creating a sustainable agricultural industry in Canada requires not only an ability to identify and alleviate environmental challenges but also to understand and deal with technological and socioeconomic changes in the agricultural sector. Accordingly, before considering human activities and impacts on agricultural environments, we

TABLE 6–1

CANADA LAND INVENTORY: SOIL CAPABILITY FOR AGRICULTURE (KM^2 OF CANADA LAND INVENTORY LANDS, BY PROVINCE[1])

Agricultural Land Capability Class[2]	NF	PE	NS	NB	PQ	ON	MB	SK	AB	BC	Canada	% of total area
1	0	0	0	0	196	21 568	1 625	9 997	7 865	211	41 461	0.05
2	0	2 616	1 663	1 605	9 071	22 177	25 306	58 744	38 371	2 355	161 908	2
3	19	1 415	9 829	11 511	12 722	29 088	24 407	94 247	61 053	6 920	25 1261	3
Total	19	4 031	11 492	13 116	21 989	72 833	51 338	162 988	107 289	9 487	454 630	5

NOTE: Figures may not add up due to rounding.

[1] The Territories (Yukon Territory, Northwest Territories, and the Territory of Nunavut) are not covered by the CLI.

[2] Land capability classes 1, 2, and 3 are Canada's prime agricultural lands containing the best soils with the highest potential to produce varied crops now and in the future. Saskatchewan possesses the largest amount of prime agricultural lands, followed by Alberta, Ontario, and Manitoba. In general, prime agricultural lands also are located where climatic conditions are favourable for farming.

SOURCE: *Human Activity and the Environment 2000,* Statistics Canada, 2000, Table 5.1.1. Reprinted by permission of Statistics Canada.

briefly identify the issues that characterize contemporary North American agriculture and review the nature of technological and socioeconomic changes encountered within Canadian agriculture.

Table 6–2 summarizes the range of economic, environmental, and social issues that characterize contemporary agriculture. The message of this table is that although we have achieved dramatic increases in short-term food production, conventional agricultural technologies have done so by increasing long-term social and environmental costs in soil degradation, loss of arable land, use of an increasingly controlled and select number of species, a greater reliance on chemicals, and increased financial debt (Alasia & Rothwell, 2003; Hoffman, 2001).

Since the 1960s, the total area of Canada's land used on farms has remained stable at about 7.4 percent (681 000 km^2). However, changes in the social, economic, and technological conditions associated with the industrialization of agriculture in Canada have been reflected in a decrease in the number, and a growth in the size, of farms in Canada. Between 1995 and 2000, for instance, the prices farmers received for their products declined 4.6 percent, while their expenses for fuel and fertilizer increased by 10 percent. As farmers have been squeezed by increasing costs and declining returns (on average, in 2000, farmers spent 87 cents in operating expenses for a return of one dollar in gross farm receipts), they have been challenged to increase their farm production in order to maintain a favourable expenses-to-receipts ratio. With expenses rising faster than revenues, the number of farms in Canada continued their decades-long decline; between 1996 and 2001, another 29 625 farms were lost (an 11 percent decline between 1996 and 2001). Farms have, however, been getting larger: from 50 hectares in 1901, they averaged 273 hectares in 2001 (Statistics Canada, 2002a, c, d). Figure 6–2 illustrates that the number of farms declined by about half, while their size increased about two and one-half times between 1941 and 2001.

Such changes have resulted in a move from mixed farming to specialized systems such as **monoculture** farming (cultivation of one species over a large area) and megafarms (consolidated farms). In terms of monoculture farming, Saskatchewan grows almost 51 percent of Canada's wheat (excluding durum). Western Canada (provinces west of Ontario) contains about 73 percent of the total hay area, while Eastern Canada (provinces east of Manitoba) accounts for about 98 percent of the soybean area and 96 percent of Canada's corn area (Statistics Canada, 2003a). One of the problems facing monoculture farming is that plants of a single species are highly susceptible to insects and diseases. Megafarms have brought another important change: in Canada, about 8 percent of farms occupy about 43 percent of all farmland, and although only 2 percent of farms have gross receipts of over $1 million, they account for nearly 35 percent of all agricultural receipts in Canada (Statistics Canada, 2003a).

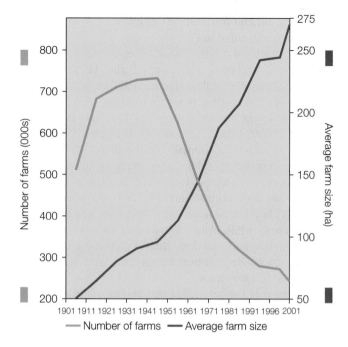

Figure 6–2

Change in number and size of farms in Canada, 1901–2001

SOURCE: *The State of Canada's Environment—1996,* Government of Canada, 1996, Ottawa: Supply and Services Canada; *Profile of Canadian Farm Operators,* Statistics Canada, Agriculture Division, 1997, Cat. no. 93-359 XPB, Ottawa: Author; *2001 Census of Agriculture: Canadian Farm Operations in the 21st Century,* Statistics Canada, 2002a, http://www.statcan.ca/Daily/English/020515/td020515.htm

These points mean that a small proportion of farmers actually determines whether sustainable practices are applied on almost half of Canada's finite agricultural lands (Government of Canada, 1996; Statistics Canada, 2000).

Since the 1960s some important changes in agricultural land use have occurred in Canada. One major change is that farmlands near urban centres frequently were converted to nonagricultural uses; land that subsequently was brought into agricultural production tended to be of lower quality. Another important change is that cropland area has continued to increase (by 4 percent from 1996 to 2001, for instance). In 2001, Canadian farmers had a total of 89.9 million acres in crops. A third major change was that the area devoted to **summerfallow** (land not sown for at least one year to conserve soil moisture and to enhance nitrogen accumulation, chiefly a Prairie practice) has continued to decline (by 25 percent from 1996 to 2001). Farmers made this change on the basis of evidence that summerfallowing contributed to soil **salinization** (see the discussion of soil quality below) (Statistics Canada, 2000). The decline in summerfallow land also reflects farmers' increasing use of no-till and other conservation seeding practices (because these practices retain soil moisture but do not require the land to remain

TABLE 6–2

ECONOMIC, ENVIRONMENTAL, AND SOCIAL ISSUES THAT CHARACTERIZE CONTEMPORARY NORTH AMERICAN AGRICULTURE

Economic	Environmental	Social
• Technologically efficient crop production	• Decline in soil productivity	• Crushing debt burden
• Increased crop specialization and reliance on monocultures	• Growing dangers to animal and human health (chemical and biological risks such as BSE)	• Income disparity between rural and urban dwellers increasingly favours urban populations
• Increased dependence on fossil fuels, chemical fertilizers/pesticides, and borrowed capital	• Increased vulnerability of plants, especially monocultures, to climatic changes, increasing frequency of extreme weather events, and new diseases	• Rapid disappearance of family farm, rural life
• Reduced profitability; farmers' average total income is between 60% and 80% of that of all tax-filers in Canada	• Ongoing requirement by corporate-based seed banks to use custom-designed fertilizers and pesticides	• Loss of agricultural land to encroaching urban development, transportation networks, airports, and industrial parks
• Increasingly mechanized approach to food production	• Reduced long-term resilience due to decreased genetic diversity in plants	• Almost 60% of land converted to urban use was formerly prime agricultural lands
• Reliance on canola oil, durum wheat, and a few others as major export crops	• Surface and groundwater contamination by pesticides	• Growing concern about pesticide safety and soil degradation
• Security of food supply and agricultural economy at risk	• Loss of wildlife habitat	• Concern about long-term effects of agricultural chemicals on human health and the environment
• Export markets subject to foreign protectionism	• Combined effects of soil erosion, acidification, compaction, salinization, and irrigation cost taxpayers almost $1.4 billion per year	• Growth of alternative farming techniques in part due to increase in public demand for organic products
• Increased reliance on imports (often of foods that could be grown here)	• Use of farm chemicals and runoff resulting in sediment damage to inland lakes and waterways, loss of recreational fishing, increased water treatment and dredging costs	• Goal of long-term stewardship of land for future generations
• Dramatic increases in input costs but low farm produce prices contributing to financial stress	• Loss of land to urbanization	
• Production of food almost totally dependent on oil and gas (to provide chemicals and machinery, and to process and distribute farm products)		

SOURCES: The Rural/Urban Divide Is Not Changing: Income Disparities Persist, by A. Alasia & N. Rothwell, 2003, *Rural and Small Town Canada Analysis Bulletin*, 4(4), 1; *Farming Facts 2002*, Statistics Canada, 2003, Ottawa: Minister of Industry; *Off Course: Restoring Balance between Canadian Society and the Environment*, by D. M. Taylor, 1994, Ottawa: International Development Research Centre. Reprinted by permission.

idle for a year). Figure 6–3 indicates changes in the use of Canada's agricultural lands from 1971 to 2001.

Farmers also are diversifying the crops they are growing. While spring wheat remains the dominant Prairie crop, oilseeds (canola, soybeans) and pulse crops (dry field peas, lentils, and beans) have increased substantially. Although Ontario continues to produce almost two-thirds of Canada's grain corn (a high-energy source of animal feed), increases in grain corn crops in Quebec and Manitoba have been driven by the significant increase in hog production in those provinces. Grain corn also is used in the production of ethanol, a crop-derived alcohol that is added to gasoline to increase octane levels and to improve engine performance (through cleaner-burning fuel). Prince Edward Island continues to farm about 25 percent of the total area (of over 418 700 acres) devoted to Canada's potato crop. Vegetables such as sweet corn, green peas, beans, tomatoes, and carrots were grown on over 330 700 acres in 2001. In Alberta in particular, and in Saskatchewan, an increasingly large area is being devoted to culinary and medicinal herbs such as basil and echinacea (Statistics Canada, 2002a). For instance, Saskatchewan grows 90 percent of the mustard that is sent to France and processed as Dijon mustard.

Blueberries have overtaken apples in terms of area devoted to fruit production; in 2001, apples were grown

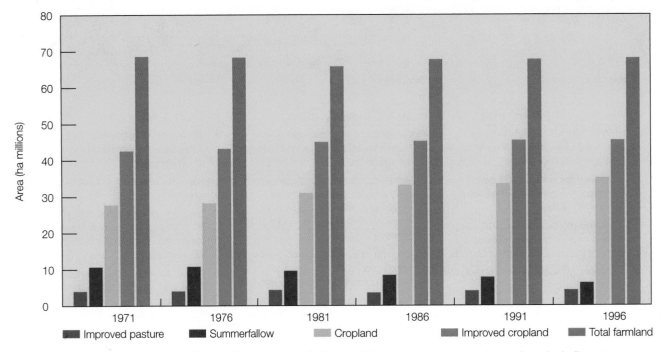

- *Improved pasture* is area improved by seeding, draining, irrigating, fertilizing, and brush or weed control, not including areas where hay, silage, or seeds are harvested.
- *Summerfallow* is area that has been left idle (not worked) for at least one year.
- *Cropland* is the total area on which field crops, fruits, vegetables, nursery products, and sod are grown; improved pasture and summerfallow are excluded.
- *Improved cropland* is the sum of cropland, summerfallow, and improved pasture.
- *Total farmland* is the total area of land operated, including improved cropland and unimproved land.

Figure 6–3

The use of farmland in Canada, 1971–1996

SOURCES: *The State of Canada's Environment—1996,* Government of Canada, 1996, Ottawa: Supply and Services Canada; *Agricultural Profile of Canada,* Statistics Canada, Agriculture Division, 1997, Cat. no. 93-356 XPB.

on 63 814 acres (a drop of 18 percent since 1996), while blueberries were grown on 108 679 acres (an increase of 21 percent since 1996). The area devoted to grapes rose 41 percent from 1996 to 2001; most of the 26 165 acres is located in southern Ontario, where about 70 percent of Canada's grapes are grown. Wineries in both Ontario and British Columbia enjoy world-class reputations, and wine tourism (where vineyards have their own wineries to sell their products and offer tours and cooking classes) helps add value to the basic grape crop. British Columbia's cranberries, accounting for over 54 percent of Canada's crop in 2001, were grown on 7453 acres; Quebec's cranberry crop accounted for 35 percent of the Canadian total. Plantings of Saskatoon berries increased almost 80 percent between 1996 and 2001, to 2937 acres. The three Prairie provinces accounted for almost 95 percent of the Saskatoon berries used to make jams and jellies (Statistics Canada, 2002a).

Eastern Canada has remained the key poultry- and egg-producing region of Canada, and the poultry sector has seen significant growth since 1996; in 2001, farmers reported raising 126.2 million hens and chickens. As of May 2001, farmers reported a record 15.6 million head of cattle on Canadian farms, and 13.9 million hogs, another record high number. Alberta accounted for 43 percent of Canada's national beef herd and 24 percent of its sheep (Ontario has about 27 percent of Canada's 1.3 million sheep). The number of dairy cows in Canada declined by almost 30 percent between 1996 and 2001, but milk production remained reasonably steady because of improved breeding and feeding techniques. Nontraditional livestock has become more popular, with goats (almost 183 000 of them) kept for meat and milk, and llamas for their meat and wool (Statistics Canada, 2002a). Also following the outbreak of mad cow disease in Europe, bison were perceived as healthier choices by markets in France, Belgium, Germany, and Britain. For information on Canada's BSE case, see Box 6–1.

As a result of technological and socioeconomic changes, the Canadian agriculture and *agri-food industry*—farmers, suppliers, processors, transporters,

BOX 6-1
BOVINE SPONGIFORM ENCEPHALOPATHY (BSE) IN CANADA

The BSE issue highlights the disadvantages of tailoring agricultural products almost exclusively to the export market. In May 2003, one eight-year-old cow from an Alberta farm tested positive for bovine spongiform encephalopathy (BSE), or mad cow disease. While not considered contagious, BSE is linked to variant Creutzfeldt-Jakob disease (vCJD) in humans. In both humans and bovine species, the disease is characterized as one of the group of TSEs (transmissible spongiform encephalopathies) that cause progressive deterioration of the nervous system by producing sponge-like changes in brain tissue (World Health Organization, 2002).

In response to the detection of BSE, 33 countries immediately closed their borders to Canadian cattle product imports. This response occurred at a time when farmers in Western Canada had moved away from crop production to cattle. In 2001, the number of cattle and calves in Canada was at a record high as farmers responded to foreign market demands (Statistics Canada 2002d). By June 2003, beef cattle producers and associated secondary industries such as trucking operations, meatpacking and processing facilities, veterinarians, feeds mills, and machinery dealers, were facing "economic disaster" (Alberta Agriculture, 2003a). In response, a short-term national compensation program, called the Canada-Alberta Bovine Spongiform Encephalopathy Recovery Program (CABSERP), was established (producers absorb 10 percent of cattle price declines and the remaining cost decline is shared on a 60 percent federal–40 percent provincial/territorial basis). Alberta's portion of the CABSERP was estimated at more than $100 million (Alberta Agriculture, 2003a). By October 2003, some international markets had reopened partially to processed Canadian beef. However, one year from when BSE was detected, no live animals were moving beyond Canada's borders.

What Is BSE?

The TSE group of diseases is suspected of being able to jump species barriers. Although the causative agent is unconventional and not fully understood, many people believe that the normal cellular prion protein (PrP) self-converts into an infectious amyloid protein or prion. An unconventional protein with no nucleic acid, prions can change normally-shaped cellular prion protein into abnormal shapes, causing changes in brain tissue (Corato, Ceroni, & Savoldi, 2000; Saborio, Permanne, & Soto, 2001; Travis & Miller, 2003; World Health Organization, 2002). Prions, however, may not be the infectious agent. Manuelidis (2003) reports that the prion protein alone is incapable of reproducing transmissible infection. Given the epidemic spread of TSEs, she points to the probable viral source of the infection.

The Beginnings of the BSE Epidemic

The BSE epidemic began in the United Kingdom in 1986, although cases may have occurred as early as the 1970s. In the U.K., over 180 000 head—mostly older dairy cattle—were known to be infected. Estimates indicate that up to one million animals may have been infected. Since the incubation period for BSE averages four to five years, during which time the animals

appear perfectly healthy, the full extent of the disease among U.K. herds was not detected immediately, given that beef cattle are slaughtered before the age of three years. Still the largest epidemic, 95 percent of known BSE cases have been detected in the U.K. (Travis & Miller, 2003).

The cause of the BSE epidemic clearly is linked to the use of cattle meat and bones, particularly the brain and spinal cord (that is, the use of recycled bovine carcasses, ground to meat and bone meal), in the preparation of cattle feed (World Health Organization, 2002). Less than one gram of brain tissue from an infected ruminant is enough to cause infection, but since BSE is not contagious (does not spread from one animal to another in a herd), it appears that the only way for cattle to become infected is to eat contaminated food (Miller & Travis, 2003; World Health Organization, 2002).

In 2003, the Canadian Food Inspection Agency published the results of an animal health risk analysis it conducted in compliance with the request from the Office International des Epizooties (World Organization for Animal Health) that all OIE member countries determine the status of BSE in their cattle populations. This risk assessment indicated a negligible probability that BSE was introduced and established in Canada (Morley, Chen, & Rheault, 2003). The World Health Organization (2002) classified North America's potential risk as unlikely, but not excluded.

Human Health and TSEs

Human health implications of the BSE epidemic rose considerably when, in 1996, 10 human cases of a new TSE similar to the well-known but very rare Creutzfeldt-Jakob disease (CJD) occurred in the U.K. Labelled vCJD to distinguish it from the classical form, the variant affects younger people (compared to people with classical CJD). Worldwide, 122 deaths have been linked to vCJD; 11 people are known to have survived the disease. While not proven conclusively, evidence suggests that humans contract vCJD by ingesting food contaminated with the BSE agent (World Health Organization, 2002).

Keeping the BSE Agent Out of the Food Chain

Even though scientific understanding of BSE and vCJD is incomplete, implementation of preventive measures can help ensure that no global BSE epidemic occurs. Protecting people from the risk of contracting vCJD means that measures must be taken, and monitored, to ensure the following:

- Food supplies are not exposed to or contaminated with the BSE agent.

- "Specified risk materials" (all tissues suspected of containing any level of infectivity, particularly beef brain, head, eyeballs, spinal cord, spine, and lymphatic materials) should be removed and destroyed, and use of these tissues for human food or rendering should be strictly prohibited (rendering processes animal byproducts and wastes by grinding and melting them down at high temperatures for a fixed time—the recovered products are used as ingredients in numerous

(continued)

BOX 6-1
CONTINUED

commercial products such as lubricants, lipstick, pharmaceuticals, soap, candles, and cement).

- Mechanically recovered meat does not contaminate the food supply. Some slaughterhouse practices strip muscle meat attached to bones and the vertebral column using wire brushes and other mechanical tools. This action can also pull out infectious nervous tissue and contaminate the recovered meat. Some experts believe BSE was transferred to humans through recovered meat products such as inexpensive hamburger, sausage meat pies, and other processed meats—the type of meats that younger people tend to eat—perhaps helping to explain why the majority of vCJD cases have occurred in younger individuals.

In terms of ruminant health, risks of contracting BSE can be reduced by putting measures in place that ensure the following:

- Meat and bone meal of ruminant origin is not fed back to ruminants.

- Recycled cattle wastes are not allowed to contaminate cattle feed (in some countries, mammalian protein-based animal feed is banned from ruminant feeds or from all feed for all farm animals).

- Infectious tissues have no opportunity to enter the food or feed chain.

For both human and ruminant protection, active surveillance systems that detect suspected cases of BSE and remove them from the food supply system provide a measure of confidence to consumers and trading partners that food safety measures are effective (Corato, Ceroni, & Savoldi, 2000; Morley, Chen, & Rheault, 2003; World Health Organization, 2002). However, because control measures appear to be straightforward, the U.K. experience was that ensuring compliance sometimes was difficult and necessitated more extensive prohibitions and aggressive enforcement as well as auditing of compliance levels (Prince et al., 2003).

Costs of BSE in Canada

Following the discovery of the isolated case of BSE, experts from Europe, the United States, and New Zealand were invited to evaluate Canada's food inspection system. Their report generally was favourable, but recommended Canada strengthen its system in terms of prohibition of specific risk material, tighter controls on nonruminant feed, improved tracking and tracing systems, improved disease testing and surveillance, and efforts to improve awareness among producers, veterinarians, and the general public (Alberta Agriculture, 2003).

Despite the recommendation given by the international panel, borders did not open immediately to Canadian beef products. Part of the reason relates to the politics of trade: in 2002 over 80 percent of Canadian beef exports (worth $4 billion) went to the United States. Mexico, Japan, South Korea, and Taiwan are the next most important export markets (Agriculture and Agri-Food Canada, 2003). Many people believe the Americans' reluctance to open their border fully was in retribution for Canada's perceived lack of support in the Iraq war. Similarly, many people believe Japan closed its border because Canada refused to accept Japan's beef for a long time following detection of BSE there. Mexico has stated they would like to import Canadian beef as long as doing so will not interfere with the reliance of their cattle business on the United States. With the December 2003 discovery of a BSE-infected cow in Washington State, international trade activity will be interesting to watch.

The stakes associated with the detection of BSE are high: lost livelihoods, damaged industries, reduced trade, weakened national economies, and human suffering from a devastating disease. As the World Health Organization (2002, p. 19) has noted, "Prevention is a responsibility shared by all those involved in the food and feed chains—from farm to fork."

SOURCES: *Fact Sheet – Trade,* Agriculture and Agri-Food Canada, 2003, http://www.agr.gc.ca/cb/trade/factsheet_e.phtml; *Alberta Commits to National Disaster BSE Relief Program,* Alberta Agriculture, Food and Rural Development [Alberta AFRD], 2003a, http://www1.agric.gov.ab.ca/$department/deptdocs.nsf/all/com6799?open; *International Panel Report on BSE Testing Released,* Alberta AFRD, 2003b, http://www1.agric.gov.ab.ca/$department/newslett.nsf/all/cotl3318?open; "Prions and Risk for Human Health," by M. Corato, M. Ceroni, & F. Savoldi, 2000, *Instituto Lombardo Accademia di Scienze e Lettere Rendiconti Scienze Chimiche e fisiche geologoche biologoche, 134*(102), 109–114; "Transmissible Encephalopathies: Speculations and Realities," by L. Manuelidis, 2003, *Viral Immunology, 16*(2), 123–139; "Assessment of the Risk Factors Related to Bovine Spongiform Encephalopathy," by R. S. Morley, S. Chen, & N. Rheault, 2003, *Revue Scientifique et Technique Office International des Epizooties, 22*(1), 157–178; "Bovine Spongiform Encephalopathy," by M. J. Prince, J. A. Bailey, P. R. Barrowman, K. J. Bishop, G. R. Campbell, & J. M. Wood, 2003, *Revue Scientifique et Technique Office International des Epizooties, 22*(1), 37–60; "Sensitive Detection of Pathological Prion Protein by Cyclic Amplification of Protein Misfolding," by G. P. Saborio, B. Permanne, & C. Soto, 2001, June 14, *Nature, 411,* 810–814; *Canadian Farm Operations in the 21st Century,* Statistics Canada, 2002d, http://www.statcan.ca/english/agcensus2001/first/farmop/01front.htm#top; "A Short Review of Transmissible Spongiform Encephalopathies and Guidelines for Managing Risks Associated with Chronic Wasting Disease in Captive Cervids in Zoos," by D. Travis & M. Miller, 2003, *Journal of Zoo and Wildlife Medicine, 34*(2), 125–133; "Understanding the BSE Threat," World Health Organization, 2002, http://www.who.int/csr/resources/publications/whocdscsreph20026/en/

grocers, and restaurant workers—has become the third-largest employer in Canada, generating about $95 billion in 2002 (Statistics Canada, 2003b). Canada's top nine agri-food exports in 2002 were non-durum wheat, slaughter cattle, fresh boneless beef, durum wheat, canola seed, fresh pork, frozen fries, frozen pork, and biscuits and crackers. Agri-food has become one of the top five industries in Canada, accounting for about 8.3 percent of the gross domestic product (Agrifood Trade Service, 2002). However, farmers have seen their net farm income decline from $2.529 billion in 1998 to $1.969 billion in 2002 (Statistics Canada, 2003c). International trade agreements such as the North American Free Trade Agreement and the World Trade Organization Agreement also have important economic and environmental influences on Canadian agriculture (see Box 6–6).

EFFECTS OF HUMAN ACTIVITIES ON AGRICULTURAL LANDS

Agriculture and land management practices affect the environmental sustainability of agroecosystems. This section identifies some of these management practices and their effects on soil resources, water resources, and biodiversity. As well, greenhouse gases and energy use are considered.

EFFECTS ON SOIL RESOURCES

In an agricultural context, soil resource quality refers to the ability of the soil to support crop growth without resulting in soil degradation or other harm to the environment (Acton & Gregorich, 1995). Soil quality is affected by land use and by land management practices. Growing a single species over a large area (monoculture cropping), leaving the land fallow, intensive row cropping (where bare soil is exposed in the spaces between crop rows), and up-and-down slope cultivation contribute to the processes that reduce soil quality. These processes result in loss of organic matter; erosion by wind, water, and tillage; changes in soil structure; salinization; and chemical contamination. Each of these concerns is discussed briefly below.

Levels of Soil Organic Matter

Organic matter (plant, animal, or microorganism matter, either living or dead) is an essential component of soil because it stores and supplies plant nutrients, retains carbon, helps water infiltrate into soil, and stabilizes soil. The amount of organic matter in agricultural soils varies

between 1 and 10 percent; how much is optimal depends on local climate, the amount of clay in the soil, and the intended use of the land.

Although there is a lack of comprehensive data regarding soil organic matter across Canada, it is known that levels of organic matter usually decline during the decade following crop cultivation on previously undisturbed forest or grassland soils. Research has shown that since initial cultivation, Canada's uneroded agricultural soils have lost between 15 and 30 percent of their organic matter. However, trends in some regions indicate that levels of organic matter are holding steady or are increasing because of improved management practices. In the Prairie provinces, for example, more farmers are replenishing organic matter taken out of the soil through cropping by adding crop residues, manure, and commercial fertilizers (note that chemical fertilizers do not replace carbon) (Gregorich et al., 1995). Still other farmers are switching to more environmentally benign tillage types (see Box 6–5). In 2001, more than 60 percent of farmers were practising conservation and zero tillage, practices that will help retain soil organic matter and reduce erosion (Statistics Canada, 2002a).

Wind, Water, and Tillage Erosion

Erosion of soil by wind and water is the most widespread soil degradation problem in Canada; these natural processes are accelerated or minimized by the types of cropping and land management practices used (Larney et al., 1998). For instance, risks of wind and water erosion can be reduced through changes in land tillage practices, including reductions in summerfallow, changes in cropping patterns, and use of erosion control measures.

Tilling (plowing) the land, particularly on rolling or hummocky land where soil moves downhill during plowing operations (i.e., *tillage erosion*), contributes to loss of topsoil. The topsoil is the soil layer best able to support life because it contains most of the soil's organic matter (the smallest, lightest soil particles contain the greatest proportion of plant nutrients and are highly susceptible to wind erosion). Wind erosion is more widespread than water erosion, even though wind erosion has a lower transport capacity. If topsoil is blown or washed away, carbon and nitrogen are lost, and the remaining soil has a reduced ability to provide the fertility required for crops as well as a lowered capacity to accept and store water. Also, if eroded topsoil is carried into water bodies, detrimental effects such as blocked waterways, buried vegetation, smothered fish, and increased costs of water treatment can occur.

Canada's agricultural lands have been classified in terms of their susceptibility to erosion. Almost all regions of the country are concerned about wind erosion, but the Prairies experience particularly extensive, damaging wind

Figure 6–4

Risk of wind and water erosion in Canada

SOURCE: *The State of Canada's Environment—1996*, Environment Canada, Ottawa: Supply and Services Canada, Figures 11.9 and 11.10. Reprinted with permission of the Minister of Public Works and Government Services Canada, 2004.

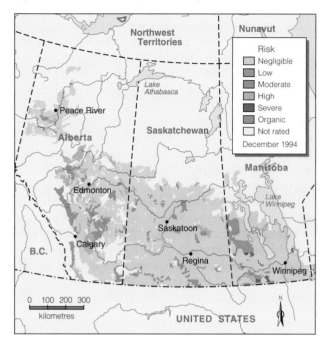

a) Risk of wind erosion in the Prairie provinces under 1991 management practices

NOTE: Management practices based on 1991 Census of Agriculture.

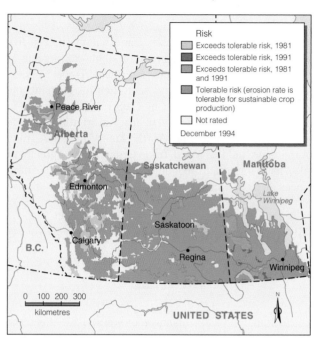

b) Risk of water erosion in the Prairie provinces, 1981 and 1991

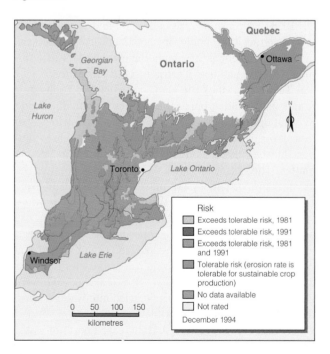

c) Risk of water erosion in southern Ontario, 1981 and 1991

NOTE: "Tolerable risk" refers to soils that are at risk of erosion at a rate that is tolerable for sustainable crop production under the most common management practices.

erosion. About 36 percent of cultivated land in the Prairies is subject to high to severe risk of wind erosion, particularly in parts of southern Manitoba and Alberta and in a large part of Saskatchewan (see Figure 6–4) (Wall et al., 1995). Soil losses to wind erosion not only lead to reduced productivity and loss of economic returns, but also cause abrasion damage to buildings, machinery, and vegetation. Airborne nutrients and pesticides eventually may degrade water quality and aggravate health problems in downwind areas. In response to such risks, between 1981 and 1991, farmers in the Prairie provinces converted their annual crops planted in sandy areas to perennial forage crops. This soil conservation measure resulted in a 7 percent decrease in risk of wind erosion (Wall et al., 1995).

Soil erosion by water varies widely across Canada, with the most severe losses of soil occurring in the Peace River area and Fraser Valley of British Columbia, and parts of Saskatchewan, southwestern Ontario, the Eastern Townships of Quebec, and Prince Edward Island. All of Canada's agricultural regions are at risk of soil erosion by water, but the risk is greatest on land that is being cultivated intensively; overall, about 20 percent of Canada's agricultural areas have a moderate to high level risk of *water erosion*. While the Prairie provinces have a low inherent risk of soil erosion by water, the same is not true for other regions. Eighty percent of cultivated lands

Photo 6–2
Water erosion is a major factor in the decline of soil quality, particularly on land under intensive cultivation.

in the Maritimes, 75 percent of British Columbia lands, and 50 percent of lands in Ontario have been identified as having high to severe risk of soil erosion by water (see Figure 6–4) (Wall et al., 1995). In British Columbia's Peace River area, erosion on summerfallow lands has been as high as 14 tonnes per hectare per year, and in the Fraser Valley, erosion rates under row crops have been as high as 30 tonnes per hectare annually. Potato lands in Atlantic Canada experience major erosion problems; in Prince Edward Island, losses of up to 20 tonnes of soil per hectare per year have occurred in the past; serious long-term problems may result if annual soil losses exceed 5 to 10 tonnes per hectare. In 2003, Prince Edward Island identified soil erosion (and associated nutrient and pesticide runoff) as a major environmental problem. Heavy rain storms and spring runoff wash soils into estuaries and ponds, destroying fish habitat and contributing to fish kills (Kelly, 2003). In southwestern Ontario, water erosion may reduce yields by as much as 40 percent. Soil eroding from agricultural areas in the Great Lakes basin has transported pesticides, such as the no-longer-used insecticide DDT, to ground and surface waters.

Not only is soil loss by water erosion costly in terms of reduced crop productivity and higher production costs, but sediment damages occur as well. These damages include reductions in channel capacity; alteration and destruction of fish habitats; accelerated plant and algal growth from excess nutrients; buildup and transportation of heavy metals, pesticides, and other toxic substances; lower recreational values; and the increased costs of ensuring water is fit for human consumption.

Soil Structure

Soils that are low in organic matter content, wet, and finely textured are most vulnerable to structural degradation, particularly if land management practices such as intensive tillage, row cropping, and short rotation periods are employed. One of the most recognized forms of structural degradation, **soil compaction** results mainly from repeated passes of heavy machinery over wet soil during tillage and harvesting, causing what was a well-aerated soil to restrict air and water movement, thus reducing the ability of plant roots to penetrate the soil and derive sufficient moisture and nutrients.

Certain crops, such as potatoes, corn, soybeans, and sugar beets, often are associated with compaction because they require long growing seasons—being planted in the early spring and harvested in the fall when the soil is frequently moist. Soil compaction not only reduces crop yields but also costs Canadian producers millions of dollars annually.

One study reported that economic losses due to soil compaction were greatest in Quebec, where it was estimated that 20 percent of the best farmland was compacted. In British Columbia, soil compaction was widespread and, in Ontario, 50 to 70 percent of clay soils were adversely affected by compaction. In Atlantic Canada, with its naturally compacted subsoils and hardpans (hardened soil layers with greatly reduced porosity), farming on moist soils causes additional compaction. Soil compaction is not a serious problem on the Prairies (Topp et al., 1995).

Soil Salinization

Salinization, an excess of salts in soils, is another factor that can reduce the capacity of soil to produce crops. Saline (salty) soils restrict the amount of water plants can withdraw from the soil, thereby reducing crop production. Unlike soil erosion, which usually occurs on the surface, salinization generally affects soils at depth. In dryland soils, such as those on the Prairies, salinization is a natural process that takes place when a high water table, high

Photo 6–3
An example of soil salinization in an agricultural area of southern Alberta.

rate of evaporation, and soluble salts all occur together (Government of Canada, 1996). About six to eight million hectares of land in western Canada contain soils that may be affected by salinization.

Prairie saline soils existed long before settlement and cultivation occurred, but land management processes that affect the soil–water balance (such as replacing natural vegetation with crops and summerfallow or applying irrigation water) may modify the extent of soil salinity. The precise role of summerfallow in soil salinization remains unclear; research has produced conflicting observations about whether summerfallow is a major cause of soil salinization (Harker et al., 1995; Statistics Canada, 2000).

Water moving from areas of saline soils to streams, lakes, or underground aquifers means that salts can degrade the quality of water available for domestic use, livestock, and irrigation. Salt water draining from saline soils may degrade the quality of neighbouring downslope areas also, suggesting that controls on drainage may help efforts to reclaim saline areas.

Other research has shown that while soil salinity is a continuing problem in some Prairie soils, only about 2 percent of Prairie agricultural land has more than 15 percent of its area affected by salinity. Most (62 percent) Prairie agricultural land has less than 1 percent of its area affected by salinity, while between 1 and 15 percent of the remaining 36 percent of land is affected (Eilers et al., 1995). Losses to farmers because of soil salinity in southern and central regions of Alberta, Saskatchewan, and scattered parts of Manitoba were estimated to have ranged between $104 and $257 million per year during the 1980s (Government of Canada, 1991).

Chemical Contamination

Fertilizers and pesticides (agrochemicals) simultaneously have helped increase food production and have had negative effects on soil and water quality and vegetation health. Some of the socioeconomic benefits of agrochemical use include a doubling of global food production since 1960, increased security of food supply, and employment opportunities in the agrochemical industry. However, use of pesticides that contain organochlorides (such as endrin, chlordane, heptachlor, mirex, and toxaphene) seriously threaten human and environmental health. Heavy metals associated with the livestock industry, and extensive use of nonrenewable fossil fuels to run agricultural equipment, also threaten soil and plant health. Agricultural pesticides are human-made chemicals used to kill unwanted plants and animals and to control insects, weeds, and crop diseases. In 1995, close to $1 billion (or $2067 per km^2) was spent on agricultural pesticides. This represents an increase of 411 percent in expenditures on pesticides from 1970 to 1995 (Statistics Canada, 2000). In 2000, sales of fertilizer in Western Canada were almost three times those in Eastern Canada (Statistics Canada, 2002b). Integrated pest management (IPM) programs are intended to help reduce the release of chemicals into the environment, and public concern about use of pesticides was a major environmental issue facing Newfoundland and Labrador during 2003 (Drover for Mercer, 2003).

IPM is a broadly based method that uses all suitable control measures to reduce pest-related losses to an acceptable level with the goal of respecting biodiversity and reducing risks to ecosystems and human health. The main components of integrated pest management programs are (1) planning and managing production systems to prevent organisms from becoming pests—this is done by controlling whether and how land is plowed, planting a diversity of crops, managing the kinds of crop rotation used, planning planting dates, and handling harvests to reduce presence of pests; (2) identifying potential pests; (3) monitoring populations of pests, beneficial organisms, and all other relevant ecological factors—this involves recognizing ecological communities and ecosystems as well as the effects one species may have on other species; (4) establishing economic, damage, and action thresholds that allow pests to exist at low, tolerable levels (because the goal is control, not extinction); (5) applying cultural, physical, biological, chemical, and behavioural control measures to maintain pest populations below threshold levels—this may include use of highly specific chemicals (used sparingly), genetically resistant stock, biological controls (natural enemies of pests including parasites, diseases such as *Bacillus thuringiensis*, and predators such as ladybugs and some species of wasps); and (6) evaluating the effects and efficiency of pest-control measures used. IPM reduces the release of toxic chemicals into the environment while enabling economically viable production of crops; the more IPM is employed, the better it is for the environment and for individual ecosystems (Government of Canada, 1996).

The suitability of soil for various uses, including food production, can be affected by **chemical contamination** from herbicides, insecticides, algicides, and fungicides. While some research shows that the contamination of agricultural soils with pesticide and nonpesticide contaminants is not a serious problem, other studies show that Canadian agricultural soil is contaminated by heavy metals such as cadmium, lead, and zinc. These metals are persistent and affect the health of plants, animals, and humans (Webber & Singh, 1995). Heavy metals enter agricultural soils mostly through atmospheric deposition as well as through fertilizers, animal manures, and sewage sludge that is applied to agricultural land as a source of organic matter and nutrients.

Acid precipitation (see Chapter 5) and the use of nitrogen fertilizers can augment the natural acidity of some soils. This *acidification* causes nutrient deficiencies and has reduced crop yields in the Maritimes and even in some Prairie areas (where soils generally are alkaline). Application of lime to reduce acidity has been a common practice among farmers. Liming has also been used in rehabilitation of land in the Sudbury area (see Chapter 10).

Farmers' efforts to increase plant production through widespread, liberal application of fertilizers can result in excess potassium, nitrogen, and phosphorus in soils, water, and plant systems. This occurs when manure and/or fertilizer, applied in excess of plant requirements, leads to a buildup of nutrients in the soil. When the capacity of the soil to retain these nutrients is surpassed, they will be lost to the atmosphere or to surface or groundwater (Chambers et al., 2001). In the lower Fraser Valley and southern Ontario, high nitrate levels and fecal coliform bacteria in wells used for drinking water have been traced to the application of fertilizers and manure to fields, as well as to leakage from septic tanks. Given the large areas involved in agriculture, and the concentrations of dairy cattle, poultry, and other livestock in these areas, heavy applications of manure may exceed the assimilative capacities of the soils and lead to declines in water quality (see the discussion "Contamination from Livestock Production Activities," below). The cumulative effects of septic systems may also result in widespread contamination of shallow groundwater.

Desertification

Desertification is a biophysical process of land degradation that occurs (in arid, semi-arid, and dry sub-humid areas of the world) as a result of complex interactions between unpredictable climatic variations and unsustainable land-use practices by people who, in their struggle for survival, overexploit agricultural, forest, and water resources. Desertification is more than just desert encroachment; it refers to degradation of dry land to the point where it is difficult to restore its former level of productivity (partly because of the loss in biological diversity). Desertification occurs in all continents when accelerated soil erosion, driven by water, wind, and salinization, causes reductions in soil quality, effective rooting depth, vegetative cover, and biomass productivity. Research has demonstrated a strong link between the desertification that affects an estimated 5.8 million hectares of land per year and the release of CO_2 from soil and vegetation to the atmosphere (Lal, 2001; Larney et al., 1998).

Worldwide, desertification directly affects about 250 million people; nearly one billion people, most of whom live in the poorer regions of the world, are at risk from desertification (Food and Agriculture Organization, 2002). African nations are the most vulnerable and the least able to combat the problem without international assistance.

Photo 6–4

Desertification is particularly problematic in Africa, where a combination of unsustainable land-use practices and climatic variations have resulted in overexploitation of the land.

During the 1970s and again in the 1980s, devastating droughts affected the West African Sahel, and thousands of people and millions of animals died. Satellite images clearly showed altered vegetation patterns, and people began to talk about expanding deserts and sand dunes on the march. Closer scientific examination revealed that much of the vegetation change reflected water shortage and not permanent loss of soil fertility or land degradation. However, the loss of soil productivity, crop failures, scarcity of fuel wood, and reduced availability of grazing lands for livestock forced many people to abandon their land and become environmental refugees. Although many African nations made plans to address the problem, little serious action was taken to combat desertification (Canadian International Development Agency, 1995; Cardy, 1994).

While the West African Sahel is the most seriously affected region in the world, vast areas in Asia as well as North, Central, and South America also are affected by desertification (see Figure 6–5). With its dry lands in the Prairie provinces, Canada is technically an affected country. Should climate changes occur as predicted, and currently dry regions become drier, Canada might well heed the lessons learned in other countries to formulate a response to this issue.

EFFECTS ON WATER RESOURCES

Both surface water and groundwater resources can be affected directly by agricultural use of land. **Surface runoff** can carry sediment, nutrients, pesticides, and bacteria from agricultural lands and contaminate surface water bodies. Groundwater resources can be contaminated by nutrients or pesticides when rainwater, irrigation water,

Figure 6–5

World drylands and desertification

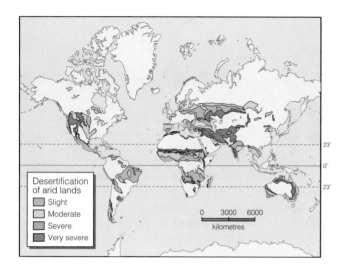

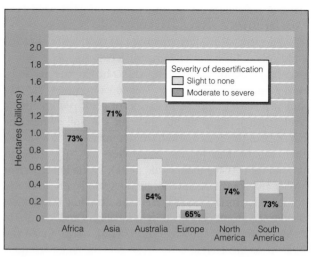

a) World status of desertification of arid lands

SOURCE: *Risk of Human-Induced Desertification,* United States Department of Agriculture—National Resources Conservation Service, 1999, http://www.nrcs.usda.gov/technical/worldsoils/mapindex/dsrtrisk.html

b) Drylands by continent

SOURCE: "Desertification," F. Cardy, 1994, *Our Planet, 6*(5), 4.

and snowmelt percolate through the soil. The discussion that follows highlights two sources of water contamination, namely crop and livestock production activities. Irrigation use and its effects also are considered.

Contamination from Crop Production Activities

The nutrients that plants need for growth are obtained from soil, water, and air. If a farmer is to attain economic yields and sustain soil fertility, it is often necessary to replace the nutrients removed when crops are harvested by adding extra nitrogen, phosphorus, and potassium in the form of manure or commercial fertilizers. Over 5 million tonnes of potash, phosphate, and nitrogen from commercial fertilizers were applied to Canadian farmlands in 2000 (Statistics Canada, 2002b). Depending on the type and intensity of crop and land management practices, soil characteristics, weather, and the type and amount of chemicals applied, that portion not used by crops or absorbed or retained in the soil can move into surface water or leach into groundwater.

Several **aquifers** in Canada show significant amounts of agrochemicals, mainly nitrates. For example, significant leaching of nitrates (which can make water unfit for human consumption) has been reported in an important aquifer near Osoyoos in the southern Okanagan Valley, British Columbia. Septic tank effluent and fertilizer use in orchards were the major suspected sources of nitrates. In

the Fraser River Valley, also in British Columbia, the Abbotsford aquifer has been highly susceptible to contamination from extensive use of high-nitrogen poultry manure on raspberry and forage crops. Leaching occurs here, too, this time in combination with climatic (heavy winter rainfall) and soil conditions. Monitoring of nitrate concentrations since 1995 suggests not only that nitrates have increased annually at a rate of 0.7 mg/L, but also that about 60 percent of the samples collected from a highly sensitive region of the aquifer exceeded the Canadian Water Quality Guidelines safe limit of 10 mg/L. Twelve different pesticides have been detected in the Abbotsford aquifer, four at levels exceeding the Canadian Freshwater Aquatic Guidelines, and four (for which there are no Canadian guidelines) that exceed Washington State water quality standards for groundwater. Studies of well water contamination by nitrates, bacteria, and herbicides, conducted in Ontario, Quebec, New Brunswick, and Nova Scotia, have shown similar findings and concerns.

Other agrochemicals also cause concerns about water quality. If phosphorus enters surface-water bodies, for example, it can lead to accelerated eutrophication (nutrient pollution). Given that about 26 percent of Canadians (mainly rural residents) rely on groundwater for domestic purposes, and that more than 85 percent of livestock consume water from underground sources, protecting water resources from agricultural contamination is essential (Government of Canada, 1996; Reynolds et al., 1995).

Water quality surveys and monitoring efforts reveal that pesticides frequently are found in concentrations below the safe limits specified in the Canadian Water Quality Guidelines, while concentrations of nutrients and bacteriological contaminants sometimes exceed acceptable limits. In terms of groundwater, although there are insufficient long-term, detailed monitoring data to provide a comprehensive understanding of the current status and trends of agrochemicals entering Canada's groundwater, one of the main impacts of agricultural activities on water quality is contamination by nitrates (from fertilizers, the use of manure instead of chemical fertilizers, and feedlots).

Pesticides sometimes are said to be less of a problem than they were 20 to 30 years ago because current pesticides are less persistent and more specialized. However, we cannot escape the fact that pesticides (particularly insecticides and herbicides) are an integral part of modern agricultural production. Public concerns remain about potential health hazards associated with water contamination, specifically such concerns as persistence and bioaccumulation (see Chapter 7). It is expected that such concerns will encourage increased use of nonchemical pest controls and decreased use of pesticides (Gregorich & Acton, 1995). Public health and safety concerns are one reason why water quality has emerged as a key environmental issue facing agriculture in the 21st century.

Contamination from Livestock Production Activities

As noted previously, meat is a prime product of Canadian agriculture. Intensive livestock operations (ILOs), confined feeding operations (CFOs), or feedlots, have become the dominant method of producing beef, hogs, and poultry in Canada. Feedlots enable livestock producers to increase their production to meet domestic and offshore demands for meat and poultry products. Feedlots also provide an ideal market for Canadian-produced grains; given the abundance of feed grains, western Canada can now produce bacon more profitably than any other region in the world (Nikiforuk, 2000). The apparent economic potential of ILO facilities has garnered government support and resulted in the rapid growth of feedlots throughout Canada, particularly on the Prairies where, in 2001, Alberta accounted for 43 percent of the national beef herd (6.7 million cattle). Such rapid expansion has generated conflict between proponents who see ILOs as positive economic operations, supporting a federal commitment to low food prices, and opponents who are concerned about contamination and health, environmental, and nuisance risks associated with feedlot facilities.

What is an ILO, and how does it generate contamination? Alberta Agriculture defines an intensive livestock operation as one where more than 300 *animal units* are confined in facilities at a density of 43 animal units per acre for more than 90 consecutive days, and where the producer has to manage the manure generated at the facilities (Alberta AFRD, 2000). Alberta Agriculture defines animal units as the number of animals of a particular category of livestock that will excrete 73 kg of total nitrogen in a 12-month period. How many animals does it take to excrete this much nitrogen per year? Alberta Agriculture (2000) determined that one beef and one bison cow or bull would do so, but that it would take 1320 calves, 60 000 broiler chickens, and 1500 sows or boars (hogs) to produce that amount of nitrogen.

Most of the negative effects of feedlots result from the volume of manure produced at these operations. Every day, one 454 kg feedlot steer produces about 27 kg of manure (9954 kg annually). In "Feedlot Alley," a small region north of Lethbridge, Alberta—Canada's largest concentration of livestock—about 500 000 cattle, 200 000 hogs, and 600 000 poultry generate a volume of manure equivalent to the waste that would be produced from a city of about 8 million. Even a single 500-sow farm producing 20 piglets per sow per year creates as much effluent as a town of 25 000 people, without a waste treatment system. Clearly, the growth of animal factories has created industrial-scale waste problems (Hasselback, 1997; Nikiforuk, 2000).

If properly applied to land, manure is an excellent slow-release fertilizer and source of organic matter. Manure application to land is particularly valuable in southern Alberta, where mineral soils (low organic content and high pH) are prevalent. However, the principal environmental hazard of ILOs is the possibility of surface and groundwater contamination from manure storage and application, including the leaching of nitrates and/or pesticides. Bacteria and parasites also contribute to contamination and to public health concerns regarding gastrointestinal illness. As residents of Walkerton, Ontario, discovered when heavy rain washed manure from farm fields into one of the wells that supplied their drinking water system, *E. coli* O157 is a deadly pathogen (see discussion of the Walkerton case in Chapter 7). Poultry ILOs generate another type of contamination. Typically, various heavy metals (such as arsenic, cobalt, copper, iron, manganese, selenium, and zinc) are added to poultry diets to improve weight gain and disease prevention. When poultry waste containing fairly high concentrations of heavy metals is dispersed repeatedly onto land, the runoff and leaching of heavy metals from the poultry waste-amended soil pose potential environmental risks to surface and groundwater (Han et al., 2000).

Air quality is a third concern associated with ILOs. Over 150 gaseous compounds may be emitted when manure is spread on land; people living in areas of dense livestock operations complain about the headaches and nausea caused by odours from these gases. As health

Photo 6–5
A farmer spreads manure on a field to replace nutrients lost through harvesting.

been applied to fruit, tobacco, and vegetable crops. Irrigated crops come with a price tag: reservoir construction floods scenic river valleys, destroys historically and culturally important sites, reduces or eliminates habitat for fish and wildlife, and alters the flow and water quality of a river, often disadvantaging downstream users. Sediments accumulate in reservoirs, reducing their

Photo 6–6a

Photo 6–6b

Photo 6–6c
Irrigation water is applied to the land using a variety of methods, including centre pivot (6–6a), oscillating sprinklers (6–6b), and irrigation canals (6–6c).

concerns have grown, and in spite of residents' appeals against ILOs, the Alberta government has indicated plans to double beef production and triple hog production by 2005.

On January 1, 2002, the Alberta government introduced a new provincial regulatory framework for its confined feeding operations. Based on the 2000 Code of Practice for Responsible Livestock Development and Manure Management, the new standards and regulations set the type and size of ILOs that must be approved, establish new water protection buffer zones for operators applying manure, and identify who should be notified of a proposed or expanding ILO (Government of Alberta, 2001). While these regulations were designed to "ensure our $5-billion livestock industry can grow and expand" (Government of Alberta, 2001), grassroots activists are concerned to ensure the long-term health of rural communities and ecosystems continues. For additional information on feedlots, manure management, public health concerns, and actions taken to resolve these issues, see Box 6–2.

Irrigation Effects

Agriculture accounted for 9 percent of the total water withdrawals in Canada in 1991 (see Figure 7–4). Of the 3991 million m^3 of water taken up for agricultural uses, 85 percent was used for irrigation and the remaining 15 percent for livestock watering (Statistics Canada, 2000). Irrigation is a highly consumptive use of water; because of the high rate of evaporation from agricultural fields, very little of the water withdrawn for irrigation is returned to its source.

In western Canada irrigation is used to increase the productive capacity of land for forage, cereal, and oilseed crops, while in eastern Canada irrigation traditionally has

Intensive livestock operations (ILOs) raise a number of environmental and community health issues, most associated with the safe, nontoxic disposal and recycling of manure. Let us examine these issues more closely.

Feedlot Operation

A typical feedlot operation consists of an enclosed animal area, a slurry pond or lagoon to catch the feedlot runoff, manure storage facilities, and surrounding agricultural land. In consuming feed or silage, livestock extract nitrogen and other elements; in turn, the majority of the nitrogen consumed is excreted either as urine or feces. The 9954 kg of manure produced annually by a single steer includes 1410 kg of solids, 56 kg of nitrogen, and 18 kg of phosphorus (Gregorash, 1997). The excreta are collected in the slurry pond, lagoon, storage tank, or trench, or removed by scraping the feedlot. The manure is then stored in piles either outside or within storage buildings and, after decomposing for six to eight months, is distributed onto fields as fertilizer for feed crops such as corn or barley. In some places, the feed grown is harvested and consumed by local livestock.

Manure Storage and Application

Storing manure (in facilities with impervious bases) allows operators to apply it to the land when the crop uptake of nutrients is most rapid and runoff from fields is least likely. Knowing when and how much manure to apply is a complex task that involves an understanding of soil characteristics, crop requirements, characteristics of the manure (moisture, salt, nutrients), nutrient releasing rates, geology, and slope of the land. Manure may be spread on any given piece of land once every three to six years. Most larger feedlot operations cannot grow all the feed grain required and must import it from other provinces or areas where the price is right. Since hauling manure beyond 18 km is uneconomical, nutrients in the manure seldom go back to the fields where the feed grain was produced (Gregorash, 1997). To avoid overloading the soil on their own land, feedlot operators make arrangements to spread manure on neighbouring fields. Since feedlots are concentrated spatially, concern has been expressed regarding increasing manure concentrations in soil.

Pollution and Human Health Problems

As the size and density of livestock feeding operations increase, so does the potential for adverse effects on soil and water quality: runoff from feeding areas, feedlots, manure storage facilities, or manured fields may contaminate surface waters with sediments, nutrients, organic matter, and bacteria. If more manure has been applied to the land than the crops can use, nitrogen, phosphorus, and salts may accumulate in the soil, and groundwater may become contaminated with nitrates. Excessive phosphate and nitrogen in water bodies promote algae, weed growth, and eutrophication.

Health researchers classify public health concerns associated with ILOs into three types: air quality, surface-water quality, and groundwater problems. Each health concern is considered briefly here.

Air Quality Problems

The public's major air quality complaint about feedlots is the smell (often considered a nuisance effect). Manure-spreading operations release gaseous compounds such as hydrogen sulphide, ammonia, carbon dioxide, and methane (Hasselback, 1997). People downwind of the manure application may complain about a variety of effects, including headaches, nausea, and aggravated asthma and respiratory problems. Air quality monitoring by Alberta Environment (2000) near livestock feeding operations in the Lethbridge area from September 1998 to July 1999 revealed that, with the exception of hydrogen sulphide, all substances detected were at levels within Alberta's ambient air quality guidelines (see Box Table 6–1).

The health implications of inhaling the mixture of gases and odours are unclear, although it is known that airborne dust (from manure spreading) can carry diseases and transmit *E. coli* (Tessier, 1998). In the Lower Fraser Valley, British Columbia, ammonia from manure is known to volatize (change from a liquid or solid to a vapour) into the atmosphere and react with particles of urban smog (industrial pollutants and vehicle emissions). When this happens, a thick band of white haze forms over the intensive poultry farming area of Abbotsford. Scientists indicate that this white haze is a rural version of urban smog, specifically associated with emissions from intensive agricultural production of poultry and other livestock manures (Environment Canada, 1999). There are concerns that rural smog will result in similar respiratory effects caused by urban smog.

Surface-Water Quality Problems

Human health problems may arise when leaching, runoff, or overflows from manure lagoons deliver increased bacterial loadings to surface-water bodies, including irrigation canals, near livestock operations. Bacteria of concern include *E. coli* O157,

(continued)

Photo 6–7
Sows in a holding pen on a hog farm.

BOX TABLE 6–1
ALBERTA'S AMBIENT AIR QUALITY GUIDELINES FOR ONE-HOUR CONCENTRATIONS OF PARAMETERS MONITORED DURING THE LETHBRIDGE AREA SURVEY (1998–1999)

Air Quality Parameter	One-Hour Guideline (ppmv)	Basis for Guideline
Hydrogen sulphide	0.01	• odour perception
Ammonia	2.0	• odour perception
Carbon monoxide	13.0	• oxygen carrying capacity of blood
Nitrogen dioxide	0.212	• odour perception
Ozone	0.082	• reduction of lung function and effects on tomatoes
Sulphur dioxide	0.172	• taste, odour perception, and effects on bluegrass

SOURCE: *Air Quality Monitoring Near Livestock Feeding Operations in the Lethbridge Area September 1998 to July 1999,* Alberta Environment, 2000, http://www.gov.ab.ca/env/air/airqual/livestock_feeding_operation_air_quality_report_June_26_2000.pdf

Campylobacter, and *Salmonella.* In Alberta, the province's highest rates of intestinal infections are found in Feedlot Alley, and many cases are associated with people in the livestock industry (Hasselback, 1997; Nikiforuk, 2000). The experience of residents of Walkerton, Ontario, in May 2000 confirms the epidemic-level of sickness (and even deaths) that *E. coli* O157 can inflict on humans when just one manure-contaminated well is connected to a drinking water supply system.

The diffusion of parasites that may be spread by cattle, such as *Cryptosporidium parvum* and *Giardia lamblia,* is a growing concern. *Cryptosporidium* has caused several large waterborne outbreaks of human disease in North America: recent outbreaks in Canada have occurred in Cranbrook and Kelowna, British Columbia, and Owen Sound, Ontario. The parasite is resistant to chlorination, but may be removed by filtration, although inadequate water treatment facilities may not remove parasite oocysts (eggs). This is of concern because spring thaw can result in the overland flow of billions of oocysts from stored manure and infected animals, and treatment facilities may have difficulty cleaning and filtering the water. Several communities in Alberta and elsewhere have been under orders to boil domestic water supplies because treatment facilities cannot guarantee parasite-free water.

Another concern is that pig manure can contain high concentrations of endocrine-disrupting chemicals, including natural estrogens. Scientists have shown that these chemicals have long-term, adverse effects on the growth, development, and reproduction of fish and wildlife (see Box 7–3). Runoff from fields treated with pig manure quickly enters adjacent streams or other water bodies and results in eutrophication or acute toxicity (Environment Canada, 1998).

Groundwater Problems

Vertical movement of liquids in the hydrological cycle may carry chemicals, including nitrates, into underlying aquifers. While humans can tolerate low levels of nitrate, a level above 10 mg/L renders water unfit for human consumption. "Blue baby syndrome" (methemoglobinemia) may affect infants who ingest high levels of nitrate (Pederson & Johnson, 1997). As a result of livestock operations and fertilization practices, nitrate contamination of shallow groundwater aquifers has occurred in many parts of North America. Generally, contamination decreases with increasing depth, but Walkerton's medical health officer has raised concerns regarding the safety of deep groundwater wells in the area. Preventing contamination of aquifers from agricultural or any other activities is vitally important because no effective technology exists to cleanse aquifers.

Community Stresses

In Alberta, stresses on long-term friend and family relationships have increased as a groundswell of concern has developed regarding ILOs and their expansion within the province. Citizen unrest, protests against feedlot expansion, and appeals regarding large-scale operations have gone largely unheeded by the Alberta government (Ahmed, 1999). Just west of Feedlot Alley, for example, the County Residents for Fair Taxation group brought attention to the subsidization of ILO operators by grain farmers and acreage owners. The County Residents group pointed out that a 65 hectare feedlot with 25 000 head of cattle paid the same annual property tax as a 65 hectare grain farm, even though the impact of the feedlot on water services and roads was up to 500 times that of the grain farm (Ahmed, 1999). Over 1200 people signed a petition urging the County of Lethbridge to introduce a business tax, which was introduced in late 1998. Other local actions have met with limited or no success. It is unlikely, however, that citizens strongly opposed to unimpeded expansion of the livestock industry will cease their pressure for reform.

Actions to Improve ILOs

In Alberta, a Code of Practice for the Safe and Economical Handling of Animal Manures has been in place since 1995. Grassroots activists note that the most important weakness of this code is that it is a guideline only and not enforceable. Without enforceable regulations, they feel, improper management of ILOs will continue. Concerns about the process by which ILO expansion could take place, and about monitoring

BOX 6-2
(CONTINUED)

and enforcing the environmental sustainability of these operations, led Alberta Agriculture to propose (in 1999) a Regulatory Framework for Livestock Feeding Operations in the province. To date the framework has not been put into effect.

Other actions to improve ILOs include efforts to reduce odour issues by altering nutritional strategies for cattle, particularly through manipulation of the type and quantity of protein in their feed (Lethbridge Research Centre, 1999). Best management practices can also be implemented to reduce the possibility of soil and water contamination from animal wastes. One example is the creation of a riparian buffer zone by lining manure lagoons with trees

and vegetation. This action provides a rich source of soil microbes that degrade and consume nutrients such as nitrate and phosphorous found in manure. Cows and Fish, a successful program aimed at helping producers protect riparian habitat and decrease manure runoff into surface water, is discussed in Box 6–4.

Other protective actions that can be taken include transporting manure to fields farther away from ILOs; testing receiving soils for nitrogen and phosphorus; and monitoring collection lagoons to prevent overflow and seepage. Even though these actions will increase operating costs, they will go a long way to preventing future pollution and public health problems.

SOURCES: "Grassroots Activism on the Prairies," A. Ahmed, 1999, *Encompass, 3*(4), 16–17; *From Conflict to Cooperation,* Alberta AFRD, 1999, http://www.agric.gov.ab.ca/c2c/index.html; *Air Quality Monitoring Near Livestock Feeding Operations in the Lethbridge Area September 1998 to July 1999,* Alberta Environment, 2000, http://www.gov.ab.ca/env/air/airqual/livestock_feeding_operation_air_quality_report_June_26_2000.pdf; "Endocrine Disruptors and Hog Manure," Environment Canada, November–December, 1998, Science and the Environment Bulletin, 4; "Manure Causing White Haze," Environment Canada, May–June, 1999, Science and the Environment Bulletin, 12; "Feedlot Manure: An Issue That Doesn't Go Away," D. Gregorash, 1997, *Encompass, 2*(2), 19–20; "Intensive Livestock Operations and Health Problems," P. Hasselback, 1997, *Encompass, 2*(2), 4–5; *Scientists Target Air Quality in New Feedlot Manure Study,* Lethbridge Research Centre, 1999, http://res2.agr.ca/lethbridge/rep1999/adva0302.htm; "When Water Kills," A. Nikiforuk, June 12, 2000, *Online Macleans,* http://www.macleans.ca/pubdoc/2000/06/12/Cover/35699.shtml; *Animal Wastes as a Source of Drinking Water Contamination,* T. L. Pederson & B. T. Johnson, 1997, ExtoxNet, http://www.ace.orst.edu/info/extoxnet/faqs/safedrink/feed.htm; *Manure Handling Strategies for Minimizing Environmental Impacts,* S. Tessier, 2001, Manitoba Agriculture, Food and Rural Initiatives, http://www.gov.mb.ca/agriculture/livestock/pork/swine/bab10s08.html

capacity, and below a dam there may be streambed erosion. These impacts associated with major infrastructure for irrigation can change flooding patterns to the detriment of some adapted species such as cottonwood trees, reduce water quality, harm fish populations, and, because of water fluctuations from reservoir operation, destroy waterfowl nests and homes of water-dwelling mammals such as beaver and muskrat.

However, in parts of the Prairies where surface waters are scarce, not all impacts of water storage areas such as impoundments and canals are negative. In addition to the water they provide for irrigation, water impoundments can be used for recreation and fishing, and may provide new types of habitat for wildlife. In the past, the brush and weeds that grew beside canals and fence lines provided excellent cover for birds and small mammals (see Box 3–1). More recently, increased chemical control of brush and weeds and the attempts to reduce water losses through canal lining and use of water pipelines have diminished the positive effects of irrigation works on wildlife.

In the Prairie provinces, irrigation water is applied to over 635 000 hectares. While the area to be irrigated has not enlarged significantly during the past decade, there has been an increase in the amount of water applied per

hectare since 1981. This trend, combined with regional climate-change scenarios that point to more frequent and prolonged droughts on the Prairies, with increased drying up of wetlands, suggests the need for improved efficiency in water use. With modern methods and technology, irrigation needs for extensive grain, oilseed, fruit and vegetable production may be reduced by 10 to 50 percent (Government of Canada, 1996).

Other environmental concerns related to irrigated land result from the increased use of pesticides and fertilizers compared with amounts farmers would usually employ under dryland conditions. When irrigation water runs off the fields, residues from these substances may enter streams or be leached into groundwater, contaminating those water bodies. Irrigation may exacerbate dryland salinization of soils by altering shallow groundwater conditions that, in turn, accelerate surface evaporation. Replacing natural vegetation with short-season cereal grains is one action that can trigger salinization. Salts also can leach from irrigated lands, sometimes as a result of unlined canals causing water tables to rise in areas where salts are present in the subsoil. About six to eight million hectares of land in western Canada contain soils that may be affected by salinization.

EFFECTS ON BIODIVERSITY

Agricultural impacts on genetic, species, and ecosystem biodiversity of native wild species have become an important environmental issue. "Generally speaking, the quantity and quality of wildlife habitat in Canada have been degraded by settlement and agricultural development. Although farmlands and rangelands do provide enhanced habitat for certain species ... others have declined as a direct result of agricultural expansion and production practices. Many species of native plants, amphibians, reptiles, fish, birds, and mammals are endangered or threatened as a consequence of habitat loss to agriculture" (Government of Canada, 1991, p. 9-9).

Agricultural development has contributed to the endangered status (threatened with imminent extirpation or extinction) of birds such as the mountain plover, sage grouse (prairie), and sage thrasher; mammals such as the swift fox; vascular plants such as the cucumber tree (see Box 9–4), American ginseng, and pink milkwort; and reptiles such as the blue racer snake (Statistics Canada, 2000).

Current agricultural land management practices have modified, and continue to modify, biodiversity in several ways. Biodiversity is reduced when agricultural production systems with little crop rotation (such as in monocultures) provide large areas of uniform habitat. Deforestation, the replacement of indigenous plants with other crops, the drying of wetlands, and the use of insecticides and herbicides have reduced the populations and areas of distribution of numerous species and also resulted in the introduction of new species. Aquatic biodiversity can be affected by water draining from agricultural fields, carrying nutrients, eroded sediments, and pesticides. Similarly, wild species composition and abundance can be affected through selective grazing of preferred forage plants that alters the vegetation composition (Mineau et al., 1994; Government of Canada, 1996).

On the prairies, much of the original habitat has been altered significantly, largely through agriculture. Less than 1 percent of the original tallgrass prairie remains, and less than 25 percent of each of the shortgrass prairie, mixedgrass prairie, and aspen parkland remain (Gauthier & Henry, 1989; Trottier, 1992). This level of modification of original ecosystems has led to a concern for their continued viability and for the survival of the species that inhabit these areas, on the prairies and beyond (Van Tighem, 1996). The following sections identify particular effects of agricultural activities on wetlands and grazing lands.

Wetland Loss and Conversion

Wetlands, along with their ecologically productive and purifying functions and their contributions to the socioeconomic well-being of Canadians (see Figure 6–6), have

Photo 6–8

The burrowing owl is just one prairie species whose existence has been endangered by the agricultural alteration of its habitat.

been disappearing since settlement began. In the Great Lakes basin, for example, wetlands have been lost to agricultural and residential development at a rate of 8100 hectares per year. Despite the massive loss of wetlands, they continue to contribute billions of dollars annually to the Canadian economy through support of commercial and sports fishing, waterfowl hunting, trapping, recreation, peatland forestry, water purification, groundwater discharge, and flood peak modification (Government of Canada, 1990).

There is potential danger to wetlands from the pesticides and nutrients used in intense agricultural activities as well. Shoreline wetlands on the Canadian side of Lake Erie are a case in point. From 1992 to 1994, researchers studying the problem showed that pesticides were transported in water and in sediments from streams and creeks surrounding the wetlands into the wetlands and downstream into Lake Erie. Alachlor, a carcinogenic and oncogenic (tumour-producing) pesticide banned in 1989, was detected at one of the sampling stations. The highest pesticide concentrations occurred between May and July, immediately after pesticides were applied to the fields and following spring precipitation. Research continues into the impacts and cumulative effects of chronic exposure of Lake Erie's marsh and lake biota to these concentrations of pesticides.

Conversion of wetlands to other uses has been significant in many regions of Canada. For instance, 70 percent of sloughs in the central prairie wetland area (including 59 percent of wetlands in the Red River Valley), 65 percent of Atlantic salt marshes, 70 percent of Pacific estuarine marshes, and 70 to 80 percent of southern Ontario and St. Lawrence Valley hardwood and shoreline swamps have been converted. Eighty-five percent of the decline in

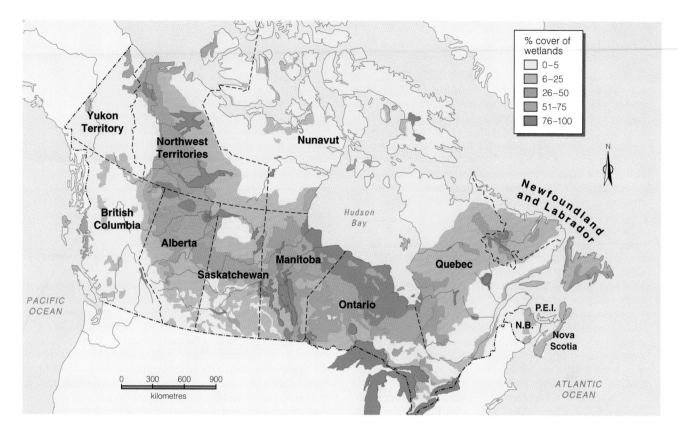

Figure 6–6

Distribution of wetlands in Canada

SOURCE: Adapted from *The State of Canada's Environment—1996,* Environment Canada, Ottawa, Figure 10.12. Reprinted with permission of the Minister of Public Works and Government Services Canada, 2004.

Canada's original wetland area is attributed to drainage for agriculture (Rubec, 1994). Severe effects such as this result in marked shifts in vegetation and animal species composition, including that of migratory species. Changes in habitat have direct effects on populations and life cycles of ducks, geese, swans, shorebirds, songbirds, and butterflies that inhabit North, Central, and parts of South America. The risk to biodiversity through habitat change is epitomized in the example of migratory songbirds and the shift from traditional shade-loving coffee plants to high-yield, sun-loving plants in modern Latin American coffee plantations (see Box 6–3).

In spite of their historical effects in Canada, some agricultural practices can enhance wildlife habitat and promote biodiversity. That is, some farmers may strive to maintain populations of pollinator species, pest predators, and soil fauna, as well as to preserve wetland habitats (which also conserves groundwater and helps protect against drought). Planting shelter belts, instituting planned grazing systems, and planting forage crops on marginal croplands are other means by which farmers may provide habitat for many forms of wildlife. Taking steps to ensure that soil organic matter is retained

(thereby providing habitat for microorganisms) and employing integrated pest management techniques can reduce the risks from pesticides to nontarget species. Furthermore, farmers can take part in habitat conservation programs such as the North America Wetlands Conservation Council (NAWCC). In partnership, the North American Waterfowl Management Plan and the NAWCC (see Chapters 7 and 12) jointly have administered $520 million into joint-venture projects. These projects have secured and enhanced over 800 000 hectares of key habitat and restored waterfowl populations and habitat (North America Wetlands Conservation Council, 2002).

Rangeland Grazing

About 40 percent of the world's land surface is used for grazing (Cardy, 1997). Most grazing takes place in arid, semi-arid, and dry subhumid climates or on land that is unsuitable for cultivated crops (because of steep slopes, for instance). Throughout the world, both domestic and wild animals forage on grasses, forbs, and shrubs, and provide people with valuable commodities such as food, fibre, and draft animals. Grazing lands also provide intangible products or values such as open space, natural beauty, and the opportunity to study natural ecosystems. Often referred to as rangelands in Canada, the value of grazing lands for wildlife and ecosystem functions has never been calculated, although their economic importance is tremendous.

Rangeland ecosystems are based on disturbance from grazing animals, fire, and drought (Trottier, 1992; Manzano & Navár, 2000). In the past, natural disturbances such as bison grazing and fire were important in stimulating new plant growth and nurturing a healthy range. Bison impacts on the prairies and foothills ranges often were severe but short-lived, because the yearly cycle of bison migration between the plains and foothills provided effective rest and regeneration periods for grazed areas (Adams & Fitch, 1995). Today, however, fire is uncommon on the range and, because of fences and roads, livestock do not emulate the periodic foraging of bison (Bork, 2000). Cattle grazing is more regular and more concentrated.

Beginning in the latter half of the 19th century, as the number of livestock on the range grew, grazing pressure resulted in the over-use of favoured plants. Heavy and very heavy grazing levels resulted in loss of important forage species (that is, a reduction in biodiversity, with a concomitant reduction in animal productivity); decrease in the protection that plants contribute to soil stability; increased erosion of exposed soils; and decreased water infiltration on hard-packed ground (Bakker, 1998; Bork, 2000; Thurow, 2000). Changes in plant biodiversity not only affected habitat for native species, including birds (Dale, 2000), but also resulted in a loss of carbon sinks, an important element in climate change (Arnalds, 2000; Janzen, 2000). Globally, the United Nations has estimated that degradation of grazing lands due to overgrazing and other agricultural practices not suited to dryland ecosystems has put the livelihoods of approximately one billion people at risk (Dowdeswell, 1997).

Livestock grazing also has been the major cause of **riparian** habitat disturbance, in which uncontrolled livestock break down stream banks, eat and trample shrubs that provide shade for stream dwellers and other wildlife, disturb stream beds, and contaminate water with fecal coliforms. For information about Cows and Fish, an Alberta initiative directed at ensuring sustainability of riparian habitats exposed to grazing livestock, see Box 6–4.

By Lorne Fitch, Provincial Riparian Specialist, Alberta Cows and Fish Program

In the United States, the use and abuse of riparian landscapes by livestock grazing has been a focal point of nearly three decades of debate. Issues about riparian use began in Alberta with a focus on fish. In the 1970s, the impact of decades of unmanaged livestock use on several high-profile trout streams in west-central Alberta became apparent through biological surveys. Those baseline surveys provided the catalyst to galvanize restoration actions designed to improve habitat conditions for trout.

Initial efforts for recovery involved fencing programs to permanently exclude livestock from variable portions of riparian areas. Exclusion fencing can provide rapid recovery and help to demonstrate a site's biological potential, often quickly; this was the case for the initial riparian management program in west-central Alberta. However, as the program to use exclusion fencing as the prime riparian management tool expanded, some issues related to the narrow focus became apparent. Initial fencing costs are high and the associated maintenance of fences in close proximity to an area prone to flood damage often exceeds the original cost. Stream-bank fencing also was perceived as a loss of abundant forage and a limitation of the opportunity for livestock water. Additionally, and contrary to disturbance process theory in ecosystem dynamics, exclusion fencing conveyed the notion that riparian areas and cattle are incompatible. As well, streams, the adjoining riparian zone, and watersheds function as units and are inseparable; exclusion fencing does not allow the opportunity to find the solution to a riparian grazing problem in the adjacent uplands and to manage on a landscape basis.

The Alberta Cows and Fish initiative began as a recognition that resolution of the impasse over riparian areas and their management would be accomplished with a range of solutions, including, but not exclusively, stream-bank fencing. In 1992 six groups and agencies sat around a rancher's kitchen table and established what would become the Cows and Fish program. This partnership between the Alberta Cattle Commission, Trout Unlimited Canada, the Canadian Cattlemen's Association, (then) Alberta Environmental Protection, Alberta Agriculture, Food and Rural Development, and Fisheries and Oceans Canada (and later PFRA) created a synergy of experience, perspective, background, and resources that broadened the approach to riparian issues.

The Cows and Fish program began (and continues) as a different way to engage with people, especially livestock producers, to move beyond suspicion, denial, and conflict to trust, acceptance, and cooperation. Engagement begins with ecological awareness, a nonthreatening, nonconfrontational extension effort to help people understand some of the ecological processes that shape the landscape they live on and make a living from. Part of that critical, initial message is that there are choices and alternatives to current management practices. As the antithesis of the top-down approach, Cows and Fish encourages the formation of local or community teams, composed of technical, producer, and other local interests, to engage with each other to "drive" the process.

The Cows and Fish program assists in the assemblage of technical advice and tools for management changes to provide alternatives to current practices. Information sources include those innovative, progressive, or practical solutions already being used by a limited and select group of landowners. Key tools, part of ecological "literacy," include demonstration sites and riparian health assessment. It is difficult to sell concepts or ideas without tangible products or examples. Demonstration sites are products, examples of changes in grazing management that people can see, review, and reflect on whether these management changes make sense for their own operations. Sites selected for demonstration purposes also represent research opportunities to test and measure riparian response to a particular grazing management option. Since many livestock producers are reluctant to experiment, at their own expense and risk, the development of demonstration sites using capital from elsewhere provides some of the first steps in a community to acceptance of other management ideas.

Riparian health assessment is a useful tool that allows people to critically observe, measure, and assess the status of ecological function on their own property or within their communities. The term "riparian health" is used to mean the ability of a riparian area to perform certain key ecological functions. These functions include sediment trapping, bank building, water storage, aquifer recharge, water filtration, flow energy dissipation, maintenance of biodiversity, and primary production. If these functions are impaired, so too will be the ability to sustain agricultural operations. Health assessment is not just an ecological "measuring stick"; it becomes a communication device to allow people with differing backgrounds and experience to "see" a riparian area and its status through the same set of eyes. Arguments about riparian condition are minimized and a much more productive discussion about how to restore damaged areas can begin. The current status of watersheds within a community can become a catalyst for action based on health assessments and forms a benchmark useful to chart progress, both on individual properties and within watersheds.

The Alberta Cows and Fish program assists in community-based conservation through a process of engagement that creates opportunity to move from conflict to cooperation. Stewardship opportunity is created through a four-stage "process" or "pathway." It begins with ecological awareness, a fundamental building block often skipped in other initiatives. The second step is assisting in the development of teams and partnerships. A network of resource professionals, landowners, and others who value riparian landscapes needs to form to solve issues and problems in a multidisciplinary fashion. Step three is the assemblage of technical advice and tools for management changes to provide options and alternatives to current practices. Part of this step includes the development and use of ecological measuring sticks to assess riparian function or "health." Those measuring sticks allow an objective review of watershed condition, link ecological status to management, help galvanize community action, and provide a monitoring framework for landowners and others. Other tools help communities link biodiversity, economics, and water quality to management actions and alternatives.

(continued)

The last step (although the process steps are often constantly repeated) is critical: it is the transfer of responsibility for action to the community, which is in the best position to make the changes and benefit from them. Riparian (and by association, watershed) actions need to be community based, locally driven, and largely voluntary. To help a community arrive at this point requires knowledge-building, motivation, acknowledgment of problems, and empowerment. The reasons for positive action may result from enhanced awareness, motivated self-interest, concern about legislation, marketing opportunity, or altruism. The net effect will be a return to a landscape that maintains critical ecological function and provides a greater measure of support for agricultural operations. Cows and Fish is about building a cumulative body of knowledge that we all should know, including how riparian systems function and link us, how watersheds work, the vital signs of landscape health, the essentials of how people need to work together, how solutions need to benefit us all, and the kinds of information that will enable us to restore or maintain natural systems and build ecologically resilient communities and economies.

[1] See the complete article on the website for this book. Each year, in conjunction with Canadian Environment Week, individuals and organizations whose actions have a positive impact on Canada's most pressing environmental issues are eligible to receive awards in the following categories: Climate Change, Conservation, Environmental Health, and Environmental Learning. The "Cows and Fish" program won the Gold Award in the Environmental Learning category for 2003 (Canadian Environment Awards, 2003, http://www.canadiangeographic.ca/CEA2003/english/press/).

Although we tend to think of rangeland as grasslands, ecosystems used for grazing in Canada include tallgrass, mixed, and fescue prairie, forested rangelands, as well as wetland, sandhill, salt flats, and valley (coulee) *complexes* (Johnson, 1995). (*Complexes* is the name given by ecologists to abrupt local changes in terrain that contain plant species adapted to local conditions; complexes are perhaps the most important wildlife habitat left on the prairies.) Before European settlement, native rangelands supported hundreds of plant species as well as grazing animals such as bison, elk, deer, and antelope. Rangelands also sustained a network of predators and scavengers including grizzly bear, prairie wolf, coyote, swift fox, eagle, magpie, and crow. Today very little native grassland remains on the Canadian prairies, and the prognosis for survival of the remaining grasslands is poor. As a result, many rangeland management efforts have focused on grasslands conservation and rehabilitation (Trottier, 1992; Johnson, 1995).

The integrity of Canada's rangeland ecosystems (and associated levels of socioeconomic well-being) has been of concern since the early 1900s. Indeed, the United Nations Environment Programme recently classified Canadian prairie soils as low to moderately degraded (in terms of soil degradation, desertification, and damaged riparian habitat) with the potential for severe degradation because our climate is cold and dry and soil recovery processes are slow (Middleton & Thomas, 1997). Rangeland restoration and management efforts require a thorough understanding of how ecosystems function and how consumers (livestock, wildlife, and people) interact with producers and decomposers, sun, climate, water, and soil (see Chapter 3). Given the understanding that rangelands are in a constant state of flux, and that an appropriate degree of disturbance provided by controlled grazing may be essential to maintaining that state, rangeland management efforts increasingly are directed toward ecosystem processes (Bakker, 1998; WallisdeVries, 1998).

In May 2004, in conjunction with the launch of the Nature Conservancy's *Campaign for Conservation: Saving Canada's Natural Masterpieces,* 50 plains bison were released into the Old Man on His Back Prairie and Heritage Conservation Area in southwestern Saskatchewan. Intended as the start of a new herd of this COSEWIC-designated threatened species (fewer than 1000 exist in Canada), the plains bison are an integral part of efforts to re-create a large prairie grassland ecosystem. By studying the effects of natural grazers such as bison on these grasslands, it may be possible to use this case as a model for other areas, not only in terms of large-scale conservation of grasslands, but also to demonstrate that a shift toward sustainable agriculture and eco-tourism will provide new economic opportunities for prairie communities (Semmens, 2004a, b).

Good stewardship of rangelands involves many considerations. Among the most important in managing disturbances are the use of *stocking rates* (the number of animal units grazed per area, depending on range quality) and grazing practices. The Society for Range Management (Holechek, 1993), for example, suggested that a conservative stocking rate of 30 to 40 percent of forage use would facilitate rangeland recovery, maintain adequate food and cover for wildlife, protect soil resources, and give the highest long-term economic returns with the least risk.

Other grazing practices that permit rest periods for plant recovery include rotation grazing, even distribution of grazing, the use of plants at an appropriate time in their life cycles, and prescribed burning (Trottier, 1992). Rehabilitation of riparian habitats, however, requires the exclusion of grazing species. This has proved a difficult issue to resolve, since producers traditionally have relied on streambed access to provide water to livestock. Restricting stream access, providing off-stream watering sites, and placing feed away from streams are measures that help protect riparian habitat (Adams & Fitch, 1995).

Balancing economic and social needs with threshold limits of rangeland ecosystems is a key challenge in rangeland sustainability. Extensive ecosystem knowledge is necessary to achieve this balance. Given the importance of rangeland health in Canada, many private individuals, ENGOs, universities, and government agencies have pooled their resources to fund the required research. Alberta's Cows and Fish program is one example; another is the Prairie Ecosystem Study (PECOS) that involved the universities of Saskatchewan and Regina, Environment Canada, and Agriculture Canada in studying regional sustainability from socioeconomic, health-risk, and land and biota health perspectives. Similarly, the International Institute for Sustainable Development's Great Plains Sustainability Study researched the relationships among economic development, societal needs, and the environment.

GREENHOUSE GASES

Given its dependence on weather and climate, agriculture will be among the sectors most affected by climate change (Manitoba Climate Change Task Force, 2001). Indeed, agriculture acts both as a sink and as a source for several atmospheric greenhouse gases thought to be responsible for climate change. Agricultural activities relate to GHG concentrations in four main ways: (1) soils are an important natural source of and reservoir for carbon; (2) methane is emitted from livestock and liquid manure; (3) nitrous oxide is released from nitrogen fertilizers; and (4) carbon dioxide is released from the burning of fossil fuels in farming activities. In combination, degradation and mismanagement of agricultural lands and deforestation have added between 2 and 9 Gt of carbon to the atmosphere annually (Hengeveld, Bush, & Edwards, 2002). In 2000, the agricultural sector (soils) provided a net CO_2 sink of 300 kt (Environment Canada, 2000a, 2000b). In 1990, soils were a source of 7300 kt of GHGs. Perhaps this change reflects the increase in Canadian farmers' use of zero and **conservation tillage**, as well as the revegetation of abandoned, marginal farmland (see Box Table 6–2; Lal, 2001).

In 2000, the agricultural sector's CO_2 equivalent greenhouse gas emissions totalled 60 Mt and contributed 8.3 per-

Photo 6–9
Given the risks of wind and water erosion of prairie lands, many farmers employ conservation tillage practices, including the use of implements such as the seed wheel.

cent of Canada's total GHGs. Agriculture accounted for 70 percent of Canada's total emissions of NO_2 (36 000 kt) and 25 percent of CH_4 emissions (24 000 kt). Carbon dioxide emissions from soils contributed 55 percent of agriculture's emissions (33.4 Mt), while enteric fermentation emissions from domestic animals accounted for 29 percent (17.7 Mt), and manure management contributed 16 percent (9.4 Mt) (Environment Canada, 2002a).

Although fertilizer use has remained relatively constant since 1985, the amount of nitrogen (N) in the total fertilizer mix increased from about 10 percent in 1960 to about 30 percent in 1985 (Government of Canada, 1994). The increased proportion of N in fertilizers is necessary to accommodate higher-yielding crops that require more N than is available in most Canadian soils. In addition to emissions from use of nitrogen fertilizers, nitrous oxides are generated from nutrient cycling in agricultural soils. The issue of atmospheric N generated from agricultural activities is receiving considerable attention, in part because losses of N from manure and fertilizers not only impact the environment but also represent potentially serious sources of economic loss for farmers (Chambers et al., 2001).

Since cultivation began, estimates are that Canada has lost between 25 and 35 percent of its total agricultural soil carbon (Smith et al., 2001). Even though it is believed that the carbon content of Canada's cultivated soil is at equilibrium (because losses during the first few years are most rapid), small fluxes of carbon into or out of soils can translate into large quantities of CO_2 when totalled across Canada (Smith et al., 2001).

Soil carbon fluxes influence the overall greenhouse gas balance for agriculture. This means that an individual farmer's actions to retain and increase levels of organic matter in soils can assist the soil to sequester carbon (because soil carbon is stored in soil organic matter) and offset CO_2 emissions from agriculture. Increasing conservation tillage and zero tillage are among the practices that are useful from the perspective of both climate change and soil quality (see Box 6–5). Additional agricultural practices that may help reduce greenhouse gas emissions include green manures (crops such as legumes grown specifically to be plowed into the soil), reduced summer-fallow area, increased forage production, improved crop yields, reduced methane emissions from farm animals (through improved feed additives and feeding technology), improved efficiency of manure use, decreased fossil fuel use, and increased use of renewable fuels such as ethanol.

In general, as the climate changes, and appropriate policy frameworks, economic incentives, and markets are established, there could be economic opportunities for Canadian (and other) farmers to plant alternative crops and to use their lands in ways that would benefit carbon sequestration. Agricultural practices such as zero tillage and agro-forestry could enhance the absorption of CO_2 from the atmosphere into soils and forests, and help in development of related agro-industries such as ethanol production (Manitoba Climate Change Task Force, 2001).

ENERGY USE

Many of today's farmers have adopted an industrial approach to agriculture: high production levels are achieved through large inputs of industrial products, including energy (Taylor, 1994; Boyd, 2003). Agricultural activities consume energy directly through the processes of tilling, harvesting, heating, and ventilation. In 1990 the pattern of agricultural energy use was 28 percent in primary production, 22 percent in processing and packaging, 18 percent in distribution, and 32 percent for storage and preparation. Fuels for transportation account for over 50 percent of the energy used in primary production on the farm, while a further 25 percent is accounted for by production and distribution of fertilizers. Approximately 3 percent of Canada's total energy consumption is used on farms to support primary agricultural production (Government of Canada, 1996). In 1996, for instance, total energy used for Prairie agriculture was 171 090 peta-joules (Agriculture and Agri-Food Canada, 2001).

A major environmental concern regarding energy use in agriculture is consumption of fossil fuels and the resultant emission of greenhouse gases. If production and use of renewable fuels such as ethanol could be expanded, environmental benefits could include lower net carbon dioxide emissions. Research at Agriculture and Agri-Food Canada in Sainte-Foy, Quebec, has been conducted on plants such as the Jerusalem artichoke, which yields up to 18 tonnes per hectare of ethanol-producing biomass. Other practices, including reduced tillage, new herbicides with lower application rates, and genetic improvements in plants such as lower fertilizer needs, point the way toward reduction in energy use and increased sustainability of agroecosystems.

RESPONSES TO ENVIRONMENTAL IMPACTS AND CHANGE

Agricultural sustainability depends on the integration of economic, social, and environmental concerns. One way in which Agriculture and Agri-Food Canada has promoted sustainability has been through development of market opportunities for Canadian agricultural and agri-food products. To that end, in 1993, Canada's agricultural industry and the federal and provincial governments set a goal of achieving $20 billion in agri-food exports by the year 2000. World trade in unprocessed grains (such as wheat), oilseeds (mostly canola), meat and meat products, and live animals enabled the agriculture sector to reach $20.7 billion in exports during 2000 and $25.88 billion during 2002. Sales to the United States comprised 67 percent of this trade, followed by 9 percent to Japan, and 4.5 percent to the European Union (Statistics Canada, 2003d).

Some people might criticize this approach for its apparent emphasis on sustainable economic growth rather than environmental sustainability. This example highlights the different views people may bring to the quest to achieve sustainability of agricultural lands. Different perspectives often generate conflict; resolution of the conflict requires responses that incorporate understanding about the cultural, economic, and ecological roles of land.

Throughout the world, many people believe that an efficient food production system involves growing crops where costs are lowest (often in developing countries) and shipping the food to markets around the world. The trend toward globalization means that food in our stores may travel an average of 2000 kilometres to get here (Olson, 1997). This model of food production is supported by international agreements such as the North American Free Trade Agreement (NAFTA) and the General Agreement on Tariffs and Trade (GATT). Some environmentalists have warned that GATT and similar negotiations regarding trade deregulation in agricultural products would spell the end of many of the world's small- and medium-scale farming operations that practise sustainable agriculture. Box 6–6 provides a brief background to the ongoing international negotiations regarding subsidies and the reform of agricultural trade rules and agreements.

BOX 6-5
CONSERVATION TILLAGE AND ZERO TILLAGE

During the past decade or so, producers, industry, and government have made concerted efforts to reduce the extent of wind and water erosion on Canada's agricultural lands through the use of soil conservation practices. An appropriate mixture of land management practices tailored to the conditions and needs of individual farms can provide multiple benefits for preserving soil health and productivity, minimizing water contamination, conserving wildlife habitat, and maintaining farm nutrient balances.

While not all farms require erosion control and some practices are applicable in some areas and not others, examples of management practices include growing forage crops in rotations or as permanent cover, growing winter cover crops, planting shelter belts, strip cropping, using buffer strips, and using conservation tillage techniques and contour cultivation.

As shown in Box Table 6–2, agricultural practices can have important effects on changes in soil carbon content.

that using conservation farming practices improved farmers' incomes by $6.42 per hectare per year in Alberta and by $32.78 per hectare per year in Manitoba (note that this figure included several additional conservation practices). "These systems appear to be the most cost-effective soil practice for general use across the country" (Government of Canada, 1996, pp. 11–20).

- *Conventional tillage:* most of the crop residue (plant material remaining after harvest) is incorporated into the soil.
- *Conservation tillage:* most of the crop residue is left on the soil surface to provide protection against erosion, reduce soil crusting, and increase the organic matter content of soils; also known as mulch tillage, minimum tillage, and reduced tillage.
- *Zero tillage:* any system where soil is not disturbed between harvesting one crop and planting the next; includes direct seeding into stubble or sod; also known as no tillage.

BOX TABLE 6–2
ESTIMATED CHANGES IN SOIL CARBON ASSOCIATED WITH AGRICULTURAL PRACTICES IN CANADA

Parameter	Change in Carbon (C)
Convert arable land to permanent cover	Sequestration: + 0.62 mg/ha/yr
Include forages in crop rotations	Sequestration: + 0.44 mg/ha/yr
Convert from conventional to zero-tillage practices	Increase in soil C = 0.13 mg/ha/yr
Reduce summerfallow to 1 in 3 years (from 1 in 2 years)	Loss reduction of 0.03 mg/ha/yr
Improve fertilizer use efficiency (by 50 percent)	Sequestration: + 0.04 mg/ha/yr

SOURCE: From "Estimated Changes in Soil Carbon Associated with Agricultural Practices in Canada," W. N. Smith, R. L. Desjardins, & B. Grant, 2001, *Canadian Journal of Soil Science, 81,* 221–227.

In 1991, conservation tillage, including zero tillage, was used on 31 percent of the land seeded; by 2001, over 60 percent of farmers practised conservation and zero tillage. Zero tillage leaves a portion or all of the crop residue on the soil surface and special equipment sows seeds for the new crop through the standing stubble from the previous crop (see Photo 6–10). This practice provides protection against erosion, reduces soil crusting, helps retain moisture by trapping snow, allows rain and snowmelt water to soak directly into the ground, and increases soil organic matter content. In addition, continuous ground cover provides better habitat for ground-nesting birds and a wide array of other wildlife. Studies in Alberta and Manitoba demonstrated

Photo 6–10
Ducks Unlimited agrologist Lee Moats, a zero till farmer, inspects a winter wheat field he seeded in August. He displays viable plants ready to spring to life in April. The old straw cover and the new wheat plants provide good duck nesting cover.

(continued)

Since the 1970s, Ducks Unlimited, a private, nonprofit charitable organization dedicated to the conservation of wetlands for the benefits of North America's waterfowl, wildlife, and people, has worked in partnership with farmers to assist them in conserving their soil and water resources while improving the environment for wildlife and people. By 1995, Ducks Unlimited had invested $1.6 million in research on and demonstration of zero tillage. Increased funding through Prairie CARE (Conservation of Agriculture, Resources and Environment), a major component of the North American Waterfowl Management Plan, allowed Ducks Unlimited to expand its role in zero tillage in all three Prairie provinces as well as in Ontario and British Columbia.

One Saskatchewan farmer began to use zero tillage in 1985 because he saw the depletion of his soil resources and wanted his farm to remain viable for his children. "Zero till addresses the long-term sustainability of the land and my family, as well as providing a better home for wildlife" (Lyseng, 1995, p. 15). An important benefit of zero tillage is that fall-seeded crops are more likely to survive over winter with the protection of straw and crop residue. For example, this farmer would plant fall rye or winter wheat following a harvest and watch the new crop grow for a month or two until freeze-up. With the spring melt the crop would be waiting for returning ducks. The cover from fall-seeded crops means ground-nesting birds of all species have a better chance to avoid predation. The absence of spring field operations means nest successes are higher than they are when conventional farming methods are used (Lyseng, 1995). Fewer compaction problems result as well.

Ontario Land CARE is another Ducks Unlimited program. One element of this program is the establishment of conserva-

tion tillage clubs in which Ducks Unlimited financially assists groups of farmers to purchase and share zero-till equipment. Conservation tillage clubs are designed to be demonstration sites for neighbouring farmers interested in learning about the practice (Kinkel & Werner, 1997).

Five farm operators from Meaford, Ontario, formed the Bighead Conservation Tillage Club. One of the members, who carries on a tradition of conservation started by his father, has added the zero-till drill to his arsenal of management techniques for growing corn silage crops. In addition to his revised tillage practices, this farmer has gained a new mindset on farm planning and land use on his property. The club agreement resulted in the securement of a 20-acre wetland on his property; the adjacent idled and no-till fields provide habitat for waterfowl and other wildlife (Kinkel & Werner, 1997).

The Bighead Club president indicated that some of his land suffered severe erosion as a result of intense row cropping. Now, having taken many years of careful management to restore the land to its full cropping potential, and with use of the no-till drill and planned crop rotation, the farmer finds that some of his fields do not need to be plowed for seven years. Such savings in time, soil, and money are lessons that many farmers in the erosion-prone area are interested in learning. By demonstrating conservation and zero-till techniques, members of the Bighead Conservation Tillage Club may impact the agricultural operations of the surrounding farming community in significant ways. At the same time, the aims of Ducks Unlimited to ensure quality wetland and wildlife habitat are being met.

SOURCES: *The State of Canada's Environment—1996,* Government of Canada, 1996, Ottawa: Supply and Services Canada; "Why Zero Till?" by R. Lyseng, 1995, *Conservator, 16*(1), 14; "Conservation Tillage … Spread the Word!" by N. Kinkel & K. Werner, 1997, *Conservator, 18*(1), 24.

The drive for efficiency does not always consider social and environmental effects, especially in developing nations. For instance, when officials in developing countries emphasize *cash crops* (crops grown for export) over food production for local people, and when people do not have access to land to grow their own food or have enough money to buy it, one result is increasing migration to cities. Few jobs are available in the cities, so poverty increases (and poor people continue to be exploited as cheap sources of labour). Also in the name of efficiency, pesticides that are banned in Canada (and elsewhere) may be used in developing nations. Other environmental degradation occurs due to nonexistent or weak legislation and regulations, and lack of enforcement.

Alternatives to the globalization of the current food system include *regional sustainability*. In this system, developing countries would be encouraged to grow food for themselves first and then crops for export. Such a

shift, however, would require either debt reduction or debt forgiveness by the developed world (Olson, 1997). Canada already provides hundreds of millions of dollars worth of food annually to countries in need; a reduction in food aid would enable Canada to shift its emphasis toward debt reduction of developing countries. Reduced debt would enable developing nations to move toward regional sustainability and decrease their reliance on developed nations.

The following section briefly considers some examples of the international initiatives that have been taken toward agricultural sustainability.

INTERNATIONAL INITIATIVES

In 1992, Agenda 21 (see Box 1–1) noted that world food production must more than double in the next 40 years to

BOX 6–6
THE WORLD TRADE ORGANIZATION AND AGRICULTURAL REFORM

The General Agreement on Tariffs and Trade (GATT) was drawn up by 23 countries and came into force in January 1948 as an international forum to encourage free trade between member states through the regulation and reduction of tariffs on traded goods. Up to 1994, the contracting parties to the GATT struggled through eight "rounds" of negotiations in efforts to reduce tariffs and produce rules to govern international trade. Typically, however, as tariffs were reduced, other nontariff barriers to trade were established. This meant that, through export and import subsidies, more developed countries (MDCs) were able to provide a protectionist advantage to their agricultural sector at the expense of less developed countries (LDCs). Agricultural trade issues are critical for LDCs since most of their GDP is connected with the agricultural sector, and the subsidies imposed by MDCs frequently prevent access to their markets by LDCs.

The most recent and most comprehensive round of negotiations, the Uruguay Round, lasted seven years (1986–1994) and established the World Trade Organization (WTO) to replace the provisional GATT. The 1994 GATT agreements included an Agriculture Agreement that, with pressure from the Cairns Group (see sidebar), established agriculture as a sector requiring trade liberalization.

The Uruguay Round of negotiations established a schedule for subsidy reduction, as did the Doha Agreement, a declaration from the Fourth WTO Ministerial Conference held in Doha, Qatar, in November 2001. The Doha Agreement also included a mandate for negotiations on agricultural trade and was geared to help poor countries—the World Bank estimated that successful negotiations could raise global income by more than $500 billion per year by 2015, with 60 percent of that gain going to poor countries.

A major purpose of the Fifth WTO Ministerial Conference was to report on the progress of the Doha Agreement. Canada and its Cairns Group partners went to the September 2003 meetings in Cancun, Mexico, seeking ambitious reforms in agricultural trade (including cessation of subsidies such as the $300 billion provided by the U.S. and the European Union to their farmers). However, talks among the 148 members of the WTO collapsed on the fourth day of the Fifth Ministerial Conference. Agriculture appeared to be the critical, divisive issue: developing countries were seeking real market-opening concessions that would require developed nations to make significant adjustments (that is, incur domestic costs) in order to achieve a "greater good" for the global economic system. In seeking genuine free trade,

The Cairns Group

The Cairns Group consists of 17 nations (including Canada) that together account for one-third of the world's agricultural exports. Formed in 1986, the group pushed for fair trade in agricultural exports and largely was responsible for reform in agricultural trade being established in the Uruguay Round (in the Agreement on Agriculture). By acting collectively, the Cairns Group has had more influence and impact on the WTO agriculture negotiations than any one individual country could have had by acting independently.

The group members seek three key reforms: (1) deep cuts to all tariffs and removal of tariff escalation, (2) elimination of all trade distortions caused by domestic subsidies, and (3) elimination of export subsidies. In addition, the Cairns Group supports the principle of special and differential treatment for developing countries. Cairns Group ministers want the WTO's framework for agricultural liberalization to support the economic and technical assistance needs of developing and small country members. The Cairns Group also is committed to achieving a fair and market-oriented agricultural trading system that places trade in agricultural goods on the same basis as trade in other goods.

Cairns member groups are Argentina, Australia, Bolivia, Brazil, Canada, Chile, Colombia, Costa Rica, Guatemala, Indonesia, Malaysia, New Zealand, Paraguay, the Philippines, South Africa, Thailand, and Uruguay.

SOURCES: *An Introduction,* Cairns Group, (n.d.), http://www.cairnsgroup.org/introduction.html

developing nations pursued removal of wealthy nations' protectionist barriers for domestic agriculture (and other items such as the pharmaceutical patent monopolies that result in "sky-high" prices for lifesaving medicines). As one commentator (Greider, 2003, p. 12) noted, "profiles in courage [were] not on the agenda at Cancun," and leaders from wealthier nations failed to offer compromises on which the trade talks depended. With the breakdown of talks at Cancun, reform of the agricultural trading system was delayed once again; rather than increasing their production, LDCs continue to face existing subsidies.

SOURCES: "Minister Vanclief in Cancun to Fight for Canada's Agriculture Sector," Agriculture and Agri-Food Canada, September 9, 2003, http://www.agr.gc.ca/cb/index_e.php?s1=n&s2=2003&page=n30909a; "$600 Million in Federal Transition Funding to be Delivered Directly to Producers," Agriculture and Agri-foods Canada, September 19, 2003, http://www.agr.gc.ca/cb/index_e.php?s1=n&s2=index&page=2003_09; *General Agreement on Tariffs and Trade,* CIESIN, (n.d.), Columbia University, http://www. http://www.economist.com/finance/PrinterFriendly.cfm?Story_ID=2071855; "The Real Cancun: WTO Heads Nowhere," W. Greider, September 22, 2003, *The Nation,* 11–17; Trade and Investment-World Trade Organization, International Institute for Sustainable Development, (n.d.), http://www.iisd.org/trade/wto/gatt.htm; "Overnight-Rich and Poor Square Off over Farm Trade in Cancun," A. Wheatley & R. Waddington, September 12, 2003, Agriculture Online, http://www.agriculture.com/worldwide/IDS/2003-09-12T063421Z_01; *Agricultural Trade: Backgrounder,* World Trade Organization, 2003a, http://www.wto.org/english/docs_e/legal_e/ursum_e.htm#aAgreement; *Negotiations, Implementation and Development: The Doha Agenda,* World Trade Organization, 2003b, http://www.wto.org/english/tratop_e/dda_e/dda_e.htm

meet the needs of a growing population, more than 80 percent of whom will live in the developing world. A key challenge then, and now, is to increase agricultural production without further degrading the environment. In 2002, the Johannesburg Summit (Earth Summit + 10) focused on the many practical steps necessary to address Earth's pressing problems of poverty and environmental degradation. Instead of producing only outcome documents (as in previous Earth Summits), delegates to the Johannesburg Summit launched over 300 voluntary partnerships among governments, NGOs, ENGOs, intergovernmental organizations, and the private sector. Tied to government responsibilities and commitments to improve implementation efforts, these partnerships are intended to ensure that established targets are met, including improving agricultural yields, expanding access to clean water and sanitation, and managing toxic chemicals (Food and Agriculture Organization, 2002; United Nations, 2002a, 2002b).

For many years, Canada's International Development Research Centre (IDRC) has promoted sustainable agriculture projects in various countries around the world. IDRC's projects have included research on indigenous knowledge systems and farming systems designed to maximize use of the marginal lands that many small-scale farmers in developing countries are obliged to cultivate, while doing the least environmental damage. IDRC has investigated the use of alternative farming practices such as integrated pest management, which reduces the need for costly chemical fertilizers (see Box 6–7 for a brief discussion of rice–fish farming in Indonesia), and agroforestry, which incorporates the use of trees for multiple purposes such as forage, firewood, windbreaks, and soil enrichment.

In many developing nations, women bear the major burden in agriculture and food production. They have extensive knowledge of local ecosystems and can help in conserving biodiversity and protecting the environment if given an opportunity to be involved equally in decision making relating to sustainable agriculture (Jowkar, 1994; Seck, 1994). See Enviro-Focus 6 for a brief discussion of the role of women in agriculture in Africa.

CANADIAN EFFORTS TO ACHIEVE SUSTAINABLE AGRICULTURE

Chapter 10 of Agenda 21 presents an integrated approach to land planning and management designed to lead toward sustainability. Based on Agenda 21, Canada identified several priorities as being relevant to our domestic land issues. These priorities emphasized provincial land use functions, namely: increased use of information systems; strengthening of federal, provincial, and territorial relations; consultations and partnerships; support of Aboriginal land-use initiatives; coordinated state-of-the-environment monitoring; and application of an ecosystem approach to land use planning and management (Government of Canada, 1996).

In practice, achieving sustainable agricultural food production systems in developed countries such as Canada is a complicated process that involves a range of socioeconomic and environmental challenges resulting from changes that industrialization and technology have brought to agriculture. Approaches to sustainable agriculture in Canada involve cooperative efforts of governments, industry, ENGOs, and farmers in developing new agricultural policies and practices.

Sustainable agriculture policies and practices need to ensure that (1) environmental quality is maintained or enhanced; (2) individuals and companies engaged in food production are rewarded adequately, both economically and socially; and (3) an adequate, accessible, and safe food supply is assured. Additionally, the Canadian agricultural industry needs to confront the possibility of adapting to potential future climate changes.

To help move Canada toward sustainability in agriculture (among other sectors), the federal government created the 5NR partnership. Since 1995, five federal departments—Agriculture and Agri-Food Canada, Environment Canada, Fisheries and Oceans Canada, Health Canada, and Natural Resources Canada—have focused their scientific expertise and technology on sustainability in our natural resources sectors. These departments have collaborated in key research areas: in agriculture, for example, they have established a national research program to assess endocrine disrupting substances (Government of Canada, 2001). The importance of science and technological innovation to achievement of sustainability lies, in part, in the generation of new or additional information that enables Canadians to make sound decisions about their effects on the environment.

Provinces also are active in efforts to attain sustainable agriculture. The following section highlights the efforts of Prince Edward Island's provincial government to meet the challenges inherent in sustainable agriculture, particularly those actions undertaken in an "institutional" context. The following sections then identify nontraditional or alternative agricultural activities, such as organic farming and game ranching, that may help promote sustainability through "individual" actions.

Examples of Prince Edward Island's Actions toward Sustainable Agriculture

Prince Edward Island is Canada's most densely settled province; 90 percent of the land is privately owned and pristine habitat is extremely limited. In the last decade, an additional 6 percent of P.E.I.'s forest has been converted to agriculture use, primarily potato production, P.E.I.'s

BOX 6 – 7
SUSTAINABLE AGRICULTURE IN INDONESIA: RICE–FISH FARMING

An excellent example of how food production can be accomplished with environmentally beneficial results can be found in a rice–fish farming project in Indonesia. Rice-growing and fish culture, carried out in the same field, demonstrate an integrated approach to farming that will help sustain productivity into the future.

Asian farmers have been raising fish in their rice fields for over 2000 years, providing farm families with important sources of carbohydrates and animal protein at the same time. Originally, wild fish bred naturally in the fields and were harvested whenever possible. Fish husbandry techniques evolved over time in many countries. But the advent of high-yielding rice varieties changed all that, because they demanded high inputs of pesticides and herbicides that were toxic to the fish. As a result many farm families were deprived of an important source of nutrition.

As the need for more efficient and sustainable farming systems increased, rice–fish farming has enjoyed a renewed interest in almost all Asian countries. A project that began in 1987 with the Sukamandi Research Institute for Food Crops and the Indonesian Research Institute for Freshwater Fisheries aimed to raise the visibility of rice–fish farming to Indonesian farmers and government policymakers.

In this project, farmers dig a small pond or trench surrounded by a protective bank of soil in a low-lying area of the rice field and introduce small fish fingerlings—carp, tilapia, catfish, or other species—into it. When the field is flooded, rice is planted as usual. The fish are let out of their pond and allowed to forage through the rice field.

What happens is that rice and fish benefit each other. Rice plants are protected because the fish feed on the pests of the rice—insects such as leafhoppers, stem borers, and aphids, plus possibly other invertebrates such as crabs and snail larvae. Fish also recycle nutrients through feeding and depositing feces in the

Photo 6–11

submerged soil, which fertilize the rice. Initial research has indicated that uptake by the rice plants of important nutrients such as phosphorus and nitrogen is significantly improved with fish in the fields.

Dr. Achmad Fagi, leader of the Indonesian project, found that "rice–fish culture with common carp actually increased the yields of commonly used rice varities." Farmers can improve family nutrition by eating fish, or they can increase their cash income by selling their fish.

Pesticides, herbicides, and, to a lesser degree, chemical fertilizers, continue to create problems for rice–fish farmers, but the Indonesian government has recognized the problem by reducing many of the import subsidies on pesticides.

SOURCE: *Agenda 21: Green Paths to the Future,* International Development Research Centre (IDRC), 1993, Ottawa: IRDC, p. 29. Reprinted with permission.

---

major crop. Potatoes have been grown successfully on P.E.I.'s relatively infertile soils through the use of agrochemicals, including fertilizers and pesticides. Unfortunately, continuous row cropping of this economically important crop has resulted in several environmental problems that the government of Prince Edward Island is trying to address to ensure long-term environmental, social, and economic sustainability.

Conversion of forest land to potato production, in addition to existing intensive crop production, has resulted in the loss of soil organic matter, making soil more prone to wind and water erosion. When Island soil, laden with nutrients and pesticides, is blown or washed

into surface waters, estuaries and ponds are silted in, fish habitat is destroyed, and fish are killed (eutrophication). Since 1994 there have been 21 suspected pesticide-related fish kills in Island rivers (Prince Edward Island Department of Fisheries, Aquaculture and Environment, 2003). Ground water quality has also been affected negatively. Over the last 15 years some surface water nitrate levels have doubled and are now higher than the "safe" level recommended in the Water Quality Guidelines established by Health Canada. These high nitrate levels are attributed to the surplus nitrogen fertilizer applied to potato crops that, subsequently, leaches through the soil profile and contaminates ground water aquifers. Some

Women and Sustainable Agriculture in Africa

Shimwaayi Muntemba, the executive director of the Environment Liaison Centre (ELC) in Nairobi, Kenya, says the search for strategies to halt environmental degradation and introduce sustainable development into Africa must begin by recognizing and legitimizing women's knowledge.

Muntemba, who coordinated much of the research of the Brundtland Commission on food security, agriculture, environment and women, works with WEDNET, a multinational and multidisciplinary project on women and natural resource management in Africa.

WEDNET's main purpose is to strengthen the role of indigenous knowledge in international development. It has involved research by women in Senegal, Burkina Faso, Mali, Ghana, Nigeria, Tanzania, and Zambia, most of it concentrated on activities such as management and conservation of livestock, water, harvesting, soil, food security, nutrition, health, and technology.

A computerized information-sharing network was established between ELC and its Canadian counterpart at York University. Muntemba hopes that the system will allow African researchers to share their knowledge with Canadians. She notes that if this had happened in the colonial past, the story of Zambia's agricultural development would have been different.

For many years, she says, the farmers of Zambia logged trees, burned the branches, and used the ash as fertilizer for the soil. That system of soil conservation was known as "citemene," and it symbolized the effective, indigenous use of soil by the African people.

"Crops rarely failed in this part of the country," Muntemba says. "Land could be used for five years before being left to rest."

But when colonial farmers came they dismissed the citemene method as backward and destructive. They promoted the use of chemical fertilizers, which acidified the soil.

"Now they have left, we must try and regenerate the soil," says Muntemba. She says that tropical soils are very fragile and require a variety of agricultural techniques.

Photo 6–12

Muntemba uses WEDNET and her position at ELC to heighten the awareness of women's indigenous knowledge in Africa. "We have come a long way since the 1970s when women began to be discussed as central to agriculture," she says. "Now there is an actual appreciation of the fact that women's economic and agricultural activities are located within the context of environmental sustainability."

The United Nations Environment Programme (UNEP) now has documented hundreds of successful attempts at controlling and remediating desertified and degraded dry lands. Descriptions of African success stories, such as the Zabré women's agro-ecological project in Burkina Faso (that achieved community land reclamation partly through empowerment of women at political, financial, and sociocultural levels), are published by the United Nations Environment Programme/International Fund for Agricultural Development.

SOURCES: *Agenda 21: Green Paths to the Future,* International Development Research Centre (IDRC), 1993, Ottawa: Author, p. 29. Reprinted by permission; *UNEP Programme on Success Stories in Land Degradation/Desertification Control,* United Nations Environment Programme/International Fund for Agricultural Development, (n.d.), http://www.unep.org/unep/envpolimp/techcoop/1.htm; *The Zabré Women's Agro-Ecological Project in Burkina Faso,* United Nations Environment Programme, 1998, http://www.unep.org/unep/envpolimp/techcoop/4.htm

ground water has nitrate levels in excess of the 10 mg/L water quality guidelines as well.

Through public engagement and partnerships with stakeholders and other levels of government, the provincial government has been trying actively to reduce the negative environmental effects of potato cropping. New and amended legislation has been employed to enhance the success of various initiatives. Education and communication with stakeholders (to gain their support and compliance) has been an important activity also. In 2003, 66 percent of P.E.I.'s farmland was managed under an environmental farm plan that supported soil stability and integrated pest management. Twelve million dollars in funding was available to farmers who implemented the plan. In 2002 the provincial government enacted the Agriculture Crop Rotation Act, the first of its kind in Canada, to require a three-year crop rotation to improve soil management. The Environmental Protection Act (EPA) was amended to provide riparian buffer zones to help reduce erosion and agrochemical contamination of waterways. The EPA permits substantial fines for noncompliance. A Drinking Water Strategy was developed that aimed to protect ground water at its source and to improve the delivery, monitoring, and reporting of water quality. Pesticide use reduction is being achieved through implementation of the mandatory riparian buffer zones and environmental farm plans as well as through cooperation with the government of New Brunswick, NGOs, the potato industry, and Health Canada.

P.E.I.'s Environment Ministry recognizes the need for public support for these new initiatives and views communication and cooperation as key components to success. "A lot of effort has been put into public and stakeholder education and providing readily understandable information" so that the public is well informed and has the information necessary to comply with the new regulations and legislation. When asked to evaluate the success of these initiatives, the ministry responded that where programs were successful "there has generally been a lot of co-operation amongst provincial government departments as well as co-operation with Environment Canada and the Department of Oceans and Fisheries" (Prince Edward Island Department of Fisheries, Aquaculture and Environment, 2003).

Nontraditional Agricultural Activities

In Canada, efforts to sustain agroecosystems have focused historically on stewardship of land and soil resources on individual farms. Individually, farmers have been searching for new ways to diversify and to achieve sustainability within their agroecosystems. Nontraditional agricultural activities have been gaining in popularity, including **organic farming** (reliance on a management system using natural soil-forming processes and crop rotation schemes rather than synthetic inputs), **alternative livestock** production (raising non-native species and domesticated native species), and **agroforestry** (combining production of trees, shrubs, agricultural plants, and/or animals in the same land area). Agricultural biotechnology efforts have resulted in biofertilizers, biofeeds, and plants with novel traits. Organic farming, game farming and ranching, and biotechnology activities are discussed briefly below.

Organic Farming Founded in 1975, Canadian Organic Growers (COG) is a national information network for organic farmers, gardeners, and consumers. Their objectives include conducting research into alternatives to traditional chemical- and energy-intensive food production practices, and endorsing practices that promote and maintain long-term soil fertility, reduce fossil fuel use, reduce pollution, recycle waste, and conserve nonrenewable resources. In addition, COG assists in educational and demonstration projects to help people understand the value and integrity of organic foods.

Organic farming is one way to promote the goals of a decentralized, bioregionally based food system that sees food produced by local farmers and consumed by local people. Reducing transportation costs, bolstering local marketing systems and economies, and promoting greater regional food self-reliance are other benefits of consuming organic (and other) foods within the region where they are produced.

The 2001 Census of Agriculture revealed that 2230 farms produced at least one category of certified organic agricultural products. Most certified organic farms (773) were located in Saskatchewan, with Ontario (405), Quebec (372), and British Columbia (319) following (Statistics Canada, 2002a). Field crops such as buckwheat, rye, and caraway comprise the principal organic crops in Ontario, Manitoba, Saskatchewan, and Alberta. Organic vegetable, fruit, and greenhouse products were most common on about 75 percent of organic farms in Atlantic Canada and British Columbia. About 40 percent of Quebec's organic production is maple syrup.

In June 1999, the government of Canada unveiled a new National Standard of Canada for Organic Agriculture. This standard outlines principles for organic agriculture that endorse production and management practices that contribute to the quality and sustainability of the environment and ensure ethical treatment of livestock. Among its provisions, the standard (1) prohibits both the use of ionizing radiation in the preservation of food and the use of genetically engineered or modified organisms; (2) promotes maximum use of recycling; and (3) encourages maximum rotation of crops and promotion of biodiversity. This standard was designed to help Canadian producers of organic foods gain greater and easier access to international markets that demand these kinds of standards (Government of Canada, 1999).

Canadian consumers could help agriculture achieve sustainability by buying organically grown or raised products. Just as in Europe, where demanding consumers increasingly have supported organic farmers and assisted in ensuring agriculture is both sustainable and competitive, Canadian consumers could help Agriculture and Agri-Food Canada to recognize organic agriculture as a viable and competitive approach to sustainable farming (Ecological Agriculture Projects, 1997).

Community shared agriculture (CSA) farms are a related development in which local people share with farmers the risk of organic food production by buying a share in the produce prior to the growing season. This enhances farmers' security in knowing that their entire harvest is sold ahead of time and reduces their debt load since they do not need a credit margin in the spring. CSA farmers also help ensure biological diversity since most grow more than 30 varieties to satisfy the needs of their sharers (Hunter, 1999). Sometimes, depending on the individual farmer, a share in the produce involves a commitment on the part of the shareholder to work in the CSA garden for a day or more. CSA farms promote local production and consumption with associated reduction in transportation needs, thus contributing in small ways toward reduced air quality problems. CSA farms also provide opportunities for urban residents to "get their hands dirty" and perhaps help people appreciate directly the value of agricultural land. (In Chapter 13 we consider urban agriculture and some of the benefits it brings to people and the planet.)

In Montevideo, Uruguay, a group of young farmers from three organic farms started Las Canastas (The

Photo 6–14
Wild elk are the basis of initial herds for game ranching and captive breeding purposes.

Baskets) in 1997. Introduced to the CSA concept by members of Lifecycles, a food security organization in Victoria, British Columbia, these Uruguayan farmers provide 70 households with "an ecologically healthy and socially just alternative to the foreign-owned chain-supermarket" (Galeano, 1999, p. 24). They also help preserve natural resources and create opportunities for small farmers in the hinterland around urban areas. However, since few individuals in Uruguay have bank accounts or savings, the concept of the consumer sharing in the risks of production is not viable. Instead, consumers pay monthly, an arrangement that still lets farmers make a living wage and customers receive their vegetables at affordable prices.

Game Farming and Ranching One type of alternative livestock production is the raising of game species. Farmers and ranchers have discovered that the pleasant-tasting, low-fat meat of the North American elk, or wapiti, makes it an attractive ranching species. The elk's velvet (nonhardened antler) is an annual crop and is valued highly in the marketplace. Although elk have been part of the ranching scene for 30 to 40 years, the industry has grown greatly since the mid-1980s. In 1990, 35 prominent elk ranchers formed the North American Elk Breeders Association (NAEBA) to promote elk ranching as an agricultural pursuit. In 1999, more than 1700 NAEBA members (including Canadians) were farming or ranching about 150 000 to 160 000 elk worldwide.

Capturing elk from the wild is illegal, and reputable elk ranchers do not take part in these activities. However, the first captive elk herd in any area is based on wild animals. In Manitoba, in 1995, the province's natural resource officers used elk baiting, particularly around Riding

Photo 6–13
The opportunity to socialize with the farmer and other shareholders is just one of the benefits of community shared agriculture farming.

Mountain National Park, to attract and then capture wild elk (Chambers, n.d.). These captured elk formed the basis of a new provincial game ranching industry.

Opposition was expressed about wild animals being kept in captivity, about the threat of the spread of disease among the unique subspecies of elk in the area, and about poaching. Like many other provinces, Manitoba already had difficulties with poaching and the trade in animal parts. Harvesting of antler velvet (called *velveting*) "to sell to the lucrative Asian folk medicine market" (Chambers, n.d.) was of great concern as a stimulus to increased poaching. According to the NAEBA, annual revenues from velveting just one mature bull elk were US$1495, and profits from velvet supplies typically would pay for feeding the entire herd year-round (Elk On Line, 1996). In Alberta in 2000, the elk herd numbered about 30 000 and was valued at $101 million, while the antlers and velvet supported a $52.5 million industry (Derworiz, 2000).

Chronic Wasting Disease (CWD) is a fatal degenerative disease of the brain that affects elk and deer. CWD is one of a group of related diseases known as transmissible spongiform encephalopathies (TSEs) that include bovine spongiform encephalopathy (BSE) or mad cow disease in cattle and Creutzfeldt-Jakob disease (CJD) in humans (see Box 6–1). While CWD is not the same as BSE, both are said to be caused by prions, abnormal proteins that accumulate in the brain. CWD has no known cure. Following the importation of a diseased elk from South Dakota in 1989, elk have tested positive for CWD on 40 farms in Saskatchewan (Alberta AFRD, 2003; Thomas, 2003).

Canada's Health of Animals Act identifies CWD as a reportable disease that falls under the Canadian Food Inspection Agency (CFIA). Since the present federal policy is to eradicate CWD from Canada, affected farms are quarantined and the infected animal(s) as well as the remaining herd(s) are "depopulated." Since 1996, the CFIA has destroyed over 7500 animals from the 40 Saskatchewan farms and one Alberta farm. The federal government has provided $33 million in compensation to the farmers for the loss and costs of disposal of their elk. In 2003, four Saskatchewan elk farmers were prohibited from growing grain or raising livestock because their land might harbour CWD organisms. Available evidence suggests that transmission of CWD may occur when the animals congregate around human-made feed and water stations and contaminate their water and feed with saliva, urine, and feces. Until it can be proven that elk (or deer) will not become reinfected with CWD, the farmers cannot grow crops, and no compensation is available to them (Alberta AFRD, 2003; Thomas, 2003).

Since the discovery of CWD, prices for farmed elk and deer have collapsed. South Korea has halted imports of elk products from North America, and the United States has stopped the sale of trophy deer and elk to American hunt farms. Saskatchewan has banned imports of male deer and elk from Alberta to its hunt farms. In August 2002, to control or prevent the further spread of CWD in Alberta, the province announced a mandatory CWD Surveillance Program. Elk and deer farmers must submit the heads from all farmed animals over one year of age that die or are slaughtered, and the product from slaughtered animals must be held at abattoirs until CWD test results are available. While there is no scientific evidence that humans can be affected by CWD, Alberta's precautions appear to heed the World Health Organization's advice that no meat source possibly infected by prions should be allowed into the human food system (Alberta AFRD, 2003).

Agricultural Biotechnology The Canadian Environmental Protection Act (1999) defines *biotechnology* as the "application of science and engineering in the direct or indirect use of living organisms or parts or products of living organisms, in their natural or modified forms." Although it is not a new discipline, biotechnology is an umbrella term that covers a broad spectrum of scientific tools from agricultural, biological, chemical, and medical sources. Biotechnology takes advantage of living organisms, or their parts, to produce products; making yogurt, cheese, and bread are said to be traditional biotechnological activities. More advanced activities include the production of antibiotics, vaccines, and enzymes.

One new aspect of biotechnology is *genetic engineering* (GE), which involves removing or transferring specific characteristics or genetic information (DNA) from one organism to another, thus altering the characteristics of these organisms. Biotechnological products are used extensively in agriculture, drug manufacturing, medical treatment, and pollution control. In Canada, the most prevalent biotechnology is **bioremediation**, used widely in resource-based industries such as oil wells, mining, and pulp and paper operations to break down or degrade hazardous substances into less hazardous or nontoxic substances (Statistics Canada, 2000).

Some of the benefits claimed for biotechnology in Canada are the production of newer and better products that may be lower in price than their traditional counterparts; more rapid diagnosis and treatment of certain diseases; and, in agriculture, the potential for superior food products and healthier agricultural plants and animals.

Proponents say that when it is combined with traditional techniques, biotechnology provides a way to develop plants, animals, and foods with novel attributes (Agriculture and Agri-Food Canada, 1996). For instance, researchers have taken a natural organic insecticide gene from a soil bacterium (*Bacillus thuringiensis*, abbreviated *Bt*) and inserted it into corn, potatoes, and cotton to enhance the bug-killing capacity of these plants. These altered plants produce toxins continuously and throughout the plant so that the corn borer, Colorado potato beetle larvae, or cotton boll weevil will be killed by "natural" means when they bite into the GE plants. By 1999, Canadian approval had been granted for insect-resistant (i.e., genetically modified) corn, potatoes, and cotton and more than 40 other genetically altered crops (Boyens, 2000; Bueckert, 2000). Commercially prepared foods that contain *Bt* include cereals, pancake mixes, corn chips, and soy formulas (Canadian Alert in Genetic Engineering, 2000).

A number of potential environmental risks are associated with commercial genetically engineered agricultural products, however. These concerns relate to the potential for organisms such as plants to spread and transfer their genetically altered material (known as outcrossing) and increase harm to nontarget species from the release of modified plants or microorganisms (Agriculture and Agri-Food Canada, 1997b). This could disrupt the balance in natural ecosystems through the replacement of a few or large numbers of species.

One example of recent research suggests this concern may be valid. Cornell University researchers reported how *Bt* corn plants (spliced with a gene from *Bacillus thuringiensis*) might represent a risk because most hybrid corn release the *Bt* toxin in pollen. When pollen from *Bt* corn was dusted on milkweed, only 56 percent of young monarch butterfly larvae survived, compared with 100 percent survival of larvae on leaves dusted with untransformed pollen or on leaves with no pollen (Losey, Rayor, & Carter, 1999). A common weed that often surrounds corn fields, milkweed is the exclusive source of food for these butterflies. This research suggests *Bt* corn pollen might be problematic for the conservation of monarch butterflies, particularly given that 50 percent of the summer monarch population is concentrated in the mid-western American "corn belt." Since the amount of *Bt* corn planted in the United States is projected to increase significantly over the next few years (suggesting that a substantial proportion of available milkweeds may be within range of corn pollen deposition), research must be conducted to evaluate the risks of this agrotechnology on the monarch butterfly. However, Prakash (2001, p. 13) reports that "the initial fear about the reported damage to monarch butterflies from *Bt* corn has not held up in additional studies." Clearly, conflicting research results suggest the need for continued investigation.

Just as some plants have been made insect-resistant, other plants have been made resistant to herbicides such as glyphosphate or glufosinate ammonium, chemicals that are capable of killing not only weeds but all plant life. In 1995, field trials of genetically engineered crops took place around the world. In 1996, the results of these trial results produced commercial crops, including two herbicide-resistant canolas grown in Canada (Boyens, 2000). By 1998, about 6.5 million acres or 50 percent of the total area of transgenic canola grown in Canada was

herbicide-tolerant (Kneen, 1999b) and, in 1999, 60 percent of Canada's canola crop was genetically engineered. As well, one in every three acres of corn, 25 percent of soybeans, and 20 percent of potatoes grown in Canada were genetically modified (Boyens, 2000).

Herbicide resistance appeals to farmers because they can reduce their chemical spraying. Canola, for instance, normally requires several passes of herbicides (to kill fox-tail, wild mustard, and other unwanted vegetation) at a cost of at least $40 per acre for the herbicides plus the labour involved in spraying. Herbicide-resistant canola requires one spraying of a herbicide such as Roundup at a cost of about $20 per acre. Multinational corporations such as Monsanto, DuPont, and Novartis claim they have created new crops that will reduce the need for agricultural chemicals. However, as critics note, these crops are designed to be used with matching herbicides manufactured by the same companies that created the crops, thus ensuring the sale of specific chemicals. There are concerns that agrotoxins eliminate biodiversity, "not only in the crop, but perhaps even more importantly, in the soil in which it is grown" (Kneen, 1999a, p. 25; Agriculture and Agri-Food Canada, 1996). Concerns for human health as well as that of other species affected by genetically engineered crops have increased over time and are among the issues considered in Box 6–8.

Given such concerns, on April 1, 1997, a new Canadian Food Inspection Agency (CFIA) took over responsibility for regulating agricultural products to see whether they are safe for humans, animals, and the environment. New regulatory requirements have been developed to address the safety of novel organisms in the environment. Before the agricultural products of biotechnology may be used, each undergoes a preregulatory review to determine if the new product is "substantially equivalent" to a product already approved (in which case it will be approved for release) or if a risk assessment will be required. If a risk assessment is necessary—say, for plants with unique traits—a series of guidelines outline the criteria that must be considered in assessing risk (Agriculture and Agri-Food Canada, 1997a).

Sometimes, potential risks can be managed by imposing conditions that reduce risks, such as limiting the release of a bioengineered product to a confined area. In the case of plants with novel traits, an environmental assessment is required for confined field trials, a second assessment is required for unconfined release, and if a plant is to be used as a food or feed, then it must undergo further safety assessments by Health Canada or CFIA before it is used in commercial production.

Canola is Canada's most genetically engineered crop, and since 1996 about 20 000 Canadian farmers have paid Monsanto Canada $15 per acre to buy herbicide-resistant canola seeds (called Roundup Ready). At an appropriate time, farmers spray their canola crop with the company's

BOX 6-8
GENETICALLY MODIFIED FOODS: ISSUES OF CONCERN

Dubbed "Frankenfoods" by the media, genetically modified foods (GMFs) are usually indistinguishable from nature's products. No labels help consumers identify these genetically engineered products, and no signs separate conventional from transgenic, or "novel," foods. How, then, are consumers to know if the gene-spliced foods on supermarket shelves are safe? We can't, at least not definitively, because very little research has been done on the effects of genetic engineering (GE) and GMFs on human health.

In Canada, and elsewhere, public opinion about the safety of GMFs varies widely. According to a recent nationwide poll, 61 percent of Canadians feel secure enough about the safety and benefits of biotechnology that they are willing to accept some long-term, unintended risks in exchange for potential health benefits (May, 2000). A 1999 poll conducted in 15 member states of the European Union indicated that only 41 percent of those polled felt biotechnology would improve their quality of life in the next 20 years. While the United States National Research Council declared that GMFs are safe (Macilwain, 2000), Britons are so concerned about the ill effects of biotechnology that a British court acquitted Greenpeace participants on a charge of destroying genetically modified, "contaminated" crops because they constituted a potential environmental threat (Chaundy, 2000). In May 1999, the 115 000-member British Medical Association called for a moratorium on GE foods and more independent research on its safety (Boyens, 2000). In 2002, the European Union determined to keep GM produce out of Europe's food system by requiring both honest labelling and a regulatory system to trace the origins of GM corn or soybeans from the supermarket back to the farm of origin (Greider, 2003).

Biotechnology applied to livestock and food crops to create GMFs has been touted as a method of securing food sources for the world's growing population (Wambugu, 2000; Coghlan, 2000). GMFs such as *Bt* plants have been created to provide potentially increased product yield; decreased pesticide use via bioinsecticides; greater resistance to bacterial, viral, and fungal plant diseases; and increased tolerance to cold, heat, and drought (Wilkinson, 1998; Pearce, 2000). From a human health perspective, edible vaccines (surgically implanted inside a food that needs no refrigeration) are being investigated to help alleviate enteric (intestinal) diseases in developing nations (Powell, 1999).

Potential negative consequences of GMFs also are widely acknowledged. Although reactions of humans to unfamiliar genetic material are uncertain, health concerns centre on the links of GMFs to toxicity, allergic reactions, antibiotic resistance, cancer, and immunosuppression, among others. Problems with unintended or secondary effects of genetic modification, such as transfer of the antibiotic resistance gene, have been raised, as have concerns about human error in creation of transgenetic material introduced into target organisms (Gasson, 1998; Powell, 1999).

Environmental concerns about GMFs focus on the unknown effects of GM crops, including the risk of a gene spreading from the transgenic organism to other organisms in the environment. This "gene escape" can occur when pollen from a transgenic plant is carried by wind or insects to a wild, nontransgenic plant.

Photo 6–15
An anti-GMF display in Munich, Germany, 2002.

If the pollen fertilizes the wild plant, the resulting hybrid will contain the transgene. If the hybrid plant survives and reproduces with other wild plants of the same species, the transgene may become firmly established in the wild plant population. While it is true that cross-pollination (outcrossing) has occurred with crop plants since the beginning of agriculture, the introduction of new genes may add new risks such as the development of "superweeds," the inadvertent production of toxins or allergens in new plants used as food, or an increased risk of resistant pests (Altieri, 2000; Arriola, 1998; Hails, 1998; Hill, 1998; Powell, 1999).

The industrialization of agriculture also has resulted in a reduction in plant and animal diversity. Of Canada's 220 livestock breeds, about 60 are rare or endangered. The greatest losses of farm animal biodiversity are occurring in the poultry and swine industries, where, for example, only three multinational companies own all of the "elite" genetic breeding stock lines to produce commercial turkeys for the world. The birds are all highly selected strains with very narrow genetic variability; because of genetic manipulation (to increase size and white breast meat), today's traditional Christmas turkeys can no longer breed naturally but are inseminated artificially. Even with high levels of disease risk with such a small gene pool, this system is considered a model for the future of other livestock industries (Chiperzak, 1999).

In evaluating GMFs, it is important to define what constitutes risk (and how much risk is acceptable) and what constitutes environmental harm (Hails, 1998). Will the impacts of GMFs be more or less harmful than existing methods of enhancing plant growth and productivity? Biotechnology holds the promise of enormous economic gain (mainly, it seems, for multinational corporations), but questions regarding environmental and social sustainability at a global scale have created heated debate among stakeholders. As governments and industry push toward acceptance of some scientific research to promote the biotechnology industry, citizens and environmental nongovernmental organizations look to other scientific documents and the precautionary principle to support

(continued)

their arguments. Opponents of GMFs, while not opposed to biotechnology per se, advocate caution and the need for decision makers to obtain more information before making irreversible decisions. Emerging from the debate is a call to science not to lose sight of the need to conduct transparent, repeatable studies to answer concerns related to the safety of GMFs.

It seems clear, however, that the future of genetic modification in agriculture will not be settled by good science and risk assessment calculations alone, but will be determined in the political and legal arenas and through regulatory regimes (Bueckert, 2000; Gray, 1998). For instance, on January 29, 2000, the United Nations Convention on Biological Diversity adopted the Cartagena Protocol (also known as the Biosafety Protocol), an international agreement regulating trade in genetically modified organisms. The Cartagena Protocol, signed by 75 countries, advocates a precautionary principle of environmental law and treats environmental issues as equal to trade-related issues. In January 2000, Canada committed to sign the Cartagena Protocol. In late September of that same year, a coalition of 80 groups including the Council of Canadians and Greenpeace Canada called on the federal government to honour this commitment. According to the coalition, Canada's failure to sign the protocol reinforced the view that government ranked trade objectives ahead of human health, the environment, and biodiversity (Bueckert, 2000).

In May 2000, the Sierra Legal Defence Fund filed a petition under the Auditor General Act concerning Canada's Federal Regulatory Framework for Biotechnology. In September 2000, the Auditor General's office responded by saying that Canada's system already provided the necessary regulatory, legal, and policy framework to evaluate the health and environmental

impacts of biotechnology (Government of Canada, 2000). Nevertheless, Agriculture Canada announced a 12-year study to assess more thoroughly the impacts of GMFs on human health and the environment (Teel, 2000).

At the Codex Alimentarius Commission, work is ongoing in Canada and internationally to examine approaches to labelling genetically modified products. In Canada, three major consultative efforts revealed strong support for mandatory labelling of foods when significant nutritional or compositional changes were made, in comparison to foods already on the market, and in cases where safety concerns such as allergenicity were identified. These consultations also determined that a voluntary labelling approach by food manufacturers or distributors to identify these foods to consumers was acceptable provided the label statement was truthful and not misleading (Canadian General Standards Board, 2000).

Designer Genes at the Dinner Table, a 1999 citizens' conference on food biotechnology, emphasized the importance of giving ordinary Canadian citizens a direct part in decision-making processes regarding the future of food. Conducted in the context of the federal government's renewal of the Canadian Biotechnology Strategy, this citizens' conference demonstrated a promising method for addressing socially controversial issues such as food biotechnology. The first of its kind in Canada, this conference enabled citizens to consult with experts as they considered issues associated with food biotechnology. In making recommendations, such as the need for the Canadian Biotechnology Advisory Committee to resolve GE food labelling issues, citizens stressed that old models of decision making were no longer viable, particularly in light of rapidly changing technology (Citizens' Panel on Food Biotechnology, 1999).

SOURCES: "Ten Reasons Why Biotechnology Will Not Ensure Food Security, Protect the Environment and Reduce Poverty in the Developing World," M. Altieri, 2000, SustainAbility Online, http://www.sustainability.com/cage/10reasons.html; "Are We Too Late?" by P. Arriola, October 8, 1998, *Nature,* http://helix.nature.com/debates/gmfoods/gmfoods_3.html; *Unnatural Harvest: How Genetic Engineering Is Altering Our Food,* I. Boyens, 2000, Toronto: Doubleday Canada; "Government Pressed to Sign Biosafety Protocol," D. Bueckert, September 28, 2000, *Calgary Herald,* p. A12; *Standard for the Voluntary Labelling of Foods Obtained or Not Obtained through Genetic Modification,* Canadian General Standards Board, 2000, http://www.pwgsc.gc.ca/cgsb/032_025/intro_e.html; "Peter Melchett: Lord of the Greens," by B. Chaundy, September 20, 2000, BBC News Online, http://news6.thdo.bbc.co.uk/hi/english/uk/newsid%5F934000/934110.stm; "Old MacDonald Had a Farm, Eee Eie Eee Eie Oh-oh," by J. Chiperzak, 1999, *Alternatives Journal, 25*(1), 15; *Citizens' Panel Final Report: Designer Genes at the Dinner Table,* Citizens' Panel on Food Biotechnology, 1999, Calgary: University of Calgary; "Judging Gene Foods," by A. Coghlan, April 15, 2000, *New Scientist,* p. 4; "Food and Drink," M. Gasson, October 8, 1998, *Nature,* http://helix.nature.com/debates/gmfoods/gmfoods_4.html; *Review of Federal Laws, Regulations, and Policies on Genetically Modified Organisms, Specifically Relating to Sustainable Development: Summary of the Response of Federal Departments to the Petition from the Sierra Legal Defence Fund,* Government of Canada, 2000, http://www.cfia-acia.agr.ca/english/ppc/biotech/enviro/sierrafse.shtml; "Be Careful What You Wish ...," by A. Gray, October 15, 1998, *Nature,* http://helix.nature.com/debates/gmfoods/gmfoods_5.html; "A High-Level Food Fight," by W. Greider, November 3, 2003, *The Nation,* p. 16; "Relative Risk," by R. Hails, October 1, 1998, *Nature,* http://helix.nature.com/debates/gmfoods/gmfoods_1.html; "Sceptically Speaking ...," by J. Hill, October 1, 1998, *Nature,* http://helix.nature.com/debates/gmfoods/gmfoods_2.html; "US Academy Study Finds GM Foods Are Safe," C. Macilwain, April 13, 2000, *Nature, 404,* 893; "Canadians Unafraid of Biotech," by K. May, July 24, 2000, *Calgary Herald,* p. A5; "Feeding Africa," by F. Pearce, May 27, 2000, *New Scientist,* 40–43; *Seminal Paper on Agricultural Biotechnology: A Summary of the Science,* D. P. Powell, 1999, Crop Protection Institute of Canada, www.plant.uguelph.ca/riskcomm/gmo/cpi/CPI-nov-99htm; "Study to Probe Altered Crops," G. Teel, August 31, 2000, *Calgary Herald,* p. A1; "Feeding Africa," by F. Wambugu, May 27, 2000, *New Scientist,* 40–44; "Benefits and Risks of Genetic Modification in Agriculture, by M. Wilkinson, October 1, 1998, *Nature,* http://www.nature.com/nature/debates/gmfoods/gmfoods_contents.html

Roundup weed killer. In 1998, a Saskatchewan farmer was sued by Monsanto Canada because he used the company's seeds without permission. The farmer, Percy Schmeiser, claimed that the genetically modified seeds blew onto his land and that he was simply following his usual practice of harvesting and reusing some of the seeds from his current crop.

After a six-year court battle, which pitted traditional farmers and environmentalists opposed to GM foods against the biotechnology industry, the Supreme Court determined that Schmeiser had infringed Monsanto Canada's patent for a genetically altered canola gene. In Canada, plants (as higher life forms) are not patentable, but in a narrow 5–4 decision, the Court ruled that Schmeiser's cultivation of a canola plant containing Monsanto Canada's patented gene without a licence deprived Monsanto Canada of the full benefits of its monopoly. A team of dissenting judges indicated that this decision indirectly allowed Monsanto Canada to acquire patent protection over whole plants, fundamentally changing Canada's patent law. The Canadian Biotechnology Advisory Committee suggested that this decision would "significantly overcompensate the biotechnology industry to the detriment of Canada's farming community" (Gold, 2004, p. A17; see also Tibbetts, 2004).

PARTNERSHIPS

Environmental issues related to agricultural use of land can be national in scope or can exhibit regional distinctiveness. National agricultural issues pertaining to the environment include greenhouse gas emissions and climate change impacts, energy use, and genetic resources. Regional concerns include soil quality, water quality, and wildlife issues. Through the agricultural component of Canada's Green Plan (which ended March 31, 1997), a number of federal–provincial–territorial agreements were established to support activities aimed at ensuring long-term sustainability of the resources that agriculture depends on and shares with other users, as well as to assist the transition to more sustainable farming practices. Activities under the Green Plan also helped Canada meet national and international environmental commitments, such as the Conventions on Climate Change and on Biological Diversity.

With funding from the sustainable agriculture component of Canada's Green Plan, and building on the National Soil Conservation Program, Ontario farmers developed the idea of environmental farm planning. Each farmer assessed his or her own farm to highlight its environmental strengths, identify areas of concern, and set realistic goals to improve environmental conditions. In addition, many farmers voluntarily organized themselves into various associations and societies, such as the Ecological Farmer's Association of Ontario, aimed specifically at environmental objectives.

By 2003, all of Canada's Ministers of Agriculture had signed or initialled the comprehensive Federal-Provincial-Territorial Framework Agreement on Agricultural and Agri-Food Policy for the Twenty-First Century (APF). The national agricultural and agri-food policy has three main goals: (1) to foster confidence in Canada's food safety and quality systems and the environment, (2) to accelerate advances in science and technology, and (3) to help farmers become more profitable by providing them with risk management and other tools they need to improve their management and technical skills (Agriculture and Agri-Food Canada, 2003a).

Part Two, Section C of the Framework Agreement sets out common environmental outcome goals that include reduction of agricultural risks to water, soils, and atmosphere (including global warming) and that ensure improved stewardship through the adoption of environmentally beneficial practices. These outcomes include increased use of appropriate manure and fertilizer management practices as well as pest and pesticide management practices, increases in zero-till or conservation tillage, improved management of riparian areas, improved practices for management of odours and particulate emissions, and protection of biodiversity (Agriculture and Agri-Food Canada, 2003a).

Under the APF, the federal government is working with the provinces and the agricultural industry to make Canada "the world leader in environmentally responsible production, food safety and food quality and innovation" (Agriculture and Agri-Food Canada, 2003b). To that end, Agriculture and Agri-Food Canada launched the Greencover Canada program in 2003. In keeping with the land and water management efforts specified in the APF (for instance, to decrease the number of bare-soil days on farmland), the five-year, $110 million Greencover program provides landowners with the technical assistance and financial incentives they need to convert lower quality cropland, or land that is severely degraded due to wind and water erosion or salinization, to perennial forage and trees (shelterbelts). Perennial forage can help protect lands that are susceptible to erosion and can help to develop riparian-area buffer strips that help protect water quality. These practices are intended to help farmers improve their grassland management practices, protect water quality, reduce GHGs, and enhance biodiversity and wildlife habitat. However, the APF continues Canada's focus on voluntary efforts rather than regulated standards to deal with environmental problems caused by agriculture (Boyd, 2003).

Cooperative research into sustainability of agroecosystems also continues, some of it in partnership with industry. Examples of these efforts include developing disease- and pest-resistant crop varieties; reducing pesticide

Photo 6–16
The Prairie Care Project helps conserve and restore wetland habitat for waterfowl and other species.

use; developing integrated approaches to pest management; improving the efficiency of animals (through breeding and nutrition), resulting in less manure and better use of forage and grains; and developing more efficient fertilizer application technology and innovative approaches to manure management, especially because these affect water quality.

FUTURE CHALLENGES

Even though there has been progress toward managing agricultural lands in a sustainable manner, difficulties remain. Farming methods have not always encouraged agriculture to depend on the natural system's heterogeneous characteristics but rather have pushed farmers to standardize procedures and technology to achieve uniform results in mass quantities (industrial agriculture). Nevertheless, the call for alternative agricultural practices—ones that are sustainable and maintain agricultural resources as renewable resources—has grown. The realization that the status quo is not sustainable means that new and creative alternatives are required. Stewardship, protection, and monitoring, as well as knowledge-building, are part of the new alternatives. Since some of the most effective agricultural innovations have been undertaken by farmers, including developments in conservation farming techniques, the ability of farmers to respond to stewardship interests and economic changes, and to adopt new management systems and technologies, needs to be incorporated in approaches to sustainable agroecosystems. The Agricultural Policy Framework (APF) referred to previously attempts to involve all stakeholders in strengthening the agriculture and agri-food

sector and has held public as well as private discussions on what is needed for the future development of this sector.

At the same time, farmers across Canada were faced with a range of pressures in 2003—from the weather to grasshoppers, and from low prices to BSE—that have resulted in severe financial pressures. Without financial health in Canada's agricultural industry and at the farm level, environmental improvements are difficult to effect. As noted earlier in this chapter, incomes have fallen and indebtedness has risen as farmers' costs for fertilizers, pesticides, and commercial feeds have risen and as stiffer competition has occurred in the global economy (MacRae & Cuddeford, 1999; Statistics Canada, 2002c). Even with $600 million in federal funding provided directly to farmers in the fall of 2003 (Agriculture and Agri-Food Canada, 2003c) the economic climate may not be favourable for environmental stewardship. The push for sustainable agriculture derives in part from its economic viability and in part from the solutions it provides to most of the erosion, contamination, and energy issues farmers face (MacRae & Cuddeford, 1999). Whether or not the APF will succeed in reaching its stated stewardship goal of developing "environmentally-beneficial agricultural production and management practices" remains to be seen (Agriculture and Agri-Food Canada, 2003a, S26.3.2).

The health and productivity of agroecosystems clearly are fundamental necessities in sustainability of agricultural land resources. However, it is difficult to monitor trends or changes in agroecosystem quality without an adequate information base. This lack also makes it difficult to determine if public and private investments in sustainability activities, such as maintaining life-support systems, preserving biological diversity, and maintaining the productive capacity of species and ecosystems, are achieving the desired ends. The APF specifies that the Implementation Agreements with each province and territory will contain targets and indicators to be used in measuring and monitoring the progress made in achieving established environmental outcome and farm environmental management goals. Here, too, the utility and effectiveness of these targets and indicators need to be demonstrated.

Among the most promising tools to monitor agroecosystems and to enhance our knowledge and database are remote sensing and geographic information systems (GISs). GIS can link together Statistics Canada's Census of Agriculture, Agriculture and Agri-Food Canada's research station experimental data, and data from farmers to improve the baseline for assessing changes. Specifically, indicators of agroecosystem health may be manipulated within a GIS, including soil degradation, soil quality, crop yield, soil cover and management, conservation practices adopted, land conversions, and nutrient balance.

Progress has been made toward preserving biological diversity and maintaining the productive capacity of species and ecosystems through the increased use of sustainable land management practices. The challenge is to continue, as appropriate, to increase conservation tillage, to decrease summerfallow, to find alternatives to agrochemicals and to reduce use of herbicides and pesticides, to remove more marginal land from crop production into forage or other uses, and to continue to restore and enhance wildlife habitat. Stewardship remains an important impetus for continued protection of agroecosystems, given that wind and water erosion, salinization, soil compaction, organic matter loss, and contamination of groundwater by nitrates, pesticides, and bacteria continue to occur.

Resolving these challenges to the sustainability of Canadian agroecosystems and land resources requires long-term commitment. Given appropriate care, the health of agricultural soils and agroecosystems can be maintained and even improved. As our understanding of environmental and other impacts on Canadian agriculture increases, and as sustainable and conservation methods continue to improve, it will become easier to avoid adverse impacts from farming operations. Achieving the goal of sustainable agriculture is a responsibility shared among farmers, the agri-food industry, government, and consumers. Cooperation and partnerships among these groups is key to ensuring the sustainability of both agriculture and the environment.

Chapter Questions

1. Identify and discuss the main ways in which soil quality of agricultural lands may be degraded. What are the sources and impacts of other human activities on agricultural lands?

2. Why should soil conservation and biodiversity of agroecosystems be a concern of every Canadian, not just farmers?

3. Why are integrated pest management approaches better for the environment than earlier approaches? How could you employ IPM principles in a small vegetable garden behind a house in a city or town?

4. Could farming lead to desertification in Canada? How might it be prevented?

5. Describe the various efforts Canada has made in attempting to achieve sustainable agriculture at international and national levels. Identify the range of efforts made (if any) by your provincial government.

6. What are some of the advantages and disadvantages of nontraditional agricultural activities (such as organic farming and biotechnology) in achieving economic and environmental sustainability of Canadian agricultural lands and plant and animal resources?

references

Acton, D. F., & Gregorich, L. J. (1995). Understanding soil health. In D. F. Acton & L. J. Gregorich (Eds.), *The health of our soils—Toward sustainable agriculture in Canada.* Ottawa: Agriculture and Agri-Food Canada. http://res2.agr.gc.ca/publications/hs/index_e.htm

Adams, B., & Fitch, L. (1995). *Caring for the Green Zone: Riparian areas and grazing management.* Lethbridge, AB: Alberta Environmental Protection.

Agriculture and Agri-Food Canada. (1996). *Biotechnology, agriculture and regulation.* http://www.aceis.agr.ca/fpi/agbiotec/geninfo.html

Agriculture and Agri-Food Canada. (1997a). *Information bulletin ... Regulating agricultural biotechnology in Canada: Environmental questions.* http://www..agr.ca/fpi/agbiotec/enviroe.html

Agriculture and Agri-Food Canada. (1997b). *Information bulletin ... Biotechnology and environmental concerns: Outcrossing.* http://www.aceis.agr.ca/fpi/agbiotec/crosse.html

Agriculture and Agri-Food Canada. (2001). *Opportunities for reduced non-renewable energy use in Canadian prairie agricultural production systems.* http://www.agr.gc.ca/spb/rad-dra/publications/reductopp/reductopp_e.php

Agriculture and Agri-Food Canada. (2003a). *Federal–Provincial–Territorial Framework Agreement on Agricultural and Agri-Food Policy for the Twenty-First Century.* http://www.agr.gc.ca/cb/apf/index_e.php?section=info&group=accord&page=accord

Agriculture and Agri-Food Canada. (2003b). *AAFC launches Greencover Canada.* http://www.agr.gc.ca/cb/index_e.php?s1=n&s2=2003&page=n30516a

Agriculture and Agri-Food Canada. (2003c). *$600 million in federal transition funding to be delivered directly to producers.* http://www.agr.gc.ca/cb/index_e.php?s1=n&s2=2003&page=n30919c

Agrifood Trade Service. (2002). *Canada's agriculture, food and beverage industry: Overview of the sector.* http://ats-sea.agr.ca/supply/e3314.htm

Alasia, A., & Rothwell, N. (2003). The rural/urban divide is not changing: Income disparities persist. *Rural and Small Town Canada Analysis Bulletin, 4*(4), 1.

Alberta Agriculture, Food and Rural Development. (2000). *Livestock regulations stakeholder advisory group.* http://www.agric.gov.ab.ca/economic/policy/ilo/index.html

Alberta Agriculture, Food and Rural Development. (2003). Chronic wasting disease (CWD) of elk and deer. *Agdex* 663-43. http://www1.agric.gov.ab.ca/$department/deptdocs.nsf/all/agdex663-43

Arnalds, A. (2000). Evolution of rangeland conservation strategies. In A. Arnalds & S. Archer (Eds.), *Rangeland desertification* (pp. 153–165). London: Kluwer Academic.

Bakker, J. P. (1998). The impact of grazing on plant communities. In M. F. WallisDeVries, J. P. Bakker, & S. E. VanWieren (Eds.), *Grazing and conservation management* (pp. 137–184). Dordrecht: Kluwer Academic.

Bork, E. (2000). Grazing can improve native plant diversity in range. Lethbridge Research Centre. wysiwyg://48/http://res2.agr.ca/lethbridge/rep2000/rep0202.htm

Boyd, D. R. (2003). *Unnatural law: Rethinking Canadian environmental law and policy.* Vancouver: UBC Press.

Boyens, I. (2000). *Unnatural harvest: How genetic engineering is altering our food.* Toronto: Doubleday Canada.

Bueckert, D. (2000, September 24). Government pressed to sign Biosafety Protocol. *Calgary Herald,* p. A12.

Canadian Alert in Genetic Engineering. (2000). *Biotechnology: Giving pollution a life of its own.* http://www.sustainability.com/cage/

Canadian International Development Agency (CIDA). (1995). *CIDA and desertification: Working to preserve our common future.* Hull, QC: Author.

Cardy, F. (1994). Desertification. *Our Planet, 6*(5), 4.

Cardy, W. F. G. (1997). Foreward. In N. Middleton & D. Thomas (Eds.), *World atlas of desertification* (2nd ed.) (p. vi). London: Arnold.

Chambers, A. (n.d.). Manitoba's elk—Just another farm animal? Sierra Club. http://www.sierraclub.ca/prairie/elk.html

Chambers, P. A., Guy, M., Roberts, E. S., Charlton, M. N., Kent, R., Gagnon, C., Grove, G., & Foster, N. (2001). *Nutrients and their impact on the Canadian environment.* Agriculture and Agri-Food Canada, Environment Canada, Fisheries an Oceans Canada, Health Canada, and Natural Resources Canada. http://www.durable.gc.ca/group/nutrients/report/index_e.phtml

Dale, B. (2000). Range management can help save prairie birds. Lethbridge Research Centre. wysiwyg://40/http://res2.agr.ca/lethbridge/rep2000/rep0217.htm

Derworiz, C. (2000, August 12). Disease may wipe out elk industry. *Calgary Herald,* pp. B1, 2.

Drover, J., for R. Mercer, Minister of Environment, Government of Newfoundland and Labrador. (2003). Response to questionnaire survey for *Our Environment: A Canadian Perspective.*

Dowdeswell, E. (1997). Preface. In N. Middleton & D. Thomas (Eds.), *World atlas of desertification* (2nd ed.) (p. iv). London: Arnold.

Ecological Agriculture Projects. (1997). *Agriculture and Agri-Food Canada's strategy for environmental sustainability.* http://eap.mcgill.ca/MagRack/EC/ec1_1_2.htm

Eilers, R. G., Eilers, W. D., Pettapiece, W. W., & Lelyk, G. (1995). Salinization of soil. In D. F. Acton & L. J. Gregorich (Eds.), *The health of our soils—Toward sustainable agriculture in Canada* (pp. 77–86). Ottawa: Agriculture and Agri-Food Canada.

Elk On Line. (1996). *Elk Breeders Home Page.* http://www.wapiti.net/

Environment Bureau. (n.d.). *Agriculture in harmony with nature: Strategy for environmentally sustainable agriculture and agri-food development in Canada.* http://aceis.agr.ca/policy/envharmon/indexe.htm

Environment Canada. (2002a). *Canada's greenhouse gas inventory 1990–2000. Factsheet overview.* http://www.ec.gc.ca/pdb/ghg/factsheet_e.cfm

Environment Canada (2002b). *Canada's greenhouse gas inventory 1990–2000. Information on greenhouse gas sources and sinks.* http://www.ec.gc.ca/pdb/ghg/query/srd_report.cfm

Food and Agriculture Organization of the United Nations. (2002). *Challenges and Opportunities for the World Summit on Sustainable Development: FAO's Perspective.* Paper prepared for the World Summit on Sustainable Development, Johannesburg, August 26–September 4. http://www.fao.org/wssd/docs/ChallengesandOpportunityfinal.doc

Galeano, P. (1999). ... and Uruguay. *Alternatives Journal, 25*(1), 24.

Gauthier, D. A., & Henry, J. D. (1989). Misunderstanding the prairies. In M. Hummel (Ed.), *Endangered spaces: The future for Canada's wilderness* (pp. 183–195). Toronto: Key Porter Books.

Gold, R. (2004, May 24). Monsanto's gain is everyone else's pain. *The Globe and Mail,* p. A17.

Government of Alberta. (2001, December 20). Confined feeding operation regulations and standards balance industry growth with environmental and health protection. News Release. http://www.gov.ab.ca/acn/200112/11733.html

Government of Canada. (1991). *The state of Canada's environment—1991.* Ottawa: Supply and Services Canada.

Government of Canada. (1994). *Canada's national report on climate change: Actions to meet commitments under the United Nations Framework Convention on Climate Change.* Ottawa: Supply and Services Canada.

Government of Canada. (1996). *The state of Canada's environment—1996.* Ottawa: Supply and Services Canada.

Government of Canada. (1999). *Canada introduces national standard for organic agriculture.* http://www.cfia-acia.agr.ca/english/corpaffr/newsrelease/19990629e.shtml

Government of Canada. (2001). *Federal science and technology for sustainable development.* http://www.durable.gc.ca/communication/long-brochure/index_e.phtml

Gregorash, D. (1997). Feedlot manure: An issue that doesn't go away. *Encompass, 2*(2), 19–20.

Gregorich, L. J., & Acton, D. F. (1995). Summary. In D. F. Acton & L. J. Gregorich (Eds.), *The health of our soils—Toward sustainable agriculture in Canada* (pp. 111–120). Ottawa: Agriculture and Agri-Food Canada.

Gregorich, E. G., Angers, D. A., Campbell, C. A., Carter, M. R., Drury, D. F., Ellert, B. H., et al. (1995). Changes in soil organic matter. In D. F. Acton & L. J. Gregorich (Eds.), *The health of our soils—Toward sustainable agriculture in Canada* (pp. 41–50). Ottawa: Agriculture and Agri-Food Canada.

Han, F. X., Kingery, W. L., Selim, H. M., and Gerard, P. D. (2000). Accumulation of heavy metals in a long-term poultry waste-amended soil. *Soil Science, 165*(3), 260–268.

Harker, D. B., Penner, L. A., Harron, W. R., & Wood, R. C. (1995). *For now we see through a glass darkly—Historical trends in dryland salinity point to future expectations under irrigation.* Saskatoon: Canadian Symposium on Remote Sensing.

Hasselback, P. (1997). Intensive livestock operations and health problems. *Encompass, 2*(2), 4–5.

Hengeveld, H. E., Bush, E., and Edwards, P. (2002). *Frequently asked questions about climate change science.* Environment Canada, Science Assessment and Policy Integration Branch.

Hoffman, N. (2001). Urban consumption of agricultural land. *Rural and Small Town Canada Analysis Bulletin, 3*(2), 1–11.

Holechek, J. L. (1993, July). Policy changes on federal rangelands. *Trail Boss News*, 1–2.

Hunter, E. (1999). Community agriculture rises in Quebec. *Alternatives Journal, 25*(1), 24.

Janzen, H. (2000). Managing range essential to maintaining soil carbon stores. Lethbridge Research Centre. wysiwyg://42/http://res2.agr.ca/lethbridge/rep2000/rep0208b.htm

Johnson, W. (Ed.). (1995). *Managing Saskatchewan Rangeland* (Rev. ed.). Regina: Economic Regional Development Agreement.

Jowkar, F. (1994). Women bear the brunt. *Our Planet, 6*(5), 16–17.

Kelly, B. (2003). Prince Edward Island, Department of Fisheries, Aquaculture and Environment, response to questionnaire survey for *Our Environment: A Canadian Perspective.*

Kneen, B. (1999a). Death science creeps onto the farm. *Alternatives Journal, 25*(1), 10–11.

Kneen, B. (1999b). *Farmageddon: Food and the culture of biotechnology.* Gabriola Island, BC: New Society.

Lal, R. (2001). Potential of desertification control to sequester carbon and mitigate the greenhouse effect. *Climatic Change, 51*(1), 35.

Larney, F. J, Bullock, M. S., Janzen, H. H., Ellert, B. H., & Olson, E. C. (1998). Wind erosion effects on nutrient redistribution and soil productivity. *Journal of Soil and Water Conservation, 53*(2), 133–138.

Losey, J. E., Rayor, L. S., & Carter, M. E. (1999, May 20). Transgenic pollen harms monarch larvae. *Nature, 399,* 214.

MacRae, R., & Cuddeford, V. (1999). *A Green Food & Agriculture Agenda for Ontario.* The Environmental Agenda for Ontario Project. http://www.cielap.org/infocent/research/agri.html

Manitoba Climate Change Task Force. (2001). *Manitoba and climate change: Investing in our future. Report of the Manitoba Climate Change Task Force.* Winnipeg: Author.

Manzano, M. G., and Navár, J. (2000). Processes of desertification by goats overgrazing in the Tamaulipan thornscrub (*matorral*) in north-eastern Mexico. *Journal of Arid Environments, 44,* 1–17.

McQuarrie, I. (1997). Agriculture and ecology. In T. Fleming (Ed.), *The environment and Canadian society.* Toronto: ITP Nelson, pp. 54–55.

Middleton, N., & Thomas, D. (Eds.). (1997). *World atlas of desertification* (2nd ed.). London: Arnold.

Mineau, P., McLaughlin, A., Boutin, C., Evenden, M., Freemark, K., Kevan, P., McLeod, G., & Tomlin, A. (1994). Effects of agriculture on biodiversity in Canada. In Environment Canada, Biodiversity Science Assessment Team, *Biodiversity in Canada: A science assessment for Environment Canada* (pp. 59–113). Ottawa: Supply and Services Canada.

Nikiforuk, A. (2000, June 12). When water kills. *Online Macleans.* http://www.macleans.ca/pubdoc/2000/06/12/Cover/35699.shtml

North America Wetlands Conservation Council. (2002). NAWCC (Canada) celebrates a decade of influencing change. http://www.terreshumidescanada.org/pubs.html

Olson, K. (1997). Agriculture and the environment. *Environment Views and Network News, 1*(1), 22.

Prakash, C. S. (2001). The genetically modified crop debate in the context of agricultural evolution. *Plant Physiology, 126,* 8–15.

Prince Edward Island Department of Fisheries, Aquaculture and Environment. (2003). *State of the environment. June 2003.* Charlottetown: Author.

Reynolds, W. D., Campbell, C. A., Chang, C., Cho, C. M., Ewanek, J., Kachanoski, R. G., et al. (1995). Agrochemical entry into groundwater. In D. F. Acton & L. J. Gregorich (Eds.), *The health of our soils—Toward sustainable agriculture in Canada* (pp. 97–109). Ottawa: Agriculture and Agri-Food Canada.

Rubec, C. D. A. (1994). Canada's federal policy on wetland conservation: A global model. In W. J. Mitsch (Ed.), *Global wetlands: Old world and new* (pp. 909–917). Amsterdam: Elsevier Science B.V.

Science Council of Canada. (1992). *Sustainable agriculture: The research challenge.* Report No. 43. Ottawa: Author.

Seck, M. (1994). Unearthing the impacts of tenure. *IDRC Reports, 22*(2), 11–12.

Semmens, G. (2004a, May 15). Threatened status may aid plains bison. *Calgary Herald,* p. A3.

Semmens, G. (2004b, May 18). Plan protects Alberta gems. *Calgary Herald,* p. A10.

Smith, W. N., Desjardins, R. L., & Grant, B. (2001). Estimated changes in soil carbon associated with agricultural practices in Canada. *Canadian Journal of Soil Science, 81,* 221–227.

Statistics Canada. (2000). *Human activity and the environment 2000.* Ottawa: Minister of Industry.

Statistics Canada. (2002a). *2001 Census of Agriculture: Canadian farm operations in the 21st century.* http://www.statcan.ca/Daily/English/020515/td020515.htm

Statistics Canada. (2002b). *Human activity and the environment: Annual statistics 2002.* Ottawa: Minister of Industry.

Statistics Canada. (2002c). *2001 Agriculture Census: Total area of farms, land tenure and land in crops, provinces.* http://www.statcan.ca/english/Pgdb/econ124a.htm

Statistics Canada. (2002d). *Canadian farm operations in the 21st century.* http://www.statcan.ca/english/agcensus2001/first/farmop/01front.htm#top

Statistics Canada. (2003a). *Farming facts 2002.* Ottawa: Minister of Industry.

Statistics Canada. (2003b). *Gross domestic product, income-based.* CANSIM Table 380-0001. http://www.statcan.ca/english/Pgdb/econ03.htm

Statistics Canada. (2003c). *Income of farm operators.* http://www.statcan.ca/english/Pgdb/prim10.htm

Statistics Canada. (2003d). *Exports—Agri-Food for January to December 2002.* http://ats.agr.ca/stats/stats-e.htm

Taylor, D. M. (1994). *Off course: Restoring balance between Canadian society and the environment.* Ottawa: International Development Research Centre.

Thomas, D. (2003, January 14). Saskatchewan: Former Sask. elk ranchers hit with total farming ban. *Edmonton Journal.* http://www.cwdinfo.org/index.php/fuseaction/news.detail/ID/208bbf5c0efa86beb752b6a515439ff6

Thurow, T. L. (2000). Hydrologic effects on rangeland degradation and restoration processes. In A. Arnalds & S. Archer (Eds.), *Rangeland desertification* (pp. 53–67). Dordrecht: Kluwer Adademic.

Tibbetts, J. (2004, May 22). Top court says farmer violated Monsanto patent. *Calgary Herald,* p. A23.

Topp, G. C., Carter, M. R., Culley, J. L. B., Holmstrom, D. A., Kay, B. D., Lafond, G. P., et al. (1995). Changes in soil structure. In D. F. Acton & L. J. Gregorich (Eds.), *The health of our soils—Toward sustainable agriculture in Canada* (pp. 51–60). Ottawa: Agriculture and Agri-Food Canada.

Trottier, G. C. (1992). *A landowner's guide: Conservation of Canada's prairie grassland.* Canadian Wildlife Service. http://www.pnr-rpn.ec.gc.ca/nature/whp/prgrass/df03s00.en.html

United Nations. (2002a). *Johannesburg Summit 2002.* http://johannesburgsummit.org/html/whats_new/feature_story41.html

United Nations. (2002b). *Frequently asked questions about the Johannesburg Summit.* http://johannesburgsummit.org/html/basic_info/faqs.html

Van Tighem, K. (1996). From wilds to weeds: Alberta's changing ecosystems. *Environment Views, 19*(5), 5–8.

Wall, G. J., Pringle, E. A., Padbury, G. A., Rees, H. W., Tajek, J., van Vliet, L. J. P., et al. (1995). Erosion. In D. F. Acton & L. J. Gregorich (Eds.), *The health of our soils—Toward sustainable agriculture in Canada.* Ottawa: Agriculture and Agri-Food Canada. http://res2.agr.gc.ca/publications/hs/index_e.htm

WallisdeVries, M. F. (1998). Large herbivores as key factors for nature conservation. In M. F. WallisDeVries, J. P. Bakker, & S. E. VanWieren (Eds.), *Grazing and conservation management* (pp. 1–20). Dordrecht: Kluwer Academic.

Webber, M. D., & Singh, S. S. (1995). Contamination of agricultural soils. In D. F. Acton & L. J. Gregorich (Eds.), *The health of our soils—Toward sustainable agriculture in Canada* (pp. 86–96). Ottawa: Agriculture and Agri-Food Canada.

"Water is a precious and finite natural resource, one which is essential to all life and vital to ecological, economic and social well-being. Yet, water is often wasted and degraded. Therefore we face both individual and collective responsibilities to use and manage water resources wisely. This will only be accomplished as we recognize the intrinsic value of water and practice conscious and committed stewardship, recognizing that this precious heritage must be safeguarded for future generations."

Canadian Water Resources Association (1994)

CHAPTER 7

Fresh Water

Chapter Contents

Chapter Objectives

After studying this chapter you should be able to

- understand the nature and distribution of Canada's freshwater resources

- identify a range of human uses of freshwater resources

- describe the impacts of human activities on fresh water and freshwater environments

- appreciate the complexity and interrelatedness of freshwater environment issues

- outline Canadian and international responses to freshwater issues

- discuss challenges to a sustainable future for freshwater resources in Canada

INTRODUCTION

Water has sustained Canadians' high quality of life for centuries, and our lakes and rivers have provided a template for the settlement patterns of the country. First Nations people located their villages along riverbanks or coastal shorelines, as did subsequent European and other settlers. Most of our major cities have grown alongside these same water bodies. Water has played and continues to play an important role in contemporary culture, including our art and music. Fresh water is also the lifeblood of the ecosphere and continues to sustain many economic and recreational activities of Canadians.

At first glance, Canadians would appear to have little cause for concern about the supply and management of fresh water. In 1999, the needs of 30.6 million Canadians were served by a renewable supply that is the envy of many other countries. Per capita, Canadians' water use is the second highest in the world; by 1999, our daily indoor household use averaged about 343 litres (Government of Canada, 2001). In rural areas of Africa, Asia, and Latin America, in contrast, average water use is between 20 and 30 litres per capita per day (Jones, 1997). European countries consume on average 150 litres per person per day (Boyd, 2003). Gleick (1996) suggests that 50 litres per capita per day would meet basic human domestic needs (Figure 7–1).

Around the world, more than one billion people do not have access to safe drinking water; pollution is rampant, and water supplies are drying up. Two billion people lack water for proper sanitation, which often leads to disease and high infant mortality rates. In situations of drought, developing nations often are prone to food shortages; in Canada, irrigation often supports agriculture in times of drought. Relative to people in other countries that belong to the Organization for Economic Cooperation and Development (OECD), Canadians use more water and pay relatively lower prices.

Given these circumstances, selling the idea that Canada has a water crisis has been difficult. Instead, many people suggest that Canada is a water-rich nation and has no problems. Despite this perception, the ability of freshwater resources to sustain Canada's ecosystems, economy, and society is under stress. Some of the stressors include the growth of large urban centres, increased industrial activity, and use of agricultural chemicals, all of which are overloading the natural ability of the hydrological cycle (see Chapter 3) to renew and purify water. Over 15 000 lakes in eastern Canada are dead because of acidic pollutants generated both within and beyond Canadian borders. Literally hundreds of chemical substances used in Canada and around the world are found in Canadian

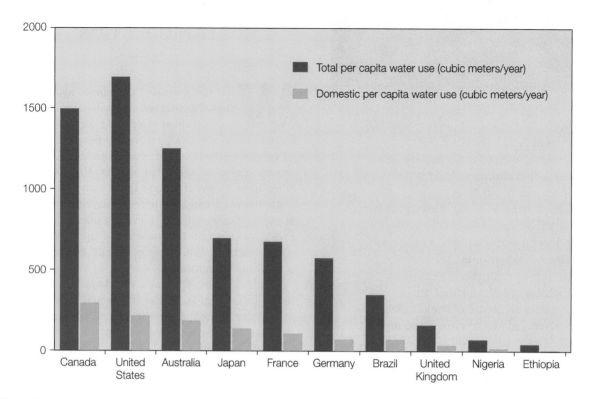

Figure 7–1

Annual and Daily per capita freshwater use, selected countries, 2000

SOURCE: Adapted from *Water Reports 23 – Review of World Water Resources by Country*, Food and Agricultural Organization of the United Nations, 2003, ftp//ftp.fao.org/agl/aglwdox/wr23e.pdf

waters, including groundwater, and some are in our food chains.

In many parts of Canada, water quality has been degraded. For years we contaminated the Fraser River with incompletely treated sewage, landfill leachates, chemicals from wood treatment and pulp and paper mills, and runoff from forestry and agricultural activities. Prairie rivers have been impaired due to agricultural runoff; the Great Lakes and St. Lawrence River have suffered from industrial and municipal pollution, urban and agricultural runoff, and atmospheric deposition; and Prince Edward Island's groundwater supplies have been threatened by agricultural pesticides. Today we face serious concerns about the ability of our water treatment plants to provide clean drinking water.

In some parts of the country such as the southern prairies and southern Ontario, high demand for water has already caused shortages. If future climate changes adversely affect the water cycle, or if our thirst for water is not curbed, current water shortages may worsen. It is also possible that pressures to divert Canadian waters southward will increase as continental economic unions (such as the North American Free Trade Agreement and the Multilateral Agreement on Investment) bring pressures to share resources, particularly water resources (Bruce & Mitchell, 1995). Although Canadian govern-

ments have repeatedly stated that Canada's water is not for sale (except in very limited quantities in bottles and other containers), Canada needs to be prepared for requests to undertake additional water diversions and for the consequences of such undertakings. (For a brief overview of two earlier water export proposals and their potential impacts, see Bocking [1987] and Schindler & Bayley [1991].)

Photo 7–1

Water quality is degraded by many human activities, including log storage and the discharge of chemicals from pulp and paper mills.

Human uses of water frequently compete with those of other species and often harm ecosystems and the people in them. For example, capturing large volumes of water behind dams to provide flood protection and to produce hydroelectricity has caused elevated levels of mercury in aquatic and terrestrial organisms in some areas. Draining land so that agriculture may be carried out has resulted in the loss of about 65 percent of Atlantic coastal marshes, close to 70 percent of southern Ontario wetlands, up to 71 percent of prairie wetlands, and 80 percent of the Fraser River delta wetlands (Government of Canada, 1991).

Clearly, reducing wasteful practices, eliminating harmful actions, preserving freshwater ecosystems, and increasing the quality and availability of Canada's fresh water are challenges that must be addressed if we are to achieve a sustainable future. In order to appreciate the need to ensure the quality of Canada's freshwater supplies for future generations, the following sections briefly describe the nature of Canada's water supply and distribution, and highlight some of the relevant concerns about the sustainability of this resource. Later in the chapter, Canadians' use of freshwater resources and some of the effects that our activities have on the quality of the freshwater supply are discussed. Selected initiatives that governments, partnerships, and local groups have undertaken to resolve freshwater issues also are considered.

WATER SUPPLY AND DISTRIBUTION

EARTH'S FRESHWATER RESOURCES

The Earth's freshwater resources have remained virtually unchanged since the beginning of time. Fresh water constitutes less than 3 percent of Earth's water supply (see Figure 7–2). The remaining 97.24 percent is in the oceans and is too salty for drinking, irrigation, and most industrial applications. Of the 2.76 percent that is fresh water, most of it is in glaciers and ice caps. Only about 0.79 percent of the Earth's fresh water is located on land and available in inland seas, lakes, rivers, soil moisture, groundwater, and other sources. Even humans' many dams, reservoirs, and irrigation works store relatively little of the Earth's fresh water. To put freshwater resources into perspective, if we imagined the world's water supply filled a gallon jug (3.78 litres), the usable freshwater supply would amount to less than one-half teaspoon (2.5 ml). Fortunately, this is a plentiful supply, continuously renewed and purified through the **hydrologic cycle** (see Figure 3–16), provided we do not overload this natural process with excessive amounts of wastes, or withdraw water from groundwater and surface water supplies faster than it is replenished.

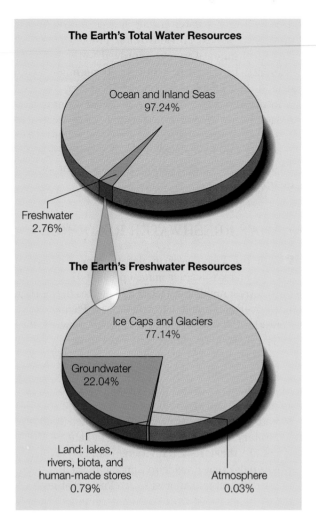

Figure 7–2

The Earth's water resources

SOURCE: U.S. Geological Survey, http://ga.water.usgs.gov/edu/earthwherewater.htm

In 1999, a comprehensive assessment of the freshwater resources of the world was submitted to the United Nations. A principal concern identified in the report was that increasing demands are being placed on Earth's renewable, but limited, freshwater resources. On a global scale, many people lack sufficient quantities of water to survive, and it is unclear how their future demands will be met. Traditionally, Canada and other nations have augmented their supplies of water to meet increased demands. Applying this strategy in the future likely will become more costly because of the greater distances and heights that water must be pumped, and because of increased social and environmental impacts. Navigation, hydro generation, and maintaining healthy ecosystems and fisheries all require adequate supplies of water. Diverting water to supply increased urban, agricultural, and industrial demands can detract from these uses.

It is unclear if a water crisis will occur in the future. We often have solved previous water problems through technological innovation and improved water resource assessments. In the future, these actions might again assist us in developing new supplies. However, a secure water future also is threatened by degraded water quality and a lack of investment in water and waste treatment systems. Future water management strategies must focus on increasing supply, reducing demand, integrating water quality and quantity, and considering water throughout the hydrologic cycle.

CANADA'S FRESHWATER RESOURCES

Climate is the key factor in determining freshwater resources. Canada's average annual precipitation is about 600 millimetres, ranging from 100 millimetres in the High Arctic to over 3500 millimetres along the Pacific Coast. Over one-third of our precipitation falls as snow, most of which is released to runoff and river flow in spring (Pearse, Bertrand, & MacLaren, 1985). Given our northern environment, 100 000 glaciers alone are estimated to contain 1.5 times the volume of surface waters (Government of Canada, 1991).

Canada's **groundwater** resources (defined as all subsurface water) continue to be a largely hidden resource. **Aquifers** are those rock formations (sands, gravels, or other materials) that provide reasonable flows and quantities of groundwater. Nationwide, we know that groundwater represents about 37 times the total amount of water contained in rivers and lakes in Canada (Government of Canada, 1991), but we know very little about its precise location, extent, quantity, and quality. This ignorance reflects the complexity of surficial and bedrock geology. As well, our abundant **surface water** resources have received much of our attention. Nevertheless, 26 percent of Canadians depend on groundwater sources for their water supplies, and 90 percent of agricultural water relies on this hidden resource.

Lakes are valuable elements of our water supply; they act as storage basins in regulating the flow of rivers to the sea, provide for water transport, and supply a rich habitat for natural life and human enjoyment. With about 7.6 percent of the country covered by fresh water in lakes and rivers, Canada probably has more lake area than any other country. For instance, Canada has 45 lakes with an area larger than 1000 square kilometres, and has (or shares with the United States) seven of the world's largest lakes.

Wetlands, primarily located in Manitoba, the Northwest Territories, and Ontario, represent 14 percent of Canada's area. These act as the kidneys for many river systems because they filter and purify water. Wetlands also moderate stream flow during floods and droughts.

Annual runoff, the flow of water in rivers, is the best measure of a nation's water supply because this flow is

TABLE 7–1
WATER SUPPLY PER CAPITA FOR SELECTED COUNTRIES

Country	Supply ($10^3 m^3$/year/person)
Canada	120
Brazil	45.2
Australia	19.7
United States	11.5
France	3.19
China	2.23
India	1.90

SOURCE: "Appraisal and Assessment of World Water Resources," I. A. Shiklomanov, 2000, *Water International*, 25(1), 11–32.

renewed by the hydrologic cycle. In contrast, only a small portion of lakes and groundwater reservoirs is renewed annually. For instance, only 1 percent of the Great Lakes is renewed through precipitation—the remaining 99 percent of the water represents meltwaters from the last ice age. This point suggests that proposals to divert water from the Great Lakes must consider the desirability and feasibility of tapping into this nonrenewable portion of the resource.

In Canada, runoff is estimated to be about 105 000 cubic metres per second (Laycock, 1987; Pearse, Bertrand, & MacLaren, 1985). That's enough water to fill 210 million backyard swimming pools per day! With 9 percent of the world's flow, this volume of water places Canada third in the world in water endowment (following Brazil at 18 percent and the former Soviet Union at 13 percent). Canada has about 0.5 percent of the world's population. When runoff per capita is calculated, we have over twice the supply per capita of Brazil (see Table 7–1). These data suggest that Canada is one of the "have" nations when it comes to an abundance of fresh water. However, there does not appear to be a direct relationship between economic growth and water availability. For instance, levels of economic development in Australia, France, and Israel are comparable to those in Canada, but availability of water is very different.

The supply of and demand for water are not evenly distributed over time and space. Average runoff rates vary across the country. About 60 percent of Canada's water drains north, while 90 percent of the population lives within 300 kilometres of the Canada–U.S. border. This means that many areas in the southern part of the country experience restricted water supplies. Five of the six water-deficient areas in the country are located in the more arid regions of Canada that, ironically, are dominated by agricultural land uses (see Figure 7–3).

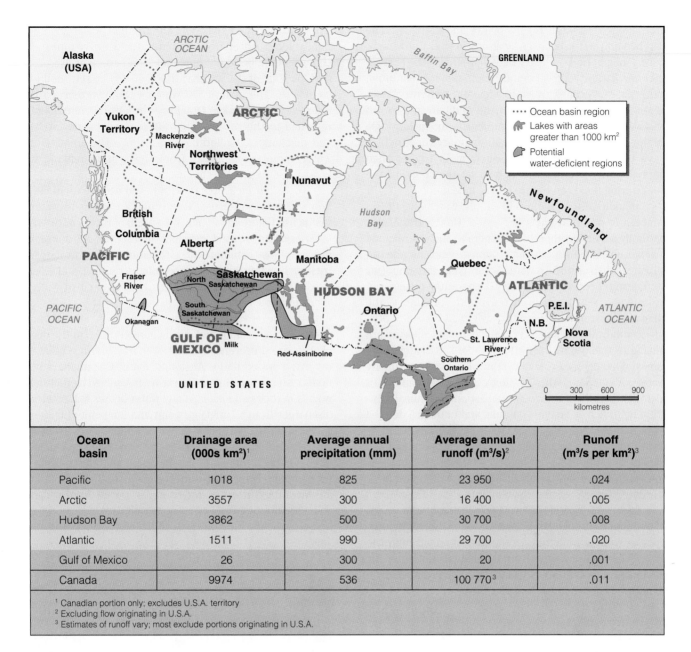

Ocean basin	Drainage area (000s km²)[1]	Average annual precipitation (mm)	Average annual runoff (m³/s)[2]	Runoff (m³/s per km²)[3]
Pacific	1018	825	23 950	.024
Arctic	3557	300	16 400	.005
Hudson Bay	3862	500	30 700	.008
Atlantic	1511	990	29 700	.020
Gulf of Mexico	26	300	20	.001
Canada	9974	536	100 770[3]	.011

[1] Canadian portion only; excludes U.S.A. territory
[2] Excluding flow originating in U.S.A.
[3] Estimates of runoff vary; most exclude portions originating in U.S.A.

Figure 7–3

Drainage regions and potential water deficiency regions in Canada

SOURCES: *Currents of Change: Final Report, Inquiry on Federal Water Policy*, P. H. Pearse, F. Bertrand, & J. W. MacLaren, 1985, Ottawa: Environment Canada; *Water is a Mainstream Issue*, Inquiry on Federal Water Policy, 1984, Ottawa; *Water: The Emerging Crisis in Canada*, H. A. Foster & W. R. D. Sewell, 1981, Toronto: James Lorimer, in association with The Canadian Institute for Economic Policy.

Water availability is a major concern even in the Great Lakes basin, the world's largest freshwater lake system, because demands for water are the highest there. Water demands of the 80 percent of Canadians who live in urban areas, which cover only 0.2 percent of Canada, are substantial. Meeting their demands and treating their sewage wastes can be a particularly significant problem when we realize that 60 percent of Canadians live in urban centres of 500 000 or more (Environment Canada, 1998).

Stormwater sewer and combined sewer (stormwater and wastewater) discharges, which can contain nutrients, biocides, fertilizers, and industrial wastes, also are urban-related water problems that add to these pressures.

Runoff rates also vary with time. During wet periods, lakes are renewed by river flow; river flow in turn is a function of the rainfall. During long droughts, little infiltration or percolation occurs to recharge groundwater storage, and there is little runoff to maintain river flows and lake levels. As a result, plants and trees become parched and water bodies may shrink in size. Conversely, during long periods of heavy rainfall, there is little absorption into saturated ground and runoff and sedimentation increase, sometimes causing lakes and rivers to flood. Major floods in June 1995 on the Oldman and South Saskatchewan rivers in southern Alberta, and in July 1996 in the Saguenay River–Lac Saint-Jean area in Quebec, have been attributed to intense rainfall. Higher than normal winter precipitation helped set the stage for the 1997 spring floods on the Red River in Manitoba.

In addition to variations in supply, water quality is a concern in different parts of Canada. Concentrations of naturally occurring impurities (minerals) such as calcium, magnesium, sodium, potassium, sulphate, and chloride can impair the use of surface water for drinking, swimming, or supporting diverse aquatic life. Groundwater quality can be affected too. In parts of the Prairie provinces, the Niagara Escarpment, New Brunswick, and Nova Scotia, the groundwater is so salty that most plant species are unable to tolerate it. In North Battleford, Saskatchewan, almost 6000 people became ill in April 2001 after drinking water contaminated with *Cryptosporidium.* Groundwater in other parts of Canada contains toxic chemical constituents (such as arsenic, fluoride, and uranium) derived from mineral deposits, while other substances (such as iron) create taste and colour problems. The quality of fish habitat can suffer as a result of these chemical changes as well as from low flow levels that reduce a river's capacity to dilute, dissolve, or absorb pollutants.

The global hydrologic cycle provides for "free trade" of pollutants through land, air, and water. Essential nutrients as well as pollutants are transported through the cycle. This explains why levels of contamination in Arctic mammals and fish are so high (see Chapter 8) even though very few humans live there.

WATER USES AND PRESSURES ON WATER QUALITY

WATER USES

Every day, at home and at work, we use water in so many situations—cooking, washing, bathing, watering lawns, carrying away the unwanted byproducts of our lives—that we are inclined to take it for granted. The two basic types of water use are instream uses and withdrawal uses. **Instream uses**, including hydroelectric power generation, transportation, waste disposal, fisheries, wildlife, heritage conservation, and recreation, occur "in the stream" (water remains in its natural setting). **Withdrawal uses**, such as municipal use, manufacturing,

irrigation, mineral extraction, and thermal power generation, remove water from its natural setting by pipes or channels for a period of time and for a particular use. All, part, or none of the water withdrawn may be returned to its source. The quantity of water withdrawn or used is referred to as **intake**. **Discharge** refers to the amount returned to the source. The difference between intake and discharge is called *consumption*—the amount of water removed or "lost" from the system making it unavailable to downstream users.

Two other measures of withdrawal use are recirculation and gross water use. **Recirculation** refers to water that is used more than once in a specific process or distribution system, or used once and then recycled to another process. Most commonly, recirculation occurs in industries such as pulp and paper, petroleum refining, and steel making. **Gross water use** is the total amount of water used (intake + recirculation).

Aquatic environments may be altered, also, if the chemical, physical, or biological characteristics of the discharge are different from the characteristics of the receiving water body. Although many Canadians now understand that pollution issues are not solved by diluting wastes in rivers and lakes, many water bodies still receive discharges from a variety of point and nonpoint pollution sources. **Point sources** discharge substances from a clearly identifiable or discrete pathway such as a pipe, ditch, channel, tunnel, or conduit from an industrial site, for example. **Nonpoint sources** discharge pollutants in an unconfined manner, for example, runoff from urban and agricultural areas.

Photo 7–2
On-farm and commercial feedlots, such as this one in Alberta, are potential nonpoint sources of water contamination.

From the perspective of supporting ecosystem species, functions, and processes, it is vital to know how much water is required to meet instream needs. In Canada, this knowledge has been growing but is still incomplete. We do, however, have a better understanding of how much water is needed for withdrawal and instream uses.

Withdrawal Uses

Thermal power production constitutes the largest withdrawal use of water in Canada, followed by manufacturing, municipal, agriculture, and mining withdrawals (see Figure 7–4). The thermal power industry, which includes both fossil fuel and nuclear electrical generating stations, was responsible for 63 percent of total water intake in 1991—four times more than the next biggest user, manufacturing. According to Statistics Canada (2000), generating one kilowatt of electricity—about enough to light a small house for one hour—requires 140 litres of water in a typical fossil fuel generating plant and 205 litres in a typical nuclear generating station. Most of this water is used for cooling purposes and is returned to its source at an elevated temperature. The manufacturing sector primarily uses water as a coolant, solvent, transport agent, and source of self-generated energy. Note that our use of thermal power has increased constantly since 1972. In contrast, the manufacturing and mining sectors have reduced their withdrawals through improved water technology efficiency. Increased recirculation of water reflects some of the declines in water use for these sectors.

In total, more than 45 billion cubic metres of water were withdrawn from Canadian sources in 1991, an 88 percent increase since 1972 (Government of Canada, 1996). This amounts to 4500 litres of water withdrawn per person per day for all uses and to about 330 litres per person per day for residential use. Urban water uses vary across the country, from a high of over 560 litres per day in Newfoundland to about 185 litres per day in Prince Edward Island (see Figure 7–5). A high intake of water produces a high volume of wastewater, which means the costs of supporting our municipal water infrastructure (water and wastewater treatment plants and pipes) are high. However, Canadian water rates are very low compared with those of other countries (see Figure 7–6). Communities generally have not paid the full cost of providing water, partly because water costs were subsidized

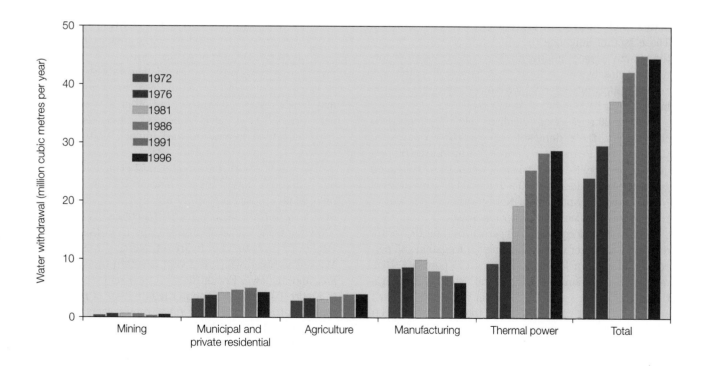

Figure 7–4

Water use in Canada, 1972–1996 (million cubic metres per year)

NOTE: Data for some sectors have been extrapolated and rounded.

SOURCE: *Water Use,* Environment Canada, http://www.ec.gc.ca/water/en/manage/use/e_use.htm. Reprinted with permission of the Minister of Public Works and Government Services Canada, 2004.

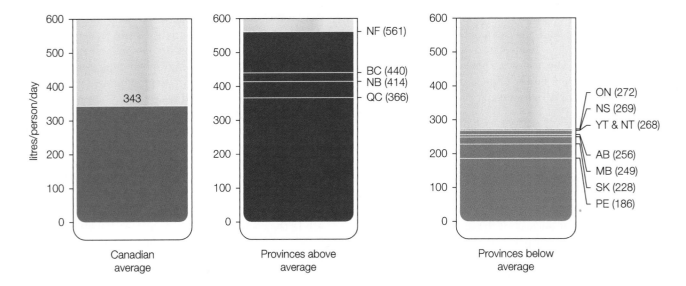

Figure 7–5

Average per capita urban water use, by province (late 1990s), litres/person/day

SOURCE: D. Scharf, Environment Canada, personal communication, 2000.

by property taxes and/or grants from provincial or federal governments and partly because major investments in such infrastructure as water treatment plants, sewage plants, and pipes were postponed or not made at all. These factors influence the efficiency of water use and the effectiveness of treatment. While low water rates correspond closely with high use, they fail to reflect the true cost of supplying water to consumers and provide little incentive for people to curb water use.

About 26 percent of Canadians rely on groundwater sources for their domestic supply (Government of Canada, 1996; Pearse et al., 1985). For instance, 100 percent of Prince Edward Island's population relies on groundwater, as does over 60 percent of the population of New Brunswick and Yukon. Nationwide, 82 percent of rural residential users rely on groundwater.

Agriculture accounted for 9 percent of total 1991 withdrawals. This water was used mainly for irrigation (85 percent), but also for livestock watering (15 percent). High rates of evaporation from agricultural fields means irrigation is the largest *consumer* of water in Canada. (Do not confuse this point with the fact that thermal power generation is the largest *withdrawal user* of water in Canada.)

Instream Uses

Instream water uses cannot be measured in the same manner as withdrawal uses since the water is not removed from lakes or streams. Instead, flow rates and water levels are the critical measurements.

In 1997, Canada obtained over 62 percent of its electricity from falling water (hydroelectricity); Newfound-land, Quebec, and Manitoba produced almost 100 percent of their electricity in this manner. With some untapped hydroelectric sites still available in Quebec, Newfoundland, Manitoba, British Columbia, and the Territories, and given Canadians' large appetite for energy (see Chapter 11), proposals for additional hydro dams may be expected in the future. However, these developments have adverse human and environmental effects, may conflict with withdrawal and other instream water uses, and likely will generate controversy (see "Hydroelectric Generation and Impacts" later in this chapter).

Water transport remains the most economical means of moving our important raw materials such as wheat, pulp, lumber, fossil fuels, and minerals. The main transportation waterways are the St. Lawrence River, which allows passage of ocean-going ships from the Atlantic Ocean deep into the heart of North America; the Mackenzie River, which is a vital northern transportation link; and the lower Fraser River on the Pacific coast (refer to Figure 7–3). Since cargo in the hundreds of millions of tonnes is transported along these routes each year, it is important to have reliable and predictable lake and river levels.

The 1996 Canadian commercial inland fishery was valued at $64 million (Statistics Canada, 2000). An even more economically important instream use, sport fishing, drew 6.5 million anglers in 1990. They spent almost $5.9 billion on sport fishing–related goods and services. Swimming, boating, and camping are additional water-based recreational activities enjoyed by millions of Canadians each year.

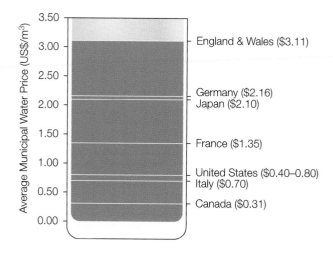

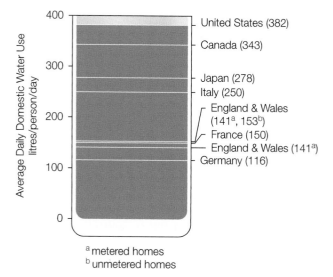

Figure 7–6

Estimated per capita residential water use and price for G7 countries in the 1990s

NOTE: Prices are calculated using a purchasing power parity method. Figures do not include the cost of waste treatment.

SOURCES: Adapted from *The Price of Water: Trends in OECD Countries,* Organization for Economic Cooperation and Development, 1999, Paris: Author; *The Poor Pay Much More for Water Use ... Use Much Less,* World Commission on Water for the 21st Century, 1999, www.worldcouncil.org. OECD source copyright © OECD, 1999.

Achieving an acceptable balance between withdrawal and instream uses is an ongoing challenge for water managers. Policies, legislation, regulations, and impact assessment procedures are some of the mechanisms frequently used to guide water and other resource management activities. Although water is becoming increasingly important as an international environmental issue and is of increasing importance to Canada's economic, social, and environmental agendas, the federal government cut budgets and staff at Environment Canada during the 1990s. In 1993, its Inland Waters Directorate (the lead federal agency for water) was eliminated; the Canada Water Act funding for collaborative activity was reduced drastically; and Environment Canada's capacity to provide other services such as hydrometric data collection and analysis was reduced (Bruce & Mitchell, 1995).

Provincial governments also have reduced their allocations to resource management agencies; neither federal nor provincial governments have made significant increases in their recent water management budgets. Water experts from across the country have provided input to Environment Canada in its efforts to identify new roles and responsibilities. All over the world, resource and environment managers are challenged to design effective institutional arrangements for water management.

PRESSURES ON WATER QUALITY

Water quality problems are seen most frequently at the local or regional level. Water managers must consider at least four aspects of water quality problems. The first aspect is the nature of the toxic or nontoxic substance, its position in the hydrologic cycle (see Chapter 3), and all the media (air, groundwater, surface water) that transport or store water. The second aspect of water quality concerns the source, amounts, types, timing, and location of substance releases. The third aspect concerns where the problem occurs, and focuses on a specific geographic region such as the Great Lakes or the Fraser River. Finally, managers must determine how to implement any remediative action. A mix of research, monitoring and analysis, standard-setting, regulations, enforcement, and penalties can help achieve the desired goals. Examples that highlight some of these aspects of water quality management are described below.

Quality issues relating to municipal use of water include increased **biological oxygen demand (BOD)** and disease-causing bacteria in the discharge of (untreated) wastewater. In the past, a large proportion of Quebec's urban population had no sewage treatment services, but noticeable progress has been made in constructing treatment plants for municipalities with populations of over 5000 (Government of Canada, 1991, 1996). Progress has been made also in providing First Nations people with potable water and sewage disposal services, even though about 17 percent of reserve dwellings still have no sewage disposal facilities (Government of Canada, 1996). In July 1996, the federal government announced a new $98.5 million program to address the most pressing water and sewage problems that posed potential health risks to community members on reserve lands. This money, in addition to $125 million allocated for water and sewer projects on reserves in 1995, provides an indication of the magnitude, as well as the cost, of rectifying water supply and treatment problems on reserves (Barnett, 1996). In 2003, the federal budget called for an additional $600 million over the next five years, including an initial investment of $200 million in the first two years, to upgrade, maintain, and monitor water and wastewater treatment facilities on reserves. (Government of Canada, 2003).

Over the past 40 years, the volume of Canadian crop and livestock production has increased considerably. Often these gains were achieved through new technologies involving mechanization, genetics, biocides, fertilizers, and irrigation. The gains in production through the use of these technologies sometimes have come at the cost of degraded water quality. Agricultural uses raise many questions, including the public cost of irrigation works, potential environmental and health problems associated with fertilizer and pesticide-laden runoff, and soil salinization and erosion problems. Since most crop irrigation takes place in the west, principally in Alberta, these issues are of particular importance there. In the South Saskatchewan River basin, for example, water returning to the rivers via irrigation channels and field runoff carries with it not only fertilizers, pesticides, salts, and sulphates in the form of suspended solids, but also dissolved solids at levels up to double what they were before irrigation (Hamilton & Wright, 1985).

Clearly, such contaminants reduce the overall quality for the next surface or groundwater user as well as for aquatic life. Leaky drainage pipes, evaporation, and seepage from irrigation canals and ditches result in inefficient use of water. In some cases less than one-third of the water withdrawn from a stream reached the intended crops (Government of Canada, 1991). In the Great Lakes–St. Lawrence region, most water withdrawals for agricultural purposes are used in stock watering.

Discharge of water used in mining (for cooling, drilling, and operating equipment) and in recovering oil from tar sands (for deep well injection) may degrade water quality by adding suspended solids, heavy metals, acids, and other dissolved substances. Tailings ponds may leak, discharging contaminated water into groundwater or surface bodies. Arctic residents, for example, are concerned about water quality deterioration from abandoned metal mines, as well as from oil and gas developments. Polluted tailings ponds at abandoned uranium mines in Saskatchewan are highly toxic and pose serious risks to the Aboriginal peoples who live near them.

Pollutants that are contained in the wastewater from the manufacturing sector range from biodegradable wastes to substances, such as PCBs and PAHs, that are toxic to fish, wildlife, and humans. Effluent from pulp and paper mills includes solid waste and chlorinated organic chemicals such as dioxins and furans, all of which may have detrimental effects on aquatic ecosystems. Residents in British Columbia, northern Alberta, northern Ontario, and the Atlantic provinces have expressed concerns about such discharges. Most thermoelectric plants use water as a coolant, returning water to the source in essentially the same quantity but at a higher temperature. This "heat pollution" can harm aquatic species, such as trout, that require cool water. Increased water temperature also can increase evaporation rates, which may raise salt concentrations to unacceptable levels (Government of Canada, 1991; Pearse et al., 1985).

Instream uses also impact water quality. Hydroelectric turbines do not consume water, but significant amounts evaporate from storage reservoirs. Damming rivers not only converts wild rivers into regulated ones but also results in loss of habitat for wildlife and fish, barriers to fish movement, and changes in downstream river flow regimes. Hydro dams have inundated valuable agricultural land in the Columbia, Kootenay, and Peace River valleys in British Columbia and have destroyed some salmon runs. In the dry Okanagan valley, irrigation sometimes has reduced tributary flows late in the season and affected fish propagation (Pearse et al., 1985).

Commercial navigation requires high water levels, which may cause bank erosion, disturb bottom sediments, and threaten beaches, while dredging to maintain depth degrades water quality. Relative to navigation and hydro-electrical generating interests, shore property owners in the Great Lakes–St. Lawrence basin prefer lower water levels because they minimize erosion and protect the owners' beaches, dock facilities, and other property. Shipping may facilitate the introduction of exotic species (such as the zebra mussel—see Box 7–1—introduced into Lake St. Clair via the ballast water of a European ship) and cause pollution, including spills of hazardous materials that may pose threats to municipal water supplies and recreation. Ice-breaking operations also pose a threat to fish and wildlife (Government of Canada, 1991; Pearse et al., 1985).

Discharging municipal and industrial wastes into water bodies has become increasingly less feasible. This is not only because growth in industrial production and population generates more wastes for discharge (often exceeding a water body's **assimilative capacity** or threshold capacity, discussed in Chapter 3), but because industrial wastes in particular contain persistent and toxic contaminants that remain in the environment. Assimilative capacity refers to the ability of a water body to accept sewage and other substances without significant harm to plants, organisms and animals, human health, or other water uses. These contaminants affect fish and wildlife and their particular habitat requirements: many species are highly sensitive to changes imposed by pollutants as well as by dams, diversions, and wetland drainage. In northern and coastal areas, fish and wildlife

Historically, pollution in the Lake Erie basin was moderated in a process by which the productive algae and fine soil particles from farmland erosion absorbed or adsorbed pollutants. As a result, Lake Erie organisms showed relatively low concentrations of toxic contaminants compared with the other Great Lakes. This may change, however, as eroded soil and nutrient levels decline and as zebra mussels deplete algal populations, thus increasing rates of bioaccumulation of contaminants.

The original aquatic community of Lake Erie was devastated by the almost total removal of native vegetation from the basin and by the exotic fish species that invaded after commercial fisheries severely exploited the native species. Along with carp, zebra mussels have impacted heavily on the recovering aquatic community. Voracious filter feeders, zebra mussels are not strongly affected by natural predators or diseases; the zebra mussel population has exploded and caused rapid changes in water quality and clarity as well as in the food web.

Zebra mussels consume large amounts of phytoplankton; their feeding caused a 77 percent increase in water transparency (clarity) between 1988 and 1991. Increased clarity of water permitted sunlight to penetrate deeper, in turn allowing rooted aquatic plants to spread into deeper water. Many organisms benefited ecologically from this change, but in some areas plant growth interfered with swimming and boating.

The mussels' eating habits, which both deplete the phytoplankton food source also used by other filter feeders and assimilate toxic contaminants, have affected the food web and may result in major changes in the future abundance of various species of fish. Mussels remove large amounts of particulates, which means that more contaminants remain in the water. The result could be higher contaminant concentrations in the remaining phytoplankton and zooplankton, as well as higher concentrations in fish and wildlife species feeding on the plankton or directly on the mussels and other bottom dwellers.

When zebra mussels arrived, they created physical problems such as clogged intake pipes and jammed machinery. We now know, however, that the invasion by this exotic species has far more complex effects. Chemical and biological methods have been proposed to control the mussels, but, mindful of other biocontrols that have proved disastrous, most of the scientific community is reluctant to take these measures. This situation highlights the point that restoring and protecting the Great Lakes ecosystem requires a commitment to achieving sustainability, fostering cooperation and coordination, and preventing pollution problems before they arise.

SOURCES: *State of the Great Lakes: 1995*, Governments of the United States of America and Canada, 1995, Ottawa: Supply and Services Canada; "Zebra Mussels: Holding Back the Tide," 1997, *Coastal Heritage, 11*(4), 10–12.

provide the major source of income and are valued as food, as integral to a way of life, and as recreational and aesthetic resources. Water-based recreational pursuits usually do not involve withdrawing or consuming water, but are affected by water body features such as surface area, depth, rate of flow, quality, temperature, and accessibility.

Canadian cities often suffer water quality problems because they rely on aging infrastructure. Sanitary sewers transport wastewater from residences and businesses to treatment plants, while storm sewers capture rainwater or snowmelt and flow into nearby water courses including natural streams, human-made channels, or pipes. Combined sewers carry both sanitary and storm drainage. During dry weather, combined sewers carry all contents to treatment plants. During wet weather, the volume of water may exceed the capacity of treatment plants. Consequently, this mix of water may flow untreated into natural water bodies (e.g., Fraser River, Lake Erie, Atlantic Ocean). Combined sewers were the standard until World War II and are still found in older parts of many Canadian cities, including Edmonton, Winnipeg, Toronto, Ottawa, Montreal, Quebec City, and Halifax

(Environment Canada, 2002).

Stormwater runoff can be a significant pollutant of municipal waterways. During wet weather, the rains flush pollutants from the streets, sometimes overloading the storm drains and flushing directly into waterways within urban ecosystems. In Vancouver, the occasional high loading of pollutants into the Fraser River from surface runoff can exceed that from sewage treatment plants. It is often difficult to trace the source of pollutants because there is no single "point source" that can be monitored and regulated. Stormwater can contain many kinds of pollutants, including trace metals that are part of gasoline, chemicals used in lawn and garden care, as well as chemicals used in commercial operations such as dry cleaners. These collect and drain into the sewage system during storm events. Factors such as land use, rainfall intensity, buildup time, and traffic intensity affect the quality of stormwater runoff and the kinds of contaminants that can enter the municipal system. In the greater Vancouver region, residential areas and roads are the largest source of contaminants, followed by industrial areas. While industrial areas have higher concentrations of pollutants, residential areas in greater Vancouver occupy the largest geographic area. On a positive note, the switch to lead-free gasoline has improved the quality of urban waterways.

On a national basis, Canadians withdraw only about 2 percent of our water resources from their natural settings. We *consume* less than 1 percent of that amount. This level of consumption might suggest that our instream resource needs are always being met. However, important regional and local problems exist in areas such as the Old Man River watershed in the southern Prairies, and individual tributary watersheds in the Great Lakes basin. Since all of these watersheds are linked to others through the hydrologic cycle, finding the right geographic scale to define and solve a water problem is crucial for successful management.

Most of Canada's serious water use problems are related to degraded water quality and to disrupted flow regimes, not to inadequate supply. First Nations people are among those whose traditional water-based activities have been most affected by deterioration from pollution and by water storage and diversion projects that manipulate lake levels and river flows. "There is hardly a major drainage system anywhere south of the Arctic which has not been affected by these pressures" (Pearse et al., 1985, p. 48). Furthermore, growing uncertainty about economic and social trends and about the impacts of human influences on climate, land use, and the distribution of water means that water management policies must support, protect, and promote a high-quality, sustainable water supply.

Photo 7–3
The purpose of this image of a fish beside a sewer lid in Montreal is to remind urbanites that the underground network of pipes is linked to "natural" waterways that have flora and fauna we seek to sustain.

THE IMPORTANCE OF WATER

Water as a Common Link

When we stop to think about our water systems, we realize how important water is to the Canadian economy and the Canadian identity—how water is a common link between citizens from every part of the country and, indeed, the world. Although water is absolutely essential in all spheres of life, Western societies have not always accorded water its full significance and have permitted its waste and degradation. This contrasts clearly with some First Nations people's view of the natural world and water's place in it (see Box 7–2). Recently, however, more and more Canadians have been demanding protection of their natural heritage, including water resources and aquatic ecosystems.

One of the ways in which rivers are being protected for future generations is through the Canadian Heritage Rivers System (CHRS). The goal is to establish a system of Canadian Heritage Rivers that reflects the diversity of Canada's river environments and celebrates the role of rivers in Canada's history and society. As well, the public is encouraged to learn about, enjoy, and appreciate Canada's rivers. On January 18, 1984, the Canadian Heritage Rivers Board was established to administer the Canadian Heritage Rivers System and to review river nominations for inclusion in the program. As a cooperative program of the federal, provincial, and territorial governments, CHRS objectives include ensuring long-term management that will conserve and protect the natural, historical, and recreational values of the best examples of Canada's river heritage (see Van Tighem, 1990).

In 1996, the French River in Ontario was the first Canadian Heritage River to be named. By 2003, a total of 39 heritage rivers across Canada had been designated for both natural and human heritage values, with management plans to address these values. Ten more were nominated, for a total (designated and nominated) of 9836 kilometres. Although only participating governments may nominate rivers to the CHRS, private citizens or groups may suggest rivers to the responsible provincial or territorial parks agencies.

Water as a Source of Conflict

As fundamental as water is to life on Earth, climate change has the potential to disrupt such water-dependent activities as agriculture and forestry. If precipitation amounts or distribution patterns change, or if temperature regimes are altered, a variety of both positive and negative impacts will be felt in these two sectors and throughout the economy.

While scientists do not agree about the specific consequences of global warming (see Chapter 5), any changes in the distribution of water resources would

BOX 7-2
A FIRST NATIONS VIEW OF THE NATURAL WORLD

The earth is central to our values. We consider ourselves part of a family that includes all of creation. We refer to the earth as Mother Earth, the giver of all life. We refer to the sun as our eldest brother and the moon as our grandmother. We consider the animal and plant life our brothers and sisters. We consider the waters of the world to be the bloodlines of Mother Earth.... We must make sure that those are always clean so that there will not be a heart attack some day to our mother.

We believe that all of creation has been given instructions by the creator. These instructions are meant to ensure that all of creation can live in harmony and peace.... The waters of the world have been instructed to quench the thirst of all life. It is said that when we drink each cold glass of water, when our throat is so dry, that there isn't a more wonderful feeling of peace and [tranquillity] than what that fresh cold glass of water can do ...

Our philosophies are based on the circle of life. To us, all life is seen as revolving in a circle and interrelated. Because movement is circular, any activity or decision made in the present will be felt in the future. In this circle, we do not see ourselves as separate or above the rest of the natural world, but an integral part of it.

Our lifestyles reflect our closeness to the natural world. We are fishers, trappers, hunters, gatherers, and farmers. These lifestyles keep us in touch with the natural world, spiritually, mentally, and physically, on a daily basis. Our close dependence on the natural world means that we must be thankful that the different parts of the natural world are fulfilling the instructions given to them by the creator. It also reminds us that we must also fulfill our instructions as well.

Because of the close relationship we have with the natural world, we cannot have healthy communities unless we have a healthy environment. Our ancestors have always understood this.... We have always recognized the importance of water. The rivers and lakes are used to transport our people from one community to the next. Fish and waterfowl have long been the major source of food for our people. The plants along its shores are the source of our medicines.

We have always followed the natural laws of the world. These laws are rooted in common sense. They say that if something you plan to do could be detrimental to the natural world, [then] don't do it. If we look at all of creation as part of our family, [then] the decisions we make must ensure that our family will come to no harm either today or in the future. Our dependence on the natural world requires us to follow these laws....

... When the Europeans first came to North America, they could not understand why First Nation people would not sell their land. For us, it was an issue of would you sell your mother? Today we cannot understand why it is okay to discharge pollutants into the waters of the world. It is akin to allowing drugs to be injected into [your] mother's blood and [then] saying it is okay because the blood will dilute it.

SOURCE: *Water Is Life,* by J. W. Ransom, September 1995, Technical Bureau Supplement to *Water News*, Canadian Water Resources Association. Reprinted by permission.

require Canadians to adapt to changing water conditions. Also, since Canada shares many water bodies with the United States, there could be increased pressure to negotiate water management agreements that satisfy the needs and desires of both countries. Depending on the scarcity or abundance and distribution of water, this could again raise the issue of large-scale water exports (Simpson, 1993/1994). This means that we need to address the goal of water sustainability for both the short and the long term, and to employ policies and programs that achieve not only ecological sustainability but also desired sociocultural outcomes.

In the global context, almost 150 of the world's 214 major river systems are shared by two countries, and another 50 are shared by three to ten nations. Many nations already have seen conflict over access to shared water resources, and the future is likely to bring more conflict, especially if global warming occurs as predicted and severe droughts become more frequent.

In the Middle East, the 1967 Arab–Israeli war was fought partly over access to water from the Jordan River basin. In the future, the issue of sharing water between countries will become even more contentious as demands grow in response to increasing population and continuing agricultural and industrial development. For example, Turkey's $20 billion Southeast Anatolian Project, "the biggest engineering feat in history," is using about 2 percent of the annual flow of the Euphrates River for irrigation and has plans to use about one-third of the Euphrates' total flow. Unless the principal states of the Tigris–Euphrates basin (Turkey, Syria, and Iraq) cooperate in determining optimal use of water, a crisis is inevitable because all three countries' plans for irrigation and hydroelectric projects would consume about one-third more than the total flow of the Euphrates (Pope, 1996; Frederick, 1996).

More recently, "water terror" has been an effective weapon in civil wars around the world. During the Bosnian conflict, Serbs discovered how "to hit their enemies where it really hurt: in the water supply." By shutting off Sarajevo's electricity, and with it the city's water pumps, city residents, including dozens of Muslims, were forced to line up at wells, making them easy targets for snipers and mortar shells (Serrill, 1997).

In the Mekong River basin, Thailand's demand for water is growing by more than 12 percent per year, Laos is considering major hydroelectric developments, Vietnam is seeking water for irrigation and industrial needs, and Cambodia is concerned about maintaining its fishing industry in light of the major dam and water diversion schemes proposed. If a series of major dams were constructed on the Mekong's main section, critics warn that rare marine species could be threatened, downstream fertility could decrease, flooding and saltwater incursions could increase, and the delta area could be threatened because of reduced silt flow (Pope, 1996).

Water as a Hazard

Sufficient water of appropriate quality is essential to a productive economy. From Canada's early days, surface and groundwater resources have supplied the demand for domestic consumption, transportation, irrigated agriculture, food processing operations, power generation, and industrial cooling. The abundance of inexpensive water and hydro-generated power has been attractive to industry. As well, the presence of water has motivated individuals to locate their homes, cottages, and other facilities on the shores of Canada's many lakes, streams, and rivers.

While communities and industries choose to locate on or close to water bodies for transportation, water supply, and recreation, the hazardous nature of water is not always fully appreciated. Dangers associated with floods and storms may compromise the safety of people and their property, and damages may entail the loss of lives and property, as well as high cleanup and repair costs. In the Saguenay–Lac Saint-Jean region floods in Quebec in July 1996, about 12 000 people were evacuated from their homes as more than 50 towns and villages were inundated by flood waters that resulted from more than 270 millimetres of rain. Ten people died, 100 homes were washed away, about 1000 homes and 20 major bridges were heavily damaged, and $800 million in damages resulted. In addition, more than 100 tonnes of toxic waste and chemicals spilled into the Saguenay River when floods damaged or destroyed industrial facilities such as pulp and paper mills. Although chemicals such as phosphoric acid, urea, and ferric chloride were dispersed in the heavy river flow, there was concern for toxic hot spots in river sediments. A $450 million relief fund was established to compensate flood victims, and many Canadians contributed to the relief effort.

The 877-kilometre-long Red River regularly floods Manitoba's valley of silty loam left behind following the retreat of Lake Agassiz about 7700 years ago. In some years the floods are more severe than others—in terms of crest levels, the 1826 flood remains the worst—it reached 11.1 metres above the river bed when it crossed the forks of the Red and Assiniboine rivers in the heart of Winnipeg. Manitobans still measure the Red River floods relative to the 1826 flood (Pindera, 1997).

In 1948 and 1950 the river crest reached 9.2 metres; 80 000 people fled from Winnipeg, 20 000 residents were evacuated from rural areas, 13 000 homes and farms were flooded, and damage estimates totalled $606 million in 1997 dollars. After those floods, a plan to "tame the Red" was implemented. Massive clay dikes and diversion dams were built in the Winnipeg area and the $63 million Red River Floodway was opened in 1968. A 47-kilometre-long channel that diverts the Red River around Winnipeg, the floodway was one of many permanent structures erected to control the flow of flood water. By the 1970s, eight towns between the U.S. border and Winnipeg were protected with permanent dikes. In 1979, a flood similar to the 1950 flood tested these defence systems; most held back the flood waters and only 7000 people fled their homes.

The history of flooding in the Red River valley reveals that floods commonly occur when the winter snowpack in the southern headwaters of the river thaws before the northern Manitoba reaches of the river are free from ice. From November 1, 1996, to April 20, 1997, record levels of snow fell on ground that had been saturated through a wet autumn. Areas south and east of Winnipeg received winter precipitation amounts that were 175 percent of the annual average, and parts of the Red River drainage basin in North Dakota received precipitation amounts of more than 200 percent of the annual average. Just days after the runoff began in early April, a major storm dumped another 50 to 70 centimetres of snow and freezing rain on top of the near-record snowpack of 250 centimetres.

The 1997 spring flood extended over 205 000 hectares, or about 5 percent of Manitoba's farmland, creating a 2000-square-kilometre "Red Sea." Had the floodway, dams, and dikes not been built, it is estimated

Photo 7–4
Some Manitoba communities were inundated by the 1997 Red River flood despite sandbagging.

that the crest of the flood, the second-highest in Manitoba history, would have measured 10.4 metres. Instead, the floodway kept the water level at 7.5 metres. Still, 28 000 Manitobans were evacuated (6000 from Winnipeg), and 2500 properties between Winnipeg and the U.S. border were damaged. Even though 45 000 laying hens and 2000 cattle were moved to safety, dairy farmers' losses were estimated at $1.3 to $2 billion. Total damages tallied more than $500 million, mostly for repairs to roads, bridges, farms, and homes. In December 2000, the International Joint Commission (2000b) released its final report on the Red River flood. Key conclusions were:

- While the 1997 event was a natural and rare event, floods of the same or greater magnitude could be expected in the future.
- The people and property of the Red River basin will remain at undue risk until comprehensive, integrated, and binational solutions are developed and implemented.
- Since there is no single solution to the problem, a mix of structural and nonstructural adjustments is required.
- Specific flood management activities are required in several Canadian and U.S. communities.
- Ecosystem considerations must be an important consideration in managing future flood damage reduction initiatives; hazardous materials must be carefully controlled and banned substances removed from flood-prone locations.
- A culture of preparedness and flood resiliency must be promoted throughout the basin.

Given the importance of ecologically healthy water bodies for habitat and ecosystem support (for all flora and fauna), and our dependence on water for life support and economic and recreational activities, the principle of sustainability must apply to the full range of water uses. That is, the value of wetlands and estuaries, of aquifers and groundwater recharge areas, and of precipitation regimes must be accounted for in the planning and management of water resources. Implementing these initiatives will require leadership at all levels of government.

Human Activities and Impacts on Freshwater Environments

As previously noted, human use of water resources—for domestic and urban, industrial, power generation, and recreational purposes—has a variety of impacts on fresh water and freshwater environments. This section covers in more detail examples of the kinds of impacts that result from each of these four types of water use.

DOMESTIC AND URBAN USES AND IMPACTS

Safe Drinking Water and Sanitation Facilities

On a global basis, one of six people lives without regular access to safe drinking water. Canada's drinking water is safer than that of many other nations. In developing countries, 80 percent of illnesses are water-related; globally, about 34 000 deaths occur daily from contaminated water and waterborne diseases (Environment Canada, 1990). These statistics reflect the facts that more than 1.5 billion people lack safe drinking water supplies, about 1.7 billion have inadequate sanitation facilities, and developing nations and the global community are unable to protect their drinking water from several threats that Canadians rarely think about.

In Canada, where municipalities have the responsibility to provide their citizens with safe drinking water, there are four major approaches to ensuring high-quality tap water. These approaches are land use planning; drinking water quality guidelines and regulations; water treatment systems; and reporting of results. The role of guidelines, regulations, and treatment systems is discussed below.

Drinking Water Guidelines and Regulations   The Guidelines for Canadian Drinking Water Quality indicate that good-quality drinking water is free from disease-causing organisms, harmful substances, and radioactive material. Good-quality drinking water also tastes good, is aesthetically appealing, and is free from objectionable odour or colour. Although they are not legally binding, these guidelines specify limits for substances and describe conditions that affect drinking water quality in Canada. Under the Canadian Environmental Protection Act, these guidelines act as "environmental yardsticks" to help assess water quality issues and concerns, establish water quality objectives at specified sites, provide targets for control and remediation programs, and provide information for state-of-the-environment reporting.

Provinces also establish laws that specify standards for treatment-plant operating procedures and how testing results will be distributed. Traditionally, publicly owned utilities operating at the municipal level have provided Canadians with drinking water. There is no single approach that utility companies use to implement drinking water quality. For instance, British Columbia, New Brunswick, Ontario (since early 2001), Quebec, and the Yukon require all samples to be tested at labs that have been certified, while the Northwest Territories provides little control. (See Sierra Legal Defence Fund [2001] and Table 14–2 for additional information on provincial approaches.)

Drinking Water Treatment Facilities   Treatment plants ensure the delivery of safe, clean water to the majority of Canadians. A typical water treatment process is described in Figure 7–7. The major goal of water treatment is to ensure, through a series of filters, that water is free of sediment, a problem frequently found in surface water sources. Filters may be so effective that waterborne pathogens such as *giardia* are removed. Disinfectants, such as chlorine, are used effectively in most water treatment systems to kill waterborne microorganisms. However, there are concerns that the use of chlorine can produce trihalomethanes (THMs), specifically chloroform, which can cause cancers of the liver and kidneys. Newer technologies include ozone and ultraviolet light. The failure to provide treatment facilities to remove pathogens and microorganisms contributes to the high incidence of disease and death in the developing world.

Canadians are now seriously questioning the ability of drinking water plants to adequately treat water. In April 2001, the residents of North Battleford, Saskatchewan, learned that the parasite *Cryptosporidium,* which caused illnesses, had entered the city's water supply. An independent judicial inquiry to investigate why the system failed considered water facilities, the role of managers and regulators, as well as government regulations and policies. The tragic consequences of improper water treatment were also experienced in Walkerton, Ontario, where seven people died and 2300 more became ill as the result of contamination of the town's water system in May 2000. The source of the contamination was identified as *E. coli* (*Escherichia coli*) O157:H7, a lethal strain of the common, usually harmless, *E. coli* bacterium found in the intestinal tract of humans and animals. Given Canadian water regulations and treatment systems, a key question became, "How did *E. coli* get in the drinking water?" Independent evaluations of both localities found human error and ineffective government regulation to be key factors in leading to these tragedies.

The inquiry into the cause of the Walkerton tragedy revealed the following:

- The contaminated groundwater well, constructed in 1978, had not been approved by the Ontario Ministry of the Environment (MOE), and nothing was done to rectify the situation.
- During 1996, the Ontario government closed MOE and Ministry of Health (MOH) water testing laboratories, actions taken without consideration of the capacity in the private sector, or among municipalities, to test municipal water supplies (Kreutzwiser, 1998).
- Private water testing laboratories were required only to report water testing results to municipalities that had requested the tests, rather than to the MOE or MOH, where water tests previously were conducted and through which MOE and medical officials could be alerted.
- *E. coli* had been present in the 1978 well prior to May 2000, and nothing had been done to rectify this.
- Walkerton's water treatment plant manager and workers lacked adequate training to effectively operate the facility; they falsified labels and test samples, and claimed the water was clean even when it was contaminated with *E. coli.*
- In May 2000, a heavy spring rainfall caused higher-than-normal runoff, some of which was contaminated with *E. coli*–infected manure and entered the 1978 well.
- At this time Walkerton's chlorinator was inoperative.
- Lab reports identifying the presence of *E. coli* were not acted on by the Public Utilities Commission (which knew the lab results), delaying for four days the order to boil water.

A follow-up study suggested that water contamination resulted from many failures in a complex system involving workers, local governments, regulatory bodies, and provincial governments. Government cutbacks and lack of regulatory enforcement contributed to both the Walkerton incident as well as the contamination of water supplies in North Battleford (Woo & Vincente, 2003). One analysis of the costs of Walkerton identified $64.5 million in hard costs and an additional $90.8 million in the less tangible costs of illnesses suffered and lives lost (Brubaker, 2002). While it is difficult to determine longer-term impacts, children under five years of age and elderly people may have other longer-lasting complications arising from the primary infection.

Some observers believe that the problems in Walkerton and North Battleford might not be isolated incidents—that increased development is depleting and degrading our water resources and putting our water supplies at risk. There is also concern about Canada's aging water and wastewater infrastructure and the need to provide significant funding to improve it.

The Sierra Legal Defence Fund (2001) compiled a report card on Canada's drinking water systems, grading each province and territory on such criteria as how it protects its drinking water sources, the quality of water treatment and testing, and ability to inform the public (Table 7–2). Alberta, Ontario (after Walkerton), and Quebec were given the best grades, while British Columbia, Yukon, Prince Edward Island, and Newfoundland received the lowest grades. Concerns were raised that no jurisdiction has developed and applied advanced standards and approaches similar to those used in the United States. Ontario has now established a Nutrient Management Act that sets high standards for activities that might introduce pollutants into watercourses. How other governments across Canada will respond to the Walkerton and North

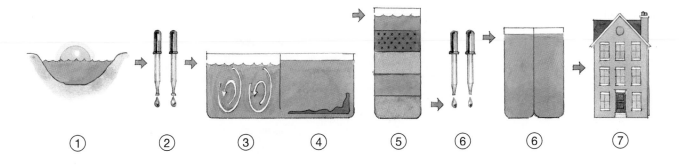

1. **Water intake pipe**

   The water intake pipe extends into the lake, river, or well that supplies water to the facility. In large lakes, these pipes may extend hundreds of metres from the facility. Where necessary, chlorine is used to keep zebra mussels from colonizing in the system.

2. **Chemical storage and feeding**

   Chlorine is added to the incoming water to kill microorganisms. Alum and lime may be added also. Alum concentrates suspended particles such as silt to aid their removal. Lime changes the pH level where required. The chemicals are mechanically mixed into the water before moving on to the flocculating basin.

3. **Flocculating basin**

   The flocculating basin stirs the water to concentrate suspended particles. The clumps of particulate that form are known as *floc*.

4. **Settling basin**

   Heavy flocs drop out of the water in the settling tank and collect along the bottom. The settled floc is removed by scrapers that move along the bottom. The cleanest water is left at the surface to be drawn off through spillways that lead to filtering basins.

5. **Rapid sand filters**

   The water is already quite clear by the time it reaches the stacked layers of fine sand, activated carbon, gravel, and rocks that form the rapid sand filters. The layer of sand removes fine bits of floc, algae, and silt from the water. The layer of activated carbon removes taste- and odour-producing chemicals.

6. **Pure water basin**

   The purified water goes into holding basins prior to distribution. Safe levels of chlorine are added to check the growth of algae and microorganisms. Lime may be added to control the pH level. A controlled pH level protects the metal components of the distribution system from corrosion.

7. The treated water is tested regularly to ensure quality. Large holding tanks store the water until it is needed in people's homes.

Figure 7–7

Typical municipal water treatment process

SOURCE: *Water and Wastewater Treatment,* Ontario Clean Water Agency, http://www.ocwa.com/frproces/htm. Reprinted with permission.

Battleford crises and the Sierra Legal Defence Fund report remains to be seen.

Wastewater Treatment Facilities Once municipal water has been used and goes down the sink or toilet, treatment of the wastewater is required. In urban centres, this takes place in sewage treatment plants (see Figure 7–8). Municipalities may offer one of three levels of sewage treatment. **Primary treatment**, the lowest degree of treatment, is a mechanical process involving removal of large solids, sediment, and some organic matter (steps 1–4 in Figure 7–8). From that point, even though it contains many pathogens, the fluid may be discharged to a receiving water body (thus the danger of contamination) or may enter a secondary treatment process. The sludge is removed and taken to a digester for further processing.

Secondary treatment employs biological processes in which bacteria and other microorganisms degrade most of the dissolved organics. After treatment, about 30 percent

TABLE 7-2
GRADING CANADA'S DRINKING WATER TREATMENT SYSTEMS

Province	Grade	Comment
Alberta	B	With Quebec and Ontario, the best of a bad lot
British Columbia	D	Rich province, poor regulations
Manitoba	C–	Not bottom of class, but ...
New Brunswick	C–	Strengths outweighed by serious weaknesses
Newfoundland	D	Should do more homework
NWT	C	No longer a regulatory deep freeze
Nova Scotia	B–	Has recently pulled up its socks
Nunavut	C	Uses NWT regulation. See above.
Ontario (pre/post revision)[1]	D/B	After a hard lesson, showing improvement
PEI	F	Bottom of the class
Quebec	B	With Alberta and Ontario, top of a lacklustre class
Saskatchewan	C	Middle province with middling to poor water protection
Yukon	D–	Frontier mentality poses a threat

[1] The first grade was assigned before Walkerton. The post-Walkerton grade is effective 2002.

SOURCE: *Waterproof: Canada's Drinking Water Report Card,* Sierra Legal Defence Fund, 2001, http://www.sierralegal.org/reports/waterproof.pdf. Reprinted with permission of the Sierra Legal Defence Fund.

of phosphates and about 50 percent of nitrates remain in the suspended solids (steps 5 and 6 in Figure 7–8).

In aeration tank digestion (activated sludge process), effluent from the primary process is mixed with a bacteria-rich slurry, air or oxygen is pumped through the mixture, and bacterial growth decomposes the organic matter. Water siphoned off the top of the tank typically is disinfected, usually by chlorination, before it is released into the environment. Sludge is removed from the bottom of the tank; some of it may be used to inoculate the incoming primary effluent, but because of its toxic content (metals, chemicals, pathogens), most of the sludge is dried and landfilled. In Calgary, sewage sludge is injected into farmers' fields as a soil conditioner. Applied at a maximum of once every six years, the injected fields are monitored carefully and consistently for contaminants such as heavy metals.

If a municipality has space, it may construct a sewage lagoon where exposure of the effluent to sunlight, algae, aquatic organisms, and air slowly degrades the organic matter (with lower energy costs). Natural or constructed wetlands also act effectively to absorb nutrients and other pollutants at low cost. Even in Canada's cold climate, wetlands are proving to be important components of municipal water treatment systems.

Tertiary treatment is a chemical process that removes phosphates, nitrates, and additional contaminants such as salts, acids, metals, and toxic organic and organochlorine compounds from the secondary effluent. Sand filters or carbon filters may also be used in advanced wastewater treatment. The treated effluent may be discharged into natural water bodies or used to irrigate agricultural lands and municipal properties such as parks and golf courses. Because tertiary treatment greatly reduces nutrients in the treated wastewater, algal blooms and eutrophication are reduced in receiving waters.

Wastewater treatment levels vary considerably across Canada since it is the provinces that enact laws. While

Photo 7–5

This solar-powered sewage treatment plant in Woodbridge, Ontario, is a biological ecosystem that simulates a natural wetland for the purposes of treating domestic sewage.

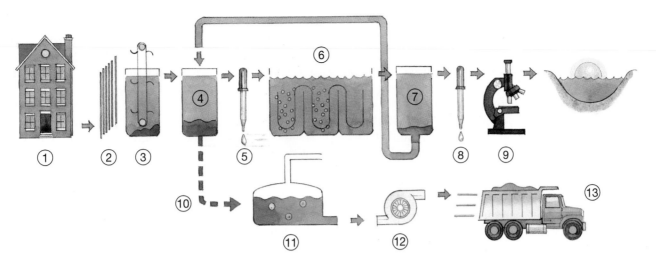

1. **Plant influent**
   Waste enters the treatment facility through the municipal sewer system.

2. **Coarse bar screen**
   Metal bars collect large debris such as rags, wood, plastics, and so on.

3. **Grit removal**
   The wastewater flows through a channel, allowing dense inorganic material to settle on the bottom. Scrapers, hoppers, and clam buckets remove the collected grits.

4. **Primary settling**
   The wastewater flows into large settling tanks, where suspended solids and organic material are allowed to sink to the bottom. The raw sludge that settles is removed through hoppers and sent through the digestion process.

5. Partially treated wastewater is drawn from the top of the settling tanks and chemicals are added to remove phosphorous.

6. **Aeration tanks**
   Large aeration tanks mix the partially treated wastewater with oxygen to support bacteria, which devour organic waste. The bacteria levels are managed to provide the most efficient removal process.

7. **Final settling**
   The cleanest wastewater is drawn from the top of the aeration tanks through spillways. By this point the water is already quite clear. Polymers may be added to concentrate any remaining material. Again, suspended particles settle to the bottom and are removed by scrapers or hoppers.

8. **Disinfection**
   The cleanest water is drawn from the surface and disinfected with chlorine or ultraviolet light to kill bacteria.

9. The treated water is tested to ensure it meets provincial standards and is returned to the original water source.

10. Sludge from the aeration and final settling tanks is drawn from the bottom of the tanks and pumped to the primary settling tank. Not only does this sludge have a high water content, but it contains oxygen and bacteria that improve the efficiency of the treatment process.

11. **Primary digest**
    Sludge removed throughout the process is pumped to digesters for processing. Anaerobic bacteria consume organic waste in the digesters. This process produces gases that are used to fuel plant boilers and heat facilities.

12. **Dewatering process**
    Vacuum filter or centrifuge systems remove water from the processed sludge to thicken it. The water removed in the process is pumped to the primary settling tank to reenter the treatment process.

13. The concentrated sludge, or bio-solid waste, is taken away for incineration or conversion into fertilizer.

Figure 7–8

Typical sewage treatment process

SOURCE: *Water and Wastewater Treatment,* Ontario Clean Water Agency, (n.d.), http://www.ocwa.com/frprocess/htm. Reprinted with permission.

almost all of the population received at least primary treatment and most received tertiary treatment of their wastewater, the highest levels of treatment exist in Ontario and the Prairie provinces. Nationwide, 6 percent of Canadians received no wastewater treatment; their wastewater was released straight into a receiving water body. In general, municipal wastewater treatment levels have been rising steadily in Canada. Although there are regional variations in treatment levels, between 1983 and 1996, the number of Canadians served with sewers who received some level of sewage treatment rose from 72 to 94 percent (Statistics Canada, 2000). During the same period, the number of people serviced by tertiary treatment rose from 28 to 41 percent.

The Sierra Legal Defence Fund confirmed these variations in sewage treatment quality when it surveyed 21 systems across the country during the 1990s (Table 7–3). Victoria, Saint John, St. John's, Halifax, and Dawson City dumped a total of 365 million litres of untreated sewage into watercourses daily. Montreal, Charlottetown, and Vancouver discharged a combined 2.4 billion litres of effluent that had been partially treated by settling tanks and skimming. At the time of the study, Calgary was judged to be the only city to have effective and environmentally sound technology. It applies ultraviolet disinfection after the wastewater has received tertiary treatment. Although there has been no update to these figures and upgrading of wastewater treatment systems is ongoing, the value of this report is to provide simple, yet effective, ways to understand the options for water treatment and the relative success of communities along a spectrum of waste treatment practices elsewhere in the country.

Water treated to a level that will protect human life may still stress aquatic ecosystems when discharged. This occurs because the lower the level of sewage treatment provided, the greater the biological oxygen demand of the

TABLE 7–3
GRADING WASTEWATER TREATMENT SYSTEMS IN CANADA: PROGRESS, 1994–1999

City	Grade		Comment
	1994	*1999*	
Victoria	F	F–	Preliminary screening, no treatment
Vancouver	D–	C–	Two plants upgraded from primary to secondary treatment
Edmonton	B–	B+	30% upgrade from secondary to tertiary, switch to UV disinfection
Calgary	A–	A	100% UV disinfection added to 100% tertiary treatment
Regina	C+	B	Switch from chlorine to UV disinfection
Saskatoon	D	C+	Upgrade from primary to secondary treatment
Brandon	C	D	Secondary treatment, no toxicity testing
Winnipeg	C–	C	Switching from chlorine to UV disinfection
Hamilton	C+	C–	Secondary and some tertiary treatment, chlorine disinfection
Toronto	B–	C/B	Secondary treatment/good plans formalized
Ottawa	B	C	Secondary treatment, chlorine disinfection
Montreal	F/C	F+	100% population now connected, no progress on treatment
Quebec City	C	C	Combined sewer overflow events reduced
Saint John	D–	E	53% of sewage not treated, some sewage receives secondary treatment
Fredericton	C	B	Switch from chlorine to UV disinfection
Charlottetown	D–	E	Primary treatment
Halifax	F	E–/C	Halifax/Dartmouth receive no treatment/good plans formalized
St. John's	F	F–	No treatment
Yellowknife	B–	B+	30% reduction in volume
Whitehorse	D	B–	From 2% secondary treatment to 100%
Dawson City	N/A	F–	Preliminary screening, no treatment

SOURCE: Adapted from *The National Sewage Report Card, Number Two,* Sierra Legal Defence Fund, 1999, http://www.sierralegal.org/reports/sewage.pdf. Reprinted by permission of the Sierra Legal Defence Fund.

effluent and the greater its impact on aquatic life. Also, unless wastewater receives secondary treatment, disease-causing bacteria may remain in the effluent. None of these levels of treatment has a proven ability to remove toxic substances, however, and persistent chemicals such as DDT, PCBs, or mirex are removed only by very advanced treatment such as that employing activated carbon. Endocrine disrupters have become a relatively recent concern in drinking water (see Box 7–3). Drinking water supplies may be affected severely by inadequately treated municipal wastewater discharges, industrial and agricultural pollution of water bodies, and leaking landfill sites (see also Chapter 8).

Demand for Water

Another urban water issue relates to domestic demand for water. At the same time as the safety and security of our water supplies are declining in some areas due to pollution, dropping water tables, and prolonged drought conditions, our demand for water is rising (Environment Canada, 1995a). Because of increasing demand, municipal water supply is becoming one of the most critical water issues in Canada. Residential water use accounts for nearly one-half of all the water used in Canadian municipalities. Each Canadian uses approximately 343 litres of water per day inside the home; the ways we use that water are identified in Figure 7–9.

Sometimes we do not realize how much water we consume in our everyday activities. For instance, 65 percent of all residential water use takes place in the bathroom; if your toilet is more than 10 years old, and you flush 4.5 times per day, over the course of one year you

5% cleaning

10% kitchen/drinking

20% laundry

30% toilets

35% showers/baths

Figure 7–9

Use of water in Canadian homes

SOURCE: *Water: No Time to Waste: A Consumer's Guide to Water Conservation,* Environment Canada, 1995, Ottawa: Supply and Services Canada.

BOX 7–3
ENDOCRINE DISRUPTERS

Research is showing that many toxic substances are capable of disrupting the endocrine system of fish, birds, reptiles, amphibians, and mammals. The endocrine system is responsible for many vital functions such as growth, development, reproduction, and the immune system. Natural hormones play a vital role in these functions. Endocrine-disrupting chemicals (EDCs), such as PCBs, dioxins, nonylphenols, lindane, dieldrin, and inorganic chemicals such as mercury and tin, mimic or inhibit hormones. These chemicals are produced and released by multiple sources, including municipal effluents, pulp and paper production, industrial processes, incineration, agricultural practices, and boat maintenance.

When released into the environment, EDCs lead to ecosystem impacts including enlarged thyroid glands in salmon and herring gulls. Cross-bill deformities, supernormal clutching, and decreased hatching success have been observed in birds from the Great Lakes. Human health also is threatened by exposure to persistent toxins from municipal sewage and industrial and agricultural chemicals. A study of Michigan women who consumed more than 0.5 kilograms of Lake Michigan fish per month during their pregnancies found that their newborns showed a significantly higher percentage of decreased birth weight, head circumference, and neurobehavioural development. Psychological tests administered to these same children when they were four years old indicated they were suffering from learning deficits.

SOURCE: *Endocrine Disrupting Substances in the Environment,* Environment Canada, 1999, http://www.ec.gc.ca/eds/fact/index.htm

Photo 7–6

Built to overcome a height difference of 99 metres between Lake Erie and Lake Ontario, the Welland Canal profoundly affected lake levels and flows.

$1.14 for 1000 litres. This amount has increased from about $0.82 per 1000 litres in 1991, and now includes a waste treatment component of about 39 percent. Because meeting increasing water demands requires the construction of new (or expansion of existing) facilities, all of which consume land, energy, and financial resources, water conservation is an important factor in improving the quality and protecting the quantity of water resources.

In fact, it is estimated that water metering reduces water consumption between 20 and 50 percent, and reduces the amount of wastewater requiring treatment by about 10 percent. For example, in 1999, Canadian households served by metered water systems used about 288 litres per person per day compared to 433 litres per day for households that paid flat rates for water. About 56 percent of Canadian municipalities use water meters, with the greatest proportion of those being in Ontario and the Prairie provinces. Water conservation programs, including restricting residential consumption during peak summer demand, are also effective in reducing consumption. Nevertheless, there is a need to address aging infrastructure. In 1996, it was estimated that over 50 percent of the urban water infrastructure was unacceptable and required an investment of $36 billion to correct the problem (Federation of Canadian Municipalities, 1996). Failure to make needed investments in water infrastructure can contribute to future water crises. In 2003, the federal budget allocated $3 billion to infrastructure improvements in Canadian cities. These funds were earmarked for all forms of infrastructure including roads and water supply facilities. Obviously, there are critical needs in urban areas that are not being met.

The Great Lakes and St. Lawrence River Basin: A Case Study

Straddling the Canada–United States border, the Great Lakes–St. Lawrence River basin contains about 18 percent of the world's fresh surface water, accounts for about 4 percent of the total length of Canada's shoreline, and is home to over 35 percent (about 9 million) of the Canadian population. About 33 million Americans also live in the Great Lakes basin. One out of three Canadians and one out of seven U.S. residents depend on the Great Lakes for their drinking water, which amounts to about 140 000 litres of water per second (Environment Canada, 1990). The Great Lakes also have played a major role in the development of both countries. The Great Lakes form part of the inland waterway for the shipment of goods into and out of the heart of the continent; are the site of industrial, commercial, agricultural, and urban development; and are a source of hydro, thermal, and nuclear energy and recreation.

Broad, long-standing issues affecting the Great Lakes–St. Lawrence River basin include deteriorating water quality through industrial and municipal uses, fluc-

will have used about 30 000 litres of water to dispose of about 650 litres of body waste. If your toilet is one of the 25 percent that leaks after flushing, you could be losing up to 200 000 litres of water in one year. Laundry accounts for 20 percent of domestic water use. If you water your lawn in summer, it could need about 100 000 litres of water during the growing season.

In 1994, 17 percent of Canadian municipalities with water supply systems reported problems with water availability. About 10 percent of Canadians on municipal water systems rely exclusively on groundwater for drinking water and other domestic uses. Larger municipalities that depend exclusively on groundwater are highly susceptible to water availability problems, particularly if the population in the municipality is growing, if the municipality permits unrestricted use of water, and if water prices are low.

The price Canadians pay for water varies significantly across the country, yet Canadians still pay the lowest rate for water among G8 nations (see Figure 7–6). A survey of municipalities in 1999 indicated that Canadians pay about

tuating water levels, flooding, and shoreline erosion. Acid precipitation, airborne toxins, depletion of wetland areas, the sale and diversion of water from the system, increased demand on shoreline recreational facilities, and climate change are among the concerns affecting the basin.

Since the early 20th century, significant changes in land use have occurred in the Great Lakes–St. Lawrence basin, including deforestation, drainage of wetlands, and urbanization. These changes have altered the runoff characteristics of the drainage basin, as have the navigational and other structures built to regulate the outflows of lakes Superior and Ontario. Other human activities have affected lake levels and the lakes themselves: examples include the Long Lac and Ogoki diversions built to bring water into Lake Superior for hydropower generation and logging; the diversion at Chicago that takes water out of Lake Michigan to support domestic, navigation, hydroelectric, and sanitation uses; and the Welland ship canal, built to bypass Niagara Falls and to provide water for power generation (Environment Canada, 1990). Dredging and channel modifications have affected lake levels and flows as well.

Compared with natural factors, human effects on lake levels are small. Nevertheless, changing Great Lakes water levels are of great concern because of their impact on the multiple and often conflicting uses of the lakes. As noted earlier, from an erosion and flood perspective, shore property owners often prefer relatively lower lake levels because low levels mean less risk to their properties. In contrast, shipping and hydroelectric companies prefer relatively higher lake levels because they permit the transport of more cargo and the generation of more hydroelectricity (Sanderson, 1993). Depending on the lake and climate change scenario employed, predictions are that lake levels may drop by 0.5 to 2.5 metres. This is a significant threat, not only to shipping and other economic concerns but also to the ecology of the Great Lakes and the lower St. Lawrence, where increases in tidal effects and saltwater intrusion could occur (Government of Canada, 1996).

Since there is an annual cycle of levels on the Great Lakes (high in summer and low in winter), as well as periodic effects from the passage of storms (on Lake Erie, for example, a major storm can cause short-term water level changes of as much as 5 metres), damage due to wave action can be extensive (Environment Canada, 1990; Sanderson, 1993). Along the Canadian shore of the Great Lakes, annual flooding and erosion damage averages about $2 million. However, for every 1 centimetre decline in Great Lakes water levels, 93 metric tonnes must be subtracted from the total load a Great Lakes boat can carry (Environment Canada, 1990). Balancing such competing interests, and improving shoreline management to achieve sustainability of water resources and related habitats, are among the challenging responsibilities of agencies such as the International Joint Commission, provincial governments, and municipal planners. Later in this chapter, efforts undertaken to restore the Great Lakes ecosystem are noted (see the section "Agreements between Canada and the United States").

Eutrophication and persistent toxic chemicals have been the focus of water quality issues in the Great Lakes basin for many years. By the late 1960s, scientists, policymakers, and the general public were aware that high levels of nutrients such as phosphorus and nitrogen were causing eutrophication of the Great Lakes. Lake Erie in particular was "dying" from uncontrolled growth of aquatic plants, lowered levels of oxygen, and conditions unfavourable to fish survival. During the same period, "the Cuyahoga River running through Cleveland was so clogged with oils and greases that it caught fire in 1969. The city had to build a fire wall and declare the river a fire hazard" (Royal Commission on the Future of the Toronto Waterfront [RCFTW], 1992). The United States/Canada Great Lakes Water Quality Agreement of 1972 initiated a program of activities that has now seen very significant declines in the concentration of nutrients.

By the late 1970s, more complex problems relating to synthetic toxic chemicals were the focus of attention. Over 360 compounds have been found in Great Lakes waters, more than one-third of which have been shown to be toxic to humans and wildlife. By 1985, 11 compounds had been identified as critical pollutants in the Great Lakes basin ecosystem (see Table 7–4) because of their **persistence**, recycling, wide dispersal, **bioaccumulation**, and **biomagnification** in the food web (RCFTW, 1992; International Joint Commission, 1992).

While levels of some critical contaminants in the Great Lakes ecosystem have been reduced, two factors combine to suggest that future improvement may be slow. The first factor is that contaminants are released continuously from sediment as the ecosystem slowly purges itself; the second factor is that the inputs of toxic substances continue. For example, although PCBs have been banned, PCB inputs continue because more than 50 percent of the PCBs that were produced are still in use, in storage, or at disposal sites. This means PCBs have the potential to enter the Great Lakes environment (International Joint Commission, 1992).

Of the many issues concerning the sustainability of the Great Lakes, pollutants remain one of the most difficult. The Canada–U.S. Strategy for the Virtual Elimination of Persistent Toxic Substances in the Great Lakes has revised deadlines and goals for many pesticides such as chlordane, aldrin, dieldrin, DDT, mirex, and toxaphene (see Table 7–4). These toxins and others are linked to a group of substances known as endocrine-disrupting chemicals (EDCs), discussed in Box 7–3.

According to the State of Great Lakes 1999 report, there have been mixed results in efforts to improve environmental conditions. On the positive side, there has been an increase in the number of wildlife species in the Great Lakes region, and 3000 hectares of wetlands and 200 kilometres of shoreline habitat have been rehabilitated. There also has been a large reduction in the levels of dioxins,

Chapter 7: Fresh Water

T A B L E   7 – 4

CRITICAL POLLUTANTS IN THE GREAT LAKES BASIN ECOSYSTEM (1991)

Pollutant	Use	Method of Entry into Great Lakes Basin
all polychlorinated biphenyls (PCBs)[a]	• insulating fluid in electrical transformers and in production of hydraulic fluids, lubricants, and inks • previously used as a vehicle for pesticide dispersal • includes 209 related chemicals of varying toxicity	• from air or in sediments
DDT and its breakdown products (including DDE)[b]	• insecticide • most uses stopped in Canada in 1970 • still used heavily for mosquito control in tropical areas on other continents	• from air or in sediments
dieldrin[b]	• insecticide once used extensively on fruits • use no longer permitted for termite control in Canada	• from air or in sediments
toxaphene[b]	• insecticide developed as a substitute for DDT • used on cotton • Canadian use virtually ceased in early 1980s	• from air or in water or in sediments
2,3,7,8-tetrachlorodibenzo-p-dioxin (TCDD) and	• chemicals created in manufacture of herbicides used in agriculture and for prairie (range) and forest management • byproduct of burning fossil fuels with chlorinated additives, wastes containing chlorine, and in pulp and paper production processes that use chlorine bleach	• from air or in water or in sediments
2,3,7,8-tetrachlorodibenzofuran (TCDF)	• created in production of pentachlorophenol (PCP) • contaminant in Agent Orange herbicide used in Vietnam • most toxic of 75 forms of dioxin (polychlorinated dibenzodioxins)	
mirex[c]	• fire retardant • pesticide to control fire ants	• from air or in sediments • residuals from manufacturing sites, spills, and landfills
mercury	• used in metallurgy • byproduct of paint, chlor-alkali, and electrical equipment manufacturing processes	• occurs naturally in soils and sediments • releases into aquatic environment may be accelerated by acidic deposition
alkylated-lead	• fuel additive • used in solder, pipes, and paint	• released when burning leaded fuel, waste, cigarettes, and from pipes, cans, and paint chips
benzo(a)pyrene	• produced when fossil fuels, wood, wastes, and charcoal are burned, including in forest fires • from automobile exhausts • one of many forms of polycyclic aromatic hydrocarbons (PAHs)	• product of incomplete combustion of fossil fuels and wood
hexachlorobenzene (HCB)	• byproduct of burning fossil fuels and wastes that contain chlorinated additives • found in manufacturing processes using chlorine • contaminant in chlorinated pesticides	• byproduct of combustion of fuels and incineration of waste

[a] manufacture and new uses prohibited in Canada and U.S.
[b] use restricted in Canada and U.S.
[c] banned for use in Canada and U.S.

SOURCES: Adapted from *Regeneration: Toronto's Waterfront and the Sustainable City: Final Report*, Royal Commission on the Future of the Toronto Waterfront, 1992, Toronto: Supply and Services Canada and Queen's Printer of Ontario, p. 105; *The State of Canada's Environment—1991*, Government of Canada, 1991, Ottawa: Minister of Supply and Services Canada, p. 18-15

furans, PCBs, and DDT. However, the report recognized that "an additional tenfold reduction may be needed to reach acceptable risk levels" (Environment Canada, 1999). Conditions in Hamilton Harbour are improving, as 12 new fish species have been observed there. Improved operations of sewage treatment plants have led to improved water quality.

While by most standards environmental quality has greatly improved in the Great Lakes area, there is consensus that the targets originally set fail to address adequately the damages that have resulted from practices of unsustainable human activities. On July 26, 2000, the International Joint Commission (IJC) released its Tenth Biennial Report on Great Lakes Water Quality. The IJC stated pollution levels were still too high and called for measures to address the problem of persistent toxic substances by strengthening the Binational Toxins Strategy. The IJC also suggested that "failure to address the challenge of restoration during this time of economic prosperity will result in future generations of Great Lakes citizens inheriting the consequences of our inaction" (International Joint Commission, 2000a). There is also concern over the emergence of a non-native fish, the round goby.

Other water quality issues exist in various locations throughout the Great Lakes–St. Lawrence basin, and the rest of the country as well. One example is Ontario's Trent–Severn Waterway, between Lake Ontario and Georgian Bay. It was conceived as a military defence route in the 1780s, but built for commerce in the 1800s (and completed in 1920). Today, nearly 200 000 recreational boats use its locks and channels. Communities that once needed the waterway as a means of bringing logs to their lumber mills now depend on tourism dollars from the boaters and tourists who patronize their marinas, stores, hotels, and restaurants. Water quality problems are among the environmental challenges that face the waterway. Phosphate pollution from agricultural lands and from lawns is causing algae and weed problems, and the growing population of cottagers has led to increased pollution from septic tanks. If these are not well maintained, they can leak disease-causing bacteria into lakes and rivers and contaminate shallow groundwater supplies. Also, as the number of residents and cottagers has grown, development has encroached into wetlands, threatening some rare plant species (Cayer, 1996). These concerns highlight the importance of land use planning to help ensure water quality throughout Canada.

INDUSTRIAL USES AND IMPACTS

Industrial sites have been located near water bodies not only for the water they supply, but also for the transportation, cooling, and effluent discharge roles they ful-

fill. Historically, ignorance about the cumulative impacts of industrial activities on the environment led to groundwater and soil contamination and other downstream effects. Today, greater awareness has reduced the severity of impacts, but human and wildlife health-related concerns from previous industrial activity continue to surface.

Industrial activities, ranging from food processing to manufacturing, rely on large amounts of clean water to produce their goods as well as water to carry away the waste and byproducts of these processes. While manufacturing withdraws the second-highest volume of water in the country (see Figure 7–4), use of more efficient water technologies has enabled this sector to withdraw significantly less water since 1981. With continued emphasis on efficiency and conservation, further reductions in withdrawal rates (perhaps as much as 40 to 50 percent) are expected to continue, with no sacrifice of economic output or quality of life.

Groundwater Contamination

Toxic industrial byproducts enter aquatic environments, including groundwater, as point sources of waste disposal or pollution. Point sources related to waste disposal sites for industrial chemicals include underground injection wells for industrial waste, waste rock and mill tailings in mining areas, and coal tar at old gasification sites. Point sources of pollution include leaks or spills from tanks or pipelines that often contain petroleum products, wood preservation facilities, and road salt storage areas. In terms of groundwater contamination, the pollution potential is greatest if disposal occurs in or near sand and gravel aquifers. Over many years in Ville Mercier, Quebec, industrial wastes had been placed in lagoons in an old gravel pit. Predictably, water supplies of thousands of residents in the region were rendered unusable, and a replacement supply had to be pumped from a well 16 kilometres away (Environment Canada, 1996a).

The overall extent of groundwater contamination from industrial sources in Canada is not known, although hundreds of individual cases have been investigated. Such cases include contamination by the pesticide aldicarb in Prince Edward Island, industrial effluents in Elmira, Ontario, and various pesticides in the Prairies, as well as creosote contamination in Calgary and other communities and industrial contamination in Vancouver. Frequently, recognition of the contamination occurs only after water users have been exposed to potential health risks and the cost of cleaning the water supply is extremely high (Environment Canada, 1996a). Cleanup often is impossible, and groundwater aquifers may remain contaminated for decades.

In addition, contaminated groundwater migrates via the hydrologic cycle to nearby rivers and lakes, creating surface water pollution that can be as serious as contamination of groundwater supplies. Given that groundwater moves slowly, it may take decades before contamination is detected; scientists expect the discovery of additional contaminated groundwater aquifers and new contaminants during the next few decades. Experience suggests prevention of contamination in the first place is by far the most practical solution to this problem.

Among the wide variety of industrial chemicals in commercial use worldwide, dense non-aqueous phase liquids (DNAPLs) are particularly troublesome. DNAPLs include dry-cleaning solvents (such as trichloroethylene and tetrachloroethylene), wood preservatives (see Box 7–4), and chemicals used in asphalt operations, automobile production and repair, aviation equipment, munitions, and electrical equipment. DNAPLs can be generated and released in accidents such as the tire fire in Hagersville, Ontario. Heavier than water, DNAPLs sink quickly and deeply into the ground, where they dissolve very slowly and then may move with the groundwater flow. These sources of contamination are very difficult to find and almost impossible to clean up (Environment Canada, 1996b). Except in large cities, drinking water is rarely tested for these contaminants, yet in 1995 a study of more than 480 municipal and communal groundwater supplies determined that almost 12 percent of them contained detectable concentrations of either trichloroethylene or tetrachloroethylene (Government of Canada, 1996). Both of these **organohalides** are toxic and are known to affect the central nervous, respiratory, and lymph systems in humans. Tetrachloroethylene is a suspected human **carcinogen**; trichloroethylene causes liver damage in experimental animals (Manahan, 1994).

Leaking underground storage tanks and piping constitute another widespread impact of industrial activities on water resources. During the past two decades there has been an increasing number of leaks of petroleum products because the 30- to 40-year-old tanks in which these products were stored had inadequate corrosion protection. (Before 1980, most tanks were made of steel and up to half of them leaked by the time they were 15 years old.) Most petroleum products have the potential to contaminate large quantities of water: for example, one litre of gasoline can contaminate one million litres of groundwater. In the Atlantic provinces, where groundwater usage is high, the problem of leaking underground storage tanks is particularly severe. Very often the problem is detected only when people start smelling or tasting gasoline in their water (Environment Canada, 1996c).

Impacts on Beluga Whales

Industrial (and agricultural) pollution poses an insidious threat to the small population of threatened beluga whales (estimated at 525 members in 1992) that inhabit the St.

Lawrence estuary ("Beluga Habitat," 1996). Research scientists are trying to confirm why the whales are not healthy. It is known that beluga calves are being poisoned by their mothers' milk, and that older beluga are dying from cancerous tumours (Béland, 1996). Pollution is one possible culprit, as the blubber and bodies of nearly all of the 15 beluga that wash ashore in an average year are heavily contaminated with more than 25 potentially toxic contaminants, likely the result of biomagnification. These include heavy metals such as mercury and lead, PCBs, DDT, and pesticides such as mirex and chlordane. Beluga carcasses are so contaminated they could qualify as hazardous waste. Other possibilities are being examined also, including in-breeding problems and marine traffic from tourists that hampers reproduction (McIlroy, 1995).

The possibility that industrial and agricultural pollution and habitat degradation are major contributors to ill health among beluga remains a prime target of investigation. For instance, the pesticide mirex, once manufactured upstream of the St. Lawrence near Niagara Falls, has been banned in Canada and the United States since 1970. There appear to be few traces of mirex in the St. Lawrence, but it is still found in Great Lakes sediments. It seems that eels migrating from the Great Lakes carry the pesticide to the St. Lawrence, where the beluga eat the eels (Norris, 1994). These substances bioaccumulate in individuals and/or biomagnify in the food chain. An aluminum smelter that once dumped waste containing a powerful carcinogen, benzo(a)pyrene, into the Saguenay River (which feeds the St. Lawrence) also is a possible cause of beluga deaths. This, in conjunction with the fact that discharges of PCBs have been eliminated and that levels of PAHs and lead have been reduced by 97 and 92 percent, respectively, may suggest that residual contamination of the river is a problem ("Beluga Habitat," 1996). (Industrial pollution of rivers is discussed in Chapter 8.)

Researchers at the University of Montreal have examined 73 of the 175 beluga cadavers found on the shores of the lower St. Lawrence since 1983. They discovered an enormous number of whales—almost 20 percent (14 of 73)—suffered from cancer. This number accounts for more than half the cancer cases found among all the dead whales in the world. Only 1 of the 1800 whales that washed ashore and were examined in the United States, for example, was found to have cancer, and there were no cases at all among Arctic beluga ("Smelters Suspected," 1995).

Acidic Deposition

Atmospheric changes, some of which result from industrial activities, are impacting freshwater systems. Acidic deposition continues to be a severe stress on freshwater ecosystems in eastern Canada. Water chemistry models predicted that up to 20 000 of the lakes in these areas would become acidic (pH under 5) under 1980 emissions

BOX 7–4
CREOSOTE WASTE PRODUCTS AND CONTAMINATED SITES

In Canada, creosote is used as a wood preservative on railway ties, bridge timbers, pilings, and large-sized lumber. Creosote itself is composed of hundreds of compounds, the largest group being the polycyclic aromatic hydrocarbons (PAHs).

Areas of contaminated soil, water, or materials result from the application, manufacture, storage, transportation, or spillage of creosote. While the total amount of waste creosote entering the Canadian environment is not known, it is estimated that 256 000 cubic metres of moderately and highly contaminated soil is associated with 11 abandoned or operating creosote application facilities in Canada. In all provinces except Prince Edward Island, waste creosote is known to be entering soils, groundwater, and surface waters at 24 sites.

In Calgary, creosote pooled beneath the Bow River near the site of a former Canada Creosote wood preservative plant has entered the river and reduced the number of benthic invertebrates (caddisflies and stoneflies) within about one kilometre downstream from the site. These species, important to the fish on which the Bow River's world-class fly fishery is based, were replaced by less sensitive snails and crane flies. Waste creosote and the PAHs found in it were detected at levels higher than those known to cause severe effects to freshwater and marine organisms.

Given that some residents downstream of Calgary drink water taken directly from the Bow River, and given that Siksika Nation people have noticed irregularities in the fish caught along their section of the river, a gravel berm was constructed as a temporary measure to stop creosote from mixing with the flow of the river. The berm was not watertight but reduced creosote seepage to the riverbed surface by up to 80 percent in the short term.

Alberta Environment and Environment Canada officials determined that a permanent, 635-metre-long, subsurface, impervious containment wall and treatment system were necessary. Costing close to $4 million, the containment project includes a pumping system to take the dirty water containing the lighter creosote oils (which float to the top of the water table in the ground) into an on-site water treatment plant prior to their discharge into

Photo 7–7
Creosote treatment site.

Calgary's sanitary sewer system. Contaminated residues from the water treatment plant are transported off-site to an approved facility for reuse or disposal. Annual costs to operate the pumping wells and water treatment are about $100 000. Since some of the creosote that soaked into the soil and rock was heavier than water, it is impossible to clean up the site totally with current technology.

Under the National Contaminated Sites Remediation Program, the Alberta and federal governments shared the cost of this project. Even though the company that owned and operated the plant from the 1920s until 1964 is known and is responsible for the contamination, the company is not paying for the cleanup because practices at the site then were standard for the time and the company probably was not contravening any legislation.

Associated with the containment project are plans for increased river and on-site monitoring of the atmosphere, land, and water to ensure public safety.

SOURCES: *Creosote-Impregnated Waste Materials,* Environment Canada & Health Canada, 1993, http://www.ec.gc.ca/library/elias/bibrec/302071B.html; *Summary Material from November 17, 1994,* Alberta Environmental Protection, 1995.

levels (Jones et al., 1990). Sulphate deposition has declined from a high of 40 kilograms per hectare per year to an average of 10–15 kilograms per hectare per year, but since the critical value is thought to be less than 8 kilograms per hectare per year, the current reduction programs in Canada and the United States (targeted at an objective of 20 kilograms per hectare per year) mean that as many as 25 percent of Atlantic Canada's lakes likely will not recover, even if reduction programs are implemented fully (Government of Canada, 1996).

It is clear that in spite of improvements in industrial emissions and reduced deposition loads, many lakes are

continuing to lose populations of fish and other freshwater species. The variable levels of improvement in lake acidity seen in monitoring studies of 202 lakes throughout southeastern Canada showed that in Ontario, 60 percent of lakes tested were improving, 7 percent were worse, the rest were stable; in Quebec, 11 percent were improving, 7 percent were worse, the rest were stable; and in the Atlantic region, 12 percent were improving, 9 percent were worse, the rest were stable (Government of Canada, 1996; Schindler & Bayley, 1991).

More recent research has suggested that the effects of acid rain, global warming, and ozone depletion are having **synergistic effects** (outcomes in which the effects of two or more substances or organisms acting together are greater than the sum of their individual effects—they are multiplicative not additive). The effects include (1) a series of reactions that reduce the likelihood that acidified lakes can ever fully recover; and (2) enhanced movement of organochlorine pollutants and mercury from warm regions to regions at high altitudes and latitudes. Scientists estimate that SO_2 emissions from the United States and Canada need to be reduced by 75 percent. However, limits for another significant airborne water pollutant, nitrous oxides (NO_x), have yet to be determined. The case of acid rain highlights the complexity of dealing with water issues. Although progress has been made in reducing sulphur emissions, synergistic effects are weakening the ability of lakes to recover, very significant reductions in sulphur are required, and there is no agreement on the amount of emissions from other substances that might have to be reduced.

Recognizing that Canada's environment must be viewed as an integrated system of ecosystems means that the 1991 Canada–United States Air Quality Agreement may be viewed partly as a response to human impacts on water. That is, because the atmosphere is an important medium for the exchange of matter and energy within and among ecosystems, actions affecting air quality are important in a water quality context also. In this instance, the focus of the Canada–United States Air Quality Agreement is on reducing emissions of SO_2 and NO_x. The link between air and water quality and acid deposition is clear. In trying to meet the need for reduced emissions, the federal government initiated a Canada-Wide Acid Rain Strategy for Post-2000. The strategy pointed out the importance of the following:

- making further SO_2 reductions in the United States
- setting new SO_2 reduction targets in eastern Canada
- establishing targets for nitrogen
- maintaining ongoing acid rain science and monitoring
- preventing pollution
- keeping existing clean areas clean
- annual reporting

HYDROELECTRIC GENERATION AND IMPACTS

Historically, Canada has been one of the world's major builders of dams and diversions. By 1991, 650 major dams had been built or were under construction in Canada; over 80 percent of the dams are for electric power generation and many of the diversions are used to concentrate flows for hydroelectric development. While substantial economic benefits result from damming and altering rivers, these activities also incur wide-ranging, long-term

ecological consequences related to the effects of impounding water in reservoirs and of altering natural patterns of stream flow (Table 7–5). Social disadvantages also occur, such as when communities are flooded out or people are forced to modify their traditional ways of life and livelihood.

From an ecological perspective, the Peace–Athabasca delta area in northern Alberta is an example of how the balance of an ecosystem that depends on flooding can be altered by construction of dams and reservoirs. The delta is a key staging and nesting area for waterfowl and is an important habitat for bison, moose, and various fish species. Since the late 1960s, when the Bennett Dam was constructed on the Peace River in British Columbia 1200 kilometres upstream from the delta, the dam has been lowering the annual flood peak levels of the Peace River. Effects on the ecosystem have been significant, particularly the successional trend from wetlands to less productive land habitats and decreased biodiversity (Environment Canada, 1993a; Government of Canada, 1991, 1996). As a remedial effort, weir construction has been undertaken to try to offset the effects of the upstream river regulation and enhance water levels for wetlands.

Other wetlands in the Saskatchewan River delta and the Atlantic provinces have been affected by the operation of upstream dams. Since the late 1960s in Atlantic Canada, hydroelectric developments have eliminated or modified extensive areas of freshwater wetlands, flooded property and wildlife habitat, altered summer water temperatures, and reduced downstream water quality during low flow periods, a factor that is particularly important for watercourses receiving pollutants (Environment Canada, n.d.a). Impassable dams on the Indian (Halifax County), Mersey, Sissiboo, and Meteghan rivers in Nova Scotia have rendered most Atlantic salmon habitat inaccessible, and inadequate flows for fish downstream of the Annapolis River dam presented serious problems in 1991.

In the late 1970s, the Smallwood Reservoir of the Churchill Falls hydroelectric project in Labrador experienced problems with high levels of methylmercury accumulation in fish downstream, but 16 years after the flooding, mercury levels dropped back to normal levels (Environment Canada, n.d.a). Concerns about the transfer of foreign species from one basin to another halted the Garrison Diversion project in North Dakota. This project could have introduced biota from the Missouri River system into the Hudson Bay watershed (Government of Canada, 1996). Cancellation of Phase 2 of the James Bay hydro project reflected concern about the ecological, social, and economic aspects of the megaproject (see Enviro-Focus 7).

It appears that the era of big dams and diversions slowed down after the 1970s in Canada. Between 1984 and 1991, only six large dams were constructed (e.g., Rafferty–Alameda, Oldman, and Laforge), some only after

TABLE 7–5

SELECTED ECOLOGICAL AND SOCIAL IMPACTS OF DAMS AND DIVERSIONS

Agent	Process/Comment	Impacts

Impounding water in reservoir	• Reservoirs enlarge existing lakes or create new lakes in former terrestrial or wetland ecosystems.
	• In Canada, hydroelectric reservoirs cover an area of 20 000 km^2.
	• Dams can lead to local and regional problems, but alternatives have environmental consequences: e.g., coal-fired thermal power plants emit more greenhouse gases and toxic substances such as mercury.
		• Displaces people.
		• Disrupts wildlife habitat.
		• Influences local climate.
		• May cause small earthquakes.
		• Affects water quality and quantity:
		  – increased evaporation, reduced flow downstream
		  – change in water temperatures
		  – decrease in dissolved oxygen level
		  – increase in nutrient loadings and eutrophication
		  – potential degradation or enhancement of fish habitat
		• Converts organic matter flooded by reservoir to toxic methylmercury that accumulates and magnifies in food webs, making fish unsafe for human consumption.
		• Causes shoreline problems due to fluctuating water levels: shoreline biological communities unable to establish; erosion and turbidity increase.

Altering natural stream-flow patterns	• Efforts to control flooding by building dikes, hardening shorelines, or regulating stream flow with dams and diversions must be balanced with the preservation of fish and wildlife habitats, ecosystem functioning, and the way of life of Aboriginal peoples.	• Reduces natural flooding and annual flood peak levels. • Causes loss of waterfowl and wildlife habitat. • Causes successional trends from wetlands to less productive terrestrial habitats, with consequent reduction in biodiversity.
Interbasin diversions	• Diversions involve the transfer of water from one river basin to another	• May result in transfer of fish, plants, parasites, bacteria, and viruses.

SOURCE: *The State of Canada's Environment—1996,* Environment Canada, Ottawa: Supply and Services Canada, chap. 10. Reprinted with permission of the Minister of Public Works and Government Services Canada, 2004.

lengthy public discussions. Long-standing plans for the Kemano River (British Columbia), Conawapa–Nelson rivers (Manitoba), and Phase 2 of the James Bay project (Quebec) have been shelved indefinitely. Combined, the changes in economics, priorities, and political influence (particularly of Aboriginal people) have made the construction of large dams and diversions less desirable and feasible.

This same trend is not evident in the developing world, where dam construction is proceeding quickly. However, as the cancellation of the Arun River dam in Nepal points out, changes in public and political thinking about the value of water resources will help protect the natural, cultural, and recreational values of rivers in the future. On the other hand, the construction of the Three Gorges dam in China, undertaken between 2000 and 2009, is the largest dam ever built both in terms of hydropower generation and the number of people requiring resettlement. Depending on the source of information, estimates suggest that between 1.2 and 1.9 million people will be moved; by its construction and resettlement plans, the dam has destroyed productive agricultural land and natural resources, and raised international protest (see Chapter 11). To compare the numbers of people moved, see Wu (1999), Li et al. (2001), and Black (2003).

RECREATIONAL USES AND IMPACTS

Many Canadians enjoy water-based recreational activities. The effects of these activities, including engine discharges,

Hydroelectric Dams in Northern Quebec: Environmental and Human Issues

Photo 7–8

On April 30, 1971, then Quebec premier Robert Bourassa unveiled plans created by Hydro-Québec to dam several rivers in the northern part of the province, thereby creating tens of thousands of jobs, a new export product (power), an enticement to investment in extractive industries, and an opportunity to increase the economic autonomy of the province. Two months later, construction of roads into the James Bay area began, though the feasibility study being conducted by Hydro-Québec had not been completed and the Cree and Inuit residents had not been informed of the plans to flood their territory. Daniel Coon Come, grand chief of the Cree, and others such as Billy Diamond, chief of the Rupert House Cree, organized quickly to protect their territory and way of life.

In May 1972, lawyers for the Indians of Quebec Association, through whom the Cree brought their protest, sought a court injunction to halt the project. On November 15, 1973, Judge Malouf granted the injunction, announcing his decision that work on the James Bay project should cease immediately because the development would damage the environment and destroy the Cree and Inuit ways of life.

The developers immediately appealed. One week later, on November 22, Judge Turgeon reversed Judge Malouf's decision, indicating that because of the investment in the development to date, the inconvenience to the James Bay Energy Corporation and the James Bay Development Corporation would be greater than the damage to the Cree if the project proceeded. Feeling powerless to stop the first phase of the James Bay project, the Cree entered into negotiations concerning Aboriginal rights, which culminated in the 1975 James Bay and Northern Quebec Agreement, giving the Cree millions of dollars in compensation and Quebec sovereign rights to the region. Ultimately, the completed Phase 1 of the James Bay project (the La Grande Rivière hydroelectric complex) flooded more than 10 000 square kilometres of land to generate more than 10 000 megawatts of power, and cost an estimated $16 billion. Work was scheduled to begin in 1992 on the addition of 5000 megawatts of power to the James Bay complex via Phase 2 in the Great Whale (La Grande Baleine) River area. This portion was expected to be completed in 1995.

The scale of Phase 2 of the James Bay project stimulated worldwide debate. Proponents of the development argued that Canadians and Americans needed large amounts of electricity for their homes and businesses, and that hydroelectric power was environmentally sound (as it was renewable and did not contribute to global warming). Opponents noted that the Great Whale project would flood over 5000 square kilometres in an area where 12 000 Cree and 5000 Inuit lived, and that caribou, snow geese, marten, beaver, black bear, polar bear, and elk were at risk of losing their habitat. Other concerns were raised about methylmercury accumulation and poisoning of the food supply for local inhabitants, and about disrupting the balance of an incompletely understood complex hydrological cycle in the area. More legal wrangling occurred, and in 1990 the National Energy Board granted Hydro-Québec a licence to export electricity to New York and Vermont, provided that there was no conflict with relevant environmental standards and that the federal government conducted an environmental impact assessment.

Groups began forming networks to lobby against Phase 2 of the project, including the James Bay Defence Coalition, the New England Energy Efficiency Coalition, coal-mining interests in the United States, and political

(continued)

and environmental activists such as the Sierra Club. More legal and political action followed. Activists in Maine and Vermont succeeded in convincing their state governments to refuse power from the Great Whale project. In 1992, the New York Power Authority cancelled a $12.6-billion contract with Hydro-Québec when the state legislature passed a law requiring the state to explore conservation and alternative energy sources before importing power. Several other states followed suit in response to pressure from environmental groups and because of a diminished need for electricity.

On August 31, 1993, Hydro-Québec released its 5000-page environmental impact statement (EIS) for the Great Whale project. This EIS indicated that the project would have impacts that were moderate and localized, that could be mitigated, and that would displace no Aboriginal communities. A joint panel of federal, provincial, and Native representatives reviewed the EIS, and on November 17, 1994, ordered Hydro-Québec to rework the study. The following day, Quebec premier Jacques Parizeau announced that the Great Whale project was no longer a priority of the provincial government and would not be constructed.

Environmental nongovernmental organizations that had worked with the Cree in their efforts to halt the project continued to work to replace megaprojects such as the Great Whale with energy efficiency and conservation. These kinds of actions may have contributed to the decline in major dam construction projects in Canada.

SOURCES: "Community-Based Observations on Sustainable Development in Southern Hudson Bay," L. Arragitainaq & B. Fleming, 1991, *Alternatives, 18*(2), 9–11; *Hydro Quebec Released Their Environmental Impact Statement,* A. A. Bennett, 1993, http://bioc09.uthscsa.edu/natnet/archive/nl/9309/0001.html; *Background on Hydro-Québec in James Bay,* D. Deocampo, 1993, http://bioc09.uthscsa.edu/natnet/archive/nl/9303/0043.html; "The Environmental and Human Issues Raised by Large Hydroelectric Dams in Northern Québec," P. Grégoire, R. Schetagne, & M. Laperle, 1995, *Technical Bureau Supplement to Water News, 14*(1); James Bay Dam, Electricity and Impacts, (n.d.), Trade and Environment Database, http://gurukul.ucc.american.edu/TED/JAMES.HTM; "Feature: Victory over Hydro-Québec at James Bay," A. Wilson, 1994, *The Planet,* http://www.sierraclub.org/planet/199412/ftr-canada.asp

RELATED SOURCE: For a summary of major environmental modifications brought about by the project, see "The James Bay Hydroelectric Project," F. Berkes, 1990, *Alternatives, 17*(3), 20.

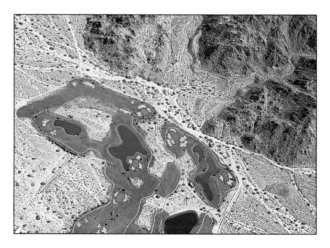

Photo 7–9

How much water is required to keep this desert golf course green?

contamination from fecal coliforms and artificial snow making, and nutrient enhancement through fertilizer and pesticide runoff, impact water resources in different ways. Second homes and seasonal residences also introduce developmental impacts such as faulty or poorly maintained septic tanks, construction of facilities on the foreshore, and higher peak demand for water.

Growing demands from ski resorts to tap nearby water bodies for snow-making purposes are likely to continue, and to increase if climate change occurs. Golf courses also require water for irrigation purposes. In light of the potential for decreased snowfall and reduced stream flow under changing climatic conditions, the use of other sources of water increases in importance. One possibility, applied on a trial basis in ski resorts in Canada, turns ski resort or household effluent into snow. Golf course irrigation technology now employs state-of-the-art systems to control the timing and amount of water applied to greens and to apply only an appropriate amount of fertilizer or pesticide to help reduce nutrient-laden runoff.

Over the past two to three decades, popular recreational areas such as the Gulf Islands in British Columbia,

Georgian Bay in Ontario, and Minnedosa in Manitoba have experienced increased population growth and related development of recreational housing and facilities. Where water is in high demand, relevant authorities in some instances have taken action to institute water restrictions during summer drought periods and to encourage water conservation efforts by offering subsidized rates on low-flush toilets, for example. Water costs (prices) to consumers also have been increased to help pay for improved supply and treatment systems.

The impacts that recreational development can have on community water supply and treatment facilities are seen clearly in the town of Banff, located in Banff National Park, Alberta. With a resident population of about 7000, the town of Banff provides the water infrastructure to accommodate up to 25 000 visitors per night and about 4 million visitors per year. Obviously, there is a large difference in the nature and cost of facilities required to service 7000 residents versus 25 000 visitors. With federal assistance, the town bears the cost of operating and maintaining water facilities for its own and for visitors' use, and levies taxes on residents and businesses to help raise funds required to do so. Up to the year 2000, the town had not instituted a direct water service charge to visitors (such as a toilet tax included in the price of a hotel room). However, since water use is now metered, implementing user pay and full cost recovery rates is an option (Draper, 1997).

RESPONSES TO ENVIRONMENTAL IMPACTS AND CHANGE

How sustainable are Canada's freshwater resources? What is needed to maintain the health of freshwater systems? How do we balance the social, environmental, and economic interests of freshwater resource users? These and many other questions highlight the importance of planning for a sustainable future, one that includes fresh water. As the following examples demonstrate, international as well as individual and cooperative efforts make a difference when it comes to achieving sustainability of freshwater resources.

INTERNATIONAL INITIATIVES

Agenda 21

Canadian representatives to the United Nations Conference on Environment and Development (UNCED) signed Agenda 21 in recognition of our international obligations to protect the quality and supply of fresh-

water resources and to manage them in an integrated fashion. Upon signing, Canada was committed as well to help developing countries and local communities link the knowledge, people, and organizations that would enhance their decisions and policies about water (International Development Research Centre [IDRC], 1993).

An example of Canada's involvement in the transfer of water quality technology took place in the communities of Split Lake, Manitoba, and Maquehue and Chol-Chol, Chile (IDRC, 1993). In the first phase of the project, Cree people from Split Lake learned how to monitor water quality and were trained to conduct appropriate tests on water samples. These skills would be useful to the Split Lake community as well as to other Aboriginal communities in the area. In a later phase of the project, the Cree and Mapuche people from Maquehue and Chol-Chol collaborated in transferring the water-testing technology to Chile. Cree technicians, for instance, were involved in the laboratory setup in Chile. This project illustrates how collaboration to build the human capacity within each community can result in gains for all parties involved. As such, capacity-building is an important component of the Agenda 21 plan of action to achieve environmentally sound and sustainable development throughout the world.

Ramsar Convention

Other international efforts to sustain freshwater resources and related environments are based on conventions and agreements that establish the objectives and obligations of those countries that signed the agreement. In 1981, Canada became a contracting party to the Convention on Wetlands of International Importance, an intergovernmental treaty that provides the framework for international cooperation for the conservation of the world's wetland habitats.

Known also as the Ramsar Convention, this agreement was first adopted in 1971 in Ramsar, Iran. There, representatives of 18 nations had gathered to determine how wetlands could be protected from drainage, land reclamation, pollution, and overuse by competing land uses, and how the economic benefits, hydrological and ecological functions, and critical habitat values of wetlands could be preserved. In 1995, with 84 contracting members, the objectives of the convention focused on stemming the loss of wetlands and ensuring their conservation and sustainable use for future generations (WetNet Project, n.d.). Canada has designated over 13 million hectares of wetlands and associated uplands in its 33 Ramsar sites.

International cooperation and partnerships are important elements in attaining conservation and sustainable management of wetland resources. In North America, international and cooperative partnerships proved to be valuable in the establishment of the North American Waterfowl Management Plan (Canada, the United States, and Mexico) and in the operation of other wetland conservation efforts including the Western Hemisphere Shorebird Reserves Network. The Canadian Wildlife Service, the lead agency in implementing the Ramsar Convention in Canada, also promotes Latin American wetland habitat initiatives. The Canadian Ramsar Network coordinates Ramsar activities within Canada.

AGREEMENTS BETWEEN CANADA AND THE UNITED STATES

Other international initiatives designed to deal with human impacts on freshwater resources and environments involve transboundary agreements between Canada and the United States. The Great Lakes Water Quality Agreements and Remedial Action Plans undertaken to restore the Great Lakes ecosystem are considered briefly here.

Great Lakes Water Quality Agreements

In 1909, the United States and Canada signed the Boundary Waters Treaty to establish the International Joint Commission (IJC). The IJC was assigned three basic responsibilities: to arbitrate disputes related to boundary waters, to conduct feasibility studies for both federal governments, and to approve applications for water diversion projects that would affect flow on either side of the border.

The first formal recognition of basin-wide water quality occurred in 1912 when the IJC was requested to investigate pollution problems in the Great Lakes. After several years of study, the IJC recommended that a new treaty between Canada and the United States was necessary to control pollution (Environment Canada et al., 1987). No agreement was reached at that time. In fact, it was not until water quality concerns escalated in the 1950s and 1960s that the signing of the first Great Lakes Water Quality Agreement (GLWQA) took place in 1972.

In addition to setting common water quality objectives, the GLWQA established cooperative research programs and called for surveillance and monitoring to identify problems and to measure progress. The main objective of the original agreement was to control eutrophication by reducing phosphorus discharges from industry and sewage treatment plants. The second GLWQA, signed in 1978, strengthened pollution standards and reinforced earlier water quality objectives. This agreement represented a move toward more holistic planning of the entire basin by calling for restoration and maintenance of the chemical, physical, and biological integrity of the Great Lakes ecosystem.

A significant step toward achieving this objective took place in 1985, when the IJC's Great Lakes Water Quality Board identified 42 Areas of Concern (AOC). An

Photo 7–10a

Photo 7–10b
Two issues associated with water contamination are diseased fish, such as this walleye with lymphosarcoma (top), and public health concerns.

AOC is a geographic area where the "beneficial use" of water or biota fails to meet the objectives of the GLWQA. In 1987, the GLWQA was revised to incorporate Remedial Action Plans as the formal process for restoring beneficial use to all the Great Lakes AOC. Of the 43 Areas of Concern identified then, 17 were in Canada. The International Joint Commission continues to monitor Great Lakes water quality.

Remedial Action Plans

A Remedial Action Plan (RAP) is a strategy that identifies the types of pollution that are present in a waterway, the geographic extent of the affected area, how water quality will be restored, and who will ensure that restoration is carried out. "RAPs are a breakthrough in cleanup programs in that they formally bring governments, businesses, industry, educators, environment groups, and individual citizens to the table on a long-term basis to discuss how to restore degraded areas in their regions to a healthy state. The aim is to focus local attention on defining problems and finding solutions, based on what residents want for the future of their harbours, bays and rivers" (Environment Canada, 1995b).

RAPs in Canada are carried out under the 1994 Canada–Ontario Agreement Respecting the Great Lakes Basin Ecosystem. This agreement sets firm targets for environmental priorities in the Great Lakes and serves as a coordinating mechanism between the federal and Ontario governments. However, involvement in the RAP process extends far beyond governments. For a RAP to be successful, local citizens, scientists, educators, the business community, and even children must be involved in cooperative learning about the specific area and in all the stages of restoring and maintaining the integrity of the Great Lakes.

Once it has been demonstrated that beneficial uses have been restored to an AOC, the AOC may be delisted. Beneficial uses are considered restored when the effectiveness of the RAP measures is confirmed by the surveillance and monitoring process. The final RAP document includes conclusions by the scientific and technical participants, the Public Advisory Committee, and the federal and provincial governments. The minister of foreign affairs is responsible for forwarding the final document to the IJC for delisting approval. In November 1994, Collingwood Harbour, a Canadian site on Lake Huron's Georgian Bay, became the first AOC officially delisted. Progress has been made in Hamilton Harbour, Bay of Quinte, and Severn Sound.

Although there has been progress in some areas, there are general concerns with the long time periods invested in the RAP process and the relatively small progress RAPs have made in addressing specific problems. On the one hand, it has been suggested that the IJC should revise and improve the RAP process (Office of Inspector General, 1998). On the other hand, the lack of progress could reflect the challenge of convincing public and private managers to adopt sustainable water management practices. It could also reflect the difficulty experienced by public agencies in implementing RAPs in the face of government cutbacks (Gallon, 2000).

CANADIAN LAW, POLICY, AND PRACTICE

If Canada is going to achieve sustainability of its water resources, ecosystem-level understanding must be incorporated into decision making about water. It is no longer sufficient to consider single aspects of the resource—quantity, quality, surface, or ground—in decision making. Now, water resource managers must collect, assess, blend, and synthesize data on the biotic and abiotic characteristics of water resources so that the health and productivity of entire aquatic ecosystems can be maintained. These requirements must be balanced with the needs of people, a central theme of sustainability. The following sections illustrate briefly some of the responses that have been undertaken to move toward sustainability.

Water Legislation and Policy Responses

There is a considerable volume of Canadian legislation pertaining to water (see Table 7–6). While the International Boundary Waters Treaty Act dates from the early 20th century, most of the water-related legislation dates from (or was revised during) the 1980s. The Canada Water Act (1970) provides for comprehensive and cooperative management of Canada's water resources, including water quality issues. The establishment in 1971 of the Department of the Environment and of the Inland Waters Directorate (now defunct) was in recognition of the need for better environmental management in general and water resources in particular.

As the agency charged wholly with the responsibility of administering the Canada Water Act, Environment Canada tries to ensure that Canada's freshwater management is undertaken in the best national interest, and promotes a partnership approach among the various levels of government and private-sector interests that contribute to and benefit from sustainable water resources.

Inquiry on Federal Water Policy   The 1984–85 Inquiry on Federal Water Policy conducted Canada-wide hearings as it worked toward development of a federal water policy. Released in 1987, the policy has the overall objective of encouraging efficient and equitable use of fresh water in a manner consistent with the social, economic, and environmental needs of present and future generations. To achieve this broad objective, governments and individual Canadians need to become aware of the true value of water in order to use it wisely. In the years since the federal water policy was established, the concept of sustainability has become prominent, prompting the Canadian Water Resources Association to develop a set of sustainability principles for water management in Canada in 1994. These principles are to practise integrated water resource management, to encourage water conservation and the protection of water quality, and to resolve water management issues (Canadian Water Resources Association, 1994).

Shared Jurisdiction   In Canada the division of responsibilities for water is complex and often is shared. Under

TABLE 7 – 6

SELECTED CANADIAN WATER LEGISLATION AND RELATED POLICIES, PROGRAMS, AND PLANS

Date	Water Legislation and Related Policies, Programs, and Plans
1909	International Boundary Waters Treaty Act (the International Joint Commission was established)
1970	Canada Water Act
1970	International Rivers Improvement Act
1971	Department of the Environment Act (DOE and Inland Waters Directorate were established)
1972	Great Lakes Water Quality Agreement
1978	Revised Great Lakes Water Quality Agreement of 1978
1979	Government Organization Act
1984–85	Inquiry on Federal Water Policy
1985	Arctic Waters Pollution Protection Act
1985	Canada Shipping Act
1985	Canada Wildlife Act
1985	Dominion Water Power Act
1985	Fisheries Act
1985	Navigable Waters Act
1987	Federal Water Policy
1987	Federal Wetlands Policy
1987	Great Lakes Health Effects Program
1987	Protocol to the Great Lakes Water Quality Agreement
1988	Canadian Environmental Protection Act
1988	St. Lawrence Action Plan
1988	Water 2020: Sustainable Use for Water in the 21st Century (Science Council of Canada)
1989	Great Lakes Action Plan (Preservation Program; Health Effects Program; Cleanup Fund)
1990	The Green Plan
1992	Canadian Environmental Assessment Act
1992	Northwest Territories Waters Act
1992	Yukon Waters Act
1993	National Roundtable on the Environment and the Economy Act
1993	St. Lawrence Vision 2000
1994	Department of Natural Resources Act
1994	Migratory Birds Convention Act
1994	Split Lake Cree First Nation Flooded Land Act
1998	Freshwater Water Strategy
2002	Nutrient Management Act (Ontario)

the Constitution Act of Canada, provinces have primary responsibility for both surface and groundwater resources and legislate flow regulation and most areas of water use. The federal government has responsibility for the northern territories, national parks, First Nations reserves, navigation and fisheries, and in areas of trade, treaty relations, taxation, and statistics. The shared responsibilities are health, agriculture, significant national water issues, and interprovincial water issues.

Thus, when rivers or lakes cross provincial or national boundaries, both the federal and provincial governments get involved. In order to provide more efficient ways of managing the environment, federal and provincial governments are trying to "harmonize" their programs. Promoted through the Canadian Council of Ministers for the Environment in 1993, the goal is to ensure that the services provided by senior governments are not duplicated.

Financial Constraints

Significant federal budget reductions to Environment Canada and other government agencies throughout the 1990s resulted in major changes in personnel and programs affecting water. Clearly, Environment Canada faces a challenge in managing water resources with fewer financial and personnel resources while at the same time responding to increasing public expectations regarding human health and water quality. These budget constraints have generated scientific and public unease about how much priority the federal government attaches to water resources.

Budget constraints also affect the provinces' ability to contribute to water research and management. For example, since 1995, the Ontario government has made unprecedented budget cuts to its lead provincial water agencies, and also has downloaded services to municipalities, cut programs, and privatized many environmental services. Almost every law pertaining to environmental protection of water resources has been weakened (Clark & Yacoumidis, 2000). In a review of the Ministry of the Environment, the Office of the Provincial Auditor of Ontario (2000) concluded that the Ministry "did not have satisfactory systems in place to administer approvals and enforce compliance with environmental regulations."

Other budgetary cuts have reduced or eliminated funding for municipal water and sewage projects that have been critical to the cleanup of the Great Lakes. It is possible that targets under the 1994 Canada–Ontario Agreement Respecting the Great Lakes Basin Ecosystem, including wetland rehabilitation and upgrading of municipal sewage treatment plants, might be scaled back because funding available through the Great Lakes Cleanup Fund may be further reduced. A generation of successful Great Lakes cleanup efforts may be threatened by these cuts. Similar effects and concerns for the future of water resources occur in other provinces and territories as environmental budgets are reduced.

Research and Application

In spite of financial constraints, research efforts to respond to effects of human activities on water resources and environments continue. The Canada Centre for Inland Waters (CCIW), for example, has a worldwide reputation as a leader in providing environmental information and knowledge about the Great Lakes (Beal, 1995). Headquartered at the CCIW is the National Water Research Institute (NWRI), where scientists conduct multidisciplinary aquatic research programs in partnership with Canadian and international researchers. For example, the NWRI conducts studies on the sustainability and remediation of groundwater resources in Canada, including how contaminants are transported in ground water and how contaminated groundwater can be restored. The new knowledge from these projects is used to support regional activities within Environment Canada such as the Great Lakes Cleanup Fund and the federal water policy (Canada Centre for Inland Waters, n.d.).

In March 2001, the federal government announced that $14.9 million would be used to establish a Canadian Water Network under the Networks of Centres of Excellence program. These funds have been used to support water research programs of 110 researchers from 29 universities, which have partners in 30 industry affiliates and 83 government agencies and other groups.

Ecological Monitoring and Assessment Network

The Ecological Monitoring and Assessment Network (EMAN) was established by Environment Canada in 1994 to improve communication and cooperation among scientists conducting ecological monitoring activities. By encouraging independent scientists and those from different jurisdictions to concentrate their work at the Ecological Science Cooperatives (ESCs) across Canada, to share information from their ecological experimentation and research, and to synthesize and integrate results, EMAN's overall objective is to understand what is changing in the environment and why.

If scientists can collect data and exchange information on environmental trends and the health of the Canadian environment with regard to climate change, water quality, air quality, and biodiversity, then a monitoring network can be used to share information and identify problems. Early identification of problems should enable scientists to establish possible consequences of environmental change, bring this to public attention, and communicate it to decision makers for action. Effectively, EMAN provides a framework for improving Canadians' understanding of ecosystem function and change at national and ecozone levels.

Scientists associated with each ESC undertake research covering all facets of ecological inquiry, including water resources. EMAN is a national coordinating network that provides the means for these scientists to link with national hydrological, weather, wildlife, forest, and agricultural networks as well as other North American and global monitoring and research networks. In addition, EMAN promotes broad distribution of results in both scientific and popular literature.

ESCs are ecozone-based and consist of a number of research and monitoring sites that, in time, will cover all the characteristic ecosystems within each ecozone. Eventually, it is intended that each of Canada's ecozones will have its own ESC, but in the meantime, some ESCs are concerned with more than one ecozone, some ESCs have sites in more than one province or territory, and most provinces and territories will have ESC sites in more than one ecozone.

Photo 7–11

In Collingwood, Ontario, Enviropark reflects cooperative efforts among industry, governments, and local groups.

Great Lakes Cleanup Fund and Great Lakes 2000
To meet its obligations under the 1987 Protocol to the Canada–United States Great Lakes Water Quality Agreement, the federal government announced its $125 million Great Lakes Action Plan in 1989. One component of the action plan was the Great Lakes Cleanup Fund, which provided funding to implement cleanup activities in the 17 Canadian Areas of Concern that existed then.

During the first term of the Great Lakes Cleanup Fund (1989–94), more than 50 ecosystem-based projects were undertaken, directly involving community members in remediation of local water quality problems. Partnerships developed among all levels of government, conservation authorities, private industry, service clubs, and public-interest groups; these partnerships contributed money, skills, and services totalling about $70 million in addition to the $35 million provided by the Great Lakes Cleanup Fund. Ultimately, the Fund enabled cost-effective remediation techniques to be tested and applied to other sites throughout the Great Lakes. These results demonstrated practical solutions to environmental problems, and provided Canadian companies with opportunities to market their environmental technologies on a global basis.

Collingwood Harbour, the first Area of Concern to be delisted, was a recipient of Great Lakes Cleanup Fund support. Fifteen partners contributed their expertise and support in upgrading Collingwood's sewage treatment plant to reduce high levels of phosphorus entering the harbour; this resulted in millions of dollars in savings for the municipality. The amount of water pumped dropped by 35 percent as a result of an environmental education program and water conservation measures ("Boundary Waters," 1995). Contaminated sediment in the harbour also was removed. Town residents were involved through

the design and construction of Enviropark, "an active learning opportunity or children's park with a difference. Play structures imitate the town's sewage network, and an obstacle course illustrates the food chain" (Environment Canada, 1993b).

In 1994, the federal government renewed its commitment to the Great Lakes Action Plan with the announcement of a seven-year, ecosystem-based Great Lakes 2000 program. The main objectives of this program are to restore degraded sites, prevent and control pollution, and conserve and protect human and ecosystem health. With the experience gained through the Great Lakes Cleanup Fund, the principles of sustainability, shared responsibility and partnerships, and pollution prevention will be important for continued success of the plan.

Northern River Basins Study The Northern River Basins Study (NRBS) was established in September 1991 by the governments of Canada, Alberta, and the Northwest Territories to examine how development affects the Peace, Slave, and Athabasca river basins in Alberta and the (then) NWT. These river basins contain two national parks and a large number of Aboriginal residents who live traditional lifestyles; the river basins are also under considerable pressure from pulp and paper, oil sand, agriculture, and municipal growth interests.

The NRBS was aimed at providing a scientifically sound information base on contaminants, nutrients, area food chains, drinking water, aquatic uses, hydrology and sediments, and traditional knowledge, as well as a synthesis or modelling of results. An important element of the study was the incorporation of members of the public, particularly those with traditional ecological knowledge. Governments will need to incorporate the new knowledge generated from the study into policies, regulations, monitoring, and research programs. It was suggested (Halliday, 1996) that there is a potential role for EMAN in the long-term management of northern river basins.

CANADIAN PARTNERSHIPS AND LOCAL ACTION

Achieving sustainability in any resource context requires an integration of effort and cooperation among nations, organizations, and individuals. Canadians have been developing the means to achieve the necessary cooperation and partnerships; the following examples demonstrate a variety of forms that actions toward sustainability have taken.

Flood Damage Reduction Program

On the national level, the Canadian government and various provinces have signed agreements regarding flood damage reduction and flood risk mapping in an effort to discourage inappropriate development and reduce flood loss and damage in flood risk areas. (Flood risk areas consist of the floodway and floodplain: see Figure 7–10.) These agreements generally sought to identify, map, and designate flood risk areas in urban communities and, through public information programs, to increase awareness of flood risk among the general public, industry, and government agencies.

Once the flood risk areas were mapped, municipalities were encouraged to adopt land use regulations that would promote only open-space uses (such as golf courses) and would prohibit construction of new buildings or other structures in the floodway. Development in the floodplain would be permitted only if certain structural controls were incorporated. No federal or provincial buildings or other structures that would be vulnerable to flood damage were to be located in the flood risk area.

After implementing appropriate land use regulations, each community would formally designate its flood risk areas under the Flood Damage Reduction Program. In the event of a flood, a community would not receive financial assistance from federal and provincial government sources for new buildings or other new structures damaged by the flood if they were located in flood risk areas. These new policies did not affect existing development in the floodway and floodplain areas. Existing property and development, however, might be *flood-proofed* (any

action or permanent protection applied to prevent flood damage, such as use of elevated pads to build above flood levels) or acquired by government.

This major policy initiative of the 1970s not only restricted building in the mapped floodplains, but also succeeded in reducing flood losses and in preserving many river valleys in high-population regions for wildlife and recreational activities. Since its development, flood damages paid by the federal government through its Disaster Financial Assistance program and by insurance companies in Canada continue to rise (see Figure 7–11).

Major flood events such as those of the Saguenay and Red rivers have raised questions about the effectiveness of existing flood protection programs. The trend of increasing flood damages might also reflect a higher incidence of intense storms and more annual rainfall, and higher losses reflecting the relatively recent accumulation of more valuable personal property and possessions by people. In response, the Insurance Bureau of Canada and Emergency Preparedness Canada are promoting the development of a National Mitigation Strategy. This government–private sector partnership is a new aspect of water management in Canada.

Watershed Planning

The planning and management of water and other resources on a **watershed** basis is an example of an ecosystem-based approach to achieving sustainability. In Ontario, Conservation Authorities have existed since the 1940s. The Conservation Authorities Act gives these agencies the power to study their watersheds in order to determine how the natural resources of the watershed may be conserved, restored, developed, and managed. Unfortunately, the act limits Conservation Authorities' powers to use water, to alter watercourses, and to fill and construct in floodplains (Mitchell & Shrubsole, 1992).

Planning for areas defined on an ecosystem basis, such as a watershed, historically has not been promoted under Ontario's Planning Act (Royal Commission on the Future of the Toronto Waterfront, 1992). However, watershed planning has become more important as a result of the Walkerton inquiry's final report. The report recommended watershed-level source water protection. As a result, in 2002 the provincial government passed the Nutrient Management Act. This act requires all farms to develop plans to deal with animal and other wastes to prevent contamination of lakes, streams, and groundwater.

In other jurisdictions, for instance in Saskatchewan and Alberta, some holistic river basin planning is being undertaken. Saskatchewan's Meewasin Valley Authority is a conservation agency for the Meewasin Valley, part of the South Saskatchewan River valley flowing through Saskatoon and the rural municipality of Corman Park. A 40-kilometre-long section of the river is the subject of a

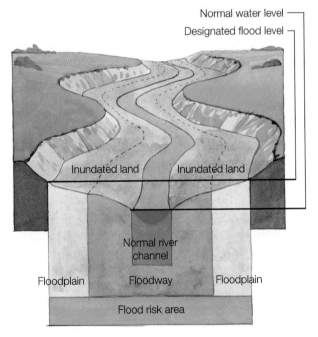

Figure 7–10

Schematic example of flood risk areas

SOURCE: Adapted from *Information Sheet: Calgary, Alberta*, Environment Canada, (n.d.), Ottawa: Canada–Alberta Flood Damage Reduction Program.

PART 3: RESOURCES FOR CANADA'S FUTURE

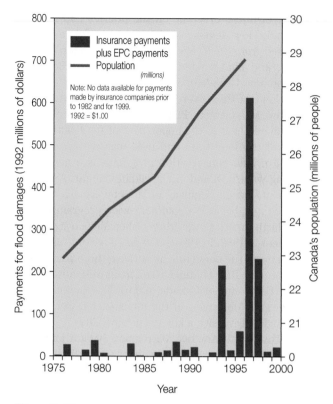

Figure 7–11

Flood damage payments in Canada, 1975–1999: Payments provided by Emergency Preparedness Canada (EPC) and the insurance industry

SOURCE: "The Cultures of Flood Management in Canada: Insights from the 1997 Red River Experience," D. Shrubsole, 2001, *Canadian Water Resources Journal, 26*(4), 461–480. Reprinted with permission.

100-year master plan that recognizes the need to preserve and enhance the valley for future generations. The plan focuses on the integrity of the natural system and employs an open decision-making process that accommodates changing ideas and needs. The river and its natural system are the basis of the plan that also addresses concerns about health, urban and rural areas, access to the river, public education, and the past, present, and future development of the valley.

A 1993 initiative, Partners FOR the Saskatchewan River Basin, has applied its mission—"to promote awareness, linkages, stewardship, knowledge and respect for the basin ecosystem and heritage that will encourage sustainable use of the basin's natural resources and nurture cultural values"—to the entire river basin (Partners FOR the Saskatchewan River Basin, n.d.). Partners FOR the Saskatchewan River Basin uses an ecosystem approach that integrates economics, environment, and society in its efforts to cooperate in managing the basin for a sustainable future.

Over 120 partners (representing education, ENGOs, First Nations, industrial users, Métis Nations, natural and

cultural heritage, tourism, and water management sectors) are participating in developing and implementing stewardship projects ranging from an eco-canoe tour project to a workshop to discuss management issues of the Saskatchewan River delta (one of only five inland deltas in the world). Their goal is to increase environmental stewardship of the basin's natural and cultural resources by developing and implementing education, information, and demonstration projects.

In Manitoba, the provincial government announced a Lake Winnipeg Stewardship Board in February 2003. Members appointed in July of that year were charged with identifying future actions to reduce nitrogen and phosphorous in Lake Winnipeg to pre-1970 levels. In addition, the Board was also charged with implementing a riparian protection action plan to address concerns in the lake as well as along the Red and Assiniboine rivers related to tillage and grazing by livestock on adjacent lands.

Fraser River Action Plan

The Fraser River produces more salmon than any other single river system in the world. The vast network of lakes and tributaries in the Fraser River basin provides spawning and rearing habitat for six salmon and 29 other fish species, as well as 87 more species of fish in the estuary. Fraser River salmon, especially sockeye, form the backbone of British Columbia's commercial fishery as well as sport and Aboriginal fisheries. Many social and cultural benefits derive from the Fraser's fisheries through recreation, tourism, and enhancement of people's way of life. (Fisheries issues are discussed in more detail in Chapter 8.)

Fraser River fish, however, are under intense pressure, mainly from degraded water quality and habitat losses due to urban and industrial development, logging impacts, water use conflicts, and overfishing of some stocks. The Fraser River Action Plan (FRAP), announced June 1, 1991, by the federal ministries of Environment and Fisheries and Oceans, aimed to improve the long-term health and productivity of the Fraser River. Specifically, the objectives of the $100 million program included assessment of the health of the river; reduction of pollution entering the river; and improvement of fish and wildlife productivity in the basin. In addition, creation of partnerships with provincial and local governments, Aboriginal and community groups, environmental organizations, industry and labour, and other stakeholders was necessary to develop a cooperative management program for the Fraser River basin based on the principles of sustainability. The role of Fisheries and Oceans in FRAP ended in March 1997, but Environment Canada continued to implement the FRAP mandate through March 1998.

In November 1998, the FRAP issued its final report. Confirmed improvements were noted in the Fraser River watershed, particularly in the reduction of industrial effluents and the improvement of riparian habitat. However, new challenges have emerged, including negative effects of climate change on this mountainous watershed, and an increase in difficult-to-detect nonpoint source pollutants from agricultural and urban sources. The Fraser Basin Council, a partnership of community stakeholders and agencies from federal, provincial, and local governments, has replaced the FRAP. Its role is to continue the management of the basin according to sustainability principles (Fraser River Action Plan, 1998).

With regard to enhancing fish and wildlife habitat, dozens of FRAP projects have been undertaken to improve conditions for rearing and spawning fish, encourage protection of wetlands and streams, and increase the amount of available habitat. To rebuild fish stocks, FRAP work included development of a comprehensive management plan for all Fraser River salmon species, and collection and analysis of scientific data on which sound management decisions could be made. This information is complemented by data on pollution point sources in the basin, and a wastewater database.

The Fraser River Action Plan alone cannot clean up pollution and restore habitats in the Fraser basin—the job is simply too large. That is why partnerships are crucial to keeping the basin healthy after the program ends. To ensure that the basin's rich water (and other) resources would be maintained for future generations, the Fraser Basin Management Board (FBMB) developed a Charter for Sustainability. When the FBMB was succeeded by the new Fraser Basin Council on April 1, 1997, the council's mandate was to promote and monitor the implementation of the charter. The Charter for Sustainability is to serve as a blueprint for people of differing opinions and experiences to work together toward the goal of long-term sustainability of the Fraser basin.

In 1998, this effort influenced the development of the Georgia Basin Ecosystem Initiative (GBEI). Pressures from growth were placing significant biological, physical, and chemical strains on the environment as well as detracting from the region's economy. The GBEI vision is to manage population growth to achieve healthy, productive, and sustainable ecosystems and communities.

North American Waterfowl Management Plan

The Canadian prairies have been bountiful producers of cereal grains and rich grazing grasses for almost a century. They also have supported abundant wildlife—almost half the North American continent's waterfowl breed in this region. After World War II, however, a combination of drought and rapid agricultural expansion involving wetland conversion from a natural state to agricultural and other uses resulted in the loss of thousands of acres of wildlife habitat and a progressive decline in waterfowl populations. Without habitat, there is no wildlife; compared with base levels of the 1970s, by the mid-1980s the number of ducks breeding on the prairies had reached alarmingly low levels. Mallard ducks, for example, decreased from 8.7 to 5.5 million, pintail from 6.3 to 2.9 million, and blue-winged teal from 5.3 to 3.8 million (Environment Canada, n.d.b).

In light of this severe decline, the North American Waterfowl Management Plan (NAWMP) was established. The plan emphasizes the importance of the health of the land over the long term and uses massive cooperative funding of land conservation programs as a method of restoring waterfowl and other wildlife habitat. A 15-year agreement, signed originally in 1986 by Canada and the United States, became a continent-wide program to conserve North America's waterbirds when Mexico signed the agreement in 1988. In 1994, an update and 10-year extension renewed the commitment of these three partners to cooperate in a fully continental effort in wetlands conservation (Environment Canada, n.d.b). The updated plan promotes the objectives of the international Convention on Biological Diversity, a key to sustainability.

Across North America, 32 regional habitat joint ventures are identified as priorities under the NAWMP agreement. The largest and most important joint venture region includes Canada's prairie potholes (so named because of the millions of wetlands dotting the prairie landscape). Most of the original grasslands in this area, as well as 40 to 70 percent of the original wetlands, have been lost to agricultural development. Because this region provides the breeding area for half of North America's waterfowl population, it is ranked as the highest priority under the NAWMP. The Prairie Habitat Joint Venture guides conservation of the Canadian prairie pothole region under the NAWMP, and aims to restore waterfowl populations to 1970 spring breeding levels of between 17 and 20 million (Environment Canada, 1995c).

The NAWMP is not an isolated program. Provincial and federal wildlife agencies work directly with landowners on habitat retention programs that offer a wide range of wildlife benefits. The Quill Lakes project in southern Saskatchewan, the Buffalo Lake project near Stettler, Alberta, and HELP in the Minnedosa region of Manitoba are examples of the commitments to prairie land and habitat conservation. The Western Hemisphere Shorebird Reserve Network is another example of cooperation among the three federal governments, which, in turn, cooperate with ENGOs such as Wetlands for the Americas and Ducks Unlimited.

Ducks Unlimited is an important partner in implementing elements of the NAWMP. Established in 1937 in the United States, Ducks Unlimited, Inc., began to raise funds to protect Canadian breeding grounds where the majority of North American ducks are produced. The ultimate goal was (and is) to ensure that hunters from the United States would have ducks to hunt. Ducks Unlimited

Canada was incorporated in 1938 as the biological and engineering branch to develop habitat projects.

Ducks Unlimited Canada works as a partner to implement NAWMP's continental conservation program, but also has delivered important wetland and wildlife habitat programs across Canada for 60 years. In Alberta, home to 20 percent of all the ducks surveyed each spring in North America, Ducks Unlimited Canada has spent more than $100 million to complete over 2800 individual habitat management projects, and restore, enhance, or preserve more than 725 000 hectares of critical waterfowl habitat. Their projects have included efforts to rehabilitate one large marsh using treated wastewater from a meat-packing plant (Haworth-Brockman & Smallwood, 1989).

Making a Difference Locally

From coast to coast, there is a growing movement composed of local residents, donors, private and public sectors to improve the quality of local streams. The Swan Lake Christmas Hill Nature Sanctuary, located just a few minutes from downtown Victoria, British Columbia, provides one example of the importance of dedicated volunteers and the differences they make to the environmental health of a community and its water resources. While the issues and events described in Box 7–5 are specific to this nature sanctuary, they are repeated—with local differences—in many parts of Canada, demonstrating the significance of local stewardship in the movement toward environmental sustainability.

The preceding examples have demonstrated a range of international as well as individual and cooperative stewardship efforts that have focused on achieving balance among social, environmental, and economic interests in freshwater resource use, and in moving toward sustainability of freshwater resources. There are challenges ahead; they are the subject of the final section of this chapter.

FUTURE CHALLENGES

Fresh water is essential to humanity and to all life. As the fundamental basis of aquatic and related ecosystems, sustainability of water is vital. Achieving sustainability of fresh water involves finding a proper balance in meeting all the competing needs for water; that is, in balancing human needs and the needs of natural ecosystems. Although Canada contains a wealth of fresh water underground and in its lakes, rivers, and wetlands, regional differences exist in the infrastructure and distribution of water, its quality, the human environments it supports, and the wildlife that depend on it.

In the prairies and interior of British Columbia, there is competition for limited water supplies between agricultural irrigation interests, growing urban demands, energy production, and the desire of residents to preserve natural ecosystems. In the most heavily populated and industrialized regions of the country, including the Windsor–Quebec corridor and parts of the Pacific and Atlantic coasts, water is plentiful, but impaired water quality and physical alterations to aquatic habitats are problematic. Persistent toxins such as PCBs in Great Lakes and St. Lawrence River sediments and the presence of DNAPLs and other contaminants in groundwater are among the unintended and poisonous byproducts of past industrial development.

In the north and other sparsely populated regions of the country, transport of contaminants (acid precipitation, toxic metals, and persistent organic compounds) from distant sites, and acid drainage from closed mines, threaten water quality. In addition, the health of numerous citizens may be at particular risk from bacteria and pathogens in well water. This is particularly important for residents of rural areas. While progress has been made in slowing the degradation of freshwater resources in Canada, uncertainties such as future climate change and its potential impacts on water supply and distribution remain as challenges for water resource managers.

As changing economic conditions have resulted in declines in government-supported programs such as water monitoring, local stewardship will become an even more important element in reaching sustainability objectives. In order to protect and conserve water resources for future generations, communities and regions will need to identify ways in which they can monitor threats to freshwater supplies and ecosystems, and judge the effectiveness of prevention and remediation measures. Just as the concerted effort of individuals associated with Prairie Habitat Joint Venture programs successfully demonstrated the value of healthy ecosystems, so too the ability of people to organize their desires for protection of a wide range of water values should empower them to devise and implement alternative systems of monitoring. While government resources must remain available to collect basic data on water quality and quantity (to support drinking water guidelines, for instance), it is not unreasonable for the community of water users to accept some increased responsibility for its sustainability.

Establishing an appropriate framework for economic, ecological, and social sustainability of water resources is a complex and difficult task. Although integrated approaches and regional plans have been implemented in major drainage basins such as the Fraser River and the Great Lakes–St. Lawrence basin, to what degree do they succeed in maintaining resilient, biologically diverse ecosystems? Do we understand yet all the principles on which Canada's water resources ought to be governed for

The transformation of Swan Lake, British Columbia, from a smelly, murky, polluted pond to a quiet nature sanctuary bordered by walking and cycling trails is a tribute to community vision and cooperation and many hours of work by local volunteers. While this story reflects events in one location, there are countless other examples of Canadians' dedication to improving local water bodies and environmental quality in general.

By the mid-1850s, the area surrounding what is now the Swan Lake Christmas Hill Nature Sanctuary had been cleared of most trees and native vegetation for agriculture. About 1860, the area's first year-round recreational resort was built. Although Swan Lake Hotel guests and local residents used the 94-hectare lake for swimming, fishing, boating, and ice-skating, it was never used for drinking water (humic and tannic acids leached from the soil gave the water a yellow-brown colour). After being destroyed by fire in 1894 and again in 1897, the hotel was not rebuilt.

In succeeding years, the water quality of Swan Lake deteriorated due to nutrient enrichment from the sulfides and yeast in effluent from two wineries; coliform bacteria contamination from sewage plant effluent discharged into waterways draining into the small lake; and biological oxygen demand problems resulting from past use of agricultural fertilizers, from cattle and waterfowl wastes, and from garden chemicals. Swan Lake became a smelly eutrophic lake that no one wanted to visit.

In the 1960s, the local municipality of Saanich began acquiring land around the lake, and, to protect the natural environment, the area was designated a nature sanctuary in 1975. Since then, the municipality has spent more than $2 million to develop facilities, and the nonprofit Swan Lake Christmas Hill Nature Sanctuary Society has raised and invested an equal amount. Provincial and federal agencies have ensured that projects meet public safety standards and habitat improvement criteria.

An interpretive centre has been built, about four kilometres of trails and boardwalks have been constructed, and a demonstration garden of native plants has been started with volunteer assistance from local contractors. In 1995, more than 15 000 school children, community group members, and casual visitors took part in the nature sanctuary's educational programs or enjoyed a quiet walk.

The society's mission is to foster understanding and appreciation of nature and to develop personal responsibility for the care of the natural environment. School programs are the main vehicle by which the society fulfills this mission, but volunteers also help maintain the sanctuary for other users. A growing appreciation of the value of this oasis in the midst of the city has contributed to

Photo 7–12

growth in volunteer hours, as well as cooperation and partnerships among citizens, community businesses, and governments.

Though no longer ugly, Swan Lake remains threatened by use of fertilizers, pesticides, and herbicides in the watershed. Chemicals enter storm drains that run into the lake, leading to algae growth and associated problems. Education programs are needed to inform homeowners and residents about the consequences of excessive fertilizer use, to promote the benefits of gardening with native plants, and to conserve water. In the future, continued funding must be established for community-based environmental education and partnership activities, including the addition of some small parcels of private land around the lake to help meet the growing public demand for meaningful natural history and outdoor experiences.

As examples of effective stewards of land and water resources, the Swan Lake Christmas Hill Nature Sanctuary Society and its governmental, industry, and citizen partners deserve recognition. They are part of the movement toward ecological sustainability and public accountability that is changing our approach to managing our shared environment.

SOURCES: "Back to Nature," M. Curtis, September 27, 1995, *Victoria Times Colonist*, p. B1; "Gardening Chemicals Causing the Algae Growth in Swan Lake," July 5, 1995, *Saanich News*, p. 24; *Partners in Stream Restoration, Winter, 1995*, Swan Lake Christmas Hill Nature Sanctuary Newsletter, p. 2; *Swan Lake Christmas Hill Nature Sanctuary: A Place to Explore; A World to Discover*, T. Roberts & T. Morrison, 1995, Victoria, BC: Swan Lake Christmas Hill Nature Sanctuary Society; *Information on Finances and Use, 1975–1995*, Swan Lake Christmas Hill Nature Sanctuary, 1995, Victoria, BC: Swan Lake Christmas Hill Nature Sanctuary Society; *1995 Annual Report*, Swan Lake Christmas Hill Nature Sanctuary, 1996, Victoria, BC: Swan Lake Christmas Hill Nature Sanctuary Society; *An Inventory of the Biota of Swan Lake with Some Basic Limnological Spects [sic] and Recommendations*, W. Zaccarelli, 1975, Victoria, BC: Swan Lake Christmas Hill Nature Centre Society.

the next century and beyond? Knowledge building—about the physical nature of freshwater resources, about how to protect water quality and aquatic ecosystems, and about human demand for water—is clearly an important dimension in achieving sustainability. Knowledge building, too, is affected by shrinking government support; therefore, new cooperative and partnership approaches to research and understanding should be examined and applied.

Chapter Questions

1. Given Canada's abundant water supply, should we be concerned about water availability in the future? Why or why not?

2. Discuss the range of pressures that humans place on water resources that lead to water quality concerns.

3. Compare and contrast the important environmental problems related to domestic and industrial uses of water.

4. We often take our water supply for granted. The following questions may challenge us to think more carefully about water in our communities:

   a. What are the major sources of your community's water supply?

   b. How is water use divided among residential, commercial, industrial, and other uses? Which sector consumes the largest volume of water?

   c. How have water prices changed in the past 20 years? Do water prices encourage conservation? Where is water being wasted?

   d. What water supply and quality problems does your community experience?

   e. What plans does your community have in place to ensure an adequate supply of safe drinking water for the future?

5. Why is it important to incorporate ecosystem understanding in the responses that Canadians make to water resource issues (at international, national, and local levels)?

6. What can you do to demonstrate stewardship with regard to water resources?

references

Anderson, J. S. (1997). Ontario's Conservation Authorities—pioneer watershed managers. *Water News, 16*(1), 4–7.

Barnett, V. (1996, July 11). Water, sewer funding on tap. *Calgary Herald,* p. A5.

Beal, S. (1995). *The Canada Centre for Inland Waters.* National Water Research Institute. http://www.cciw.ca

Béland, P. (1996). The book of the dead. *Canadian Geographic, 116*(3), 46–47.

Beluga habitat in St. Lawrence improving, but still 12 years from full recovery. (1996). *Environment Policy and Law, 6*(12), 352.

Black, J. K. (2003). Three Gorges gates close on Chinese history: China builds the world's largest dam. *The Z Magazine, 16,* 7–8.

Bocking, R. C. (1987). Canadian water: A commodity for export? In M. C. Healy & R. R. Wallace (Eds.), *Canadian Aquatic Resources* (pp. 105–135). Canadian Bulletin of Fisheries and Aquatic Sciences, 215. Ottawa: Department of Fisheries and Oceans.

Boundary waters a shared resource. (1995). *Global Agenda, 3*(1). http://www.dfait-maeci.gc.ca/english/news/newsletr/global/

Boyd, D. (2003). Unnatural law: Rethinking Canadian environmental law and policy. Vancouver: UBC Press.

Brubaker, E. (2002, April 1). Lessons from Walkerton. *Fraser Forum.* www.environmentprobe.org

Bruce, J., & Mitchell, B. (1995). *Broadening perspectives on water issues.* Canadian Global Change Program Incidental Report Series No. IR95-1. Ottawa: The Royal Society of Canada.

Canada Centre for Inland Waters. (n.d.). Groundwater remediation project, Aquatic Ecosystem Restoration Branch, National Water Research Institute. http://gwrp.cciw.ca/gwrp/

Canadian Water Resources Association. (1994). *Sustainability principles for water management in Canada.* Cambridge, ON: Canadian Water Resources Association.

Cayer, S. (1996). Lakes and ladders: Up and down the locks and channels of Ontario's Trent–Severn Waterway. *Canadian Geographic, 116*(4), 32–47.

Clark, K. L., and Yacoumidis, J. (2000). *Ontario's environment and the Common Sense Revolution: A fifth year report.* Toronto: Canadian Institute of Environmental Law and Policy.

Draper, D. (1997). Touristic development and water sustainability in Banff and Canmore, Alberta, Canada. *Journal of Sustainable Tourism, 5*(3), 183–212.

Environment Canada. (n.d.a). *State of the environment in the Atlantic region.* http://www.ns.ec.gc.ca/soe/cha4.html

Environment Canada. (n.d.b). *The North American waterfowl management plan.* http://www.doe.ca/tandi/NAWMP/bkgd_e.html

Environment Canada. (n.d.c). *Information sheet: Calgary, Alberta.* Ottawa: Canada–Alberta Flood Damage Reduction Program.

Environment Canada. (1990). *A primer on water: Questions and answers.* Ottawa: Supply and Services Canada.

Environment Canada. (1993a). *Water works!* Freshwater Series A-4. Ottawa: Supply and Services Canada.

Environment Canada. (1993b). *Project highlights: Great Lakes Cleanup Fund.* Ottawa: Supply and Services Canada.

Environment Canada. (1995a). *Water: No time to waste: A consumer's guide to water conservation.* Ottawa: Supply and Services Canada.

Environment Canada. (1995b). *Canadian Great Lakes Remedial Action Plan update.* http://www.cciw.ca.glimr/rep-test/overview.html

Environment Canada. (1995c). *Prairie Habitat Joint Venture: Conserving an international resource.* http://www.mb.doe.ca/ENGLISH/LIFE/WHP/PHJV/phjv.home.html

Environment Canada. (1996a). *How we contaminate groundwater.* http://www.doe.water/water/en/nature/grdwtr/e_howweg.htm

Environment Canada. (1996b). *DNAPLs.* http://www.cciw.ca/glimr/data/water-fact-sheets/facta5-e.html

Environment Canada. (1996c). *Leaking underground storage tanks and piping.* http://www.cciw.ca/glimr/data/water-fact-sheets/facta5-e.html

Environment Canada. (1998). *Canada and freshwater: Experiences and practices.* Monograph #6. Ottawa: Environment Canada. www.ec.gc.ca/agenbda21/98/Default.htm

Environment Canada. (1999). *State of the Great Lakes report.* Ottawa: Environment Canada.

Environment Canada. (2002). Science. *Environment bulletin: Tackling urban water pollution.* http://www.ec.gc.ca/science/sandenov02/article2_e.html

Environment Canada, United States Environmental Protection Agency, Brock University, and Northwestern University. (1987). *The Great Lakes: An environmental atlas and resource book.* Toronto: Environment Canada.

Federation of Canadian Municipalities. (1996). *Report on the state of municipal infrastructure.* Ottawa: Author.

Fraser River Action Plan. (1998). *Fraser River Action Plan Final Report.* http://www.pyr.ec.gc.ca/ec/frap/ press.htm

Frederick, K. D. (1996). Water as a source of international conflict. *Resources, 123* (Spring), 9–12.

Gallon, G. (2000, June 2). Ontario government cuts beaches protection funding by 70 percent. *The Gallon Environment Letter, 4*(20). Canadian Institute for Business and the Environment, Montreal, Quebec. E-mail: ggallon@pcstarnet.com

Gleick, P. H. (1996). Basic water requirements for human activities: Meeting basic needs. *Water International, 21,* 83–92.

Government of Canada. (1991). *The state of Canada's environment—1991.* Ottawa: Supply and Services Canada.

Government of Canada. (1996). *The state of Canada's environment—1996.* Ottawa: Supply and Services Canada.

Government of Canada. (2001). Urban water use: Municipal water use and wastewater treatment. *State of the Environment Bulletin* No. 2001-1. National Environmental Indicator Series. Ottawa: Environment Canada.

Government of Canada. (2003, Feb. 18). *Highlights of Budget 2003.* Ottawa: Department of Finance.

Halliday, R. A. (1996). *Northern river basins study.* Paper presented at the Ecological Monitoring and Assessment Network, Second National Science Meeting, Halifax, Nova Scotia, January 17–20.

Hamilton, H., & Wright, A. (1985). *Water quality and quantity model for irrigation in the South Saskatchewan River Basin.* Paper presented at the 38th Canadian Water Resources Association Conference, Lethbridge, AB.

Haworth-Brockman, M., & Smallwood, S. (1989). Cool, clear, refreshing ... effluent. *Conservator, 10*(3), 10–14.

Heming, L., Waley, P., & Rees, P. (2001). Reservoir resettlement in China: Past experience and the Three Gorges Dam. *The Geographical Journal, 167*(3), 195–212.

International Development Research Centre (IDRC). (1993). *Agenda 21: Green paths to the future.* Ottawa: Author.

International Joint Commission. (1992). *Sixth biennial report under the Great Lakes Water Quality Agreement of 1978 to the governments of the United States and Canada and the state and provincial governments of the Great Lakes basin.* Ottawa: Author.

International Joint Commission. (2000a, July 26). *Tenth biennial report on Great Lakes water quality.* http://www.ijc.org/ijcweb-e.html

International Joint Commission. (2000b). *Living with the Red.* Windsor: Author. http://www.ijc.org

Jones, J. A. (1997). *Global hydrology: Processes, resources and environmental management.* London: Longman.

Jones, M. L., Minns, C. K., Marmorek, D. R., & Elder. P. C. (1990). Assessing the potential extent of damage to inland lakes in eastern Canada due to acidic deposition: II. Application of the regional model. *Canadian Journal of Fisheries and Aquatic Sciences, 47,* 67–80.

Kreutzwiser, R. (1998). Water resources management: The changing landscape in Ontario. In R. D. Needham (Ed.), *Coping with the world around us: Changing approaches to land use, resources and the environment* (pp. 135–148). Department of Geography Publication Series No. 50. Waterloo, ON: University of Waterloo.

Laycock, A. (1987). The amount of Canadian water and its distribution. In M. C. Healey & R. R. Wallace (Eds.), *Canadian aquatic resources* (pp. 13–41). Canadian Bulletin of Fisheries and Aquatic Sciences, 215. Ottawa: Supply and Services Canada.

Li, H., Waley, P., & Rees, P. (2001). Reservoir resettlement in China: Past experience and the Three Gorges Dam. *The Geographical Journal 167*(3), 195–212.

Manahan, S. E. (1994). *Environmental chemistry* (6th ed.). Boca Raton, FL: Lewis.

McIlroy, A. (1995, September 9). Beluga behavior beguiles scientists. *Calgary Herald,* p. B4.

Mitchell, B., & Shrubsole, D. (1992). *Ontario conservation authorities: Myth and reality.* Waterloo: Department of Geography Publication Series Number 35, University of Waterloo.

Norris, K. S. (1994). Beluga: White whale of the north. *National Geographic, 185*(6), 2–31.

Office of Inspector General (U.S.). (1998). *Canada–U.S. strategy for the virtual elimination of persistent toxic substances in the Great Lakes.* www.epa.gov/glnpo/p2/bns.html

Office of the Provincial Auditor of Ontario. (2000). *Special report on accountability and value for money.* Toronto: Government of Ontario.

Organization for Economic Cooperation and Development. (1999). *The price of water: Trends in OECD countries.* Paris: OECD.

Partners FOR the Saskatchewan River Basin. (n.d.). http://www.saskriverbasin.ca

Pearse, P. H., Bertrand, F., & MacLaren, J. W. (1985). *Currents of change: Final report, inquiry on federal water policy.* Ottawa: Environment Canada.

Pindera, G. (1997). Red River dance. *Canadian Geographic, 117*(4), 52–62.

Pope, H. (1996). Sharing the rivers. *People & the Planet, 5*(1), 5.

Royal Commission on the Future of the Toronto Waterfront. (1992). *Regeneration: Toronto's waterfront and the sustainable city, final report.* Toronto: Supply and Services Canada and Queen's Printer of Ontario.

Sanderson, M. (1993). Climate change and the Great Lakes. In P. L. Lawrence & J. G. Nelson (Eds.), *Managing the Great Lakes shoreline: Experiences and opportunities* (pp. 181–193). Waterloo, ON: Heritage Resources Centre, University of Waterloo.

Schindler, D. W., & Bayley, S. E. (1991). Fresh waters in cycle. In C. Mungall & D. J. McLaren (Eds.), *Planet under stress: The challenge of global change* (pp. 149–167). Toronto: Oxford University Press.

Serrill, M. S. (1997, November). Wells running dry. *Time, 150*(17A), 16–21.

Shrubsole, D. (2001). The cultures of flood management in Canada: Insights from the 1997 Red River flood experience. *Canadian Water Resources Journal, 26*(4), 461–480.

Sierra Legal Defence Fund. (2001). *Waterproof: Canada's drinking water report card.* Vancouver: Sierra Legal Defence Fund. www.sierralegal.org/reports/html

Simpson, B. (1993/1994, December/January). Liquid gold: Parts of the United States are drying up, and Americans are looking north for salvation. Will Canada eventually have to sell water south? *Earthkeeper: Canada's Environmental Magazine,* 14–19.

Smelters suspected in belugas' cancers. (1995, June 11). *Calgary Herald,* p. A9.

Statistics Canada. (2000). *Human activity in the environment 2000.* Ottawa: Author.

Statistics Canada. (2003). *Population by sex and age group.* http://www.statcan.ca/english/Pgdb/demo31a.htm

Van Tighem, K. (1990). Heritage rivers: Preserving the spirit of Canada. *Canadian Geographic, 110,* 2.

WetNet Project. (n.d.). *Canada and the Ramsar Convention.* http://www.wetlands.ca/wetcentre/wetcanada/RAMSAR/booklet/booklet.html

Woo, D. M., & Vicente, K. J. (2003). Sociotechnical systems, risk management, and public health: Comparing the North Battleford and Walkerton outbreaks. *Reliability Engineering & System Safety, 80,* 253–269.

Wu, M. (1999). *Resettlement problems of the Three Gorges Dam.* International Rivers Network. http://www.irn.org/programs/threeg/resettle.html#one

"The ability of future generations to live in harmony with the ocean environment and enjoy the fruits of its resources depends on the decisions taken now to protect and cherish our marine areas."

Brian Tobin (Fisheries and Oceans Canada, 1994)

Oceans and Fisheries

Chapter Contents

Chapter Objectives

After studying this chapter you should be able to

- identify a range of human activities occurring in the marine environment
- discuss the effects of human activities on the marine environment
- describe the complexity and interrelatedness of marine environment issues
- specify Canadian and international responses to oceans and fisheries issues
- discuss challenges to a sustainable future for oceans and fisheries resources

INTRODUCTION

As a maritime nation, bordered by the Arctic, Atlantic, and Pacific oceans (see Figure 8–1a), Canada has the longest coastline (about 244 000 kilometres), the longest inland waterway, the largest archipelago, and the second-largest continental shelf (about 3.7 million square kilometres) of any country in the world. Canada also holds jurisdiction over an almost 5-million-square-kilometre **exclusive economic zone**, which we claimed in 1977 and was later enshrined in the United Nations Convention on the Law of the Sea (UNCLOS).

In spite of the geographical and historical importance of oceans, few Canadians realize how extensive and varied our ocean environments really are; few understand the wealth of natural resources that oceans contribute to the subsistence, social, economic, and cultural needs of the nation's people; and fewer still appreciate that Canada has international legal obligations and economic incentives to protect oceans from degradation.

In the sections that follow, a brief overview of selected biophysical characteristics and threats to ecosystem integrity in the Arctic, Atlantic, and Pacific oceans is provided. In each ocean and on each coast the discussion identifies important sociocultural, economic, ecological, and sustainability issues and challenges facing oceans and fisheries in Canada and internationally. Discussion continues about human activities and their effects on marine environments. In addition to these issues, coastal development, overharvesting, climate change, and ozone depletion affect the living and non-living resources of oceans and coasts in Canada and the world over. International, national, and local initiatives to regulate human behaviour toward coastlines and oceans are discussed later in the chapter.

CANADA'S MARINE ENVIRONMENTS

CANADA'S ARCTIC OCEAN ENVIRONMENT

Major Characteristics

Canada's coldest ocean area, the Arctic, contains about 173 000 kilometres of coastline (twice that of the east and west coasts combined) and over one million square kilometres of continental shelf waters that provide most of the country food for Canadian Inuit. The majority of this

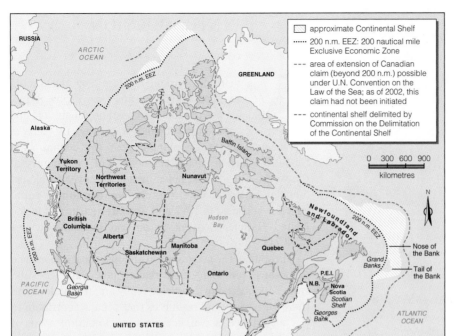

Figure 8–1a

Canada's marine environment

Figure 8–1b

Selected characteristics of Canada's
Arctic Ocean environment

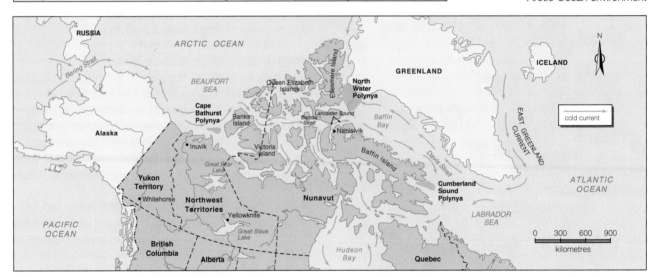

Figure 8–1c

Selected characteristics
of Canada's Pacific
Ocean environment

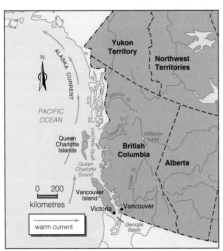

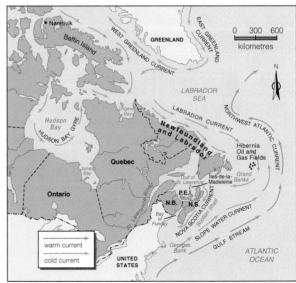

Figure 8–1d

Selected characteristics
of Canada's Atlantic
Ocean environment

SOURCE: Adapted from *A Vision for Ocean Management*, B. Tobin,
1994, Ottawa: Fisheries and Oceans, p. 1.

area is covered seasonally by ice one to two metres thick. Maximum sunlight enters the water column in July, after the ice breaks up, limiting phytoplankton production to the late summer. As a result, Arctic waters produce only about one-quarter of the organic biomass per unit area compared with that produced over the east and west coast continental shelves.

Major currents influencing Arctic waters include the flow of polar water southeast through the archipelago, the West Baffin Current, and the Hudson Bay gyre, a circular current that creates a small, nearly closed circulation system characterized by nutrient enrichment and high levels of biological productivity (see Figures 8–1b and d). Locally, currents help maintain open water areas called **polynyas** for much or all of the year. The largest of these polynyas, the North Water in north Baffin Bay (see Figure 8–1b), allows an early and persistent phytoplankton bloom. A biological hotspot, the North Water and other smaller polynyas serve as winter refuges for marine mammals such as polar bears, seals, whales, and sea birds that overwinter in the north (Welch, 1995; National Advisory Board on Science and Technology, 1994). The ice edges that exist at polynyas and along floe edges (between landfast and drifting ice) are very important feeding and staging grounds for marine mammals and sea birds, particularly during spring and early summer migrations.

While subsistence hunting and fishing dominate the human use patterns in Arctic waters, some commercial fishing does take place. True polar Canadian waters have only limited commercial potential. The small arctic char commercial fishery, for example, produced goods worth about $1.1 million in 1996, and arctic char in northern Labrador are part of the commercial fishery that caught approximately 27 000 fish during the 2000 season (Fisheries and Oceans Canada, 2001b). There is concern, however, that most of the major arctic char populations have been harvested at or above maximum **sustainable yield** levels (Welch, 1995). The status of the species is unknown, as no subsistence or recreational catch data exist (Fisheries and Oceans Canada, 2001b). Other species such as redfish, round-nosed grenadier, Greenland sharks, clams, shrimp, and scallops may support small commercial operations. Although its commercial potential has not been assessed, there is a large biomass of kelp in nearshore shallow waters throughout the eastern Arctic.

Marine mammals and sea birds are the main biological components of the Canadian Arctic Ocean ecosystem. Generally, both sea bird and marine mammal populations have been stable through the last three decades. During the past few years scientists from several countries have been monitoring a warming trend that has important implications for the size and location of the pack ice. Many bird and mammal species are closely dependent on the ice edge for foraging and nesting areas. While most research on the causes and effects of Arctic Ocean changes is not

Photo 8–1
Narwhals are an important source of country food for indigenous communities in the Canadian Arctic.

conclusive, scientists are increasing their scrutiny of sea ice conditions, water temperature, and other elements of the Canadian Arctic marine environment.

With regard to marine mammals such as whales, population data are inadequate to detect any but the largest changes. Between 1990 and 2001, scattered, low-density Inuvialuit communities such as Aklavik, Inuvik, and Tuktoyaktuk, which depend on marine mammals for food and cultural continuity, landed an average of 111 beluga whales from the Eastern Beaufort Sea stock. During the 1992–2001 period, the community of Pangnirtung on Baffin Island hunted and landed an annual average of 36.5 belugas from the Cumberland Sound stock. The Nunavut Wildlife Management Board co-manages the hunt with Fisheries and Oceans Canada (DFO) and, for the 2002–2003 season, increased the annual quota from 35 to 41 belugas. This change reflects the beluga population's recovery from commercial hunting during the 1920s through the 1960s. However, beluga catches in both the Eastern Beaufort and Cumberland Sound areas consistently have exceeded quota limits, a continuing concern for the Canadian Committee on the Status of Endangered Wildlife in Canada (COSEWIC), who designated the Cumberland Sound stock as endangered in 1990.

Since 1996 there has been a limited subsistence hunt for endangered bowhead whales in northwestern Hudson Bay. While the International Whaling Commission, ENGOs, and many scientists have criticized the Canadian bowhead hunt, Inuit people in Nunavut do not view this stock of bowheads as endangered, and they support a hunt because of its social and cultural significance. Balancing the needs of Arctic hunting cultures with the desire to maintain endangered species and biological diversity will require resource managers, scientists, and the Canadian public to become keenly attuned to the wildlife and people of the North.

Figure 8–2
Biomagnification of PCBs in the Arctic

SOURCE: *The State of Canada's Environment—1996,* Environment Canada, Ottawa: Supply and Services Canada. Used with permission of the Minister of Public Works and Government Services Canada, 2004.

Arctic ecosystems are particularly vulnerable to contamination by chemical compounds that have been transported over long distances (in gaseous or vapour form in the atmosphere, for instance). These compounds are persistent and lipophilic, which means they tend to concentrate in fatty or protein-rich tissues. Compounds such as PCBs and DDT **bioaccumulate** readily in fatty tissues of long-lived animals at the tops of food chains such as marine mammals (polar bears, seals, whales), as well as terrestrial animals (caribou), seabirds (gulls), fish (arctic charr), and hunters and their families (see Figure 8–2).

The long and complex food web (with five trophic levels) in Arctic marine systems allows persistent contaminants to become highly concentrated in top predators. This occurs because predators at each higher level of the food chain accumulate the total chemical contaminant burden of all their prey. For instance, the bowhead whale sweeps up millions of copepods in a single day's foraging and, in doing so, accumulates the small amounts of chemical present in each individual prey item. The cumulative increase in the concentration of a chemical in successively higher levels of a food web is known as **biomagnification**. This is the reason why PCB levels in marine mammal blubber may be a million or more times higher than those in the Arctic ocean water. The issue of bioaccumulation is revisited briefly later in this chapter (see the pollution section of "Human Activities and Impacts on Marine Environments").

Threats to Ecosystem Integrity

Most of the major threats to Arctic ocean waters impact the higher trophic levels, including marine mammals, sea birds, and polar bears. This is important not only because these species are of the most interest to people, both local people as well as tourists, but also because the roles of the top predators are important in the structure and function of the ecosystem. If top predators are impacted by one or more of the threats described briefly below, feedback may alter the structure and function of the ecosystems.

Hydroelectric Development Hydroelectric dams and developments in Quebec, Ontario, and Manitoba will continue to affect rivers flowing to the Arctic primarily by increasing winter flows and decreasing summer flows. In turn, altered flows may cause changes in ocean currents, nearshore ice conditions, nutrient availability, the timing and magnitude of ice algal and phytoplankton production, the use of estuaries by marine mammals and **anadromous** fish that migrate into fresh water to spawn, and the ways in which coastal residents use the land and the sea. Although research programs are continuing, knowledge of many Arctic ecosystems is so poor it is possible only to guess at long-term changes. Caution is necessary since, once put in place, hydroelectric developments are irreversible in the sense of meaningful human time horizons.

Long-Range Transport of Pollutants Ecosystem integrity is threatened by the long-range transport of pollutants (LRTP) into the Canadian Arctic from agricultural and industrial activities elsewhere in the northern hemisphere. POPs, or persistent organic pollutants (halogenated compounds), and heavy metals are the most important groups of LRTP contaminants. These substances are difficult to metabolize and excrete, so they bioaccumulate along the food chain, culminating in top Arctic predators.

PCBs evaporate from cropland, dumpsites, lands, and waters to the south. While they may enter the Arctic via ocean currents, their main pathway is atmospheric transport. Winter air flowing from Europe and Asia over the North Pole deposits thousands of tonnes of soil, fly ash particles, and associated pollutants on land and in the sea. When snowmelt occurs in June, these substances enter the oceans, are concentrated by algae and phytoplankton, and passed on up the food chain. The long length of Arctic food chains concentrates (biomagnifies) contaminants to levels that are high enough to constitute potential health hazards for the animals and humans higher up the chain.

Mercury occurs in natural ecosystems and is concentrated heavily in long-lived top predators such as seals, whales, and polar bears. Research has revealed that concentrations of mercury in the hair of humans and seals currently inhabiting Greenland are three to four times higher than they were in the hair of preindustrial humans and seals. While we do not know what physiological and behavioural effects mercury concentration in top marine predators might be, we can surmise they are at risk since they cannot switch to an alternate food source as humans might do. LRTP contaminants may pose the most important threat to the integrity of the Arctic marine ecosystem; the cultural and economic costs of the loss of polar bears, for example, are almost incalculable. Taking the required global actions to clean up these pollutants is difficult; Canadians can regulate directly only what happens within our own boundaries. However, we can help persuade political leaders to follow up with their commitments to ban POPs through such international agreements as the 2000 Stockholm Convention.

Climate Change The scientific consensus is that the effects of climate change will be most evident at high latitudes in the form of increases in winter temperature and snowfall as well as reductions in the extent and thickness of sea ice. Gradually, arctic will be transformed into subarctic; animals, ice edges, and other boundaries will shift. Analyses of changes in both body condition and cub production by polar bears in western Hudson Bay over the past two decades, in relation to the timing of sea ice breakup, suggest that climate change already may be affecting this population (Lunn & Stirling, 2001).

Climate change also presents potentially serious security concerns. If the Northwest Passage were to melt, as predicted, Canada's undefended Arctic coast would be open to sea-going vessels, water piracy, waste dumping, and oil spills. An open Passage also could threaten Canadian sovereignty (Canadian Arctic Resources Committee, 2002). Increased UV radiation as a result of ozone depletion over the poles may damage phytoplankton at high latitudes, but perhaps not enough to affect ecosystem integrity. Much remains unknown, however, and climate change remains a concern for Arctic seas.

Nonrenewable Resource Extraction Specific geographic sites such as the Polaris and Nanisivik base metal mines (see Figure 8–1b) and the Bent Horn oil field in the High Arctic appear to pose minor threats to local ecological integrity. Oil and gas exploration in the Beaufort Sea has been a source of hydrocarbon contamination (from drilling muds and fuel spills), but greater threats will occur during the production phase, perhaps during the next 20 years. Oil spills from wells, tankers, and pipelines around the world demonstrate that, in spite of stringent regulations, the Arctic will not be exempt from disasters. Depending on the time of year a massive oil spill might occur, Arctic marine ecosystem integrity would receive moderate to extremely severe damage.

Industrial growth in the Arctic has increased demands for more roads and port facilities. In 2001, a road and port project in Bathurst Inlet, Nunavut, was under review through the Nunavut Land Claims Agreement. Proposed to connect base metal and diamond mines with a new coastal port, several organizations felt this project would have potentially severe effects on people and wildlife. Not only would the proposed road overlap with the calving grounds of the important Bathurst caribou herd, port development would result in heavy ship traffic through summer beluga whale concentrations. These organizations requested rigorous (federal) reviews of the project.

CANADA'S PACIFIC OCEAN ENVIRONMENT

Major Characteristics

Of Canada's marine environments, the Pacific shoreline is the shortest (27 000 kilometres) and has the narrowest continental shelf (typically 16 to 32 kilometres wide), as well as the warmest waters (8 to 14°C at the surface) (see Figures 8–1a and c). The fusion of offshore currents (including the northward-flowing Alaska Coastal current, the southward-flowing California current, weak coastal currents over the continental shelf, such as the Vancouver Island Coastal current, tidal streams, and upwellings) create a complex oceanographic system that mixes the waters, giving the region a very high level of productivity.

Over 300 finfish species occur in this region, including anadromous species such as Pacific salmon (five species), steelhead, cutthroat, and Dolly Varden trout; **catadromous** eel family members; and other marine species such as Pacific cod, rockfish, flounder, lingcod, and herring. Numerous species of shellfish such as shrimps, scallops, crabs, and clams also proliferate in Pacific coastal waters. Millions of sea birds, both coastal and offshore species, and large numbers of marine mammals, such as whales, porpoises, dolphins, sea otters, seals, and sea lions, live and feed in or migrate through British Columbia waters.

The coastline is rocky, and only a few small estuaries exist. Mudflats are found at the head of fjords, especially in the southern part of the province, and deltaic deposits occur at the mouths of major rivers such as the Fraser, Skeena, and Stikine (see Figure 8-1c).

In 2001, British Columbia's commercial fishing sector continued to decline, employing about 5400 people and generating $358 million in revenue. Commercial landings of wild salmon were valued at over $33 million, while recreational fishing (that supported about 8900 jobs) generated revenues of $675 million. One of British Columbia's fastest-growing industries, aquacultural production had a wholesale value of more than $318 million (British Columbia Ministry of Agriculture, Food and Fisheries, 2003).

Threats to Ecosystem Integrity

Global Change Not all is well in Canada's Pacific Ocean marine environment. Scientific evidence suggests that ocean surface temperatures are increasing on the Pacific coast, although clear answers as to why this is happening and what effects they will have on marine environments are not yet available. We do know that global environmental changes such as the El Niño–Southern Oscillation (see Chapter 5) not only reduce upwelling and lower productivity but also influence fish behaviour. In an El Niño year, salmon swim out away from the Alaska coast and closer to the British Columbia coast, resulting in record catches off British Columbia, and a scarcity for Alaska. These processes also have been implicated in very poor salmon returns in 1998, 1999, and 2000, possibly due partly to high ocean mortality from northerly movement of predatory mackerel.

Marine Pollution There are growing threats to the sustainability of the natural resource base, particularly in areas such as the Georgia Basin, where growing human coastal populations generate increasing waste disposal, municipal wastewater, urban and agricultural runoff, and industrial, oil, and chemical discharges and spills. In part because of immigration, population numbers in Vancouver and Victoria have grown; in 2002, more than 50 percent of British Columbia's 4.069 million inhabitants lived in these two cities. As of February 2002, Vancouver's Iona Island sewage plant still provided only primary treatment of the more than 416 billion litres of liquid waste that are disposed of annually. Both the Iona Island and Lions Gate sewage treatment plants failed bioassay toxicity tests, following which British Columbia's Minister of Water, Land and Air Protection required both plants to provide full secondary treatment by 2020 and 2030 respectively.

Both Vancouver and Victoria discharge large amounts of effluent into fish habitats. Since 1962, Vancouver's effluent has contributed to the closure of shellfish harvesting in Boundary Bay. As of February 1995, over 120 200 hectares of coastal marine habitat were closed to shellfish harvesting due primarily to sewage contamination, as well as contamination by toxic substances such as dioxins and furans (Department of Fisheries and Oceans, 2002a).

CANADA'S ATLANTIC OCEAN ENVIRONMENT

Since John Cabot's small ship arrived off the coast of Newfoundland in 1497, the Grand Banks have been hailed as the world's richest hunting ground for Atlantic (northern) cod. The cod stocks, at one time so thick that ships were said to be slowed by them, seemed inexhaustible. But, in 1992, in light of unrestricted exploitation, stock declines, and fish plant closures, the Canadian Department of Fisheries and Oceans placed a two-year moratorium on cod fishing in an attempt to avert ecological disaster. The moratorium remained in effect more than ten years later.

Cod stocks off Newfoundland and Labrador have declined by 97 percent during the past 30 years, while stocks in the northern Gulf of St. Lawrence have declined by 81 percent. These statistics provide clear evidence of how a once bountiful resource can be decimated through ecosystem changes, human ignorance, and disregard of precautionary conservation principles. The declining biomass is the prime reason why, in May 2003, COSEWIC listed the Newfoundland and Labrador population of Atlantic cod as an endangered species and the northern Gulf of St. Lawrence cod stock as threatened (COSEWIC, 2003).

Major Characteristics

The eastern coastline of Canada is about 40 000 kilometres long, but the dominant physical feature of the east coast marine environment is the large, submerged continental shelf (see Figures 8–1a and d). Characterized by raised offshore areas of the seabed known as banks, with shallow water depths often of 50 metres or less, these areas of the continental shelf are associated with high levels of biological productivity and marine life. At 250 000 square kilometres, the Grand Banks is the largest bank in the northwest Atlantic.

Three interconnected ocean currents—the Labrador Current, Gulf Stream, and Nova Scotia Current—mix and exchange coastal and deeper ocean waters, causing upwelling and bringing stored nutrients from the sea bottom up through the water column. This results in increased biological productivity, particularly along the edge of the Scotian shelf and the southern Grand Banks. The Labrador Current brings cold Arctic waters down to the eastern margin of the Grand Banks, while the warm Gulf Stream flows up the east coast of the United States and over the southern Grand Banks. The Nova Scotia Current is a smaller coastal movement of cool water from the Gulf of St. Lawrence along the Scotian Shelf to the Gulf of Maine. The complex mixing of these major currents, local gyres, eddies, and tides, as well as spring discharge from northern rivers and melting ice from the Arctic, combine to increase biological productivity and marine life diversity.

Temperature and salinity also have consequences for marine environments and climates. Atlantic Canada's offshore seawater is relatively cold, but a slight change in temperature can have great impacts on biological production. For instance, because each fish species has a range of tolerance to water depth and temperature, a change of 1°C can affect the distribution of a species, particularly if the species is living at the extreme limits of its distribution. A recent decline in water temperature may be a contributing factor in the collapse of the groundfish fishery. Similarly, declines in salinity resulting from increases in Arctic snowmelt may have contributed to the decline in cod and other groundfish (Meltzer, 1995). However, scientific knowledge of the effects of changes in temperature and salinity is incomplete, and further research is needed.

For centuries, the spring phytoplankton bloom on the wide Atlantic continental shelf has supported several major fisheries. In 1990, for example, cod and other groundfish such as haddock, plaice, flounder, and halibut accounted for about 80 percent of total Canadian landings in weight and were valued at over $375 million. In 1998, groundfish accounted for about 18 percent of Canada's total landed value ($1573 million). Catch values of pelagic species, such as herring, mackerel, tuna, salmon, and capelin, reached over $147 million in 1998. Invertebrate fisheries for shrimp, lobster, crab, scallops, and clams netted more than $475 million in 1990. The 1998 invertebrate catch from the Atlantic coast accounted for over 91 percent of Canada's total landings by value (Statistics Canada, 2003).

In 1995, then federal Fisheries minister Brian Tobin announced that cod stocks may have declined to the point of commercial extinction. The dramatic drop in the 400-year abundance of northern cod and other groundfish species has continued. Although there are some "good news" stories associated with the recovery of two localized fish stocks, the

Fisheries Resource Conservation Council report (2003) noted with alarm the sustained failure of most groundfish stocks in the Scotian Shelf and Bay of Fundy areas. Fisheries scientists and inshore fishers cite predation by seals, reduced capelin populations, ongoing problems with by-catches, unaccounted mortality (such as dumping and discarding, and landing unreported catches), changes in ocean temperatures, seismic testing, food chain disruptions, and low biomass thresholds as causes for the continued decline in stocks.

The Fisheries Resource Conservation Council (FRCC, 2003, p. 5) has recommended that every conservation recommendation should be explored—"we owe it to the fish to do so." COSEWIC's listing of Atlantic cod as an endangered species may be viewed as an appropriate conservation action because it forces the federal Cabinet to consider listing the cod as legally protected under the Species at Risk Act. Once listed legally, it would be a crime to kill a cod fish. Such an action would preclude any possibility of reopening the fishery; angry Newfoundland fishers already have begun protesting (COSEWIC, 2003; Jaimet, 2003).

East coast waters support the feeding and breeding of many marine mammals and sea birds. While whales are not harvested commercially, they are vulnerable to fishing gear entanglement and ship collisions. As whale watching and other associated nature-based tourism activities have become more popular, the overall need to protect whales has been acknowledged, and the Department of Fisheries and Oceans has established several whale conservation areas off New Brunswick and Nova Scotia. Seals, on the other hand, have been a controversial and politicized issue on the east coast for almost 20 years. Landsmen in Newfoundland and residents of the Îles de la Madeleine currently undertake commercial fisheries for harp and hood seals. Public and scientific controversy still rages over the role played by rebounding seal populations in the biomass decline in groundfish.

Millions of sea birds also nest and breed on Canada's east coast. However, changes in oceanic conditions and currents as well as overfishing have affected capelin and herring, important food fish for the sea birds. As capelin populations have declined—capelin are the key link between upper and lower levels of the food web, and are the principal food source for larger fish, sea birds, and marine mammals—scientists have documented declines in sea bird populations (Carscadden, Frank, & Miller, 1989).

Productivity of the Atlantic marine ecosystem is a function of diverse geographic and physiographic conditions such as shallow, rich estuaries and bays that provide ideal habitat for shellfish, and the oceanic currents, tidal flows, gyres, and upwellings on the continental shelf that influence the interactions of fish, mammalian, invertebrate, and plant communities. There are few comprehensive productivity studies, however, and limits to our knowledge have contributed to overfishing, marine degradation, and other problems.

In addition to its biological productivity, sedimentary formations underlying the continental shelf contain a variety of mineral resources, including petroleum, sand and gravel, silica sands, and precious metals. Commercial production of crude oil began in June 1992 in the Cohasset-Panuke oil field located 256 kilometres southeast of Halifax. The giant Hibernia project 312 kilometres east of St. John's on the northeast Grand Banks began pumping oil at a rate of 20 000 barrels per day in 1997. In 2002, Hibernia's average daily production was 180 000 barrels (about 66 million barrels per year) from an estimated recoverable reserve of 750 million barrels. Together, Terra Nova and Hibernia produce about 121 million barrels of oil per year.

Threats to Ecosystem Integrity

Lack of Knowledge   Much remains unknown about the highly variable yet interconnected Atlantic marine environment, and much research is required if scientists are going to be able to predict and prevent the negative consequences of human activities on marine ecosystems. The same is true if scientists are to understand the nature and implications of environmental changes. Fundamentally, this lack of knowledge underscores the importance of a precautionary approach toward use of ocean resources.

Anthropogenic Impacts and Marine Pollution
Although the coastal population of the Atlantic provinces is widely dispersed and there are few areas of industrial development, there are some hot spots of pervasive marine pollution including the St. Lawrence River and estuary, Halifax Harbour, and St. John's Harbour. Hot spots are likely to grow in number and magnitude with increasing population, economic growth, and industrial development in coastal areas. Increasing competition for ocean space and limited resources also is anticipated, highlighting the need to have a coordinated approach to management of marine resources in the region.

Many productive shellfish areas have been contaminated by municipal or industrial effluents such as sewage, heavy metals, and polycyclic aromatic hydrocarbons. About 500 000 m^3 of municipal sewage (50 percent untreated) is discharged daily into coastal waters. Since 1940, the number of shellfish area closures has increased steadily until, in 1997 for instance, over 33 percent (more than 200 000 hectares) of the area classified as suitable for direct harvesting of shellfish in Atlantic Canada was closed (Menon, 1998). Similar conditions cause closures of recreational beaches, restrictions on siting of aquaculture operations, and limits on development options in general.

Other point sources of pollution occur adjacent to Canada's Atlantic Ocean environment, including effluents from pulp and paper mills, mines, and mineral processing plants; food processing plant discharges; and oil and hazardous chemical spills. Nonpoint sources of pollution

Photo 8–2
Although a "nonconsumptive" use of marine life, whale watching may have negative effects on some species.

include pesticides and other chemicals from agricultural and urban runoff, and atmospheric acid deposition.

Commercial Fishing Clearly, our ability to set fishing harvest levels in harmony with the ocean's level of productivity is difficult. Incomplete knowledge of the impacts of fishing and fishing practices on the natural marine ecosystems and the role of other ocean changes means the effects of overfishing are not always clearly identifiable. Nevertheless, during 2000, **total allowable catches** for most stocks continued to fall. Efforts to regain spawning biomass have met with only limited success.

Other factors contributing to stock decline include discarded fishing nets and gear that continue to catch fish, marine mammals, and sea birds without human supervision (**ghostfishing**), and the dumping of plastics and other refuse. The use of mobile fishing gear, such as trawls, rakes, and dredges, has caused extensive structural destruction of ocean floor ecosystems. The trawl fishery drags nets and gear weighing more than one tonne along the ocean floor, removing important components of the marine habitat and affecting most fish and invertebrates that live and feed at the ocean bottom. For instance, critical damage has been done to the continental shelf off Nova Scotia, at depths below about 200 metres, where 500-year-old seafan coral groves have been "clearcut" by trawlers. Their recovery will take centuries. Such damage means the area's biodiversity is reduced and fish catches decline (McCallister, 2000; Willison, 2002).

Sea-Level Rise

Some scientists predict that global warming will have serious implications for Atlantic Canada, particularly with regard to sea-level rise. Under current climate change scenarios, much of the coasts of Prince Edward Island, New Brunswick, and Nova Scotia would face increased rates of bluff erosion, beach erosion, and destabilization of coastal dunes. While some parts of the coast would be submerged permanently, in other places new beaches, spits, and barriers could form. With an increase in sea level, many communities could be damaged by storm surges. More research is needed here to identify specific impacts and to design appropriate protection strategies.

SUMMARY OF CONCERNS FACING CANADA'S OCEAN REGIONS

Canada's coat of arms bears the motto *A Mari usque ad Mare*—from sea to sea—a clear designation of the significance of Canada's oceans to the life of the country. For centuries, different cultures in Canada have depended on the bounty of the oceans to support their traditional ways of life. From the Inuit groups in the Arctic and the First Nations peoples on the Pacific and Atlantic coasts, to the one thousand or so communities of fishers bordering the Atlantic, the ability to harvest various species of fish and shellfish, marine mammals, and sea birds to support both subsistence needs and commercial opportunities has been fundamental to the continuation of the lifestyles of these people.

Until relatively recently, these lifestyles were supported by generally stable, healthy ocean ecosystems that permitted growth in both fishing and processing industries based on what were thought to be renewable resources. Today, however, as humans have exploited fish stocks and other marine mammal and sea bird populations in order to earn a living, not only has the renewability of these resources come into question, but also their sustainability. In 2000, 12 countries accounted for about 66 percent of the volume of total world catch, mostly Peruvian anchoveta, Alaskan pollock, Atlantic herring, and skipjack tuna (Figure 8–3a). After increasing steadily for more than 30 years, the world catch of ocean fish dropped sharply in 1990 and appeared to have peaked at between 85 and 90 million tonnes per year (Figure 8–3b). In 2000, however, the world fisheries catch reached 94.8 million tonnes, the highest level ever. If China's catch is excluded (because their catch statistics likely are inaccurate), world catches in 2000 were similar to early 1990s levels of about 77 to 78 million tonnes (Food and Agriculture Organization, 2002).

The United Nations Food and Agriculture Organization (FAO) has estimated that catches of 70 percent of marine species have reached or exceeded sustainable levels. Given the need to conserve aquatic resources for the future, the FAO adopted the formal, global Code of

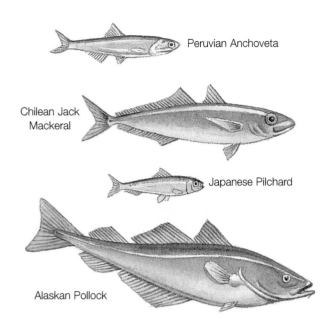

Figure 8–3a

World's most-fished species, 2002

SOURCE: *The State of World Fisheries and Aquaculture,* Food and Agriculture Organization, 2002, http://www.fao.org/docrep/005/y7300e/y7300e.00.htm

Conduct for Responsible Fisheries in 1995. This non-mandatory code indicated that the right to fish carried with it the obligation to do so in a responsible manner in order to ensure effective conservation and management of living aquatic resources. Another important principle contained in this code is the admonition to all fisheries management organizations to apply a precautionary approach to conservation, management, and exploitation of aquatic resources and environments (Food and Agriculture Organization, 1997).

In addition to continuing ecological stresses such as growth in fishing pressure and stock or species collapses, a number of new human stresses as well as climatic and natural uncertainties have added to the difficulties in managing and protecting marine environmental resources for the future. For instance, the pace of coastal development, including tourism and recreational growth, has exacerbated problems relating to use of environmentally sensitive sites, water pollution, and waste disposal, while industrial growth has added new toxic chemicals such as dioxins and furans to effluent discharges. These stresses have ecological and economic implications such as bio-magnification of contaminants in fish, wildlife, and humans, reduced fish reproduction, fish die-offs, shell-fishery closures, beach closures, and tourism losses.

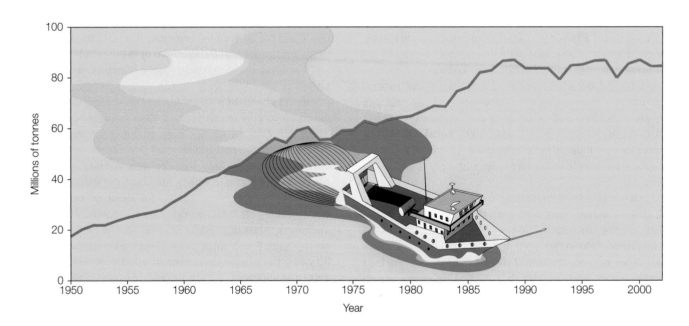

Figure 8–3b

World total fish production in marine waters, 1950–2002

SOURCE: *Yearbook of Fishery Statistics: Summary Tables,* Food and Agriculture Organization, 2002, http://www.fao.org/fi/satist/statist.asp; *Yearbooks of Fishery Statistics: Capture production by countries and areas,* Table A2, Food and Agriculture Organization, Fisheries Department, 2002, http://www.fao.org/fi/statist/statist.asp; *Yearbooks of Fishery Statistics: World capture production,* Table A1, Food and Agriculture Organization, Fisheries Department, 2000, http://www.fao.org/fi/statist/statist.asp

NOTE: Data used to plot line from 1993 to 1995 were calculated values, ±1 million tonnes. Statistics reflect changes in data collected between 1950 and 2002 as understanding of fisheries dynamics improved.

Additional human stresses relate to energy supplies; as Canadian land-based supplies of petroleum resources decline, offshore oil and gas finds become more viable economically. Exploration and production activities may alter or contaminate habitats temporarily, whereas operational or accidental releases of toxic substances may degrade habitat for long periods. Marine transport safety and pollution issues continue to affect Canadian as well as international waters.

The volume of marine traffic carrying toxic or harmful substances is a major concern for ports and their approaches. The port of Vancouver, for example, annually ships millions of tonnes of chemicals and fuels to offshore markets. While marine accidents can cause spills of large volumes of oil and other hazardous substances, most spills from oil tankers result from routine operations such as loading and discharging. While on the high seas, some captains deliberately decide to pump out bilge water contaminated with leaked fuel oil or lubricating oil, or to discharge washings from fuel or cargo tanks over the ship's side. These actions, and the perception that oceans have an infinite capacity to absorb waste, need to be changed if long-term sustainability of ocean environments is to be assured. Canada's National Defence aerial surveillance activity has resulted in the successful prosecution of offshore vessel captains who illegally discharge pollutants. Such enforcement activities help ensure environmentally responsible

Photo 8–3
The high volume of traffic carrying toxic cargo through the port of Vancouver is a continuing concern.

commercial marine operations in Canadian waters.

Global atmospheric, climatic, and sea-level changes are difficult to predict with certainty, but will affect all three Canadian coasts, particularly the Arctic. Inundation of coastal communities and wetland habitats, for example, could displace coastal populations and affect marine species distribution patterns, potentially leading to major economic restructuring.

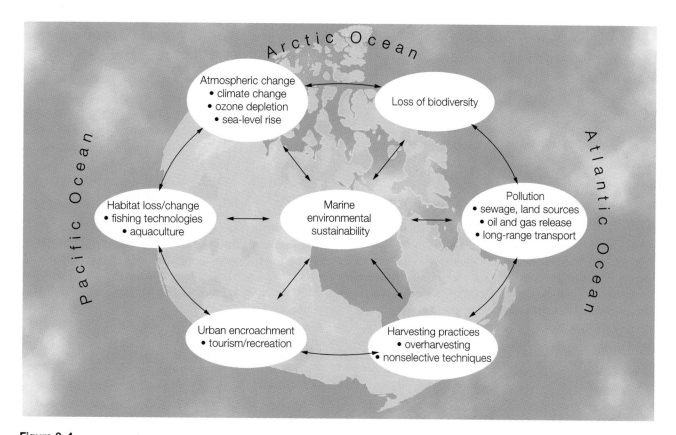

Figure 8–4
Common threats to Canada's marine environmental sustainability

CHAPTER 8: OCEANS AND FISHERIES

The effects of overharvesting both within and beyond Canada's 200-nautical-mile exclusive economic zone (EEZ) have been most evident on the Atlantic coast, where thousands of jobs were lost during the 1990s in Newfoundland and Labrador (see Coward, Ommer, & Pitcher, 2000). In both Atlantic Canada and British Columbia, the issues of excessive fishing capability and the power of fishing technologies have compounded the fundamental problem of too many fishers chasing too few fish. In the Arctic, there is concern about the potential consequences of opening a commercial char fishery when there is a lack of knowledge about the links between arctic char and other species (Beckmann, 1995). This is a useful illustration of the importance of making decisions at an ecosystem, rather than species, level.

Six major interconnected and strikingly similar threats to Canada's marine environmental sustainability exist (to varying degrees) on all three coasts: harvesting practices, urban encroachment, pollution, habitat loss, atmospheric (climatic) change, and loss of biodiversity (see Figure 8–4). A selection of these threats is discussed in detail in the following section.

Human Activities and Impacts on Marine Environments

FISHERIES

The 1995 Canada–Spain Turbot Dispute

> The unthinkable has come to pass: The wealth of oceans, once deemed inexhaustible, has proven finite, and fish, once dubbed "the poor man's pro-tein," have become a resource coveted—and fought over—by nations. (Parfit, 1995, p. 2)

The right of nations to exploit their coastal waters is well established in customary law and formal treaties, but it is only recently that international oceanic law has come to grapple with the question of whether a coastal nation has a right to demand protection of migrating species important to coastal economies even while these species are in international waters. Some nations that have large coastal fish stocks, such as Canada and the United States, argue for extending jurisdiction, while other nations that depend on fishing in distant waters argue for limitations to control overfishing beyond certain boundaries. The 1995 Canada–Spain conflict over turbot (or Greenland halibut—*Reinhardtius hippoglossoides*) is one incident that highlights the issue of transboundary stock protection.

As commercially valuable stocks of the bigger, slower-growing species have declined, commercial fishing fleets have turned to "fishing down the food chain," targeting increasingly large quantities of smaller species of fish with less commercial value. In the case of the Spanish factory trawler fleet, their arrival on the Grand Banks in the mid-1980s was a result of their "fishing down African hake" and being "diverted to the Northwest Atlantic to find new fishing possibilities" (Department of Fisheries and Oceans, 1995c).

Canadian fishers have harvested turbot in our coastal waters for many years but, in 1986, European Union (EU) vessels, mostly from Spain and Portugal, began seriously to overfish groundfish stocks managed by the Northwest Atlantic Fisheries Organization (NAFO), an agency of the United Nations Food and Agricultural Organization. Some of these stocks were located on the "nose" and "tail" of the Grand Banks just outside Canada's 200-mile limit. Turbot catches remained at about 20 000 tonnes from 1985 to 1988, then grew to 47 400 tonnes in 1989 and to 63 000 tonnes in 1992 (Department of Environment, n.d.; Department of Fisheries and Oceans, 1995a; Gomes, 1995; Revel, n.d.).

In September 1994, NAFO adopted a 1995 total allowable catch (TAC) for turbot of 27 000 tonnes, less than half of what had been caught in 1993 and 1994. In January 1995, the minister of Fisheries and Oceans (Brian Tobin) attended a special NAFO meeting in Brussels to determine how the turbot that migrated across the 200-mile limit should be shared. At the close of that meeting, Tobin claimed victory for Canada's conservation focus regarding turbot stocks, noting that most of NAFO's 15 major fishing nation members had voted for the allocation of 16 300 tonnes to Canada (a 60 percent share), and 3400 tonnes to the EU (a 12 percent share). The remainder of the TAC was allocated to Russia (3200 tonnes), Japan (2600 tonnes), and others (1500 tonnes) (Department of Fisheries and Oceans, 1995b, 1995c).

Under the provisions of a NAFO objection procedure, however, the EU could reject the NAFO decision, set its own quotas, and allow EU vessels to fish virtually without restriction in the northwest Atlantic. Objection procedures seriously hamper the effectiveness of international ocean treaties, yet in some cases are the only way to activate treaties and bring all the resource users to the same table. As expected, the EU filed its objection to the turbot allocation in March 1995, and announced that its members would catch 18 000 tonnes of turbot, unilaterally setting the quota at 69 percent of the catch. Essentially, quota allocations sparked the confrontation that evolved into the Canada–Spain "turbot war."

In a move designed to permit Canada to conserve the turbot and other groundfish stocks that straddle the 200-mile limit, coastal fisheries protection regulations were amended in 1994 and in 1995. These regulations allowed

Canada to seize Spanish and Portuguese vessels (and other ships flying flags of convenience) if they were caught fishing outside the 200-mile limit in defiance of NAFO quotas. This step was necessary because many European Union vessels registered with non-NAFO countries such as Panama and then fished without quotas, thus exacerbating the effects of EU overfishing. Also, according to the Department of the Environment (n.d.), foreign fisheries on the nose and tail of the Grand Banks outside 200 miles were characterized by the harvest of prespawning juveniles (immature fish), damaging future recruitment of the turbot stocks.

Diplomatic efforts to resolve these problems had continued, but it was under the coastal fisheries protection regulations that on March 9, 1995, Canadian authorities boarded and seized the Spanish fishing trawler *Estai* after a chase at sea ended when warning shots were fired across its bow. The *Estai* captain was arrested and the ship was escorted into St. John's Harbour by Fisheries and Coast Guard vessels. Subsequently, Spain accused Canada of breaking international law when it seized the *Estai* in international waters. That case was put before the International Court of Justice; however, the case was dropped when Canada refused to let the court have jurisdiction (a requirement for this court to proceed). Some observers feel this was because, in strictly legal terms, Canada likely was in the wrong.

Inspection of the *Estai* revealed that 79 percent of its catch was undersized turbot and that 25 tonnes of American plaice, an endangered species that NAFO had put under moratorium, were stored behind false bulkheads. The *Estai*'s net, which its crew had cut deliberately during the chase at sea, was recovered from the Grand Banks—its mesh measured 115 millimetres, although the smallest mesh size mandated for turbot by NAFO was 130 millimetres. Furthermore, the 115 millimetre net had an 80 millimetre mesh liner, to take even smaller fish (Bryden, 1995a). International regulations that have been adopted into the United Nations Convention on the Law of the Sea decree that fish cannot be caught before reaching a certain spawning size, to ensure the survival of the species. Canada claimed that this illegal net would account for the undersized turbot catch aboard the *Estai*, and was proof of Spanish overfishing and violation of international rules intended to preserve endangered fish species.

The *Estai* was released on March 15, 1995, on payment of a $500 000 bond, and discussions with the EU to settle the dispute resumed. About two weeks later, then Fisheries and Oceans minister Tobin was in New York City on a public relations mission. As diplomats meeting at the UN Conference on Straddling Fish Stocks and Highly Migratory Fish Stocks were considering how to enforce fisheries rules on the high seas, Tobin displayed the 5.5 tonne net from the *Estai* on a barge on the East River

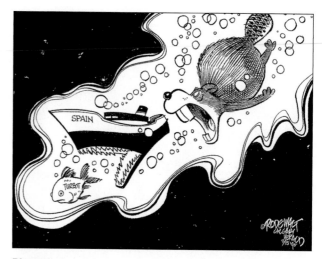

Photo 8–4

Calgary Herald, March 15, 1995, p. A4. Drawing by Vance Rodewalt.

across from the United Nations buildings. Holding up a tiny turbot in front of the illegal net, Tobin told "an army of international reporters and television crews" that "no baby fish can escape that monstrosity, that ecological madness. That's vacuuming the ocean floor and that's destroying and killing everything there" (Bryden, 1995b, 1995c). This was the day "a clever publicity stunt turned the tide in Canada's favor and the Spanish were ... branded around the world as despoilers of the high seas unscrupulously vacuuming baby fish from the ocean floor" (Gessell, 1995).

On April 15, 1995, Canada and the EU reached an agreement on the conservation and management of transboundary stocks. The Canada–European Union Control and Enforcement agreement was a bilateral commitment to provide better rules, effective enforcement, and more severe penalties governing all Canadian and EU vessels fishing in specific areas regulated by NAFO. Among the major components of the new enforcement agreement were placement of independent, full-time observers on board vessels at all times; increased satellite surveillance and tracking; increased inspections and more rapid reporting of infractions; verification of gear and catch records; significant penalties to deter violations; and new minimum fish size limits. Also under the agreement, new catch quotas for both Canada and the EU were put in place for the remainder of 1995 (Canada and Spain would each receive 10 000 tonnes of the 27 000 tonne quota).

In conjunction with their bilateral negotiations with the European Union, Canadians had been working internationally toward a binding United Nations convention regarding conservation of migratory or straddling fish stocks on the high seas. This effort continued a major thrust Canada made in the original UNCLOS negotiations a decade earlier. A United Nations agreement relating to

CHAPTER 8: OCEANS AND FISHERIES

Photo 8–5
Fisheries minister Brian Tobin displays an undersized turbot during his public relations mission in New York.

the conservation and management of straddling and highly migratory fish stocks was finalized and adopted without a vote by the UN General Assembly in August 1995. The agreement came into force in December 2001 (United Nations, 2003).

Turbot hostilities with Spain, from March 3 to April 15, 1995, were estimated to have cost the Canadian taxpayer over $3.24 million. This includes over $954 000 to operate ships that monitored the Spanish fishing boats, $231 000 for air surveillance, more than $752 000 in salaries for the RCMP and other law enforcement officers, and over $89 000 for Fisheries and Oceans minister Tobin's New York publicity campaign (Gessell, 1995).

Did Canada win the "turbot war"? A reduction in numbers of major violations of NAFO regulations (from 25 in 1994 to one in 1995) suggested that Spain and other members of the EU were cooperating with NAFO and Canada in regulation of the Grand Banks turbot fishery (Cox, 1996). The ability to cooperate in order to regulate catches and to conserve stocks is an important step toward sustainability of the resource. However, if sustainability is to be achieved, it is critical that all parties act responsibly and look to the best long-term interests not only of all participants but also the fishery resources themselves. Responsible fishing includes reduction in waste, as described in Enviro-Focus 8.

In 2002, however, given evidence of continuing irresponsible fishing practices, Fisheries minister Robert Thibault announced that Canada would close its ports to fishing vessels that were believed to have committed

serious violations of NAFO conservation and enforcement measures. In March and April, Canada closed its ports to vessels from the Faroe Islands and Estonia, respectively, for overfishing and misrepresentation of shrimp catches. In December 2002, ports were reopened to Estonian vessels following passage of special decrees to comply with Canadian and NAFO fishing regulations. Port closures had a negative economic impact on two Newfoundland communities that previously had supplied commercial cold storage and stevedoring operations. Minister Thibault indicated that residents in these communities understood that the long-term benefits of port closures would help to conserve fish stocks for future generations (Fisheries and Oceans Canada, 2002).

The Northern Cod Moratorium

The unsustainable harvest of fish stocks is an increasingly important issue in Canada, perhaps nowhere felt more keenly than in outport Newfoundland. The northern cod (*Gadus morhua*) was the focal point of the distinctive culture of the small, once isolated communities scattered around the coast adjacent to fishing grounds. Peopled mostly by those of English and Irish descent, Newfoundland outports have persisted for over 300 years based on small-scale, seasonal fisheries production. Although historical records indicate that northern cod have experienced general and localized cycles of abundance and severe decline, the fish always came back, and outport people adapted to these fish failures in a variety of ways (Coward, Ommer, & Pitcher, 2000; Ommer, 2002).

Change began in the 1930s with the decline of the saltfish trade, linked to the Great Depression among other things. When markets rebounded, the frozen-fish trade became the centre of development ideals, and traditional systems of merchant credit and old-fashioned processing methods began to disappear. A new scale of change began in the 1960s, led by the arrival on the Grand Banks of foreign trawlers built to withstand the icy winter storms of the North Atlantic. By 1968 their size and fishing power had boosted northern cod catches to over 800 000 tonnes, four times their historical average of 200 000 tonnes. This increase in catch clearly was the result of actions of the offshore foreign fleet and it came at the expense of the inshore Newfoundland fishers (see Figure 8–5). The 1968 catch, labelled "the killer spike," had two important impacts on the fishing industry in Newfoundland. Not only did the huge harvest in 1968 remove a large number of the northern cod population, it also appears to have reduced the resiliency of the stock to rebound from fishing mortality and changing environmental conditions, including ocean temperature changes (Department of Environment, n.d.). From a sustainability perspective, the 1980s expectation that northern cod stocks could sustain

50 Million Meals Dumped at Sea

Volume of Fish Wasted (Alaska)

1992	227 million kilograms
1993	336 million kilograms
1994	340 million kilograms

In 1994, a record 340 million kilograms of edible fish were dumped back into the ocean off Alaska because they were too big, too small, or the wrong sex—that's about 50 million meals wasted by discarding fish! The Alaska Department of Fish and Game reported that among the seafood estimated to have been dumped by large factory trawlers were 7.7 million kilograms of halibut, 1.8 million kilograms of herring, 200 000 salmon, 360 000 king crabs, and 15 million tanner crabs. Fifteen percent of the total bottomfish catch of 2.2 billion kilograms, including pollock, cod, and sole, was wasted. Some of the discarded fish originated in British Columbia streams flowing through the Alaska Panhandle.

The U.S. National Marine Fisheries Service calculates these volumes of waste based on observational data, which means that the actual amount of wasted fish could be much higher. If the statistics shown here represent the waste occurring off Alaskan coasts alone, how much fish is wasted worldwide?

Factory trawlers routinely haul in 10 to 30 tonnes of fish in a single net. If the fish are not of the right species or size to process aboard ship, or if only female fish are required in order to harvest their eggs (a delicacy in Japan that fetches top prices), they are simply thrown back in the water. Usually dead or dying, these fish are a deplorable display of waste. Alaska state officials want the North Pacific Fishery Management Council to endorse a plan requiring offshore fishers to keep everything they catch. The Canadian Fisheries minister could impose a similar rule on ships fishing in Canadian waters.

Responsible behaviour in the fishing industry, including controls on waste, is a necessity for sustainability in global fisheries, particularly in light of depleted stocks and declining catches around the world. Unless stocks are conserved, fish could become a luxury only the wealthy can afford.

SOURCES: "Millions of Kilos of Seafood Dumped in Sea," Canadian Press and Associated Press, December 3, 1995, *Victoria Times Colonist*, p. D8; "Stop High Seas Plunder," December 12, 1995, *Victoria Times Colonist*, p. A4.

---

annual catches of at least 400 000 tonnes, double historical averages, was highly significant.

In 1977, in the context of negotiations at the UN Convention on the Law of the Sea, Canada declared a 200-mile exclusive economic zone (EEZ) (see Figure 8–1a). That zone excluded about 10 percent of the Grand Banks—the nose and tail—where important stocks of cod, flounder, and redfish moved between Canadian and international waters and were fished commercially, outside Canada's control. In 1979, NAFO assumed responsibility for conservation of 10 northwest Atlantic fish stocks (including cod) outside Canada's 200-mile limit. Within the EEZ, Canada imposed strict controls: foreign fishing was phased out, the offshore fishery became a Canadian fishery, and the federal government developed a science-based system of fisheries management (Department of Fisheries and Oceans, 1995d).

As a consequence of the extension of fisheries jurisdiction to 200 miles, there was a perception of increased resource abundance and economic opportunity. Canadian fish-catching and fish-processing capacity expanded rapidly. After an initial period of increased catches in the late 1970s and early 1980s, however, groundfish stocks within and outside the EEZ declined drastically. By the mid-1980s, excess capacity and overcapitalization were evident in, for example, the development of jumbo draggers with larger hold capacities, bigger nets, more powerful engines, and higher price tags than conventional draggers.

About this time, the Department of Fisheries and Oceans (DFO) discovered serious inaccuracies in their stock assessments. In response, quotas were lowered, but it was not until 1992 that the magnitude of the problem was accepted—the fish had not come back, and maybe

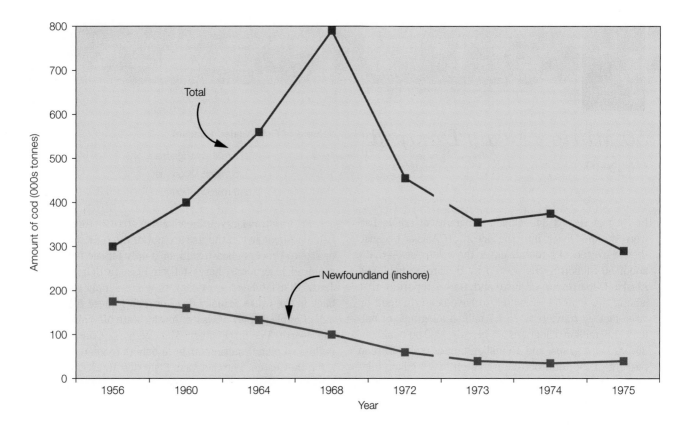

Figure 8–5

Newfoundland cod harvests showing the 1968 "killer spike" and impact on inshore fishers' catches

NOTE: These cod harvests took place in non-NAFO areas.

SOURCE: Adapted from *The Political Ecology of Crisis and Institutional Change: The Case of the Northern Cod,* B. J. McCay & A. C. Finlayson, 1995, paper presented at the Annual Meeting of the American Anthropological Association, Washington, DC, http://www.lib.unconn.edu/ArcticCircle/NatResources/cod/mckay.html

never would. By July 1992 it was estimated that the biomass of northern cod, Atlantic Canada's most important commercial fish stock, was about one-third its 30-year average and on the verge of commercial extinction. On this basis the Canadian government declared a two-year moratorium on the entire $700 million northern cod fishery: the intent of the moratorium was to allow the stock to rebuild. (Table 8–1 identifies selected events related to the moratorium.)

Over 40 000 fishers and fish plant workers ultimately lost their jobs as a result of the moratorium; their families, businesses, and community organizations dependent on their work also were affected. Government assistance for the planned two-year closure was provided through programs such as the Northern Cod Adjustment and Recovery Program. In 1994 the northern cod stock was reassessed and results indicated a continued decline and concern for biological extinction of the stock. The moratorium continued, and with closure of the food fishery in the summer of 1994, it became illegal for Newfoundlanders to jig for cod to feed their families. The recreational (food) fishery

reopened on a limited basis in 1996. Fishery-dependent workers and families in Atlantic Canada and Quebec were promised $1.9 million in income replacement and retraining assistance until the end of May 1999 (Department of Environment, n.d; Sinclair & Page, n.d.). However, the support program ended in August 1998 because it ran out of money. The moratorium remains in place, perhaps indefinitely, given the endangered status of certain key cod stocks.

The collapse of the groundfish fishery is a crisis of historic proportions for Newfoundland and the rest of Atlantic Canada as the displaced fishers and plant workers have lost their traditional livelihoods, independence, and way of life, and all levels of society and the economy have suffered (Blades, 1995; Coward, Ommer, & Pitcher, 2000; Davis, 2000). As understanding of the socioeconomic effects of the northern cod closure has grown (Ommer, 2002), it is clear how difficult it is to achieve intergenerational equity and to find a balance between protection of fish stocks and their habitat and provision of fish and fishing opportunities for Canadians

TABLE 8−1
SELECTED EVENTS LEADING TO THE 1992 NORTHERN COD MORATORIUM IN NEWFOUNDLAND

Settlement (1600s)	• Many coastal communities settled because of the abundance of cod. • Early boats (known as shallops) were 9 to 12 metres in length and carried up to five men fishing with hooks and lines.
Late 1800s	• Trawler fishing off the Grand Banks began to replace traditional methods. • Dragging in rich spawning areas of the Grand Banks signalled a potential for a resource collapse. • Calls to stop trawling were ignored due to the push for competitiveness. • Opponents of trawling were forced to adopt the practice to remain competitive. • The division between inshore and offshore fishers became firmly entrenched.
1955	• The town of Grand Bank opened a fish plant with three company-owned trawlers.
1963	• Northern offshore cod stocks were estimated at 1.6 million tonnes.
1977	• Canada extended its fishing boundary from 12 to 200 nautical miles with the expectation of rebuilding the resource to provide increased catch rates for Canadian fishers. • Fishing companies expanded rapidly.
1983	• The east coast fishery was restructured to rescue five financially troubled fish companies (National Sea Products, H. B. Nickersons, Fishery Products Ltd., Lake Group, and Connors Bros.). • Two large companies emerged: Fishery Products International and National Sea Products. • The result was a further rift between inshore and offshore fishers: inshore continued to be fished by small, independently owned boats, while the offshore fishery was dominated by corporate-owned, large-volume fishing trawlers.
Mid-1980s	• To gain a competitive edge, National Sea Products introduced The Cape North, a factory freezer trawler that could stay at sea for up to three months.
1988	• The total allowable catch (TAC) of northern cod was 266 000 tonnes. • 95 percent of fishers worked the inshore waters; they were allotted 15 percent of the TAC.
1990	• The TAC of northern cod was 197 000 tonnes. • The actual catch of northern cod was 127 000 tonnes.
March 1992	• The TAC for 1992 was reduced from 180 000 tonnes to 120 000 tonnes.
July 1992	• A two-year moratorium on the northern cod fishery was announced by federal Fisheries minister John Crosbie.
1994	• The moratorium was extended "indefinitely."
September 1996	• A limited recreational fishery was announced (cod could be caught for personal use).
1997	• A small commercial fishery was permitted on the south and west coasts.
2000	• A recreational food fishery, index fisheries, and sentinel fisheries to collect scientific data and occasional, small commercial catches are all that remain of the northern cod fishery.
2003	• Cod stocks are not recovering despite fishery closures; COSEWIC lists the Newfoundland and Labrador population of the Atlantic cod as endangered and the northern Gulf of St. Lawrence cod stock as threatened.

SOURCES: *Two Atlantic Cod Populations Designated at Risk,* Committee on the Status of Endangered Wildlife in Canada, 2003, http://www.cosewic.gc.ca/eng/sct7/sct7_3_1_e.cfm; "When the Future Died," J. Demont, July 13, 1992, *Maclean's, 105*(28), 15–16; "When All the Fish Were Gone," S. Frazer, May 1992, *Canadian Forum, 71*(809), 14–17; "A Fine Kettle of Fish," D. Gillmor, July/August 1990, *Equinox,* 66–75.

now and in the future. The causes of the crisis are many, complex, and interrelated, and include domestic and foreign overfishing, predation by seals, ghostfishing and **driftnetting**, seismic testing, government policies, corporate interests, the failures of international management, environmental and climatic factors, and the errors

and uncertainties of science. Clearly, the knowledge base for fisheries decision making must be improved.

Critical gaps in knowledge and their role in the collapse of the northern cod stock were recognized by a broad spectrum of fishery stakeholders. For instance, some inshore fishers were among the first to identify the

CHAPTER 8: OCEANS AND FISHERIES

Photo 8–6

Many of Atlantic Canada's fishing communities were affected economically and socially by the decline of fish stocks.

inaccuracy of the Department of Fisheries and Oceans' science-based stock estimates, but the fishers' warnings went unheeded. Academics and other analysts were among those who called for an internal reevaluation of scientific stock assessment and its methods. Scientists, too, who reassessed their estimates of stock sizes and showed that their earlier claims had overestimated the stock's abundance by as much as 100 percent and underestimated fishing mortality by about 50 percent, concurred that lack of knowledge had played a role in overfishing and, ultimately, the moratorium. In addition, it was recognized that the scientific side of fisheries decision making often was subject to considerable political pressure and intervention. Thus, when changes in the process of assessing and allocating the resource were initiated, there was agreement that it was necessary to open up the decision-making process.

The value of a more participatory and inclusive decision-making process was reflected in the 1993 decision of the minister of Fisheries and Oceans to create the Fisheries Resource Conservation Council (FRCC). Comprising 14 members, including scientists, academics, fishing industry groups, and other experts outside of the Department of Fisheries and Oceans, the FRCC was given the final authority for resource assessments and for recommendations to the minister regarding quotas and other conservation strategies. The spectrum of interests and expertise represented on the FRCC adds important dimensions to decision making by providing more broadly based views and public input into the process. Unfortunately, the only good news in the FRCC's 2003 groundfish recommendations was that Scotian Shelf haddock were showing signs of recovery, and one Newfoundland cod stock was increasing. Other cod stocks

had collapsed or were well below historical averages on indicators of stock health. Overall, the FRCC recommended continued tough conservation measures for groundfish stocks.

As cod stocks collapsed, other groundfish, including hake, haddock, pollock, halibut, and redfish, became economically more important as substitutes for cod. However, despite increased measures to protect fish biomass, almost all stocks have continued to decline. Although no one seems able to explain fully why the decline continues, one possibility to consider is whether the threshold effect has occurred. Is the lack of recovery the result of gross ecosystem imbalance? At least part of the solution lies in good scientific research (conducted with sufficient funding) to increase understanding of the Atlantic Ocean ecosystem. Another part of the solution may lie in establishment of a national sustainable fisheries policy (Auditor General of Canada, 1997). Discussion such as that associated with the Atlantic Fisheries Policy Review (2001) could contribute to the future of Canada's fisheries based on conservation, orderly management, and shared stewardship objectives.

Pacific Herring and Salmon Stocks

The "turbot war" and the northern cod moratorium are just two examples of human effects on fisheries in the northwest Atlantic where harvesting practices have had critical impacts on fish stocks, their habitat, and biodiversity, as well as significant social and economic effects on fishing communities and regions. If we shift our attention to the west coast fisheries, we find similar long-standing issues. Evidence is growing that Pacific salmon and herring stocks are being overfished and spawning habitats are being impacted severely. Managers have responded with quota restrictions, gear restrictions, and area closures. More recently they have been able to reduce fleet capacity, an essential element of any program designed to ensure economic returns to Aboriginal, commercial, and recreational fishers. Here, too, scientists have been grappling with their ability to assess stocks accurately and set TACs, to model ecosystem interrelationships, and to deal with the uncertainty of the ocean environment.

Pacific Herring Pacific herring are the most abundant fish species on Canada's west coast, providing employment for up to 6000 people. Herring harvests contribute millions of dollars to the provincial and national economies (wholesale value of herring products in 2001 was $119 million).

Just like capelin, herring are central to the marine food web: they are a key fish in the summer diets of Chinook salmon, Pacific cod, lingcod, and harbour seals, and herring eggs are important in the diet of migrating sea birds and grey whales (Environment Canada, 1994). Pacific herring require abundant kelp beds and uncontaminated waters in which to spawn, but herring spawning habitat is threatened by coastal development. In addition, Pacific herring are sensitive to natural fluctuations in ocean climate and ecology, including ocean temperature changes and predators such as the Pacific hake. Waters off the west coast of Vancouver Island undergo alternating periods of cool and warm water that have been intensified by strong El Niño events (see Figure 8–6). Young herring survival is reduced during warm events because Pacific hake are abundant and also because large numbers of Pacific mackerel migrate north into British Columbia waters and feed on herring and other species during the summer. On average, the eight most abundant predatory fish off Vancouver Island's west coast consume an estimated 45 000 tonnes of herring annually (six times more than is harvested there each year). The spawning biomass declines because fewer young herring survive to join the spawning stock. Conversely, survival and growth are relatively strong when the summer biomass of hake is low and the annual water temperature is cool, about 10°C (Environment Canada, 1994).

Until the late 1960s, herring were harvested and reduced into low-value products such as fish meal and oil. Very large quantities of herring—up to 250 000 tonnes in 1962—were caught in this reduction fishery, exceeding the biomass that was left alive to spawn (Figure 8–7). By the mid-1960s, the commercial herring fishery could not be sustained as most of the older spawning fish had been removed from the populations. Coast-wide, only 15 000 tonnes of herring were left to spawn in 1965, and in 1967, the federal government closed all herring fishing, except traditional food and bait fisheries, for four years.

Fortunately, Pacific herring are among a group of fish species that can recover dramatically from a reduced pop-

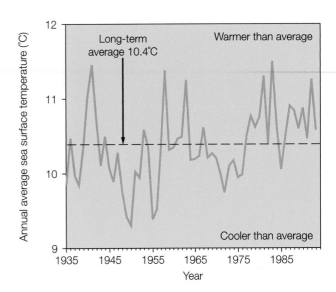

Figure 8–6

Variations in sea surface temperature off the west coast of Vancouver Island, 1935–1993

SOURCE: *Sustaining Marine Resources: Pacific Herring Fish Stocks,* Environment Canada, 1994, SOE Bulletin, No. 94-5, Ottawa: Author, p. 6. Reprinted with permission of the Minister of Public Works and Government Services Canada, 2004.

ulation size; by 1993 the stocks had rebuilt to at least 100 000 tonnes and reached an estimated 165 000 tonnes in 2001. The fishery is managed by setting a fixed quota based on a 20 percent harvest rate of the forecast mature stock biomass. To meet conservation objectives, a minimum spawning stock biomass also is enforced. If the forecast biomass falls below the cutoff threshold (that varies by region), the commercial fishery is closed to allow stock recovery. One-year closures have occurred in 1985, 1986, and 2001 for west coast Vancouver Island stocks, and in 1988, 1994, and 2001 for the Queen Charlotte Islands roe fishery, all because of low biomass. Fisheries and Oceans Canada admits that very little is known about the factors that affect recruitment in the different herring stocks, making it difficult to forecast future stock trends.

In 1972, a new fishery began to harvest British Columbia herring for its high-quality roe. Known as kazunoko, herring roe is a traditional delicacy that, until the economic downturn in the Japanese economy, sold for $120 to $150 per kilogram in Japan. The highly controlled roe fishery removes about 31 000 tonnes per year (average for 1990–99). Even though only about one-tenth the amount of herring is taken compared with the 1960s reduction fishery, the landed value of the roe fishery averaged about $37.5 million annually between 1997 and 2001 (see Figure 8–7). The processed value of the catch is two to three times the landed value. First Nations people, for whom herring has been a traditional food, have been active participants in the food and commercial roe fishery.

Photo 8–7

Herring spawning off the west coast of Vancouver Island.

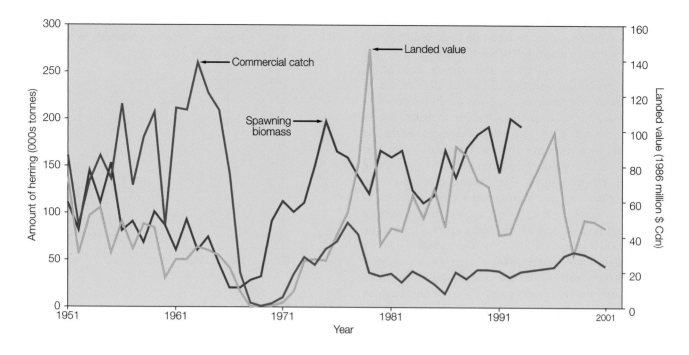

Figure 8–7

Spawning biomass, commercial catch, and landed value of Pacific herring, 1951– 2001

SOURCES: *Sustaining Marine Resources: Pacific Herring Fish Stocks,* Environment Canada, 1994, SOE Bulletin, No. 94-5, Ottawa: Author, pp. 3, 4; *British Columbia Seafood Industry in Review,* B.C. Minister of Agriculture, Food and Fisheries, 1998, 1999, 2001, http://www.agf.gov.bc.ca/fish_stats/statistics.htm#StatPubList; *Stock Status Reports: Pacific Region 2002,* Pacific Scientific Advice Review Committee, Fisheries and Oceans Canada, 2002, B6-01, B6-02, B6-03, B6-04, B6-05 http://www.pac.dfo-mpo.gc.ca/sci/psarc/SSRs/pelagic_srrs_e.htm; *B.C. Herring Production,* Ministry of Agriculture, Food & Fisheries, Government of British Columbia, 2002, http://www.agf.gov.bc.ca/fish_stats/Herring.htm

They have maintained their presence in the herring fishery for the past decade, owning about 25 percent of the herring roe fleet vessels and almost 75 percent of the licences to harvest roe spawned on kelp (Fisheries and Oceans, 2001a).

In 1991, 1200 boats participated in the lucrative roe fishery; herring contributed $214 500 to the gross income earned on each boat (in comparison, salmon accounted for $146 500 per boat) (Environment Canada, 1994). A combination of high financial stakes and stringent control efforts by the Department of Fisheries and Oceans to ensure that quotas are not exceeded means the roe fishery is "an aggressive event where fishermen attempt to net as much herring as they can in the few minutes an opening is declared" (Meissner, 1995). In 1994, for example, the Strait of Georgia roe herring fishery for seine vessels was open for only 30 minutes. In 1995, herring roe openings on different parts of the coast ranged from 5 minutes to 24 hours, 50 minutes. In the latter case, in Barkley Sound (on the west coast of Vancouver Island), a pool system was set up to take the low quota of 1394 tonnes. Because of limited stocks, only 3 of the 23 seine boats were permitted to set their nets for the herring, but

the value of the catch was to be split evenly among the 23 vessels. This decision to fish on a pool basis was made in an effort to come as close as possible to meeting the Department of Fisheries and Oceans' quota. This style of management has been used with relatively high success for roe herring since that time.

Pacific herring play an important and complicated role in ocean food webs. The herring, their prey, and predators also interact with natural ocean change and human-induced changes. The corresponding high variability in spawning biomass between years makes this species a management priority for the Department of Fisheries and Oceans. In addition, according to the Supreme Court of Canada, DFO must maintain conservation values, followed by the interests of First Nations in the fishery.

Coastal Salmon Salmon stocks pose different kinds of problems, including the need for conservation of stocks through enhancement programs; improvement of environmental conditions through installation of sewage treatment systems; and political negotiations to apportion catch levels between Canadian and American fishers. Due

Photo 8–8
Herring punts used by First Nations fishers in the herring roe fishery.

to a combination of factors—*ocean regimes* (alternating periods of cool and warm water that are relatively stable for long periods of time, then shift abruptly, with dramatic effects on fish stocks), harvesting practices, and spawning habitat destruction due to forestry practices— wild Chinook salmon stocks and more recently coho salmon stocks on the west coast of Vancouver Island and in the Gulf of Georgia have been severely depressed, some risking extinction. In 1992 and 1993, for example, young salmon migrating out to sea suffered elevated predation by increases in mackerel and Pacific hake (related to El Niño–induced changes in ocean water temperature). These events meant that, in 1995 and 1996, numbers of Chinook returning to spawn in the 68 wild Chinook streams where they were hatched on Vancouver Island were forecast to be as low as one-tenth of recent averages. That is, a wild stock that would usually have 100 spawners could have as few as 10. If this were to occur, many Chinook stocks could suffer a major loss of genetic diversity, which would hinder stock rebuilding ("Saving the Wild Chinook," 1995).

To enable as many spawning adults as possible to get back to their streams, the Department of Fisheries and Oceans reduced harvest rates by up to 50 percent and also took eggs from wild Chinook stocks to be raised in Vancouver Island hatcheries for later release into their home streams. In 1996, in order to reduce the harvest rate by over 90 percent, commercial catch restrictions planned for west coast Vancouver Island Chinook included closure of some area fisheries, time limits on others, and strict enforcement measures. Recreational and Aboriginal fisheries also were constrained in efforts to contribute to conservation of threatened stocks (Department of Fisheries and Oceans, 1996). Total closures on coho fishing by all users were implemented in 1999 and 2000. In 1999 and again in 2000 Fraser River sockeye stocks collapsed and the commercial fishery was closed.

Predictably, reaction was not all positive, particularly among recreational fishing industry operators who were affected by the closure of fisheries on the west coast of Vancouver Island, the Gulf of Georgia, and off the Queen Charlotte Islands. When the Department of Fisheries and Oceans closed the fisheries, many of these operators' clients demanded refunds for fishing trips they had booked months in advance (Masterman, 1996). Major advertising campaigns were required to prevent the collapse of the recreational charter industry in 1999 and 2000.

The reasons for the demise of the west coast salmon fishery are similar to those encountered in the collapse of the east coast fishery: overfishing, transboundary and international conflicts, stakeholder conflicts, mismanagement of stocks, and lack of information. Although the current status of the west coast fishery is in flux, it is clear that the commercial salmon fishery is severely restricted, sometimes closed (depending on gear type, location, species, and fishing method). Recreational fishing is permitted, subject to specific regional regulations that change frequently.

Wild salmon species continue to decline. In the Yukon, for example, all but one species reached all-time lows in 1998. Six salmon species demonstrated largely downward trends that continued into 2000, causing the Department of Fisheries and Oceans to continue with precautionary-based management. Of particular concern are the coho stocks that remain dangerously low, despite DFO's zero mortality policy, likely due to poor marine survival and successive El Niño events (Pacific Fisheries Resource Conservation Council, 2000). Salmon scientists indicate an ocean regime shift may have occurred in 1998 and suggest additional research is needed to track salmon survival in the ocean phase of their life cycle (Yukon Department of Renewable Resources, 2000).

Destruction of stream spawning beds has been cited as one of the primary causes of declining Pacific salmon stocks. In an effort to involve local communities in restoration and protection of this important salmon habitat, DFO, with funds from the Fraser River Action Plan, initiated the Streamkeepers Program. The objectives of the program are to provide training and support to volunteers interested in restoring and protecting local aquatic environments, to provide public education on the value of watershed resources, and to facilitate communication and cooperative efforts in watershed management. Since its inception in 1993, the Streamkeepers Program has supported local programs throughout British Columbia and the Yukon. Volunteers have removed dams, stabilized streambanks, improved salmon habitat, and created partnerships with local businesses with the goal of improving associated stream habitat (Pacific Streamkeepers Federation, 2001; Qualicum Beach Streamkeepers, 2001).

With the recognition that "Pacific fisheries of the future must be based on real conservation and a sustainable fishery," conservation has become the Department of Fisheries and Oceans' top priority ("Pacific Fisheries for Tomorrow," 1995). Recent research shows that salmon are a keystone species (see Chapter 3) as a food resource for such wilderness icons as bears and bald eagles. Healthy salmon runs, especially those genetically distinct salmon in smaller streams, are linked intimately to the well-being and survival of everything from small Aboriginal coastal communities, whose entire existence may depend on salmon, to whole forest systems and the provincial economy. Conservation of salmon is vital to preserving ecologically significant links between oceanic, freshwater, and terrestrial ecosystems (Robinson, 1995; Willson, Gende, & Marston, 1998).

The Pacific salmon fisheries ran into additional difficulties when the United States began a regime of serious overfishing of Fraser River sockeye, doubling their allocation under the 1985 Pacific Salmon Treaty between Canada and the United States. Canada was restrained from intercepting American-origin fish, mainly from endangered runs in Washington and Oregon, but American fishers freely intercepted increasing numbers of Canadian-bound Chinook in Alaska, and pink and sockeye in the Fraser River. Quota negotiations and the treaty itself collapsed by 1994 (McDorman, 1995).

Then Fisheries and Oceans minister Brian Tobin took up the political and legal challenges of Chinook conserva-

Photo 8–9
Fish hatcheries are one of several artificial means of increasing spawning production.

tion in 1995 when Alaskan officials unilaterally announced increased harvest rates on Chinook salmon, far exceeding historical levels and reasonable stock strength estimates. Among other things, the Pacific Salmon Treaty promoted shared Canada–U.S. responsibility for conservation of salmon stocks and provided for equitable distribution of catch between Americans and Canadians. When Alaskan officials refused to stop the commercial catch of Chinook salmon returning through Alaskan waters to spawn in B.C. and U.S. streams, U.S. Pacific Northwest tribal groups along with Washington and Oregon state officials initiated legal proceedings to halt the Alaskan fishery. Canada joined the action as an "amicus curiae" (friend of the court), pointing out that Canada had imposed severe restrictions on domestic harvest in the name of conservation and rebuilding of Chinook stocks. Also, Canadian fishers felt there was an imbalance in the number of Canadian-spawned salmon caught by U.S. fishers and the number of U.S.-spawned salmon caught by Canadians (which translated into a $60 to $70 million loss for Canadian fishers).

In early September 1995, a U.S. district court judge decided to halt the commercial Chinook harvest. Tobin was pleased, noting that Canadian efforts to conserve and rebuild Chinook stocks would not be frustrated by excessive harvesting in Alaska. Canada and the United States also agreed on the appointment of a mediator to help reach an agreement on the long-standing differences in interpretation of the equity principles of the Pacific Salmon Treaty (Department of Fisheries and Oceans, 1995e; Kenny, 1995).

As is frequently the case, however, it is difficult to obtain agreement: as of June and July 1996, Canadians again were criticizing the United States' Chinook management plans, which proposed catch levels two and one-half times what Canadian scientists were recommending as necessary for conservation. Canada had proposed a 1996 catch limit of 60 000 Chinook for all fisheries in southeast Alaska, where 60 percent of the Chinook caught originate in Canadian waters. Canadian fishers have greatly reduced their catch to achieve conservation, including, in 1996, complete closure of all directed commercial fisheries on Chinook and all sport and Aboriginal fisheries of west coast Vancouver Island Chinook stocks.

The inherent ecological difficulty in maintaining the health and sustainability of Chinook stocks is compounded by the economic pull of potential catches. This is one fisheries instance where it might well be said that "the force of economics overpowers conscience" (Parfit, 1995, p. 11). In June 1997, despite year-long consultations, negotiations between the United States and Canada collapsed when the two sides could not agree on how small a quota of endangered United States–bound coho salmon should be intercepted by Canadian fishers, and how much the United States should compensate Canada by taking fewer Canadian-bound fish entering the Fraser River. For the fourth consecutive year the two nations failed to agree on quotas; each ended up setting its own maximum quotas, as if there were no treaty.

Canada's "vigorous but not aggressive" quotas were designed with a specific political component in mind: to demonstrate that U.S. fishers would be better off with a joint management plan and shared quotas (Howard, 1997). Under this plan to take up to 24 million fish, Canadians could intercept up to 78 percent of southbound sockeye salmon before they reached U.S. waters. Americans, in turn, announced quotas similarly favourable to their fishers.

Scientists and others expressed concern that Canada's key conservation objectives could be ignored in retaliatory actions, and that the level of fishing possible under unilateral quotas could signal the demise of some stocks. High-level officials, including the more recent federal Fisheries minister David Anderson, suggested Canada should seek international arbitration (a proposal the United States previously has rejected). The El Niño advisory issued for 1998, and the knowledge that there would be "lots of mackerel munching on the salmon" (Murphy, 1997), highlighted the continuing need for long-term solutions to sustainability of living marine resources.

The collapse of coho and sockeye stocks and the dangerously low level of many Chinook stocks evident in 2003 are the end result of a series of resource management mistakes. The ceaseless rounds of finger pointing to single causes, such as poor land use practices, overfishing, ocean change, seals and sea lions, fishers from different nations and from different sectors within the fisheries, aquaculture practices, and Aboriginal fisheries, clearly indicate the harmful effects of crisis-response management. The use of artificial means of increasing spawning production, such as hatcheries, artificial spawning streams, and fish ladders, coupled with advanced fishing techniques to increase the selective capture of target stocks, have only served as stopgap measures against falling stocks.

Fortunately, some of the confrontation among stakeholders that has existed in the past is giving way to new ways of dealing with complex and sensitive fisheries management issues. We examine some of these initiatives and actions later in the "Responses to Environmental Impacts and Change" section of this chapter.

AQUACULTURE

Aquaculture is the farming of aquatic organisms, such as fish, shellfish, and plants, in controlled environments. Commercial aquaculture began in Canada in the 1950s, focused on trout in Ontario and British Columbia, and oysters in British Columbia, New Brunswick, and Prince Edward Island. It was not until the 1980s that the industry expanded to every province and the Yukon and the species base expanded to include Atlantic and Pacific salmon, rainbow and steelhead trout, mussels, clams, scallops, char, and marine plants. In 1999, 8000 people were employed directly in the aquaculture industry (Canadian Aquaculture Industry Alliance, 2000).

British Columbia and New Brunswick dominate the aquaculture industry; in both provinces, Atlantic salmon is the leading agri-export. In 2001, salmon (105.3 tonnes), trout (6.5 tonnes), and steelhead (4.7 tonnes) were the major finfish species produced in Canada. Total value of this production was over $538.2 million, with salmon accounting for over $468.9 million. The value of shellfish production in 2001 was $58.1 million, and mussels (21 666 tonnes) accounted for $30.5 million of that total (Fisheries and Oceans, 2002a; Fisheries and Oceans, 2002b).

Aquaculture has grown significantly in Canada (and globally), partly in response to declines in wild fish stocks. With the collapse of local commercial fisheries, many people consider aquaculture to be a legitimate alternative that helps meet the demand for fish (for food) and reduces the economic hardship (including the need to relocate) on those who relied on wild stocks. By 2030, the UN Food and Agriculture Organization anticipates that more than 50 percent of fish for human consumption will come from aquaculture. Anxious to take part in this growing industry, the federal government has made sustainable aquaculture a priority and, in support, established the Office of the Commissioner for Aquaculture Development (OCAD).

Although socioeconomic benefits from expansion of the aquaculture industry have been important, particularly in coastal regions, concerns exist about possible effects of aquaculture operations on coastal ecosystems. Such concerns include potential introduction and transmission of fish disease and its impact on wild stocks, escapes of fish (including genetically engineered fish) from net pens and genetic interactions among wild and cultured stocks, fouling of the benthic environment under the pens, and aesthetic effects. These concerns are not unfounded, as the following examples illustrate.

On the Atlantic coast, where aquaculture cages often are located in shallow water (about 12 metres deep), research undertaken in the early 1990s identified impacts from deposition of waste feed and fecal matter in 77 percent of the sites studied (Thonney & Garnier, 1992). Since then, the industry has responded to many of these concerns. Rainbow trout, likely escapees from aquaculture operations in Cape Breton and New Brunswick, are breeding in streams in Newfoundland's Gros Morne National Park. DFO is concerned these non-native rainbows will out-compete wild Atlantic salmon for food resources (Canadian Broadcasting Corporation, 2003). Sea lice infestations in wild pink salmon have been linked with salmon farms in British Columbia (see Box 8–1).

Since 1987, the Atlantic Salmon Watch Program has recorded 1085 escaped adult Atlantic salmon in 80 British Columbia rivers. Two points of view exist concerning this issue. Some people suggest that the presence of small numbers of escaped salmon over 15 years does not constitute colonization (the establishment of self-sustaining or "feral" runs). Others note that the widespread presence of even small numbers of escaped fish suggests the possibility of colonization and subsequent threats to viability of native stocks (Gardner & Peterson, 2003). It is unclear which perspective is accurate, principally because significant gaps in knowledge remain.

In an effort to assess the concerns noted previously, two major independent inquiries have been held in British Columbia. Both the Salmon Aquaculture Review in 1997, and the joint British Columbia Environmental Assessment Office–Pacific Fisheries Resource Conservation Council investigation in 2002, reached a similar conclusion: that overall, given current knowledge, practices, and production levels, British Columbia's salmon aquaculture industry represented low risk to the environment. However, both inquiries indicated the need for increased protection against effects of aquaculture operations on coastal species and environments, and both inquiries emphasized the need to invoke the precautionary principle.

Public concern about the effects of a rapidly growing, unregulated industry on coastal environments, and associated health and safety concerns, led the B.C. government to place a seven-year moratorium on salmon aquaculture in the province in 1995. DFO, however, aggressively promoted fish farming and "presid[ed] over a 300-per-cent [sic] increase in open-net-pen fish farms" during the moratorium—hardly a precautionary approach (Anderson, 2002). Although the moratorium was lifted in 2002, many members of the public remain concerned, particularly because evidence from government and independent agencies frequently has been contradictory (or unavailable).

Many agencies are working proactively to increase knowledge and decrease public concerns about aquaculture in Canada. For instance, the B.C. Aboriginal Fisheries Commission (BCAFC) works in partnership with DFO, British Columbia's Ministry of Agriculture, Food and Fisheries, the RCMP, other ENGOs and First Nations groups, and industry to create a safe, productive, knowledgeable, and sustainable fishery, and to monitor fish stocks. In 2002, the BCAFC signed a protocol with Pacific National Aquaculture, a Norwegian salmon tenure holder, to work together to develop a sustainable finfish aquaculture business (British Columbia Aboriginal Fisheries Commission, 2002).

DFO has been active also; in 1999, the Ministry of Fisheries and Oceans developed an Aquaculture Partnership Program to promote improved collaboration among key corporate and government players. This project was to fulfill OCAD's mandate of bringing together all appropriate federal government resources, leading necessary regulatory reforms, and working with the provinces to develop an environmentally sustainable aquaculture industry. Unfortunately, by 2001, lack of funding had terminated the project. In 2000, DFO embarked on its $75 million Program for Sustainable Aquaculture. This five-year program is intended to improve the federal government's capacity to assess and mitigate the impacts of aquaculture on aquatic ecosystems; maintain consumer and market confidence in the safety and quality of aquaculture products; and enhance the DFO legislation, regulations, and policies that govern aquaculture (Fisheries and Oceans Canada, 2003).

Not everyone is satisfied with DFO's handling of oceans and fisheries resources, however. Criticism has been levelled at DFO for its general failure to protect Canada's fish stocks, and for its failure to implement the integrated management plans and marine protected areas possible under the 1996 Oceans Act. For instance, the Sierra Legal Defence Fund (SLDF) is a Canadian organization dedicated to enforcing and strengthening the laws that safeguard our environment, wildlife, and public health. This organization has criticized the federal government's routine licensing of boats that use destructive fishing gear in areas of sensitive and important habitat— even though destruction of fish habitat is against Canadian law, and even though DFO has acknowledged that dragging is destructive to fisheries and fish habitat. According to the SLDF, by not basing management of ocean resources on the local ecology and economy, and by not employing the precautionary principle, DFO has failed to support the type of fishing that is ecologically sustainable and that generates local employment.

POLLUTION

Coastlines are the primary habitat of the human species, probably because coastal ecosystems possess an extraordinary endowment of natural resources from which everyone can benefit (Olsen, 1996). Unfortunately, the litany of human effects on the marine environment is lengthy. Bacterial/viral contamination, oxygen depletion, toxicity, bioaccumulation, habitat loss or degradation, depletion of biota, and degradation of aesthetic values are among the notable effects of human-induced change in the coastal zone.

Contamination of Canada's oceans and coastlines is principally the result of human activities, including discharges of municipal sewage and wastewater; industrial effluents such as discharges from pulp and paper mills and other industrial sources; urban and agricultural runoff; solid waste and litter such as plastics, logs, and nets; ocean dumping; coastal developments such as housing, harbours, causeways, marinas, aquaculture farms; and marine shipping.

Photo 8–10

Sea lice find ideal breeding conditions on high-density fish farms.

In November 2002, the Pacific Fisheries Resource Conservation Council[1] (PFRCC) issued a warning to Canadian and British Columbia Fisheries ministers that there were dramatic decreases in the number of spawning pink salmon in the Broughton Archipelago. Located near Campbell River, a well-known salmon sport fishing centre on Vancouver Island, the Broughton Archipelago is a group of islands that contains a high concentration of fish farms. The decline in numbers of spawning fish in 2002 (from over 3.6 million to 147 000) raised PFRCC's concerns about the potential impact of salmon aquaculture and sea lice on wild stocks. This concern led the PFRCC to recommend that Canada and B.C. act to maximize the passage of fish through the archipelago during April 2003.

The PFRCC indicated that while there was scientific uncertainty regarding the cause of the decline, European research had indicated that sea lice infestations can be associated with salmon farming. Since the decline in spawning salmon was confined to the Broughton Archipelago, and since the pink juveniles in this area were infested with sea lice (a condition virtually unknown among juvenile pink salmon in the natural environment elsewhere), the conclusion was that the sea lice (and fish farming) were associated with the observed decline.

In advocating the precautionary approach in this situation, the Chair of the PFRCC noted that in the absence of any evidence of some cause other than the sea lice, action was warranted. Among the options suggested were fallowing of all salmon farms in the Broughton Archipelago and implementation of rigorous sea lice control measures on the salmon farms to protect wild fish. The fallowing option would involve temporarily removing all salmon from the sea pens six weeks prior to when pink salmon entered the marine environment (as early as mid-April). Cooperative development of a Broughton Archipelago sea lice management plan could involve a variety of potential actions, including application of chemotherapeutants by all Broughton fish farms to kill (not just shed) the sea lice.

The PRFCC believed that fallowing was the lower risk option and had the greatest likelihood of improving conditions for passage of juvenile pink salmon to the sea. In addition to noting that more research into the ecology and life history of sea lice was required, the PFRCC advised that monitoring of the environment, sea lice levels, and juvenile pink salmon was required immediately.

Strongly opposed to fish farming, the Union of British Columbia Indian Chiefs demanded (in a January 2003 letter to the Regional Director General of DFO) that the federal government curb harmful activities of fish farms in the Broughton Archipelago. They asked DFO to close (fallow) the fish farms from the end of February through mid-June to allow the wild juvenile fish to pass safely out to sea without the risk of infection from sea lice. Fisheries and Oceans Canada's response was to indicate that they were working on an Action Plan to address potential risks to wild pink salmon stocks. This Action Plan was to include freshwater and marine monitoring programs, an approach to salmon farm management, a long-term research plan, and a public consultation/dialogue process.

DFO is both enabler and regulator of the aquaculture industry; as such, DFO indicated it is "committed to the protection of wild salmon resources in British Columbia" (Fisheries and Oceans Canada, 2003). But, as the B.C. Chiefs asked, did DFO fail to enforce the Fisheries Act and thus allow salmon farming to infect wild salmon stocks? Did DFO fail to take appropriate steps to minimize the possibility of a sea lice outbreak, given similar experiences in Norway, Ireland, and Scotland? Did DFO's approval of fish farm tenures on salmon migration routes reflect the lack of a scientific basis for siting criteria? Did DFO have a formal plan to manage the risks of salmon farming? These and other questions have been raised with respect to the sea lice outbreak in the Broughton Archipelago, and challenge the ability of a single agency (DFO) to promote both industry development and resource protection.

[1]Established in 1988, the role of the Pacific Fisheries Resource Conservation Council is to provide independent, strategic advice and relevant information to the Fisheries Ministers of both Canada and British Columbia, and to inform the Canadian public about the status and long-term sustainable use of wild salmon stocks and their freshwater and ocean habitats.

SOURCES: *Status Report on Sea Lice Monitoring in the Broughton Archipelago,* British Columbia Ministry of Agriculture, Food and Fisheries, 2003, http://www.agf.gov.bc.ca/fisheries/health/sealice_monitoring.htm; "Fish Farm Flap," *Disclosure,* Canadian Broadcasting Corporation, February 4, 2003, http://www.cbc.ca/disclosure/archives/030204_salmon/report.html; *News Release: A Collaborative Action Plan for Sea Lice to Be Implemented by Fisheries and Oceans in British Columbia,* Fisheries and Oceans Canada, 2003, http://www.dfo-mpo.gc.ca/media/newsrel/2003/hq-ac02_e.htm; *Summary of the Sea Lice Outbreak in the Broughton Archipelago,* Georgia Strait Alliance, 2001, http://www.georgiastrait.org/Articles2001/sealice2.php; *Communiqué: Pink Salmon in Broughton Archipelago in Crisis: Report,* Pacific Fisheries Resource Conservation Council, 2002, http://www.fish.bc.ca/html/fish3011.htm; *Impact of Fish Farms on the Salmon Runs of the Broughton Archipelago,* Union of British Columbia Indian Chiefs (letter), 2003.

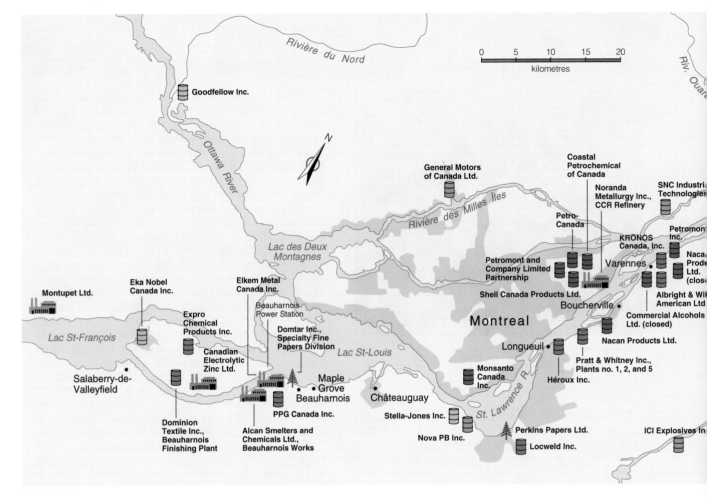

Figure 8–8

Montreal area factories involved in the St. Lawrence River cleanup plans

SOURCE: "It's Not Easy Being Green," A. Brocklehurst, 1996, *Canadian Geographic, 116*(3), 40–41. Reprinted by permission of The Royal Canadian Geographic Society.

Industrial and Chemical Effluents

Industrial effluent is an example of a point source pollutant (see Chapter 7), as are those from municipal sewage, mining and ore-processing operations, discrete oil and chemical spills, and food processing plants. Nonpoint sources of pollution (see Chapter 7) include urban and agricultural runoff as well as atmospheric deposition of pollutants. Both point-source and nonpoint-source releases of pollution to inland freshwater receiving bodies can be carried along rivers to the oceans where they can have effects that are just as devastating to the marine environment as they are to the shorelines where they are discharged.

The St. Lawrence River channels large quantities of runoff and industrial wastewater past many communities on its way to the Atlantic Ocean. In 1988, when the fed-eral and provincial governments announced cleanup plans for the St. Lawrence River (St. Lawrence Action Plan), 50 of the largest factories in the Montreal area were selected to cut their discharges into the river (see Figure 8–8). Measurements at that time showed that every day, the 50 companies involved were discharging 572 000 kilograms of suspended solids, 455 000 kilograms of organic matter, 1830 kilograms of oils and greases, 995 kilograms of heavy metals, and 74 100 kilograms of other metals into the St. Lawrence.

Since then, initiatives of the St. Lawrence Action Plan (1988–93) and the St. Lawrence Vision 2000 program (1995–2000) have resulted in declining levels of mercury, PCBs, and other heavy metals, and improved health of fish. Quebec government tests show large perch caught in 1993 near Maple Grove contained average mercury levels of 0.42 milligrams per kilogram (compared with 1.02 mil-

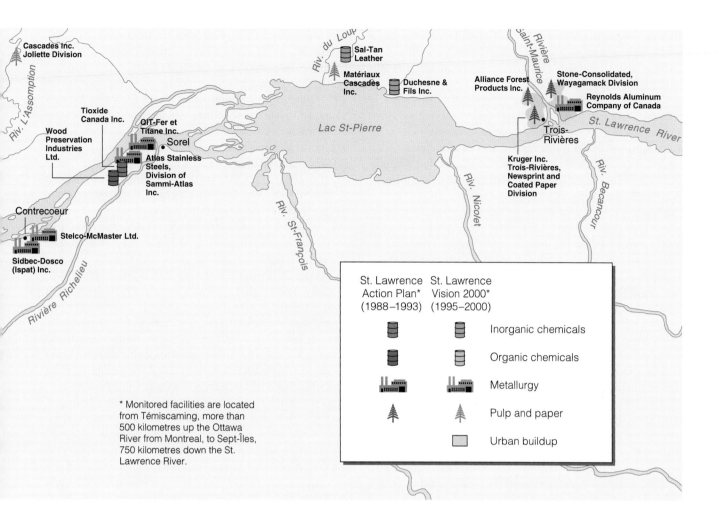

* Monitored facilities are located
from Témiscaming, more than
500 kilometres up the Ottawa
River from Montreal, to Sept-Îles,
750 kilometres down the St.
Lawrence River.

ligrams per kilogram in 1985) and undetectable levels of
PCBs (below 0.02 milligrams per kilogram compared with
0.08 milligrams per kilogram in 1985). Emissions from the
original 50 factories continued to be measured regularly,
and another 56 companies were targeted to reduce their
levels of toxic discharges (Brocklehurst, 1996). By 1998
even better results from pollution control efforts became
evident, when the original 50 industries reduced effluents
by 96 percent. These positive indications of success must
be interpreted with caution because they do not measure
the impact that historical pollutant levels may yet have on
the river ecosystem (St. Lawrence Vision, 1999, 2000).

In 2003, the St. Lawrence Ecosystem Monitoring
Committee (established in 1999) produced a report on
the improved state of the river. Using 21 environmental
indicators as their basis, the committee noted that toxic
contaminants had been reduced, animal populations
restored (northern gannet and great blue heron in particular), consumption of seafood and freshwater fish no
longer posed a health risk, and water quality was adequate for most recreational activities (but not swimming).
Some problems remained, including increased bacteriological contamination due to wastewater and storm sewer

overflows, and the potential for resuspension (by storms
or dredging) of contaminants remaining in deeper sediments (St. Lawrence Vision, 2003).

Although lower levels of industrial pollution now
reach the St. Lawrence River and Atlantic Ocean, not all
companies in Atlantic Canada (or the rest of the country)
comply with current pollution prevention regulations. For
instance, in May 1996, a Newfoundland provincial judge
ordered Corner Brook Pulp and Paper Limited to pay a
$500 000 fine (for violations of the regulations) and
$250 000 penalty (to ensure compliance with those same
regulations) for pollution violations under the 1992 federal Pulp and Paper Effluent Regulations issued under the
federal Fisheries Act. These regulations are designed to
control industrial discharges from pulp and paper mills in
order to protect fish and their habitat from suspended
solids, biochemical oxygen demanding matter, and lethal
effluent. This fine, the largest paid to that date in Canada,
was levied because the company had been out of compliance with the Pulp and Paper Effluent Regulations for
several months by failing to complete construction of a
new effluent treatment system. Corner Brook Pulp and
Paper Limited, which was discharging an estimated

31 463 289 cubic metres of untreated effluent per year, was charged with depositing mill effluent into Humber Arm coastal waters (Environment Canada, 1996; Environment Canada, n.d.a).

Mill effluent has three major types of constituents: dissolved organic compounds, suspended solids, and an inorganic component. Each of these effluents, often lethal to fish, can degrade water quality and aquatic habitat in rivers, estuaries, and coastal waters. Suspended solids such as cellulose fibres, wood particles, small pieces of bark, and lime mud can settle out of suspension near mill outflows, smothering the benthic habitats of species such as shellfish, lobsters, and polychaete worms, and interfere with species such as salmon that use gravel on river bottoms to spawn or feed. Large areas can be affected by suspended solids—wood fibre has been deposited on the bottom of Humber Arm for up to two kilometres north and northeast of the outfall of the Corner Brook Pulp and Paper Limited mill (Environment Canada, n.d.a).

The proceeds from the $250 000 penalty were distributed among three local organizations with environmental interests. Scholarship funds were set up for environmental programs at the local West Viking College and Sir Wilfred Grenfell College, and funding was given to the Corner Brook Stream Development Corporation for its efforts to conserve the Corner Brook stream area (Environment Canada, 1996).

Other industrial activities in Canada such as mining and manufacturing are monitored for their environmental effects. In New Brunswick, for example, the No. 12 lead–zinc mine of the Brunswick Mining and Smelting Company began production in the mid-1960s and, until about 1982, discharged acidic, heavy-metal–contaminated wastewater into the Little River, which empties into Bathurst Basin. Great improvement in the effluent quality was achieved in the early 1980s when the company built a separate wastewater treatment system for mine water. However, this created a new and more difficult problem: partially oxidized sulphur compounds (thiosalts) generated in the milling process at the mine were chemically and biologically oxidized when released to the environment and generated large quantities of sulphuric acid. This acid depressed the pH of Little River, making it uninhabitable for fish and other aquatic organisms; only the bacteria that oxidize the thiosalts remain alive in the river. Efforts to treat and remove the thiosalts from the mine water have proved expensive and not very effective (Environment Canada, n.d.a) (see Chapter 10 for more information on mine closure and reclamation).

Abandoned mines also cause environmental problems; hundreds of derelict mines in Nova Scotia, New Brunswick, and Newfoundland continue to release contaminants such as arsenic, mercury, lead, and zinc for many years after their closure. In Elsa, Yukon, the United Keno Hill Mine (which operated from 1946 until 1989 when it closed for economic reasons) has been the subject of considerable attention from Environment Canada. Environmental impact assessments have been conducted on the hardrock mining operation every five years since 1974–75 to ensure compliance with the Fisheries Act and to ensure adequate measures are in place to protect the receiving waters. The 1995 impact assessment report noted a significant increase since 1985 in heavy metals in stream sediments, as well as a reduction in the diversity of the benthic invertebrate community downstream (Environment Canada, 1995). Should the owners of the property decide to reopen the mine, the series of environmental impact assessments will provide valuable information on which to develop sound advice for its future development.

As noted previously, Arctic marine systems encourage biomagnification of pollutants such as PCBs and DDTs (see Figure 8–2). From 1985 to 1987, research on Broughton Island, Northwest Territories, that assessed the possible risk to the health of Arctic residents consuming country foods (seal, caribou, narwhal, beluga, fish, walrus) found that 15.4 percent of male and 8.8 percent of female residents ingested more than the Canadian tolerable daily intake of PCBs. Also, PCB concentrations in blood samples exceeded tolerable guidelines in 63 percent of children under 15 years of age, 39 percent of females aged 15 to 44 years, 6 percent of males 15 years and older, and 29 percent of women 45 years and older. One-quarter of the breast milk samples analyzed also exceeded the tolerable PCB level (Kinloch, Kuhnlein, & Muir, 1988).

The level of concern over contamination in the food chain, and the importance of complete information to those who are affected by contamination and have to make decisions about their food intake, is illustrated by the case of one Inuk mother. When she brought her sickly baby to the local nursing station, she indicated concern that her breast milk might be contaminated because she ate country foods. She had attempted to protect her baby by feeding the child Coffee Mate mixed with water (National Advisory Board on Science and Technology, 1994).

Although there may be risks to health associated with the presence of PCBs in the traditional Inuit diet, country foods are nutritionally superior to "southern" foods. Switching from a country foods diet to a southern Canadian diet may lead to nutritional deficiencies and associated known risks to health such as obesity, diabetes, cardiovascular disease, and cancer. The research on Broughton Island revealed that the benefits of country foods and of breast-feeding were greater than the risks from the PCBs present. Clearly, however, contamination must be controlled, and levels monitored so local communities can be assured that country foods are safe to eat (Government of Canada, 1991; National Advisory Board on Science and Technology, 1994).

Municipal Sewage

Municipal sewage, a point source of pollution, continues to be an issue in Canada. Although there has been progress in increasing the level of treatment and number of people serviced by wastewater treatment systems (see Chapter 7), concerns remain about environmental and human health effects of municipal sewage. Among the concerns are continuing eutrophication of lakes and rivers, contamination of domestic water supplies, shellfish harvesting closures due to fecal coliform contamination, heavy-metal contamination of sediments, risks associated with swimming in water contaminated with human fecal material, and aesthetic objections to the visible signs of sewage discharge.

In the Arctic, the extremely cold climate severely restricts the rate at which wastes break down in the environment. With local populations growing rapidly, it is not clear that traditional methods of sewage waste disposal alone will be sufficient to avoid contamination of coastal waters in the long term.

Population pressures also affect municipal sewage discharges in the lower Fraser River basin. Currently, municipalities release half of the wastewater entering the lower Fraser River, much of it untreated or primary-treated sewage. If current trends continue, the population of Vancouver and its suburbs will rise from about 1.7 million to about 3 million in the next 25 years ("Population: A Growing Problem," n.d.). Knowing this in advance, it should be possible to plan population and city growth while keeping in mind the sustainability of ocean waters receiving municipal discharges.

Sewage discharges into the marine environment have adverse impacts on shellfish fisheries and tourism but do not pollute the groundwater and surface waters used for domestic water supplies. In the past, since protection of public health (and not the environment) was the prime motivation for treating municipal wastewater, discharges to marine environments were considered as not too serious. In Atlantic Canada, where many communities are located on or near marine waters, the proportion of the population served by waste treatment facilities is lower than in the rest of Canada. Prince Edward Island is the exception: to protect its important coastal-oriented tourism industry and shellfish resources and to avoid groundwater contamination, almost all municipal waste is treated.

Halifax and Dartmouth, Nova Scotia, have been dumping raw sewage into Halifax Harbour for nearly 250 years. By 1999, 40 outfalls from the two cities were discharging 187 million litres (or roughly enough to fill 40 Olympic-size swimming pools) per day of untreated industrial and urban sewage into the harbour. Beaches on the harbour and the Northwest Arm have been closed to swimmers periodically during the summers (and year-

Photo 8–11
Hypodermic needles and other medical waste are among the debris found on beaches.

round for shellfishing) because of bacterial contamination. Public criticism also has grown regarding aesthetic concerns (odours and floating debris).

Sewage treatment had been studied for at least two decades, but it was not until 1987 that federal–provincial negotiations established some funding for the cleanup and for construction of a central primary treatment plant (Environment Canada, n.d.a). The plan for a central treatment mega-plant was scrapped in 1999 and replaced with a more environmentally friendly design for five smaller primary treatment facilities. By September 2002, Nova Scotia and Ottawa had pledged a total of $62 million of the $315 million required for the harbour cleanup. As in the case of the Fraser River basin, growing awareness of municipal sewage impacts on ocean waters suggests there is opportunity to take corrective action and move toward sustainability.

MARINE SHIPPING, OCEAN DUMPING, AND PLASTICS

Marine Shipping

As noted previously, shipping activity contributes to degradation of marine environments. From contaminated bilge water to major oil spills, most impacts and problems are caused by human error; many can be avoided. As low temperatures and limited species diversity make Canada's North exceptionally sensitive to pollution, northern Canadians are concerned about accidental pollution of Arctic waters, which could occur through shipping incidents or during hydrocarbon exploration. Furthermore, effective spill cleanup in ice-covered areas may be impossible, particularly since the effects of oil on the Arctic system are largely unknown. As early as 1972 the government recognized the need for special consideration of the region with the passage of the Arctic Waters Pollution Prevention Act, and again pushed for special consideration in the Convention on the Law of the Sea for ice-covered areas.

Aquaculture operations also may be particularly vulnerable to marine shipping, as the 1993 stranding of the oil tanker *Braer* on the Shetland Islands, Scotland, demonstrated. Heading to Canada when it grounded on Sumburgh Head, the *Braer* spilled 96 million gallons of light crude oil, threatening salmon farms 80 kilometres away from the accident site. The Shetland Salmon Farmers Association determined that approximately 2.5 million fish at 16 aquaculture sites were tainted by the oil spill, resulting in a loss of $67.5 million (Golden, 1993).

Oil spills may kill farmed fish and shellfish by direct toxicity or by smothering them, and can damage them by tainting their flesh. Even the hint of oil contamination may cause consumer uncertainty and affect world fish markets. Given that the salmon (aquaculture) industry in the Bay of Fundy is similar in scope to that of the Shetland Islands, fish farmers in both areas were concerned that the environmental quality around their operations could be destroyed by a tanker accident. With about 350 tankers passing through the Bay of Fundy annually, a serious oil spill might be difficult to contain, particularly if the very high tides in the area were combined with high seas. Such a potential threat emphasizes the need for protection of unpolluted environments in which to raise salmon (Golden, 1993) and other marine species.

Ocean Dumping

Ocean dumping is defined under the Canadian Environmental Protection Act as deliberate disposal at sea from ships, aircraft, platforms, and other human-made structures. Dredged material, fish waste, scrap metal, decommissioned vessels, and uncontaminated organic material of natural origin are among those inert, nonhazardous substances permitted to be dumped. Permits for ocean dumping are granted after an evaluation has been conducted regarding the type of material to be dumped, the intended locations for the loading and disposal, potential environmental impacts, and alternatives to ocean disposal. A permit is not issued if practical opportunities are available for recycling, reuse, or treatment of the waste. Furthermore, Canada banned the disposal of industrial and radioactive wastes at sea in September 1994.

From 1987 to 1997, Environment Canada issued 1781 permits for disposal at sea of excavated earth and material from dredging of waterways, fish processing waste, retired vessels, and "other" wastes including scrap metal. More than 98 percent of wastes disposed at sea were from dredging of harbours and land excavation activities, while in Atlantic Canada, most permits were issued for disposal of fish waste (Statistics Canada, 2000). In 1998–99, two permits to dispose of vessels were issued, one of which was an artificial reef-building project (see Box 8–2).

A permit is issued only if there are no practical alternatives to disposal at sea. In March 1997, a new fee of $470 per thousand m³ (of material to be disposed) came into effect as an incentive to companies to reduce the amount of wastes they disposed of at sea. While the number of permits declined from a high of 223 in 1991, to 91 in 1997, the World Wildlife Fund of Canada (WWF) has charged that fines levied against ships that discharge bilge water at sea are less than the cost of legally disposing of bilge water in port. This suggests that it "pays to pollute"—and foreign ships take advantage of that situation, dumping oil in Canadian waters before returning to their home ports. Instead of tolerating illegal oil discharges that kill 300 000 birds a year, the WWF noted that stricter laws and better air, ship, and satellite surveillance; ship traffic monitoring; and in-port inspections would help reduce the amount of oil dumped into the oceans. "These are … major crimes against nature, and there should be major fines to stop it" (Chandarana, 2002).

Plastics

Plastics in the marine environment are an increasing problem. In high demand because of their durability, light weight, and relatively low cost, plastics do not break down readily and tend to remain in the marine environment for three to five years or longer. In 1990, volunteers with the British Columbia Coastal Cleanup Campaign found more than 1000 pieces of debris per kilometre on some beaches; most of the debris they catalogued was plastic and foam. As the amount of plastic released into the environment grows each year, plastic in the marine environment is accumulating faster than it can break down; tides, winds, and storms deposit plastic all over the world's coastline and seabeds. One study on Sable Island,

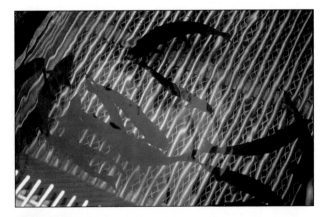

Photo 8–12

Aquaculture operations may be highly sensitive to any degradation in the marine environment.

BOX 8-2
DECOMMISSIONED NAVAL SHIP BECOMES ARTIFICIAL REEF

With a loud bang and clouds of brown and white smoke, the decommissioned Canadian destroyer escort *Mackenzie* sank 18 metres below the waters off Sidney, B.C., on September 17, 1995, ending a 34-year naval career in just over four minutes. A popular new tradition for old ships, the Mackenzie was the third ship sunk in B.C. waters by the Artificial Reef Society of British Columbia. Within two years of settling on the bottom, over 120 species of marine life were expected to be found in and on the new artificial reef and diving attraction.

The sinking of the *Mackenzie* had been preceded by the scuttling of HMCS *Chaudière* off Sechelt, B.C., in 1992 and the HMCS *Saguenay* in Nova Scotia in 1994. The *Matthew Atlantic* followed, sunk near Port Mouton, Nova Scotia, in 1998 to enhance an underwater diving park.

A special ocean disposal activity permitted by Environment Canada under the Canadian Environmental Protection Act, preparing the *Mackenzie* or any other ship to become an artificial reef requires removal of potentially polluting materials such as PCBs, oil, and gauges containing radioactive materials. In addition, holes are cut between compartments and decks to allow safe access for divers. Before the ship is sent to the bottom, Environment Canada officials inspect the ship for environmental readiness.

Turning the 2370-tonne destroyer HMCS *Chaudière* into an artificial diving reef north of Vancouver was a proving ground for this activity. Environmentalists and First Nations people objected to its sinking because of potential pollution fears. However,

Photo 8-13

Environment Canada instituted a program to observe any effects at the disposal site, and in 1993 data showed the old destroyer was home to a wide variety of marine life. In 1994, further videos showed a rich abundance of marine life completely covering the vessel, and no evidence of chemical contamination in the water and nearby sediments has been found.

SOURCES: "Former Destroyer to be Sunk off Sidney Today," J. Bell, September 16, 1995, *Victoria Times Colonist,* p. B3; "That Sinking Feeling," M. Eggen, 1997, *Alternatives, 23*(1), 7; "Ocean Disposal: Waste Management and Remediation, the Atlantic Region," Environment Canada, 2002, http://www.pyr.ec.gc.ca/EN/ocean-disposal/English/fact2_e.htm; "Mackenzie Goes Below as Artificial Reef, N. Gidney, September 17, 1995, *Victoria Times Colonist,* p. A7; *Canadian Environmental Protection Act: Report for the Period April 1994 to March 1995,* Minister of Supply and Services, 1996, Ottawa: Author.

160 kilometres east of Nova Scotia, estimated eight tonnes of debris washed up on the island each year, 94 percent of which was plastic.

Marine sensitivity to plastic is high: marine birds and other creatures are hurt or killed when they mistakenly eat or become entangled in it. Plastic can kill by blocking a digestive tract, by releasing toxins as a byproduct during digestion, or through starvation by giving a false sense of being full. Entanglements in plastic often lead to starvation, exhaustion, infection from wounds, and drowning. People, too, are affected by plastic debris when it gets caught in boat propellers, clogs water intakes, or blocks pumping systems. Repairs, lost fishing opportunities, and rising insurance claims cost individuals and the fishing industry both time and money. Communities also may face increasing costs for litter collection (Environment Canada, n.d.b).

Marine plastic debris comes from many sources, including careless boaters, beach users, and tourists;

cargo vessels, passenger ships, and commercial fishing vessels discharging garbage or accidentally losing cargo; workers at construction sites or other industrial sites who thoughtlessly dispose of waste into marine waters; and poor management practices at landfills (if uncovered, materials may blow into the ocean) and municipal sewage outlets (people dispose of plastic wastes in sewer systems). Rope, containers, grocery and garbage bags, cups and cup lids, and foam pieces are some of the most prevalent forms of plastic debris encountered in the marine environment.

A National Marine Plastic Debris research, information, and education program was established by Environment Canada under the Green Plan to improve efforts to deal with the problem. Debris pollution is one of those problems that requires international cooperation because the oceans carry lightweight plastics over long distances. While Canada harmonizes its own legislation with international standards, nations that do not regulate effectively and ships that fly flags of convenience and exist outside of most national and international regulation can dump plastic debris that lands on Canada's coasts.

COASTAL DEVELOPMENT

Growing human populations in coastal areas are posing increasing problems to marine ecosystems as a result of sewage, municipal and industrial wastes, litter, and urban runoff. As well, habitat loss through construction activities and physical alteration of coastal environments is an issue of concern in several parts of the country. One of the main attractions of living on the coasts is their aesthetic appeal, which can be lost through poorly planned development.

Urban Runoff

Urban runoff is one of the most significant examples of nonpoint-source (NPS) pollution. Not only is NPS pollution hard to track, it is also hard to control using conventional regulations. Nothing short of a change in our overall behaviour as "urban humans" will combat this problem. The Lower Fraser Valley, with its urban, industrial, and agricultural growth, is perhaps Canada's best example of this problem's complexity.

When rainwater runs off farm land in the lower Fraser River valley into ditches or washes city streets and industrial sites before draining away, it picks up all kinds of contaminants that eventually end up in the river. Manure is one major potential pollutant in agricultural runoff, especially in the Fraser Valley, which supports high livestock densities. If improperly applied on fields, pesticides can be problematic if they enter the drainage systems and reach the river untreated or reach groundwater aquifers. Wastewater from poorly maintained septic systems can have the same potential effects. All of these nonpoint sources of polluted runoff are diffuse and hard to quantify, but can have serious effects (as in Walkerton, Ontario; see Chapter 7).

It is known, however, that runoff from urban areas collects sediments and chemical pollutants such as trace metals, PCBs, and hydrocarbons (from cars and trucks). The volume of urban runoff for the Fraser River basin as a whole amounts to 500 million cubic metres per year—enough to fill B.C. Place Stadium 250 times (Fraser River Action Plan, n.d.). That volume is more than the annual discharge from municipal and pulp and paper sources combined (see Figure 8–9). Because of its large population and extensive urban areas, the lower Fraser River basin contributes the largest amounts of urban runoff in British Columbia, including almost 55 000 tonnes of suspended solids that enter the Fraser basin each year.

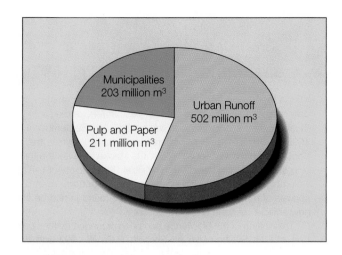

Figure 8–9
Urban runoff in the Fraser River basin

SOURCE: *Fact Sheet 2: Pollution in the Fraser,* Fraser River Action Plan, (n.d.).

The changes that growing urban pressures in the Fraser Valley and expansion of the Roberts Bank Superport area (such as the high-tech container port, Deltaport) bring to Fraser delta shorebirds are worrisome to researchers. According to current research, environmental changes in the Fraser delta could affect migrating shorebirds profoundly. If an oil spill or any other mishap reduced the birds' feeding, many would arrive too late at their Arctic breeding grounds to nest and produce a brood. In other words, habitat degradation on the Fraser River and delta could hinder the shorebirds' success on a breeding ground 3000 kilometres away (Obee, 1996). Such interconnections between human activities and environmental impacts on migrating birds give important meaning to the adage "think globally, act locally."

Urban runoff and other sources of pollution are being studied with a view to cleaning up the Fraser River and restoring its salmon population, among other efforts. Some of these initiatives are considered later in this chapter in the "Responses to Environmental Impacts and Change" section.

Physical Alterations

Physical restructuring of coastal environments can have impacts on the biotic as well as the abiotic components of an area. The construction of a fixed-link bridge between Prince Edward Island and New Brunswick by the consortium Strait Crossing raised numerous such concerns. As noted in Box 8–3, an ENGO, Friends of the Island, raised awareness of and concern about the potential harm that ice buildup around the bridge might have on lobster and scallop stocks.

When it opened to the public on May 31, 1997, Confederation Bridge was the longest continuous multispan marine bridge in the world. Crossing Northumberland Strait at its narrowest point, the 12.9-kilometre, $840 million bridge has replaced ferry service between Cape Tormentine, New Brunswick, and Borden, Prince Edward Island. The bridge is just over 11 metres wide. The typical bridge elevation off the water is 40 metres, while the navigation span (250 metres wide to accommodate the largest commercial vessel in the region) is 60 metres above the water. All vehicles permitted to travel on the Trans-Canada Highway are able to use the two-lane toll bridge, and a shuttle service accommodates pedestrians and cyclists.

Northumberland Strait is a channel of water about 300 kilometres long, between 13 and 55 kilometres wide, and covered with ice from January to April. The strait is one of the richest lobster-fishing areas in Atlantic Canada, and lobster provides about 75 percent of the value of shellfish landed in the area. Friends of the Island was a coalition of fishers, ferry workers, farmers, environmentalists, and Islanders who challenged the construction of the fixed-link bridge on environmental grounds. Supported by the Canadian Environmental Defence Fund, the group attempted to overturn a Federal Court ruling that the Government of Canada had conducted a proper assessment of the project's environmental impacts.

The coalition was concerned that the $100-million-a-year Northumberland Strait fishery might be jeopardized by the bridge. The group claimed ice buildup in the strait at the bridge could harm valuable lobster and scallop stocks. Of primary concern to the Friends of the Island was that the potential environmental impacts were determined by consultants retained by the consortium building the bridge.

In June 1995, the Federal Court of Canada ruled that the Government of Canada had taken all the necessary steps before concluding that the megaproject posed no harm to the environment.

Bridge construction began with the consortium using "proven technology" in its design and construction methods. The bridge was built to carry two lanes of traffic, 24 hours per day, year-round. To prevent any adverse effects of the bridge on the marine environment, the consortium's environmental management and environmental protection plans permit only environmentally safe de-icers to be used on the roadway surface during winter. Similar limitations prevail for other seasons and other issues, from wind monitoring to storm drainage systems.

Environmental effects monitoring programs required of Strait Crossing call for repeated measurements of environmental variables over time in order to detect changes caused by external

Photo 8–14

Opening celebrations for Confederation Bridge included an opportunity for runners and walkers to cross the bridge unimpeded by vehicular traffic.

influences, verify impact predictions, and evaluate the effectiveness of mitigation measures. Among the biophysical components studied are terrestrial wildlife areas, physical oceanography (currents, tides, sediment modelling), fisheries resources, and ice climate and effects of ice scouring in the Northumberland Strait. It will be interesting to see how effective Strait Crossing's lobster enhancement program (which dumped dredged material in nonproductive habitats to establish new lobster terrain) will be, and to determine how accurate environmental impact predictions were.

SOURCES: *Crossing That Bridge: A Critical Look at the P.E.I. Fixed Link,* L. Begley, (Ed.), 1993, Charlottetown: Ragweed Press; Northumberland Strait Crossing Project: Information Package, Public Works and Government Services Canada and Strait Crossing Development Inc. (n.d.); *Northumberland Strait Crossing Project: A Link to the Future,* Public Works and Government Services Canada, 1997, http://cycor.ca/nscp/index-e.html; *The Northumberland Strait Bridge,* Strait Crossing Team, (n.d.), http://wwwpeinet.pe.ca/SCI; "Steps across the Strait," H. Thurston, March–April 1997, *Canadian Geographic,* 52–60.

OFFSHORE HYDROCARBON DEVELOPMENT

Marine ecosystems may be impacted directly or indirectly and for the short or long term by hydrocarbon exploration and production activities. Our still-limited abilities to deal with iceberg collisions with production rigs or with spills, well blowouts, and containment in the Atlantic and Arctic ocean environments are key issues related to offshore development. In 1972, the British Columbia government imposed a moratorium on offshore exploration, and on the east coast, in response to fishing industry lobbying, exploration activity was banned by the federal and Nova Scotia governments in the Georges Bank area until the year 2000. Even under growing public and industry pressure, neither of these moratoria had been lifted as of 2001.

In early 2003, however, a dispute concerning a reversal of British Columbia's ban on offshore drilling arose between the then federal ministers of Environment, David Anderson, and Natural Resources, Herb Dhaliwal. British Columbia Premier Gordon Campbell's 2003 throne speech suggested an offshore oil and gas industry could be creating jobs and wealth by 2010. Dhaliwal indicated the moratorium could end by July 2003 and suggested a panel be appointed to advise government on that decision. Anderson, who had lobbied for the original ban, remained opposed, citing potential impacts on the already compromised fishing industry and the $200 million cruise ship industry. He asked: what was the likelihood of an Exxon *Valdez* spill? Could even the sight of oil rigs be a disincentive to tourist cruises?

These are not unimportant questions, and they expose the challenges faced by managers in making appropriate decisions for resource sustainability. British Columbia would value the coastal employment opportunities offshore oil development could bring to replace jobs lost in both fishery and forestry sectors. However, investment in aquaculture activity appears to have been made without adequate research, regulatory, or enforcement support. Anderson warned that the oil industry would require many years and $100–120 million to study a variety of environmental and related issues prior to any lifting of the ban. Aboriginal land claims and jurisdictional issues also would need to be considered prior to any decision to lift the ban.

Exploration for petroleum hydrocarbons began in the Arctic in 1973 and, by 2000, 134 wells had been drilled off the Arctic coast; as well, 346 wells were drilled off the Atlantic coast. The Hibernia project off Newfoundland, smaller projects on the Scotian Shelf, and the gas field near Sable Island are in production. These projects provide economic development and expansion opportunities for the region, but also pose potentially significant threats to local marine life on the Grand Banks and to grey and harbour seals on Sable Island breeding grounds. With the passage in 2003 of the federal Species at Risk Act (SARA), it will be interesting to see how government responds. For instance, in 2002 COSEWIC uplisted (to endangered) the status of the Scotian Shelf population of bottlenose whale. This species is threatened by the loud noises associated with oil and gas exploration around Sable Island. How might the application of SARA affect seismic testing or drilling operations in this and other regions?

Another issue of concern in the drilling of a well is that drilling muds and drill cuttings (typically in the order of 200 to 500 cubic metres) are discharged directly into the ocean. Mud additives used in this process are of particular concern because they may contain a variety of contaminants, including heavy metals, hydrocarbons, and biocides.

Legislation establishing the current drilling moratorium on Georges Bank expired on January 1, 2000; subsequently, a panel conducted a public review of the environmental and socioeconomic impacts of hydrocarbon drilling. An important source of new environmental information for the panel was a multidisciplinary research program on the effects of drilling wastes conducted by researchers at the Bedford Institute of Oceanography in Nova Scotia. Scientists studied the physical oceanography and sedimentology of Georges Bank, the flocculation behaviour of drilling wastes, drilling waste dispersion around an active rig on Sable Island Bank, and the sub-lethal effects of drilling wastes on sea scallops, the most important commercial species on Georges Bank. Particular attention was paid to water circulation so that the horizontal dispersion and transport of particulate drilling wastes could be understood, particularly in the benthic boundary layer. That layer of water, just above the sea floor, is where the scallops filter their food particles. Scientists used the models they developed to explore the potential impacts of specific hypothetical drilling scenarios on Georges Bank and conveyed that information to the review panel for their consideration ("Drilling on Georges Bank," n.d.). This and other socioeconomic consultation by the panel allowed them to submit a recommendation to extend the moratorium until the end of 2012, which was accepted jointly by the governments of Canada and Nova Scotia.

ATMOSPHERIC CHANGE

Scientists agree that the rate and patterns of climate change over the past century point toward a discernible human influence on global climate. As major components of the ecosphere, oceans and the atmosphere are inextricably linked, primarily through exchange of gases and heat. Oceans absorb CO_2 from the atmosphere and are the largest reservoir of carbon in the global carbon cycle. Important changes in the properties and composition of the atmosphere are occurring, including changes in greenhouse gas concentrations, general warming of the atmosphere near the Earth's surface, and a decline in stratospheric ozone concentrations. All of these changes have implications for the world's oceans, as do changes in sea level, which threaten loss of wetlands and other fish and wildlife habitat, flooding of property, shoreline erosion, contamination of coastal water supplies, reduced viability of ports, and the further disruption of established fisheries.

Although research is continuing, detailed knowledge of all the processes involved in ocean–atmosphere interactions is lacking. Increases in the amount of CO_2 in solution, changes in the temperature of ocean waters, rising sea levels, and ozone depletion potentially have negative impacts on marine environments and marine life. In anticipation of climate change, Canadian and international scientists have been working independently and collaboratively to improve the knowledge base regarding oceans.

RESPONSES TO ENVIRONMENTAL IMPACTS AND CHANGE

Are Canada's commercial fisheries sustainable? How can the health of major aquatic ecosystems be maintained? How do we balance social, environmental, and economic interests in areas that are important for fisheries, shipping, and urban and industrial development? These questions and many others like them reveal the worries that people have regarding the present effects of human activities and the need to plan and manage resources for a sustainable future. While we know that we need greater ecological knowledge to answer such questions and to move toward sustainable ocean resource use, there have been efforts at all levels, from international to individual, to try to make a difference. This section outlines some of these efforts.

INTERNATIONAL INITIATIVES

A variety of international agreements and programs are in place to safeguard the oceans and their resources for present and future generations. The Organization for Economic Cooperation and Development (OECD) and the United Nations Environment Programme (UNEP) run oceans and coastal area programs. UNEP's Regional Seas Programme, for example, emphasizes integrated coastal zone management, pollution control measures, and climate change issues. Earlier in the chapter we noted NAFO's difficulties with the lack of enforceable international law to preserve high-seas turbot for future generations (Schram & Polunin, 1995). The United Nations Convention on the Law of the Sea and Agenda 21,

described briefly below, are important international initiatives.

United Nations Convention on the Law of the Sea

On November 16, 1994, the United Nations Convention on the Law of the Sea (UNCLOS) came into force, concluding a process that had begun in the mid-1930s. Prior to World War I, it had been recognized that the world needed to develop a legal order for the oceans that would "promote the peaceful uses of the seas and oceans, the equitable and efficient utilization of their resources, the conservation of their living resources, and the study, protection and preservation of the marine environment" (UNCLOS, 1982, cited in Alexandrowicz, 1995).

While the history of the United Nations Convention on the Law of the Sea is both lengthy and complex, the legal regime was developed through international political negotiations during three Law of the Sea (LOS) Conferences. The Third LOS Conference took 10 years (until 1982) to negotiate a final agreement, which, in turn, took 12 years to come into force (in 1994). The resulting United Nations Convention on the Law of the Sea (UNCLOS) is the centrepiece of the international regime for managing the world's oceans. UNCLOS reconciled the widely divergent interests of 150 independent states, and also established the ideal for a new equity in use of oceans and their resources. UNCLOS is not the only body setting out rules, however, as the Food and Agriculture Organization also has established fishing areas for management purposes (see Figure 8–10).

Among other items in UNCLOS agreements, exclusive economic zones (EEZs) were established that gave coastal states legal power and international obligation to apply sound principles of resource management to oceans. In addition, UNCLOS declared that more than 45 percent of the seabed area and its resources were "the 'common heritage of mankind,' a concept that represents a milestone in the realm of international cooperation" (World Commission on Environment and Development, 1987, p. 273). This concept means that the world's oceans are inextricably linked and that Canada, as every other maritime nation, has an international responsibility to manage the oceans as a shared global resource.

Having ratified the convention in 2003, Canada is a strong supporter of the UNCLOS process and is a major beneficiary of its provisions (UN Convention on the Law of the Sea, 2004).

Agenda 21

Chapter 17 of Agenda 21 of the United Nations Conference on Environment and Development (UNCED) deals with the marine environment. Linked directly with

UNCLOS, Agenda 21 sets out the international basis on which protection and sustainable development of marine and coastal environments and their resources are pursued. Agenda 21 calls for new approaches to marine and coastal area management and development that integrate knowledge from all sources and that are both precautionary and anticipatory (Gerges, 1994).

Agenda 21 focuses on a number of areas such as integrated management and sustainable development of coastal areas, including EEZs, marine environmental protection, climate change, and strengthening regional and international cooperation and coordination in dealing with oceans issues. Given the continued growth of human settlement along the world's coasts, increased coastal recreation, the concentration of industrial development in coastal areas, and the wealth of exploitable, living marine resources, emphasis is placed on protecting the health of coastal waters. Particular attention is paid to the land-based sources of marine degradation that contribute 70 percent of ocean pollution (International Development Research Centre, n.d.). The real challenge, of course, is to implement these approaches effectively in order to achieve the desired goals.

Photo 8–15
Marinas expand as recreational demand grows. Marinas may cause wetland destruction through dredging for access to sheltered water.

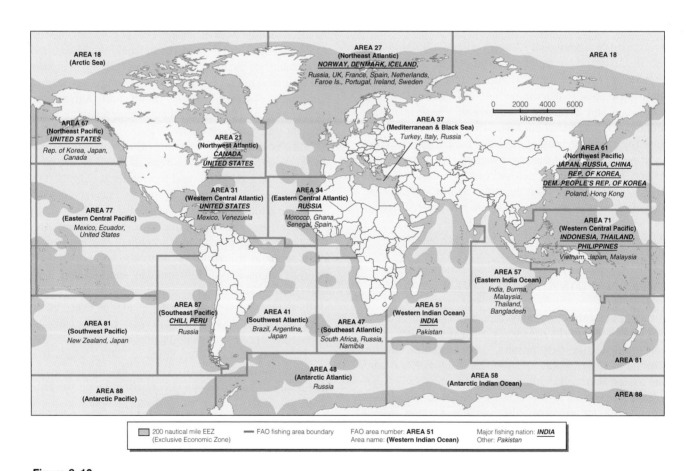

Figure 8–10
Food and Agriculture Organization fishing area boundaries

SOURCE: *WWF Atlas of the Environment,* G. Leon, D. Hinrichsen, & A. Markham, 1990, New York: Prentice Hall, pp. 158–159. Reprinted by permission of Prentice Hall.

CANADIAN LAW, POLICY, AND PRACTICE

Revised Oceans and Fisheries Legislation

Canada's history of governance of marine environments has reflected both federal and provincial responsibilities and jurisdictions. The federal government has responsibility for all matters in waters below the mean high-water mark (except in aquaculture, where certain responsibilities have been delegated to provinces through specific memoranda of understanding). Provincial jurisdiction covers provincial lands, shorelines, and freshwater resources (except navigable waters, inland fisheries, and federal lands), and certain areas of the seabed. Federal and provincial jurisdictions overlap in matters of species and habitat conservation, and provincial and municipal jurisdictions include many of the land-based activities that affect the marine environment. With the conclusion of some major land claims agreements in the Arctic, Aboriginal governments also are involved in the management of human activities in marine environments.

The main federal responsibility for the marine environment and economy lies with Fisheries and Oceans Canada (DFO), which includes Coast Guard functions. Primarily, DFO has managed commercially harvestable marine species, but it also is responsible for all marine species (except sea birds) and marine mammals. Other key departments and their major responsibilities are identified in Table 8–2.

The multiple and overlapping jurisdictions that characterize Canadian government in general give rise to fragmentation, duplication, and lack of coordination in marine environmental management and decision making (Beckmann, 1996). Aware that fragmentation tends to cause environmental considerations to be lost among the competing, more powerful economic interests, the federal government has been generating policies and guidelines to deal with that problem. Among the legislation and policies developed are the National Marine Conservation Areas Act (2002) and the Canadian Arctic Marine Conservation Strategy.

Despite some excellent plans for marine conservation, pollution control, and species and habitat protection, collapse of numerous fish stocks indicates that Canada has had difficulty managing the commercial fishery in its own waters (Beckmann, 1995). Despite the government's commitment to environmentally sustainable development, reasons for Canada's failure to prevent degradation of marine waters and species include political inertia and fragmentation of responsibility.

In the early 1990s, four key events looked very promising for the advancement of marine conservation. The first event was the 1994 release by the National

---

TABLE 8–2
FEDERAL DEPARTMENTS WITH SIGNIFICANT MARINE RESPONSIBILITIES

Department	Selected Major Responsibilities
Fisheries and Oceans	• management of commercially harvestable marine species, and all marine species (except sea birds) and marine mammals; safeguarding of Canada's oceans; Coast Guard functions
National Defence	• border patrol
Natural Resources	• extraction of marine resources such as oil, gas, minerals, and aggregates south of 60°N latitude
Environment	• protection of marine environmental quality; pollution prevention; sea bird and sea bird habitat protection
Parks Canada	• establishment of National Marine Conservation Areas and National Parks through Parks Canada
Indian Affairs and Northern Development	• environmental issues and offshore oil, gas, and mineral development north of 60°N latitude
Foreign Affairs and International Trade	• marine-related negotiations with other countries (as in the case of turbot negotiations with Spain, and the Pacific Salmon Treaty)
Emergency Preparedness Canada	• response to natural and human-made disasters such as oil spills (on land and at sea)

SOURCES: *Seas the Day: Towards a National Marine Conservation Strategy for Canada,* L. Beckmann, 1996, Ottawa: Canadian Arctic Resources Committee/Canadian Nature Federation; *Overview of the East Coast Marine Environment,* E. Meltzer, 1995, Ottawa: Canadian Arctic Resources Committee/Canadian Nature Federation.

Advisory Board on Science and Technology (which reports directly to the prime minister) of its report on oceans and coasts (National Advisory Board on Science and Technology, 1994). Recommendations made by the advisory board included a comprehensive marine environment protection system that would safeguard Canada's oceans for the health, enjoyment, and economic welfare of future generations. The second event was Parks Canada's release of its revised "Guiding Principles and Operational Policies," which included a section on establishing marine conservation areas and which recognized the need to design these areas to accommodate varying levels of human activity within them. A conservation areas program that originated from the Guiding Principles document has resulted in establishment of three marine conservation areas. These are Gwaii Haanas national marine conservation area off the Queen Charlotte Islands, British Columbia; Fathom Five National Marine Park adjacent to Bruce Peninsula National Park in Georgian Bay, Ontario; and Saguenay–St. Lawrence Marine Park near Tadoussac, Quebec. The need for marine protected areas is great; this is highlighted, perhaps, by knowing that COSEWIC has identified more than 100 British Columbia marine ecosystems species that need protection.

Amendments to the Canada Wildlife Act, which enabled the establishment of wildlife areas out to 200 nautical miles, was the third event. This is an important means of extending environmental regulation to protect marine mammals and their habitat. The fourth event was DFO's release of *A Vision for Ocean Management,* a document that not only outlined the essentials of a new oceans management strategy for Canada but also recommitted DFO to the creation of a Canada Oceans Act.

The Canada Oceans Act was passed as Bill C-26 by the House of Commons on October 21, 1996, and given royal assent in December 1996. The Canada Oceans Act recognized Canada's jurisdiction over its ocean areas by declaring a contiguous zone and an EEZ, and provided the legislative basis for an oceans management strategy based on the principles of shared stewardship, sustainable development of ocean resources, and the precautionary approach. There were high hopes for the Canada Oceans Act because it was the basis on which cooperative work with the provinces could begin, and through which all Canadians interested in the marine environment and economy could help build an oceans management strategy. However, since the Oceans Act was "enabling legislation" only and required designation of either management plans or marine protected areas to be effective, DFO was criticized for its failure to follow through on implementation of the act (Sierra Legal Defence Fund, 2002).

In 2002, when Canada's Oceans Strategy was released, the creation of Marine Protected Areas (MPAs) became possible. MPAs are areas of the ocean designated for special management measures under the Oceans Act. This means that enforceable regulations can help protect the area and its marine organisms. In March 2003, Canada's first MPA was designated: the Endeavour Hydrothermal Vents Area, southwest of Vancouver Island, British Columbia, contains 12 species of marine life that exist nowhere else in the world. Currently, 13 other areas are being considered for designation as MPAs, including an area in the southern Beaufort Sea as an MPA for critical beluga whale habitat.

Prior to the passing of the Canada Oceans Act, the provinces had been active. For example, Nova Scotia and New Brunswick had formulated coastal zone management policies and plans. Other cooperative initiatives were undertaken as well, such as the Burrard Inlet Environmental Action Plan (Environment Canada–B.C. Ministry of Environment, Lands and Parks) to reduce toxic loadings in the inlet. The Inuvialuit Final Agreement (for the western Arctic) and the Nunavut Final Agreement (for the eastern Arctic) provided for a joint federal–territorial–Aboriginal management system that fosters integrated decision making. (These co-management initiatives are noted briefly in the section "Canadian Partnerships and Local Actions" later in this chapter.) Municipalities, with their solid waste and sewage management responsibilities, also have important roles to play in improving the health of Canada's marine systems through careful urban planning and appropriate waste management practices.

In December 1995, then Fisheries and Oceans minister Brian Tobin tabled the first major rewrite of the Fisheries Act since 1868, updating the legal basis for conservation and fisheries management. In particular, these changes provided greater opportunity for shared management of the resource through partnerships, more effective enforcement, and more flexible regulations—the kinds of changes that had been advocated for a long time by various analysts. Tobin indicated that changes to the Fisheries Act were designed to allow government and industry to move forward into the "fishery of the future."

Six guiding principles inform the "fishery of the future": (1) conservation comes first; (2) industry capacity must be balanced with resource capacity; (3) the fishery must be conducted by professionals; (4) access to the resource should be through multi-licensed enterprises, while recognizing the reality of certain specialized fleets; (5) government and industry must operate in partnership, with binding agreements signed by both parties; and (6) Aboriginal rights must be respected (Department of Fisheries and Oceans, 1995f).

Coastal Zone Management Efforts

As the examples above illustrate, Canada has made progress in revising existing legislation and in creating new legislation pertaining to the marine environment. Efforts to implement coastal zone management in Canada, however, continue to lag.

To mitigate the negative aspects of increasing stress on coastal environments, a number of countries have adopted national coastal zone management (CZM) programs. Coastal zone management may be defined as "the process of implementing a plan designed to resolve conflicts among a variety of coastal users, to determine the most appropriate use of coastal resources, and to allocate uses and resources among legitimate stakeholders" (Hildebrand, 1989, p. 9). Canada does not have a national coastal zone management program; however, regional efforts exist in parts of the country where resource allocation conflicts and problems of environmental degradation are severe.

In 1978, in an effort to establish a national CZM program, the Canadian Council of Resource and Environment Ministers (CCREM) sponsored a national seminar on coastal zone issues. Although a national strategy did not emerge from this symposium, 10 principles were developed to guide local and regional coastal management efforts. The principles, along with the early initiative for CZM, disappeared before the end of the decade, leaving coastal management to be implemented through a variety of regional initiatives such as the Fraser River Estuary Management Program and the Atlantic Coastal Action Program.

The rapidly urbanizing lower Fraser River basin is biologically productive and highly desirable for agriculture, industrialization, and marine transportation. In 1985, intense allocation conflicts among resource users in the area prompted establishment of the Fraser River Estuary Management Program (FREMP) under a federal–provincial agreement. FREMP was developed according to the recommendations of the CCREM symposium and was intended as a model for regional coastal zone management programs (Day & Gamble, 1990).

FREMP administered programs and activities for the estuary by coordinating the efforts of more than 30 agencies, including representatives from all levels of government, First Nations, and interest groups. Programs for water quality, waste management, recreation, fish and wildlife habitat, port and industrial development, and navigation and dredging were coordinated successfully under FREMP.

With an improved understanding of the interrelationships of ecosystem components came the realization that, to properly manage the Fraser River estuary, comprehensive planning and management of the entire Fraser River basin was required. In 1991, the Fraser River Action Plan (FRAP) was established under Canada's Green Plan with the objective of developing a management plan for the remainder of the Fraser basin (see Chapter 7). Having partnered with the Burrard Inlet Environmental Action Program (BIEAP), the two programs have moved to develop and implement regional management plans, monitoring programs, and other sustainability initiatives (Burrard Inlet Environmental Action Program and Fraser River Estuary Management Program, 2002). Coordinated initiatives such as these offer promise for promoting sound environmental planning and management in other coastal regions of Canada where development and allocation conflicts exist.

Growing demand from the public to become involved in environmental decision making and increasing concern about the quality of Atlantic Canada's waters provided the stimulus for the Atlantic Coastal Action Program (ACAP). Established in 1991 as a six-year, $10 million project under Canada's Green Plan, ACAP called on local community initiatives to achieve its central objective of ensuring a sustainable future. All ACAP projects were community owned and organized, with all stakeholders, from governments to interested citizens, participating as equal partners. The federal government provided seed money of up to $50 000 per year to start and maintain each local ACAP community initiative. The enthusiasm of 13 local communities for the ACAP process was witnessed by the rapid start-up of initiatives, the diverse number of participating stakeholders, and the pace at which plans were developed (Hildebrand, 1993).

The ACAP mandate has continued, with local action remaining the key to its achievements in developing coastal zone management plans. Between 1997 and 2002, Environment Canada invested over $1.07 million, funding 95 scientific projects within ACAP communities and strengthening the links between science and community. Partnerships undertaken by Environment Canada and the 14 ACAP communities involved have included river and harbour cleanup projects and the Sable Island Preservation Trust (Environment Canada, 2002).

CANADIAN PARTNERSHIPS AND LOCAL ACTIONS

In the early 1990s, the towns of Chatham and Newcastle continued to discharge both treated and untreated sewage into the Miramichi River in northeastern New Brunswick. In addition, the Repap Miramichi Pulp and Paper mill discharged its treated industrial and domestic effluent directly into the river at the mill site. Among other effects, these sewage discharges caused ongoing shellfish harvesting closures in the lower estuary of the river.

Historically, local residents had encountered problems in determining who was responsible for dealing with their environmental concerns about the Miramichi River. Wanting to do something about this situation, the Miramichi River Environmental Assessment Committee (MREAC) became an ACAP community group. One of the first things MREAC did was to establish a River Watch program, which, by 1993, included a public phone-in service called the River Watch Line. By simply calling 1-800-RIVER, callers were put in touch with the proper government department or organization and received the information they required. With ACAP financial assistance, MREAC provided this toll-free service to give local people easy access to environmental information and to share their expertise with residents. Water quality assessments were added to the River Watch program in 1994 as MREAC trained and equipped volunteers to patrol the Miramichi River by boat and to monitor water quality. A seasonal Swim Watch program monitored and informed the public of the water quality at key swimming locations (Environment Canada, n.d.c).

These actions from the government–community partnership demonstrate the effort not only to support development of local knowledge and application of environmental, social, and economic information to decision making in the Miramichi watershed, but also efforts to improve the health and sustainability of the Miramichi watershed and receiving ocean waters for the benefit of future generations.

The Islands Trust in British Columbia provides another example of the importance of local residents' involvement in planning and development to help achieve sustainability in a marine environment. The islands in the Strait of Georgia and Howe Sound are recognized nationally for their beauty, tranquillity, and unique natural environments. In 1974, the government of British Columbia enacted special protective legislation entitled the Islands Trust Act, which recognized these special qualities. The act states that the objective of the Islands Trust is to "preserve and protect the trust area [of approximately 5178 square kilometres] and its unique amenities and environment for the benefit of the residents of the trust area and of the Province generally, in cooperation with municipalities, and organizations, and the government of the Province" (Islands Trust, 1995). On April 1, 1990, the act was amended to establish the Islands Trust as an autonomous local government agency with land use planning and regulatory authority.

Given its mandate to "preserve and protect ... for the benefit of the residents," the Islands Trust spent more than two years developing its policy statement based on public forums and public comment on drafts of the policy statement. Public forums identified six themes that captured the concerns of Island Trust area residents: fresh water, resources and environment, growth and development, community, local control, and government.

Throughout the policy development and public consultation process used to formulate the policy statement, residents and trust members alike were educated about the social and environmental challenges facing the area and were encouraged to consider environment and development issues simultaneously. This process stimulated public thinking about issues such as marine pollution, about how to encourage stewardship, and about the meaning and use of ecosystem approaches to achieve protection of the Islands Trust area (Crance, 1995). As a coordinating agency that combines regulatory and non-regulatory functions to achieve its objectives, the Islands Trust is a good example of effective agency–community partnership in action at the local level.

Two large land claims agreements in the Arctic have demonstrated a federal–territorial–Aboriginal partnership system that fosters integrated decision making in marine management. Conservation and sustainability are promoted by the use of co-management (cooperative management) in both the Inuvialuit and Nunavut final agreements, which together cover more than half the Arctic coastline. This joint-management approach to marine conservation and sustainable development offers important keys to success in co-management on the east and west coasts, should fishers and managers there agree on its value.

Canadian partnerships have involved governments, agencies, First Nations, coastal communities, and industrial interests. One example is the first fish habitat bank completed in 1992 on the North Fraser River, a joint initiative of the Department of Fisheries and Oceans and the North Fraser Harbour Commission ("Fish Habitat Bank," 1994). The 1988 North Fraser Harbour Environmental Management Plan helps ensure that environmental considerations are an integral part of waterfront development in the harbour. Specifically, construction of compensation habitat is required prior to approval of a project, thus complementing the no-net-loss principle of the Department of Fisheries and Oceans' national habitat policy.

The importance of nutrient-rich marshes for juvenile salmon on their way to sea is well known, and the North Fraser River was lacking in such environments. In this project, almost two kilometres of what was once heavily industrialized foreshore of the Fraser River was reworked to build marsh habitat that would protect fish habitat and river bank vegetation. Development of the Fraser Lands Riverfront Park, which also included boardwalks and a park for people, involved installing a low rock berm parallel to the shore and backfilling with sandy soil to create an intertidal beach to support the growing marsh. About 14 000 plants were planted in the marsh, and native plants and grasses were used in creating the fish habitat and park area.

Even though the working aspect of the Fraser River is giving way to urban uses, the logging industry still floats its logs and log booms down the river; wave action from other boat traffic erodes the river banks, and debris smothers the marshes. Historically, the Department of Fisheries and Oceans did not support storing logs next to marshes because bark settled into the marshes and the logs themselves grounded on the marsh. However, log storage farther away in the river actually helps break the waves down and keeps debris off the marsh. By 1994, two years into its operation, monitoring showed a healthy, successful marsh project ("Fish Habitat Bank," 1994).

Canadian partnerships for marine environmental protection have extended to other nations as well. Ocean Voice International is an Ottawa-based ENGO with international and Canadian activities. Their use of geographic information systems (GIS) is an important element in enhancing data collection as well as the effectiveness of marine conservation efforts.

Drawing on experience from their land-based Yellowstone to Yukon (Y2Y) initiative (see Chapter 12), the Canadian Parks and Wilderness Society (CPAWS) recently proposed an equivalent multinational, marine protected area. Using the migratory route of the grey whale to define the marine area to be protected, the Baja to Bering (B2B) project sees CPAWS working in partnership with other ENGOs, Parks Canada, and federal agencies in Mexico and the United States. Given the great need for marine protected areas, this project represents an important precedent for cooperative action on the international level.

FUTURE CHALLENGES

What would characterize a sustainable future for Canadian oceans and fisheries? A number of points may be identified, including deliberate application of a sound ecological approach to management, ethical principles, including interpersonal and intergenerational equity, appropriate technology, and an improved science with clear and objective data. Other features would include achievement of a tolerable level of human pressure on natural resources and ecosystems, an acceptable quality of life (which recognizes the interdependence of humans and their environment), the maintenance of diversity, and people striving cooperatively together toward agreed-on goals that meet environmental, social, and economic objectives.

As the examples in this chapter have illustrated, human development activities in and relating to marine environments have resulted in a wide variety of impacts on coastal waters and species. International, national, regional, and local agreements and actions are helping to resolve some of these problems. In general, however, challenges for the future regarding sustainability of Canada's oceans, fisheries, and environments lie in ongoing efforts to improve our stewardship, protection and monitoring, and knowledge-building activities.

Stewardship, the management of oceans and fisheries resources so that they are conserved for future genera-

tions, requires an active, shared awareness of both ethical and ecological principles on which safeguarding the future of marine environments depends. Equity and respect for all other users of oceans and fisheries resources, and a holistic understanding of ecosystems and their interdependencies at all levels, are among the characteristics required to develop attitudes and practices that will help achieve a sustainable future. This is as true for professional fishers as it is for recreational users. One related challenge arising here is to educate everyone who uses or impacts marine environments to include care and respect not only for themselves, but also for other people, and to be aware of the interactions that may occur from their decisions.

New agreements, such as Canada's Oceans Act, that are intended to work toward and achieve sustainability, and new fishing management agreements that are intended to protect particular fish species and stocks, may assist improved stewardship practices. Follow-through is required, however, to ensure the agreements are put in place and enforced. Increasingly—and this applies to all environments, not just oceans and fisheries resources—stewardship implies that the necessary attitudinal characteristics on the part of individuals, agencies, corporations, and governments should include awareness of human effects, receptivity to change, accountability for decisions taken, and acceptance of responsibility to rectify any negative impacts.

Sustainability depends on many related factors, some of which are little understood or poorly defined, and most of which are very difficult to predict. If Canada's oceans and fisheries resources and environments are going to be protected for the future, monitoring efforts need to be incorporated for measuring the environmental, social, and economic parameters of our development activities. In addition, monitoring effects of variables ranging from effluent discharges to climate change may, over time, help us overcome the problem of inadequate data on which to base management decisions (although there is no guarantee that the acquired knowledge will satisfy the questions that need to be answered). In the meantime, as Canada and other countries present at the 1992 Earth Summit recognized, when the "weight of evidence" suggests that action to protect the environment should be taken to prevent serious or irreversible damage, then the lack of full scientific "proof" is not a reason to postpone measures to prevent that degradation. In light of this precautionary principle, perhaps the Fisheries minister's actions in regard to protection of turbot stocks were appropriate.

Observer programs for fisheries (as in the case of European Union vessels and others who fish stocks that cross the Canadian 200-mile limit), and the establishment and application of standards for various types of coastal development activity including housing construction, port expansion, and other infrastructure facilities, are means by which marine protection can be promoted. A shared sense of the responsibility for establishing and effectively implementing programs and regulatory mechanisms to ensure continued ecological functioning of marine environments should enable all parties to cooperate and achieve desired common goals, including sustainability. An important part of any cooperative effort is the ability to anticipate what our future needs might be, what impacts climate change might have on marine environments, for instance, and the ability to develop appropriately sensitive environmental protection and monitoring approaches and devices. Protection measures also can be applied to those marine environments that have been restored or rehabilitated.

Given the lack of knowledge of fundamental as well as more sophisticated elements of marine environments and species functions, a key current and future challenge is to enhance research programs and data collection opportunities to build the necessary knowledge. Partnerships at all levels are appropriate here, and are particularly valuable when they involve First Nations' traditional ecological knowledge (TEK), and the knowledge of fishers and others with long-term experience in marine environ-

ments. Opening up the research process to include TEK and the environmental understanding and contributions of people with extensive fishing experience should enable a fuller appreciation of stock status and regeneration methods, for example, as well as an earlier identification of critical gaps in knowledge. Additionally, support for and participation of those affected by decisions may encourage meaningful domestic (and perhaps international) partnerships in identifying alternative economic and employment opportunities.

Rachel Carson, Jacques Cousteau, and many other observers since the 1970s have pointed out that the oceans do not hold the key to humanity's burgeoning needs and wants. The collapse of the northern cod fishery and the more recent downward trends in Pacific salmon stocks should be very clear and ominous calls for the attention of all Canadians. During the heyday of those fisheries, few believed they would collapse with such harsh echoes resounding over the now-quieted fishing harbours of maritime Canada. As discussed above, governments are moving in new directions, but they require the support and critique of an informed public to create a better future for Canada's oceans.

Chapter Questions

1. Most biological productivity in the oceans occurs near continents, in areas of upwelling currents, on continental shelves, or near river estuaries. For each of the Arctic, Atlantic, and Pacific oceans, identify the kinds of threats that Canadians and their activities (on land and sea) pose to these areas of biological richness.

2. Fish do not respect political or administrative boundaries such as the 200-nautical-mile limit to Canada's exclusive economic zone. What difficulties does this characteristic of fish bring to efforts to protect transboundary fish stocks from overharvesting? What are the implications for sustainability of fish stocks?

3. Discuss the effects that the El Niño phenomenon can have on the west coast fisheries and ocean environment.

4. In what ways can scientific research help us make decisions that will enhance the sustainability of oceans and their resources?

5. Debate the value of the precautionary principle or approach in dealing with international and national oceans and fisheries issues.

6. If you live in a coastal area, identify one or more examples where human activities have impacted or could impact negatively on the sustainability of oceans or fishery resources. What kinds of action were taken (or could be taken) to address the issue(s)? What could you do to make a difference in a current issue?

references

Alexandrowicz, G. (1995). *Law of the sea: A Canadian practitioner's handbook.* Kingston, ON: Faculty of Law, Queen's University. http://qsilver.queensu.ca/law.seati.htm

Anderson, M. (2002, August 14). There is a way to save our oceans—but is there a will? *The Globe and Mail,* p. A11.

Atlantic Fisheries Policy Review. (2001). *The management of fisheries on Canada's Atlantic coast.* http://www.dfo-mpo.gc.ca/afpr-rppa

Auditor General of Canada. (1997). Chapter 14. In *Report of the Auditor General of Canada.* http://www.oag-bvg.gc.ca/domino/reports.nsf/html/ch9714e.html

Beckmann, L. (1995). Marine conservation—keeping the Arctic Ocean on the agenda. *Northern Perspectives, 23*(1), 1–2.

Beckmann, L. (1996). *Seas the day: Towards a national marine conservation strategy for Canada.* Ottawa: Canadian Arctic Resources Committee/Canadian Nature Federation.

Blades, K. (1995). *Net destruction: The death of Atlantic Canada's fishery.* Halifax: Nimbus Publishing.

British Columbia Aboriginal Fisheries Commission. (2002). Are fish farming practices contaminating shellfish beds? *Aboriginal Fisheries Journal, 8*(2), 7.

British Columbia Fisheries. (2000). British Columbia aquaculture harvests and values 1997–1999. http://www.bcfisheries.gov.bc.ca/stats/statistics-aqua.html

British Columbia Ministry of Agriculture, Food and Fisheries. (2003). *Fisheries statistics 2001.* http://www.agf.gov.bc.ca/fishstats/statistics.htm and http://www.agf.gov.bc.ca/fishstats/statistics-aqua.htm

Brocklehurst, A. (1996). It's not easy being green. *Canadian Geographic, 116*(3), 40, 41.

Bryden, J. (1995a, March 14). Ship loaded with tiny turbot. *Calgary Herald,* p. A2.

Bryden, J. (1995b, March 29). Canada delivers proof of "ecological madness." *Calgary Herald,* p. A3.

Bryden, J. (1995c, March 29). Lonely, unloved turbot clinging by its fingernails, Tobin says. http://www.southam.com/nmc/waves/depth/fishery/turbot032295.html

Burrard Inlet Environmental Action Program and Fraser River Estuary Management Program. (2002). *Annual Report.*

Burrard Inlet Environmental Action Program (BIEAP) and Fraser River Estuary Management Program (FREMP). (2002). *Annual report 2001/02.* Burnaby, B.C.: Authors.

Canadian Aquaculture Industry Alliance. (2000). *Canadian aquaculture industry profile.* http://www.aquaculture.ca/English/TheIndustry/CAIA IndustryProfile.html

Canadian Arctic Resources Committee. (2002). Action on climate change. *Compass,* p. 7.

Canadian Broadcasting Corporation. (2003, January 30). Rainbow trout may push out Atlantic salmon. *CBC Radio News.* http://cbc.ca/cgi-bin/templates/print.cgi/2003.01.30/salmon trout

Carscadden, J. E., Frank, K. T., & Miller, D. S. (1989). Capelin (Mallotus villosus) spawning on the southeast shoal: Influence of physical factors past and present. *Canadian Journal of Fisheries and Aquatic Sciences, 46,* 1743–1754.

Chandarana, R. (2002, September 25). Ships dumping bilge are slaughtering birds off Canada. Reuters. http://www.enn.com/news/wire-stories/2002/09/09252002/reu_48512.asp

Committee on the Status of Endangered Wildlife in Canada. (2003). *Two Atlantic cod populations designated at risk.* http://www.cosewic.gc.ca/eng/sct7/sct7_3_1_e.cfm

Coward, H., Ommer, K., & Pitcher, T. (Eds.). (2000). *Just fish: Ethics and Canadian marine fisheries.* St. John's, Newfoundland: Institute of Social and Economic Research.

Cox, K. (1996, March 16). Who won the great turbot war? *The Globe and Mail.* http://www.docuweb.ca/~pardos/globe.html

Crance, C. (1995). *Government coordinating agencies in Canadian coastal planning and management: The Islands Trust and the Waterfront Regeneration Trust.* Unpublished M.A. thesis, Wilfrid Laurier University.

Davis, D. (2000). Gendered cultures and conflict and discontent: Living "the crisis" in a Newfoundland community. *Women's Studies International Forum, 23*(3), 343–352.

Day, J. C., & Gamble, D. B. (1990). Coastal zone management in British Columbia: An institutional comparison with Washington, Oregon, and California. *Coastal Management, 18,* 115–141.

Department of Environment. (n.d.). Sustainability for commercial fisheries. http://www.ns.doe.ca/soe/ch6-43.html

Department of Fisheries and Oceans. (1995a, January 27). *Tobin says NAFO must decide equitable sharing arrangement for Greenland halibut.* News release. http://www.ncr.dfo.ca/communic/newsrel/1995/HQ08E.htm

Department of Fisheries and Oceans. (1995b, February 2). *Canada wins critical vote on turbot at NAFO.* News release. http://www.ncr.dfo.ca/communic/newsrel/1995/HQ10E.htm

Department of Fisheries and Oceans. (1995c, February 15). *Tobin says Canada will not let the EU devastate turbot.* News release. http://www.ncr.dfo.ca/communic/newsrel/1995/HQ08E.htm

Department of Fisheries and Oceans. (1995d, July). *The fisheries crisis in the northwest Atlantic.* Backgrounder. http://www.ncr.dfo.ca/communic/backgrou/1995/HQ16e.htm

Department of Fisheries and Oceans. (1995e, September 12). *Tobin comments on U.S. District Court decision on Alaskan chinook fishery.* News release. http://www.ncr.dfo.ca/communic/newsrel/1995/HQ107E.htm

Department of Fisheries and Oceans. (1995f, December 11). *Tobin tables Fisheries Act amendments.* News release. http://www.ncr.dfo.ca/communic/newsrel/1995/HQ140E.htm

Department of Fisheries and Oceans. (1996). *Commercial catch restrictions and Chinook conservation.* Backgrounder. http://www.ncr.dfo.ca/communic/backgrou/1966/pr11e.htm

Department of Fisheries and Oceans. (2001a). *Herring and minor finfish. Pacific region roe herring fishery. Overview of the fishery.* http://www.pac.dfo-mpo.gc.ca/ops/fm/Herring/ROE/roe.htm

Department of Fisheries and Oceans. (2001b). *North Labrador arctic char.* DFO Science Stock Status Report D2-07.

Department of Foreign Affairs and International Trade. (1999). Canada and the Oceans. http://www.dfait-maeci.gc.ca/sustain/EnvironIssu/canOcean/oceans-e.asp

Drilling on Georges Bank. (n.d.) http://biome.bio.dfo.ca/science/drilling.html

Environment Canada. (n.d.a). *State of the environment in the Atlantic region.* http://www.ns.ec.gc.ca/soe/cha4.html

Environment Canada. (n.d.b). *Marine plastics debris.* http://www.ns.ec.gc.ca/udo/cry.html

Environment Canada. (n.d.c). *ACAP communities in action.* Halifax: Author.

Environment Canada. (1994). Sustaining marine resources: Pacific herring fish stocks. *SOE Bulletin,* No. 94-5. Ottawa: Environment Canada.

Environment Canada. (1995). *United Keno Hill Mines impact assessment report (Released).* http://yvrwww1.pwc.bc.doe.ca/ep/programs/eppy/yukon/ukhm.html

Environment Canada. (1996). *Corner Brook Pulp and Paper Limited pays $750 000 for pollution violations.* http://www.doe.ca/enforce/cor2_p_e.htm

Environment Canada. (2002). *What is the Atlantic Coastal Action Program (ACAP)?* http://www.ns.ec.gc.ca/community/acap/index_e.html

Fisheries and Oceans Canada. (2002a). *Canada reopens ports to Estonian fishing fleet.* News Release. http://www.dfo-mpo.gc.ca/ media/newsrel/2002/hq-ac149_e.htm

Fisheries and Oceans Canada. (2002b). *Aquaculture production statistics.* http://www.dfo-mpo.gc.ca/communic/statistics/aquacult/AQUA01.htm and http://www.dfo-mpo.gc.ca/communic/statistics/aquacult/aqua00.htm

Fisheries and Oceans Canada. 2003. *Sustainable aquaculture: DFO's Aquaculture Action Plan.* http://dfo-mpo.gc.ca/aquaculture/response_details_program.htm

Fisheries and Oceans Canada. 1994. *A vision for ocean management.* (B. Tobin.) Ottawa: Minister of Fisheries and Oceans.

Fisheries and Oceans Canada, Pacific Region. (2002). *Shellfish contamination closures – Pacific Region.* http://www.ops-pac.dfo-mpo.gc.ca/fm/shellfish/closures/default.htm

Fisheries Resource Conservation Council. (2003). 2003/2004 Conservation requirements for groundfish stocks on the Scotian Shelf and in the Bay of Fundy (4VWX5Z), in Subareas 0, 2 + 3 and redfish stocks. http://www.frcc.ca/2003/sf2003.pdf

Fish habitat bank completed on the North Fraser. (1994). *Pacific Tidings, 7*(2), 4–5.

Food and Agriculture Organization. United Nations. (1997). *Code of conduct for responsible fisheries.* http://www.fao.org/waicent/faoinfo/fishing/agreem/codecond/codecon.htm

Food and Agriculture Organization. (2002). *The state of world fisheries and aquaculture.* http://www.fao.org/docrep/005/y7300e/y7300e.00.htm

Fraser River Action Plan. (n.d.). *Fact Sheet 2: Pollution in the Fraser.* http://yvrwww1.pwc.bc.doe.ca/ec/frap/fr-fs2.html

Gardner, J., and Peterson, D. (2003). *Making sense of the salmon aquaculture debate: Analysis of issues related to netcage salmon farming and wild salmon in B.C.* Vancouver: Pacific Fisheries Resource Conservation Council.

Gerges, M. A. (1994). Marine pollution monitoring, assessment and control: UNEPs approach and strategy. *Marine Pollution Bulletin, 28*(4), 199–210.

Gessell, P. (1995, June 22). *Turbot war has high net cost for Canada.* http://www.southam.com/nmc/waves/depth/fishery/turbot062295.html

Golden, S. (1993). Shetland tanker spill worries Canadian fish farmers. *Alternatives, 19*(4), 13.

Gomes, M. C. (1995, March 20). Turbot affair: The EC vs Canada. Message posted on fish-ecology mailing list of the Bedford Institute of Oceanography. http://hed.bio.ns.ca/lists/war/msg00050.html

Government of Canada. (1991). *The state of Canada's environment—1991.* Ottawa: Supply and Services Canada.

Hildebrand, L. P. (1989). *Canada's experience with coastal zone management.* Halifax: Oceans Institute of Canada.

Hildebrand, L. P. (1993). Coastal zone management in Canada—the next generation. In P. L. Lawrence & J. G. Nelson (Eds.), *Managing the Great Lakes shoreline: Experiences and opportunities* (pp. 13–30). Waterloo, ON: University of Waterloo Heritage Resources Centre.

Howard, R. (1997, June 28). "Vigorous but not aggressive" quotas set for salmon catch. *The Globe and Mail,* p. A5.

International Development Research Centre. (n.d.). *Ocean facts: Land-based sources of ocean pollution.* Ottawa: Author.

Islands Trust. (1995). *An introduction to the Islands Trust.* Victoria: Islands Trust. http://www.civicnet.gov.bc.ca/muni/istrust.html

Jaimet, K. (2003, May 3). Scientists add Atlantic cod to endangered list. *Calgary Herald,* p. A15.

Kenny, E. (1995, July 11). East, west disputes oceans apart. *Victoria Times Colonist,* p. A5.

Kinloch, D., Kuhnlein, H., & Muir, D. (1988). Assessment of PCBs in Arctic foods and diet: A pilot study in Broughton Island, NWT, Canada. *Arctic Medical Research, 47* (Supplement 1), 159–162.

Lunn, N., & Stirling, I. (2001). Climate change and polar bears: Long-term ecological trends observed at Wapusk National Park. *Research Links, 9*(1), 1, 6–7.

Masterman, B. (1996, June 3). Rules anger industry. *Victoria Times Colonist,* p. C7.

McAllister, D. E. (2002). Marine biodiversity of Canada: Threats and conservation solutions. *Biodiversity, 3*(1), 11–16.

McDorman, T. L. (1995). The west coast salmon dispute: A Canadian view of the breakdown of the 1985 treaty and the transit license measure. *International and Comparative Law Journal, 17*(3), 477–506.

Meissner, D. (1995, March 1). Boat owners set to throw book at roe protesters. *Victoria Times Colonist,* p. B2.

Meltzer, E. (1995). *Overview of the east coast marine environment.* Ottawa: Canadian Arctic Resources Committee/Canadian Nature Federation.

Menon, A. (1998). Shellfish water quality protection program. Department of Fisheries and Oceans. http://www.mar.dfo-mpo.gc.ca/science/review/1996/AmarMenon/Menon_e.html

Murphy, P. (1997, July 9). El Niño's coming and it's going to be hot. *Victoria Times Colonist,* p. A2.

National Advisory Board on Science and Technology. (1994). *Opportunities from our oceans. Report of the Committee on Oceans and Coasts.* Ottawa: Author.

Obee, B. (1996). Fragile havens for millions of shorebirds. *Beautiful British Columbia, 38*(2), 24–29.

Olsen, S. (1996, February 29). The primary habitat of our species. *Providence Journal-Bulletin.* http://brooktrout.gso.uri.edu/ProJoEd.html

Ommer, R. (Ed.). (2002). *The resilient outport: Ecology, economy, and society in rural Newfoundland.* St. John's, Newfoundland: Institute of Social and Economic Research.

Pacific fisheries for tomorrow. (1995). *Pacific Tidings, 7*(3), 5–6.

Pacific Fisheries Resource-Conservation Council. (2000). *Annual report: 1999–2000.* http://www.fish.bc.ca/reports/annual_2000/chap2_e.html

Pacific Streamkeepers Federation. (2001). *The Streamkeepers Program.* http://www-heb.pac.dfo-mpo.gc.ca/PSkF/id/program.html

Parfit, M. (1995). Diminishing returns: Exploiting the ocean's bounty. *National Geographic, 188*(5), 2–37.

Population: A growing problem for the Fraser Basin. (n.d.). Fraser River Action Plan Fact Sheet. http://yvrwww1.pwc.bc.doe.ca/ec/frap/fr-fs2.html

Qualicum Beach Streamkeepers. (2001). *Projects in 1999–2000.* http://www.mvihes.bc.ca/projects/qualicum.html

Revel, B. (n.d.). *The fish: The Greenland halibut, or turbot.* http://www.sfu.ca/~revela/thefish.htm

Robinson, B. (1995, October 15). Healthy ecology rides on the salmon. *Victoria Times Colonist,* p. F2.

St. Lawrence Vision 2000. (1999). *Five-year report: 1993–1998.* http://www.slv2000.qc.ec.gc.ca/slv2000/english/library/report/quin9398/anglais.Pdf

St. Lawrence Vision. (2000). *Study tracking contaminants in the St. Lawrence River released.* http://www.slv2000.qc.ec.gc.ca/communiques/phase3/1998_1510_etatstlaurent_a.htm

St. Lawrence Vision. (2003). Monitoring the state of the St. Lawrence. *Le Fleuve, 13*(6), 1–6. http://www.slv2000.qc.ca/bibliotheque/lefleuve/vol13no6/accueil_a.htm

Saving the wild Chinook. (1995). *Pacific Tidings, 8*(2), 3–4.

Schram, G. G., & Polunin, N. (1995). The high seas "commons": Imperative regulation of half our planet's surface. *Environment Conservation, 22*(1), 3.

Sinclair, M., & Page, F. (n.d.). *Cod fishery collapses and North Atlantic GLOBEC.* http://www.usglobec.berkeley.edu/usglobec/news/news8/news8sinclair.html

Statistics Canada. (2000). *Human activity and the environment 2000.* Ottawa: Minister of Industry.

Statistics Canada. (2003). *Landed value of fish by species.* http://www.statcan.ca/english/Pgdb/prim70.htm

Thonney, J.-P., and Garnier, E. (1992). *Bay of Fundy salmon aquaculture monitoring program, 1991–1992.* Fredericton: New Brunswick Department of the Environment.

United Nations. (2003). *The United Nations agreement for the implementation of the provisions of the United Nations Convention on the Law of the Sea of 10 December 1982 relating to the conservation and management of straddling fish stocks and highly migratory fish stocks (in force as from 11 December 2001): Overview.* http://www.un.org/Depts/los/convention_overview_fish_stocks.htm

United Nations Convention on the Law of the Sea. (2004). *Chronological list of ratifications of, accessions and successions to the Convention and the related Agreements as at 16 July 2004.* http://www.un.org/Depts/los/reference_files/chronological_lists_of_ratifications.htm

Welch, H. E. (1995). Marine conservation in the Canadian Arctic: A regional overview. *Northern Perspectives, 25*(1), 5–17.

Willison, M. (2002). Science and policy for marine sanctuaries. *Biodiversity, 3*(2), 15–20.

Willson, M. F., Gende, S. M., & Marston, B. H. (1998). Fishes and the Forest. *Bioscience, 48*(6), 455–462.

World Commission on Environment and Development. (1987). *Our common future.* Oxford: Oxford University Press.

Yukon Department of Renewable Resources. (2000). *Yukon state of the environment report 1999.* Whitehorse, YT: Author.

> "[Canada's] goal is to maintain and enhance the long-term health of our forest ecosystems, for the benefit of all living things both nationally and globally, while providing environmental, economic, social and cultural opportunities for the benefit of present and future generations."
>
> Canadian Council of Forest Ministers (1992b)

Forests

Chapter Contents

Chapter Objectives

After studying this chapter you should be able
to

- understand the nature and distribution of
  Canada's forest resources

- identify a range of human uses of forest
  resources

- describe the impacts of human activities on
  forests and forest environments

- appreciate the complexity and interrelatedness of forest environment issues

- outline Canadian and international
  responses to forest issues

- discuss challenges to a sustainable future
  for forest resources in Canada

INTRODUCTION

As biologically diverse as the people who live here, Canada's forests are a symbol of our national heritage. From the lofty Douglas fir and Sitka spruce in the old-growth temperate rain forests of British Columbia, to the rare sassafras and endangered cucumber trees in the Carolinian forests of southern Ontario, to the ground-hugging black spruce, jack pine, and tamarack of the Boreal forest that drapes "like a great green scarf across the shoulders of North America," forests continue to enrich the lives of all Canadians (Natural Resources Canada, 1996a).

More than 330 communities and 870 000 people are directly and indirectly supported economically by Canada's forests (Natural Resources Canada, 1996b, c; Canada's Forest Network, 2000). Moreover, forest products are the biggest net contributor to Canada's visible trade balance, to a considerable degree defining Canada's global economic role. Capital investment in the forest industries is massive (Natural Resources Canada, 2000a, b). Yet, employment, trade, and investment figures are only part of the story of the values, products, and services associated with forests. Forests are more than a source of timber and fibre for newsprint and other industrial commodities. Left standing, forests are complex systems that provide many important ecological services such as moderating climate (carbon storage), improving air quality, stabilizing soil, regulating water flow, protecting aquatic ecosystems (in rivers and streams), and providing habitats for plants, fish, and wildlife, including nesting and breeding grounds for many migratory bird species (Environment Canada, 1995; McKibben, 1996; Sierra Club of Canada, 1996).

Recreation and tourism are increasingly important activities in forestry environments. The spiritual and cultural values associated with forests, long important to First Nations peoples, are also of great interest to many Canadians and visitors seeking solitude and sanctuary from urban lifestyles (Davidson, 1996). Specific life forms such as large trees and marbled murrelets also are important forest values. For more information on forest ecosystem–based values, see Table 9–1.

TABLE 9-1
FOREST ECOSYSTEM–BASED VALUES

Forest Value	Comment
Air quality	Most life on earth depends on a unique chemical reaction—photosynthesis—that happens inside the cells of green plants. The green pigment chlorophyll combines carbon dioxide gas from the air with water from the soil to produce carbohydrates and oxygen. Since plants began to photosynthesize, almost all life has relied on this reaction to produce food, generate oxygen, and remove carbon dioxide. The oxygen people breathe comes from green plants; large forests are major producers of oxygen and also filter pollutants from the air.
Water and soil	Forests act like massive pumps, helping to recycle water, making it repeatedly available for plant growth. Through this action and their extensive rooting systems, forests also help to maintain a regular pattern of water flow in streams and reduce erosion, thus helping to maintain soils and their nutrients. In doing so they help maintain stream conditions favourable for fish and other species.
Climate	Forests capture carbon dioxide and store vast amounts of carbon that might otherwise accumulate in the atmosphere and contribute to global warming. By producing oxygen and absorbing carbon dioxide, forests provide a vital air-conditioning service to the planet.
Biodiversity	Natural (unmanaged) forests are remarkably rich in species. Survival of many species depends on the structural complexity and variety of habitats found in old, natural forests. Managed forests are deliberately simplified to make management easier. This simplification alters resident biodiversity, sometimes dramatically.
Scenic values	People experience scenery over a large area. Thus, to understand scenic resources, it is necessary to look at broad patterns in the landscape. For residents, scenery provides a backdrop to their lives and reflects on their lifestyles. For tourists, scenic resources often provide the context for a trip or recreational activity. Forests are part of many of the world's most highly valued landscapes. To many people, removal of the forest reduces scenic resource values.
Cultural and spiritual values	Forests have values that go beyond specific resource attributes, such as the presence of large trees or deer. They provide traditional foods, materials, and medicinal plants important to indigenous cultures. As systems, they provide a context in which physical and spiritual events take place. Because of their longevity and many values, forests often form part of the cultural identity of the people who inhabit or live near them.
Economic values	Forests provide many goods, such as wood and its diverse products, fish, wildlife, and water—all of which support human society. The sale of forest products and forest-based experiences generates funds that support health, education, and other social services.
Intergenerational values	Many forest trees, especially those in the Pacific Northwest, are potentially long-lived, some reaching ages greater than 1000 years. Thus, the values associated with any individual forest can benefit several human generations. Values attributed to forests have changed over human history, and it is reasonable to expect that they will continue to change. The obligation of current generations is to sustain forest systems without damaging their potential value for future generations.

SOURCE: Adapted from *A Vision and Its Context: Global Context for Forest Practices in Clayoquot Sound,* Clayoquot Sound Scientific Panel, 1995, Report 4 of the Scientific Panel for Sustainable Forest Practices in Clayoquot Sound, Victoria: B.C. Ministry of Forests, p. 4. Copyright © 2000 Province of British Columbia. All rights reserved. Reprinted with permission of the Province of British Columbia.

If Canadian forests are to continue to provide jobs, recreation, places of spiritual expression, and wildlife habitat, then a central concern is how to maintain the biological diversity on which the multiple benefits and roles of forests depend. In turn, biodiversity conservation requires forest management practices that sustain the health and productivity of forest ecosystems. Within the past two to three decades, a major challenge—internationally as well as in Canada—has been to develop an understanding of the complex environmental, economic, social, and political dimensions of forests. Both national and provincial governments in Canada have responded to the growing public concern regarding forests and the environment, and have taken action intended to shift forest management from its historical focus on sustaining output levels for specific forest products (for example, timber) toward sustainability of forest ecosystems that protect both timber and nontimber values of Canada's forests.

THE EARTH'S FORESTS

GLOBAL DISTRIBUTION, PRODUCTS, AND DEMAND

About 40 percent of the Earth's land surface supports trees or shrub cover. Although estimates vary considerably (Mather, 1990), one conservative suggestion indicates that forests occupy about 3.4 billion hectares (27 percent) of the world's land area, and open woodland and mixed vegetation occupy almost 1.7 billion hectares. The former USSR contains the largest concentration of forests and wooded areas (942 million hectares), followed by North America (749 million hectares), Europe (195 million hectares), and the Pacific nations of Australia, Japan, and New Zealand (178 million hectares). Combined, the forests of Brazil, Russia, the United States, and Canada contain more than 50 percent of the Earth's forests (Natural Resources Canada, 1996a).

Globally, forests may be classified as temperate, boreal, or tropical (see Figure 9–1). Tropical forests, found between the tropics of Cancer and Capricorn, form the most species-diverse ecosystem in the world, containing more than 50 percent of all living species on this planet (Natural Resources Canada, 1996b). In 1990, tropical forests occupied about 1.79 billion hectares, while boreal and temperate forests occupied about 1.67 billion hectares. The predominantly coniferous boreal forests cover 920 million hectares and are located between the Arctic tundra and the temperate zone.

During the 1980s, as dramatic satellite imagery showed, the area of tropical forests declined by an average of 15.4 million hectares annually, mainly as a result of the clearing of land for agricultural use in developing nations. Subsequently, according to the Food and Agriculture Organization (FAO) (1999), total forest area in developing countries continued to decline, by another 56.3 million hectares between 1990 and 1995 (or 1.5 percent of the forest cover annually). Although direct clearing of forest land appears to be the most immediate cause, many interrelated factors contribute to the decline. Forest fires in 1997–98 were the worst recorded in recent times, causing losses of millions of hectares worldwide, with the greatest destruction in southeast Asia. Pollution is also a threat to forests in developed countries. However, forest cover is increasing among developed countries, particularly the United States, as former agricultural lands are converted to forest (Mather, 1990). The federal department Natural Resources Canada is examining the feasibility of increasing forest cover on agricultural lands to help Canada meet its commitments to reducing carbon emissions (see Chapter 5). It is not clear, however, whether the amount of land that could be returned to forest cover would be sufficient to make a significant contribution to these commitments.

The estimated value of fuel wood and wood-based products to the world economy is US$400 billion (Natural Resources Canada, 1996b). Worldwide, the forest sector provides subsistence and wage employment equivalent to 60 million work years, 80 percent of which is in developing countries. In addition, forests provide an extremely wide range of products, including fuel, oils, medicinal plants, and household furniture, as well as all products conventionally classified with the pulp and paper and wood-processing industries. Many forests also are a direct source of food, including fruits, honey, and mushrooms. Increasing awareness of the multiple roles that forests fulfill and growing scientific knowledge of forests as complex ecosystems have meant that forest issues have transcended political and sectoral boundaries, becoming a priority in general international debates regarding the future of the Earth's environment and its expanding population (Natural Resources Canada, 1996b).

As the human population continues to expand, global demand for wood and wood products continues to increase. Based on the world's current annual wood consumption of 0.7 cubic metres per person, it has been estimated that the demand for wood could increase by as much as 70 million cubic metres annually (Natural Resources Canada, 1995). This is about as much wood as British Columbia harvests annually. The FAO (1999) predicts that demand for wood will increase 1.7 percent annually between 1999 and 2010. However, there is not enough commercial forest in the world today to meet this demand. Does this mean the world will run out of wood? According to the Canadian Forest Service, this scenario is unlikely. If wood becomes scarcer, "[p]rices for wood products will rise, recycling and use of wood waste will increase, more substitutes for wood will be used, technologies will improve, and new wood products will be developed" (Natural Resources Canada, 1996b, p. 42). With the rapid growth of plantations and a greater commitment to sustainability in some parts of the world, the world as a whole is likely not facing a biological shortage of wood. However, severe regional problems exist, the survival of old-growth timber is an issue, and the projected increase in rates of human consumption and global population are a significant challenge to forest managers throughout the world.

Since the 1970s, the international community has engaged in debates concerning the future of the world's forests, particularly in relation to deforestation in tropical forests. By 1990, deforestation was recognized as symptomatic of a fundamental conflict between human needs and the environment. Subsequently, international discussions broadened in scope to balance consideration of environmental, social, and economic factors in forest

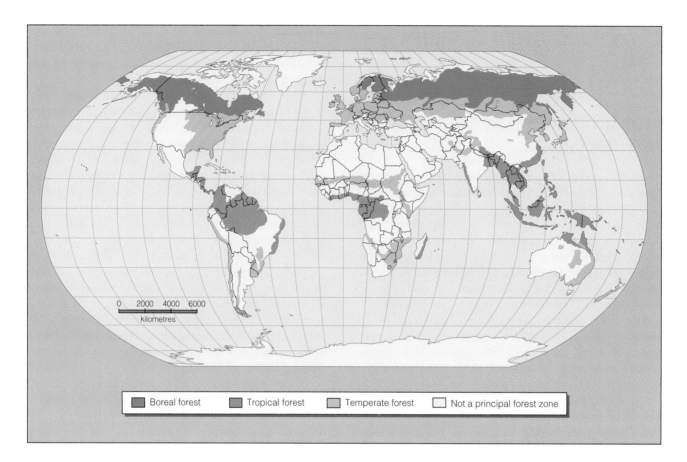

Figure 9–1

Principal forest zones of the world

SOURCE: *The State of Canada's Forests: Sustaining Forests at Home and Abroad, 1996,* Natural Resources Canada, Ottawa: Natural Resources Canada, p. 25. Reprinted with permission of the Minister of Public Works and Government Services Canada, 2004.

development. As is described later in the chapter, efforts to promote sustainable forest management have been part of the focus of the "global forest dialogue" since then (Natural Resources Canada, 1996b).

Because Canada's forests are of benefit to more than Canadians (forests contribute to air quality and biodiversity, for instance), other nations want Canada to nurture our forests wisely so they will continue to provide benefits. And, because Canada relies so heavily on exporting its wood products, other countries have powerful means of encouraging good stewardship. Canada is highly sensitive to international actions such as consumer boycotts or "eco-certification" of wood products (discussed later in the chapter).

FORESTS IN CANADA

Among the 12 regions classified in Figure 9–2, there are eight major forest regions in Canada, covering approximately 45 percent of our land base. The three types of boreal forests occupy more than one-third of Canada,

constituting the country's largest biome and providing direct employment for an estimated 254 400 people. British Columbia's high-volume coastal (temperate) forests, however, produce a relatively large share of Canada's forest products and, because of the large size of the trees, account for about 46 percent of the annual volume cut in Canada. The other major forest regions shown in Figure 9–2 are the Columbian, Deciduous, Great Lakes–St. Lawrence, and Acadian. With almost 418 million hectares of forested land, Canada is caretaker of about 10 percent of the world's forests, including about 15 percent of the world's softwood supply (see Table 9–2 for additional Canadian forest facts).

Unlike most nations, the vast majority (94 percent) of Canada's forests are publicly owned. Forests are managed on behalf of each one of us (the public) by provincial governments (71 percent of forests) and federal and territorial governments (23 percent), while the remaining 6 percent of Canada's forest lands are owned privately (see Figure 9–3). This pattern can be explained by our history of settlement.

Prior to the 20th century, the Crown allocated private property rights to land and resource owners. But at the

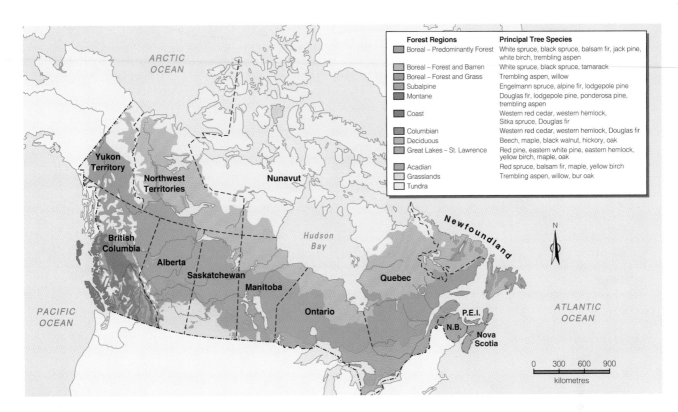

Figure 9–2
Forest regions of Canada

SOURCE: *The State of Canada's Forests: A Balancing Act*, 1995, Natural Resources, Ottawa. Reprinted with permission of the Minister of Public Works and Government Services Canada, 2004.

turn of the 20th century, all federal and provincial governments in Canada turned their backs on the policy of granting complete or outright title to land and resources, except for urban and agricultural purposes. For other resources, such as timber, minerals, and water, governments developed rights in the form of leases, licences, and permits that they issued to private parties, giving private individuals or companies access to resources while keeping the title in the Crown. Thus, comprehensive private ownership is more prevalent in the regions that were settled first. This means that private ownership of forest lands is greatest in the Atlantic provinces and generally declines toward the more recently settled West. In British Columbia, private ownership of resources—including timber—is very small, while in the northern territories it is almost nonexistent.

Today, the majority of forest land is provincial Crown land. However, in Prince Edward Island (92 percent), Nova Scotia (69 percent), and New Brunswick (51 percent), the majority of forested land is held privately, often in private woodlots. Given that more than half of Maritime forests are privately owned, largely by small woodlot owners who are not subject to provincial regulations, the implications for sustainable management are different than in the rest of Canada, where Crown land dominates. Forest management practices on woodlots in the Maritimes are highly variable,

and, in New Brunswick, logging continues at a faster rate than replanting on most woodlots (see Box 9–1). Numerous initiatives recently have been promoted to enhance forest stewardship among woodlot owners (National Round Table on the Environment and the Economy, 1997). Bearing in mind the deep criticisms of corporate forestry in the rest of Canada, the problems facing small woodlots in the Maritimes underline the complex challenge of achieving sustainable forestry practices that meet a balanced range of goals.

Forest management in Canada is a matter of provincial jurisdiction; each province as well as the Northwest Territories and Nunavut has its own legislation, policies, and regulations to govern forest activities within its boundaries. The Yukon also is seeking to gain control over its forest resources from the federal government. In addition to overseeing management of Yukon forest lands, the federal government focuses on trade and investment, national statistics, forest science and technology, Aboriginal affairs, environmental regulations, and international relations (Natural Resources Canada, 1996b).

Forest products continue to be the single largest contributor to Canada's balance of trade (Natural Resources Canada, 2000b). In 2001, for instance, Canada's forest products exports were valued at approximately $44.1 billion; softwood lumber, paper and paperboard, newsprint,

TABLE 9–2

SELECTED FACTS ABOUT FORESTS IN CANADA

Forested Lands

- Forests cover 417.6 million hectares (45 percent) of Canada's land base (921.5 million hectares).
- 56 percent, or 234.5 million hectares, are considered commercial forests.
- 38 percent, or 156.2 million hectares, are open forests consisting of muskeg, marshes, and sparse tree cover.
- Most (94 percent) are publicly owned: provincial governments manage 71 percent, federal and territorial governments manage 23 percent, and the remaining 6 percent is private property of 425 000 landowners.

Commercial Forests

- 50.2 percent, or 118.9 million hectares, are managed for timber production.
- Over 12 percent, or 50 million hectares, are protected from harvesting by legislation (heritage forests) or policy (protection forests).
- Annually, in recent years, about 0.4 percent of the accessible commercial forest is harvested, removing an average of 174.5 million m^3 of wood and contributing about \$95 per m^3 to Canada's gross domestic product.

Forest Regions (see Figure 9–2)

- Boreal
- Subalpine
- Montane
- Coast

- Columbian
- Deciduous
- Great Lakes–St. Lawrence
- Acadian

Forest Type (2000)

- Softwood (e.g., pine, spruce)   62 percent
- Hardwood (e.g., poplar, maple)   16 percent
- Mixed wood   22 percent

Tree Species

- approximately 180

Total Annual Allowable Cut Estimates (millions of cubic metres)

Year	Total	Softwoods	Hardwoods
1990	253	192	61
1991	253	190	63
1992	247	185	62
1993	228	172	56
1994	230	172	58
1995	233	174	59
1996	234		
1997	237		
1998	241		
1999	225.3		
2000	232.9		

SOURCES: *Compendium of Canadian Forestry Statistics 1996,* Canadian Council of Forest Ministers, 1997, Ottawa; "Sustaining Canada's Forests: Timber Harvesting," Environment Canada, 1995, *Overview SOE Bulletin,* no. 95-4 (Summer); *The State Of Canada's Forests 1995–1996: Sustaining Forests at Home and Abroad,* Natural Resources Canada, Canadian Forest Service, 1996, Ottawa: Author; *The State of Canada's Forests 1996–1997: Learning from History,* Natural Resources Canada, Canadian Forest Service, 1997, Ottawa: Author; *The State of Canada's Forests 1997–1998: The People's Forests,* Natural Resources Canada, Canadian Forest Service, 1998, Ottawa: Author; *The State of Canada's Forests 1998–1999: Globally Competitive through Innovation,* Natural Resources Canada, Canadian Forest Service, 1999, Ottawa: Author; *The State of Canada's Forests 1999–2000: Advancing into the New Millennium,* Natural Resources Canada, Canadian Forest Service, 2000, Ottawa: Author; *Natural Resources Fact Sheet,* Natural Resources Canada, 2000, http://www.nrcan.gc.ca/statistics/factsheet.htm; *The State of Canada's Forests 2000–2001: Sustainable Forestry: A Reality in Canada,* Natural Resources Canada, Canadian Forest Service, 2001a, Ottawa: Author; *The State of Canada's Forests 2002–2003: Looking Ahead,* Natural Resources Canada, Canadian Forest Service, 2004, Ottawa: Author.

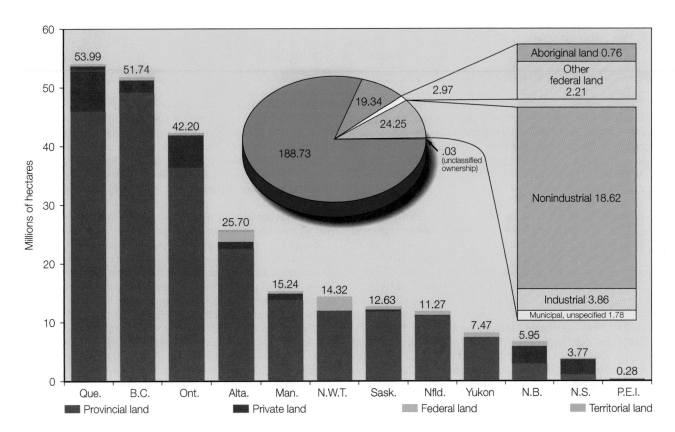

Figure 9–3

Ownership of timber productive forest lands

SOURCES: *Compendium of Canadian Forestry Statistics, 1996,* Canadian Council of Forest Ministers, 1997, Ottawa. Figure 1, p. 1; *The State of Canada's Forests 2001–2002,* Natural Resources Canada, Canadian Forest Service, 2002, Ottawa: Author; *Canada's National Forest Inventory* (CanFI 92, V. 94), Natural Resources Canada, Canadian Forest Service, 2002, http://www.pfc.cfs.nrcan.gc.ca/montioring/inventory/facts/facts-table2_e.html

and wood pulp account for most of these exports (Natural Resources Canada, 2001b). In 2001, the forest sector contributed $28.5 billion to Canada's gross domestic product (or 2.9 percent of the total for all industries) and $34.4 billion to Canada's balance of trade (Natural Resources Canada, 2001b). The forestry sector is the principal reason for Canada's positive visible trade balance in 8 of the last 10 years. One in every 15 Canadian jobs is in the wood and paper products or related industries (Canadian Forest Service, 1996a).

On a worldwide basis, the amount of forest is decreasing. Most of the loss of forests is recent, 60 percent having occurred since the Industrial Revolution. In North America, 64 million hectares were cleared for settlement and agriculture between 1860 and 1978 (Clayoquot Sound Scientific Panel, 1995). As noted, since the 1900s in the United States, forest area has increased. In Canada, between 1979 and 1997, the average area harvested (logged) was almost 1 million hectares (or 0.4 percent of Canada's productive forest) per year, and the average annual volume of wood harvested was 163 million cubic metres. These statistics appear to be trending

upward: from 1995 through 1998, the average area harvested was 1.03 million hectares, from which an average of almost 179.5 million cubic metres was harvested (Natural Resources Canada, 1996b, 1997, 1998, 1999, 2000; Canadian Council of Forest Ministers, 1997). British Columbia's share of this volume of wood produced is about 70–80 million cubic metres per year (including about 10 percent from private lands), enough wood so that each year a stack of lumber one metre high and one metre wide would circle the equator twice (M'Gonigle & Parfitt, 1994). Overall in British Columbia, 50 percent of the volume logged has been cut since 1972. Coastal old-growth forests are being clear-cut at a rate of about 200 000 hectares per year (Sierra Legal Defence Fund, 2000).

About 200 000 hectares of boreal forests have been opened up for exploitation since the 1980s, doubling the amount of boreal forest harvested in Canada since the 1920s. Provincial governments have opened public lands to multinationals from North America, Europe, and Asia (especially Japan), supporting their plans for pulp, paper, and saw mills, simulated plywood plants, and chopstick

The fact that more than half of Maritime forests are in the hands of private owners has different implications for sustainable forest management than in the rest of Canada, where the vast majority of forests are on Crown land. In the Maritimes, private woodlots are not subject to provincial forestry regulations but are considered freehold property where owners may use the resource as they choose.

There are more than 422 000 private forest owners in Canada (National Round Table on the Environment and the Economy [NRTEE], 1997). While some of this land is owned and operated by large forestry corporations, most is classified as private woodlots, private property used for, or capable of, growing trees, and with an average size of 40 hectares (NRTEE, 1997). Because they represent such a large percentage of forested lands in the Maritimes, private woodlots are critically important for all the economic, social, cultural, and ecological reasons discussed in this chapter. However, there is serious concern about the sustainability of woodlot forestry in the Maritimes. Stakeholders at meetings in 1997 agreed that a combination of overharvesting, lack of forest stewardship practices, and a perception that there are no apparent economic incentives to manage woodlot forests sustainably could create a forest industry collapse in the Maritimes similar to the collapse of the ground fishery in Atlantic Canada (see Chapter 8).

The sense of impending crisis in the woodlot industry prompted a request to the National Round Table on the Environment and the Economy (NRTEE) to explore problems and make recommendations for achieving a sustainable woodlot industry. Members of the NRTEE are appointed by the prime minister and are charged with improving economic and environmental quality by providing decision makers with information to make decisions about stewardship of Canada's natural resources (NRTEE, 1997). Stakeholders involved in the process included federal and provincial government representatives, the forest industry, woodlot owners, environmentalists, and First Nations people. Although there is consensus that current activity is not sustainable, no one agrees about the best way to manage the resources of private woodlots. The role of regulation, administration, market demand, value systems, and science and technology are all contentious issues with respect to sustainable woodlot management. Recognizing that many unsustainable forestry practices are economically driven, the NRTEE recommendations seek to promote sustainability through methods that make sustainable forestry practices more affordable through, for example, voluntary measures, tax incentives, and industry-supported trust funds, rather than through legislation (NRTEE, 1997).

Responses to NRTEE's 1997 recommendations have been mixed. For example, in New Brunswick, logging is still continuing in woodlots faster than owners are replanting (Natural Resources Canada, 2000). On a more positive note, the Prince Edward Island Department of Agriculture and Forestry is conducting an education program for woodlot owners. In Nova Scotia, the recognition that current woodlot harvesting practices are not sustainable has resulted in a regulation that puts the onus on buyers of large quantities of primary forest products to undertake some of the responsibility toward sustainable forestry. On the industry side, one Maritime pulp and paper company pays a bonus for fibre from private woodlots and is providing private owners with nominally priced seedlings to encourage replanting (Natural Resources Canada, 2000c).

SOURCES: *State of the Debate on the Environment and the Economy: Private Woodlot Management in the Maritimes,* National Round Table on the Environment and Economy (NRTEE), 1997, Ottawa: Author; *The State of Canada's Forests, 1992–2000. Forests in the New Millennium. Species at Risk,* Natural Resources Canada, 2000, Ottawa: Author, http://www.nrcan.gc.ca/cfs/proj/ppiab/sof/sof00/spart3.html

---

factories, often with subsidies on the order of millions of Canadian taxpayer dollars.

Large-scale exploitation of timber in Alberta's boreal forest began in 1986; the Alberta Pacific pulp mill (45 percent owned by Mitsubishi of Japan, 14 percent by Canadian shareholders) consumes 3.2 million cubic metres of timber per year, and the Daishowa-Marubeni bleached kraft pulpmill consumes 1.8 million cubic metres of boreal forest annually (Marchak, 1995; Pratt & Urquhart, 1994). These two multinational firms have leased 15 percent of Alberta's boreal forest land base.

Every day, Saskatchewan's giant Prince Albert pulp and paper complex turns the equivalent of 30 football fields of forest into pulp. The company's 20-year lease covers about 34 000 square kilometres of boreal forest land. In Manitoba, the provincial government has granted the right to about 17 percent of the province (77 percent of its prime boreal forest lands) to one firm, Repap Enterprises Limited.

Ontario and Quebec contain the bulk of Canada's timber-productive boreal forest. Expansion of older mills is occurring throughout these provinces, as the boreal forest is expected to provide 60 years of cutting opportunities (staving off the day when Brazil and Indonesia will dominate pulp markets). In both provinces, as logging has moved north, timber harvesting has conflicted with indigenous people's traditional uses of forest land. In Quebec, for instance, harvesting in the southern fringe of the James Bay region has consumed 600 square kilometres annually, the size of a hunting territory that supported about 30 Cree people (McLaren, 1993). Similarly, in Newfoundland and Labrador, there have been proposals to cut as much as 1.7 million hectares of the sparse, widely separated stands of boreal forests on land claimed by the Innu Nation (Innu Nation, 1995).

Large multinationals also lease much of Nova Scotia and New Brunswick's Acadian forest lands. About 80 percent of Nova Scotia's forest production is in pulp, cut by companies such as STORA Forest Industries from Sweden. In New Brunswick, Repap leases about 25 percent of the province's Crown land, mostly for pulp production. The challenge is that forests that are ecologically productive are also economically desirable, pitting these two interests against one another in land use debates. Overall, nearly 100 percent of Canada's most ecologically productive boreal forest, including several provincial parks and wildlife reserves, is committed to forest companies in 20-year leases and is available for logging (McLaren, 1993).

In light of the globally increasing demands for wood, Canada's important role as a provider of wood products, and the significant contribution of forest products to our nation's economy, questions have been raised about whether Canada can maintain, let alone increase, its supply of timber to provide products for world consumption. Still other questions have been raised about whether current rates of production are too high to permit forests to replenish themselves and to promote forest sustainability (Hammond, 1991; M'Gonigle & Parfitt, 1994; Sierra Club of Canada, 1996). Related to questions about the rate of wood cutting in Canada are concerns about the impacts of harvesting systems and the "timber bias." A brief discussion of these concerns follows.

Harvesting Systems

Clear-cutting remains the dominant method of harvesting in Canada. Traditionally, about 90 percent of Canada's timber harvest has been clear-cut, and about 7 percent of timber harvested is done using selection cutting. However, it is not always easy to distinguish logging types. These statistics probably include other forms of logging referred to in Table 9–3 (see also Figure 9–4), such as patch cutting, as well as other variations such as partial retention. There are also indications that clear-cutting is declining slightly, in 1995 accounting for 85.7 percent of Canada's harvest. Timber companies prefer clear-cutting because it is normally the most cost-effective way to harvest trees. It is also the safest system for loggers, and foresters have argued that on many sites it is ecologically the most appropriate form of logging (Kimmins, 1992).

Clear-cutting is sometimes defended as a method of logging that mimics natural disturbances. To some degree this may be true in boreal forests, where fire means renewal (for example, seeds in cones of black spruce and lodgepole pine are released by fire). Clear-cutting, like

Photo 9–1
This clear-cut area shows signs of initial regrowth.

fire, may establish site conditions conducive to regeneration (Hebert et al., 1995). In ancient temperate rain forests, however, there is no forest-fire dependent cycle and clear-cutting may have numerous negative effects, including landslides, water pollution, loss of plants and animals, and soil degradation. Replanting after clear-cutting can be extremely difficult.

In practice, clear-cutting throughout Canada (and elsewhere) typically focused first on the most accessible, high-quality forests, leaving less-accessible, often lower-quality forest areas for later harvest. This is a practice known as **highgrading**. While highgrading typically yields high short-term economic returns, it may have long-term ecological consequences. Since accessibility and timber quality often overlap (if by no means perfectly), many of the highest quality forests across Canada have long since been logged. Consequently, some nonindustrial values, such as high-quality seed sources (with genetic codes adapted precisely to sites), have been lost forever.

In Canada, highgrading rarely means "cherry picking" the very best individual trees from a forest. Rather, highgrading typically occurs within the system of clear-cutting, where all trees are removed from large continuous tracts and from entire watersheds. The removal rate may occur in a short time (five years) or over a longer period of 20 or 30 years, but the impacts that occur are similar.

Until recently, clear-cuts across Canada were extremely large (hundreds of hectares), and the practice

T A B L E   9 – 3
FOREST HARVESTING METHODS

Method	Characteristics
Selective (or selection) cutting	• The original forest is of varied species and ages. • Only the most valuable species of trees or only trees of prescribed size or quality are cut. • The forest is left to regenerate naturally. • Trees are cut individually or in small clusters. • About 7 percent of Canada's timber has been cut using selective techniques.
Shelterwood cutting	• The original forest may be evenly or unevenly aged. • All mature trees are removed in a series of cuttings, stretched out over about 10 years. • First cut removes most canopy trees, unwanted tree species, and diseased, defective, and dying trees. • After a decade or so, when enough seedlings have taken hold, a second cut removes more canopy trees but leaves some of the best mature trees to shelter the young trees. • After perhaps another decade, a third cut removes the remaining mature trees and the remaining uniformly aged stand of young trees grows to maturity.
Seed tree cutting	• The original forest is harvested in one cutting, but a few uniformly distributed trees are left to provide seeds for regeneration. • After the new trees have become established, the seed trees may be cut. • About 4 percent of Canada's timber is cut using the shelterwood cutting and seed tree methods.
Clear-cutting	• Clear-cutting removes all the trees from an original forest at the same time. • Trees may be cut as whole stands, as strips, or as patches. • After the trees are cut, the forest may be left to regenerate naturally or may be replanted. • Clear-cut areas have varied in size (from small patches to thousands of hectares). • Almost 90 percent of Canada's timber is cut using clear-cutting.
Patch (clear) cutting	• Patch clear-cutting leaves small-scale clear-cut areas, perhaps 100 to 200 hectares in size. • Several clear-cut patches adjacent to one another can result in a continuous clear-cut.
Strip cutting	• Strip cutting involves clear-cutting narrow rows of forest, perhaps 80 metres wide, leaving wooded corridors that may serve as seed sources. • After regeneration, another strip is cut above the first, and so on, allowing the forest area to be clear-cut in narrow strips over several decades.
Whole tree harvesting	• Machines harvest entire trees (including roots, leaves, bark, small branches) and cut them into small chips to be used as pulpwood or fuelwood products. • Another variation of clear-cutting, this method deprives soil of plant nutrients and can support replanted trees only if they are fertilized.

of **continuous clear-cutting**—in which cut blocks are located adjacent to each other in successive years—rapidly laid bare much larger areas. Since 1995 in British Columbia, continuous clear-cutting has been banned and the size of permissible clear-cuts greatly reduced as part of efforts to create a more environmentally sensitive forest sector (Hayter, 2000).

Clear-cutting results in—and is defined by—altered microclimates (Kimmins, 1992). Clear-cutting removes cover needed by wildlife and new plant growth, degrades the soil, damages fish habitat and water quality, and destroys the forest diversity needed to sustain all forest uses, including future timber supplies (Hammond, 1991). With very large clear-cuts, especially in association with continuous clear-cutting, these problems become more serious. The tendency to cut the most accessible and the best first also means that every year both the quality and the value of the remaining forest declines (Clapp, 1998).

(a) Selection cutting

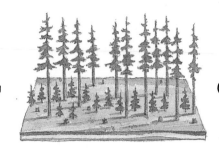
(b) Shelterwood cutting

(c) Clear-cutting

(d) Patch cutting

Figure 9–4
Major systems of tree harvesting

SOURCE: *The State of Canada's Forests: A Balancing Act,* 1995, Natural Resources Canada. Reprinted with permission of the Minister of Public Works and Government Services Canada, 2004.

The timber now standing in many provinces is less valuable per tree and costs more to log than the average quality timber of a few decades ago.

As clear-cutting–highgrading depletes high-volume, high-value old-growth forests, it also results in a decline in the average volume of timber per hectare in remaining old-growth forests. This means that for the same land area, the volume and the quality of the timber removed is less. If timber managers seek to "sustain" the volume or the value of their cut, they must log a larger forest area each year. As British Columbia's coastal old-growth forests were clear-cut, for instance, timber companies moved into the interior of the province, to higher elevations, and to remoter parts of the coast where the trees are smaller, requiring more land area to be cleared to supply the volume of wood needed to feed local mills. Elsewhere in Canada, after a century or more of clear-cutting, the loss of the Atlantic hardwood forests and almost 90 percent of the Carolinian forests has forced companies in the eastern and central provinces to reach into more remote regions of their provinces to supply societal demands. In New Brunswick, for instance, clear-cutting of the remote Christmas Mountains region, the only old-growth forest left in the province, began in 1992.

Yet, alternatives to clear-cut logging also pose challenges for sustainability. At least four issues arise, if we criticize clear-cut logging. First, under contemporary arrangements, clear-cutting is undoubtedly the least expensive means of logging. Other strategies such as patch-cutting or selective logging may require specialized equipment (e.g., helicopters) that is very expensive to operate, or the construction of more access roads.

Second, where selective forms of harvesting require more patches to be protected from harvest, both economic and ecological concerns arise due to the cost of building access roads. Road building is the most ecologically damaging activity related to timber harvesting, often resulting in undesirable environmental effects such as soil compaction and erosion, disturbance of natural water courses, and loss of soil cover and nutrients. With the exception of helicopter logging (which is the most expensive option), the smaller the cut area, the larger the number of access roads required, and the greater the ecological impact.

Third, safety of forest workers is also a key issue. Logging is a dangerous occupation and clear-cutting remains the safest method of felling trees. In other methods of selective logging where trees are left standing within an active logging area, loggers themselves are at high risk for fallen trees and other "debris" in the area. In some cases, worker compensation boards require that clear-cutting be undertaken in order to protect the safety of forest workers.

Fourth, where logging has historically been undertaken under a particular regime (e.g., clear-cut logging), a change to new harvesting methods may result in job lay-offs, increased training requirements, and the shedding of a long-standing workforce. In many cases, loggers have built their livelihood and identity on skills they have learned over many years of work in the woods. They believe that they make a key contribution to society by providing a product that is in heavy demand by urbanites and others who build homes, city infrastructure, and paper products from timber. Changes in logging practices may threaten both the incomes and long-held beliefs of woods workers related to their value in contemporary society. Politicians who are sensitive to their constituents may be reluctant to make policy decisions about harvesting if they believe such decisions would be vehemently opposed by their electorate. These issues illustrate the complexity of sustainability. As we consider changes that we believe are ecologically responsible, we face challenges related to social and economic elements of sustainability. The choices are not always clear.

Tree Plantations

After clear-cutting, forest firms engage in some form of **silviculture** or managed forestry, defined as the "theory and practice of controlling the establishment, composition, growth and quality of forest stands. Basic silviculture means planting and seeding; intensive silviculture includes site rehabilitation, spacing and fertilization" (Natural Resources Canada, 2000a, p. 110). Traditionally, the most common approach to re-creating a forest, albeit in much simplified form, first emphasized slash burning, where debris (natural supplies of woody material on the ground and in the soil) was removed by fire. The conventional wisdom was that slash burning facilitated regeneration, including natural regeneration. Moreover, as commitment to planting trees in slash-burned areas increased after the 1960s, there was a strong tendency to plant one or two species only (known as monoculture), to be grown and cut again in short cycles of 60 to 120 years. In a third silvicultural step, brush may be removed, often with chemical pesticides, as forest managers attempt to maximize timber growth by controlling or eliminating competing forms of life from the forest.

This kind of three-step basic silviculture—slash-burning, monoculture, and brush removal—poses problems, however. Slash burning increases greenhouse gases, increases the likelihood of soil erosion, decreases the availability of soil nutrients, and creates human health problems. Respiratory ailments are of particular concern as slash-burning smoke contains two carcinogens, namely formaldehyde and polynuclear aromatic hydrocarbons.

In addition, replacement of diverse old-growth forests with monoculture stands brings potential prob-

Photo 9–2

Most tree plantations are characterized by stark rows of one or two species of similar age. Underbrush is absent.

lems, including rapid spread of wildfires in evenly aged stands, blowdown due to poor root formation in planted trees, and loss in the number and variety of mycorrhizal fungi needed for seedling growth (Hammond, 1991). Finally, brush provides shade for young trees and cover for animals, enriches the soil, and repels unwanted insects. In natural forests, brush plays an important role in sustaining the whole forest (as does every other organism). While economically viable from a forest industry perspective, monoculture and development of plantation forests potentially reduce both economic and biological diversity (M'Gonigle & Parfitt, 1994; Taylor, 1994). Tree plantations are not the same as old-growth forests.

Until recently, **deforestation** was a term applied mostly to tropical rain forests, but deforestation also occurs in Canada. In regard to our timber-cutting practices, during the 1990s, environmental activists referred to Canada—particularly British Columbia—as the "Brazil of the North" (Hammond, 1991; McCrory, 1995). The timber industry and the provincial government denied the comparison, pointing out that Amazon rain forests are being cleared for agricultural uses, whereas Canada's forests are being regenerated for timber. However, some critics estimate that until the 1990s, about 50 percent of the logged area in Canada did not regenerate to productive species within five years of cutting (Diem, 1992). In addition, during the period 1979 to 1993, only 36 percent of the area harvested was replanted or seeded; the remainder was left to "regenerate naturally," in an effort to "help maintain the natural diversity of the forest *and reduce costs*" (Natural Resources Canada, 1996b, p. 88; emphasis added). Nevertheless, basic silviculture has been significantly extended, especially in British Columbia and Quebec (Hayter, 1996, p. 105). In British Columbia, for many years now, the number of trees planted has exceeded the number of trees harvested, planting frequently has involved a variety of species to suit site-specific conditions, and slash burning has been reduced greatly.

The Timber Bias

For many decades, a great number of Canadians and their governments have viewed trees as timber and forests as log supply centres; Canada's institutional approach to forest management has evolved out of that perspective. Typically, politicians legislated for timber cutting and pro-

duction, and not for forest protection and the stewardship of diverse ecosystems. The "timber bias" prevalent in our use of forests has resulted in a number of ecological, economic, and sociocultural impacts visible on the landscape and in communities.

Early timber industry barons viewed forests as a limitless timber supply. As evidence accumulated that this was not the case, government and industry embraced the idea of **sustained-yield** forest management. The commonly held interpretation of sustained yield is that timber should be cut no faster than new trees can grow so that an even flow of timber in perpetuity may be obtained. In effect, this concept means that, if yield is to be maintained, all the trees cut down should be replaced each year with new trees that are allowed to mature to a size comparable to the original trees on the site before they are logged. This simple concept becomes obscured in practice, however.

Tree growth data show that the average annual accumulation of wood *volume* (not wood *quality*) peaks before the old-growth stage (see Figure 9–5). Timber managers want to grow trees to the point where average timber volume production is greatest (this is called the culmination age), and then clear-cut the trees to produce the greatest timber volume over time. Timber managers consider trees at culmination age—60 to 120 years old—

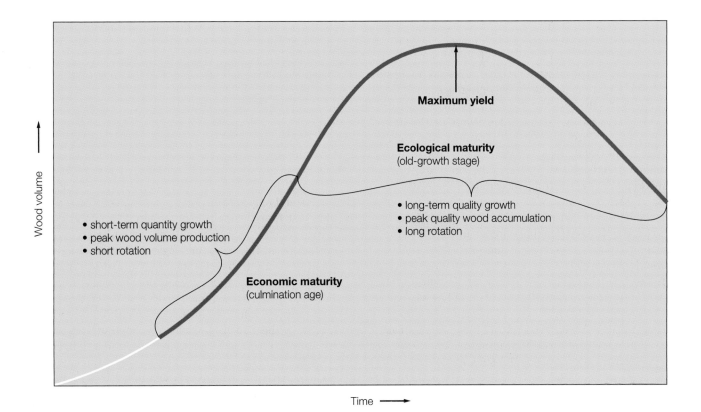

Maximum yield

Ecological maturity
(old-growth stage)

- long-term quality growth
- peak quality wood accumulation
- long rotation

- short-term quantity growth
- peak wood volume production
- short rotation

Economic maturity
(culmination age)

Wood volume

Time ➞

Figure 9–5
Timber volume production

to be "mature"; they rationalize that old-growth or "deca-dent" timber should be cut down because it grows timber volume too slowly.

The word "mature" is used differently by foresters and by scientists. Foresters use the term to mean the *economic maturity* of a tree or forest stand—the youngest age at which the trees can be cut and sold for a profit. On the other hand, scientists call wood mature when it has developed its maximum fibre length; that is, the *natural or ecological maturity* at which trees develop their strongest and highest value wood. A tree is not ecologically mature until it has passed through all forest stages from shrub to old growth. For short-lived tree species, this may be 200 to 300 years, but in long-lived species, trees may take up to 1500 years to reach ecological maturity. Trees and forests may live for centuries after reaching maturity, during which time they perform vital ecological functions and accumulate the highest quality wood of their lifetimes (Franklin, 1984). Clearly, economic maturity is reached far sooner than ecological maturity.

Cutting trees at economic maturity versus ecological maturity, however, is a little like picking green tomatoes instead of waiting for ripe tomatoes. When green, toma-toes have nearly the same volume as red, ripe tomatoes, but the ripe fruit of the tomato is far more desirable and far more valuable. Just as ripe tomatoes sell for a lot more money than green tomatoes, high-quality, old-growth wood can be sold for a lot more than economically mature wood (Hammond, 1991).

In order to calculate how much timber should be cut in any one year (called the annual allowable cut, or AAC), current Canadian forestry practice relies on a measure called the rotation period. For any given species, its rota-tion period is the length of time required to grow a tree to economic maturity. Actual calculations of the annual allowable cut are complicated, however; the rotation period, the total available natural old-growth timber volume, and the annual growth of the forest are the most important factors affecting calculation of the AAC.

A serious issue with respect to forestry is the declining number of people who are required to under-take it. In British Columbia, for example, total forest sector employees (in logging and processing) declined from 1.32 per 1000 cubic metres harvested in 1963 to 0.85 in 1995 (Marchak et al., 1999). This decline suggests that to maintain the workforce in forestry, either more wood must be cut or better use must be made of it. It is often argued that secondary wood processing—such as making cabinets, mouldings, panelling, furniture, and guitar tops from lumber—adds value and results in higher levels of employment generated from each tree cut. In fact, value can be added by capital and technology, as well as labour, and greater efficiency requires less labour. Nevertheless, in British Columbia, which has the highest quality and highest volume of timber per hectare of any province in Canada, the value added to wood products in 1984 was only about half that added in the rest of Canada. In spite of its large timber production, for every one job per 1000 cubic metres of timber (about 33 truckloads of logs) pro-duced in British Columbia in 1984, 2.2 jobs were produced from the same amount of timber in the rest of Canada (Hammond, 1991; Hayter, 2000; Travers, 1993). This dis-crepancy suggests that there are unrealized value-added potentials in British Columbia.

These issues are of great interest to many people concerned about the sustainability of British Columbia's and Canada's forests. With the increasing global scarcity of old-growth timber (Food and Agriculture Organization, 1997), the province could tap into the economic advan-tage that exists in the luxury market potential that old-growth trees embody. But to date it appears as if "corporate, government and union policies remain hooked on a volume economy of mass commodity production.... As if in a process of reverse alchemy, we still convert our old-growth gold into the dull lead of ... two-by-fours" (M'Gonigle & Parfitt, 1994, p. 44).

The pursuit of economic productivity through high-volume, low-labour logging has resulted in declining forests, closure of community mills, and loss of jobs throughout the country (Hammond, 1991; Marchak, 1995; M'Gonigle & Parfitt, 1994; Taylor, 1994). As the most accessible and highest quality (usually old-growth) forests have been clear-cut (often with considerable wood waste) to achieve short-term maximum profit, younger and smaller trees have been logged, shortening the rota-tional period. These practices have contributed to the "falldown effect" and to the potential loss of future eco-nomic and ecological diversity.

The Falldown Effect

Diversity is the cornerstone of a stable, healthy forest and of a stable, healthy economy. Given that forest diversity is achieved through forest protection, and that we do not understand fully how forests work, it would seem reason-able that forest management activities should give pri-ority to, and be consistent with, protection and ecological sustainability. However, timber managers are aware that their monoculture tree plantations and managed natural stands will never produce the amount or quality of timber provided by the old natural forests they are replacing.

Foresters call the reduction in volume of production caused by the shift from exploiting old-growth forests to producing managed (monoculture or more diverse) forests *falldown*. Government projections of falldown estimate that future harvests will be 20 to 30 percent below allowable annual cuts; other forecasts are highly variable. Falldown occurs because old-growth forests have typically had a long time to produce the fibre for which they are renowned. Although definitions vary (see

Kimmins, 1992), old-growth forests often are thought of as forests that contain trees 200 to 1000 or more years old and that have not been seriously changed by human action. Yet, when these forests are harvested, they are typically placed on a rotation schedule of approximately 60 to 120 years. That means that an old-growth tree that may be cut after 200 years would be replaced by a tree that is "ready to harvest" in 60 to 120 years. The shorter period between harvests means there is less time to build the biomass; hence the resulting reduction in harvest volume. If falldown is to be avoided—both biologically and economically—new growth forests would have be cut on 200- to 400-year cycles or rotation periods (Hammond, 1991). (See also the section "The Need for Protection" later in this chapter.)

Falldown and loss of biological diversity have clear economic impacts. Once the historical diversity of tree species and sizes is gone, employment levels in traditional timber cutting and wood manufacturing will decline, and the wood products industry may be unable to adjust rapidly to local, national, and international demands. In addition to reduced quantities of timber, reductions in quality of timber produced in the second-growth forests are of concern. Old-growth wood from slow-growing trees is clear, soft, fine-grained, strong, and easy to work with. In contrast, second-growth wood is frequently knotty, hard, coarse-grained, weak, and more difficult to work with. Old-growth trees contain about 80 percent mature wood, while 60-year-old second-growth trees contain about 50 percent juvenile wood. Long-fibre, mature wood is not formed on the stem of the tree until branches die (natural pruning), while juvenile, short-fibre wood is produced on the stem as long as wood is growing around green or living branches.

Forintek, a Canadian industry–government research cooperative, analyzed the wood fibre of second-growth Douglas fir and determined that it had low structural strength and problems with warping and workability. Almost no high-quality lumber could be milled from the short rotation second-growth trees (Kellogg, 1989). This decline in quality (strength and workability) is due largely to the shorter length of fibres that make up the new-growth wood. Even pulp products such as paper also are stronger when produced from old-growth timber. Forintek's research revealed that second-growth Douglas fir trees grown on average or better sites needed to reach the age of 90 years and be pruned to increase wood quality in order to have positive dollar value. This finding suggests that annual cutting rates need to be reduced by at least one-third to accommodate a lengthening of rotation periods from 60 to more than 90 years so that wood will have sufficient value to realize a profit in the long term.

Some industry analysts suggest that future timber management in Canada can be based on a "fibre" or "volume" economy. This means that technology can be used to provide us with the ability to fabricate construction materials from bonded composites of chips or from pulp obtained from plantation-grown trees or natural regeneration trees grown on short rotations. The flaws in this plan include the fact that high-quality wood equates to high-value wood, both today and in the future. A study in Washington state, for example, revealed that Douglas fir aged 160 or more years was 56 percent more valuable than wood of the same species aged less than 100 years (Wigg & Boulton, 1989). High-quality wood produced by older trees will always be in demand for manufacture of fine wood products, for quality pulp and paper, and other specialty products.

Much of the high-quality, long-fibre, old-growth timber left in the world is located in western Canada. This supply can be liquidated and sold cheaply or it can be cut more slowly to make high-quality wood products and command premium prices. From an economic standpoint, scarcity favours the latter strategy. World demand for high-quality old-growth wood suggests that the costs of careful stewardship and labour-intensive practices could be recovered. However, such a strategy would reverse historical practices and require a fundamental rethinking of land tenure arrangements, fees and royalties paid by private operators to provincial governments, and management practices required of forest companies. To date, no government has demonstrated the political will to dismantle and replace the current system of land allocation and forest management.

It has been argued that sustainable cutting of old growth and manufacturing of high-quality wood products can provide more jobs in both the short and long term than can plantation forestry and pulp or wood fibre or volume economies (Travers, 1993). Selective cutting systems in natural forests require the highest labour per volume logged of all conventional logging systems. Timber products from old-growth forests can feed a diverse sustainable industry, including large log sawmills, small log sawmills, pulp mills, paper mills, cabinet shops, millwork plants, furniture factories, beam and truss lamination plants, and so on. Similarly, labour-intensive operations such as commercial thinning in older diverse forests provide growing space for old trees and furnish intermediate timber products for wood manufacturing. In addition, if planned sensitively, selective harvesting can protect forest-based recreation, although the latter itself can cause environmental damage.

Tree plantations, where trees of uniform size grow in straight rows and are logged by machines, tend not to be labour intensive. Plantations that produce products (such as disposable chopsticks) made from single-species trees of a uniform size require much less labour because production systems are highly mechanized and the range of products is very limited. The wood that comes from monoculture forests can be made into pulp and composite products such as fibre boards. In a fibre or volume economy, more and more timber is required to account for fewer and fewer jobs. Consequently, over time, as natural forests are reduced to fibre farms, both tree sizes and employment will decline (Hammond, 1991; M'Gonigle & Parfitt, 1994). In addition, tourism and recreational values and opportunities that depend on more diverse forest values may be reduced or even destroyed. These social impacts of falldown and loss of biological diversity are in addition to the loss of the forests' function to buffer climate change, because large carbon sinks are gone (see Chapter 5), a lower quality of water is produced (due to possible contamination from pesticides used in the clearing of brush), and species are eliminated.

Although hard to define, loss of forests and their diversity results, too, in the loss of their spiritual values, not only for First Nations but for all people. First Nations have also been affected significantly by logging taking place on lands over which they assert Aboriginal title. While trying to reach a just settlement of the land question in court, logging has continued. Consequently, First Nations people have sought injunctions to stop logging on these lands so that if they become recognized, then First Nations people themselves will have the right to decide how they will be utilized.

We know how to practise sustainable use of our forests and how to avert loss of biodiversity. Achieving sustainability requires a reduction in cutting rates, protection of old-growth forests, design of timber management systems that perpetuate natural forest diversity, and the ability to provide for a variety of forest uses. Even though single-species plantations are not forests, there may be a place for some plantations under particular circumstances. There is the possibility that plantations of hybrid cottonwoods or genetically uniform Douglas fir—if not grown on good forest land and not displacing natural forests or their regeneration—could become sources of significant employment and regional income. Some new towns could be based on scientific research related to plantation forestry as well as exploitation of these unnatural forests (Schoonmaker, von Hagen, & Wolf, 1997).

The shift away from the old-forest economy, based on volume cutting and unsustainable economics of corporate and bureaucratic growth, toward a value-based, forest economy brings many opportunities. One caveat to this perspective is that the costs such a strategy entails will have to be paid for by revenues generated from hoped-for value-added activity. This is an assumption and should not be taken for granted (Hayter, 2000), and the performance of woodlots in New Brunswick should be kept in mind in this regard. But old-growth forests offer nonindustrial values as well. Part of the challenge in learning to value forests differently and in considering strategies for conservation-based development is to understand why old-

growth forests are so important ecologically. Another part of the challenge will be to figure out how forest workers can be retained and retrained for new kinds of jobs in a value-based forest economy.

THE ECOLOGICAL IMPORTANCE OF OLD-GROWTH FORESTS

Old-growth or ancient forests share similar characteristics (in terms of life cycle, carbon storage, and biological diversity) that make them important legacies for the future of Canadian forests. These characteristics, and the need for their protection, are considered briefly below.

THE LIFE CYCLE IN THE OLD-GROWTH FOREST

Standing Live Trees, Snags, and Fallen Trees

Large, old living trees are virtual forest communities in themselves and provide an excellent example of the connectivity of species within ecosystems described in Chapter 3. For example, the foliage of a single old-growth Douglas fir may have a surface area of over 2800 square metres that, because of the microclimatic differences created by the lean of the tree, may attract epiphytic (aerial) plants such as mosses and lichens to the moist cool portions of the crown and lichens to the drier portions of the canopy. Up to 1500 species of insects have been found in the irregularly shaped branches of a single stand of old-growth forest (Hammond, 1991). Small mammals such as squirrels and tree voles depend on the large accumulations of organic matter in the crowns of individual old-growth trees for their food and shelter. Certain birds such as the northern spotted owl and the marbled murrelet need old-growth canopies for nesting areas and as habitat to rear their young.

No tree lives forever. Within a dying tree, the processes of matter recycling and energy flows, along with the roles of producers, consumers, and decomposers, become apparent (see also Chapter 3). However, when large, old trees die, their extensive root systems enable them to remain upright for 50 to 75 years in the case of Douglas fir, and up to 125 years in the case of western red cedar. Large snags are a distinct feature of old-growth and unmanaged forests. They provide habitat for many nesting birds and mammals such as woodpeckers, flickers, and marten. Many birds prefer the large, old-tree snags that require centuries of tree growth, growth that is not part of the life cycle in the managed forest (see Figure 3–11). Managed forests and young forests simply do not provide the diversity of habitat options available in old-growth forests over 200 years of age (because snag density increases as forests age). A reduction in the number of snags means a reduction in the numbers of hole-nesting birds that assist in balancing insect populations.

When old snags finally fall to the forest floor, they are perhaps even more valuable in ensuring continuation of tomorrow's forests than they were when they were standing. A fallen tree is literally the soil for future generations of forests. Carpenter ants (which eat insect eggs and larvae, and help keep the defoliating spruce budworm in check), bark beetles, wood borers, and mites invade the wood of the fallen tree, using it as a home and contributing to its decay. As the decaying wood gets wet, it acts like a giant sponge, holding water and slowly releasing water and plant nutrients to the forest. The mycorrhizal fungi that assist forest plants in obtaining the elements needed for growth thrive in this medium (see Box 9–3; Ricklefs, 1993, p. 146).

A healthy, living, large tree in an old-growth forest may have 30 to 40 species of mycorrhizal fungi attached to its roots, providing a rich source of nitrogen to the host tree. European researchers found that in intensely managed forests without old-growth decaying wood in the soil, only three to five mycorrhizal fungi were present. Thus, old-growth decaying wood is an extremely important legacy to future forests. The presence of mycorrhizal fungi enables a fallen tree to become a nurse tree, an ideal place for the germination and growth of the next generation of trees and many other plants. Nurse trees are important for the regeneration of conifers in river or riparian zones; the elevated surface of a fallen tree may offer one of the few places safe from flooding where a young tree may germinate and grow (Hammond, 1991). If a nurse tree is broken or spread apart, its moisture-holding and nutrient-cycling functions are greatly reduced or destroyed.

Over the 250 years it would take a 400-year-old Douglas fir to decompose (or 400 years if the fir were 800 years old), the fallen tree performs other functions, including maintaining the stability of forest slopes. When large living trees or snags fall across the slope, a natural retaining wall is created, holding organic material and soil behind the terrace or barrier, and preventing slumps and erosion. Also, the water and nutrients collected here provide rich conditions for plant growth. Animal habitat is created along the sides of (and often underneath) large fallen trees, providing travel routes for squirrels, mice, and rabbits, and their predators.

Trees that fall into or across streams act as dams or breakwaters to slow the erosive forces of the stream and add diversity to the stream channel. They provide a variety of spawning and rearing habitats for fish, including sites for aquatic plants and insects needed for fish food.

Photo 9–3
Various stages of growth are evident in this part of a B.C. rain forest.

Fallen trees that land in streams help stabilize sediment transport by acting as barriers for movement of debris. Over time as the tree decomposes, parts of it may move down the stream and become embedded in a bank, contributing to the formation of a new forest community.

CARBON STORAGE

As we saw in Chapter 3, maintaining the 0.03 percent carbon dioxide in the Earth's atmosphere is crucial to maintaining life as we know it. Old-growth forests are the planet's most important land-based storage systems for carbon. Huge amounts of carbon are stored in branches, trunks, roots, fallen trees, and soil organic matter, and the older the tree, the more it can store. For instance, a 450-year-old Douglas fir forest stores more than double the total amount of carbon stored in a 60-year-old Douglas fir forest. There is evidence that the northern hemisphere's land-based carbon sinks are more important for carbon storage than the oceans. This means that Canada's old-growth forests—particularly the vast northern boreal forests—are immensely important in regulating Earth's carbon dioxide levels (Jardine, 1994). With this realization, the Canadian government has initiated two programs, the "Shelterbelt Enhancement Program" and a "Feasibility Assessment of Afforestation for Carbon Sequestration," to determine the potential for the large-scale creation of new forests on lands currently used for agriculture.

KEYS TO DIVERSITY

The keys to diversity in an old-growth forest are its multiple canopy layers, canopy gaps, and understory patchiness. Just as the diverse canopy provides varied microhabitats for plants and animals, so too do the canopy gaps and understory patches. The Pacific yew, for example, needs the shade, cool temperatures, and high humidity provided by old-growth canopies (see Enviro-Focus 3). Specialized mammals such as the northern flying squirrel and fisher require the massive spreading branches, deformed tops, hollow trees, and open spaces of the old-growth canopy for their homes. Ungulates such as mule deer, moose, and Roosevelt elk use the dense forest patches for winter shelter and browse the vegetation within the canopy gaps. In fact, of the native mammals on Vancouver Island, 85 percent reproduce in ancient forests (Hammond, 1991).

In contrast, the closed canopies of young forests result in forest simplicity. When a forest is growing, most of the forest's energy is diverted to growing trees, resulting in a uniform, closed canopy that blocks almost all usable light from reaching the ground. Understory trees, shrubs, and herbs are very limited or absent, and populations and species diversity of many life forms decrease during this period of forest development. Natural diversity rebounds when the forest canopy begins to open due to individual tree mortality in the early old-growth phase.

Biological Diversity

Humans have identified only a small number of the organisms in an old-growth forest, and we know even less about plant, animal, fungi, and bacteria functions than we do about the number of species. We do know, however, that the biomass (total amount of living matter) in a Canadian northwest old-growth rain forest is three to eight times as great as the biomass of a tropical rain forest. And if plants, animals, and microorganisms above, at, and below soil level are included, a northwest old-growth temperate rain forest may be more biologically diverse than a tropical rain forest (Kelly & Braasch, 1988).

Our ignorance of the functioning of Canadian forests is immense, in part because of this diversity. In contrast to tropical rain forests, where the diversity and functioning depend on about 500 different tree species and less than 10 mycorrhizal fungi, the boreal forest contains about a dozen tree species and may depend on as many as 5000 species of mycorrhizal fungi to sustain their integrity. We know that each of these mycorrhizal fungi has a specific role in growth and development of boreal forests, but for the most part we do not know the nature of these functions (Hammond, 1991).

The ongoing existence of forests depends on the biological diversity or richness of old-growth forests. If measured by numbers of species only, recently disturbed forests have the greatest variety of species. The species that colonize openings created by fire, wind, or the falling of a single tree are called *aggressive generalists,* as these species grow in harsh environments. Since disturbances are created constantly by nature and by humans, neither these species nor their habitats are at risk of loss.

In contrast, old-growth species are specialist species that require the ancient forests to survive. Humans cannot create old growth: "[L]ichens which fix nitrogen, small mammals which move mycorrhizal fungi around, and predator insects which eat foliage-consuming insects are some of the gifts of old growth forests which benefit all phases of a forest" (Hammond, 1991, p. 32). While some species require old growth, others may need it only for certain periods in their lives. Grizzly bears need the hiding cover provided by the large crowns of old-growth trees, and they utilize the berry supplies of old-growth canopy gaps. Salmon, required by grizzlies, thrive on the quality of water and habitat supplied only by old-growth forests. In turn, the old trees benefit mutualistically from nutrients provided by salmon carcasses, left partially eaten at their bases, by grizzly and other bears. Scientists are just beginning to understand the extensive land–sea food web and its importance in forest and fisheries management.

In unmanaged forests, every organism is different genetically, allowing each organism to adapt to its particular environment today and to meet the uncertainties in tomorrow's environments. Maintaining genetic diversity is a natural process in a healthy forest. Just as the complexity and diversity of each individual's contribution within human societies enables societies to continue and thrive, so genetic diversity enables forests to survive. This is why managed forests, such as the genetically identical clones of Douglas fir or white spruce planted to replace logged, old-growth forests, are not adequate substitutes for the complexity and diversity found in old-growth, natural (unmanaged) forests. Because of their variety of species and long lives, old-growth forests are a vital storehouse of genetic material, from soil microorganisms to giant trees.

Old-growth forests exhibit extremely large ecosystem diversity in their multilayered canopies and patchy undergrowth. Yet the old-growth soil with its thousands of organisms is the most biologically rich part of the old-growth forest. If this biological legacy of old-growth forests were to be lost, it could lead eventually to critical losses in all forests. Without old-growth forests, essential parts required to maintain forests through time are lost. This is perhaps most evident in water and forest relationships. The canopies of standing giant trees catch snow while their root systems and the organic, rich soil filter, purify, and slowly release the water. Fallen trees serve as reservoirs for water and as buffers for stream flows so that most parts of the forest get the water they need, seldom too much or too little. Old growth is a necessary part of any forest ecosystem, a necessary phase in the life of any forest.

FOCAL POINT: CLAYOQUOT SOUND

In British Columbia, and internationally, the Clayoquot (pronounced *klak-wot*) Sound area became an important focal point in the conflict over forest values and issues of environmental and economic sustainability. Located on the west coast of Vancouver Island, British Columbia (see Figure 9–6), the Clayoquot Sound area is rich in forest values, including spectacular natural, unmanaged forests; a long history of First Nations' settlement; world-class scenic and tourism resources; and major commercial fishery and timber industries.

Over 60 percent of the Clayoquot Sound area is assessed as commercially productive forest land and about 20 percent of that has been clear-cut logged. In 1995, about 70 percent (or 90 400 hectares) of the remaining unlogged, merchantable forest was primary or old-growth forest (Clayoquot Sound Scientific Panel, 1995).

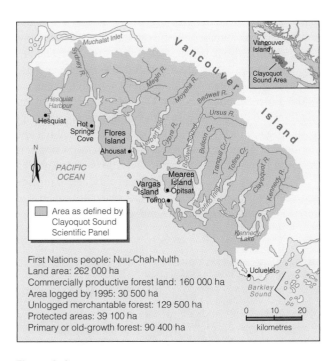

Figure 9–6

Clayoquot Sound area: location and selected statistics, 1995

SOURCE: *A Vision and Its Context: Global Context for Forest Practices in Clayoquot Sound,* Clayoquot Sound Scientific Panel, 1995, Report 4 of the Scientific Panel for Sustainable Forest Practices in Clayoquot Sound, Victoria: B.C. Ministry of Forests, pp. i, 7. Copyright © 2000 Province of British Columbia. All rights reserved. Reprinted with permission of the Province of British Columbia.

Photo 9–4

A researcher examines a 320-year-old Douglas fir in Canada's temperate rain forest.

In the summer of 1993, thousands of concerned citizens and environmentalists from British Columbia and from around the world congregated at Clayoquot Sound to protest the British Columbia government's decision to allow major timber companies to clear-cut log more than two-thirds of the Sound's rain forests. In their attempts to stop the clear-cutting, over 850 people were arrested (some jailed and later fined) for defying court orders that outlawed the blockading of logging roads (Curtis, 1996; Morell, 1994). Counter-demonstrations by forest workers and their families underlined the conflict.

To address the controversy, in 1993 the provincial government created a Clayoquot Sound Scientific Panel whose mandate was to help "make forest practices in the Clayoquot not only the best in the province, but the best in the world" (Clayoquot Sound Scientific Panel, 1995, p. 1). Upon seeing the recommendations of the committee, in 1996, the major corporation active in the area negotiated a joint venture, controlled by local First Nations, known as Iisaak Forest Resources. This company is now undertaking logging operations in the region, using criteria associated with sustainable forestry (Hayter, 2000).

In 2000, Clayoquot Sound was designated a United Nations Biosphere Reserve (see Chapter 12). Biosphere reserves are designated when representatives of local, provincial, and national interests can come together to protect the cultural and ecological integrity of a region, while allowing for its sustainable use. In Clayoquot Sound, the significance of old-growth forests as a bank for genetic variability, as a wildlife sanctuary, and as a place of cultural and spiritual importance, among other values, were rationales for the designation.

A large part of the local and international concern about Clayoquot Sound arose because of its significance within the coastal temperate rain forest biome. Although they are found around the world in places such as Chile, Norway, and Tasmania, coastal temperate rain forests are a rare forest type, originally covering less than 0.2 percent of the Earth's land surface. British Columbia forests contain an estimated 18 to 25 percent of the world's coastal temperate rain forests (Kellogg, 1992). About 60 percent of the world's unlogged coastal temperate rain forests and over 95 percent of the unlogged coastal temperate rain forests in the Pacific Northwest occur in British Columbia and Alaska (Clayoquot Sound Scientific Panel, 1995). (The Pacific Northwest is the geographical region including southeast Alaska, British Columbia, Washington, Oregon, Idaho, western Montana, and northern California.)

In North America, the distribution of these globally important, vigorously growing forests is centred on Vancouver Island and "attains its most dramatic expression around Clayoquot Sound" (Clayoquot Sound Scientific Panel, 1995, p. 8). Tall trees are evidence of this "dramatic expression"—for example, the two tallest western red cedars in British Columbia (59.2 and 56.4 metres), the tallest Sitka spruce (95.7 metres), the tallest Douglas fir (82.9 metres), the tallest western hemlock (75.6 metres), and the two tallest yellow cedars (45.4 and 44 metres) are located near Clayoquot Sound (Clayoquot Sound Scientific Panel, 1995). In addition, as scientists have begun to study the complexity of forest ecology, including the canopies of coastal rain forests, they have discovered hundreds of new species that previously were unknown to science (Moffett, 1997).

The controversy over the use and management of forests in Clayoquot Sound reflected the growing international concern for sustainability of forests and their ecosystem-based values. As we shall see in the following sections, awareness of the importance of forests in the global environment has created new obligations for Canada in helping to find solutions to forest issues abroad and at home. This chapter also reviews the state of Canada's forests, identifies the kinds and effects of human uses of and activities on forest resources, and notes the efforts of governments, partnerships, and local groups to resolve outstanding forestry and land use issues.

THE NEED FOR PROTECTION

Sustainable human communities require healthy forests. One way to maintain healthy forests and healthy communities is to apply the principles of ecological responsibility and balanced use (Kimmins, 1992). Ecological responsibility means that all human activities within forest landscapes must be carried out in a way that protects and maintains necessary forest structures and functions. Balanced use means that all forest users, both human and nonhuman, are entitled to a fair and protected land base on which to carry out their various activities. In turn, this means that sufficient natural forest reserves must be protected from all but the gentlest of human uses. Protection of natural areas is required to maintain landscape connections, to maintain a reservoir of species and genetic components, and to provide natural benchmarks so that we can evaluate impacts of forest activities and restore degraded forests.

Photo 9–5a   **Photo 9–5b**   **Photo 9–5c**   **Photo 9–5d**   **Photo 9–5e**

Among the tallest trees in Canada's west coast forests are (from left) the western red cedar, Sitka spruce, Douglas fir, western hemlock, and yellow cedar.

Part of the controversy over logging in the Temagami region of Ontario during the 1990s derived from a concern that not enough of the old-growth white and red pine was being protected. Since 1992, it has been known that less than 1 percent of the old-growth pine forest remains in North America, and less than 1 percent of Ontario's original white pine forest remains. Temagami has the highest concentration of old-growth white and red pine forest ecosystems remaining in the world. It is official Ministry of Environment and Energy policy that old-growth white and red pine forest should be protected. Despite this policy and the knowledge that these are endangered ecosystems, both the Ontario Ministry of Natural Resources (MNR) and the Temagami Comprehensive Planning Council allowed approximately 50 percent of the old-growth pine in Temagami to be logged beginning in 1996. Timber harvests were permitted to supply the needs of people for wood products. Earthroots (1996) estimated that in 1996 each old-growth tree was worth approximately $600 in the sawlog market, but governments in Ontario only received approximately $15 stumpage for each tree (Earthroots, 1996).

In 1997, the Sierra Legal Defence Fund (SLDF) launched a court action against the Ontario Ministry of Natural Resources charging that logging approvals in the Temagami Forest region violated the government's own Crown Forest Sustainability Act. In 1998, decisions by two courts, the Ontario Divisional Court and the Ontario Court of Appeal, agreed with the SLDF (Sierra Legal Defence Fund, 1998a). As a result, the Temagami logging plan was struck down, and the government requested a two-year grace period to amend the plan (Sierra Legal Defence Fund, 1998b).

In the case of the 1400-hectare Owain Lake stand, the third-largest old-growth white pine stand in North America, logging was permitted because the MNR assumed that another area containing old-growth white pine (Lake Obabika) was representative of that ecosystem. This assumption was based on a comparison of surficial geology only. Essentially a claim that the stand had no ecological representation value, this assumption permitted the Temagami Planning Council and MNR to indicate the Owain Lake stand did not need protection. To determine accurately if an area is ecologically representative, however, the standard practice is to describe community types (flora and fauna), providing lists of plant species and plant communities as well. Although the MNR had produced such information in the past for other similar situations, it did not undertake this detailed work for the Owain Lake pine stand (Quinby, 1996).

Independent botanical field studies show that the Owain Lake old-growth pine stand has many ecological features—including 10 plant species, four that are rare regionally and three that are rare locally—that make it different from the other old-growth pine stands (Quinby, 1996). For instance, the rare plant species found at the Owain Lake stand were absent at the Lake Obabika stand. Failure to adequately assess the ecological representation for the Owain Lake stand meant that the value of protecting more than one example of a community type—especially an endangered one—from total loss due to natural disturbances was eliminated.

In effect, the acknowledgment of the importance of all life forms provides impetus to fully protect biological diversity in order to maintain healthy forests and healthy communities. Sustaining biological diversity means protecting the integrity of the forest—the biological diversity of forests—at the species, genetic, and ecosystem levels. Because diversity works in interconnected and interdependent ways, biological diversity must be protected in at least two ways. One way is by limiting the fragmentation of the forest so that the patterns of, and connections between, the various clusters or stands that compose a forest landscape are maintained. Another way requires the protection of the structure and composition of the individual parts that constitute the forest landscape, including the fallen trees, the soil, the old growth, the riparian and upland corridors, and the natural flows of water, nutrients, and other forms of energy throughout the forest (Hammond, 1991; Ness & Cooperrider, 1994; Quinby, 1996). Ultimately, protection requires managers to act on the understanding that each forest stand requires old trees, snags, and fallen trees, that brush or competing vegetation is biologically necessary for forest integrity, and that insects and disease are essential parts of a fully functioning forest.

HUMAN ACTIVITIES AND IMPACTS ON FOREST ENVIRONMENTS

Human activity inevitably affects forest environments. While technologically sophisticated and specialized activities extract great socioeconomic benefits, these same activities exert the greatest impacts on forest ecosystems. Herein lies the dilemma of sustainable forest management.

Canadians depend on their forests for a wide range of services, products, and values. The key to sustaining all of these uses and values—in order that the forest needs of both present and future generations of Canadians may be met—lies in maintaining the health, diversity, and productive capacity of forest ecosystems. This means that forests in Canada and everywhere must be managed so that their sustainability is assured. However, recognition of this challenge is relatively recent, and we continue to see a wide range of problems resulting from practices that did not, or do not, promote sustainability. This section provides a brief historical context of human activities in Canadian forests and illustrates the effects of these activities.

A BRIEF HISTORICAL OVERVIEW OF THE FOREST INDUSTRY

First Nations and European Settlers

Before this country became a nation, many of Canada's Aboriginal peoples derived their basic means of survival—food, clothing, shelter, and tools—from the forests. Forests also were a fundamental dimension in their cultural and spiritual lives. Among the wealthiest nonagricultural (traditional) societies ever known, the First Nations people on the west coast based their lives on the use of the abundant cedar trees and the salmon that spawned in the coastal rain forest streams. These people carved wooden totem poles and masks, highly notable elements of their culture, that represented and expressed the integration of the forces of natural and spiritual realms that surrounded them (Knudtson & Suzuki, 1992). The wealth of the coastal forests enabled these First Nations to develop the potlatch economy where masks and other goods were redistributed according to social status.

Photo 9–6

First Nations people from Canada's west coast used wood to carve artworks in addition to a great number of more utilitarian items.

The European settlers, who challenged the original forest inhabitants for possession of lands, had a different conception of, and relationship with, nature and the forest. From the 17th to the 19th centuries, the vast majority of European immigrants wanted to establish farms. The forests of butternut, oak, white pine, walnut, maple, and black cherry trees, as well as the Aboriginal people, were seen as obstacles, something to be conquered and to be driven back to permit homesteading to proceed. With more powerful technology, Europeans used force, treaties, or both to remove Aboriginal peoples from lands the newcomers wanted to settle. With axes and oxen, European settlers slowly but surely began removing the forest cover.

Timber Exports and Government Ownership

At the beginning of the 19th century, Britain and the United States began to be interested in extracting Canadian forest products for large-scale export (Lower, 1973). The Royal Navy turned to Canada for the white pine wood required for shipbuilding, and the forests in New Brunswick, Quebec, and the Ottawa Valley were cut and exported as square timber to Britain. The dramatic increase in demand soon attracted British businessmen to Canada, where they established thriving commercial interests based on small operators out in the bush where the trees were being cut. "The prevailing ethic was simple. Cut it down and get it out in as great a quantity and as fast as possible" (Swift, 1983, pp. 34, 35).

By 1826, authorities recognized that forest exploitation could bring in revenues, so they issued licences to cut timber in specific areas. People cutting wood on Crown land had to pay a licence fee as well as duty on the wood actually cut. These actions effectively established the principle of government ownership and control over timberlands that continues to be applied today.

Changing Market Demands, Changing Industry

By the mid-1800s, the Canadian forest industry began to change when sawmills were set up to meet changing market demands for rough-sawn chunks of lumber. The capital required for such facilities led to concentration of control as powerful industrial enterprises emerged. In its rush to secure all available supplies of timber, the forest business became more centralized and integrated—the timber barons ran the bush camps, the sawmills, and the plants that produced shingles, laths, and doors.

Following Confederation, from 1867 to 1906, the governments of Quebec and Ontario received millions of dollars through the liquidation of their forests, collecting various rents and fees from the timber industry. For instance, between 1867 and 1899, forest-generated income contributed 28 cents out of every dollar collected by the Ontario treasury (Swift, 1983).

The Shifting Frontier and Conservation Concerns

Beginning in the 1840s, the United States began to surpass Britain as the most important market for Canadian forest products. Railroad construction and city building stimulated increased demand for wood that was met by the spread of logging in Ontario from the Ottawa Valley across the Canadian Shield to Georgian Bay and the Lakehead. These forests helped supply Prairie markets until the British Columbia forest industry hit its stride after 1900. In Quebec, the second half of the 1800s saw exports to the United States steadily gaining ground, while in New Brunswick, the majority of the good pine had been removed from the forests by the 1870s.

Growth in Pulp and Paper and Increased Concentration

Even though the conservation movement in Canada argued for forest conservation, through the decades of the early 1900s the pulp and paper industry was growing, based on the northern expanses of spruce trees too small to be turned into lumber. New pulping technologies (making pulp not from rags but from alternative tree species such as black spruce), new market opportunities that accompanied the growth of literacy, the increased importance of newspapers, and the growth in consumer spending in Canada during the early 1900s created the impetus for a new pulp and paper industry. The newsprint industry grew rapidly after 1920, as did the chemical pulp industry after 1945. In contrast to sawmilling operations, where small entrepreneurs were still numerous, pulp and paper mills were invariably large and capital-intensive, much more under the control of large companies. Moreover, these companies, with the help of favourable government policies, were able to gain large timber concessions that assured access to adequate supplies of pulpwood and the necessary loans. Most of the 20th century was marked by increasing concentration of control over timber harvesting by a small number of large companies. In British Columbia, by the 1940s, 58 firms held about 52 percent of the 4 million acres under timber licence. By 1973, 54 percent of the harvesting rights were held by 10 large companies that were multinational in scope with integrated forest management operations around the world. "In 1975, the four largest firms accounted for 22 percent of timber production, but 75 percent of plywood, 52 percent of market pulp, and 94 percent of newsprint production" (Hayter, 2000, p. 55). In the 1990s, this level of concentration remained.

Sustained-Yield Focus

The turn by government to large-scale companies was a deliberate policy of the 1950s and 1960s. Governments of the day believed that a sustained or steady harvest of timber would lead to socially desirable objectives such as community stability (Byron, 1978), and that large-scale companies were in the most favourable position to meet them (Marchak, 1983). Yet, a steady supply of timber did not account for changing market demand and fluctuating prices, while large companies did not provide more stable employment for workers than smaller companies. Instead, larger companies had significant political influence in setting policy priorities related to payments made to governments, allocation of harvesting rights, and management practices (Williston & Keller, 1997). As the history of labour strikes and continuous shedding of employment indicates, larger companies did not provide a high level of stability for forestry workers and ultimately their actions increased dependency and vulnerability of workers and forestry communities to globalizing forces (Marchak et al., 1999; M'Gonigle, 1997; Burda and M'Gonigle, 1998).

By the end of World War II, no adequate inventory of forest resources existed. Without knowledge of how much timber was left, how fast it was growing or being cut or being depleted by fires and pests, and how well it was being utilized, forest management was almost impossible. Nevertheless, pressure on timber reserves continued to escalate, particularly for housing for veterans returning to civilian life. Wartime overcutting, with even greater postwar demands, meant high-grade wood was becoming harder to find in central and eastern Canada. Logging moved up from the valley bottoms onto the mountainsides, visible evidence of the decline in timber quality and accessibility.

In 1976, a royal commission on forest tenure in British Columbia was led by Dr. Peter Pearse, a resource economist. Pearse called the concentration of corporate control a matter of "urgent public concern" because it meant, in the communities where these corporations operated, that competition in the logging business had been largely eliminated. Large companies did not sell their timber on the open market where it would be available to the most efficient mills, and in some cases, companies even threatened to overwhelm the smaller resident business community (Drushka et al., 1993). In other words, experience with British Columbia's system of land allocation and management has not provided the conditions to ensure that forested landscapes are ecologically, economically, or socially functional and productive.

TIMBER PRODUCTION ACTIVITIES AND THEIR EFFECTS

Historical timber production activities in Canada have shifted from small-scale, low-impact use of forests by small numbers of Aboriginal people to increasingly large-

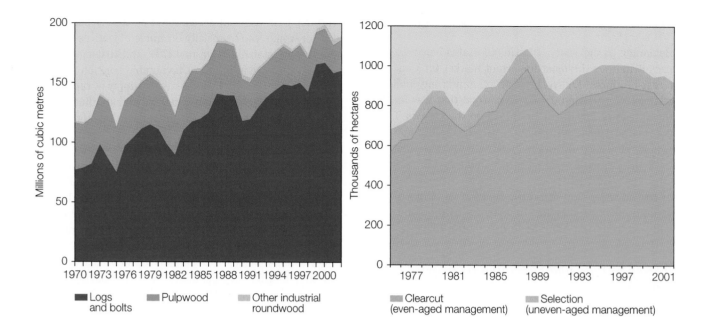

Figure 9–7

Annual volume and area of timber harvest in Canada

NOTES:

- *Roundwood* is the term applied to the major types of products harvested from Canadian forests. Roundwood includes sections of tree stems (with or without bark), logs, bolts (short logs to be sawn for lumber or peeled for veneer), pulpwood, posts, pilings, and other products "still in the round." Industrial roundwood also includes fuelwood for industrial or institutional needs, and firewood for household or recreational needs.
- Regional harvest trends vary substantially from this national picture.

SOURCE: *Compendium of Canadian Forestry Statistics 2004,* Canadian Council of Forest Ministers, 2004, Ottawa. Reprinted with permission of the Canadian Council of Forest Ministers, http://www.ccfm.org

scale, high-impact forest cutting by settlers, speculators, entrepreneurs, and, ultimately, by large corporations on Crown land. Basically, in Canada, forest policy has been equated with industrial policy, and clear-cutting has become the dominant form of harvesting (Figure 9–7). Although the fundamental problem of failure to protect forests for the future was recognized more than a century ago, it is only recently that actions toward sustainability have begun to be developed. Before examining these actions and efforts in more detail, a number of examples of the impacts of timber production on selected forest elements, as well as other effects of human activities on forest ecosystems, are identified.

Habitat, Wildlife, and Life-Support Effects

Timber cutting, especially clear-cutting, can have significant negative effects on habitats for all species, on wildlife, on humans, as well as on ecological life-support systems. From their place in the world's carbon storage system to their reservoir of genetic diversity, old-growth forests across Canada provide a range of immensely important global functions. After so much of the country's old-growth timber has been clear-cut, or committed to being logged, it is only recently that the search for appropriate levels of forest use has begun. Only in recent years, as well, have increased levels of knowledge provided some basic premises on which to make more informed choices about management efforts.

Canada's forest regions are home to more than two-thirds of all species found in Canada, including about 76 percent of our land-dwelling mammals and 60 percent of breeding bird species (Natural Resources Canada, 1996b). Of the roughly 200 000 species that depend on forest habitats, 85 have been identified as species at risk (see Table 9–4). The decline in numbers of many of these forest-dwelling species may be attributed partly to habitat loss due to timber harvesting.

The very rare Queen Charlotte goshawk, for example, was assigned "vulnerable" status in 1995 by COSEWIC—the Committee on the Status of Endangered Wildlife in Canada (see Tables 12–3 and 12–4). Only three nests have been reported on the Queen Charlotte Islands and six have been reported on Vancouver Island. The goshawk

TABLE 9–4

A SAMPLE OF FOREST-DWELLING SPECIES AT RISK IN CANADA (2000–2002)

	Mammals	Birds	Plants	Reptiles & Amphibians

Endangered	American marten *(NFL)*
	Vancouver Island
	  marmot *(BC)*
	Wolverine *(QC, NFL)*
	Woodland caribou *(QC)*
		Acadian flycatcher *(ON)*
		Kirtland's warbler *(ON)*
		Northern spotted owl *(BC)*
		Screech owl sub.
		  macfarlanei (BC)
		Prothonotary warbler *(ON)*
			American ginseng *(ON, QC)*
			Blunt-lobed woodsia *(ON, QC)*
			Cucumber tree *(ON)*
			Deltoid balsamroot *(BC)*
			Drooping trillium *(ON)*
			Few-flowered club rush *(ON)*
			Heart-leaved plantain *(ON)*
			Large whorled pogonia *(ON)*
			Nodding pogonia *(ON)*
			Prairie lupine *(BC)*
			Purple twayblade *(ON)*
			Red mulberry *(ON)*
			Seaside centipede lichen *(BC)*
			Small whorled pogonia *(ON)*
			Spotted Wintergreen *(ON)*
			Wood Poppy *(ON)*
				Blue racer (snake) *(ON)*
				Tailed frog *(BC)*

Threatened	Ermine haidarum
	  subsp. *(BC)*
	Grey fox *(MB, ON)*
	Pallid bat *(BC)*
	Wood bison
	  (AB, BC, NT, YT)
	Woodland caribou
	  (NT, BC, AB, SK,
	  MB, ON, QC, NL)
	  Boreal pop.
	Woodland caribou
	  (BC, AB) Southern
	  Mountain pop.
		Hooded warbler *(ON)*
		Marbled murrelet *(BC)*
		White-headed woodpecker
		  (BC)
		Yellow-breasted chat *(BC)*
		Queen Charlotte
		  goshawk *(BC)*
			Goldenseal *(ON)*
			American chestnut *(ON)*
			Bird's-foot violet *(ON)*
			Blue ash *(ON)*
			Boreal felt lichen *(NB, NS)*
			Brunt's jacob's ladder *(QC)*
			Crooked-stemmed aster *(ON)*
			Deerberry *(ON)*
			Kentucky coffee tree *(ON)*
			Lyall's mariposa lily *(BC)*
			Phantom orchid *(BC)*
			Pocket moss *(BC)*
			Purple sanicle *(BC)*
			Round-leaved greenbrier *(ON)*
			Wild hyacinth *(ON)*
			White wood aster *(ON, QC)*
			White-top aster *(BC)*
			Yellow montaine violet *(BC)*
				Black rat snake *(ON)*
				Blanding's turtle *(NS)*
				Eastern Massasauga
				  rattlesnake *(ON)*
				Jefferson salamander
				  (ON)
				Mountain dusky
				  salamander *(QC)*
				Rocky Mountain northern
				  cricket frog *(ON)*
				Northern leopard
				  frog *(BC)*
				Northern ribbon snake
				  (NS) Atlantic pop.
				Pacific giant
				  salamander *(BC)*
				Tiger salamander *(BC)*
				Western toad *(YT, NT,*
				  BC, AB)

Special concern	Ermine *(BC)*
	Fringed myotis
	  (bat) *(BC)*
	Gaspé shrew
	  (NB, NS, QC)
	Grizzly bear *(AB, BC,*
	  NT, YT, NU)
	Keen's long-eared
	  bat *(BC)*
	Mountain beaver *(BC)*
	Nuttall's cottontail *(BC)*
	Southern flying squirrel
	  (NB, NS, ON, QC)
	Spotted bat *(BC)*
	Wolverine *(AB, BC, MB,*
	  NT, ON, SK, YT, NU)
	Woodland caribou *(AB,*
	  BC, MB, NT, ON, SK)
	Woodland vole *(ON, QC)*
		Bicknell's thrush *(NB,*
		  NS, QC)
		Cerulean warbler *(ON, QC)*
		Flammulated owl *(BC)*
		Lewis's woodpecker *(BC)*
		Louisiana waterthrush
		  (ON, QC)
		Red-headed woodpecker
		  (MB, ON, QC, SK)
		Yellow-breasted chat *(ON)*
			American columbo *(ON)*
			Broad beech fern *(ON, QC)*
			Coastal wood fern *(BC)*
			Crooked-stemmed aster *(ON)*
			Cryptic paw lichen *(BC)*
			Dwarf hackberry *(ON)*
			False rue-anemone *(ON)*
			Green dragon *(ON, QC)*
			Hop tree *(ON, QC)*
			Oldgrowth specklebelly
			  lichen *(BC)*
			Seaside bone lichen *(BC)*
			Shumard oak *(ON)*
				Coeur d'Alène
				  salamander *(BC)*
				Five-lined skink *(ON)*
				Mountain dusky
				  salamander *(QC)*
				Northern red-legged
				  frog *(BC)*
				Pacific giant
				  salamander *(BC)*
				Wood turtle *(NB, NS,*
				  ON, QC)

SOURCE: *The State of Canada's Forests 2002–2003: Looking Ahead,* Natural Resources Canada, Canadian Forest Service, 2004, Ottawa: Author.

prefers to nest in large, contiguous stands of mature forests with closed canopy cover. The dense vegetation provides ideal breeding habitat, with cover and protection from predators. In addition to threats from poaching and pesticides, the Queen Charlotte goshawk has lost suitable nesting trees and foraging habitat due to timber harvesting (Natural Resources Canada, 1996b). In recent years, the situation has not improved; the goshawk is now declared to be "threatened," that is, "a species likely to become endangered if limiting factors are not reversed" (COSEWIC, 2003).

As trees disappear through logging, so do other living things. The woodland caribou, for instance, lives in mature forests, where it feeds on lichen. While some populations are not at risk, others—for example, in the boreal habitat in northwestern Ontario and parts of British Columbia—are declining because of logging, road and pipeline construction, and hunting (Natural Resources Canada, 1995). The Queen Charlotte woodland caribou population is extinct and the Atlantic-Gaspésie population in Quebec is in the endangered category (COSEWIC, 2003). Migratory birds such as the Cape May warbler, whose summer range includes Ontario's boreal forests, are affected by the loss of habitat caused by logging. As recent research findings have shown, loss of neotropical migratory birds from forests has important implications for control of insect outbreaks (see Box 9–2). In other parts of the world, similar wild species losses occur. Australia has lost 18 of its native mammals in the recent past, many from Western Australia, where 95 percent of the natural woodland has been cleared (Pimm, 1996).

Other habitat losses occur as soil and debris from clear-cut slopes erode into rivers and streams, suffocating aquatic species. Rainfall can exacerbate erosion problems: in the Clayoquot Sound area in January 1996, for instance, heavy rain generated over 250 landslides, mostly in clear-cut logged areas (Rainforest Action Network, 1996). Companies responsible for other damage to British Columbia streams and fish habitat allegedly caused by logging roads and operations have been difficult to prosecute (Friends of Clayoquot Sound, 1996).

Such effects are found worldwide; for instance, in the northern Philippines, logging roads were found to have caused erosion more than 200 times greater than on undisturbed sites (Ryan, 1990). Millions of dollars of damage to fisheries and coral reefs caused by logging-induced sedimentation has been documented. Near Palawan in the Philippines, fisheries in Bacuit Bay were depleted after logging began on surrounding hillsides. Sediment entering the bay smothered up to half of the living coral that supported the fishery, depriving local people of their source of protein (Ryan, 1990). In the mid-1960s, harvesting $14 million worth of timber from the watershed of the South Fork of the Salmon River in central Idaho caused an estimated $100 million damage to the

river's Chinook salmon fishery. That industry still has not recovered.

A variety of plants are threatened by timber operations, too. The white wood aster, a perennial herb that is known to grow in only five sites in Quebec and three in the Carolinian forests of Ontario, has declined and was designated by COSEWIC as a threatened species in 1995 (see Table 9–4). This aster is threatened largely because of habitat loss resulting from human and natural changes in the forest ecosystem (Natural Resources Canada, 1996a). In Alberta's boreal forests alone, about 100 plant species are known to grow only in the boreal forest and about half of these are rare already (Acharya, 1995).

In the Pacific Northwest, existing wild mushroom harvests likely would be destroyed should logging occur in the old-growth forests where the fungi grow (see Box 9–3). The case of the "worthless" Pacific yew tree (see Enviro-Focus 3) also highlights the need for enhanced awareness of the importance and protection of what may appear to be insignificant species in our forests. If people are to gain a full range of benefits from forest products and avoid the irrecoverable losses of biodiversity and ecological function that can occur under certain harvesting regimes, the precautionary principle (see Chapter 1) is applicable.

Degradation and Deforestation of Tropical Forests

Although tropical rain forests occupy only about 6 percent of the world's land area near the equator in Latin America, Africa, and Asia, they provide habitat for at least one-half of the Earth's plant and animal species, and homes and livelihoods for about 100 million people. The global importance of tropical forests relates to their ecological functions as well as to the products produced from their plants—nuts, fruits, chocolate, gums, coffee, wood, rubber, pesticides, fibres, and dyes. In particular, people with high blood pressure, Parkinson's or Hodgkin's disease, multiple sclerosis, and leukemia have been treated with drugs made from tropical plants. Scientists believe many more plants with medicinal values remain to be discovered.

In the developing parts of the world, chronic poverty plays a role in the massive destruction and degradation of tropical forests. Reasons for the deforestation of tropical forests vary regionally: in Latin America, rain forests often are converted to pasture land; in sub-Saharan Africa, they are cleared to meet increasing demands for farmland and firewood; and in Southeast Asia, rain forests provide hardwood products for export to industrialized countries. In many cases, forests that are converted to agricultural or pasture land produce food products for export to industrialized countries like Canada.

Neotropical migratory birds such as the rufous hummingbird, the western tanager, or the several warbler species that inhabit the Pacific Northwest and boreal forests spend only one-third of their lives in Canada or the United States. The remainder of their time is spent in tropical regions such as Mexico, the Caribbean, and Central and South America (or areas in between their summer and winter homes). Of the nearly 200 species of neotropical birds, some travel thousands of miles twice a year.

These neotropical birds, as well as resident birds, such as juncos, thrushes, pine siskins, chickadees, nuthatches, and some woodpeckers, play an important role in helping to keep forest trees healthy. While not all birds eat insects, the majority do: as many as 300 insects per day during the summer months. A pair of breeding evening grosbeaks, for example, can devour 25 000 to 50 000 caterpillars just in the period it takes them to raise their brood.

Recently, United States Forest Service biologists in the Pacific Northwest learned that 35 species of birds, including 24 neotropical migrants, feed on the western spruce budworm and the Douglas fir tussock moth, which are the two most destructive defoliating insects there. When the caterpillars (or larvae) of the western spruce budworm and Douglas fir tussock moth eat the needles of these trees, they weaken them, making them vulnerable to attack by other insects or to diseases that ultimately kill the trees. Severe outbreaks of these insects can result in the loss of millions of trees over thousands of hectares of forested land.

One of the responses to these insect outbreaks has been to spray insecticides to kill the budworm and tussock moth. However, this partially effective action kills beneficial insects and spiders along with the pests. Killing the beneficial insects affects forest birds and animals that depend on all types of invertebrates for food.

Knowing that both neotropical and resident birds feed on budworm and tussock moth larvae, the Forest Service scientists planned an experiment to see just how effective the birds were in eating budworms off fir trees. The scientists enclosed entire, 30-foot-tall fir trees in cages of PVC pipe and plastic netting to prevent birds from feeding on the caterpillars on those trees. The results showed that six times more budworms survived on the caged trees than on uncaged trees. This means that the birds were eating five of every six caterpillars, and were making a tremendous difference to the health of the forests.

Unfortunately, the populations of dozens of species of neotropical migratory birds that spend their summers in the forests of the Pacific Northwest and the boreal forests of northern Canada have declined during the past two decades. Biologists believe these declines are the result of changes in nesting or wintering habitat (or both). Activities such as logging and land clearing result in less available food, fewer nesting sites, and less protection from natural enemies for these birds. Logging and land clearing also affect streamside (or riparian) habitats, which are favourite nesting or foraging sites for many birds species. Building roads and allowing grazing near streams also have caused problems for birds.

If healthy forests are to be maintained, responsible managers need to realize that essential forest-dwelling birds need suitable nesting and foraging areas. This implies that managers need to plan for forest diversity with a variety of tree species of different ages, including standing snags and fallen logs. One international effort to help stem the decline in populations of neotropical birds is the Partners in Flight–Aves de las Americas Neotropical Migratory Bird Conservation Program.

Initiated in 1990, the Partners in Flight program brings public and private partners, including Canadian, American, Mexican, Caribbean, and Latin American conservation organizations, together in efforts to conserve migratory songbirds. The basic principles of their strategy, known as the Flight Plan, include promotion of conservation of habitats in breeding, migration, and wintering areas when it should be done—before species and ecosystems become endangered. Conservation based on sound science, such as that described here, is also an important principle of the Flight Plan. Saving our birds is indeed a way to save our forests.

SOURCES: *Partners in Flight—Canada,* Environment Canada, 1996, http://www.ec.gc.ca/cws-scf/canbird/pif/p_title.htm; *Neotropical Migratory Bird Conservation,* National Fish and Wildlife Foundation, (n.d.), http://www.nfwf.org/nfwfne.htm; *National Fish and Wildlife Foundation,* National Fish and Wildlife Foundation, 1996, http://www.bev.net/education/SeaWorld/conservation/nfw.html; *Save Our Birds—Save Our Forests,* T. R. Torgersen & A. S. Torgersen, 1995, Portland, OR: United States Department of Agriculture, Forest Service; *Partners in Flight* home page, United States Geological Survey, (n.d.), http://www.pwrc.nbs.gov/pif/

Cash crops displace forests in many parts of the world. The African nation of Ethiopia, for example, once 40 percent forest covered, now uses about 60 percent of its land to grow cotton and a further 22 percent to grow sugar cane for export to pay its heavy debt load. Only about 3 percent remains in forests (Bequette, 1994; Mungall & McLaren, 1990). In other countries, trees are cut to provide land for large plantations on which to grow banana, tea, and coffee crops, mostly destined for the more developed countries. Cultivation of crops such as cacao, rubber, marijuana, and cocaine-yielding coca also contributes to destruction of tropical forests.

The high rate of removal of trees in the Brazilian rain forest in the 1980s—the catalyst for much of the global concern regarding loss of tropical rain forests—was based mostly on changes occurring in the state of Rondonia as the World Bank–financed highway (BR 364) was constructed. Undertaken without environmental impact

BOX 9-3
FORESTS AND WILD MUSHROOMS

The Pacific Northwest has been recognized for a long time for its rich mycota—such as mushrooms, truffles, conks, puffballs, and cup fungi—that are found in conjunction with old-growth forests. The richness of the mycota is related directly to the region's expansive forest communities, diversity of tree species, and weather patterns. Under these conditions, forest tree species form beneficial root symbioses (mycorrhizae) with specialized fungi that obtain their carbohydrate nutrition via the roots of the host trees and, in return, provide access to nutrients to much larger tree roots (Brady & Weil, 1999). The wood of both live and dead trees and the abundance of other organic debris on the forest floor provide rich resources for numerous fungi. In turn, these fungi are important contributors to the dynamic functioning of forest ecosystems, providing food for organisms from microbes to mammals and contributing to the overall resiliency and diversity of forest ecosystems. "If we are to succeed at managing forest ecosystems in their entirety, we must integrate the biological and functional diversity of forest fungi into future management plans" (Pilz & Molina, 1996, p. 1).

Pushing through the mossy floor surrounding 100- to 200-year-old lodgepole pine, Douglas fir, and western hemlock, the firm, white to pale brown pine mushrooms (*Tricholoma magnivelare*) are found along the coast and interior mountain ranges of western North America. In Canada, pine mushrooms also are found in the eastern Maritimes and throughout the boreal forests of Manitoba and Saskatchewan. Under ideal conditions pine mushrooms can grow to more than five pounds each!

Of the more than 30 species of wild edible mushrooms that are harvested in British Columbia, the pine mushroom is one of the few that is picked commercially on a large scale. Pickers who harvest top-grade pine mushrooms can command anywhere from $8 to $300 per pound for these fungi.

The pine fungi are closely related to the very popular Japanese *Tricholoma matsutake* mushroom, which, because of the medicinal qualities it is believed to contain, has been an integral part of the Japanese diet for centuries. In the 1980s, however, Japanese *matsutake* crops declined as a result of "the ailing health of its red pine forests, the symbiotic partner of the matsutake" (Welland, 1997, p. 66). (Of note are similar declines in European fungi that have been linked to pollution such as acid precipitation and sulphur dioxide [Amaranthus & Pilz, 1996].) At the same time as supplies of the pine mushroom declined, the demand for it grew; by the early 1990s, Canada was the world's fourth leading exporter of pine mushrooms to Japan.

The pine mushrooms are "fussy fungi, needing an undisturbed forest floor to propagate; any pawing or raking of the moss covering can set future growth back for years" (Welland, 1997, p. 64). Extensive harvesting of several species of edible forest mushrooms during the past decade has heightened

Photo 9–7
Pine mushroom of the Pacific Northwest.

awareness and concern for forest fungi on the part of the public and resource managers. While economic benefits are sizable—such as the $40 million that wild mushroom picking contributed to the economies of Oregon, Washington, and Idaho in 1992, or the almost $4 million earned by harvesters in 1994 in the Nass Valley in northern British Columbia—so are the concerns about the potential overharvesting of wild mushrooms and the impact of mushroom harvesting on forest ecosystems. Clear-cut logging, scheduled to occur in the Nass Valley soon, also is a concern.

In the United States, legislation and permit systems to regulate the commercial harvest of wild mushrooms on public lands have been instituted. As well, then-president Clinton's 1993 Forest Ecosystem Management Assessment Team report identified the need to study and protect fungi throughout Pacific Northwest forests and to integrate these forest fungi into ecosystem management plans.

In British Columbia, regulation of the mushroom harvesting industry has been resisted. The Nisga'a First Nation's recent land claims settlement may signal a change, however. As they now own all forest resources within some 2000 square kilometres of the Nass Valley, the Nisga'a patrolled the valley in 1996 to contain the harvesters within main camping centres. As well, through their company, Nass Valley Resources, Inc., they hope to develop a food processing facility in the area for mushrooms, fish, and other local foods. Obviously, none of these actions will result in the opportunity to study or protect the pine mushroom should logging occur. Thus, an adaptive management process is recommended in which, as new information is acquired, it may be incorporated continuously into revised management plans.

SOURCES: "Productivity and Sustainable Harvest of Wild Mushrooms," M. Amaranthus & D. Pilz, 1996, in D. Pilz & R. Molina (Eds.), *Managing Forest Ecosystems to Conserve Fungus Diversity and Sustain Wild Mushroom Harvests* (pp. 42–61), Portland, OR: United States Department of Agriculture, Forest Service, Pacific Northwest Research Station: General Technical Report PNW-GTR-371; *The Nature and Property of Soils*, N. C. Brady & R. R. Weil, 1999, Englewood Cliffs, NJ: Prentice-Hall; "Introduction," D. Pilz & R. Molina (Eds.), 1996, *Managing Forest Ecosystems to Conserve Fungus Diversity and Sustain Wild Mushroom Harvests* (pp. 1–4), Portland, OR: United States Department of Agriculture, Forest Service, Pacific Northwest Research Station: General Technical Report PNW-GTR-371; "Mushroom Madness," F. Welland, 1997, *Canadian Geographic, 117*(1), 62–68.

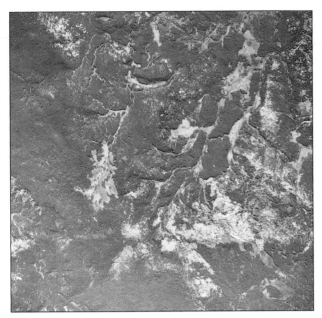

Photo 9–8

The central Swan Hills area, 35 kilometres north of Whitecourt, Alberta, September 27, 1949 (9.1 × 9.1 km). North is to the top of the image, as it is in the two accompanying photos.

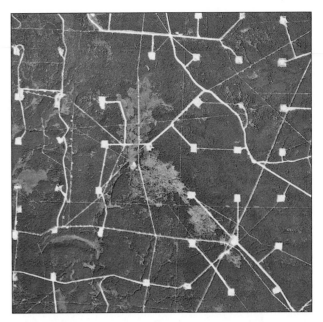

Photo 9–9

The central Swan Hills area, July 2, 1964 (7.2 × 7.2 km). The grey lines are roads, pipelines, and transmission lines; small squares are well sites; large patches are clear-cuts.

Photo 9–10

The central Swan Hills area, October 7, 1991 (13.6 × 13.6 km). In just 42 years a wilderness area is transformed into an intensely fragmented, ecologically dysfunctional landscape.

assessments, the highway was used to serve to funnel urban migrants (who were looking for ways to improve their quality of life) into the forests. Forest lands were cleared to build homes and grow crops. Much of the forest was burned and soils were depleted of nutrients. The result was small-scale farming and livestock ranching operations that destroyed large tracts of forest. In other regions, such as Central America, about two-thirds of tropical rain forests have been lost to livestock raising (Bequette, 1994). Underlying the pattern of forest loss are conditions of grinding poverty and an international debt load that impel countries and individuals to seek out livelihoods in the short term that provide some form of income. Conditions for loan repayment are typically not set by the countries themselves, whose nationals are forced to liquidate natural assets to provide some measure of immediate return. Unless these issues (poverty and debt) are addressed, it is unlikely that the pattern of natural capital liquidation is likely to be reversed.

Firewood consumption adds to the problem of tree loss in developing countries. Half of the world's wood is used as fuel for cooking and heating, and yet in 1985, about one in three persons on Earth was not able to get enough fuelwood to meet basic needs or was forced to meet those needs by consuming wood faster than it was being replenished. Wood scarcity creates considerable hardship for poor families as buying fuelwood or charcoal can take 40 percent of a family's small income. Women, especially, often have to walk long distances to gather fuel. If they cannot get enough fuelwood, poor families often burn dried animal dung and crop residues. This means that these natural fertilizers are not returned to the soil, and, as cropland productivity declines, hunger and malnutrition increase. In addition, cutting trees for fuelwood often results in increased erosion and may even lead to permanent lowering of water tables.

Photo 9–11
People in developing nations rely heavily on fuelwood cutting, which threatens renewability of forests and associated resources.

The burden of debt carried by many tropical and developing countries has placed pressures on them to over-exploit their resources. Many of the countries with the highest rates of deforestation in the 1980s were also the largest debtors at that time. Furthermore, structural adjustment programs, designed by the World Bank and the International Monetary Fund to address the indebtedness of many developing countries, encouraged countries to engage in export-led economic growth. The result was a massive liquidation of natural resources. For example, Ghana lost 75 percent of its forest area during its economic adjustment in the 1980s (Elliott, 1999). Furthermore, austerity measures by government to provide funds for debt servicing often required reductions in programs and agencies that addressed environmental management, health provision, and poverty alleviation. If these countries were able to reduce their payments toward debt relief, they may be able to reduce pressures on environmental resources and improve the infrastructure to protect both their natural and social systems. However, decisions about how debts may be reduced have frequently been made by the World Bank, the International Monetary Fund, and/or individual

lending countries that have not had environmental protection, poverty alleviation, or improved social welfare in these countries as primary criteria when deciding how best to restructure a country's economy. These examples illustrate how environmental sustainability cannot be achieved without action on economic and social fronts as well.

The timber industry is another factor in deforestation. Since 1950, the consumption of tropical hardwoods has increased by a factor of 15, satisfying markets in Japan (which consumes about 60 percent of annual tropical timber production), the United States, and Great Britain. Developing countries often sell off their forests to pay their debts and to create jobs. Ninety percent covered by virgin rain forest, Surinam (on the northeast coast of South America) granted large timber concessions to an Indonesian logging company in 1994. This decision was based on the fact that children in Surinam were dying of hunger and cutting forests would provide much-needed jobs and an improved ability to purchase required foodstuffs (Bequette, 1994). The Bahinemos of Papua New Guinea, whose ancestral home is the Hunstein forest, faced similar pressures to gain cash for the education of their children and to acquire Western goods. However, foreign logging activities, no matter how careful, would cause the loss of large birds such as cassowaries and impact dramatically on the Bahinemo culture and landscape (Bakker, 1994).

Effects of such decisions can be extreme: countries that once exported timber—Nigeria and the Philippines, for example—now import it. Other Southeast Asian and Central American nations have almost totally deforested their lands: Haiti has lost 98 percent of its original forest cover, the Philippines 97 percent, and Madagascar 84 percent. In Asia, in Thailand, the Philippines, Papua New Guinea, Myanmar, Malaysia, Indonesia, India, Fiji, and Cambodia, log production declined by over 20 percent from 1995 to 1999 (International Tropical Timber Organization, 1999), and depletion there will encourage more cutting in Latin America and Africa. African mahogany often is logged selectively, perhaps one tree per hectare, but in order to reach logging sites, trails as much as 100 kilometres long must be cut, opening the way for land-hungry farmers and subsequent degradation of forest lands (Bequette, 1994). Most cleared tropical forests are not replanted, in part because timber companies are held responsible for few of the costs of environmental degradation, and in part because of corrupt, deeply entangled local practices (Dauvergne, 1997).

POLLUTION

Pollution is a widespread problem for Canada's forests. Pulping operations release toxic organochlorines, such as dioxins, and other substances into water bodies.

Sometimes disastrous effects on aquatic life occur, and often damaging effects on the health of First Nations people are observed. Examples abound: in Ontario, long-term discharge of mill effluent threatened local people with Minamata disease (mercury poisoning) from eating contaminated fish from the Wabigoon River system. In the case of the largest bleached kraft mill in the world, Mitsubishi's Alberta-Pacific mill located on the Athabasca River, chlorine is used to bleach 1500 metric tonnes of pulp per day (Acharya, 1995). Completed in 1993, this mill releases dioxins and other toxins that require long-term monitoring to prevent cumulative toxic damage. Throughout British Columbia, the health of indigenous people has been threatened for many years by the eating of fish and shellfish contaminated by pulp mill effluents; however, increasing understanding of the causes of these health problems and improvements in technology have reduced contemporary releases of such substances.

Some of the world's most serious cases of air pollution causing damage to forests are found in Russia. Russian mining and smelting operations produce among the world's highest sulphur emissions, which, in turn, have killed off entire Russian forests and damaged forests outside of Russia. During the 1950s and 1960s, huge smelters were built on the mineral-rich Kola Peninsula bordering Norway and Finland. These plants use obsolete technology, and produce nearly as many tonnes of sulphur emissions as they do metals and minerals. In 1988, for instance, the Severo-Nickel smelter on the Kola Peninsula produced 243 000 tonnes of nickel, copper, and cobalt, and 212 000 tonnes of sulphur dioxide (Acharya, 1995).

These sulphur emissions have killed forests outright on 40 000 to 100 000 hectares and damaged 3 million hectares of forest in Russia alone. The damage extends into Norway and Finland; both countries regarded the pollution as a major foreign policy problem and offered to help upgrade the Russian plants. The offer was declined, and the plants continued to operate using imported ores that contained even higher sulphur content than the exhausted local ores.

Elsewhere in Russia, such as at Krasnoyarsk in the south-central part of the country, fluorine emissions from aluminum smelters are identified as the prime cause of 3.2 million hectares of dead and dying forests. Southern Sweden and Norway are impacted by the long-range transport of acid precipitation and other forms of air pollution from this area. Deposition of sulphur dioxide and nitrous oxides has affected soils so radically that some of them may not be able to support another generation of trees. Radioactive pollution is another problem affecting forests in parts of the former Soviet Union, including about 1 million hectares around Chernobyl. As fires in contaminated forests release dust and smoke, radioactive pollution drifts on air currents for hundreds of kilometres (Acharya, 1995).

Photo 9–12
Acid rain has had a detrimental effect on the world's forests.

In Germany, the word *Waldsterben* (meaning "forest death") describes the forest decline that has affected more than one-third of the country's forests, as well as up to 25 percent of the fir and spruce trees in Switzerland and more than 1.2 million acres of forest in Poland and the former Czechoslovakia (Hammond, 1991). The primary cause of this forest decline is the long-range transport of airborne pollutants (LRTAP) from combustion of fossil fuels in vehicles and from many industrial processes, including thermal power generation and smelting.

The acid deposition that results from this LRTAP causes direct damage and loss of foliage and needles to trees by leaching out nutrients, causing thinning and per-haps fatal damage to the tree's crown. As well, indirect damage through soil acidification (in which heavy metals are released in forms taken up by tree roots) occurs. In Canada, acid precipitation contributes to, and may well be the major cause of, forest decline in southwestern British Columbia as well as the loss of sugar maple trees in Quebec. Slash burning and the pulp manufacturing process also contribute to the problem of forest decline in Canada.

Slowing the rate at which boreal, temperate, and tropical forests are being destroyed requires changes in government policies as well as in consumer and corporate awareness, attitudes, and behaviour. Evidence that some of these changes are occurring is presented in the "Responses to Environmental Impacts and Change" section of this chapter.

TOURISM AND RECREATION

Natural forest landscapes are an important factor in the health of the tourism industry across Canada. Valued at billions of dollars per year, forest-related tourism occurs in national and provincial parks and in provincial forests through activities such as hunting and fishing, wilderness hiking, river rafting and canoe tripping, skiing, and camping and picnicking. While these activities disperse revenues in the communities where the tourism occurs to some degree, many goods and services used by tourists are supplied by major urban centres. This is a concern, as is the seasonal nature of tourism and the often low wages paid to tourism workers. Nevertheless, the overall scale of tourist activities occurring in, or related to, forest areas is substantial.

Wildlife viewing—from waterways, forest trails, and viewing stations—is a multimillion-dollar, "nonconsumptive" forest use and activity in Canada. Provided that forest-based tourism occurs in forest areas that are protected and used wisely, long-term annual benefits can accrue to communities. Old-growth forests are prime wildlife habitat and provide superb scenic backdrops for visitors who have come to the forest for their interest in birds, mammals, fish, and plants. Even the trees themselves are visitor attractions. Cutting old-growth forests for their timber values potentially precludes the economic benefits that come from sustainable, nontimber values such as wildlife viewing, even though logging has been important for facilitating access for recreationalists. Even though it is difficult to calculate a precise dollar value, trees are worth a lot to the tourism industry.

As a growth industry in Canada, tourism and recreational services will continue to depend on healthy, nat-

ural amenities, including unlogged forests. In 1995, foresters and environmentalists clashed over plans to clear-cut much of the 700 square kilometres of the Algoma Highlands, 150 kilometres northeast of Sault Ste. Marie, Ontario. A 60-square-kilometre conservation reserve was excluded from harvesting, as was a large (temporary) area around Megasin Lake in the centre of the highlands. Wilderness outfitters in that area argued that logging was incompatible with their successful remote tourism operations, and forced the provincial government to undertake its first ever environmental assessment of a forest management plan. The environmental assessment, however, focused only on the impacts of logging on tourism, and did not consider other environmental impacts (Leahy, 1995).

A proposed "view-scape cut," where narrow bands of trees would be left to ring the major lakes and streams used by tourists, may hide clear-cut slopes beyond the tree ring, but does not provide the healthy, natural forest amenities that are part of the expected quality experience for wilderness visitors. Outfitters worry that this lack of a quality experience could lead to a decrease in repeat customers that, in the longer term, could cause a reversal in the economic viability of their operations.

Yet, possibilities for complementary relationships between the forest industry and ecotourism exist, as the experience of Fiddlehead farm near Powell River, British Columbia, reveals. Fiddlehead Farm is surrounded by extensive tree farm licences and could have been threatened by logging slated for the adjacent Giovanno Valley. Visitors to the 32 hectare wilderness hostel arrive by boat and then walk in through two kilometres of second-growth, mixed-conifer forest, including Douglas fir, hemlock, and cedar. The forest company's logging plan anticipated clear-cuts in the valley adjacent to the farm and a mainline logging road across the walk-in access route (McPhedran, 1997). Indeed, since 1997 more roads have been built and more cut blocks have been created, including a logged site that extends to the shore of Giovanno Lake. Nevertheless, despite the farm owner's initial fears regarding noise and visual impact of summer logging, both the farm and logging co-exist.

The manager of the forestry company, which annually puts about $30 million into Powell River's economy, indicated that because the industry has to exist somewhere, the option of leaving the valley untouched was not negotiable. However, the manner in which the logging was conducted was negotiable. Fiddlehead Farm was most concerned with trail protection from the farm to Giovanno Lake and the forest company agreed not to clear-cut in a way that would compromise this trail and to log between October and March when the farm was closed. Fiddlehead Farm, the former wilderness hostel, is now in a clear cut. However, Weyerhaeuser and the Powell River Parks and Wilderness Society have the approval of government to

Photo 9–13
Canada's forests provide economic returns through recreation and tourism as well as commercial forestry.

relocate the trail to avoid the logged area. The project is slated for 2005–2006, subject to funding. This example illustrates both the popularity of wilderness tourism and its continuous vulnerability to being granted secondary status to other user groups with more secure property rights to forest ecosystems.

Admittedly, recreational activities that access public or private forests themselves impose risks and problems. For instance, human-started fires occur, and demands for improved infrastructure (such as toilets, roads, and parking lots) may arise, especially as numbers escalate. Forest values can be destroyed by recreational activities, most obviously in the case of skiing, snowmobiling, and golfing but also from hiking, cycling, and visiting. Backcountry hiking may introduce exotic species into previously inaccessible regions. Indeed, where tourism becomes a successful industry, the large-scale presence of visitors may reduce the values they came to appreciate. In Squamish, British Columbia, the popularity of the winter bird count coincided with temporary declines in returning birds. Local and provincial officials took strong measures (including designation of a protected area and provision of public information at the site) to ensure the safety of bald eagles who congregate there by the hundreds in the winter. Caution must be exercised in interpreting such activities as "nonconsumptive." In addition, governments and private landowners may face increased liability if natural forest events such as wildfires or falling trees damage equipment or injure or kill visitors. These realities, however, do not diminish the "multiple values" of natural forests.

SOCIOCULTURAL DIMENSIONS

Groups in Canadian society whose interests in the forest are based on long-term (rather than short-term) economic considerations include Aboriginal people, nature-based tourism operators, rural water users, ranchers, trappers, small-business owners in forest-based communities, wilderness users, scientists, artists, educators, and future generations of Canadians. Governments are expected to manage public forests and forest lands for the long-term benefit of all these people. These groups have been challenging governments to change the ways forests are used, to move away from the old volume economy, and to develop a new forest economy, one based on value-added industries and on community, rather than corporate, control of forests.

Indigenous peoples lived as part of the North American forests for centuries before Europeans arrived in this land. Hammond (1993, p. 134) claims that their sovereignty "has never been diminished through conquest, prior discovery, purchase, or fair treaties by the various governments of European descendants which have controlled Canada for the last two centuries." Displacement of indigenous peoples from their traditional homelands by European settlers and the subsequent suppression and sometimes abuse of Aboriginal cultures have diminished the opportunity to learn about sustaining the forests from those people who once were part of them. Throughout Canada, land questions remain largely unresolved. From Alberta to the East Coast, treaties were signed, however, disputes still arise over their interpretation. For most of British Columbia, treaties have not yet been signed. Nevertheless, access to resources has been given to private individuals and companies whose "third-party" interests must be considered as governments and Aboriginal peoples come to the negotiating table.

For Aboriginal peoples, pursuing the land claims process embodies an enigma: Why must something be claimed that was never given up? Even though the land claims process is far from complete, Aboriginal influence over British Columbia's forests is growing. A concern is that logging means not only that these people will be denied timber sale royalties, but also that they will lose traditional hunting and gathering territories. While struggles are ongoing to address Aboriginal title adequately and to settle land questions justly, the timber industry is said to be "swiftly foreclosing on the options that indigenous people will have following any agreement" (Hammond, 1991, p. 135). On the other hand, in recent years, timber companies in British Columbia have been extremely active in negotiating joint ventures with Aboriginal peoples (Hayter, 2000). For example, Iisaak is a company formed as a joint venture between Weyerhaeuser and the Mamook (Aboriginal) Development corporation, with First Nations holding controlling interest. It now has primary rights of access to timber in Clayoquot Sound, and in 2001 it was recognized for its outstanding environmental and social commitment to the forests by the World Wildlife Fund.

In other parts of Canada as well, Aboriginal peoples are active in forestry. In Saskatchewan, Mistik Management is a joint venture between the Meadow Lake Tribal Council, Norsask Forest Products (jointly owned by Aboriginal peoples and sawmill workers), and Millar Western. Mistik has been in operation for over 20 years, gradually increasing the numbers of First Nations peoples it employs and developing practices that are consistent with ecological aspects of sustainability. Given the large proportion of Aboriginal peoples living in northern Saskatchewan (where most of the timber is harvested), in the late 1990s the Government of Saskatchewan declared that any new forest management agreements must be operated as a co-management arrangement between Aboriginal and industry partners. In Labrador and Quebec, in the summer of 1994, the Innu living in Nitassinan were involved in land rights negotiations with Canada and Newfoundland when the Newfoundland

Forest Service constructed new access roads in Forestry Management District 20, despite legislation requiring an approved forestry management plan before any new activity took place (Innu Nation, 1995). While the forests were not exploited actively by large-scale industrial forest operations at that time, the wood deficit forecast for the island of Newfoundland meant that large pulp interests were looking to Nitassinan to fill the gap. Subsequently, a land claims agreement has been reached in principle (Department of Indian Affairs and Northern Development, 1999).

Indigenous people are not the only ones attempting to gain more control over timber operations so that forestry-based lifestyles may continue. In many communities, people are working toward environmentally sustainable, socially acceptable, economically feasible, and community-based forestry practices. Some of these efforts are documented in the section "Local Partnerships and Responses."

HUMAN ACTIVITIES AND ONTARIO'S CAROLINIAN FOREST

Clearing of forest lands for agriculture and grazing, and to accommodate community and urban development, has been ongoing in Canada almost since settlement began. Perhaps nowhere are the effects of these human activities seen more clearly than in the Carolinian forest in Ontario. Located on the fertile plain north of Lake Erie, less than 10 percent of the area remains in forest cover, the remainder having been cleared since the time of the European settlers. More than 95 percent of the Carolinian forest is privately owned, and less than 1 percent is contained in national and provincial parks. The nature of this forest, its importance, and an indication of efforts being undertaken to conserve what remains of it are discussed in Box 9–4.

RESPONSES TO ENVIRONMENTAL IMPACTS AND CHANGE

Many countries that produce and consume forest products are trying to come to grips with the need to balance economic and environmental requirements in order to ensure healthy, vigorous forests that can meet the needs of today as well as tomorrow. Over time there has been a continued focus on balancing economic development and conservation, although the relative emphasis placed on development and conservation has varied. A relatively

early example is the 1983 International Tropical Timber Agreement (ITTA). In this agreement, most attention was focused on promoting economic expansion, increasing timber processing, and diversifying international trade in tropical timber in order to promote industrialization and improve export earnings. The concept of sustainable use and conservation of tropical forests and their genetic resources was included as the final objective of the ITTA.

Much of the recent international dialogue on the future of the world's forests took place during the two-year process of preparing for the 1992 United Nations Conference on Environment and Development (UNCED). The Canadian delegation "played a lead role in structuring the debate and engaging the world forest community in the deliberations" (Forestry Canada, 1993, p. 68). The following sections identify some important international and Canadian initiatives that have been taken in response to the challenges relating to forest sustainability.

INTERNATIONAL INITIATIVES

UNCED Forest Principles

In recognition that environmental considerations were becoming increasingly important in the international forest trade, one of Canada's priorities prior to the Earth Summit (UNCED) was to establish an internationally accepted definition and measurement of sustainable forest development. The thinking was that scientifically based international criteria for sustainable forest development would create a level playing field for competitors in forest products markets and would encourage more producers to practise sustainable forest management (Forestry Canada, 1993).

Initially, Canada was aiming to develop criteria for sustainable forest development within a legally binding international agreement. That did not happen, partly because of the different perspectives that nations hold regarding the social, economic, and environmental importance of forests. For instance, developed nations were concerned about protecting tropical rain forests as storehouses of biodiversity and as greenhouse gas sinks, while developing nations felt their tropical forests could be exploited for their timber as well as for potential farmland and as a free source of fuel (Taylor, 1994). By the end of the meetings, however, consensus was achieved and expressed as the UNCED Forest Principles (see Table 9–5).

Even though they are not legally binding, these Forest Principles represent an important international breakthrough in agreement about what constitutes sound forest management. Attaining a legally binding agreement will be challenging, however, as the Forest Principles document contains contradictory elements. On the one hand, for instance, the document indicates that forests should be managed sustainably to meet the social, economic,

BOX 9-4
CANADA'S CAROLINIAN LIFE ZONE

Canada's Carolinian forest covers the southernmost part of Ontario, stretching from the Rouge River Valley in Toronto to the shore of Lake Erie. The forest represents the northern extreme of the eastern deciduous forest region that covers a vast area south of the Great Lakes. The Carolinian forest supports natural habitats and species found nowhere else in Canada. Sixteen endangered plants and animals are native to the Carolinian region, and at least one-third of our country's rare, threatened, or endangered species depend on its natural habitats. More than half of Canada's bird species are found in the Carolinian forest; so is the highest representation of reptiles in Canada. The Carolinian forest is home to the opossum, North America's only marsupial. The forest area is one of the few places in North America where the American chestnut has not been eradicated by chestnut blight. Other Carolinian tree species include the tulip, Kentucky coffee, black gum, cucumber, sycamore, and sassafras.

Early settlers were attracted to the region's relatively warm climate and highly productive agricultural lands. Today more than 20 percent of Canada's population lives in the Carolinian zone. As a result of extensive urbanization in the Toronto to Windsor corridor, less than 10 percent of the land has any forest cover. In fact, all of the counties within the Carolinian zone have less than 20 percent natural forest cover remaining, while some townships are approaching zero cover. Clearly, southern Ontario's Carolinian forests have been and are influenced greatly by human activities.

Although large tracts of Carolinian forest containing similar habitats and range of species remain in the United States, Canada's Carolinian zone is particularly worthy of conservation. Ontario's small patch is considered crucial because species living near the limit of their range have unique characteristics—such as exceptional hardiness—that can be passed on to and strengthen those species over their entire range (see the section in Chapter 3 entitled "Tolerance Ranges of Species").

Protecting Carolinian habitat is especially challenging because most forest remaining in Ontario is fragmented. A variety of complex issues, such as understanding the problem of natural succession, must be considered when managing fragmented natural areas. As a forest matures, new species take over from older ones in a cycle of replacement. In large wilderness areas, most plant species normally survive because occasional natural disturbances, such as fires, storms, and insect infestations, merely slow or reduce the natural succession of the area. While an isolated forest fire may be healthy to a large forest, the same size fire could be devastating to a forest remnant. Conversely, if the remnant is left alone, succession by new species may replace existing ones. The possibility of Carolinian species being replaced by more northerly species is a significant threat in Ontario's Carolinian forests.

One of the best remaining examples of mature Carolinian forest in Canada is the 265-hectare Backus Woods, located a few kilometres from Lake Erie. The site was owned by the Backus family, operators of a flour mill and sawmill beginning in 1798. Backus Woods remained largely intact because the family

Photo 9-14
Cucumber tree.

recognized the value of preserving their forest to protect the watershed for the creek that powered their mills. Backus Woods presently is owned by the Long Point Region Conservation Authority and managed by an advisory committee. In the Backus Woods management plan, half the forest is designated as a natural zone, where no human interference is allowed. The remaining half is a conservation zone where action may be taken to stop natural succession if Carolinian species are threatened.

Both conservationists and forest managers agree that Carolinian forests must be further protected and expanded in Ontario. By expanding Carolinian forests, genetic exchange between adjacent forest remnants can be encouraged. Genetic exchange is important to ensure that plants can build up the tolerance needed to respond to environmental changes such as global warming. Organizations such as the World Wildlife Fund and Long Point Region Conservation Authority are working toward enlarging the woods by buying adjacent properties and encouraging voluntary stewardship initiatives.

Stewardship initiatives in Canada's Carolinian zone began in 1983 with the Carolinian Canada project. The initiative was conceived by the Natural Heritage League (a network of 38 private organizations and public agencies linked by mutual interest and involvement in the identification, protection, and management of Ontario's natural heritage), the Nature Conservancy of Canada, and the World Wildlife Fund.

(continued)

BOX 9–4

(CONTINUED)

Photo 9–15
Tulip tree.

Between 1984 and 1994, 38 priority unprotected Carolinian forest sites were identified, two-thirds (9800 hectares) of which were privately owned. A landowner contact program was established to target these landowners and encourage their willingness to enter into stewardship agreements. Voluntary agreements were then negotiated by the league's Natural Heritage Stewardship Program based at the University of Guelph. By 1988, 347 verbal agreements had been negotiated in Carolinian Canada sites, and as of 1996, more than 1000 landowners had been contacted by members of the Natural Heritage Stewardship Program. These efforts resulted in private landowners agreeing to conserve over 6000 hectares of land and another 800 hectares acquired by conservation groups.

In 1996, Carolinian Canada developed a conservation strategy for a regional approach to environmental education and conservation. It relies on its member organizations to undertake initiatives to protect specific areas and species.

SOURCES: *What is Carolinian Canada?* Federation of Ontario Naturalists, (n.d.), (brochure); "The Enchanted Woodland," P. Gorrie, 1994, *Canadian Geographic, 114*(2), 32–42; *The State of Canada's Forests 1993: Forests, a Global Resource,* Natural Resources Canada, Canadian Forest Service, 1994; "Natural Heritage Protection: Voluntary Stewardship or Planning Control?" by M. Van Patter & S. Hilts, 1990, *Plan Canada, 30*(5), 20–28; *What is the big picture?* www.carolinian.org/ConservationPrograms.htm

ecological, cultural, and spiritual human needs of present and future generations. On the other hand, the document underscores the sovereign rights of nations to exploit their forests in ways that continue to endanger the world's forests. Whether Canada's efforts to change international policy will be successful remains to be seen.

In the absence of a binding agreement, Canada began to implement these Forest Principles through action plans established under the national forest strategy (discussed more fully in the section "Canadian Policy, Practice, and Partnerships," below).

Agenda 21

The 1992 Earth Summit produced a second document of relevance to forests, *Agenda 21.* The massive (but not legally binding) Agenda 21 document set an international agenda for development and the environment in the 21st century, focusing particularly on the needs of developing nations. Agenda 21 recognized the impacts human activities have had on forests: "Forests worldwide have been and are being threatened by uncontrolled degradation and conversion to other types of lands uses ... and environmentally harmful mismanagement including ... unsustainable commercial logging ... and the impacts of loss and degradation of forests are in the form of soil erosion, loss of biological

diversity, damage to wildlife habitats and degradation of watershed areas" (cited in Taylor, 1994, p. 104).

While Agenda 21 called for national action and international cooperation to achieve sustainable development, it exhibited serious weaknesses. For instance, the chapter on forestry contained no recommended policy for sustainable forest management. However, Canada viewed Agenda 21 as a basis for cooperation with, and assistance to, developing nations. Canada's Official Development Assistance Program, managed primarily by the Canadian International Development Research Centre (IDRC), was chosen to implement Agenda 21. Established in 1970, IDRC—which already was one of the world's largest donors of development assistance in forestry—provided increased funding for forest research programs.

UNCED Conventions and Other Responses

Global conventions on biodiversity and climate change also were signed by heads of state at the Earth Summit. The Convention on Biological Diversity, a legally binding agreement, commits Canada to prepare and adhere to a national biodiversity strategy. Part of this commitment is to ensure that a representative sample of Canada's forests is protected, research and education on biodiversity are supported, and Canadian forest management does not

TABLE 9–5
UNCED FOREST PRINCIPLES

The Non-Legally Binding Authoritative Statement of Principles for a Global Consensus on the Management, Conservation, and Sustainable Development of All Types of Forests

- Establish national guidelines and scientifically based international criteria for the conservation, management, and sustainable development of forests.

- Perceive forests as integrated ecosystems with a whole range of diverse values (such as timber, culture, wildlife, and soil conservation).

- Promote public participation in decision making, and, in particular, ensure the participation of women and Aboriginal peoples.

- Develop the skills, education, knowledge, and institutions needed to support forest conservation and sustainable development.

- Strengthen international cooperation and assistance for forests in developing countries.

- Identify and deal with pressures placed on forest ecosystems from outside the forest sector.

- Develop policies to ensure the conservation and sustainable development of forests.

- Encourage fair international trade in forest products.

SOURCE: Adapted from *The State of Canada's Forests 1992: Third Report to Parliament,* Forestry Canada, 1993, p. 69.

affect biodiversity adversely. The Canadian Biodiversity Strategy was completed in 1996 and is to be implemented by federal, provincial, and territorial governments in association with other stakeholders (Environment Canada, 2000). The Convention on Climate Change, which seeks to stabilize greenhouse gases in the atmosphere to prevent additional threats to climate and to forests, requires Canada to adopt national policies and measures on climate change, limit emissions of greenhouse gases, and report regularly on progress in maintaining emissions levels at early 1990 levels (Forestry Canada, 1993).

While the Earth Summit efforts regarding sustainable forestry have been criticized, international efforts continue to promote sustainable management of forests. In 1995, for example, the United Nations Food and Agriculture Organization's Committee on Forestry confirmed the need for a holistic approach to forests that balances their environmental and developmental functions. In particular, the committee highlighted the need to develop and apply criteria and indicators for sustainable management of all types of forests. Another United Nations agency, the Commission on Sustainable Development, is the body mandated to review and pro-

mote implementation of UNCED's decisions in the field of forests.

Canada has chosen to honour its Earth Summit forestry commitments through involvement in two initiatives aimed at defining criteria and indicators for sustainable forest management. These two initiatives are known as the Montreal Process (for boreal and temperate forests) and the Helsinki Process (for forests in Europe). Both are working to establish internationally accepted principles and standards of forest management. Also as part of its recognized responsibility to assist other nations in the sustainable management of their forests, Canada has instituted a number of model forests. Discussion follows regarding these and other Canadian efforts to contribute to global understanding of the role of forests in sustaining planetary health.

United Nations Forum on Forests and the Montreal Process

The United Nations Forum on Forests (UNFF) was established in October 2000 with the main objective of facilitating and promoting implementation of the proposals for action of the Intergovernmental Panel and Forum on Forests (IPF/IFF). UNFF was also charged with considering how to develop a legal framework for all types of forests and devising approaches for financial and technology transfer to support sustainable forest management. In 2001, countries participating in the UNFF (of which Canada is one) developed a structure for the sessions to follow. It was decided that each would include a multistakeholder dialogue and discussion on progress toward sustainable forest management under specific themes. In 2002, UNFF focused on issues related to conservation and protection, deforestation and forest degradation, poverty, food security, and access to water. In 2003, UNFF dealt with the economic aspects of forests, forest health and productivity, and maintaining forest cover to meet present and future needs. The 2004 session addressed scientific and traditional knowledge; social and cultural aspects of forests; and monitoring, including setting appropriate criteria and indicators of sustainable forest management. In 2005, the UNFF plans to review the effectiveness of its work and consider future actions.

In addition, in an initiative entitled the Montreal Process Reports on Criteria and Indicators, 12 countries (Argentina, Australia, Canada, Chile, China, Japan, Korea, Mexico, New Zealand, Russian Federation, the United States of America, and Uruguay) representing 90 percent of all temperate and boreal forests (and 45 percent of all trade in forest products) have been working together to develop and implement criteria and indicators for the conservation and sustainable management of temperate and boreal forests. This science-based initiative is aimed at preparing a common reporting framework to exchange information, identify trends in forest use, and cooperate and enhance national capacity to measure and report on forest management practices. Its first report was released at the 12th World Forestry Congress in 2003; its current efforts are to increase reporting capacity, consistency, cooperation, and effectiveness for policymakers around the world.

CANADIAN POLICY, PRACTICE, AND PARTNERSHIPS

As the predominant natural resource department of Canada, Natural Resources Canada (NRC) is mandated to promote sustainable development and responsible use of Canada's forests and other resources. One of five NRC sectors, the Canadian Forest Service (CFS) provides leading-edge forest science and expertise. The CFS also brought stakeholders together to develop the National Forest Strategy, the Canadian and International Model Forest programs, and criteria and indicators for sustainable forest management. As well, the CFS co-funds research partnerships with other institutes and councils.

The following information about Canada's National Forest Strategy, Model Forest Program, and development of criteria and indicators of sustainable forest management highlights important elements of current forestry policy, practice, and partnerships in Canada. One might observe that these efforts at the federal level are mainly directed toward research and negotiations of international treaties and protocols. The provincial governments have the primary *regulatory* responsibility for forest management. Therefore, it is provincial governments that set the rules for allocating land for timber harvesting, for determining the rates paid by private industry for timber, and for determining the required management and stewardship practices for companies to retain their rights of harvest. As provincial concerns grew about managing forest resources effectively for Canadians, the provinces and territories also initiated new approaches to forest management, expanded protected areas, enacted tougher environmental regulations, and invested in forest renewal.

Canada's National Forest Strategy

After a year of discussion with provincial and territorial governments and organizations representing the interests of naturalists, wildlife, First Nations, foresters, labour, private forest landowners, academics, and forest industries, the Canadian Council of Forest Ministers (CCFM) (1992a) released a national forest strategy document. This commitment to sustainable forest management was ratified by the signing of the first Canada Forest Accord on March 4, 1992. Together, these documents represented broad new directions for forest management in Canada.

Photo 9–16
Road construction through rugged terrain often increases the risk of landslide.

Nine strategic priorities were identified in the strategy (see Table 9–6); they were meant to ensure that in addition to protecting ecological integrity and biodiversity, Canada's approach to sustainable forest management included a range of timber and nontimber values. A key goal of the forest strategy was to develop forestry practices that respected a range of values while maintaining the health of forest ecosystems. One of the key commitments Canada made was to implement a national network of model forests where sustainable development principles could be tested and applied (Forestry Canada, 1993).

Canada's Model Forest Program

The Model Forest Program began in 1991 with the selection of 10 working-scale forests between 100 000 and 2 500 000 hectares in size. Since this time, changes have been made in the composition of the Model Forest Network. The network now includes 11 forests: McGregor MF (British Columbia), Foothills MF (Alberta), Prince Albert MF (Saskatchewan), Manitoba MF (Manitoba), Lake Abitibi MF (Ontario), Eastern Ontario MF (Ontario), Waswanipi Cree MF (Quebec), Bas-Saint-Laurent MF (Quebec), Nova Forest Alliance MF (Nova Scotia), Fundy MF (New Brunswick), and Western Newfoundland MF (Newfoundland). Each model forest is managed by a partnership of organizations and interested individuals that determine, on a consensus basis, particular management objectives for their forest. Recent developments include the setting of local level indicators, the expansion of involvement from Aboriginal groups, and the implementation of programs targeted at private woodlot owners (Canadian Model Forest Network, 2003).

Each model forest has a different number of partners, including provincial governments, forest industries, First Nations, recreational users, community organizations, private landowners, government agencies, environmental and conservation organizations, and academics. The idea behind the partnerships is that full discussion of different viewpoints (as well as any conflicts or tradeoffs) in the earliest stage of forest management planning should lead to an understanding of the important dimensions of the forest and its uses, as well as agreed-on solutions to problems (Forestry Canada, 1993).

The lands involved in model forest projects include national parks, private lands, and First Nations lands, as well as the predominant provincial lands. In each model forest, sustainability initiatives reflect the issues relevant to each region. For example, wildlife concerns have been addressed in the Western Newfoundland Model Forest, where forest management programs focused on the conservation and protection of the pine marten's mature forest habitat. In the Foothills, Manitoba, and Lake Abitibi Model Forests, the fate of the woodland caribou has been an important concern. Deer monitoring programs (to

determine their preferred winter habitat) are under way in the Fundy Model Forest. In Prince Albert, Saskatchewan, research has been undertaken on how to involve local residents in determining social criteria and indicators of sustainability for their communities. In addition, the Model Forest program is tackling the broad issue of how prepared forest-based communities are to adapt to changing global conditions such as climate change.

At the Earth Summit, then prime minister Brian Mulroney announced Canada's intention to expand the concept of model forests to the global level by seeking international partners. With a $10 million commitment from Canada, Mexico became the first international partner (with three sites) in 1993, and Russia became the second in 1994. Malaysia and the United States (with three sites) joined the model forest network in 1995. Other participating countries now include Argentina, China, Japan, and Vietnam. Canada's contribution to forest management through the Model Forest Program offers promise for building strong international partnerships for sustainable management of the world's forests.

Criteria and Indicators of Sustainable Forest Management

Each of Canada's National Forest Strategy, Agenda 21, and UNCED Forest Principles documents recognized the need to formulate scientifically based, internationally accepted criteria and indicators of sustainable forest management. If criteria and indicators are in place, it should be possible to monitor progress toward the goals established in the National Forest Strategy (as well as in international agreements).

In 1993, officials and scientists from provincial, territorial, and federal governments, academics, and representatives from First Nations, industry, and ENGOs contributed to the Canadian Council of Forest Ministers report on criteria and indicators for sustainable forest management in Canada (Canadian Council of Forest Ministers, 1995). The four ecological and two socioeconomic criteria that were identified provide a common understanding of what is meant by sustainable forest management in Canada (see Table 9–7). Obtaining data for these criteria is not straightforward, and by the year 2000 only the socioeconomic indicators had been measured (Canadian Council of Forest Ministers, 2000). Potentially, however, these criteria and indicators can help identify gains made in implementing sustainable forestry (Aplet et al., 1993; Maser, 1994) and reveal elements of the forest ecosystem that must be sustained or enhanced, thus contributing to improved information for decision makers and the public (Canadian Council of Forest Ministers, 1995).

As criteria and indicators are implemented in assessing sustainable forest management in Canada, and experience informs our understanding of which elements

TABLE 9–7

ECOLOGICAL AND SOCIOECONOMIC CRITERIA, INDICATORS, AND CRITICAL ELEMENTS OF SUSTAINABLE FOREST MANAGEMENT IN CANADA

ECOLOGICAL

Conservation of biological diversity

Biological diversity is conserved by maintaining the variability of living organisms and the complexes of which they are part.

- Ecosystem diversity is conserved if the variety and landscape-level patterns of communities and ecosystems that naturally occur on the defined forest area are maintained through time.
- Species diversity is conserved if all native species found on the defined forest area prosper through time.
- Genetic diversity is conserved if the variation of genes within species is maintained.

Maintenance and enhancement of forest ecosystem condition and productivity

Forest ecosystem condition and productivity are conserved if the health, vitality, and rates of biological production are maintained.

- Forest health is conserved if biotic (including anthropogenic) and abiotic disturbances and stresses maintain both ecosystem processes and ecosystem conditions within a range of natural variability.
- Ecosystem resilience is conserved if ecosystem processes and the range of ecosystem conditions allow ecosystems to persist, absorb change, and recover from disturbances.
- Ecosystem productivity is conserved if ecosystem conditions are capable of supporting all naturally occurring species.

Conservation of soil and water resources

Soil and water resources and physical environments are conserved if the quantity and quality of soil and water within forest ecosystems are maintained.

- Physical environments are conserved if the permanent loss of forest area to other uses of factors is minimized, and if rare physical environments are protected.
- Soil resources are conserved if the ability of soils to sustain forest productivity is maintained within characteristic ranges of variation.
- Water resources are conserved if water quality and quantity are maintained.

Forest ecosystem contributions to global ecological cycles

Forest conditions and management activities contribute to the health of global ecological cycles. This contribution is maintained if

- The processes that are responsible for recycling water, carbon, nitrogen, and other life-sustaining elements are maintained.
- Utilization and rejuvenation are balanced and sustained.
- Forests are protected from sustained deforestation or conversion to other uses.

SOCIOECONOMIC

Multiple benefits to society

Forests provide a sustained flow of benefits for current and future generations if multiple goods and services are provided over the long term. Multiple benefits are maintained if

- Extraction rates are within the long-term productive capacity of the resource base.
- Resource businesses exist within a fair and competitive investment and operating climate.
- Forests provide a mix of market and nonmarket goods and services.

Accepting society's responsibility for sustainable development

Fair, equitable, effective, and just resource management choices are in the best interests of present and future generations, including those of particular cultural and/or socioeconomic communities. To achieve these kinds of management choices,

- Aboriginal and treaty rights should be respected in sustainable forest management.
- All forest stakeholders and Aboriginal communities need to participate and cooperate in achieving sustainable forest management.
- All members of society have an obligation and responsibility to understand forest sustainability issues and the positions of others on forest issues.

SOURCE: "Demanding Good Wood," M. von Mirbach, 1997, *Alternatives, 23*(3), 10–17. Reprinted courtesy of Alternatives Journal: Environmental Thought, Policy and Actions.

of this approach are effective, it is likely that there will be a further evolution in both the criteria and indicators and in approaches toward managing Canada's forests as social and ecological systems.

Local Partnerships and Responses

Slow to react at first, over the past decade or so, the forest industry has become increasingly conscious of the need to respond to environmental and social imperatives. A more systemic indication of this trend is the growing number of forest firms that are seeking ISO 14 000 registration and various forms of eco-certification (see Box 9–5). In most cases, these attempts involve various kinds of formal and informal partnerships with local community groups and with ENGOs.

Although often considered implacable foes, partnerships between ENGOs and forest firms have become a discernible trend. A noteworthy example is provided by the accord between the World Wildlife Fund and Tembec Inc., an innovative Quebec-based corporation that also has extensive interests in southeast British Columbia, to promote sustainable forestry practices (World Wildlife Fund, 2001). In its British Columbia operations, centred at Crestbrook, for example, Tembec has obtained ISO 14 001 certification at two mills; developed ecosystem land use plans; shifted away from sole reliance on clear-cutting while reducing the size of remaining clear-cuts; practised more intensive silviculture; and participated in various cooperative efforts with local institutions.

Additionally, First Nations and private forest companies have formed joint business ventures to undertake

---

BOX 9–5
SETTING STANDARDS FOR SUSTAINABLE FOREST MANAGEMENT

As public concerns about forests and forest practices have grown, and environmental groups have had increasing public recognition, there has been a corresponding growth in industry support for independent assessments of forest operations to ensure continued or enhanced market access. The objective is to provide purchasers with a form of guarantee that the forest products they buy were managed according to sustainable forest management principles.

The Forest Stewardship Council (FSC), an international nongovernmental organization, was established in 1993 with the support of the World Wildlife Fund. Members of the council include representatives of environmental groups, indigenous peoples, certification organizations, and other nongovernmental groups from 25 countries. The council's goal is to provide consumers with information about forest products and their sources through certification.

There are important similarities between the objectives and approaches of the Canadian Standards Association (CSA) and the FSC. For instance, both organizations promote better forest management and require third-party audits. However, there are also important differences in their processes. The FSC focuses on product labelling and tracking of forest products to their origin (the "chain of custody"). Wood that is guaranteed to have come from environmentally well-managed forests is sometimes called "certified wood" (Polson, 1996). The CSA's registration program does not involve tracing products but assesses a company's ability to manage in an environmentally sound manner and includes performance indicators tailored to specific sites. The CSA process is not an eco-labelling program because the standards do not apply to consumer products at the retail level.

Another difference is that the FSC's certification is based on a series of principles that they developed independently to apply to all forest types. The CSA's Sustainable Forest Management system is based on both the Canadian criteria and the ISO 14 000 system, which deal with the quality of the management systems, not the quality of the products themselves. CSA's Technical Committee on Sustainable Forest Management has met with FSC representatives to discuss the potential for aligning the two processes.

In March 2000, two small woodlot operators became the first two forests in western Canada to have been certified under the FSC certification system. In 2001, Mistik Management, a joint venture in Saskatchewan, was granted certification. As of April 2001, in Canada, about 37 percent of Canada's 119 million hectares of managed forest land was certified under one or more of the four main certification systems: the Canadian Standards Association (CSA), Forest Stewardship Council (FSC), Organization for International Standardization (ISO), and the Sustainable Forestry Initiative (SFI). This area is almost three times larger than that of the same time 2000 (Natural Resources Canada, 2002). Products acquired from certified forest operations are in high demand, fuelled by the purchasing policies of large retailers such as Home Depot and IKEA that give preference to certified wood (Vasbinder & Brewer, 2001). It is anticipated that the demand for certified products will continue to grow, increasing pressure on producers to identify and implement approaches to certification that are consistent and clear for consumers.

SOURCES: *The State of Canada's Forests 1995–1996: Sustaining Forests at Home and Abroad,* Natural Resources Canada, Canadian Forest Service, 1996; "Cutting with Conscience," S. Polson, 1996, *E Magazine, 7*(3), 42–43; "The Promise of Forest Certification," W. Vasbinder & C. Brewer, 2001, *Encompass, 5*(3), 8–9, 28; "Demanding Good Wood," M. von Mirbach, 1997, *Alternatives, 23*(3), 10–17.

---

sustainable forestry that employ Aboriginal peoples. In Saskatchewan, for example, forest companies can no longer have access to a forest management agreement unless they are partnered with a First Nation. Mistik Management is a company jointly owned by Millar Western and NorSask Forest Products, the latter of which is owned by the Meadow Lake Tribal Council. Since the late 1980s, Mistik has attempted to conduct sustainable forestry, working with nine community-based advisory boards to guide its operations, ensuring that most of its contractors are owned by First Nations or Métis community members, and engaging in practices and research to assure the long-term sustainability of the resource base.

First Nations have become important participants in forestry across the nation. In British Columbia, efforts have been made to increase First Nations' participation in the forest sector through "interim measures agreements" that have been signed prior to resolution of the land question through treaties. In Alberta, the Aboriginal Apprenticeship Project, announced in September 2002, encourages Aboriginal people to take up careers in many trades, including forestry jobs. In October 2002, the Government of Newfoundland and Labrador and the Labrador Métis First Nation signed a memorandum of understanding to facilitate Métis participation in forest management. The Government of Canada renewed the First Nations Forestry Program from 2003–2008 to provide financial assistance to activities that encourage opportunities for First Nations people to participate in Canada's forest sector.

The many local responses to forest concerns provide a useful counterpoint to the formal partnerships established to develop national or provincial elements of sustainable forestry management. For example, Saskatchewan developed a new policy framework for managing wildland fire and forest insects and diseases following extensive public consultation. A few examples that focus on the need to preserve animal habitats are identified below to demonstrate the value of individual actions and partnerships in achieving viable responses to forest-related issues on the local scale.

Wildlife and Forestry Activity   Grizzly bear populations in Alberta have declined from an estimated historic population of 6000 to about 800 today, and are at risk of extinction if their needs are not integrated into land use planning and hunting quotas. The Eastern Slopes Grizzly Bear Research Project began in 1993 as a partnership between researchers at the University of Calgary and about 30 other conservation groups, resource users, developers such as the cattle and the oil and gas industries, and government agencies. Researchers are attempting to understand grizzly bear habitat as it relates to all human activities. If this understanding can be developed, then it may be possible to design an enduring land use system that would include movement corridors for

the endangered bears and enable them to repopulate their former home ranges in Banff National Park and the Kananaskis Country areas.

There are no regulations requiring the Alberta forest industry to demonstrate concern for grizzly bear habitat. However, an Alberta forest products company, Spray Lakes Sawmills Ltd., became a partner in the Eastern Slopes Grizzly Bear Research Project in 1996, donating $10 000 to support the conservation project ("Grizzly Study Finds Forestry Friend," 1996). The woodlands manager, aware that logging can be made compatible with other forest values, entered the project with the willingness to develop creative ways to change the company's forest management practices. Committed to learn from researchers about habitat and movement corridors that are important for grizzlies in areas they propose to log, the company gained crucial information for making environmentally responsible decisions.

Clearly, being a partner in this research project helped the company identify how to manage the impact of its forestry operations on sensitive wildlife species such as the grizzly bear. In addition, researchers gained an important opportunity to learn how to integrate grizzly bears' needs into the design of future forest harvesting plans for the eastern slopes of the Rocky Mountains. By protecting grizzly bears and their habitat, this partnership also helped maintain habitat for many other species in healthy ecosystems. The Eastern Slopes Grizzly Bear Research Project is an important demonstration of the win–win situation that can evolve through sharing information about cumulative impacts and being willing to change management practices.

White-tailed deer reach the northern limits of their natural habitat in Quebec. Their survival is influenced strongly by availability of winter habitat that provides shelter and food. When establishing their winter habitat, deer seek out mature coniferous stands that provide shelter from the cold and wind. Coniferous stands also facilitate the animals' movements because the trees permit only minor accumulations of snow on the ground. Since twigs constitute the deers' basic winter diet, an abundance of young broad-leaved trees also is necessary. The greater the diversity of food and shelter opportunities within an area, and the shorter the distance deer have to travel during harsh winters, the lower will be their vulnerability to winter weather and predators (Fondation de la faune du Québec, n.d.; Natural Resources Canada, 1995).

To secure this important winter habitat, the forest industry, Quebec government departments, Natural Resources Canada, Wildlife Habitat Canada, and the Fondation de la faune du Québec developed and funded the Deer Yard Program. The program works in cooperation with the forest industry by increasing woodlot owners' awareness of their land's wildlife potential and by providing landowners with technical and financial assistance to plan timber harvesting and tree-planting regimes that are suitable to the habitat needs of white-tailed deer. The more severe the winter, the more local landowner yards become essential to the survival of the species. The Deer Yard Program initiative is helping to ensure the continued sustainability of Quebec's deer population.

Setting Codes of Practice The Forest Practices Code of British Columbia Act (1995) was a major attempt to balance the full range of industrial and nonindustrial values within a framework of sustainable forest management. Achieving such a balance has been difficult, however (Cashore et al., 2001). The government acknowledged public concern for the protection of landscape values and wildlife habitat, and the Forest Practices Code was an attempt to regulate industrial forestry in a way that would satisfy these desires while preserving jobs (B.C. Ministry of Forests, 1997). During the late 1990s, the code regulated all aspects of forestry and rangeland management on Crown lands and some private lands in British Columbia, including all aspects of logging and road construction, silviculture policies, fire protection, and the safeguarding of environmental and recreational values. Unlike most previous forestry-related regulations, which were contractual, the code was statutory, enforceable by law.

Nevertheless, environmental groups and the industry continued to criticize the code based on whether it worked to improve management practices or whether it simply added costs to industry (Hayter, 2000). In 2001, government revised the code, in part to reduce what was considered a bureaucratic burden on industry. Environmental groups were alarmed at what they saw as a weakening of an already too weak and inadequately enforced code. In 2002, the provincial government introduced a new Forest and Range Practices Act. The act came into effect in January 2004 and will be fully operational by 2006, replacing the Forest Practices Code. Government policymakers suggested that the code was highly prescriptive, rigid, and expensive to enforce. In its place, government introduced a results-oriented system in which government sets objectives to enforce forest values, and forest companies are granted greater flexibility to determine how best to meet those objectives. Monitoring and enforcement remain part of the new act (see British Columbia Ministry of Forests, 2003). Environmental groups have argued that the new legislation further weakens government commitment to protection of forest values, while forest industry advocates disagree. Thus, while the Forest Practices Code was considered a radical policy innovation, it was short-lived and ultimately failed to reconcile competing values in British Columbia's forests.

Beyond these initiatives, there is a growing interest in providing conditions for communities to be more directly involved in undertaking and regulating forestry. There are many ways in which local communities might be directly involved in forestry. In Quebec, for example, the government has been testing an approach known as the *Forêt Habitée* ("inhabited forest"). This is a concept of joint forest management that allows diverse users to make management decisions. In Manitoba, a government report released in 2002 outlined ways for government, industry, and First Nations to work together to enhance forest stewardship and to ensure that both scientific and traditional knowledge are considered to help promote a sustainable forest economy. In Toronto, the Urban Forestry and Natural Environment and Horticulture Section of Toronto's Parks and Recreation Division has undertaken several projects related to urban forestry and biodiversity protection that demonstrate the benefits of community stewardship in ecosystem protection. On northern Vancouver Island, the North Island Woodlot Association has promoted a type of grassroots, small-scale forestry that involves local people in forest management. In the late 1990s, the association spearheaded the Comox Valley Community Forest. As of 2004, this is one of 10 places in British Columbia that holds a community forest licence. Today, it is negotiating with First Nations to determine how the community forest can continue while respecting the rights of Aboriginal people in the region. Community forests, in all their diversity, are small but important examples of people working toward the social, economic, and environmental objectives of sustainability.

FUTURE CHALLENGES

In March 1997, the World Resources Institute reported that only 20 percent of the world's major virgin forests remained, mostly in remote or "frontier" areas in Canada's far north, in Russia, and in the Amazonia region of Brazil. Only in these areas are the forests large enough to support indigenous species and to survive indefinitely without human intervention—if protection and responsible forest management are put in place now. The deforestation of tropical forests for agriculture and for fuel and industrial wood continues: the Food and Agriculture Organization (1999) estimates that they are disappearing at a rate of 133 000 km^2/year. At the same time, just 3 percent of virgin temperate forests remain. Combining this knowledge with the fact that in North America we use as much wood, by weight, as all metals, plastics, and cement combined (Black & Guthrie, 1994), it is clear that Canada's forests require careful stewardship. Balancing the demands on Canada's forests to attain sustainability requires that the full range of forest values—ecological, economic, and social—should be integrated into decision making. And, in a time when scarcity of forests is an issue, protecting forest ecosystems is paramount.

One way to protect forests and forest ecosystems is to identify an alternative source of fibre, such as hemp (see Enviro-Focus 9). New drug-free strains of hemp are revitalizing what was once a major source of fibre for paper, textiles, and other composite materials such as fibreboard. Since hemp is an agricultural crop grown on farms, it has significant implications for providing high-quality fibre with lower costs and fewer environmental impacts, potentially reducing the pressure on Canada's remaining old-growth forests. Other nonwood fibres may be poised for a comeback also (Rosmarin, 1997).

International commitments to protection and monitoring have been implemented also, and new initiatives continue to be developed. For instance, the Canadian Sustainable Forestry Coalition is promoting the Canadian Standards Association's (CSA) development of an international system of forestry certification using the ISO 14 000 series of Environmental Management Systems standards. The International Organization for Standardization (ISO) focuses on developing systems standards so that quality products of a particular type (such as pharmaceuticals or chemicals) are known to have similar or identical manufacturing histories. With regard to forestry, the ISO emphasizes whether a logging

ENVIRO-FOCUS 9

Hemp: Fibre of the Future?

Humans have used hemp for fibre for thousands of years; in fact, in about 150 B.C., the Chinese made the world's first paper out of hemp. Until about 1850, hemp was used for making textiles, fishing nets, sails, rope, and the caulking between ship planks. More recently, hemp has been identified as an excellent source of biomass for alternative fuels, as a substitute for petrochemicals in the manufacture of some plastics, and in the manufacture of particle board, fibreboard, and other composites used in the construction industry.

Better known as the marijuana plant, hemp used to be grown throughout the western and central provinces of Canada as a textile crop. By 1937, hemp was extremely profitable as the hemp combine and other new machinery had simplified harvesting and made production more cost-effective. Manufacturers became interested in the byproducts of hemp, including seed oil for paint and lacquer, and "hurds" (the woody inner portion of the stalk) for paper. However, in September 1937, the United States government banned hemp production totally. In spite of the benefits of the plant for industrial uses, and established markets for paper, textiles, and medicine derived from hemp, Canada followed suit, banning production under the Opium and Narcotics Act in 1938. The plant "disappeared" from cultivation for over 55 years.

The U.S. government's reasons for banning such a beneficial plant were purely economic. Timber baron William Randolph Hearst (who controlled large tracts of forested land for pulp and paper), and the multinational

Photo 9–17
Textiles made from hemp have multiple uses.

DuPont (which owned the patents on new sulphate/sulphite processes for making paper out of wood), stood to lose billions of dollars if low-cost hemp became widely used. Across the United States, Hearst used the newspapers he owned to create a new perception of hemp as "the assassin of youth." Hearst's tactics resulted in the criminalization of hemp.

Although it made a brief comeback during World War II, hemp might never have been heard of again, except that a new strain of the plant was engineered in France. This plant contains very low levels of THC (the active ingredient responsible for the marijuana "high"). The new "drug-free" hemp was legalized in many parts of Europe and crops were growing again by 1993. In June 1996, the Canadian government amended the Narcotic Control Act to make it legal to cultivate industrial grades of hemp.

There are enormous implications of switching from wood to hemp fibre. Fibreboard industry representatives claim that anything that can be made out of a tree can be made out of hemp, more cost-effectively and with less negative impact on the environment. Since hemp is an annual crop that can be grown on existing farmland, there is no need to build and reclaim expensive logging roads, and no need to clear-cut vast areas of forest using heavy equipment that damages the interconnected elements of the ecosystem and reduces its biodiversity. Hemp fibres are ideal for producing superior quality paper; long and light-coloured, hemp fibres require less bleaching than wood pulp, resulting in the production of lower levels of organochlorines.

If hemp becomes an internationally significant alternative fibre source, it could reduce harvesting pressure on remaining old-growth forests in Canada and elsewhere. The forest industry could then focus on sustainable forest management, value-added production, and community health and sustainability. The viability of the concept of sustainability could become crystal clear to consumers; it makes much more sense to use plant material that grows in 100 days to build a house that lasts 50 years than it does to use plant material that takes between 200 and 500 years to grow. Hemp may be one of the fibres that helps achieve stewardship (conservation) of Canada's forests for the future.

SOURCES: "Back to the Future: Hemp Returns," S. Black & A. Guthrie, 1994, *Earthkeeper, 4*(3), 18–25; *Hemp and the Marijuana Conspiracy: The Emperor Wears No Clothes,* J. Herer, 1991, Van Nuys, CA: Hemp Publishing; "Harvesting Opportunity," P. Marck, June 23, 1997, *Calgary Herald,* p. C4.

company, for instance, has an adequate forest management planning process in place; the ISO does not focus on their actual on-the-ground performance. (Note, however, that the sustainable forest management system is based on the six Canadian criteria for sustainable forest management outlined in Table 9–7.)

The ISO 14 000 system has been criticized for not requiring a clearly documented and verifiable "chain of custody" (see Box 9–5) back to the forest from which the wood originated. Although participants in this system must define a "designated forest area" where their management system will apply, there is no requirement that the entire output of a specific mill, for instance, originate from that area. That is, there is no clear link between the products sold by a company and its forest management system (von Mirbach, 1997).

Consumer pressure, especially in Europe, has been driving efforts to establish certification standards or methods to identify forest products that have been produced in ways that do not degrade the environment. As a result, "certified wood" is gaining popularity (Box 9–5). Canadian companies and governments have responded. For example, in November 2002, the Ontario Ministry of Natural Resources signed a memorandum of understanding with the Standards Council of Canada (SCC) to recognize each other's requirements for forest certification. Federal model forests of Eastern Ontario and Bas-Saint-Laurent have gained resource manager forest certification in accordance with the principles and criteria of the Forest Stewardship Council. The first boreal certification in Canada was granted to Gordon Cosens Forest, an area of two million hectares located in northern Ontario managed by Tembec. Because Canada exports so much of its wood products, companies and governments must respond to international pressures to provide these products in a manner that is ecologically and socially sustainable.

If Canadians wish to have sustainable forests, sustainable forest industries, and sustainable environments, there is a great deal that we have yet to know and predict about forests and forest ecosystems. We must also gain an understanding of the social effects that will result from changes in land allocation and management strategies. Knowledge building remains a critical part of developing sustainable forests, sustainable forest communities, and sustainable forest management. This is one reason why Canada's old-growth forests, boreal and temperate, must be considered as living laboratories. By considering the forests as laboratories, we can test new systems of harvest, new silvicultural methods, and new ways to gain additional value from timber. We can also try new methods and participatory processes to learn how to respect other cultural and non-timber values of the forests, and how to address concerns over displacement of forest workers as we move toward a value-based forest economy.

Knowledge-building technology, such as satellite imagery and GIS, can be used effectively to augment our understanding of the impacts of human activities on forest systems. For instance, satellite imagery of the world's forests, evaluated by experts around the world, enabled the World Resources Institute to specify what percentage of natural forests remained on the planet. There is a need, also, to provide and expand these technical forms of assistance to developing nations to enable them to determine what constitutes sustainable forest management.

On a different level, knowledge-building processes help facilitate community understanding of forest uses and values that, ultimately, find expression in community-based land use planning decisions. Knowledge of ways to create sustainable forests and forest-based communities is improved, too, when partnership experiences and successes are shared with others. Overall, there is a critical need to combine natural sciences, social sciences, and even intuition and spirituality in our quest for a more sensitive and sustainable relationship between people and forests. Finally, we should recognize that across Canada, there has been a "re-regulation" of forest policy (Hayter, 2000). The last decade has witnessed a remarkable experimentation in forest legislation that is desperately seeking a more balanced use of the forest. As Canadian and international public interests in forests are likely to continue, the need to experiment with new harvest practices and management strategies, driven by a heightened commitment to protecting ecological, social, and cultural integrity, is likely to continue.

Chapter Questions

1. What is falldown? In what ways are the falldown effect and the life cycle in old-growth forests inter-related?

2. Outline the range of impacts that human activities have on forests and associated resources. If you live in a community or area in which forestry activities are important, which types of impacts are most visible? Can you classify impacts as economic, environmental, social, or cultural? Which kinds of impacts are most important? Why?

3. Comment on the following statement: clear-cutting is appropriate and necessary for forest management. In what ways would your comments be different for the statement, "clear-cutting is appropriate and necessary for forest sustainability"?

4. If clear-cutting were banned today, what would be the implications for people who work in the forests and live in forestry communities? What would be the impacts on government revenues? What recommendations might you make to address these impacts?

5. What cultural and social values do you hold in relation to forests? Of these, what values do you think are consistent with the aims of environmental, social, cultural, or economic sustainability?

6. If all the world's remaining tropical forests were to be destroyed, in what ways might your life change? What difference could the loss of all old-growth forests in Canada have on your life and on the lives of your descendants?

7. Discuss the value of the Canadian Council of Forest Ministers criteria and critical elements (outlined in Table 9–7) to achieve sustainability of Canada's forests. Are there additional elements that you would add to the list?

Acharya, A. (1995, May–June). Plundering the boreal forests. *World Watch,* 21–29.

Aplet, G. H., Johnson, N., Olson, J. T., & Sample, A. V. (Eds.). (1993). *Defining sustainable forestry.* Washington, DC: Island Press.

Bakker, E. (1994). Return to Hunstein forest. *National Geographic, 185*(2), 40–63.

Bequette, F. (1994, November). Greenwatch: Red alert for the Earth's green belt. *The Unesco Courier,* 41–43.

Black, S., & Guthrie, A. (1994). Back to the future: Hemp returns. *Earthkeeper, 4*(3), 18–25.

British Columbia Ministry of Forests. (1997). *BC's new Forest Practices Code: A living process.* FPC 1. www.for.gov.bc.ca/pab,publctns/fpcliv/process.htm#begin

British Columbia Ministry of Forests. (2003). *British Columbia's forests and their management.* Victoria: Author.

Burda, C., Gale, F., and M'Gonigle, M. (1998). Eco-forestry versus the state(us) quo. *BC Studies, 199:* 45–72.

Byron, R. N. (1978). Community stability and forest policy in British Columbia. *Canadian Journal of Forestry Resarch, 8,* 61–66.

Canada's Forest Network. (2000). *The importance of Canada's forests.* http://www.forest.ca/details.php.3

Canadian Council of Forest Ministers. (1992a). *Canada Forest Accord.* Ottawa: Author.

Canadian Council of Forest Ministers. (1992b). *Sustainable forests: A Canadian commitment.* Ottawa: Author.

Canadian Council of Forest Ministers. (1995). *Defining sustainable forest management: A Canadian approach to criteria and indicators.* Ottawa: Author.

Canadian Council of Forest Ministers. (1997). *Compendium of Canadian forestry statistics, 1996.* Ottawa: Author.

Canadian Council of Forest Ministers. (2000). *Criteria and indicators of sustainable forest management in Canada. National status, 2000.* http://www.ccfm.org/pi/4

Canadian Forest Service. (1996a). *About the CFS: Industry initiatives.* http://www.nrcan.gc.ca/cfs/mandat/facts/ii_e.html

Canadian Forest Service. (1996b). *Model forest program.* http://mf.ncr.forestry.ca/

Canadian Model Forest Network. (2003). *Network news.* http://www.modelforest.net/e/what_/networke.html

Cashore, B., Howlett, M., Wilson, J., Hoberg, G., & Rayer, J. (2001). *In search of sustainability: British Columbia forest policy in the 1990s.* Vancouver, BC: University of British Columbia Press.

Clapp, R. A. (1998). The resource cycle in forestry and fishing. *The Canadian Geographer, 42,* 129–144.

Clayoquot Rainforest Coalition. (1995). *Clayoquot Rainforest Coalition overview.* http://www.ran.org/ran/ran-campaigns/rain-wood/bc-overview.html

Clayoquot Sound Scientific Panel. (1995). *A vision and its context: Global context for forest practices in Clayoquot Sound.* Report 4 of the Scientific Panel for Sustainable Forest Practices in Clayoquot Sound. Victoria: B.C. Ministry of Forests.

Committee on the Status of Endangered Wildlife in Canada (COSEWIC). (2003). Canadian Species at Risk. Ottawa: Author.

Curtis, M. (1996, August 27). The fight for woods "is over." *Victoria Times Colonist,* pp. A1, A6.

Dauvergne, P. (1997). *Shadows in the forest: Japan and the politics of timber in Southeast Asia.* Cambridge, MA: MIT Press.

Davidson, B. (1996). Forests: A symbol of national heritage. *World Conservation, 3* (October), 16.

Department of Indian Affairs and Northern Development (DIAND). (1999). *Agreement in principle reached between the Innu nation, Canada and Newfoundland.* November 24. http://www.inac.gc.ca./nr/prs/s-d1999.1~99167.html

Diem, A. (1992). Clearcutting British Columbia. *The Ecologist, 22*(6), 261–266.

Drushka, K., Nixon, B., & Travers, R. (1993). *Touch wood: B.C. forests at the crossroads.* Madeira Park, BC: Harbour Publishing.

Earthroots. (1996). Temagami. http://www.sll.fi/TRN/index2.html

Elliott, J. (1999). *An introduction to sustainable development.* London and New York: Routledge.

Environment Canada. (1995). Sustaining Canada's forests: Overview. *Overview SOE Bulletin,* no. 95-4 (Summer).

Environment Canada. (2000). *Biosafety protocol negotiations.* January 24–28. http://www.bco.ec.gc.ca/ProjectsBiosafe03_e.cfm

Fondation de la faune du Québec. (n.d.). *Deer yard program.* Sainte-Foy, PQ: Fondation de la faune du Québec. (Brochure).

Food and Agriculture Organization. (1997). *Factfile: Where have all the forests gone?* http://www.fao.org/news/FACTFILE/FF9704-E.HTM

Food and Agriculture Organization. (1999). *State of the world's forests, 1999.* http://www.fao.org/forestry/FO/SOFO/sofo-e.htm

Forestry Canada. (1993). *The state of Canada's forests 1992: Third Report to Parliament.* Ottawa.

Franklin, J. F. (1984). Characteristics of old-growth Douglas-fir forests. In *New forests—forests for a changing world.* Proceedings of the 1983 Society of American Foresters Conventions, Bethesda, MD.

Friends of Clayoquot Sound. (1996, October 2). *Province drops prosecution of MacMillan Bloedel in Clayoquot Sound.* News release. http://www.island.net/~focs/nr100296.htm

Grizzly study finds forestry friend. (1996). *University of Calgary Gazette, 26*(10), 6.

Hammond, H. (1991). *Seeing the forest among the trees: The case for wholistic forest use.* Vancouver: Polestar Press.

Hammond, H. (1993). Forest practices: Putting wholistic forest use into practice. In K. Drushka, B. Nixon, & R. Travers (Eds.), *Touch wood: B.C. forests at the crossroads* (pp. 96–136). Madeira Park, BC: Harbour Publishing.

Hayter, R. (1996). Technological imperatives in resource sectors: Forest products. In J. N. H. Britton (Ed.), *Canada and the global economy* (pp. 101–122). Montreal: McGill–Queen's UP.

Hayter, R. (2000). *Flexible crossroads: The restructuring of British Columbia's forest economy.* Vancouver: UBC Press.

Hebert, D. M., Sklar, D., Wasel, S., Ghostkeeper, E., & Daniels, T. (1995). *Accomplishing partnerships in the boreal mixed wood forests of northeastern Alberta.* Transactions of the 60th North American Wildlife and Natural Resources Conference, pp. 433–438.

Innu Nation. (1995). *Adaptive mismanagement proposed for Nitassinan forests.* http://www.web.net/~innu/adaptivemm.html

International Tropical Timber Organization (ITTO). (1999). *Annual review and assessment of world timber situation.* Prepared by the Division of Economic Information and Market Intelligence, International Tropical Timber Organization. Yokohama, Japan. http://www.itto.or.jp/inside/review1999/index.html

Jardine, K. (1994). Finger on the carbon pulse: Climate change and the boreal forests. *The Ecologist, 24*(6), 220–223.

Kellogg, E. (Ed.). (1992). *Coastal temperate rain forests: Ecological characteristics, status and distribution worldwide.* Occasional Paper No. 1. Portland, OR: Ecotrust/Conservation International.

Kellogg, R. M. (Ed.). (1989). *Second growth Douglas fir: Its management and conversion for value.* A report of the Douglas-fir Task Force. Special Publication No. SP-32. Vancouver: Forintek Canada Corporation.

Kelly, D., & Braasch, G. (1988). *Secrets of the old growth forest.* Layton, UT: Gibbs Smith.

Kimmins, H. (1992). *Balancing act: Environmental issues in forestry.* Vancouver: UBC Press.

Knudtson, P., & Suzuki, D. (1992). *Wisdom of the elders.* Toronto: Stoddart.

Leahy, S. (1995, September/October). Clayoquot Sound East. *Equinox, 83,* 14.

Lower, A. R. M. (1973). *Great Britain's woodyard: Britain, America and the Timber Trade, 1763–1867.* Montreal: McGill-Queen's UP.

Marchak, M. P. (1983). *Green gold: The forest industry in British Columbia.* Vancouver: UBC Press.

Marchak, M. P. (1995). *Logging the globe.* Montreal: McGill-Queen's UP.

Marchak, M. P., Aycock, S. L., & Herbert, D. M. (1999). *Falldown: Forest policy in British Columbia.* Vancouver: David Suzuki Foundation and Ecotrust Canada.

Maser, C. (1994). *Sustainable forestry: Philosophy, science and economics.* Delray Beach: St. Lucie Press.

Mather, A. S. (1990). *Global forest resources.* London: Bellhaven.

McCrory, C. (1995, June). Canada's forests still "Brazil of the north." *Taiga News, 14.* http://www.sll.fi/TRN/TaigaNews/News14/CanadaUpdatel.html

McKibben, B. (1996). What good is a forest? *Audubon, 98*(3), 54–63.

McLaren, C. (1993). Heartwood. In T. Leighton, *Canadian regional environmental issues manual* (pp. 118–125). Orlando, FL: Harcourt Brace.

McPhedran, K. (1997, Spring). Fiddlehead Farm—out of tune? *Beautiful British Columbia Traveller, 7.*

M'Gonigle, M. (1997). Reinventing British Columbia: Towards a new political economy in the forest. In T. Barnes and R. Haytor (Eds.), *Troubles in the Rainforest: British Columbia's forest economy in transition* (pp. 15–35). Canadian Western Geographical Series no. 33. Victoria: Western Geographical Press.

M'Gonigle, M., & Parfitt, B. (1994). *Forestopia: A practical guide to the new forest economy.* Madeira Park, BC: Harbour Publishing.

Moffett, M. W. (1997). Climbing an ecological frontier: Tree giants of North America. *National Geographic, 191*(1), 44–61.

Morell, V. (1994, February). New hope for old growth? *Equinox, 73,* 99.

Mungall, C., & McLaren, D. J. (Eds.). (1990). *Planet under stress: The challenge of global change.* Toronto: Oxford UP.

National Round Table on the Environment and the Economy (NRTEE). (1997). *State of the debate on the environment and the economy: Private woodlot management in the Maritimes.* Ottawa: Author.

Natural Resources Canada, Canadian Forest Service. (1995). *The state of Canada's forests 1994: A balancing act.* Ottawa: Author.

Natural Resources Canada, Canadian Forest Service. (1996a). *The boreal forest* (poster-map). Ottawa: Author.

Natural Resources Canada, Canadian Forest Service. (1996b). *The state of Canada's forests 1995–1996: Sustaining forests at home and abroad.* Ottawa: Author.

Natural Resources Canada, Canadian Forest Service. (1996c). *National Forest Week, May 5–11, 1996.* http://nrcan.gc.ca/cfs/nfw/nfw_e.html

Natural Resources Canada, Canadian Forest Service. (1997). *The state of Canada's forests 1996–1997: Learning from history.* Ottawa: Author.

Natural Resources Canada, Canadian Forest Service. (1998). *The state of Canada's forests 1997–1998: The people's forests.* Ottawa: Author.

Natural Resources Canada, Canadian Forest Service. (1999). *The state of Canada's forests 1998–1999: Globally competitive through innovation.* Ottawa: Author.

Natural Resources Canada, Canadian Forest Service. (2000a). *The state of Canada's forests, 1999–2000: Forests in the new millennium.* Ottawa: Author.

Natural Resources Canada, Canadian Forest Service. (2000b). *Natural resources fact sheet.* Ottawa: Author. http://www.nrcan.gc.ca/statistics/factsheet.htm

Natural Resources Canada, Canadian Forest Service. (2000c). *The state of Canada's forests, 1999–2000. Forests in the new millennium. Species at risk.* Ottawa: Author.

Natural Resources Canada, Canadian Forest Service. (2001a). *The state of Canada's forests, 2000-2001. Sustainable Forestry: A Reality in Canada.* Ottawa: Author.

Natural Resources Canada, Canadian Forest Service. (2001b). *Natural resources fact sheet.* Ottawa: Author. http://www.nrcan.gc.ca/statistics/factsheet.htm

Natural Resources Canada, Canadian Forest Service. (2002). *The state of Canada's forests 2001–2002: Reflections of a Decade.* Ottawa: Author.

Ness, R., & Cooperrider, A. Y. (1994). *Saving nature's legacy: Protecting and restoring biodiversity.* Covelo, CA: Island Press.

Pimm, S. (1996). The lonely earth. *World Conservation, 27*(1), 8–9.

Pratt, L., & Urquhart, I. (1994). *The last great forest: Japanese multinationals and Alberta's northern forests.* Edmonton: NewWest Press.

Quinby, P. A. (1996). *A critique of the proposed management of old-growth white and red pine forest in Temagami, Ontario resulting from the comprehensive planning process of 1996 with a case study analysis of the Owain Lake old-growth pine stand as a representative ecosystem.* http://www.sll.fi/TRN/index2.html

Rainforest Action Network. (1996, February). Clayoquot Sound landslides add to mountain of evidence against rainforest clearcut. *Rainforest Action News.* http://www.ran.org/ran/info_center/press_release/landslide.html

Ricklefs, R. E. (1993). *The economy of nature.* New York: W. H. Freeman.

Rosmarin, H. (1997). Rethinking paper: Non-wood fibres poised for comeback. *Global Biodiversity, 7*(2), 33–36.

Ryan, J. S. (1990, July/August). Timber's last stand. *World Watch,* 27–34.

Schoonmaker, P. K., von Hagen, B., & Wolf, E. C. (Eds.). (1997). *The rain forests of home: Profile of a North American bioregion.* Washington, DC: Island Press.

Sierra Club of Canada. (1996). *Canadian forests fact sheet.* http://www.sierraclub.ca/national/forests/forests-fact-sheet-1996.html

Sierra Legal Defence Fund. (1998a, May 21). *Ontario court rules budget cuts no excuse for MNR's continued non-compliance with forestry laws.* (Press release.) http://www.sierralegal.org/

Sierra Legal Defence Fund. (1998b, October 27). *Ontario Court of Appeal upholds environmental groups' win in landmark forestry case against government.* (Press release.) http://www.sierralegal.org/

Sierra Legal Defence Fund. (2000). *Forest facts.* http://www.sierralegal.org/issue/forest_facts.html

Swift, J. (1983). *Cut and run: The assault on Canada's forests.* Toronto: Between the Lines.

Taylor, D. M. (1994). *Off course: Restoring balance between Canadian society and the environment.* Ottawa: International Development Research Centre.

Travers, O. R. (1993). Forest policy: Rhetoric and reality. In K. Drushka, B. Nixon, & R. Travers (Eds.), *Touch wood: B.C. forests at the crossroads* (pp. 171–224). Madeira Park, BC: Harbour Publishing.

von Mirbach, M. (1997). Demanding good wood. *Alternatives, 23*(3), 10–17.

Western Canada Wilderness Committee. (1994). Scientific support increases for preserving Clayoquot's magnificent ancient rainforests. *Western Canada Wilderness Committee Educational Report, 13*(5). http://www.wildernesscommittee.org

Western Canada Wilderness Committee. (2000, February 11). *WCWC thrilled to see Clayoquot Sound designated a UN biosphere reserve.* http://www.wildernesscommittee.org/clayoquot/clayoquot-atlas.htm

Wigg, M., & Boulton, A. (1989). Quality wood, sustainable forests. *Forest Watch, 9*(7), 7–12.

Williston, E., & Keller, B. (1997). *Forests, power and policy: The legacy of Ray Williston.* Prince George, BC: Caitlin Press.

World Wildlife Fund. (2001). *World Wildlife Fund and Temebc Inc. reach historic accord to promote long-term sustainability of Canadian forestry.* http://www.wwfcanada.org/news-room/tembec_forestry.htm

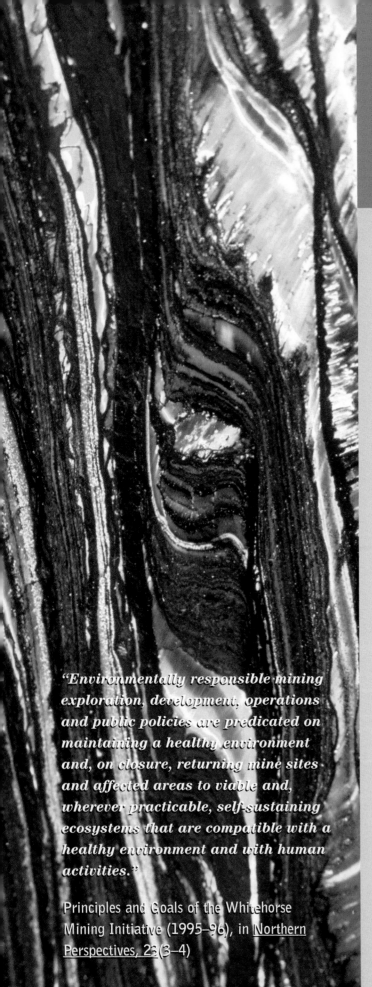

Mining

Chapter Contents

"Environmentally responsible mining exploration, development, operations and public policies are predicated on maintaining a healthy environment and, on closure, returning mine sites and affected areas to viable and, wherever practicable, self-sustaining ecosystems that are compatible with a healthy environment and with human activities."

Principles and Goals of the Whitehorse Mining Initiative (1995–96), in <u>Northern Perspectives</u>, 23(3–4)

Chapter Objectives

After studying this chapter you should be able to

• understand the nature and distribution of Canada's mineral resources

• describe the impacts of mining on natural environments

• outline Canadian and international responses to mining issues

• discuss definitions of and challenges to sustainable mining in Canada

Earth's mineral resources, including sand, gravel, clay, rock, minerals, and fossil fuels, touch almost every aspect of our lives. Mineral products are essential in the construction of our homes and workplaces (concrete, bricks, tiles, and structural steel; saw, hammer, and nails); in the provision and distribution of energy and water (coal and uranium, electrical wires and copper pipes); in our transportation (roads, gasoline, trains, and bicycles); in many luxury goods (televisions, stereos, telephones, and computers; gold and diamond jewellery; CDs; aluminum baseball bats); and in medicines, vitamins, and other products to keep us healthy (including zinc, an essential ingredient in sunscreen). Mining is not something we could readily do without!

The mining industry is a vital contributor to the Canadian economy—every province and territory in Canada supports mining and oil drilling activities. In 2001, the mining and mineral processing industries directly employed approximately 361 000 Canadians; 47 000 were employed in mining, 52 000 in smelting and refining, and 262 000 in the manufacture of mineral and metal products (Natural Resources Canada, 2002). In 2001, the value of minerals produced in Canada reached $77 billion, and the mining and mineral processing industries contributed $36.1 billion to the Canadian economy, or 3.7 percent of the national gross domestic product. About 80 percent of our mineral and metal production is exported.

Although Canada does not have accurate data on the nature and extent of land used by the mining industry, in 1982 it was estimated that less than 0.03 percent (279 477 hectares) of Canada's land area was disturbed, used, or alienated by mining activities since metal mining began more than 150 years ago (Government of Canada, 1996a). This intensive use of a relatively small area (less than half the size of Prince Edward Island) produces all the minerals we use every day. However, there is concern that habitat destruction, increased access to natural areas resulting from construction of infrastructure (including mine access roads), and effects of acidifying mine waste runoff on fish, wildlife, and water quality are among the most important effects of mining. Mineral staking rushes, too, such as in the cases of the Northwest Territories diamond fields and Voisey's Bay nickel deposit, can affect or block land uses.

Even though mineral resources are key components of national and global economies, their extraction, transportation, and processing often result in environmental harm and disruption. In Western economies, heightened environmental policies and regulations have prompted the mining industry to improve its practices, to invest in research and development, and to develop new environmental technology. In its activities in developing nations, however, the industry does not always display the same level of concern for sustainability as it does at home (Johnson, 1999; Holden, 2003).

With growing worldwide demand for minerals and mineral products, Canada is faced with the challenge of developing our mineral resources in a way that fosters our economy without compromising the sustainability of our environment. This chapter examines mineral development in Canada and provides an overview of the environmental hazards and disturbances that result from mining activity. Strategies that governments, industries, and non-governmental organizations use to reduce the environmental and social impacts of mining are also discussed.

HUMAN ACTIVITIES AND IMPACTS ON NATURAL ENVIRONMENTS

HISTORICAL OVERVIEW

Many communities have strong ties to mineral extraction; in fact, much of Canada's regional and infrastructural development has proceeded in parallel with the development of natural resources. We all have heard tales of the Klondike gold rush where thousands of people from around the world flocked to the Yukon in the hopes of striking it rich. Despite treacherous conditions, prospectors made their way over icy mountain passes and down the Yukon River to Dawson City. By 1898, three years after the first discovery of gold by George Carmack, more than 40 000 people had set up camp in Dawson.

Then referred to as the Paris of the North, Dawson became a thriving community, creating wealth for both prospectors and the Canadian government. Although few of the men made their fortunes (because the rich gold claims had been staked before most prospectors arrived), the tax earnings from gold and alcohol sales prompted the federal government to make the Yukon a separate territory in 1898. Major infrastructure projects, such as the White Pass and Yukon Route Railway (opened in 1900 between Skagway, Alaska, and Whitehorse, Yukon), were constructed to accommodate the growing population.

Not everyone appreciated the economic boom generated by the gold rush. Encroaching populations disrupted traditional ways of life of First Nations people. Miners and other new residents joined in the hunt for game, leaving Aboriginal hunters to travel farther for food. While some First Nations people elected to earn wages packing supplies for miners or cutting fuelwood for steamships, others were forced to leave their lands to escape the growing mining towns.

Photo 10–1
In the rush to extract gold, 19th-century mining communities often were erected hastily. This photo of Barkerville, B.C., was taken the day before it was razed by fire in September 1868.

By 1928, about $200 million worth of gold had been produced in the region, mostly by individual placer mines. For years, gold dredges had operated in the creeks near Dawson City, removing gold until the level of recovery declined sufficiently that it was no longer profitable to run the dredges. Despite the end of the gold rush in the early 1900s, and the depopulation of Dawson City, mining activity remains a valuable component of the Yukon economy.

The Yukon gold rush illustrates the boom–bust cycle associated with the mining industry and the history of mining in Canada. Like all sectors of the economy, the mining industry is affected by recession and growth periods within the business cycle. However, mining is especially vulnerable to short-term changes in the supply, demand, and price of an individual commodity. These market conditions have a great deal of influence over the regions and communities where mining takes place.

We know that there is a significant imbalance between the distribution of people (south) and the distribution of mineral resources (north) in Canada. This distribution gave rise to a large number of single-resource or **one-industry towns**, established to provide a pool of labour to operate and service a mine effectively. When a town's mine is booming, additional workers and support services are required, drawing people into the community. During these good times, one-industry towns are very successful economically, with low unemployment and relatively high personal income levels. In 1999, for example, average weekly earnings in the mining, quarries, and oil wells industry were $1113, one of the highest levels of any industry in the Canadian economy (Natural Resources Canada, 2000). However, once there is a downturn in the

Photo 10–2

Many of Canada's communities, such as Sudbury, Ontario, built their wealth on the extraction of mineral resources.

commodity price or reserves are exhausted, mines close, workers leave, and the community is left with little or no economic base. Quite often, entire communities are abandoned, leaving usable facilities and infrastructure behind. The costs to establish and abandon towns can be very high.

The problems associated with one-industry towns are not as prevalent today as they were in the past. Improvements in air transportation and communication have fostered a regime of **long-distance commuting**, in which miners fly in to a mine to work for a designated period and then are flown back to their homes in larger communities for another period. Workers are provided with food and temporary lodging, but no expensive infrastructure or support services are constructed at the mine site. Changes in the regulatory environment have also led to fewer one-industry towns being established.

In the past, little attention was paid to community planning; often, the urban environment in single-resource towns was of poor quality. Today, in response to concerns of governments and miners' families, new towns are subject to impact assessment processes, structured planning efforts, and substantial infrastructure investments (Shrimpton & Storey, 1988). Tumbler Ridge, British

Columbia, for example, was created in the early 1980s within a strong planning and governance framework. Nevertheless, the town was still subject to cycles in mineral production and the population has declined since the 1990s. Planning and infrastructure requirements for new towns, combined with companies' striving to increase productivity, reduce costs, and rationalize unproductive operations, have led to a movement away from one-industry towns. Even so, mining is the mainstay of employment in over 128 Canadian communities, mostly in rural and remote areas (Natural Resources Canada, 2000).

From the coal mines of Nova Scotia to the asbestos mines of Quebec, over the years mineral exploration, development, and processing have taken many lives and affected the health of countless mine workers. In the early days of mining, health and safety concerns were of minimal importance to mine operators. Beyond the obvious threats from collapsed mineshafts, equipment failures, and site explosions, little thought was given to the long-term effects of exposure to hazardous substances, mine dust, and other emissions. Black lung disease, silicosis, asbestosis, and cancer are among the common diseases that miners and other workers have contracted.

Today we recognize these threats; however, workers continue to be employed in conditions that place their health at risk. The mining industry is concerned about worker health and safety, and most mine sites now have programs to monitor exposure to hazardous substances, assess noise impacts, and protect respiratory health. Even so, a key problem confronting the industry is the uncertainty surrounding the long-term, cumulative effects of exposure to mining operations. Governments and industry must continue to strive for continual enhancement of workplace safety through reliable monitoring programs, technical innovation, and enforceable regulatory measures.

PRODUCTION, VALUE, AND DISTRIBUTION OF MINERAL RESOURCES IN CANADA

Canada's mineral resources comprise both mineral fuels and nonfuel minerals. **Mineral fuels**—crude oil and equivalents, natural gas, coal, and natural gas byproducts—accounted for about 77 percent of the total value of Canada's mineral production in 2002. **Nonfuel minerals** are categorized as metallic (e.g., iron ore, gold), nonmetallic (potash, asbestos), and structural (lime, sand, and gravel). In 2002, nonfuel minerals accounted for the remaining 23 percent of the total value of Canada's mineral production, or about $18.0 billion (see Table 10–1). There were 190 metal, nonmetal, and coal mines; about 3000 stone quarries and sand and gravel pits; and 50 nonferrous smelters, refineries, and steel mills operating in Canada in 1999 (Natural Resources Canada, 2002). Canada's principal metal and mineral mining regions are identified in Figure 10–1.

Canada produces more than 30 metallic minerals; five of these—copper, gold, iron ore, nickel, and zinc— accounted for 79 percent of the total value of metal mining production in 2002. Canada produced just over 62 044 351

TABLE 10–1
PRODUCTION VOLUME AND VALUE OF CANADA'S LEADING MINERALS (2002)*

Category/Commodity	Production (000 tonnes unless noted otherwise)	Value ($ millions)
METALS		
Gold (kg)	147 866.2	2 292.4
Copper	577.0	1 418.9
Iron ore	30 969.3	1 391.7
Zinc	891.9	1 089.9
Nickel	178.3	1 883.3
Uranium (tU)	13 055.9	608.4
Silver (t)	1 344.4	314.5
Platinum group (kg)	21 829.3	449.5
Cobalt	2.0	49.0
Lead	99.1	70.7
Other metals	n.a.	656.7
Total metals	n.a.	10 225.1
NONMETALS		
Potash (K$_2$O)	8 969.0	1 667.0
Cement	13 201.3	1 387.5
Sand and gravel	229 535.0	1 047.4
Stone	119 113.5	971.8
Diamonds (000 ct)	4 984.0	801.5
Salt	13 191.8	399.5
Clay products	n.a.	235.2
Lime	2 237.3	220.5
Peat	1 127.3	169.7
Gypsum	8 847.0	112.7
Asbestos	240.5	98.0
Other nonmetals	n.a.	667.3
Total nonmetals	n.a.	7 751.8
FUELS		
Crude oil and equivalent (000m^3)	137 356.5	30 794.5
Natural gas (million m^3)	171 347.8	23 295.9
Natural gas byproducts (000 m^3)	29 541.7	3 290.8
Coal	66 822.0	1 593.1
Total fuels	n.a.	58 974.3

* Numbers for production and value are preliminary.

SOURCE: Information Bulletin: Mineral Production, Canadian Mineral Production Demonstrates Continued Strength in 2002, Natural Resources Canada, 2003b, http://mmsd1.mms.nrcan.gc.ca/mmsd/production/default_e.asp

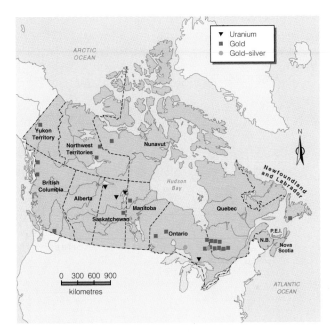

Figure 10–1a

Principal mining regions of Canada: uranium and precious metals mines

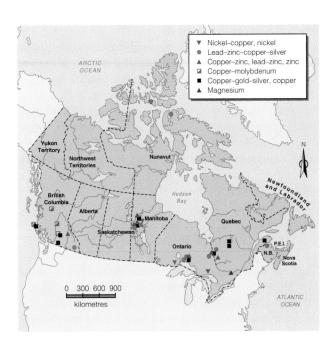

Figure 10–1b

Principal mining regions of Canada: base metal mines

Figure 10–1c

Principal mining regions of Canada: ferrous metal mines

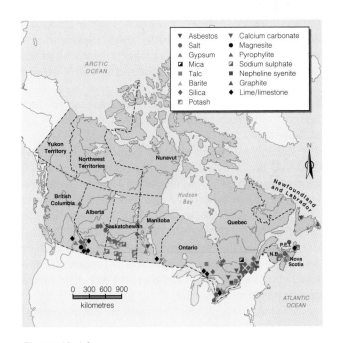

Figure 10–1d

Principal mining regions of Canada: industrial mineral mines

SOURCE: *The State of Canada's Environment—1996,* Environment Canada, 1996, Figure 11.27. Reprinted with permission of the Minister of Public Works and Government Services Canada, 2004.

tonnes of copper, nickel, lead, zinc, iron ore, gold, potash, salt, and gypsum in 2002, ranking first in the world for production of potash and aluminum, and second in production of nickel and zinc. In terms of 2002 production value, the four most important metals were gold ($2.29 billion), nickel ($1.88 billion), iron ore ($1.39 billion), and zinc

($1.09 billion). Potash, the most important commodity in the nonmetals category, was valued at $1.6 billion in 2002 (Natural Resources Canada, 2003b).

About 60 percent of expenditures on mineral exploration for the years 1998 to 2000 were focused in the Northwest Territories, Quebec, and Ontario. Typically,

precious metals, particularly gold, account for the largest share of total Canadian exploration expenditures. However, one of the largest staking rushes in recent Canadian history was prompted by the 1991 discovery of diamonds in the Northwest Territories. In 2002, exploration for diamonds contributed 25 percent of total minerals expenditures for a total of $133 million, while diamond production resulted in a value of $802 million. By 2003, diamonds had become Canada's tenth largest non-fuel mineral commodity in terms of value of production (Natural Resources Canada, 2003a).

CANADA'S FIRST DIAMOND MINE

Since the discovery of diamonds, the central region of the Northwest Territories (known as the Slave Geological Province) has attracted intense mineral exploration and development. Encompassing an area approximately one-third the size of Alberta, the Slave Geological Province extends north from Great Slave Lake to Coronation Gulf on the Arctic coast. Between 1993 and 1996, the diamond boom accounted for 20 percent of Canada's total exploration expenditures, or $560 million (Natural Resources Canada, 1996a). In these three years, roughly 22 million hectares of land were staked by 200 companies—this compares with less than 4 million hectares staked in the previous decade (Department of Indian and Northern Affairs, 1996a). By 1996, close to 60 companies were active in diamond exploration, attempting to find reserves as lucrative as the BHP mine located northeast of Yellowknife.

The Broken Hill Proprietary Billiton (BHP Billiton) diamonds project, located in the Lac de Gras region 300 kilometres north of Yellowknife, is the largest mineral project in the Northwest Territories and the first diamond mine in North America (see Figure 10–2). Its creation has been followed by the opening of the Diavik diamond project in 2003, and an anticipated mine to open in 2006 or 2007. Thus, efforts to determine the environmental and social effects of the BHP Billiton project will affect future developments in the north.

On June 21, 1996, a Canadian Environmental Assessment Review Panel (EARP) approved the BHP Billiton diamonds project. Twenty-nine recommendations were made in the EARP report regarding the ecological and social impacts of the mine. Most of these concerns were expected to be satisfied through the terms and conditions of the project's water licence, land lease, and land use permits. However, a number of EARP recommendations fell outside the scope of these regulatory instruments. To satisfy these concerns, the federal government required BHP Billiton to enter into an environmental agreement and to negotiate **impact-benefit agreements** with the Treaty 11 Dogrib, the Yellowknives Dene,

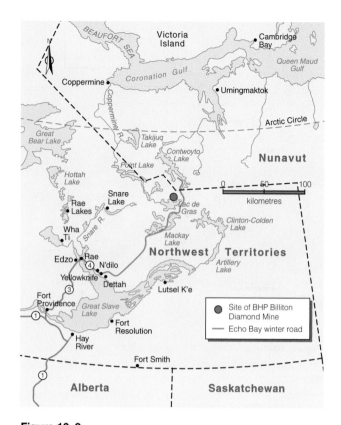

Figure 10–2

Great Slave Lake area and location of BHP Billiton diamond mine

SOURCE: Adapted from "The Nasty Game," S. Wismer, 1996, *Alternatives*, 22(4), 12. Reprinted courtesy of Alternatives Journal: Environmental Thought, Policy and Action.

the Inuit of Coppermine, and the Métis Nation who live in various parts of the remote region. Impact-benefit agreements are negotiated between a private company, various levels of government, and local communities to enhance local benefits from resource production. Normally, these agreements are limited to Aboriginal communities where issues such as employment and training; economic development and business opportunities; social, cultural, and community support; implementation and coordination; and funding of various initiatives may be addressed (Kennett, 1999). The federal government has used similar agreements in the north for several mines in the Yukon as well as for the Norman Wells Pipeline.

In consultation with the government of the Northwest Territories and Aboriginal groups, the federal government negotiated the required environmental agreement with BHP Billiton. The agreement obliged BHP Billiton to establish an environmental advisory group in addition to developing monitoring and management plans for birds and caribou. Annually, BHP Billiton was to submit public monitoring reports on social and environmental effects that, in turn, would be used in preparation of longer-term monitoring reports every three to five years. Water quality issues that were not included in the water licence also were to be addressed (Department of Indian and Northern Affairs, 1996b). However, even though there was a commitment to monitor social effects as well environmental ones, only the environmental effects have actually been monitored under the independent environmental monitoring agency. Social effects of the mine operation have received very little formal attention. Yet, as mining proceeds, it is clear that numerous social and environmental issues remain.

At the time of the environmental assessment, numerous criticisms were raised about the way in which the environmental assessment of the diamond project was conducted. Intervenor groups, for instance, noted that BHP Billiton's environmental impact statement was deficient in traditional knowledge, monitoring, mitigation, community impacts, and handling of issues related to land claims. Given this and other northern EARP experiences, questions were raised about the ability of Canada's environmental assessment process to ensure fair, effective, and efficient decision making (Wismer, 1996).

Since the BHP Billiton mine was opened, several other diamond mines have been or are being developed. The first mine site located at Lac de Gras opened in October 1998. BHP Billiton's operation contains seven diamondiferous kimberlite pipes scheduled for production at the Ekati diamond mining complex between 1998 and 2008. In 2003, production from Ekati and Diavik mines totalled $1.72 billion, while over 25 years these two mines plus a third mine will contribute an estimated $25.7 billion to the GDP of the Northwest Territories.

In 2002, about 900 employees were employed directly by the Ekati mine and 1947 people worked for mine contractors. Just over 55 percent were northern residents. At Diavik, 2003 employment was estimated at 625, of which 73 percent are northern residents and 37 percent are Aboriginal. A third project at Snap Lake will likely begin production in 2007 and is expected to employ 525 people during operations (Government of the NWT, 2004).

According to original estimates by BHP Billiton, two-thirds of the workforce would be composed of northern residents and about half of those would be Aboriginal. The company's preference in hiring and giving on-site training to Aboriginal people was estimated to result in a 10 percent reduction—from 40 to 30 percent—in the regional unemployment rate in Aboriginal communities (Department of Indian and Northern Affairs, 1996c). By January 2003, 78 percent of all workers were northerners, and of these 38 percent were Aboriginal. Approximately 30 percent of the total employees are Aboriginal; therefore, employment targets have yet to be reached. Furthermore, in the construction phase, the mine operated on a two weeks in/two weeks out, fly-in/fly-out, basis. Although BHP Billiton indicated this schedule suited employee needs, there were no provisions for workers to take time off on a seasonal basis. The costs of social and

Photo 10–3
A diamond mine northeast of Yellowknife, N.W.T.

family disruption and loss of opportunity to participate adequately in community life arising from such a schedule were not addressed (Wismer, 1996). The Canadian Arctic Resources Committee (CARC), a nongovernmental organization, continues to advocate for mining practices that will allow Aboriginal peoples to maintain their cultural and social values while participating in the economic activities of the north (see Box 10–1).

Beyond the social effects described above, the World Wildlife Fund (Canada) expressed public concerns over the rapid pace of the approvals process. The organization filed for a judicial review of the EARP's procedures in an effort to obtain commitments for action on protected areas in the region. The international conservation organization was concerned that the diamond mining area is located at the centre of the migration route of the Bathurst caribou herd. After the governments of Canada and the Northwest Territories committed to produce a Protected Areas Strategy (PAS) for the entire Northwest Territories by 1998, the World Wildlife Fund withdrew the court action (World Wildlife Fund, 1997). On September 30, 1999, the Northwest Territories Protected Areas Strategy (PAS) was signed by the federal minister of Indian Affairs and Northern Development and the territorial minister of Resources, Wildlife and Economic Development. Although the PAS is a consensus-built strategy and is intended to provide greater certainty to industry interested in northern investment, concerns about the effectiveness of such protected areas strategies remain. One indication that such strategies might work occurred in the case of Tuktut Nogait National Park, established in 1996 to protect the tundra hills natural region and the calving grounds of the Bluenose caribou herd. The mining industry wanted park boundaries amended to permit mining on 415 square kilometres of land. In 1998, Parliament refused, perhaps an indication that approaches such as the PAS will achieve their objectives (McNamee, 1999).

Aboriginal groups also expressed concern over the BHP Billiton mining project. The site falls within the traditional hunting and trapping grounds asserted by the Yellowknives Dene and Dogrib Nation. As the mine project was undergoing the approval process, both the Yellowknives Dene and the Dogrib were in the process of negotiating land claims agreements with the federal government. The Yellowknives Dene entered into treaty land entitlement negotiations, while the Dogrib were negotiating a comprehensive land claim and self-government agreement. Both groups were concerned that mining development would compromise the government's ability to conduct negotiations (Department of Indian and Northern Affairs, 1996d).

The Dene were concerned that without a land agreement they risked becoming an "embittered minority" as mining development would attract many new residents to the area (Freeman, 1996). Although BHP Billiton negotiated impact-benefit agreements with Aboriginal groups, there is concern about the enforcement of such agreements because no provision for enforcement was set out in the EARP recommendations. Attention to equity issues may have been insufficient to ensure that northern people and their communities can remain healthy and sustainable (Wismer, 1996).

Aboriginal leaders also were skeptical of BHP Billiton because of the company's environmental performance record, including its problems with the Ok Tedi copper mine in Papua New Guinea. There, BHP Billiton operated without a mine tailings system for more than 10 years, dumping 80 000 tonnes of waste rock per day into the Ok Tedi and Fly rivers, and rendering 70 kilometres of the Ok Tedi River almost biologically dead (Papua New Guinea, 1996). Aboriginal leaders claimed that although BHP Billiton might offer monetary compensation for damaging the waters and the land, money does not replace the natural values and opportunities the land provides. In particular, the Yellowknives Dene were concerned about the fuel oil, arsenic, and cyanide that would be hauled on winter roads across their hunting grounds (Wismer, 1996). What troubles many Aboriginal groups and environmentalists is the increased potential for accidents and spills as additional mining companies show interest in developing the area.

The environmental impact statement prepared by the project's proponents outlined the methods and techniques that were to be used to mitigate environmental damage. Although diamond production avoids the use of toxic chemicals, land and water resources are severely altered by diamond mining. For instance, the mine will have to dig through 6 tonnes of granite for every 1 tonne of kimberlite diamond ore that will be processed. The mine is expected to process about 9000 tonnes of ore per day; that will yield about 2 kilograms of diamonds per day, enough to fill a coffee can (Weber, 1997). In addition, the BHP Billiton project will drain six lakes and use another for tailings storage. Over the planned 25-year operation of the diamond processing plant, 133 million tonnes of tailings will be impounded at the storage lake (BHP Billiton Minerals Canada Ltd., 1995). At full capacity, the mine is expected to supply 3 Mct annually, or about 5 percent of the world's diamonds (Natural Resources Canada, 2000).

As noted above, five new mines may open in the region where BHP Billiton is developing its Panda mine. In early March 2000, the Government of the Northwest Territories signed off on the environmental agreement with the second diamond mine, that of Diavik Diamond Mines Inc. This agreement, too, addressed environmental provisions identified in the Diavik Comprehensive Study Report that would not be covered by existing regulatory instruments, such as a water licence or land permit (Government of the Northwest Territories, 2000). The mine opened in 2003.

Diamond mining poses important challenges. Notwithstanding the industrial uses of some diamonds, most diamonds are produced for a luxury market. Yet, their production also provides employment in a region where steady jobs are hard to come by. Canadians might ask: Can mining make a contribution to the longer-term health and to the sustainability of northern communities? This question is considered in Box 10–1.

ENVIRONMENTAL IMPACTS OF MINING

Mineral deposits often are located in areas desirable for other land uses, such as forestry, agriculture, and recreation. Therefore, mining is not always an activity located in remote regions. In the case of sand and gravel mining, these deposits are frequently located near urban areas. Using ecosystem-based approaches to make environmental decisions helps us realize that the environmental effects of a mine operation extend far beyond the mine site. Mining has a considerable influence on land surrounding the mine, and the range of interaction between mining and other land uses has become increasingly complex. This influence, referred to as the *shadow effect,* includes all the indirect activities associated with mining such as the construction of roads, rail links, and power facilities (Marshall, 1982). Both the direct uses at the mine site and the accompanying indirect uses in the shadow zone have the potential to conflict with other land use activities (see Figure 10–3). As a result, a mining company's responsibility for environmental protection extends beyond its working operations to include neighbouring lands and watersheds.

The Brundtland Commission described sustainable development as "development that meets the needs of the present without compromising the ability of future generations to meet their own needs." Can we consider sustainability in light of human demands to extract nonrenewable resources? Recently, industry, government agencies, nongovernmental organizations, and Aboriginal peoples in Canada have been thinking about how sustainability might apply to the mining industry. While these concerns only deal with the production of minerals and fossil fuels rather than their consumption, these efforts do provide an opportunity to consider whether sustainability and mining are compatible. Here are some efforts to put these two together.

In 1994, the Whitehorse Mining Initiative Leadership Council Accord was signed by members of the industry, senior governments, labour unions, Aboriginal peoples, and the environmental communities. This accord was aimed at maintaining healthy and diverse ecosystems in Canada, and for sharing opportunities with Aboriginal peoples. In 1995–96, the northern watchdog organization Canadian Arctic Resources Committee (CARC) stated that the Initiative Accord was not fully effective and suggested that several issues still needed to be addressed in relation to northern mineral exploration and development. Although neither the accord nor CARC explicitly used the language of "sustainable mining," the issues raised are consistent with later efforts to consider sustainability and mining. The outstanding issues included

- Ensuring notification, consultation, and consent for exploration and development by affected Aboriginal organizations and communities.
- Ensuring regulation of the impacts of exploration to protect environmental as well as social and cultural effects.
- Using traditional knowledge to create regulatory requirements for resource development.
- Negotiating fair impact and benefits agreements.
- Developing regional baseline studies using traditional knowledge and appropriately directed science.
- Providing Aboriginal people with opportunities to become familiar with mining practices so that they can become involved in identifying and assessing potential impacts.
- Providing opportunities for cross-cultural awareness programs for members of the mining industry (including some in government), possibly modelled on programs developed for the petroleum industry in the Western Arctic.
- Ensuring that all stages of mineral development focus on protecting cultural, economic, and environmental concerns; that is, infusing resource development decisions with the customs, traditions, and values of Aboriginal peoples. Because women, in particular, give testimony to these cultural issues, some believe that women can and should play more central roles in development decision making.

In the 1990s, the environmental impact assessment panel investigating the impacts of the Voisey's Bay nickel project off the coast of Newfoundland required the proponent to include sustainability objectives in its application. The Environmental Impact Assessment panel noted that to contribute to sustainability, the mine must

- Not impair ecosystem integrity or biodiversity
- Not significantly damage local and regional ecosystem functions
- Not reduce the capacity of renewable resources to support present and future generations
- Deliver durable and equitable social and economic benefits (even after mine closure), with special attention to the needs of Aboriginal peoples
- Proceed in a manner compatible with stewardship of nonrenewable resources
- Respect Aboriginal rights and land claims agreements

Based on the panel's review and recommendations, the rate of extraction was slowed to extend the life of the mine operation and to promote longer social and economic benefits for the local community. Would these recommendations "fit" your definition of sustainability?

In November 2001, the Mining Association of Canada developed a set of guiding principles to demonstrate the commitment of its members to sustainable development. Part of the principles state,

In all aspects of our business and operations, we will:

- respect human rights and treat those with whom we deal fairly and with dignity
- respect the cultures, customs and values of people with whom our operations interact
- obtain and maintain business through ethical conduct
- comply with all laws and regulations in each country where we operate and apply the standards reflecting our adherence to these Guiding Principles and our adherence to best international practices
- support the capability of communities to participate in opportunities provided by new mining projects and existing operations
- be responsive to community priorities, needs and interests through all stages of mining exploration, development, operations and closure
- provide lasting benefits to local communities through self-sustaining programs to enhance the economic, social, educational and healthcare standards they enjoy. (Mining Association of Canada, 2004)

In May 2002, a workshop was held involving geological scientists across Canada. Participants recognized that resource development may be a key contributor to future economic inde-

(continued)

pendence and stability (for Canada's northern regions), but preservation of the social and environmental values must also be paramount to any development decision. They defined sustainable mining as mining that ensured that its economic benefits are shared fairly with affected communities and are not offset by long-term negative impacts to the environment.

What considerations do you think are important if mining is to be considered under the heading of sustainable development?

SOURCES: "Aboriginal Communities and Mining in Northern Canada," R. F. Keith, 1995/96, *Northern Perspectives, 23*, 3–4. http://www.carc.org/pubs/v2n3-4/mining2.htm; *Towards Sustainable Mining: Guiding Principles,* Mining Association of Canada, April 2004, http://www.mining.ca/english/tsm/principles~eng.pdf; Personal communication, B. Noble, 2003; *Whitehorse Mining Initiative,* Natural Resources Canada, 2003, http://www.nrcan.gc.ca/mms/poli/wmi_e.htm; "Sustainable Mining in the 21st Century," (SUM 21), Workshop for Geoscientists, Conference Report, (n.d.), http://www.cim.org/geosoc/sum21_report.cfm

Increasingly, mineral claims have been staked in areas designated for environmental protection; this action has sparked some intense land use conflicts. For example, a mining development proposed by a subsidiary of the Canadian multinational Noranda Inc., adjacent to Yellowstone National Park, was halted by U.S. president Bill Clinton following a lengthy period of public opposition (see Box 10–2). In exchange for all rights to the mining claim, the U.S. government compensated the mine proponents with US$65 million worth of federal lands. Similarly, a proposal to develop a copper mine on British Columbia's Windy Craggy Mountain, in the Tatshenshini watershed, was denied and a land swap compensation package was negotiated in 1993. The region around the Tatshenshini River now is preserved as the Tatshenshini-Alsek Wilderness Park and is permanently closed to mining (Newcott, 1994). In 1994, the area was designated as a World Heritage Site.

Most mining operations follow a four-stage sequence of development: exploration, development and extraction, processing, and closure and reclamation (each stage is described briefly in the following sections). The environmental impact at each stage of development varies according to the mineral type, consistency, location, and the form in which the final product is delivered. Each stage of the mineral production process—from prospecting and exploration to mine development and extraction to refining and processing—introduces potentially disruptive environmental impacts. Figure 10–4 summarizes the waste impacts and potential hazards of mining at each stage of the production process, while Table 10–2 highlights a broad range of impact issues associated with the mining activities illustrated in Figure 10–4.

The extent of environmental impact from a mine depends on a range of factors, from the type of mineral and its chemical properties to the local ecology, geology, and climate characteristics at the mine site. For example, most metal (gold, copper, zinc, and nickel) mines in Canada contain sulphide materials. When exposed to air and water, the sulphide materials oxidize and generate sulphuric acid, resulting in **acid mine** (or acid rock) **drainage**. Acid mine drainage is the most serious environmental problem facing the mining industry (see the discussion under "Mine Closure and Reclamation"). Climatic variables, such as strong winds and precipitation, may further compound problems of acidic drainage and increase the risk of freshwater contamination.

Smelting, refining, and fabrication processes use chemicals that may leak or be discharged into the environment. Furthermore, these processing activities require a large amount of energy—the mining industry accounts for about 13 percent of Canadian industrial energy demand (Natural Resources Canada, 2000). Generally, energy is provided by burning fossil fuels, which releases carbon dioxide and other gases into the atmosphere.

Mineral Exploration

Mineral exploration involves finding geological, geophysical, or geochemical conditions that differ from those of their surroundings (Marshall, 1982). Discovering such anomalies in the landscape may signal the presence of significant mineral deposits. Even though there are extensive geological mineral records compiled over the past century in Canada, actually detecting an anomaly can be like finding a needle in a haystack. A company's expenditures are high during the exploration stage, and there is no guarantee of turning a discovery into an economically feasible mine.

During exploration, construction of access roads, trenches, pits, and drill pads disturbs the land surface, and may interfere with wildlife and local drainage. In some instances, where vegetation is stripped to accommodate testing activities, soil erosion and sedimentation (and possible disruption of fish habitat) follow. The shadow effects of constructing roads in areas previously devoid of them include opening up access to potentially sensitive areas and permitting hunters, wilderness

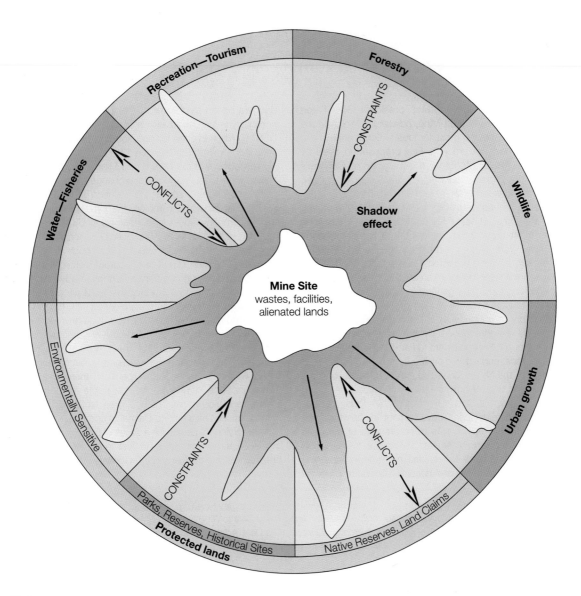

Figure 10–3

Conceptual land use conflicts and constraints for mining

SOURCE: Adapted from *Mining, Land Use and the Environment: A Canadian Overview,* I. B. Marshall, 1982, Ottawa: Environment Canada. Land Use in Canada series. No. 22–23, p. 194. Reprinted with permission of the Minister of Public Works and Government Services Canada, 2004.

tourists, guides, outfitters, and others to cause potentially significant impacts, ranging from noise to harassment of wildlife. Companies involved in exploration are required to follow guidelines aimed at reducing the disruptive environmental effects of their activities. In addition, most provinces require companies to have reclamation plans and adequate financing to rehabilitate exploration sites.

Having discovered a mineral deposit, a company must first assess the technical and economic requirements of bringing a mine into operation. Costs of extraction, transportation, and processing are considered, as well as costs of environmental controls and reclamation procedures. A company's commitment to operate within an acceptable

environmental standard must be demonstrated before a project is approved. As with proposals for many large-scale developments, mining proposals are subject to environmental review. Decisions to proceed with, modify, or restrict development normally are reached through the **environmental impact assessment** (EIA) process.

An EIA aims to provide decision makers with scientifically researched and documented evidence to identify the likely consequences of undertaking new developments and changing natural systems (Wiesner, 1995). The magnitude of a review and the requirements to be satisfied vary from project to project and from province to province. A key component of the overall EIA process is the preparation of

On August 12, 1996, United States president Bill Clinton announced that the New World Mine, adjacent to Yellowstone National Park (the first and most famous national park in the United States), would not proceed. First proposed in 1989 by Crown Butte Mines of Montana, a subsidiary of Canada's Noranda Inc., the project would have extracted gold, copper, and silver worth between US$600 and $800 million.

Conservation groups were adamantly opposed to the project because of the threat of acid mine pollution to the Yellowstone ecosystem, most notably the Clarks Fork of the Yellowstone River. The New World Mine would require construction of a 28-hectare reservoir to store a projected 11 million tons of potentially toxic mine tailings. Geologists expressed concern about the reservoir cracking (and thus generated questions about safe storage of the tailings) because of the high elevation, severe climate, and history of earthquakes in the region. For three consecutive years, dangers posed by the mine prompted the conservation group American Rivers to name the Clarks Fork the United States' most endangered river.

Opposition to the New World Mine extended far beyond the states of Montana and Wyoming. Major newspapers, including the New York Times, the Denver Post, and USA Today, repeatedly editorialized against mine development in the vicinity of Yellowstone National Park. For its editorials against the New World Mine, the New York Times was awarded a 1996 Pulitzer Prize. Conservationists credit the national media for enlightening the public, as well as regional and national officials, to potential dangers posed by the mine. Significantly, Americans saw the nationality of the Canadian firm as a key issue; Noranda had paid US$1.9 million in fines for pollution and health and safety violations in the United States since 1981.

The owners of the New World Mine did not walk away from Yellowstone empty handed. In exchange for abandoning mining plans and cleaning up historic pollution at the site (at an estimated cost of US$22.5 million, the U.S. government gave the mining company federal land worth up to US$65 million.

The Yellowstone victory has given U.S. conservationists some momentum in pressing for reform of the country's 1872 Hardrock Mining Act. Written in part to encourage settlement of the west, the act gives mining companies economic advantages such as the purchase of federal land for the 1872 price of $5 per acre. Canadian conservationists were pleased with the land swap as well, noting that Yellowstone is part of a patchwork of wilderness areas that extends north into Alberta—encompassing Banff and Jasper national parks—and all the way to Yukon. Large carnivores such as bears and wolves migrate great distances between these wilderness areas and depend on them for their survival. What happens in Yellowstone has ripple effects in Banff and Jasper (and vice versa). The U.S. president's decision to halt the mine was seen as a help to Canadian and U.S. environmentalists who are trying to reestablish a series of protected areas along the whole eastern slope of the Rockies in a major project known as the Yellowstone-to-Yukon (Y2Y) Conservation Initiative.

SOURCES: "Yellowstone Deal Elates Naturalists," J. Adams, August 12, 1996, Calgary Herald, p. A1; President Clinton Stops New World Mine, American Rivers, 1996, http://www.igc.apc.org; President Clinton Signs Deal to Stop New World Mine, Greater Yellowstone Coalition, 1996, http://www.desktop.org/gyc; Two Weeks after Proposed Headwaters Deal, Administration Offers Similar Land Swap to Mining Company, Redbud News, 1996, http://www.redbud.com/news/paper.html

an **environmental impact statement** (EIS). In Canada, it is the responsibility of the project proponent (individual or company proposing the project) to prepare the EIS and ensure that all provincial and federal policy requirements are satisfied. A balanced EIA should consider the scope of a project from a systems perspective. Table 10–3 outlines the procedural elements involved in the EIA process and highlights basic features of an EIS.

While project proponents are required to publish terms of reference for their EISs, and opponents are entitled to respond to these terms, the nature and complexity of the EIA process often leads to frustration for all interested stakeholders. In spite of the possibility of review under the federal Environmental Assessment Act, many people are critical of impact statements because they are commissioned by the developer and prepared by consultants and researchers hired by the developer. The EIS for the BHP Billiton diamond mine in the Northwest Territories, for example, was described as superficial and totally inadequate in assessing the effects of the mine on the environment (Freeman, 1996).

Opponents of development projects are not the only ones critical of environmental assessment procedures. In 1996, the Cheslatta Carrier Nation of northern British Columbia obtained documents that revealed "extraordinary corporate pressure to obtain federal approval for the proposed Huckleberry mine near Houston, BC" (Nelson, 1996, p. A5). Japanese investors in the mine opposed "unreasonable delays" in the environmental assessment process and threatened to withdraw from the project and reduce further mining investment in Canada. Such corporate pressure undermines the federal environmental assessment process and may have led to the report entitled "Streamlining Environmental Regulation for Mining" tabled in the House of Commons in November 1996. In the report, the Standing Committee on Natural Resources made 11 recommendations for reforming the federal environmental regulatory regime for mining.

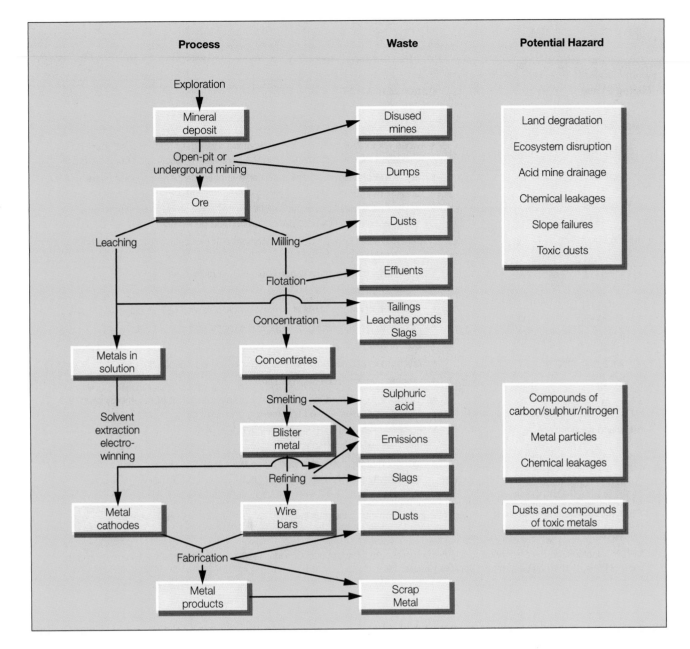

Figure 10–4

The mining process and the environment

NOTE: Please refer to Table 10–2 for information on potential environmental impacts of mining phases illustrated in this figure.

SOURCE: Adapted from *Environmental Degradation from Mining and Mineral Processing in Developing Countries: Corporate Responses and National Policies,* A. Warhurst, 1994, Paris: Organization for Economic Cooperation and Development, p. 14.

Mine Development and Mineral Extraction

Mining development may have a double impact on the environment, not only as a result of the mine site development but also because of the infrastructure put in place to service the mine. As noted above, roads increase access to remote areas and result in additional environmental pressure from nonmining activities such as hunting, fishing, and recreation. In the case of sand and gravel pits, increased traffic on existing roads brings increased levels of dust, noise, and maintenance costs that must be borne by local residents and municipalities. Although these activities are beyond the direct control of mining companies, it would be difficult or impossible to undertake them without the mine development.

Effective transportation links are essential for servicing mining operations. Although a large degree of processing occurs at the mine site, large shipments of

TABLE 10-2
POTENTIAL ENVIRONMENTAL IMPACTS OF MINING PHASES

Mining Phase	Potential Impacts
Exploration	• generally low or no impact • when exploration stage requires trenching, drilling, or road access, there is increased habitat disturbance and the potential for discharge of contaminants
Extraction and processing	• acid mine drainage containing contaminants is released to surface water and groundwater; there are particular concerns relating to:
• mining and milling	– heavy metals that originate in the ore and tailings (their release may be accelerated by naturally occurring acid generation) – organic compounds that originate in the chemical reagents used in the milling process – cyanide, particularly from gold milling processes – ammonia • alienation of land as a result of waste rock piles and tailings disposal areas • increased erosion; silting of lakes and streams • dust and noise
• smelting and refining	• discharge to air of contaminants, including heavy metals, organics, and SO_2 • alienation of land as a result of slag • indirect impacts as a result of energy production (most of the energy used in mining processes is used for smelting and refining)
Closure and reclamation[a]	• continuing discharge of contaminants to groundwater and surface water (particularly heavy metals when naturally occurring acid generation exists) • alienation of land and one-time pulse discharge of contaminants and sediment to water as a result of dam failure

NOTE: A particular concern centres on the responsibility for orphaned mine sites; liability falls to society through the government.

[a] Does not apply everywhere.

SOURCE: Adapted from *The State of Canada's Environment—1996,* Environment Canada, 1996, Ottawa. Reprinted with permission of the Minister of Public Works and Government Services Canada, 2004.

minerals are transported to smelters, refineries, and other processing locations. In 1998, for instance, Canadian railways earned 55 percent of their total freight revenue from transporting mineral products, and almost 70 percent of the volume of all products loaded at Canadian ports in international trade was from mineral-related products (Natural Resources Canada, 2000). Environmentalists, and others, are concerned about spills of minerals, mine wastes, and processing chemicals en route to or from mine sites. Roads constructed beside rivers are of special concern as an accident could release toxic substances into the watercourse and threaten aquatic habitat. Mine sites located in areas of seismic activity also are of concern due to the threat of an earthquake rupturing tailings ponds and subsequently washing out roads and bridges.

Until the early 1900s, mining activity was concentrated underground. To extract the desired materials, shafts were sunk or tunnels were driven on a slope or horizontally into the ore zone (Figure 10–5). Then the ore was drilled, blasted, and collected for transport to the surface by ore hoists or wheeled haulage vehicles.

Development of surface mining allowed lower-grade ore bodies to be mined over a wider, more dispersed land area. Surface mining involves two basic techniques. In **open-pit mining**, ore is extracted from deposits by making a progressively larger and deeper pit from which overburden and waste rock are removed. In **strip mining**, material that lies relatively flat and is not buried too deeply is exposed by shovels or draglines, and the waste materials are thrown back into the previous cut made where the ore or coal was extracted. Placer mining is a form of surface mining, such as panning for gold, that occurs mainly in the Yukon and involves mining river- and streambeds for eroded particles of minerals. In 1992, over 70 percent of mineral production in Canada was from surface mining operations.

TABLE 10–3
THE ENVIRONMENTAL IMPACT ASSESSMENT PROCESS

Procedural Elements in the Environmental Impact Assessment Process

- The developer calls together consultants, regulatory bodies, and other organizations.
- An EIS is published and used as a basis for consultation involving regulatory authorities, industry, local residents, and the general public.
- The findings of the consultation process are presented to the competent authority reviewing the project.
- Any mitigating measures and claims for compensation are considered.
- Progressive and postproject monitoring of environmental consequences arising from implementation of the project are set in place.

Basic Features of an Environmental Impact Statement

1. Description of main project characteristics
2. Provision of an analysis of the aspects of the environment likely to be affected by the project
3. Description of the measures envisaged to reduce the harmful effects
4. Listing of alternatives to the proposed project and reasons for their rejection
5. Assessment of the compatibility of the project with environmental regulations and land use plans
6. Systems to be set in place for ongoing and postproject monitoring
7. Nontechnical summary

SOURCE: Adapted from *The Environmental Impact Assessment Process,* D. Wiesner, 1995, Dorset: Prism Press. Used by permission.

Photo 10–4
Gold, coal, potash, and salt are among the minerals extracted in underground mining.

Photo 10–5
Open-pit mines may have environmental impacts beyond land disturbance and aesthetic degradation, including altered surface drainage patterns and the release of harmful trace elements.

Open-pit mining has allowed mineral companies to exploit economies of scale by using larger mining equipment and large-scale extraction techniques (Warhurst, 1994). However, extensive surface excavations of overburden material may clog streams, create excessive dust, and disturb habitat. In 1989, surface mines produced eight times as much waste per tonne of ore as underground mines (Warhurst, 1994). Land disruption from surface mining also reduces land available for alternative uses, such as forestry, agriculture, or recreation.

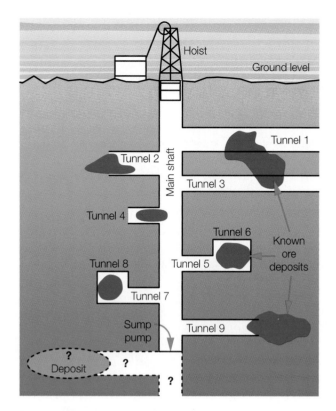

Figure 10–5

Cross-section of a mineshaft

NOTE: Sump pumps keep the mine dry, and sophisticated ventilation systems, usually requiring a second shaft, must be installed in most underground mines to remove explosive and radioactive gases and provide fresh air.

SOURCE: Adapted from *Conservation of Natural Resources: A Resource Management Approach,* D. C. Castillon, 1992, Dubuque, IA: Brown, p. 325.

Processing of Minerals

The processing of minerals through milling, smelting, and refining is less land intensive than exploration and extraction; however, the environmental impacts are more significant and long-term. The **milling** process involves the crushing and grinding of ores to separate the useful materials from the nonuseful ones. **Tailings**, the nonuseful materials, are removed from the mill after the recoverable minerals have been extracted. Generally, the amount of concentrate produced per tonne of ore is small in comparison with the amount of rock waste.

Most metal ores require crushing and grinding, plus additional treatment with chemical or biological reagents to extract the desired minerals. Base-metal milling commonly uses a flotation process that mixes the ore with chemicals (such as kerosene, organic agents, and sulphuric acid) and water to produce a fine mineral concentrate (subsequently shipped to a smelter for the next stage of recovery). Following the flotation process, tail-

Photo 10–6

Usually, pure minerals constitute a tiny fraction of the material extracted to obtain them.

Photo 10–7

A potash slag heap at Vanscoy, Saskatchewan.

ings are normally filtered and washed to remove most of the reagents. Once tailing solids have settled, the effluent is discharged to natural water bodies providing it meets regulatory standards.

The milling of gold uses cyanide, which is lethal to fish at concentrations as low as 0.04 milligrams per litre (Government of Canada, 1996b). Once the milling of gold is complete, cyanide must be treated because the retention time in tailing ponds normally is not long enough for cyanide to break down naturally. Various processes are

Photo 10–8

As the demand for mineral resources continues to increase, the recycling of metals remains an important means to reduce waste. Aluminum products such as these beverage cans are among the most commonly recycled metals. However, recycling often requires large quantities of fossil fuel energy, which does not resolve the issue of how much we consume in the first place.

used to treat cyanide in gold mill effluents and tailings pond waters where cyanide concentrations are high. Most of the cyanide and metallocyanide complexes can be destroyed or recovered. However, some metallocyanides are more stable and difficult to treat; they may end up in the aquatic environment and harm fish. Provincial regulations specify limits on the amount of cyanide that may be released through effluent discharges.

Usually, mine wastes are treated and stored on site in tailing retention facilities, such as dams and ponds. Effluents containing metals are treated with lime during retention to precipitate the dissolved metals as hydroxides. Such treatment generally removes up to 99 percent of metals and suspended sediments from mine and mill effluent (Government of Canada, 1996a). Serious aquatic damage may result if tailings retention facilities leak or rupture, as was the case with tailings dike failures at Canadian-owned or operated mines in Spain (Los Frailes) and Kyrgyzstan during 1998. These failures were preceded by the failure of a drainage tunnel at the Marcopper mine in the Philippines in 1996, and by the crash of a truck carrying sodium cyanide into a local river near the Kumtor mine in Guyana during 1995 (Johnson, 1999). Each of these accidents caused severe contamination of surrounding lands and rivers. There have been past instances in Canada, too, where leaking or ruptured tailings ponds

have affected local water bodies and contaminated community water supplies, such as when the tailings pond at Western Mines discharged into Buttle Lake on Vancouver Island, British Columbia, about 30 years ago. Mines that have long been abandoned pose enormous environmental costs today. Toxic substances deposited into Howe Sound, British Columbia, from the Britannia mines continue to destroy fish habitat, while no government or private company is prepared to take responsibility for its cleanup. Even today, many sites are highly toxic.

Production of most base metals (copper, lead, zinc, and nickel), all ferrous metals (iron and steel), and aluminum occurs in smelters and blast furnaces that operate at high temperatures (pyrometallurgy) and emit various pollutants into the atmosphere. Particulate matter, nitrogen oxides, sulphur dioxide, metals, and organic compounds may be deposited locally or transported over long distances. Since the 1970s, stricter emission control standards, new smelting technologies and processes, and voluntary pollution prevention efforts by the mining industry have helped to reduce sulphur dioxide and other emissions (see Figure 10–6). However, a substantial portion of the Boreal Shield ecozone continues to receive elevated levels of acidic deposition (Government of Canada, 1996b).

Tailings and waste rock that are naturally high in sulphide materials and are stored at working and abandoned mines can result in acid mine drainage. Acid mine drainage, the most serious control problem facing the Canadian and global mining industry, is discussed in the following section.

Mine Closure and Reclamation

In common with other nonrenewable resources, mineral reserves are finite. Once deposits become depleted or extraction becomes uneconomical, a mine site will close. Efforts to extend the life of a mine include introducing exploration programs to find nearby ore deposits, increasing recovery efficiencies, and providing financial incentives. Regardless of the efforts employed, the eventual closure of a mine is inevitable.

In one-industry towns, mine closures can devastate an entire community. In regions dependent on mining, the closure of a mine can lead to the rapid deterioration of social and physical infrastructure. People move to find employment elsewhere and leave behind a small population base with high levels of unemployment and minimal opportunities for diversification. Unless a secondary employer is found early in the life of a mining community, there is little chance of avoiding a rapid decline of the community once a mine closes. In most instances, it is too late to attract a secondary employer once it is known that most of the population is going to disappear soon (Province of Manitoba, n.d.). Governments, and more recently companies, may provide relocation allowances

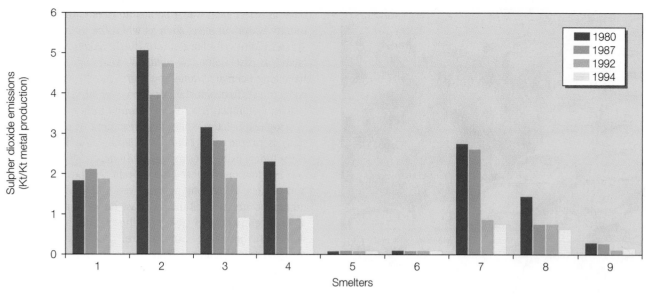

Smelter/location

1. Hudson Bay Mining and Smelting, Flin Flon, Man.[b]
2. Inco Ltd., Thompson, Man.
3. Inco Ltd., Sudbury, Ont.
4. Falconbridge Ltd., Sudbury, Ont.
5. Falconbridge Ltd., Kidd Creek Division, Timmins, Ont.
6. Canadian Electrolytic Zinc (Noranda), Valleyfield, Que.
7. Noranda Minerals Inc., Rouyn-Noranda, Que.
8. Noranda Minerals Inc., Murdochville, Que.
9. Brunswick Mining and Smelting Ltd., Belledune, N.B.

Figure 10–6

Decline in sulphur dioxide emissions per kilotonne of metal production at selected smelters in eastern Canada for selected years since 1980[a]

[a] In this figure, smelting refers to pyrometallurgical processes used in the production of metals from sulphide concentrates, including the roasting of zinc from concentrates.

[b] The increase in SO_2 per kilotonne of metal production from 1980 to 1987 is attributable to variability in the sulphur content of concentrates handled at this custom smelter.

SOURCE: *The State of Canada's Environment—1996,* Environment Canada, Figure 11.32. Reprinted with permission of the Minister of Public Works and Government Services Canada, 2004.

and retraining programs to help residents of single-industry towns.

Some communities have managed to retain a healthy economic base after the closure of a mine. In 1996, Island Copper Mine on northern Vancouver Island was closed after 24 years in operation. Most employees lived in Port Hardy. Despite the fact that hundreds of employees lost jobs at the site, its economic impact was considered small because of the long lead time to closure, the ability of the company to find alternative employment for most of the workers, the transitional provision of services negotiated between the company and the local governments, and the ability of Port Hardy to draw on its local advantages to diversify its economic base. Elliot Lake, Ontario, developed an economic diversification strategy to promote the community for retirement living and outdoor recreation. With the closure of the area's last uranium mine in 1996, Elliot Lake also hoped to position itself as an international centre for research on tailings management (Canadian Environmental Assessment Agency, 1996). When the coal mines (and timber mills) closed near Chemainus, British

Columbia, tourists came to view the large heritage murals painted on community buildings and to enjoy shopping and socializing over tea or coffee in the many small businesses that were established following the success of the murals. Nevertheless, those people involved in the tourist trade were not the same as those who had been previously employed in primary resource industry. The process of transition is often wrenching for resource workers, leaving distinct "winners" and "losers" during the process.

Mine closure not only affects the social fabric of a community but can have far-reaching negative effects on the environment. In the past, in keeping with environmental and mining practices of the day, mine sites were abandoned with minimal or no rehabilitation work. Little, if any, concern was shown for land restoration, environmental stability, or even public safety. Today, most mining projects will be denied approval if the proponents fail to provide a detailed closure plan or do not have the financial security to ensure site rehabilitation. To be effective, these plans must include a long-term monitoring program to safeguard against future environmental damage.

Photo 10–9
Recent reclamation efforts have reintroduced vegetation and improved air and water quality in the area of Sudbury, Ontario, where mining slag heaps had historically been left unattended.

In Canada, jurisdiction over mineral resources is assigned to the provinces. As a result, **reclamation** standards are determined by each province. The federal government has direct responsibility for reclamation in Yukon and the Northwest Territories and in relation to uranium. As well, the federal government influences mine reclamation issues at the national level through the Fisheries Act, the Canadian Environmental Assessment Act, and through tax policies, and science and technology activities.

Mine reclamation seeks to rehabilitate a mine site to a viable and, wherever practicable, self-sustaining ecosystem that is compatible with a healthy environment and other human activities (Government of Canada, 1996b). Returning land disturbed by mining activity to a condition that is safe, stable, and compatible with adjoining lands requires a well-planned series of activities that incorporate sustainability objectives at all stages of mineral production. In other words, rehabilitation should be a continual activity that occurs over the life of the mine. In addition to initial studies, such as an EIA and clo-

sure plan, rehabilitation reports should be prepared annually and the site should be continuously monitored to identify ecological and land use changes.

Mines are no longer abandoned; they are closed following legislated federal or provincial procedures and, when maintained according to these procedures, have few downstream impacts. Mines were abandoned in the past, however, and Canada still has many abandoned mining sites scattered across the landscape that have never been reclaimed. Abandoned metal mines potentially introduce the most serious mining-related environmental problems. The tendency of sulphide-bearing rocks to oxidize and generate acidic effluents was noted earlier in the chapter. Every year, the Canadian mining industry generates approximately 650 million tonnes of tailings and waste rock, about half of which are from sulphide ore operations. In the absence of naturally occurring (or applications of) acid-neutralizing materials such as calcite or limestone, toxic metals from surface tailings and mine wastes may leach into nearby watercourses in the form of acid mine drainage. When tailings contain high levels of sulphide material, the potential for acid generation is severe. Untreated acidic wastes also carry toxic concentrations of metals and high levels of dissolved salts. Rainfall and melting snow flush the toxic solutions from waste sites to the surrounding watershed, contaminating watercourses and groundwater.

Although the process of acid mine drainage can be slowed, and the acidic effluent can be treated, it is difficult to prevent completely. Current treatment facilities at mine sites are effective in preventing environmental contamination downstream provided they are well maintained and operated. Unfortunately, acid generation may continue for hundreds of years following mine closure, making treatment plants neither financially nor operationally viable. Furthermore, conventional treatment

Photo 10–10
Tourism allows communities affected by mine closures, such as Elliot Lake, Ontario, to maintain their viability.

using lime offers limited long-term benefit as the volume of sludge produced in the treatment process would exceed the volume of tailings in only a few decades.

In 1994, over 12 500 hectares of tailings and 740 million tonnes of waste mine rock were generating acid drainage in Canada. According to these figures, acidic drainage liability in Canada was estimated at between $2 billion and $5 billion. Promoting the growth of vegetation on tailings and waste rock was expected to alleviate acid drainage so that mining companies could abandon sites without future liability. However, it was discovered that the presence of vegetation did little to improve the quality of water drainage. The Canadian mining industry and governments realized that new reclamation technology needed to be developed and in 1986 established a task force to conduct research on acidic drainage.

Recommendations from the task force led to the creation of the Mine Environmental Neutral Drainage (MEND) program in 1989 (see Box 10–3). The MEND program has advanced research and led to improved technology for reducing problems associated with acid mine drainage. Many mine operators favour underwater disposal of mine tailings as part of a decommissioning program. An anaerobic (without oxygen) environment prohibits the production of sulphuric acid in tailings kept underwater. In the long term, natural sedimentation will cover the tailings, preventing their contact with oxygen.

There are more than 10 100 abandoned mines in Canada (see Table 10–4), including eight Atomic Energy Control Board uranium mines (Mackasey, 2000). The cost to clean up abandoned mine sites across Canada was estimated in 1996 to be $6 billion (Young, 1996). A large portion of this total can be attributed to degradation resulting from past practices that are no longer permitted. Abandoned mine sites that represent an unacceptable risk to the environment or human health and safety need to be rehabilitated. Most provincial and territorial governments have surveyed abandoned (no longer operating, but owners known) or orphaned (no longer operating, owners unknown) mine sites to identify the number involved and to assess the level of degradation and cleanup efforts required. Responsibility for cleaning up many of these abandoned sites is often left with provincial governments, as the previous owner or operator of the property can no longer be identified, is insolvent, or is otherwise unable to pay. However, governments are reluctant to take responsibility and frequently have more immediate demands on public funds.

It is important to note that governments and nongovernmental organizations do not believe that the list in Table 10–4 is complete (Mackasey, 2000). The numbers likely grossly underestimate the number of abandoned mines in Canada today.

Each of the provinces and territories has legislation in place to administer abandoned mines and mine reclamation (Mackasey, 2000). Mining-related acts are employed in British Columbia, Newfoundland and Labrador, Nova Scotia, New Brunswick, Quebec, Ontario, and Yukon.

BOX 10–3
CANADA'S MEND PROGRAM

Canada's initial Mine Environmental Neutral Drainage (MEND) program was a cooperative program financed and administered by the Canadian mining industry, several federal government agencies (including Natural Resources Canada, Environment Canada, and Indian and Northern Affairs), and provincial governments.

The program was a collaborative effort to research and develop technologies that would prevent or substantially reduce environmental problems caused by acid mine drainage and the financial liabilities that accrued to public agencies at abandoned mine sites. For tailings and waste rock piles of many existing and orphaned mines, however, the best that can be expected is long-term containment and treatment to neutralize acid drainage and remove dissolved metal contaminants. The MEND program confirmed that prevention is the best solution for acidic drainage.

Up to 1997, more than $18 million was spent on the MEND program, resulting in such research advances as:

- development of precise methods to predict and measure the extent of acidic drainage before it occurs
- subaqueous tailings disposal (such as at the Polaris Mine)
- layered earth covers
- engineered wetlands (such as in Elliot Lake)
- use of solid covers and wet barriers to prevent and control acid runoff
- use of biotechnology to treat small acidic seeps

MEND 2000 was a three-year program (1998–2000) designed to transfer the knowledge gained from MEND and other related acidic drainage research projects to users via workshops, reports, and on-line services. Funded equally by the Mining Association of Canada and Natural Resources Canada, the program also initiated and is monitoring long-term field scale and other projects designed to further reduce the environmental liability associated with acidic drainage.

SOURCES: *The State of Canada's Environment—1996,* Government of Canada, 1996a, Ottawa: Supply and Services Canada; *Mine Environmental Neutral Drainage Program,* Natural Resources Canada, 1996, http://www.emr.ca/mets/mend/; *MEND 2000 Mission,* Natural Resources Canada, 1999, http://mend2000.nrcan.gc.ca/mission_e.htm)

TABLE 10-4
ABANDONED MINES IDENTIFIED AND ON FILE IN CANADA (2000)

Jurisdiction	Number of Abandoned Mines
Alberta	2100
British Columbia	20 (estimated)
Manitoba	Not available
New Brunswick	60
Newfoundland and Labrador	39
Nova Scotia	300
Ontario	6015
Prince Edward Island	No mining in province
Quebec	1000
Saskatchewan	505+
Yukon	120
Nunavut	3
Northwest Territories	37
Atomic Energy Control Board	8

SOURCE: *Abandoned Mines in Canada,* W. O. Mackasey, 2000, http://www.miningwatch.ca/publications/Mackasey_ abandoned_mines.html. Reprinted by permission of WOM Geological Associates Inc.

Environment-related acts are used in Saskatchewan, Northwest Territories, Nunavut, and Yukon, while Alberta uses the Coal Conservation Act. Manitoba has a policy document in place. Environment-related acts also apply in Alberta, British Columbia, and Newfoundland and Labrador. Federally, acts that apply to abandoned mines and reclamation include Environmental Assessment, Environmental Protection, Fisheries, and the Atomic Energy Control Act. In May 2000, the Atomic Energy Control Board became the Canadian Nuclear Safety Commission with a mandate to be more active in regulating and licensing sites with significant radioactive substances resulting from nuclear operations (see also Chapter 11).

Mined-out shafts do not necessarily have to be closed and capped as is commonly assumed. For example, a mined-out chamber of a zinc–copper mine in Manitoba has been converted successfully to a thriving garden supporting a wide variety of plant species (see Enviro-Focus 10). Reclamation of gravel mine sites is now quite common as, in order to obtain permits to extract gravel, mining companies are required to establish remediation strategies that may include planting new vegetation and providing recreational facilities (e.g., small lakes, picnic areas) on site after the gravel has been removed. Across the country, abandoned underground mines have been considered possible locations for mine tailings and/or for disposal of solid and nuclear wastes generated elsewhere. Proposals for such uses are often controversial as they require safe transportation, disposal, and maintenance of wastes over considerable periods of time. No one knows, for example, if we can guarantee the safety of nuclear wastes for thousands of years. The use of mine sites for these purposes, however, draws attention to the increased awareness that sustainable mining must consider the entire life cycle of the materials that are extracted and of the sites where extraction takes place.

In a very curious example of remediation of toxic sites, one location in the United States has been called the "World's Most Ironic Nature Park." The Rocky Mountain Arsenal, located just outside Denver, Colorado, was the location of a major chemical weapons manufacturing facility of the U.S. Department of Defence during World War II. After the war, the land was leased to companies that made commercial pesticides. For nearly 40 years, millions of gallons of highly poisonous chemicals were deposited in landfills and waste basins on the site. The chemicals also contaminated soils and groundwater. Yet, there was no urban sprawl from Denver in this direction, and in 1986, scientists discovered that the Arsenal was home to a large population of wintering eagles. Since then, other wild flora and fauna have been observed, so that the site now boasts wildlife populations that are more diverse and abundant than anywhere else in the central Rockies. In 1992, Congress passed an act to designate the site a future wildlife refuge. The Rocky Mountain Arsenal National Wildlife Refuge is now managed by the U.S. Fish and Wildlife Service in association with the Army, Shell Oil Company, and regulatory agencies. Staff members at the refuge now work with biologists and other scientists to make it one of the largest urban national wildlife refuges in the United States. Efforts are being placed on restoring the landscape to the prairie ecosystems that characterized the region prior to World War II. The refuge also has a large public education campaign to provide environmental education for visitors who come to enjoy its "natural" beauty. From efforts at this site, we might learn a great deal about how best to remediate abandoned mine sites that have also experienced severe biophysical and chemical degradation.

Beyond mining activities directly, the recycling of minerals and metals has become an important economic venture. Reuse and recycling practices, combined with the long life span of minerals and metals, can help increase the stock of extracted minerals and metals and reduce the need for primary extraction. Because Canada has developed a large metal smelting and refining capacity, and because scrap and recycled metals follow the same metallurgical pathway through smelters and refineries as do primary metals, Canada has an excellent base for metal recycling.

Most metal products produced in Canada are made from a mixture of primary and recycled metals. For

Precious Metals, or Petals?

Exhausted mineshafts and tunnels do not necessarily have to end up abandoned. In a fully operational copper mine in Flin Flon, Manitoba, a rich diversity of plant species flourishes 365 metres below the surface. And with advances in biotechnology, the value of underground gardening is just beginning to take root.

Canada's adventure in underground gardening began in the late 1970s when Inco began growing market vegetables at one of the company's mines in Creighton, Ontario. In 1984, the company switched to growing pine seedlings, replacement trees for those damaged by mine activity. Looking to make use of its kilometres of exhausted mine tunnels, the Hudson Bay Mining and Smelting Co. Ltd. approached Saskatchewan-based Prairie Plant Systems to evaluate the potential of using the spent tunnels for biotechnology research. The mining company also was interested in the possibility of growing plant material for use in mine site reclamation. Despite initial concerns, the underground garden has been extremely successful, with some varieties growing almost three times as fast as they would above ground.

The spectacular level of growth may be attributed to the reduced amount of stress plants endure in mine tunnels. Conditions underground—everything from lighting and temperature to moisture and fertilizer—are close to ideal. In the mine, aboveground stresses such as drought, wind, excess sun and water, insects, and disease are avoided. The president of Prairie Plant Systems believes that with conditions such as these, a plant's genetic system is able to direct all its energy toward growth.

The opportunity to expand underground botanical activities may be limited only by the availability of mine sites. Experience in Flin Flon indicates that a mine acts

Photo 10–11
Underground gardening in Flin Flon.

like an artificial growth chamber. Not only do mines speed up the growing process, they also provide an environment that produces good-quality plants and trees. Farmers and pharmaceutical companies are attracted by the prospect of perfect plants. For instance, Saskatoon berry bushes have become a crop of choice for diversifying western farmers. Although there is a shortage of good-quality Saskatoon berry bushes, the mine has been able to produce perfect plants that are hardy and produce berries that are uniform in size, flavour, and time of ripening.

Drug companies also are intrigued by the potential of mines to provide high-quality, rapidly growing plants for medicinal purposes. The Pacific yew, for instance, is in high demand because its bark contains taxol, used in the treatment of ovarian cancer (see Enviro-Focus 3). As a result, the Flin Flon mine is cultivating yew trees with the intention of easing pressure on the wild Pacific yew. The federal government has also been using the mine site in Flin Flon to grow marijuana for medicinal purposes.

SOURCE: "Roses from Rock," B. Ryan, January–February 1995, *Equinox, 79,* 50–55. Reprinted by permission of the author.

instance, about 50 percent of the 15 million tonnes of iron and steel produced annually in Canada comes from recycled iron and steel scrap. More than 90 percent of the lead consumed in Canada can be recycled economically; in 1995, about 40 percent of Canada's total refined lead production came from secondary lead sources. On a global basis, scrap copper accounts for about 40 percent of the raw material input of refined copper production and consumption. The automotive industry is responsible for about 80 percent of the end use of secondary aluminum. In 1971, an average of about 35 kilograms of recycled aluminum was used in the production of each new car; by 1994, that average figure had climbed to 68 kilograms per vehicle (Government of Canada, 1996a). Currently, over 1000 scrap metal recycling companies operate in Canada, providing direct employment to about 20 000 people (Natural Resources Canada, 2003). They handle over 11 million tonnes and $3 billion worth of metals annually. In 1994, the Canadian trade in recyclable metals exceeded 4 million tonnes and was valued at over $2 billion (Government of Canada, 1996a).

Metal recycling has significant environmental benefits when compared with primary production of metals. The reduced demand for raw material means more efficient use of minerals and increased energy conservation. Energy savings are gained through the reduced quantity of fossil fuels used to generate electricity or operate smelters. Reduced use of fossil fuels means reduced levels of air emissions such as carbon dioxide, sulphur dioxide, and nitrogen oxides, as well as reductions in effluent discharges. Typical energy savings realized by recycling are 95 percent for aluminum, 85 percent for copper, 65 percent for lead, and 60 percent for zinc. Producing steel from recycled materials results in energy savings of 74 percent when compared with the energy used in primary production.

RESPONSES TO ENVIRONMENTAL IMPACTS AND CHANGE

MARKET FORCES

Canada is proud of its position as a global leader in mining and is making great efforts to ensure that its high ranking continues. Both federally and provincially, policies and guidelines have been adopted to promote the growth of the mining industry. Although sustainability is recognized as an important component, the quest to grow and to achieve international competitiveness often seems to take precedence. For example, sustainable development objectives of Natural Resources Canada's 1996 Minerals

and Metals Policy include ensuring the international competitiveness of Canada's minerals and metals industry in the context of an open and liberal global trade and investment framework.

Unfortunately, environmental policies related to mining are not uniform throughout the world. As a result, countries that have higher environmental standards are often left at a disadvantage when it comes to attracting investment and remaining competitive. In Chile, for example, monitoring and enforcement of environmental laws are rare, obtaining access to information on environment and development impacts is difficult or impossible, and health and safety standards are inadequate (Environmental Mining Council of British Columbia, 1996). Industry leaders, such as Canada's Barrick Gold, praise the attitude of Chile toward mining companies, while others, including the World Bank, are critical of the absence of regulations and administrative obstacles in Chile that make it difficult to assess or enforce environmental standards. The environmental and worker safety conditions that exist at many South American mine sites would not be tolerated in Canada. However, with operating costs averaging about two-thirds that of Canada, South America is gaining a distinct competitive advantage for copper production (Environmental Mining Council of British Columbia, 1996; Holden, 2003).

PARTNERSHIPS FOR ENVIRONMENTAL SUSTAINABILITY

Major players in Canada's mining industry realize that they have to work hard to overcome a long history of environmental neglect. Abandoned operations, high emissions, and dangerous human health conditions have given the mining industry a reputation that is less socially desirable than that of other forms of economic and land use activities. Government regulations, at both the federal and provincial levels, have prompted mining companies to invest in research and environmental technologies to improve their operations. Environmental nongovernmental organizations (ENGOs), such as the Mining Council of British Columbia, act as environmental watchdogs over industrial practices at home and abroad. In the past decade or so, a variety of cooperative programs and initiatives have been working toward fostering sustainable development of mining and mineral processing activities.

The Environmental Mining Council of British Columbia (EMCBC) is a coalition of environmental groups working toward environmentally sound mining laws and practices. Organizations included in the EMCBC are the Canadian Nature Federation, the Canadian Parks and Wilderness Society, the West Coast Environmental Law Society, British Columbia Spaces for Nature, and the

Yukon Conservation Society. Formed in an effort to advance a coordinated strategy on mining, the EMCBC acts as a clearinghouse for a variety of technical and strategic information on mining. Member groups acknowledged that while British Columbia's environmental community had a great deal of depth regarding forestry and wilderness issues, there was a significant gap in their ability to respond effectively to the technically complex and politically distinct world of mineral development. They acknowledged, also, that more work needed to be done to increase awareness of the threats of mining pollution and habitat destruction (British Columbia Spaces for Nature, n.d.).

In response to growing public concern about provincial mining regulations and assessment processes, the EMCBC initiated an education and advocacy program called BC Mining Watch. The program aimed to build the capacity of activists to respond to environmental threats posed by mineral development, from exploration to mine abandonment. The BC Mining Watch project carried out its central objective in three main ways, by: (1) supporting effective documentation of environmental mining conflicts; (2) supporting communication and alliance between ENGOs, First Nations, and labour groups on environmental mining issues; and (3) coordinating and communicating a clear agenda for environmentally appropriate mining regulations and practices (British Columbia Spaces for Nature, n.d.).

One of most successful advances in reducing environmental impacts of mineral processing involves the initiatives that have been taken to lower sulphur dioxide emissions. Ontario's Countdown Acid Rain Program, initiated in 1985, was a key motivator in prompting companies to attain the required reductions in sulphur dioxide emissions. At Inco's nickel smelter in Sudbury, Ontario, the program prompted a 90 percent reduction in emissions. This positive event was tempered by intensification of continued emissions, however (see Box 10–4).

Efforts to reduce emissions in other parts of Canada have also generated positive results. For instance, at Flin Flon, Manitoba, the Hudson Bay Mining and Smelting zinc smelter implemented a new hydrometallurgical pressure process rather than conventional ore roasting. This process leached toxic metals and sulphur ore out of wastes in solution. Installation of a new zinc plant resulted in significant emission reductions—up to 98 percent of the sulphur contained in the ore was captured. In addition, use of Gore-Tex fabric for filtering purposes greatly reduced particulate emissions and allowed mine operators to aim for particulate emissions as low as 16 percent of the allowable limit (Government of Canada, 1996b).

The Accelerated Reduction/Elimination of Toxics (ARET) program is another initiative that helps companies reduce emissions of particulate matter and sulphur dioxide from industrial smelters. As part of the ARET program, 17 base-metal production companies committed to reduce emissions of arsenic, cadmium, lead, mercury, and nickel by 80 percent by 2008, from 1988 levels (see Table 10–5). Between 1988 and 1993, the companies reduced releases by 43 percent, and by 1998 had achieved a reduction of 73 percent (5824 tonnes) from 1988 levels. Among the companies contributing to major reductions in zinc, lead, arsenic, and cadmium emissions were Hudson Bay Mining and Smelting and Noranda's Horne smelter. Increases in lead and copper emissions were reported at Inco. (Other air quality initiatives are discussed in Chapter 5.)

Cooperative efforts in the minerals sector extend beyond atmospheric antipollution programs. In 1993, the Aquamin program was initiated to assess the impacts of mining on aquatic environments. Aquamin's multistakeholder group included representatives from various federal government departments, eight provincial governments, the Mining Association of Canada, Aboriginal groups, and ENGOs. The final report of Aquamin made a number of key recommendations, including improving federal effluent regulations and updating the Environmental Code of Practice for Mines.

Canada's role in fostering technological development in the mining and metals industry is an important one. Research activities of federal agencies such as the Canada Centre for Mineral and Energy Technology (CANMET) include the pursuit of sustainable development objectives in advancing mineral science. The federal government also supports stronger links between the scientific community and policy organizations to advance sustainability objectives (Government of Canada, 1996b).

TABLE 10–5
MINING AND SMELTING EMISSIONS, 1998

Substance	Percent of Emissions (total: 2112 tonnes)
Lead	26
Zinc	25
Copper	22
Nickel	15
Arsenic	8
Cadmium	2
Other	2

SOURCE: *Environmental Leaders 3 Update: Voluntary Action on Toxic Substances,* Accelerated Reduction/Elimination of Toxics, 2000, http://www.ec.gc.ca/ARET/el3u/aret_elsu_e.pdf

BOX 10-4
THE GREEN TRANSFORMATION OF SUDBURY

Over a century of nickel smelting by Inco and Falconbridge near Sudbury, Ontario, devastated the surrounding landscape and waters. Acidic smelter emissions fouled the air, and, as species' tolerance limits were exceeded, hundreds of square kilometres of once vast forests essentially disappeared along with the aquatic life of the region's lakes. Much of the worst damage was done in the first 50 years of operation, when ore was smelted over huge, open wood fires.

Since the 1970s, new smelting processes and the construction of high smokestacks (including the 381 metre superstack built in 1972) have reduced the local and overall impacts of emissions. In Sudbury, for instance, the level of sulphur dioxide in the air dropped 83 percent between 1970 and 1988. (If we recall the law of matter, however, we know that emissions from stacks do not simply disappear but appear somewhere else. In this case, air pollution problems shifted to areas downwind, sometimes as far away as Quebec.)

Between 1965 and 1985, Inco's Sudbury operations reduced emissions of sulphur dioxide by 70 percent, the largest decrease in emission tonnage among North American smelters. In 1985, the Ontario government unveiled its Countdown Acid Rain Program, which required Inco to increase its containment of sulphur dioxide to 90 percent. During the 1980s, Inco invested $530 million on related research and developed the necessary new technology. Inco's sulphur dioxide abatement project turned into a wide-ranging facility improvement plan that also generated cost savings of over $50 million per year and reduced carbon dioxide, nitrogen oxides, and particulate emissions, as well as sulphur dioxide emissions.

On January 1, 1994, the company met the government's deadline for reducing sulphur dioxide emissions to a maximum annual level of 265 000 tonnes, setting new standards of environmental stewardship for other companies to follow. This is a considerable achievement, but its effectiveness has been offset somewhat by the fact that sulphur dioxide emissions continue, and their effects have been intensified by other acidifying emissions such as nitrogen oxides.

In the 1970s and early 1980s, competition from foreign nickel mining operations resulted in substantial declines in mining and smelting employment in Sudbury. However, public-sector jobs and other white-collar employment opportunities developed in their place, partly as a result of interventions of provincial and federal governments, and partly as a result of local initiatives. This employment shift not only eased the social effects of the downturn in the nickel industry, it also transformed Sudbury into a regional service and hospitality centre.

As Sudbury's air quality improved, and as lake water quality also improved, extensive land and aquatic restoration efforts began. Beginning as early as 1973, through the efforts of an interested (and empowered) citizenry and with the cooperation and contributions of private-sector businesses and governmental agencies, a reclamation strategy was developed. Careful research, including a baseline inventory, and field trials were undertaken as the first steps toward reclamation (Bagatto & Shorthouse, 1999).

From 1981 through 1998, thousands of students and others helped reclaim 4700 hectares of heavily degraded land. Limed, fertilized, and planted with grasses, legumes, and millions of shrubs, conifers, and hardwood trees, Sudbury's landscape is recovering. Insects, birds, and mammals are coming back. An unexpected ecological benefit—the decline in lake and river acidity—has prompted efforts to restore aquatic systems as well. The care and concern of individuals, and the support of the community at large, have helped restore degraded ecosystems and provided for continued business and tourism growth in the local area.

Photo 10–12a   **Photo 10–12b**

Facility improvements at the Copper Cliff smelter in Sudbury have reduced atmospheric emissions.

SOURCES: Adapted from *The State of Canada's Environment—1996,* Government of Canada, 1996; "Biotic and Abiotic Characteristics of Ecosystems on Acid Metalliferous Mine Tailings near Sudbury, Ontario," G. Bagatto & J. D. Shorthouse, 1999, *Canadian Journal of Botany, 77,* 410–425; "Beyond Repair," D. Havinga, 1999, *Alternatives Journal, 25*(2), 14–17; "The Transformation of Sudbury," N. Richardson, 1999, *Alternatives Journal, 25*(2), 15.

The Ottawa-based International Council on Metals and the Environment (ICME) represents major nonferrous and precious-metal producers from five continents. Canadian companies help ICME foster environmentally sustainable economic development by defining environmental management systems for the mining industry, including the ISO 14 000 series of international standards.

The Whitehorse Mining Initiative (WMI) was established in 1992 in order to find solutions to economic and environmental realities in the Canadian and global mining industry. Initiated by the Mining Association of Canada, the WMI accord is perceived as an important key to the future of the Canadian mining industry. The accord advocates change toward a sustainable mining industry within the context of a commitment to social and environmental goals and within the framework of an evolving and sustainable Canadian society (Natural Resources Canada, 1996b) (see also Box 10–1).

The vision of the WMI is one of a socially, economically, and environmentally sustainable and prosperous mining industry, underpinned by political and community consensus. In addition to this vision, the accord contains 16 principles, 70 goals, and a statement of commitment to follow-up action. The principles and goals have been grouped into six main categories: (1) addressing business needs, such as streamlining and harmonizing regulatory and tax regimes; (2) maintaining a healthy environment, by adopting sound environmental practices and establishing an ecologically based system of protected areas; (3) resolving land use issues, beginning by recognizing and respecting Aboriginal treaty rights and guaranteeing stakeholder participation where the public interest is affected; (4) ensuring the welfare of workers and communities by providing workers with healthy and safe environments and a continued high standard of living; (5) meeting Aboriginal concerns by ensuring the participation of Aboriginal peoples in all aspects of mining; and (6) improving decisions through the creation of a climate for innovative and effective responses to change (Natural Resources Canada, 1996b; "Principles and Goals," 1995–96).

The WMI's vision is complex, and its successful implementation requires the full commitment of all stakeholder groups. The WMI is an important initiative for balancing economic and environmental goals and achieving sustainability of Canada's mineral and ecological resources.

FUTURE CHALLENGES

STEWARDSHIP

The cooperation demonstrated by the various parties in preparing the Whitehorse Mining Initiative is an impor-

tant step toward the stewardship of Canadian mineral resources. Successful implementation of the WMI accord would demonstrate to the world Canada's commitment to environmental sustainability through sound economic, social, and ecological frameworks and policies. Efforts made toward improving the stewardship of minerals in Canada, and other parts of the developed world, should be transferable to the developing world.

Developing nations, often desperate for revenue, frequently fail to legislate (or enforce) sustainable mining policies and regulations. As a result, mining companies, particularly multinationals, must assume a high degree of corporate responsibility when undertaking projects in countries with less stringent environmental regulations. Although it may be economical to cut corners, the risk of accidents, environmental disasters, and injury to mine workers increases if adequate environmental and safety conditions are not followed. In the long term, the initial savings gained by failing to implement proper environmental safeguards may be outweighed by the high costs of accident cleanup and reclamation efforts at a later stage. Corporations that demonstrate effective stewardship of natural resources should be recognized by individuals and governments for their accomplishments.

The stewardship of mineral resources, particularly metals, is demonstrated by efforts to reduce, reuse, and recycle. Recycling extends the efficient use of metals, reduces pressure on landfills and incinerators, and results in energy savings relative to the level of energy inputs required to produce metals from primary sources. In fact, many minerals and metals can be reused indefinitely because of their value, chemical properties, and durability.

With these kinds of characteristics in mind, Germany instituted an extended producer responsibility (EPR) policy through its Packaging Ordinance of 1991. Subsequently debated (and frequently copied)

Photo 10–13
Recycling plant in Burlington, Ontario.

throughout the world, this approach extends "the responsibility of producers for the environmental impacts of their products to the entire product life cycle, and especially for their take-back, recycling, and disposal" (Fishbein, 1998). That is, environmental costs are incorporated into product costs, and the costs of collecting, sorting, and recycling used car parts, computers, televisions, or refrigerators, following their use by consumers, are shifted from municipal governments to private industry. The idea is that making producers (not the public) pay for waste management would give them an incentive to make less wasteful and more economically recyclable products. In addition, EPR was expected to stimulate new recycling technologies and enhance Germany's competitive position as a major exporter of environmental technologies.

Industry responded to the challenge of designing and implementing its own system to take back and recycle packaging waste (the first sector targeted for EPR) by initiating the Dual, or Green Dot, System. In this system, households have two bins, one for regular garbage that they pay their municipality to collect, and one for packaging that the Duales System Deutschland (DSD) collects for free (DSD is the nonprofit company established to operate the recycling system). Between 1991 and 1999, packaging consumption per person decreased from 95.6 kilograms in 1991 to 82.5 kilograms in 1999 (almost 14 percent). Between 1993 and 1996, packaging recycling increased from 52 to 84 percent. By 1999, 52 million tonnes of packaging had been forwarded for recycling. In 2002, approximately 6.3 million tonnes of garbage were collected from households and containers. Of this total, 5.3 million tonnes of sales packaging were forwarded for recycling. It was estimated that used sales packaging saved 67.5 billion megajoules of primary energy in 2002 while it prevented the emission of 1.5 million tonnes of greenhouse gases. This is the equivalent to the exhaust emissions that would be caused by city buses driving 1.2 billion kilometres (*Der Grüne Punkt*, 2003). At least 28 countries now have packaging take-back laws. The Packaging Ordinance also stimulated development of high-tech sorting and recycling technologies using infrared and laser beams. Germany already is licensing some of its new technologies in Japan and expects to expand its exports of environmental technology within Europe and Asia (Fishbein, 1998).

Vehicles are an excellent example of how extended producer responsibility can impact product design. European vehicle producers have been redesigning their cars for disassembly and recycling since the early 1990s. They have increased the recycled content, reduced the number of plastic resins, labelled plastics, marked parts to permit draining of fluids (to avoid contamination of recycling feedstock), and used fasteners that facilitate disassembly. In the United States, where no EPR policies are in

Photo 10–14
Plastic containers are recycled into durable picnic tables and benches.

place for end-of-life vehicles (ELVs), members of the voluntary Vehicle Recycling Partnership, including Chrysler, Ford, and General Motors, are working on design changes to make it easier to recycle discarded vehicles.

As experience has shown, EPR policies can result in reduced consumption of energy and materials, lower toxicity of products, fewer negative environmental effects of manufacturing, and improved efficiency in resource use. Canada likely has something to gain by considering the contributions EPR policies can make to stewardship as well as to sustainability.

PROTECTION AND MONITORING

With increased global competition in the minerals and metals sector, Canada is striving to maintain its role as a leading mineral producer. To attract foreign investment and retain a competitive edge, domestic policy- and decision-making processes must be responsive to international factors.

Federal initiatives, such as the streamlining of environmental regulations for mining, are intended to promote a positive investment climate. For the mining industry to remain an important component of the Canadian economy, new mineral deposits must be discovered. Because mineral reserves are finite, exploration will advance into more remote regions of the country and conflicts will arise over key wilderness areas. The Cheviot Mine proposal in Alberta is a case in point (see Box 10–5). It is important that the exploration and development of these new deposits be conducted with ecosystem sustainability in mind. Canada must identify and protect areas of significant terrestrial and marine habitat from mineral and other forms of development. Protected areas strategies should be coordinated with the provinces to ensure critical regions are safeguarded. Where mining is permitted, monitoring strategies should be implemented to assess the effect of development on ecosystem health.

KNOWLEDGE BUILDING

As land-based mines become exhausted and their extraction becomes less cost effective, we may turn to the oceans for many of our mineral requirements. The extraction of mineral fuels from the ocean floor has become a major economic activity; however, deep-sea mining of nonfuels remains in its infancy. As mineral exploration of the seabed increases, we must consider the diverse physical and ecological processes that occur in our oceans and proceed with development on a sustainable basis. Development decisions, both on land and in the ocean, should be based on the precautionary principle.

Research efforts in the mining field focus on all aspects of mineral production, from exploration to processing and site decommissioning. Recently, we have seen important scientific advances in controlling toxic emissions and acid mine drainage. However, much more work needs to be done in the areas of waste management, aquatic effects monitoring, and reduced energy consumption. For instance, there are over 10 000 active, abandoned, and orphaned tailings sites in Canada (see Table 10–4). There is no comprehensive inventory of the risks these sites pose. This is one of the significant gaps in knowledge about the environmental effects of mining that needs to be filled.

Coordination of research efforts through national partnerships may increase technical innovation and allow programs to be delivered with maximum efficiency. International collaboration and the sharing of expertise are essential for meeting the challenge of sustainable development, particularly in the developing world. As a leader in mining, Canada has an important role to play in fostering the global sustainability of mining operations through our research, technology, and policy efforts.

BOX 10–5
THE CHEVIOT MINE: STRIP MINING PARADISE?

Jasper National Park is part of a UNESCO World Heritage Site, meaning its wilderness and wildlife are of outstanding global significance. Despite this designation, Cardinal River Coals (CRC) received permission to dig an open-pit coal mine less than two kilometres from the park boundary. The proposed mine consists of a chain of 26 deep pits extending for 22 kilometres and 1 to 3 kilometres wide. Over a 20-year period, CRC plans to extract about 90 million tonnes of coal, 65 million of which is expected to be marketable, for a total of about US$3.4 billion. About 450 people would be employed directly in the mine.

The huge mine would lie across the headwaters of the McLeod and Cardinal rivers (the Cardinal Divide), near treeline, at the foot of the front ranges. This location—as environmentally sensitive as it is beautiful—is 10 040 hectares of gently rolling subalpine forest owned by the Crown (public land). The mine's processing plant and some of its pits would be located in the old coal mining area of Mountain Park, which was damaged by mining in the first half of the century and continues to recover. Much of the new mining would occur in undamaged lands.

CRC is a joint-venture company consisting equally of Canadian Luscar Ltd. and US Consolidated Coal (the largest coal mining company in the United States). CRC's mine application was reviewed at a joint federal–provincial public hearing in January 1997; two of the three-person panel were members of the Alberta Energy and Utilities Board, and the third person was appointed by the Canadian Environmental Assessment Agency. CRC justified the Cheviot Mine on the basis that the company's existing Luscar coking-coal mine would exhaust its supply in about four years, and a new mine would be needed to supply the Japanese market.

At the hearing, Parks Canada stated that the Cheviot Mine proposal had the potential to adversely affect the ecological integrity of Jasper National Park. Concerns related specifically to the loss or alienation of habitat, to impacts on essential wildlife travel corridors that link Jasper National Park and the high-quality habitat in adjacent provincial lands, and to increased wildlife mortalities. Over time, the cumulative impacts of this project and other planned or proposed activities such as timber harvesting, access, and oil and gas exploration would result in the extirpation of grizzly bears, wolves, wolverines, and cougars from the region. Expert witnesses from both CRC and Parks Canada concluded that the mine would result in the direct loss of quality habitat and wildlife travel routes for at least 100 years.

The Cardinal Divide is a hot spot of biological diversity, an unglaciated area that is home to 76 species of birds, 40 species of mammals, and 1 species of amphibian. There are many rare, disjunct, and threatened species in the proposed mine area; some species have been found nowhere else on Earth. The Upper McLeod River system supports the largest concentration of breeding harlequin ducks known in western Canada. Testimony at the hearing showed harlequin ducks would be affected negatively, as would Alberta's threatened bull trout. Species richness and diversity of songbirds is "as high as it gets in North America."

The Cheviot mine area was identified as part of the habitat needed for the recovery of Alberta's grizzlies, but Alberta has no endangered species legislation to protect them. In addition, the provincial government had identified the proposed mine area as a critical wildlife zone, and scientific assessments recommended the entire Cardinal Divide area be designated as a Natural Area

BOX 10-5

CONTINUED

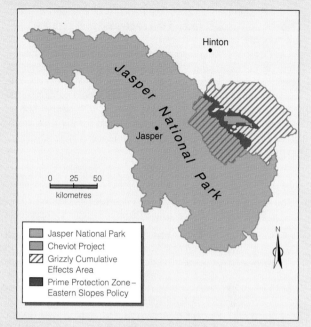

SOURCE: "Coalminer's Fodder," S. Legault, 1997, *Canadian Geographic, 117*(2), 20. Reprinted by permission of Onjo Graphics.

of Canadian Significance. Western science has pointed to the irreplaceable value of the area, and so has the traditional knowledge of First Nations people, who have used medicinal plants from the proposed mine area—plants not found elsewhere.

Reclamation plans call for 14 of the 26 pits to be left partly open to fill with water and function as trout ponds; this mitigation measure does not re-create the diversity of life found in natural streams and along their banks. Neither the bull trout nor the harlequin duck live in lakes. Open-pit lakes do not meet the Canadian government policy that requires no net loss of productive fish habitat, nor do they support the international commitments Canada has made to preserve biodiversity.

Alberta's land use planning process identified the need to diversify the region's resource extraction–based economy and highlighted tourism as an alternative. The wildlands of the proposed mine area were identified as one of the two key tourism assets in the entire region. Furthermore, environmental organizations pointed out that there were alternative mining locations that would not threaten the World Heritage Site or destroy the Cardinal Divide area. The Cheviot Mine will foreclose other options unless (international) opinion can be mobilized to persuade government officials to reject the $250 million project. U.S. environmental groups have been involved in lobbying because of Jasper's status as a World Heritage Site and because of the similarities of the Cheviot mine development to the proposed goldmine next to Yellowstone National Park that U.S. president Clinton turned down in 1996 (see Box 10–2).

Concerns about the consequences of the Cheviot Mine continued to build following the October 1997 federal Cabinet deci-

sion to approve the Cheviot Mine (subject to authorization from the federal Department of Fisheries and Oceans). In November 1997, the AWA Coalition (a group of five national and provincial ENGOs)[1] filed an application for a judicial review of the Joint Review Panel Report, challenging both the authorization itself and the environmental assessment on which the decision was based. The director of UNESCO also expressed concerns about the mine and asked the federal government to consider alternative sites.

In December 1998, finding that the Panel's original report was not in compliance with the Canadian Environmental Assessment Act, the Federal Court quashed the authorizations for the project to proceed. The three original members of the Joint Review Panel were reappointed to hear additional evidence regarding cumulative impacts and to consider alternatives to open-pit mining.

The supplemental hearing was conducted in spring 2000. Once again, the Joint Review Panel decided that approval of the Cheviot Mine was in the public interest, noting:

- any permanent harm was justifiable

- most environmental and cumulative impacts were insignificant

- any significant impacts could be mitigated effectively after development starts

- compensation would be appropriate if mitigation were not possible

(continued)

BOX 10-5
(CONTINUED)

On September 12, 2000, the Panel recommended that the Cheviot Coal project receive regulatory approval from the government of Canada.

In October 2000, Luscar Ltd. announced that the Cheviot Mine project was being shelved because the lengthy process to obtain necessary approvals had resulted in the loss of potential markets. As well, Luscar announced that it would be closing its existing mine near Hinton, Alberta, putting 400 people out of work. In December 2003, provincial permits for the mine were obtained, although environmental organizations were stepping up their opposition campaigns.

The startup of the $50 million Cheviot Mine was scheduled for November 2004. By August 2004, the Sierra Legal Defence Fund (SLDF) had filed an application in the Federal Court of Canada seeking a new environmental impact assessment (EIA) of the project. The SLDF felt a new EIA could consider (1) how the addition of a 20 km haul road increased the cumulative effects of the mine, and (2) updated scientific information about the importance of the grizzly bear population in the area to maintaining bears in southern Alberta. While the coal company acknowledged the SLDF filing could delay the project, they expected that final approval from the Department of Fisheries and Oceans would be granted.

NOTE:

[1] The AWA Coalition included Alberta Wilderness Association, Canadian Nature Federation, Canadian Parks and Wilderness Association, Jasper Wilderness Association, and Pembina Institute for Appropriate Development.

SOURCES: *Panel Findings,* Alberta Energy and Utilities Board, September 12, 2000, www.eub.gov.ab.ca/CyberDOCS30/Libraries/Default_Library/Common/frameset.asp; *Legal Action over the Proposed Cheviot Mine,* Alberta Wilderness Association, (n.d.), http://www.web.net/~awa/cheviot/cheviot.htm; *Group Wins Appeal over Cheviot Coal Mine,* Alberta Wilderness Association, December 1, 1998, http://www.web.net/~awa/alert/120198_1.htm; *Coal Mine Recommendation a Tragic Set-Back for Jasper Park and Neighboring Community,* AWA Coalition, 2000, http://www.albertawilderness.ab.ca/News/Press/20000913w.htm; Between Alberta Wilderness Association, Canadian Nature Federation, Canadian Parks and Wilderness Association, Jasper Environmental Association, Pembina Institute for Appropriate Development, Applicants, and Cardinal River Coals Ltd., Respondent. Reasons for Order. Federal Court of Canada. Docket T-1790-98, J. Campbell, April 8, 1999, http://www.fja.gc.ca/en/cf/1999/orig/html./1999fca24281.o.en.html; *10:00 a.m. news* [Radio One broadcast], Canadian Broadcasting Corporation, October 24, 2000; *Cheviot Review Panel Issues Report,* Canadian Environmental Assessment Agency, September 12, 2000, http://www.ceaa.gc.ca/panel/cheviot/rel000912_e.htm; "A Hard-nosed Look at the Proposed Cheviot Mine," B. Gadd, 1997, *Wild Lands Advocate, 5*(2), 4–5; "Coalminer's Fodder," S. Legault, 1997, *Canadian Geographic, 117*(2), 20; "Strip Mining Paradise," D. Pachal, 1997, *Encompass, 1*(2), 12–14; "Cheviot Mine Hit by Legal Snag," G. Teel, August 17, 2004, *Calgary Herald,* p. D1.

Chapter Questions

1. If mining operations occur on only 0.03 percent of Canada's land surface, why is there so much environmental concern about them?

2. Which of the four phases of mineral development do you think has the greatest environmental impact? Why?

3. What mineral resources are extracted in your local area and in your region? What mining methods are used? What laws and regulations require restoration of the landscape after mining is completed? How stringently are these laws and regulations enforced? What parallels are there between the impacts and regulation of urban development and those of mining?

4. Discuss several harmful environmental effects that mining and processing minerals have on atmospheric, aquatic, and land environments.

5. Reusing and recycling mineral and metal resources is one way to extend the availability of these nonrenewable resources. What kinds of mineral and metal recycling occur in your educational institution? In your community? What more could be done in your community or region to encourage reuse and recycling of mineral resources?

6. What are the similarities between the proposed goldmine located near Yellowstone National Park in the United States and the proposed coal mine located near Canada's Jasper National Park? Do you think Canadian public opinion could have a similar effect on the outcome of the political decision to approve the Cheviot Mine as U.S. public opinion did regarding the New World Mine?

7. Mining contributes to economic well-being, while mining activities, both directly and indirectly, affect economies, society, and environments. How do you balance these three elements in your definition of sustainable mining?

references

BHP Billiton Minerals Canada Ltd. and DIA Met Minerals. (1995). *NWT diamonds project: Environmental impact statement/BHP; DIA MET.* Vancouver: BHP Diamonds.

British Columbia Spaces for Nature. (n.d.). http://www.sunshine.net/www/0/sn0004/bc-miningwatch/

Canadian Environmental Assessment Agency. (1996). *Decommissioning of uranium mine tailings management areas in the Elliot Lake area.* Ottawa.

Department of Indian and Northern Affairs. (1996a). *Canada's diamond mine project: West Kitikmeot Slave study—a regional study of development impacts.* Backgrounder #5. http://www.INAC.ca

Department of Indian and Northern Affairs. (1996b). *Canada's diamond mine project: Environmental agreement.* Backgrounder #7. http://www.INAC.ca

Department of Indian and Northern Affairs. (1996c). *Canada's diamond mine project: Canada's gross domestic product to grow by $6.2 billion.* Backgrounder #3. http://www.INAC.ca

Department of Indian and Northern Affairs. (1996d). *Canada's diamond mine project: Land claims near the BHP site.* Backgrounder #2. http://www.INAC.ca

Der Grüne Punkt. (2003). *Duales System Deutschland AG.* http://www.gruener-punkt.de

Environmental Mining Council of British Columbia. (1996). *The real story of mining in Chile.* http://www.sunshine.net/www/0/sn0004/

Fishbein, B. K. (1998). *EPR: What does it mean? Where is it headed?* http://www.informinc.org/eprarticle.htm

Freeman, A. (1996). Government approves largest diamond mine in North America. http://www.igc.apc.org

Government of Canada. (1996a). *The state of Canada's environment—1996.* Ottawa: Supply and Services Canada.

Government of Canada. (1996b). *The minerals and metals policy of the Government of Canada.* Ottawa.

Government of the Northwest Territories. (2000). *GNWT signs off on Diavik Environmental Agreement.* http://www.gov.nt.ca/RWED/99news.htm#diaviksignsoff

Government of the Northwest Territories, Investment and Economic Analysis. (2004). *NWT economic trends,* Issue 2, 2nd quarter. http://www.gov.nt.ca/RWED/iea/index.htm

Holden, W. N. (2003). *The role of environmental protection as a locational determinant of the nonferrous metals mining industry: A synecdochic investigation of a stylized fact.* Unpublished PhD dissertation, University of Calgary.

Johnson, E. (1999). Canadian mining companies build a shaky reputation abroad. *Alternatives Journal, 25*(4), 29.

Kennett, S. A. (1999). *A guide to impact benefits agreements.* Canadian Institute of Resources Law, University of Calgary, Calgary, AB.

Mackasey, W.O. (2000). *Abandoned mines in Canada.* http://www.miningwatch.ca/publications/Mackasey_abandoned_mines.html

Marshall, I. B. (1982). *Mining, land use and the environment: A Canadian overview.* Ottawa: Environment Canada. Land use in Canada series, p. 22.

McNamee, K. (1999). Undermining wilderness: The Canadian mining industry is abandoning its support for a national network of protected areas. *Alternatives Journal, 25*(4), 24–31.

Natural Resources Canada. (1996a). *Canadian mining facts.* http://www.nrcan.gc.ca

Natural Resources Canada. (1996b). *Whitehorse mining initiative.* http://www.emr.ca/ms/sdev/wmi_e.htm

Natural Resources Canada. (2000). *Canadian mining facts.* http://www.nrcan.gc.ca/mms/efab/mmsd/facts/canada.htm

Natural Resources Canada. (2002). *Minerals and metals sector, year in review.* http://www.nrcan.gc.ca/mms/cmy/2002revu/con_e.htm

Natural Resources Canada. (2003a). *Canadian mining facts.* http://mmsdl.mms.nrcan.gc.ca/mmsd/facts/canFact

Natural Resources Canada. (2003b). *Canadian mineral production demonstrates continued strength in 2002.* Information Bulletin, Mineral Production. http://mmsd1.mms.nrcan.gc.ca/mmsd/production/default_e.asp

Nelson, J. (1996, March 14). Environmental review process put under pressure. *Victoria Times Colonist,* p. A5.

Newcott, W. R. (1994). Tatshenshini-Alsek Wilderness Park: Rivers of conflict. *National Geographic, 185*(2), 122–134.

Papua New Guinea: BHP agrees to settlement for Ok Tedi spill. (1996). http://www.igc.apc.org

Principles and goals of the Whitehorse Mining Initiative. (1995–96). Cited in *Northern Perspectives, 23*(3–4) (Fall/Winter), 9–11.

Province of Manitoba. (n.d.). *Sustainable development: Provincial mineral policies and their application.* Sustainability Manitoba series, MG 3605.

Shrimpton, S., & Storey, K. (1988). *The urban miner: Long distance commuting to work in the mining sector and its implications for the Canadian north.* Paper presented to the Canadian Urban and Housing Studies Conference. Institute of Urban Studies, University of Winnipeg.

Warhurst, A. (1994). *Environmental degradation from mining and mineral processing in developing countries: Corporate responsibilities and national policies.* Paris: Organisation for Economic Cooperation and Development.

Weber, R. (1997, July 18). Diamond mine rises from tundra. *Calgary Herald,* p. A8.

Wiesner, D. (1995). *The environmental impact assessment process.* Dorset: Prism Press.

Wismer, S. (1996). The nasty game. *Alternatives, 22*(4), 10–17.

World Wildlife Fund. (1997, January 13). *WWF Canada withdraws court action on BHP diamond mine.* News release.

Young, A. (1996, October 22). *Achieving investor security through environmentally sustainable mining.* Speech presented to the Fraser Institute.

CHAPTER 11

Energy

"The continuing prosperity and high quality of life of Canadians depend on a secure supply of affordable energy, produced with minimal impact on the natural environment. Our ability to sustain this supply will depend on two factors. First, there must be a critical mass of people with the right knowledge and a passion for achieving this goal. Second, these people must be able to become experts at collaboration, applying their diverse skills, perspectives, and interests to the creation of technological breakthroughs.....
When it comes to production of this energy, 'business as usual' is not an option."

Bolger & Isaacs (2003, pp. 58, 61)

Chapter Contents

Chapter Objectives

After studying this chapter you should be able to

- understand the supply and demand of energy resources in Canada
- identify a range of human uses of energy resources
- describe the effects of human activities related to the production and use of Canada's energy resources
- appreciate the complexity and interrelatedness of energy issues
- outline Canadian and international responses to energy issues
- discuss challenges to sustainable energy production and use in Canada

CANADIANS AND ENERGY

Canadians use energy every day when we heat our homes, travel to work, or cook our meals. Business and industry depend on abundant, reliable sources of energy to produce goods and deliver services. Energy has been important to the economy throughout Canadian history. While we may change our preferences for the products we consume, we cannot choose to abandon the use of energy. In fact, many of us take our use of energy for granted. For example, when we flip on a light switch, we do not think about the coal or hydro power that may have been used to produce electricity. However, this is not the case in many parts of the world where electricity is considered a luxury and much thought is given to how to obtain light (see Box 11–1).

While our need for energy may be constant, our sources of energy are variable. Canada is fortunate to have an abundant supply of fossil fuels, and these have provided us with most of our energy requirements over the last century. Although a driving force behind our economy, the production and consumption of fossil fuels affects our environment. Concern over climate change and air pollution (see Chapter 5) provides impetus for rethinking how energy is delivered. Consumer demand for "green" energy, such as wind power, is beginning to lead business to invest in renewable energy. This change is noticeable in the way petroleum companies have begun to market themselves as "energy companies." While this may not mean the end of fossil fuels as our dominant energy source, it does signal a potential shift toward alternative energy sources.

Photo 11–1
If electricity for electric vehicles (EVs) is produced by wind and solar technologies, these vehicles are pollution-free. Cars such as the University of Waterloo's *Midnight Sun 2* (above) competed in a long-distance race to demonstrate the potential of solar technology in transportation. Recently, gasoline–electric hybrid vehicles from several automobile manufacturers have become available commercially.

BOX 11-1

LIGHT UP THE WORLD PROJECT

The Light Up the World (LUTW) Foundation was founded by Dr. David Irvine-Halliday, a professor of Electrical and Computer Engineering at the University of Calgary, following a 1997 visit to rural Nepal. Alerted to the fact that schools and homes lacked adequate lighting, Dr. Irvine-Halliday set out to find a simple, safe, healthy, reliable, and economic source of light for the villagers.

In many parts of the world, rural homes are lit by kerosene wick lamps and, to a lesser extent, by candles or resin-soaked twigs—if they are lit at all. Each of these forms of lighting constitutes a fire hazard and leads to serious and well-documented health problems in poorly ventilated homes. Dr. Irvine-Halliday recognized the potential, afforded through the development of white light–emitting diodes (WLEDs), to bring a safe, simple, and affordable source of light to homes that otherwise would not be able to acquire traditional lighting. Batteries that supply power to the WLED lamps may be recharged using forms of renewable energy, such as human-powered pedal generators, solar photovoltaic panels, and Pico hydro generators. A child is able to read a book with the light emitted by a single 0.1 watt WLED lamp, and a room can be lit to a very acceptable level using a 1 watt WLED lamp (life expectancy of 40 years). Since electrical energy is very costly in the developing world, LUTW's objective is not to light up entire rooms or homes to North American standards of illumination but to provide adequate light to those areas where it is most useful. As a community development tool, the Solid State Lighting (WLED lamp) has had an enormous impact on the social, economic, and physical well-being of many people who live in poverty-stricken parts of the world. The ability to read and study after dark, and to operate cottage industries at night, using either fixed or mobile task lighting helps people improve their education and earn a modest living.

From the initial project in Nepal, LUTW's activities have grown to include lighting programs in India, Sri Lanka, the Philippines, Afghanistan, Mexico, the Galapagos Islands, the Dominican Republic, Angola, and Pakistan. In 2004, LUTW implemented its first pilot project in Canada to install and evaluate the potential of the Foundation's WLED lamps in the off-grid First Nation community of Xeni Gwet'in (located in the Chilcotin region of British Columbia). Participants in this project replaced their flashlights and propane lamps with WLED lamps, saving an estimated $220 per household annually in disposable battery costs alone.

Working with humanitarian organizations around the world, and through the support of interested individuals, host country

Photo 11–2
Children reading by the light of a 6-WLED lamp at Child Haven Orphanage, Bhaktapur, Nepal.

organizations, and international foundations and grants, the Foundation's goal is to install more than 200 000 WLED systems by the end of 2005. The LUTW Foundation believes that with a virtual *one-time* cost of US$60 for the WLED lighting system, compared to the US$40 to $100 *annual* cost of kerosene, the prospect of providing almost permanent lighting systems for the world's two billion people who do not have electricity is a challenging, yet reachable, goal.

The successful efforts of the LUTW Foundation to bring light to the developing world has earned it a number of prestigious international awards. In 2003, these awards included the Rolex Award for Enterprise (Geneva), the Tech Museum Award for Equality (San Jose), and the Saatchi and Saatchi Award for Innovation (London and New York). Since the LUTW continues to work to provide the capacity for host communities to continue, independently, after initial projects are completed, the Foundation is helping to ensure that lighting, literacy, equality, and economic and social progress are inseparable allies.

SOURCES: Adapted from text provided by Dr. David Irvine-Halliday, the Light Up the World website (www.lutw.org), and the LUTW's newsletter, *Enlightenment*. Courtesy of David Irvine-Halliday.

This chapter examines some of the obstacles in adopting new energy sources into our energy use mix. Our reliance on nonrenewable fossil fuels is discussed, as are other energy supplies such as nuclear fission and hydroelectric power. The environmental effects of energy use are investigated along with steps taken to improve energy efficiency. We begin by considering Canada as a producer and consumer of energy.

HUMAN ACTIVITIES AND IMPACTS ON NATURAL ENVIRONMENTS

Photo 11–3

In Canada, long corridors of transmission lines carry power from generating facilities to consumer markets.

ENERGY SUPPLY AND DEMAND IN CANADA

Canada's high standard of living is attributed partly to a reliable and relatively low-cost supply of energy. Because of such factors as our cold climate, vast distances (which encourage car use), and an energy-intensive industrial base, Canadians are the third largest per capita consumers of energy in the world and consume five times the world average (Boyd, 2001). In 2000, Canada's energy consumption was 351.5 gigajoules per person (Statistics Canada, 2002). Canadians spend more than $89 billion per year to heat and cool their homes and offices and to operate their appliances, cars, and industrial processes

(Natural Resources Canada, 2000a). Employing nearly 280 000 people, the energy sector plays a major role in the Canadian economy. Consistently during the past four decades, the energy sector constituted 6 to 10 percent of Canada's gross domestic product (GDP). Figure 11–1 illustrates Canada's primary energy consumption from 1958 to 2000 as well as growth in Canada's GDP during those years. Recently, our energy consumption per dollar of GDP has been declining. The decline is a result of two factors: (1) energy conservation practices and improved energy-efficient technologies have reduced the energy intensity of the economy, and (2) energy consumption has increased at a slower rate than the GDP, causing a decline in the energy consumption per dollar ratio (Statistics Canada, 2002).

Throughout Canada's history, different types of energy have dominated. About the time of Confederation, wood accounted for almost 90 percent of Canada's energy market, while today wood accounts for less than 5 percent. The use of coal for energy peaked about 1920, when it commanded 75 percent of the market. By 1996, the market share for coal had declined to roughly 12 percent. With industrialization and the widespread adoption of the combustion engine, oil (with a market share of 60 percent) became the dominant energy source during the 1960s. With the surge in natural gas, the market share of oil dropped to about 39 percent while natural gas accounted for 36 percent of the energy market in 2000. Figures 11–2 and 11–3 illustrate changes over time in production and consumption of primary types of energy in Canada. Alternative energy sources, such as solar and wind power, also are slowly increasing their market share as the technologies become more viable.

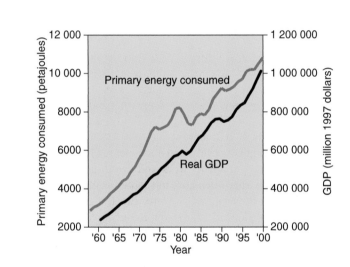

Figure 11–1

Primary energy consumption in Canada and GDP, 1958 to 2000

[1] Energy units: A joule is the international unit of measure for energy. The energy content of a 40-litre tank of gasoline is about 1.36 billion joules. One gigajoule (GJ) is 10^9 (1 000 000 000) times larger than one joule, one petajoule (PJ) is 10^{15} (1 000 000 000 000 000) times larger than one joule, and one exajoule (EJ) is 10^{18} (1 000 000 000 000 000 000) times larger than one joule. One exajoule is roughly equivalent to 28 billion litres of motor gasoline.

[2] These data exclude the use of wood and wastes as energy sources.

SOURCE: *Human Activity and the Environment: Annual Statistics 2002,* Statistics Canada, 2002. Reprinted by permission of Statistics Canada.

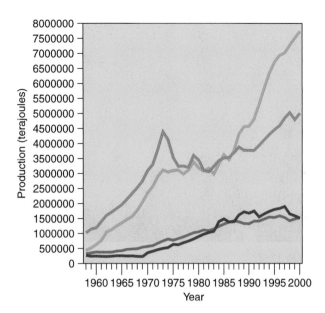

Figure 11–2

Canadian primary energy production by source, 1958–2000

SOURCE: *Human Activity and the Environment: Annual Statistics 2002,* Statistics Canada, 2002. Reprinted by permission of Statistics Canada.

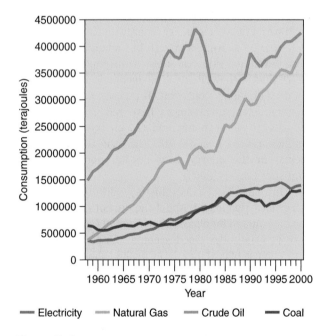

— Electricity   — Natural Gas   — Crude Oil   — Coal

Figure 11–3

Changes in primary types of energy consumed in Canada, 1958–2000

Note: One terajoule is 10^{12} (1 000 000 000 000) times larger than one joule.

SOURCE: *Human Activity and the Environment: Annual Statistics 2002,* Statistics Canada, 2002. Reprinted by permission of Statistics Canada.

HOW DO WE USE ENERGY?

Of the energy consumed in Canada, it is estimated that transportation uses 30 percent, industry uses 30 percent, and the remaining 40 percent is used in agriculture, homes, and businesses (Statistics Canada, 2003). Six industries—pulp and paper, metal smelting, steel making, mining, cement manufacturing, and petrochemicals—account for about 81 percent of all energy used by industry.

Energy use can be described as primary or secondary. **Primary energy use** represents the total requirements for all uses of energy, including energy used by the final consumer, energy in transforming one energy form to another (e.g., coal to electricity), and energy used by suppliers in providing energy to the market. **Secondary energy use** is energy used by final consumers for residential, agricultural, commercial, industrial, and transportation purposes. From 1990 to 1997, total primary energy use increased 15.3 percent from 9500 **petajoules** to 10 955 petajoules, and demand is projected to grow 27 percent between 1997 and 2020 (Natural Resources Canada, 2000b).

Net Useful Energy

It takes energy to produce energy and, as described in Chapter 3, converting energy sources into useful energy products leads to waste. This waste is divided almost evenly between degradation of energy to low-quality energy (the second law of energy) and waste brought about by avoidable human practices.

Society unnecessarily wastes a considerable amount of energy. Many Canadians work and live in poorly insulated buildings, drive gas-guzzling motor vehicles, leave the lights on when leaving a room, run the dishwasher when it is only half full, and so on. Reducing the amount of energy we waste is paramount in reducing the environmental effects of our energy consumption. For example, taking the bus, walking to the corner store, or adding an extra blanket instead of turning up the thermostat help to reduce our energy consumption and reduce energy waste.

Improving the efficiency of our energy consumption also reduces the impacts of energy use on the environment. Improving energy efficiency, for example, reduces the amount of low-quality, unusable energy that is lost during conversion. Upgrading our appliances to more energy-efficient models, ensuring our homes are properly insulated, and using a hot water heater cover are examples of ways to improve energy efficiency.

Net useful energy is the usable amount of energy available from an energy source over its lifetime. To determine the net useful energy of an energy resource, all losses are subtracted, including those automatically wasted (second energy law) and those wasted in the dis-

Photo 11–4a
Thirty cars carry 40 people . . .

Photo 11–4b

covery, processing, and transportation phases. For example, if 10 units of coal energy are required to supply 15 units of electricity, the net useful energy gain is 5 units of energy. If it takes 15 units of coal energy to produce 10 units of electricity, there is a net energy loss of 5 units over the lifetime of the system. Presently, oil has a relatively high net useful energy as it is readily accessible and easily transported; however, as deposits decline and their locations become more remote, net useful energy of oil will decrease. The concept of net useful energy is important in understanding why many alternative technologies, such as hydrogen fuel cells, are being explored at the present time.

Environmental concern over energy use is reflected throughout the life cycle of the energy source. During exploration and production, there is often disturbance to the land, conflict with other land uses, and the risk of spills or accidents. Delivering energy to the user requires some form of transport (such as that of oil through pipelines or electricity through power lines) that may be intrusive to natural ecosystems. In addition, the risk of accidents or spills increases as the distance or number of transfers increases. At the point of consumption, burning fossil fuels or wood releases emissions that affect air quality, contribute to acid precipitation, and influence global climate. At the end of the life cycle, issues surrounding waste disposal affect our environment. For example, finding a safe long-term disposal solution for radioactive waste is one of the most controversial issues in energy decision making today.

ENERGY RESOURCES

Canada's economic well-being is tied to the energy sector. Increasingly, however, Canadians realize that our well-being is achieved with considerable social, environmental, and economic cost. The construction of large energy

Photo 11–4c
. . . but one bus carries 40 passengers and takes up a lot less space.

Photo 11–4d

developments, or megaprojects, has been heralded by government and industry as critical in securing Canada's national and international economic success. Government-supported ventures, such as the Hibernia oil project off the coast of Newfoundland (see Box 11–2), oil sands development in northern Alberta, and the James Bay hydroelectric project in Quebec, offer the promise of jobs and economic security for residents of the respective regions.

BOX 11-2
IN QUEST OF OIL: THE HIBERNIA PROJECT

On December 11, 1979, Chevron Canada Resources Ltd. announced a major offshore oil discovery at the Hibernia P-15 well, located 315 kilometres east of St. John's, Newfoundland, in 80 metres of water. Later, tests on the drilling cores identified three principal oil zones with a producing capability of more than 20 000 barrels per day of light, high-grade crude. Since that day on the Grand Banks, approximately 3 billion barrels of light oil and 3.5 trillion cubic feet of natural gas have been discovered in the Jeanne d'Arc Basin, one of five identified oil-prone basins off Canada's east coast.

To put these finds in perspective, the discovered resources in the Jeanne d'Arc Basin are equal to about 45 percent of established conventional crude oil reserves in western Canada, while the undiscovered potential in the Jeanne d'Arc Basin alone is

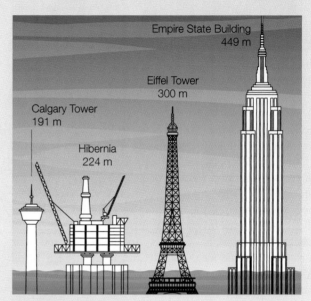

Box Figure 11–1
Relative size of Hibernia drilling platform.

Photo 11–5
The Hibernia drilling platform.

equal to the remaining light-oil potential in western Canada (the other basins have yet to be evaluated fully). Hibernia, the fifth-largest oil discovery ever in Canada, began producing oil on November 17, 1997. Production from Hibernia's estimated recoverable 750 million barrels is expected to reach peak production of about 180 000 to 220 000 barrels per day. In 2001, Hibernia averaged close to 150 000 barrels of light conventional crude oil per day (in December 2001, Hibernia produced 173 000 barrels per day) (Canadian Association of Petroleum Producers, 2002).

The major owners and players in the development of the $5.8 billion Hibernia project are Mobil Oil Canada (33 percent), Chevron Canada Resources Ltd. (27 percent), and Petro-Canada (20 percent), while the federal government (9 percent), Murphy Oil Co. Ltd. of Calgary (7 percent), and Norway's Norsk Hydro ASA (5 percent) hold smaller ownership stakes in the project. An illustration of the size of the capital investment required to undertake and develop the Hibernia megaproject, and of the cooperation required among industry and government partners, was the $150 million collaboration of Mobil, Murphy, and Chevron on the design, construction, and financing of one of the two tankers needed to move oil from Hibernia to a transshipment point on Newfoundland's Avalon Peninsula. The ships, which will hold up to 850 000 barrels, were designed to operate in the often stormy seas of the Grand Banks.

BOX 11–2
CONTINUED

Geographical conditions on the Grand Banks also meant the partners had to design a production platform that would be ice-berg resistant. Research showed that, between February and June each year, a few icebergs weighing up to several million tonnes would enter the Grand Banks in the vicinity of the oil field. Iceberg management involves two approaches: resistance or avoidance. At Hibernia, as elsewhere, avoidance is preferred.

The $5.8 million production platform sits on top of a gravity base structure (GBS) made of concrete and steel. The 1.2-mil-lion-tonne GBS, its base surrounded by a 1.4-metre-thick wall featuring 16 force-dispersing teeth, was towed out to the drilling site in June 1997 and ballasted to the ocean floor using iron ore. If tugboats are unable to tow an iceberg off a path that would lead to a collision with the production facility, the GBS was designed to withstand the impact of a 6-million-tonne iceberg travelling at 1 knot. The production system uses a floating plat-form that looks like a large ship. If necessary, the lines from the drilling rig to the production platform can be disconnected and the platform can sail out of danger if an iceberg collision appears imminent.

Although the beginning of the project was marred by delays and cost overruns, partners in the project expect about a 10 per-cent return on their investment (depending on oil prices) over the expected 20-year life of Hibernia. As of 1996, Canadian tax-payers had provided $1.1 billion in grants and $1.8 billion in loan guarantees to the Hibernia project partners. At the end of 1998, the first full year of production, Hibernia employed 705 people, 83 percent of whom were from Newfoundland. The Newfoundland Department of Finance estimated that for 1998 the province's GDP was $626 million higher and total employment was 3100 greater as a result of Hibernia. Hopes continue to be strong that the economic impacts of oil and gas development will equal or exceed those of the fishing industry.

SOURCES: "Special Report: One of a Kind," B. Bergman, March 3, 1997, *Maclean's 110*(9), 30–31; "It's a Win-Win Project," A. Boras, July 10, 1996, *Calgary Herald,* p. C2; *Offshore Eastcoast,* Canadian Association of Petroleum Producers, (n.d.), http://www.capp.ca; *Production,* Canadian Association of Petroleum Producers, 2002, http://www.capp.ca; *About Hibernia,* Hibernia Website, 2002, http://www.hibernia.ca/index.html; "Hibernia Ahead of Schedule," M. MacAfee, July 8, 1997, *Calgary Herald,* p. C8; "Last Hope," D. Martin, August 30, 1995, *Calgary Herald,* p. C1; "Drilling Starts on First Hibernia Well," D. Martin, July 29, 1997, *Calgary Herald,* p. E2.

Although large-scale energy projects have become less common overall, the oil and gas industry continues to be involved in energy megaprojects. The Terra Nova off-shore oil project is the second-largest oilfield on Canada's east coast, located about 35 kilometres east of Hibernia. A Petro-Canada–led venture, the $4.5-billion Terra Nova project commenced production in 2002. With potential reserves estimated at 470 million barrels, the project is expect to yield 129 000 barrels per day (Petro-Canada, 1999). Unlike Hibernia, which produces oil from its fields using a concrete platform base with extended reach drilling capabilities, Terra Nova oil is pumped from the seabed into a vessel capable of processing and storing hundreds of thousands of barrels per day.

As oil prices increase and Aboriginal land claims are settled, exploration and development in the Arctic are expected to grow considerably. The North is particularly attractive because large land tracts are available and fed-eral royalty schemes have subsidized companies oper-ating in high-cost environments.

Canada's ability to provide relatively inexpensive and abundant energy is paramount to the Canadian way of life. But in recent years, management difficulties, cost over-runs, and safety issues have troubled the energy industry. The following section reviews some of the societal bene-fits and environmental challenges of Canada's primary energy resources: fossil fuels (oil, heavy oil, coal, and nat-ural gas), hydroelectricity, and nuclear power.

Fossil Fuels

Oil   Oil is a mixture of hydrocarbon compounds, those containing hydrogen, carbon, and other elements. Most oil is found in sedimentary rock located deep below the surface of the land and sea floor; Canada's most important sedimentary basins are illustrated in Figure 11–4. Most **hydrocarbons** are the remains of prehistoric animals, forests, and sea-floor life, hence the name **fossil fuels**. Buried in layers of sediment, these plants and animals decomposed very slowly and eventually were converted into crude oil.

Canadians consume roughly 1 665 000 barrels of oil every day. Compared with other energy sources, oil has a high energy value per unit of volume, making it the world's most important traded commodity. In addition, oil is relatively inexpensive (at present) and is transported easily. Nature, however, does not view oil as favourably as our economy does. As we know, the combustion of fossil fuels releases gases that contribute to climate change,

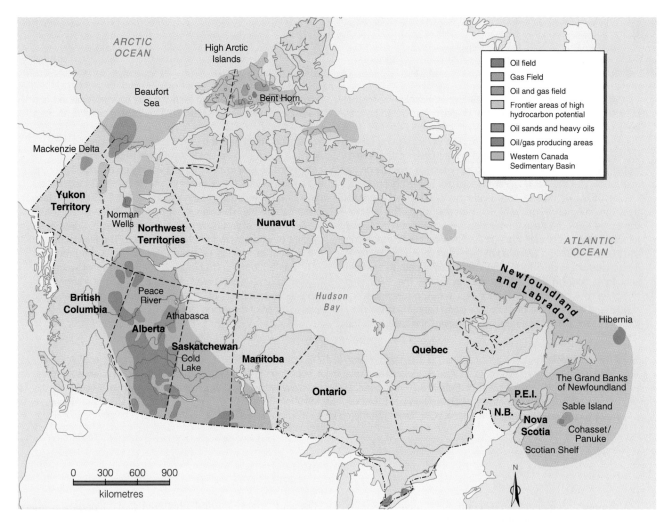

Figure 11–4

Canada's important oil-producing basins

SOURCE: *The State of Canada's Environment—1996,* Environment Canada, 1996, Figure 11.43. Reprinted with the permission of the Minister of Public Works and Government Services Canada, 2004.

acid precipitation, and photochemical smog. Furthermore, crude oil and other petroleum products often are toxic to wildlife and in some cases can result in drastic changes to wildlife ecosystems. It has been estimated that more than 10 000 cubic metres of oil enter our environment each year through spills and oil well blowouts (Environment Canada, 1996). An oil spill can occur at any point from production to consumption; spills are most likely, however, during transportation (see Box 11–3).

Between 1958 and 1997 our total energy consumption increased by 313 percent (Green Lane, 1997). Canadian consumption of energy resources raises concerns about economic and environmental sustainability. We recognize that creating a sustainable economy requires us to use nonrenewable energy resources at rates that do not exceed our capacity to create substitutes for

them, and to use renewable energy resources at rates that do not exceed their capacity to renew themselves. We recognize also that full-cost accounting must be in place to ensure that environmental, economic, and social costs are accounted for in decisions relating to land, resource use, species depletion, and economics (Taylor, 1994).

These principles suggest that market prices of oil should reflect accurately the actual costs of production, distribution, consumption, and environmental impacts. If market prices were to reflect true costs, it is likely that alternative energy sources such as wind power would become more viable. Currently such renewable sources of energy have limited markets, while nonrenewable sources such as oil receive government incentives for megaprojects (see Box 11–2). If these incentives were removed, would oil resources continue to dominate our energy industry?

BOX 11-3
OIL SPILLS AND TRANSPORTATION SAFETY

Oil tanker spills—such as the 1989 Exxon *Valdez* spill off the Alaska shoreline or the 1996 *Sea Empress* spill off the Welsh coast—garner a substantial amount of media attention because spills are visible and the potential for environmental damage is clearly evident. However, it has been estimated that tanker accidents account for only 6 percent of the oil released into marine environments (Chivers, 1996) and that 92 percent of all tanker spills occur at the terminal when oil is being transferred (Environment Canada, 1996). Considering that petroleum is transported by ocean tankers, trains, pipelines, and tanker trucks, the more transfers that occur, the greater is the risk of a spill. In fact, delivering petroleum from the source to the consumer may require up to 15 transfers.

Transportation-related accidents and spills account for only a small portion of the total oil that enters the marine environment each year. Oil may enter rivers, lakes, and oceans in many ways, including from oil seeps, offshore oil and gas production, municipal and industrial waste runoff, and atmospheric fallout. Besides human-induced discharges, oil enters the environment naturally. Worldwide, approximately 200 natural underwater oil seeps have been identified. In Canada, natural seepage has been observed off the north coast of Baffin Island and off the coast of Labrador. All sources considered, municipal and industrial waste runoff are the major sources of oil input into Canada's environment.

Tanker spills are serious because of the large volume of oil that often is spilled and the uncertainty surrounding the nature and extent of environmental damage that may result both from the spill itself and from cleanup efforts. Predicting the damage from a spill is difficult because of the number of factors that must be considered. In addition to the quantity spilled, the location (open sea, close to shore, estuary), weather (temperature, wind speed), tides and currents, grade of oil (crude or refined), and the surrounding environment all play a part in determining the environmental damage. Consequences of cleanup efforts are unclear, as well. Evidence from the Exxon *Valdez* remediation efforts suggests that some measures imposed short- and long-term damage to shoreline plants and animals. Having the longest coastline in the world, Canada is aware of the risk of oil spills and the responsibility for ensuring safe and effective cleanup. Although it is impossible to eliminate spills completely, Canada supports international initiatives to make transportation safer.

The size of tankers has increased dramatically since the 1950s. For example, ships of 30 000 deadweight tons (dwt) were considered very large in the 1950s. Today, most tankers exceed 250 000 dwt, and some are up to 500 000 dwt. A number of safety measures have been instituted to improve the safety of these larger vessels. For example, individual holding tanks are limited in size to reduce the volume of spill if one tank is ruptured. Technological advances, such as electronic charting and computer imagery, are making it safer for crews to navigate around hazards. In addition, tanker ships built after 1993 are required to have double hulls (bottoms) to reduce the risk of spills if one layer

Photo 11–6

is damaged. At present, however, there is minimal global use of double-hulled tankers. For example, more than 90 percent of oil entering American ports is transported in single-hulled vessels. The Natural Resources Defense Council estimates that by the year 2005, only 25 percent of the world's tankers will be double-hulled (Chivers, 1996).

Even with improvements in shipping safety regulations, operational requirements, and global conventions and standards, the safety of shipping has not improved since World War II (Frankel, 1995). Major changes in shipping safety standards normally occur following a major disaster. The Exxon *Valdez* spill, for instance, dumped 50 million litres of crude oil into Alaska's marine ecosystem and resulted in the United States introducing the Oil Protection Act in 1990. The act requires the oil industry to pay damage costs resulting from supertanker spills but has been criticized for prompting a transportation shift from tankers to the less regulated tugboat–barge combination. In Canada, the party responsible for the spill is liable for the cleanup costs as well as any economic losses that result from environmental damage. One of the most significant criticisms of the oil transport industry concerns the industry's preoccupation with reducing the amount and impact of accidental discharge and not with preventing accidents in the first place (Frankel, 1995).

SOURCES: "Troubled Waters," C. Chivers, 1996, *E, The Environment Magazine*, *7*(1), 14–15; *Oil, Water and Chocolate Mousse*, Environment Canada, 1996, http://www.ec.gc.ca/chocolate/en/Default.htm; *Ocean Environmental Management: A Primer on the Role of the Oceans and How to Maintain Their Contribution to Life on Earth*, E. Frankel, 1995, Englewood Cliffs, NJ: Prentice-Hall.

Heavy Oil   Oil deposits, in the form of shale and oil sand, are found close to the Earth's surface. **Oil shale** is rock that contains a solid mixture of hydrocarbon compounds called kerogen. Once crushed and heated, kerogen vapour is condensed to form heavy, slow-flowing shale oil. Shale oil is more expensive and difficult to extract and process than conventional oil, and therefore its net useful energy yield is lower. Because oil shale is extracted on land, many of the same problems associated with above-ground mining are evident with shale processing. In addition to altering the landscape and interfering with wildlife, salts and toxic compounds from processed shale can leach into watercourses and contaminate groundwater.

Oil sand is a combination of clay, sand, water, and **bitumen**, a black oil rich in sulphur. Removed by surface mining, oil sands are heated and treated with steam to separate the bitumen from other compounds. The bitumen is then treated and chemically upgraded into synthetic crude oil. The net useful energy yield is lower than for conventional oil because more energy is required to extract and process the bitumen. Deriving oil from oil sands introduces significant landscape changes, notably the creation of large waste disposal ponds. The process also releases large quantities of sulphur dioxide.

Canada is home to the largest known oil sands deposits in the world. Comprising 58 percent of the world's bitumen resources, and covering an area about the size of New Brunswick, the Athabasca tar sands lie in the boreal forest zone of northern Alberta (Bolger & Isaacs, 2003). Alberta's Energy and Utilities Board estimates that about 315 billion barrels of bitumen are recoverable from the oil sands—several hundred years' supply at current production rates (Canadian Association of Petroleum Producers, n.d.). However, there are environmental barriers to realizing the full value of this resource, including CO_2 emissions, landscape destruction, and volume of water used in production of bitumen.

Investment in northern Alberta oil sands projects began in the late 1960s and, since then, industry and government have poured billions of dollars into oil sands research and development. Originally, analysts thought that upgrading the bitumen into usable oil would be profitable only if the price of oil were US$50 per barrel (Boras, 1995). The two major players in Canada's oil sands, Suncor and Syncrude, produce one of every five barrels of oil in Canada. In 1992, it cost the pair of companies $19 per barrel to dig, process, and blend the synthetic oil (Boras, 1995). Since 1999, operating costs have been in the range of $14 per barrel (Suncor Energy Inc., 2003).

Burgeoning interest in secure oil supplies, particularly by the United States, has markedly increased the demand for oil from the tar sands. As demand increases, environmentalists are expressing more concern about Canada's oil sands developments. As of 1995, about $5 bil-

Photo 11–7
Companies extracting bitumen from Alberta's oil sands rely heavily on technology, including massive equipment such as this 80-ton shovel, shown here loading a 240-ton truck.

lion had been allocated for oil sands expansion in northern Alberta, and the industry plans to spend over $24 billion on additional development by 2007 (Natural Resources Canada, 2000b). A large portion of these funds is being directed at "in-situ" production methods. In-situ production uses steam to separate bitumen from underground oil sands deposits, some buried as deeply as 760 metres. With in-situ technology, recoverable oil sands reserves that are not accessible through conventional surface mining are becoming more economically viable. Environmentalists are concerned, given projected expenditures and the scope of proposed developments, that weak environmental review processes will fail to safeguard Alberta's boreal forest ecosystem from the cumulative effects of oil sands projects.

Although the federal government has exclusive authority over fish and migratory bird habitat, a key component of forest ecosystems, the federal government's role in assessing the environmental impacts of oil sands development has been limited. Under the Canadian Environmental Assessment Act, federal environmental assessments are required when the federal government exercises authority over a project. Federal authority is triggered when a federal department or agency proposes a project, or provides land or money to facilitate a project. To varying degrees, provinces also require environmental reviews. Under recent "harmonization" agreements, where two or more governments are required by law to assess the same project, a single assessment and review process takes place.

In the absence of federal reviews, companies prepare environmental impact assessments (EIAs) and submit them to the province for review before a project licence is issued. Ideally, an EIA should consider all potential environmental impacts from a project, including the cumulative effects of other developments, and outline measures to minimize negative effects. According to many environmental groups, company environmental reviews are often narrow and lacking in depth. Consequently, ENGOs are challenging company-sponsored environmental reviews and approval decisions of provincial governments.

For instance, the Oil Sands Environmental Coalition (OSEC) opposed approval of the Christina Lake thermal oil sands project. The OSEC maintained that the EIA for the PanCanadian Resources project, the first of as many as 18 in-situ projects in the region, was narrowly defined and unable to adequately address the potential cumulative effects of the project on the region. Of particular concern is the increase in habitat fragmentation for wildlife through the construction of extensive seismic lines during exploration, roads and wells during development, and pipelines for product delivery (Smith & MacCrimmon, 1999). Other concerns, including greenhouse gas emissions, flaring, land reclamation, and effects on traditional lifestyles of the Chipewyan community, need to be addressed also (Laird, 2001). The group called for a temporary moratorium on in-situ oil sands developments in the boreal forests of Alberta. Although the Alberta Energy and Utilities Board declined the request, the OSEC felt a moratorium would have allowed time to assess comprehensively the cumulative impacts of widespread in-situ development (Pembina Institute, 2000).

The Pembina Institute, a watchdog on energy and development issues, is critical of oil sands expansion because of the high concentration of greenhouse gases released during upgrading and processing activities. The Institute claims that expansion of production in Alberta's oil sands will make it even harder for Canada to meet its current and future commitments to combat climate change (Chambers, 1996). Attaining international emission targets may be possible in spite of increased production, however. Suncor, for example, has made a commitment to lower its net greenhouse gas emissions by 6 percent below 1990 levels by 2010 even in light of the company's aggressive growth plans (Suncor Energy Inc., 2000).

With the recognition that world energy markets will be dominated by fossil fuels for some time, the large reserves of oil sands in northern Alberta should provide an important impetus for the Canadian economy. However, further reducing toxic emissions and greenhouse gases and successfully reclaiming strip-mined lands are two of the challenges that remain in reducing the environmental effects of this energy megaproject.

Coal **Coal** is the most abundant fossil fuel in the world, with reserves four to five times that of oil and gas combined. Coal is burned to generate 20 percent of Canada's electricity, and roughly 45 percent of the world's electricity. Coal has a relatively high net useful energy yield and is highly effective for providing industrial heat. As a result, 75 percent of the world's steel is produced from coal energy. Globally, 4 billion tonnes of coal is consumed per year, 75 million tonnes of which is produced in Canada's 24 operating mines (Natural Resources Canada, 2000b). The combined impact of coal mining, coal transportation, and coal-fired electricity on the Canadian economy is 73 000 jobs and over $5.8 billion in GDP.

Of all the fossil fuels, coal burning produces the most carbon dioxide and air pollution per unit of energy. Concern over climate change and acid precipitation has led the coal industry to develop advanced pollution control technologies. Even so, in the United States alone, air pollutants from coal burning kill between 5000 and 200 000 people, cause 50 000 cases of respiratory disease, and account for several billion dollars in property damage every year. Extracting coal from underground mines endangers human lives directly through accidents (explosions, shaft collapses) and prolonged exposure to coal dust (black lung disease). Surface coal mining severely alters the landscape, causes soil erosion, and can pollute nearby water supplies. Despite coal's drawbacks, energy analysts forecast that the world demand for coal will increase significantly over the next 20 to 50 years. To improve the environmental acceptability of coal, technologies to reduce emissions and improve combustion efficiency are being developed. The challenge is to commercialize clean coal technologies while keeping a competitive price.

For nearly a century, the Sydney steel plant (Cape Breton Island, Nova Scotia) was the lifeblood of the community, providing employment for 1500 workers in its prime. But virtually every day since 1905, pollution from the plant's coke (coal) ovens poured into the air and untreated effluent from the coking operations was dumped into Muggah Creek. Today, the 34 hectares of the Sydney Tar Ponds, which once were the tidal flats of the Muggah Creek watershed, contain an estimated 500 000 tonnes of toxic coal-tar deposits. Among the contaminants are PAHs, PCBs, and heavy metals (Gjertson, 1997; Hamilton, 1997). With each tide, contaminants are flushed into Sydney Harbour; scientists estimate that about 800 kilograms of PAHs are released annually. In 1980, Environment Canada closed the fishery in the harbour's South Arm due to contamination—PAH levels in lobsters were 26 times the norm.

In 2000, the Acting Minister of Nova Scotia's Department of Environment stated that managing contaminated sites and prioritizing them for remediation was a first priority for his Ministry. He cited the Sydney Tar Ponds as an example of this problem and called it Canada's worst contaminated site (Baker, personal communication, 2000).

More than 15 years of concerted efforts and an enormous financial investment from three government jurisdictions (civic, provincial, and federal) have failed to provide a solution to the cleanup and remediation of the Sydney Tar Ponds. A recent effort to address the tar ponds problem involved a community-based joint action group (JAG), which conducted an extensive public participation process to chart a preferred method for the cleanup. The process targeted two problem areas: tar ponds decontamination and coke oven cleanup. Citizen participants marginally rejected capping and containment of the site in favour of soil washing, bioremediation, incineration, and containment to remediate the ponds. Incineration was the method of choice to clean the coke ovens. The estimated cost for the remediation effort was between $210 million and $450 million. However, the Nova Scotia government was not inclined to accept the previously rejected recommendation to incinerate; as a result, remediation efforts were stalled once again. Over the last 20 years, the federal government alone has spent $250 million on the decontamination effort (Commissioner of the Environment and Sustainable Development, 2002; Joint Action Group for the Environmental Clean-up of the Muggah Creek Watershed, 2003; "The Sydney Tar Ponds," 2003). In the spring of 2004, the governments of Canada and Nova Scotia jointly committed $400 million to the cleanup of the Tar Ponds. The solution involves high-temperature incineration of the most serious contaminants and bioremediation and containment of the rest. The cleanup is expected to take 10 years and 2500 person years of employment (Sydney Tar Ponds Agency, 2004).

Natural Gas Conventional, or associated, **natural gas** is located underground above most reserves of crude oil and is a gaseous hydrocarbon mixture of methane combined with smaller amounts of propane and butane. When found on its own in dry wells, natural gas is called nonassociated or unconventional natural gas. Approximately 72 percent of world reserves of natural gas are of the nonassociated type and the remaining 28 percent are the associated type. Typically, it has been more economical to extract associated reserves; however, advances in extraction technology have improved the cost-effectiveness of nonassociated production. Natural gas production had been rising faster than oil, making it increasingly more important to the Canadian economy. By 2000, natural gas represented almost 30 percent of Canada's total primary energy production (International Energy Agency, 2003a). However, some experts estimate that conventional natural gas production has peaked and by 2006 will start to decline (Bolger & Isaacs, 2003). Canada produces virtually all of the natural gas used by Canadians and produces large quantities for export to the United States.

Many analysts view natural gas as the bridge from "dirty" hydrocarbon-based energy to cleaner renewable energy. Natural gas burns more efficiently than oil, and produces one-third less carbon dioxide per unit of heat energy and fewer pollutants overall. Simply substituting

Photo 11–8
Installation of the Trans-Canada pipeline near North Bay, Ontario.

natural gas for coal in electrical generation facilities could reduce carbon emissions by 50 to 70 percent, depending on efficiency of individual facilities. However, because methane is 30 times more powerful than carbon dioxide as a greenhouse gas, it is critical that turbines and associated machinery in natural gas–burning facilities be sealed against leaks. If only 3 to 4 percent of the methane finds its way to the atmosphere, the lower emission benefit of burning natural gas is nullified (Hill, O'Keefe, & Snape, 1995). Natural gas generally is less expensive than oil and transports easily over land through pipelines (see Figure 11–5). Pipeline construction, however, can impact sensitive aquatic, grassland, and other environments (see Wallis, Klimek, & Adams, 1996).

The process of **flaring** also is a source of considerable concern. Flaring is a way of disposing of unwanted, unprocessed natural gas; the industry burns this gas to release hydrogen sulphide (sour gas) and to avoid the buildup of potentially explosive levels of gas at work sites. In Alberta, for instance, only about 1.4 percent of the natural gas processed in the province is flared, but that amount translates into more than two billion cubic metres of gas burned off each year (Francis, 1997). For almost two decades, people living near flares have expressed their concern about the impacts of flaring emissions.

Recent research has revealed that emissions from many flares are greater than was assumed previously; that even though flaring destroys many sulphur compounds, others are created; and that the risks to cattle and humans from exposure to many of the emissions from flaring are not well understood. In 1996, the Canadian Cattle Commission noted that flaring eventually should be eliminated, given that flares are Alberta's single largest source of a variety of volatile compounds including PAHs, VOCs (volatile organic compounds), carbon monoxide, carbon dioxide, nitrous oxides, and heavy metals. Following a five-year study, in 1996 the Alberta Research

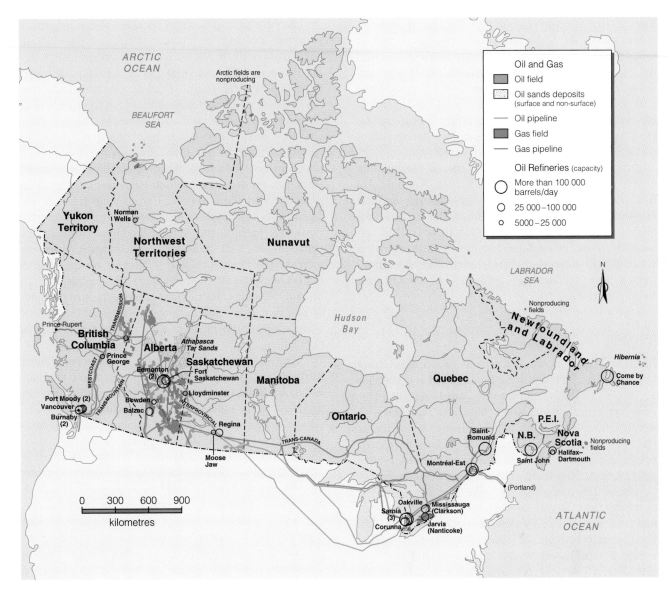

Figure 11–5

Oil and natural gas fields and pipelines in Canada

SOURCE: Adapted from *Canadian Oxford World Atlas: New Edition*, Q. H. Stanford (Ed.), 1992, Toronto: Oxford UP, p. 250. Used by permission of Oxford University Press, Inc.

Council identified more than 200 different chemical compounds produced by flaring, including more than 30 varieties of cancer-causing benzene (Francis, 1997).

Public frustration over flaring escalated in the late 1990s when two residents of Alberta were convicted following acts of "eco-terrorism" against an oil company. The two citizens, concerned about the health of their families and livestock, bombed an oil well shed in northern Alberta. Mounting evidence of the health risks of flaring continued into the new century. In July 2000, claiming that his herd of cattle had experienced reproductive failure, a central Alberta rancher accepted a settlement from a major oil company after an eight-year legal battle (Lau, 2000).

Although governments have been slow to react, in 2000, an advisory committee appointed by the Alberta Energy and Utilities Board recommended a series of measures to develop a better understanding of the effects of sour gas on human health and improve the public consultation process (Public Safety and Sour Gas, 2000). Alberta also introduced regulations requiring petroleum companies to reduce flaring by 25 percent from 1996 levels, by the end of 2001 (Alberta Energy and Utilities Board, 2000). By 2001, the industry had reduced flaring in Alberta by 53 percent (Canadian Association of Petroleum Producers, 2001). Given the potential seriousness of chronic exposure to risks from flaring, including increased costs of medical care, precautionary approaches continue to be warranted.

The 1970s Energy Crisis In 1973, a then little-known cartel, the Organization of Petroleum Exporting Countries (OPEC), disrupted global supplies of imported oil. The OPEC embargo, combined with the U.S. energy policy, resulted in an energy crisis in the United States that had far-reaching economic, social, and political effects. In addition to high prices and long lines at gas stations, the "oil shock" raised concerns over energy policy, particularly U.S. dependence on imported petroleum.

The energy crisis prompted critical thinking on energy futures in areas such as estimates of global fossil fuel supplies, ways to improve energy efficiency, and use of alternative energy sources. In hindsight, the oil shock helped the North American economy to realize the volatility of energy markets. The energy crisis has been credited with improving our understanding of environmental and societal consequences of energy use (Feldman, 1995).

Biomass

Biomass, organic matter that includes wood, agricultural wastes and manure, and some types of garbage, supplies 15 percent of the world's energy. Canada relies on biomass—principally waste wood chips from pulp mills—to supply about 6 percent of our energy requirements (Natural Resources Canada, 2000b). Less-developed countries are much more dependent on biomass, relying on it for up to 50 percent of their energy requirements.

When trees and plants are replaced at a level equal to or greater than the rate at which they are harvested, biomass is considered a renewable energy resource. In addition, no net increase in atmospheric carbon dioxide occurs if replacement equals harvest. Sound land use management practices must be in place to avoid problems associated with land clearing, such as soil erosion, water pollution, flooding, and habitat loss. Unfortunately, in many parts of the world, harvesting practices are not sustainable and fuelwood shortages occur. Many forested lands in developing countries are disappearing to accommodate agriculture and urbanization, not to provide fuelwood to meet energy needs and demands.

Increasingly, developing countries are turning to agroforestry to meet their energy needs. Agroforestry is a multipurpose land use system that combines indigenous trees (to provide erosion protection and fuelwood) in areas used to plant crops and graze animals. Land use activities reflect local sociocultural values. Agroforestry initiatives, such as the Kenyan Woodfuel Development Program, have been successful in encouraging farmers to increase the amount of woody biomass on their farms (Hill et al., 1995). The introduction of similar initiatives in other developing countries may improve the sustainability of biomass fuels, enabling people to use wood for fuel while meeting the increasing energy demands of local populations.

Biomass also can be converted into gaseous or liquid **biofuels (ethanol)** and used to power motor vehicles. Ethanol is used directly for fuel but is more commonly used as an octane-enhancing gasoline additive. In Canada, ethanol is available as a blended gasoline with concentrations of between 5 and 10 percent. Ethanol-blended gasoline in concentrations of 10 percent or less can be used in all gasoline-powered automobiles without engine or carburetor modification. Ethanol blends have been found to reduce carbon monoxide emissions by up to 30 percent, especially during the winter months when emissions increase due to colder temperatures. Many U.S. cities mandate use of ethanol-blended gasoline during winter as a measure to reduce carbon monoxide emissions. As a result, use of ethanol-blended gasoline is more widespread in the United States than it is in Canada.

In 1975, Brazil introduced the ProAlcohol program to promote use of ethanol in automobiles. The program was initiated to counter the oil price shocks in the early 1970s and to promote self-efficiency. Unlike in Europe, the United States, and Canada, in Brazil many automobiles operate on pure ethanol or higher blended concentrations (20 to 22 percent ethanol). In 1989, five million cars in Brazil operated on pure ethanol and another nine million ran on a 20–80 ethanol-gasoline blend (Hill et al., 1995). However, as gasoline prices fell, it became difficult to sustain the program and subsidies were reduced. By 1997, less than 1 percent of new cars sold were fuelled by ethanol (Energy Information Administration, 1999). With ethanol being more environmentally benign than oil, and with opportunities to create employment in the sugar cane sector, Brazilian regulators again invested in ethanol. In 1999, government programs encouraged 200 000 taxis and 80 000 government vehicles to be replaced with ones operating on

Photo 11–9

In Los Angeles in 1979, gasoline rationing resulted in long lineups at local gas stations.

100 percent ethanol (Energy Information Administration, 1999). In 1999, about 41 percent of Brazil's transportation fuel demand was met with ethanol. Brazil is capitalizing on its success with ethanol development by exporting this technology to developing countries.

Hydroelectricity

Hydroelectric power supplies roughly 5 percent of the world's total commercial energy and about 20 percent of the world's electricity. Canadian consumption of hydropower is above the world average; about 12 percent of our total commercial energy comes from hydroelectric sources (see Figure 11–6). Consumption of hydropower increased a marginal 1 percent between 1958 and 1992,

indicating low growth potential for further hydroelectric development. Indeed, there are few suitable sites remaining for large-scale hydro projects in Canada, and those that are suitable would introduce massive land use changes and generate strong opposition from resident communities.

Hydroelectric projects are expensive to build but have low operating and maintenance costs. In addition, their life spans are 2 to 10 times greater than those of coal or nuclear plants. Hydroelectric stations emit no air pollutants or greenhouse gases and help regulate downstream irrigation. Unfortunately, construction of dams to provide hydroelectric power introduces far-reaching landscape changes that displace people and wildlife, destroy cropland and forests, and interfere with aquatic ecosystems.

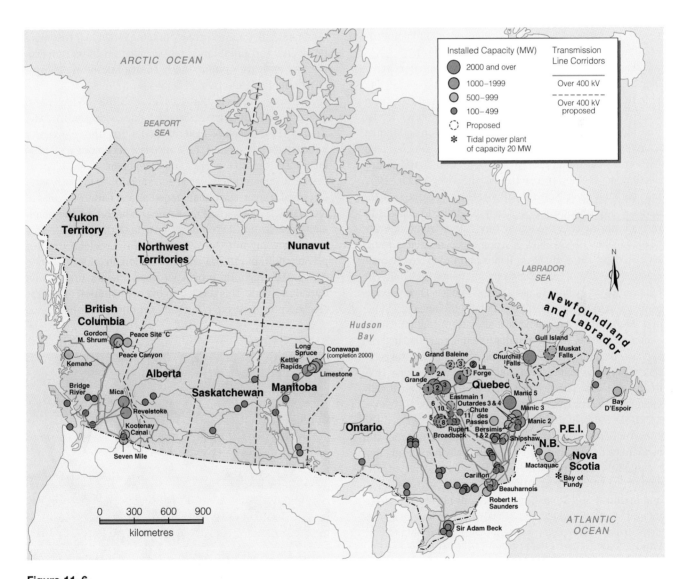

Figure 11–6
Major hydroelectric dams and transmission line corridors in Canada

SOURCE: Adapted from *Canadian Oxford World Atlas: New Edition,* Q. H. Stanford (Ed.), 1992, Toronto: Oxford UP, p. 23. Used by permission of Oxford University Press, Inc.

The James Bay hydroelectric project in northern Quebec is an example of the irreversible environmental damage that can result from the construction of a megaproject (see Enviro-Focus 7).

On November 9, 1997, engineers finished dumping 60 000 cubic metres of rockfill and cement to dam the Yangtze River, irrevocably changing the landscape of this part of China. In building the world's largest dam across the world's third-largest river, China displaced 1.2 million people from fertile farmlands along the river to less productive, steeper hillsides. The dam flooded farmland that had produced 40 percent of China's grain crops and 70 percent of its rice crops. The lives of about one-quarter of the Chinese population (400 million people) who lived along its banks were affected. By the time the US$25 billion Three Gorges project is completed in 2009, 140 towns and 326 villages will be submerged under 670 square kilometres of water, along with 657 factories, 953 kilometres of highway, and 139 power stations (Sly, 1997).

The purpose of the dam is to provide energy for China's increasingly industrial economy. Proponents of the dam note that it will produce 18.2 million kilowatts of clean electricity (the equivalent of 18 nuclear power stations) needed for China's economic growth and modernization and reduce China's reliance on coal-fired electric power, thus reducing GHG emissions from China. In addition, the dam will control flooding in the lower reaches of the river (Zich, 1997; International River Network, 1998). Critics charge that the dam's ability to control floods may be limited because the river's lower reaches are fed by three other tributaries that also contribute to flooding. A string of smaller dams would have had the same effect on floods and produced as much electricity without as many environmental side effects. Other impacts include the loss of the famous Three Gorges as a landmark; they will be flooded and the ancient (Stone and Bronze Age) archaeological sites they contain will be submerged. Also, the river supports numerous endangered species, including the Yangtze dolphin, Chinese sturgeon, finless porpoise, giant panda, and Siberian white crane. How many of these species will survive in the dramatically changed environment is uncertain.

Both national and international sources opposed the construction of this dam. Fifty-three senior engineers and academics from China, and internationally honoured environmental and social activist Dai Qing, warned the Chinese government that the dam was rife with technical and financial problems. Concerns about negative environmental, human rights, and social impacts made investment from Western sources more difficult to obtain. The World Bank denied financial support to the project and so-called responsible investment funds have led campaigns to discourage other investors (International River Network, 1998).

The energy conservation movement and advancements in energy-efficient technologies have reduced the demand for large power-producing utility projects in the Western world. Supporters of utility megaprojects maintain they provide economies of scale necessary to meet the energy requirements of future populations. In reality, however, many hydroelectric megaprojects across North America have fallen far short of expectations.

One Canadian ENGO whose objectives include stopping unnecessary hydroelectricity plants is the Energy Probe Research Foundation (EPRF). Incorporated in 1980 (but originating in 1970 as the Energy Team of Pollution Probe), EPRF is financially independent of governments and corporations, and receives the majority of its funding from the general public. The goals and objectives of EPRF include educating Canadians about the benefits of sound resource use (Energy Probe Research Foundation, 1995). Often viewed as a maverick because of its stand against "ill-advised projects," EPRF promotes democratic processes by encouraging individual responsibility and accountability, and provides business, governments, and the public with information on energy, environmental, and related issues.

Energy Probe is the division of EPRF dedicated to research and advocacy in Canada's energy sector. In 1980, Energy Probe established a model for the electricity industry that separated the transmission component of the business from the generation component. This allowed electrical generation to become a competitive industry with electrical transmission maintained as a separate, regulated monopoly. The Energy Probe model is becoming the dominant organizational structure throughout the world.

Following an extensive study of hydroelectric water pricing, both in Canada and internationally, Energy Probe concluded that our provincial governments undervalue Canada's rivers. This has led to overexploitation of hydroelectric power through such megaprojects as the James

Photo 11–10
One standard incandescent bulb (left) uses as much energy as four compact fluorescent bulbs (right).

Bay project in Quebec and the Peace River project in British Columbia. Energy Probe also is working to promote competition in the Canadian electricity sector in an effort to stop unnecessary electricity plants. The organization suggests that an increasingly competitive system would eliminate new coal and nuclear generation facilities and hydro dams and favour conservation, renewable resources, and high-efficiency gas technologies while improving the economy at the same time.

Nuclear Energy

Nuclear energy generates approximately 75 million megawatt-hours of electricity in Canada each year, equivalent to about 13 percent of the total electricity consumed in Canada. Forty percent of Ontario's electricity is supplied by nuclear energy. The nuclear industry is an important aspect of Canada's economy, contributing more than $2 billion dollars to Canada's GDP annually in uranium production, nuclear energy generation, reactor sales and service, fuel and isotope production, research and development, and in providing direct employment for over 20 000 people. Of the 22 nuclear reactors in Canada, 15 are operating: 13 in Ontario, one in Quebec, and one in New Brunswick. The other seven reactors are either laid up for refurbishment or have been decommissioned. Canada has developed a strong competitive advantage in the nuclear technology sector, primarily through development and export of the CANDU reactor. There are CANDU reactors operating in Argentina, China, India, Romania, and South Korea. Atomic Energy of Canada Limited (AECL), a Crown corporation, generates about 60 percent of its revenue through export of nuclear expertise, technology, and services to foreign governments and utilities. AECL currently is developing a new prototype called the Advanced CANDU Reactor (Canadian Energy Research Institute, 2003; Stothart, 1996).

Nuclear energy emits one-sixth the amount of CO_2 per unit of electricity than does coal. Excluding subsidies, generation of nuclear energy is less expensive than energy produced from fossil fuels and creates minimal land disturbance. Once regarded as the energy of the future, the capacity of nuclear fission to provide the world with energy has not lived up to expectations. In the 1950s, researchers predicted that 1800 nuclear power plants would supply 21 percent of the world's commercial energy by the end of the century. In March 2003, there were 437 commercial reactors in 31 countries with 360 gigawatts capacity (Macfarlane, 2003) producing about 5 percent of global energy. Government subsidies, cost overruns, accidents, and public concerns about safety have affected growth of the nuclear industry.

Despite Canada's investment in nuclear energy, its future is uncertain. Two catastrophic accidents raised serious concerns about the safety of nuclear energy. In the

United States in 1979, a series of human errors and mechanical failures allowed unknown quantities of radioactive material to move through an open valve to the containment building inside the Three Mile Island reactor. Subsequently, radioactive material escaped to the atmosphere through leaky pipes in the building's exhaust system. Two days after the accident, 200 000 people within a 40-kilometre radius were evacuated from the area. Officially, no deaths were linked to the Three Mile Island accident (although some contradictory claims exist). This incident sparked the anti-nuclear movement and severely reduced public trust in nuclear power. Then, in 1986, the world's worst commercial nuclear accident occurred at the Chernobyl plant in the former Soviet Union (now Ukraine). Two explosions occurred at the nuclear power plant as a result of human error. The first explosion released fission products to the atmosphere. A second explosion ejected fragments of burning fuel and graphite from the core and allowed air to rush in, causing a graphite fire that burned for nine days. Forty-two exposed workers died immediately, childhood thyroid cancers increased 25 times in nearby Belarus, and an estimated 6500 more cancer deaths likely will occur in nearby residents (Macfarlane, 2003). Winds carried radiation particles thousands of kilometres from the site. The accident went unreported in Russia for three days, and another two days elapsed before the rest of

Photo 11–11
A concrete sarcophagus encloses the damaged reactor at Chernobyl.

the world was alerted. Local residents were unaware that they had been exposed to an estimated 30 to 40 times the radioactivity of the bombs dropped on Hiroshima and Nagasaki during World War II.

Although Canada's nuclear industry has been operating for 50 years without a serious accident, Canadians' concern over nuclear safety was reaffirmed in 1997, when an independent review charged that "performance of Ontario's stations was well below that being achieved by the world's best nuclear stations" (Epp, Barnes, & Jeffrey, 2003). Acting on the report, Ontario Power Generation (then Ontario Hydro) temporarily decommissioned four operating units at Pickering A and three at Bruce A. A budget of $780 million was approved for work to be completed by June 2000 on the four Pickering A units. After countless delays and an actual cost of $1.25 billion to upgrade the one unit that returned to service in September 2003, the Government of Ontario ordered a review panel to investigate the delays. The conclusion of the report was that management of the project was seriously flawed from the outset (Epp, Barnes, & Jeffrey, 2003).

The findings of the review panel provided impetus for opponents to further their campaigns against nuclear energy. Two national ENGOs, the Sierra Club of Canada and Energy Probe, lobbied for a nationwide phase-out of nuclear energy. Their arguments centred on three main issues: cost, safety, and disposal of radioactive waste. The Sierra Club of Canada estimates that over the 50 years Canada has supported the nuclear industry, the federal government has provided subsidies of $17.5 billion. That figure includes an annual $211 million subsidy to AECL. Estimates are that returning the three remaining Pickering units to operation, plus refitting currently operating plants, is expected to cost more than $2 billion (Martin, 2003a).

Safe operation of nuclear power plants also is a concern. Although nuclear reactors are constructed with multiple safety systems designed to keep highly radioactive materials inside the reactor core, it was human error and mismanagement, not failure of the safety systems, that caused the Three Mile Island and Chernobyl disasters (as well as the initial shut-down of the seven Ontario plants). Finding suitable facilities to store **radioactive wastes** remains problematic. Depending on their **half-life**, some wastes emit radiation for tens of thousands of years; no wonder few communities volunteer to act as landfill sites for radioactive waste storage!

Nuclear fuel waste from Canadian reactors is stored in water-filled pools or dry storage concrete canisters at reactor sites. In an effort to find a more suitable, long-term solution, AECL has conducted extensive research into the concept of deep geological disposal (DGD). The proposed solution calls for disposing nuclear waste 500 to 1000 metres below the surface in the granite of the

Photo 11–12
The nuclear power station at Pickering, Ontario.

Canadian Shield. The waste would be placed in a disposal vault, which would be a network of horizontal tunnels and disposal rooms with vertical shafts extending from the surface to the tunnels. After more than 10 years of analysis, and a year of public hearings, an Environmental Review Panel released its assessment of the concept in 1998. The panel concluded that while the concept is technically safe, in its current form it does not have the required level of social acceptability to be adopted as Canada's approach for managing nuclear fuel wastes (Canadian Environmental Assessment Agency, 1998).

The federal government responded to the panel report in December 1998, indicating its expectation that producers and owners of nuclear fuel waste will establish a waste management organization to develop long-term management and disposal solutions. Regardless of the final decision for DGD in Canada, energy experts concur that rising stocks of nuclear waste from civilian energy programs are a great challenge for the future. Not only is there concern about passing on to future generations the need to look after radioactive waste, but there is also the potential that nuclear waste could become a resource as technology continues to identify alternatives to terminal disposal solutions.

Uncertainties about Canada's nuclear energy future evolve from the decision regarding whether the advantages of clean, available energy outweigh the economic costs and safety risks associated with the industry. Regardless of the disadvantages, nuclear energy represents a significant proportion of Ontario's energy supply. During the summer of 2003, large portions of southern Ontario and several midwestern and northeastern states experienced a complete power blackout. A joint U.S.–Canada Task Force was charged with determining why an estimated 50 million people were left without electrical power for up to two days. The task force identified the reasons as inadequate appreciation of the difficulties at a power generation company in Ohio, and inadequate support from the interconnected power grid, (U.S.–Canada Power System Outage Task Force, 2003). The blackout reminded Canadians just how reliant we are on electricity and reemphasized how desperately we need abundant, reliable power.

The nuclear issue is a good example of the many, often conflicting, opinions that must be resolved when moving toward energy sustainability. Coal-generated power has been replacing nuclear-generated electricity, but some jurisdictions, such as the Ontario government, have committed to shutting coal plants down as part of their greenhouse gas reduction plans. Ontario government officials have included nuclear energy and an additional hydroelectric station in their long-term power generation plans (McLearn, 2003). Hydroelectric facilities supply clean energy but are devastating to ecosystems. Proponents of nuclear energy remind us that nuclear energy emits comparatively small amounts of greenhouse gases and represents an option to fossil fuel–generated energy. Obstacles to nuclear power expansion include cost (high capital investment in construction of nuclear plants), safety, nuclear waste, and security against sabotage and theft. Some ENGOs recommend phasing out coal plants and nuclear power plants, suggesting that our long-term energy requirements can be met by increasing energy efficiency, implementing conservation programs, instituting pricing policies that reflect the actual cost of electricity, and developing decentralized green energy sources (Martin, 2003b; Sierra Club, 2003). Still others believe that the combination of conservation efforts and development of renewable energy sources will not meet domestic and global energy needs for the next 40 to 50 years (Bolger & Isaacs, 2003). Clearly, considerable challenges remain with respect to energy sustainability in Canada and elsewhere.

RESPONSES TO ENVIRONMENTAL IMPACTS AND CHANGE

EMERGING ENERGY RESOURCES

If Canada continues to rely on hydrocarbon-based and nuclear energy, we may damage our long-term competitiveness in the world energy market. In other parts of the globe, **alternative energy** is price-competitive with fossil fuels. For example, California wind turbines produce electricity for the same cost as nearby coal-fired plants, and

India is on target to be the world's fastest-growing installer of wind turbines. More than 40 countries are estimated to employ geothermal power, including Nicaragua and the Philippines, where presently 25 percent of their power is generated from geothermal sources.

Although they are renewable resources, wood and other forms of biomass often are exploited to produce energy without concern for the pollutants released through burning, the low net useful energy yield, or the long-term sustainability of the resources. Except for microhydro opportunities, expanded use of hydroelectricity in North America is unlikely as there are few ideal sites remaining, construction costs are high, and alteration of landscapes introduces significant ecological and social concerns. Work of researchers and scientists continues to improve the feasibility of alternative energy technologies. Increasingly, environmental issues, such as air pollution and climate change, are sparking government interest in using renewable energy as a way to decrease greenhouse gases and other pollutants. However, despite all the very good reasons for expanding our access to renewable energy, the contribution of renewable energy to Canada's total primary energy supply declined marginally between 1990 and 2001 (International Energy Agency, 2003b). The following section discusses some of the research advances in developing alternative technologies as well as barriers to their adoption.

Solar and Wind Energy

The sun provides Earth with close to 99 percent of its heat energy requirements. Just imagine the benefits possible if we could harness the sun's energy to provide our commercial energy needs. If we were able to combine **solar energy** with other renewable energy sources, such as wind, flowing water, and biomass, our energy production would cause much less environmental damage and pollution per unit of energy than any fossil fuel.

Solar power gained most of its strength at the height of the energy crisis in the late 1970s. At that time, it was anticipated that solar energy would provide the United States with 2 to 5 percent of its energy requirements; by 1994, only 0.5 percent came from solar sources. However, in 2001, wind developers in the United States installed 1700 megawatts of wind turbines, almost doubling America's windpower capacity. In Canada, although solar and wind power provide a minor proportion of our energy consumption, wind power grew by 47 percent, to 215 megawatts in 2001 (Fairley, 2003). Advances are being made in wind turbine technology to make them operate effectively in Canada's climate—in the Northwest Territories, residents of Cambridge Bay have found that their 80 kilowatt wind turbine reduces their costly reliance on diesel power. Speculators have envisioned massive wind farms off the coast of Vancouver Island and the Queen Charlotte Islands.

Limited subsidies are helping to generate investment in wind and solar power, but if the economics of renewable energy are to become more attractive, the Canadian government (and others) will need to ensure that conventional power producers pay for the pollution they generate. Although wind power continues to cost more to produce than coal-fired electricity, wind power advocates maintain that when the costs of environmental degradation and health problems are factored into energy pricing, wind power is actually cheaper than coal-fired electricity. If Canada were to charge energy utility companies for the environmental and health effects associated with coal-fired power (that is, carbon emissions), power costs would increase substantially. What clean energy advocates want is the price of fossil-fuelled power to reflect its environmental cost (Fairley, 2003; Newman & Brooks, 2004). This idea often is resisted because of the significant budgetary implications that fossil fuel sales have on national economies. The Pembina Institute for Appropriate Development indicates that this lack of "eco-pricing" is the most significant barrier facing renewable energy in Canada (Pape-Salmon et al., 2003).

With consumers looking for clean energy sources, the quest to improve the viability of renewable energy is again gaining momentum. Canada appears to be in support of developing renewable energy resources despite our economic ties to fossil fuels. In 1996, Canada's energy minister unveiled a renewable energy strategy that included tax breaks and incentives aimed at encouraging research and development of renewable energy technology. A key aspect of the federal government's environmental strategy is to increase market demand for "green power" by showing leadership. In its 2000 budget, the federal government committed to purchase $15 million in renewable energy over 10 years for its facilities in Saskatchewan and Prince Edward Island. Also in 2000, the federal government established a Sustainable Development Technology Fund to stimulate the development and demonstration of environmental technologies, particularly those aimed at reducing greenhouse gas emissions. However, it was not until 2002 that the federal government provided its first credits for wind power production.

In 2003, Canada's 449 wind turbines produced 0.2 percent of Canada's energy needs, amounting to over 300 million kilowatt hours of electricity—enough to supply about 38 000 homes (Canadian Wind Energy Association, 2003). Alberta is becoming a leader in producing and marketing wind energy. One hundred and fourteen turbines (windmills) near McBride Lake constitute Canada's largest wind plant, contribute more than 75 000 kilowatts to the power grid, and supply (to the grid) the equivalent of all the power needed to run Calgary's light-rail transit commuter train (Canadian Wind Energy Association, 2003).

Growth in the Alberta wind power industry has continued at a steady pace since the opening of the Cowley Ridge Windplant in 1993 (see Photo 3–1b). The Canadian Wind Energy Association has set an ambitious target of producing 10 000 megawatts of installed wind capacity by 2010. Meeting this objective would secure wind power's position as the fastest-growing source of energy (Butt, 2000a). However, when the federal government announced investments to implement its Climate Change Plan for Canada in the fall of 2003, designated funding was inadequate to stimulate investment in even the 1000 megawatts of new wind power announced in the 2001 federal budget (Hornung, 2003). Market access remains the major hurdle to achieving rapid growth in wind power.

Despite the lack of market access, utility customers are interested in buying green power (Scanlan, 2001). In 1998, Enmax, a subsidiary of the City of Calgary, started Canada's first green electricity program. Enmax offered its customers a variety of options to integrate wind power into their energy bill. Customers could add $5, $10, or $15 to their electricity bill every month and Enmax would ensure that up to 250 kilowatt hours (kWh) of wind-generated power entered the Alberta power pool. Each 250 kWh is equivalent to about 45 percent of the average home's monthly power consumption. Over a year, a $15-per-month purchase would prevent the air pollution equivalent to driving a car for approximately 15 500 kilometres, or planting about 400 fully grown trees, or keeping more than 1600 kilograms of coal in the ground (Butt, 2000b; Enmax, 2001). As beneficial as this program is, the financial cost may prove a barrier to individual consumers who might wish to support the use of green power but who are unable to afford the additional charge on their utility bills. Table 11–1 provides data on the results of Enmax's Green Power residential program from 1998 to 2003.

Corporations also are investing more money in research and development of energy alternatives. For example, a University of Calgary professor teamed up with a manufacturer in Regina, Saskatchewan, to export Canadian wind turbine technology to Africa. And in 1998, Le Nordais, a 134-turbine wind farm on the shore of the St. Lawrence River in Quebec, began producing electricity for sale to Hydro-Québec.

Solar energy has three major applications: photovoltaics, passive solar, and active solar. Although none of these applications is widespread, photovoltaics are the most widely used, primarily in off-grid situations. Investment in **photovoltaics (PV)** is growing. The photovoltaic effect, the direct conversion of sunlight into electricity, was first observed in 1839. However, only recently has the technology become cost-effective for commercial use. Since 1995, the use of PV in Canada has tripled, largely because of its ability to provide power in remote areas away from power grids and its compatibility with hybrid technologies. Applications include water pumping and purification, navigational lighting and beacon systems, telecommunications, and handheld calculators. The largest individual PV system user in Canada

TABLE 11-1

RESULTS OF CALGARY'S GREENMAX RESIDENTIAL PROGRAM, 1998–2003

Year	Greenpower Produced (kWh)	CO_2 Not Released (tonnes)	Equivalent Km Not Driven	Equivalent Fully Grown Trees
1998	326 000	326	1 418 100	35 452
1999	1 571 250	1 571	6 834 937	170 873
2000	2 377 726	2 377	10 343 108	258 578
2001	3 876 773	3 877	16 863 962	421 599
2002	5 254 550	5 255	22 857 293	571 432
2003*	11 414 828	11 415	49 654 502	1 242 363
Total	24 821 127	24 821	107 971 902	2 699 297

* To July 30, 2003

SOURCE: *Greenmax Newsletter,* August 2003, p. 2.

is the Canadian Coast Guard. An estimated 7000 navigational buoys, beacons, and lighthouses use PV (Dignard-Bailey, 2000; Natural Resources Canada, 2000b).

Passive and active solar energy are two other solar energy technologies. Active solar energy involves the conversion of solar energy into other energy forms through energy transfer fluids such as water or air. Air and water heating are the major uses of active solar energy. Passive solar energy utilizes high-performance windows, increased insulation, and draft-free buildings to maximize heat gains and decrease the need for other forms of heat energy (Natural Resources Canada, 2003a).

Hydrogen

Hydrogen is a colourless, odourless gas that makes up 75 percent of the mass of the universe. Industry uses hydrogen to manufacture ammonia and to refine petroleum. Hydrogen is an important component of the NASA space program; space shuttles are fuelled with hydrogen and fuel cells convert hydrogen to electricity to provide astronauts with heat, light, and drinking water. The commercial use of hydrogen as a power source would provide us with a virtually inexhaustible source of energy while reducing our dependency on nonrenewable fossil fuels. So why is this nonpolluting and abundant element not being used to fuel our vehicles, power our airplanes, and provide electricity to our homes and offices?

A number of complexities, both technical and socioeconomic, are working against the mass commercialization of hydrogen as an energy source. Despite its abundance, hydrogen is found only in combination with other elements, primarily oxygen, carbon, and nitrogen. Hydrogen must be separated from these elements to be a useful energy source. A substantial amount of energy is required to isolate hydrogen; if this energy comes from fossil fuels or nuclear power, the environmental benefit of using hydrogen is negated. However, the production efficiency of hydrogen improves when renewable energy sources (solar and wind power) are used, and fewer stresses are placed on the environment. Currently, producing hydrogen from these sources is expensive and the production efficiency, while improving rapidly, is not yet commercially viable.

Other technical obstacles in moving toward a hydrogen-based economy involve concern over the safe storage of hydrogen and the lack of infrastructure to support its widespread use. Hydrogen usually is stored as a compressed gas or cryogenic liquid. The volatility of hydrogen gas is prompting researchers to develop storage systems that not only improve safety but also increase capacity per unit volume. For example, researchers at the National Renewable Energy Laboratory (a division of the United States Department of Energy) are developing a solid-state storage system that binds hydrogen to microscopic carbon tubes. The hydrogen attaches to the surface of the carbon and is released by changing temperature and pressure levels. In another experiment, engineers at Lawrence Livermore National Laboratory in California have found that under pressurized conditions, small glass bubbles absorb hydrogen then release it when crushed or heated (Frenay, 1996).

To achieve a commercial market for hydrogen, effective infrastructure and distribution networks are required. Establishing these networks will take time and may involve lengthy transition periods. Government and business leaders need to consider carefully the alternative methods of supplying hydrogen to meet growing consumer demands. For example, the extensive natural gas pipeline network throughout North America could be converted to carry hydrogen. At the other end of the scale is localized production, where automobile owners produce their own hydrogen using home reformers or on-board hydrogen converters (Frenay, 1996). Developing safe and economical production, infrastructure, and storage systems will help bring hydrogen one step closer to commercial use. However, once these technical challenges have been overcome, the most restrictive barriers remain, those of economics and political will.

The economic success of many nations depends on the production and delivery of fossil fuels. As the world's most traded commodity, oil has given the petroleum industry considerable economic and political influence worldwide. Convincing oil companies to invest in hydrogen research is difficult because most consider hydrogen as a serious threat to their short-term economic well-being. Furthermore, many oil companies are opposed to large-scale government-funded hydrogen research. As a result, government-supported hydrogen research is lower in nations whose economies are strongly tied to the petroleum sector. For example, compared with Japan and Germany, the United States allocates about one-seventh of the funding those countries do to hydrogen research. Additional reasons working to slow investment in hydrogen research include uncertainty surrounding the short-term prospects for its widespread use and the estimated lengthy transition period to convert the population to **hydrogen power**.

Despite uncertainty and objections from major petroleum companies, hydrogen research is drawing the attention of governments and the private sector. Although there were across-the-board funding cuts to budgets for solar and other renewable energy research, U.S. federal funding for hydrogen research increased during the mid-1990s. A "hydrogen corridor" is taking shape in the desert basin east of Los Angeles, where Xerox operates a $2.5 million hydrogen production complex (Frenay, 1996). The facility in El Segundo, California, is North America's largest solar converter of water into hydrogen. The use of hydrogen-powered golf carts on local California streets has been legalized, and the community of Palm Desert is looking into building two fuelling stations that will use hydrogen produced from a local wind farm.

In the near term, the transportation sector likely will benefit most from hydrogen power. Current technology enables combustion engines to be fuelled directly with pure hydrogen or hydrogen blended with natural gas and to achieve efficiencies up to three times greater than gasoline-powered engines. Automobile manufacturers are focusing on hydrogen fuel cell technology to power clean-air vehicles. The city of Los Angeles tested a vehicle certified by the California Air Resources Board and the United States Environmental Protection Agency as a Zero Emission

Photo 11–13
The 275-hp Ballard fuel cell engine fits into the same space as the diesel engine; these buses meet performance requirements of transit authorities without emitting any pollution.

Vehicle (Honda Corporation, 2003). A Canadian company, Ballard Power Systems, is "front and centre" in developing hydrogen as a fuel source for commercial transportation.

In 1995, Ballard supplied the city of Chicago with three transit buses powered by hydrogen fuel cells. During a public ceremony, Mayor Daley enjoyed a "glass of exhaust" collected from the tailpipe of an idling Ballard bus. The taste test was performed to illustrate the environmental cleanliness of the exhaust—nothing but steam and condensed water. With the success of the $5.8 million test program, final trials of the "Xcellsis" fuel cell bus began in California and Vancouver in 2000. Xcellsis, a developer of fuel cell engines, is a partnership between Ballard, DaimlerChrysler, and Ford that is in the process of commercializing its fuel cell engines (Ballard Power Systems, 2004).

Ballard uses a type of fuel cell known as a proton exchange membrane (PEM). The PEM fuel cell is one of five fuel cell types currently available that produces energy from a hydrogen reaction. With a PEM cell, hydrogen is directed through one side of a central core that is divided by a proton-permeable membrane. Oxygen is directed into the other side of the core and attracts the hydrogen protons to move through the membrane. The hydrogen protons then bond with the oxygen to form water. With the absence of protons on the hydrogen side, the hydrogen electrons are drawn to a metal electrode, creating a charge that provides energy. The PEM cell operates at temperatures below boiling and is presently the most suitable choice to power a motor vehicle with hydrogen. In conjunction with DaimlerChrysler and Ford Motor Companies, Ballard has produced a concept Jeep Commander, a sedan, and a sport utility vehicle that use Ballard fuel cells (Ballard Power Systems, 2000a).

Through Natural Resources Canada, the federal government supports projects aimed at making fuel cell technology a practical energy option. In addition to funding a project to improve the cost effectiveness of the Ballard fuel cell, Natural Resources Canada facilitated a utility demonstration program between Western Economic Development (a federal funding agency), the Government of British Columbia, and Ballard to develop a fuel cell electricity-generating plant (Natural Resources Canada, 1996a).

BARRIERS TO THE ADOPTION OF ALTERNATIVE TECHNOLOGIES

Advances in solar, wind, hydrogen, and other renewable energy sources are occurring rapidly. In some locations, freeway signs are being illuminated by hydrogen fuel cells and photovoltaics, entire communities are powered by wind farms, and a Canadian-built solar car successfully travelled from coast to coast in the summer of 2000 (see Box 11–4).

Despite these advances, renewable technologies are often considered small-scale alternatives that are incapable of providing large populations with affordable energy. Localized renewable energy provision also is contrary to contemporary corporate mega-thinking and the quest to achieve economies of scale (through pipelines, hydroelectric plants, nuclear power stations, and so on). Utility companies generally prefer large-scale projects because of the perceived benefits of lower costs and more customers. In addition, megaprojects often have fewer players, and therefore the utilities and other large companies have more control over a region and its energy. However, as Canada and the world come to accept the need to reduce greenhouse gas emissions, alternative energy technologies are receiving more support.

With respect to wind energy, perceptions of protectionism and the lack of a level playing field for new electrical generation projects may be seen as barriers to the encouragement of energy sustainability. For renewable energy resources. To be accepted by the public, prices must be comparable and quality of service must parallel current sources.

The implementation of alternative energy sources likely will reduce the dominant market position of many large utilities. For this reason it is difficult to convince these companies to invest in alternative energy sources. Instead, utilities have worked to keep their costs down and promote energy efficiency to keep their product more affordable than a renewable alternative. Even so, renewable energy resources will be developed to serve an increasing number of people as they become cost-effective.

The potential to capitalize on alternative energy offers developing nations a double opportunity. Besides reducing dependency on oil imports, developing nations

solar energy systems, thus creating jobs and expanding their industrial base. In both developed and developing nations, the world's major oil, utility, and automotive companies bring strong political and economic influences to bear on decision-making processes; these influences may jeopardize widespread research and implementation of alternative energy technologies.

IMPROVING ENERGY EFFICIENCY

Although roadblocks stand in the way of adopting renewable energy sources for widespread use, Canadians increasingly are conscious of the environmental and cost-saving benefits of using energy wisely. The implementation of energy-efficient technology is helping to reduce the environmental burden of energy use in our transportation, industrial, and residential requirements. Without the energy-efficient technologies introduced in the last decade or so (1990–2001), energy use in Canada would have grown by 28 percent, rather than 14 percent. The difference means Canadians have achieved a 10 percent increase in energy efficiency (Environment Canada, 2003).

Transportation Efficiency

The transportation sector consistently accounts for about 30 percent of secondary energy demand in Canada (Natural Resources Canada, 2003b). Passenger travel, including road, rail, and air transportation, represents 60 percent of the energy used in transportation. With road travel dominating, the best way to improve energy efficiency in transportation is to increase the fuel efficiency of motor vehicles. For the period 1973 to 1985, the average fuel efficiency of all new domestic cars doubled, and the efficiency of imports improved by 54 percent. Since that time, however, only slight gains in fuel efficiency have been observed.

Although small cars represent the largest percentage of vehicles on the road, light trucks have been the fastest-growing vehicle type on the road. This growth is attributed at least partly to the popularity of sport utility vehicles (SUVs). Classified as light trucks, SUVs are not subject to the same fuel consumption and tailpipe emission standards as passenger cars. SUVs average 20 centimetres taller, are 33 percent less fuel efficient, and produce 30 percent more carbon monoxide emissions than the average passenger vehicle. Public concern over the inefficiencies of SUVs is beginning to effect change in regulatory and manufacturing standards. In 2004, the state of California will begin the phased-in implementation of new tailpipe emission standards that will require SUVs, light trucks, and minivans to reduce heat-trapping emissions (carbon dioxide, carbon monoxide, and nitrous oxides) from tailpipes by 30 percent over 1998 levels. In spring 2004, Ford Motor Company introduced its first fuel-hybrid SUV.

Even with the controversy over SUVs, advances in vehicle technology promise to improve environmental efficiency in transportation. For example, the Honda Insight achieved a 3.9 L/100 km city (3.2 L/100 km highway) fuel rating, making it Canada's most fuel-efficient vehicle in 2004 (Office of Energy Efficiency, 2004). The Insight combines a gasoline engine with an electric motor to achieve its high-efficiency rating. Known as a hybrid vehicle, the Insight relies on gasoline; however, its electric motor assists by supplying extra power when needed. The electric batteries are recharged during braking by capturing the energy generated from forward momentum, eliminating the requirement for an external power supply (Honda Canada, 2001).

Another way to improve energy efficiency in transportation is to examine the types of vehicles we use to ship goods and products. For instance, transporting durable goods by planes and transport trucks wastes a considerable amount of energy. Increasing the use of trains and ships, as well as improving the fuel efficiency of transport trucks, promises rapid gains in energy savings to the transport sector.

One example of innovative thinking was RoadRailer, a truck–train freight service that cut transportation costs by moving truck trailers on rails. Until recently, moving freight short distances via rail was not economical. RoadRailer trailers are hauled on and off railroad tracks quickly by truck tractors and do not require overhead cranes. Solutions such as RoadRailer help railroads compete in the short and medium corridors by lowering costs and improving efficiency. In Canada, Canadian National operates RoadRailer service between Montreal, Toronto, and Chicago.

Industrial Efficiency

Industrial uses account for almost one-third of all energy consumed in Canada. Unfortunately, industry wastes a large percentage of the energy it uses through inefficient motors and boilers, improper use of lighting, and machine operation at higher settings than required. There are a number of ways industries can reduce their consumption and waste of energy at the plant level. In addition to installing energy-efficient lighting and equipment, industries can use computer-controlled energy management systems to turn off equipment and lighting when not required.

District heating through **co-generation** is another option industries and governments can explore to maximize energy efficiency. Under normal operating conditions, fossil fuel–powered electrical generating stations produce more waste energy than electricity. For example, the Tufts Cove Generating Station in Dartmouth, Nova Scotia, uses only about 35 percent of the energy that is available from fossil fuels to produce usable electrical energy. The remainder is lost to the environment as waste heat in the form of hot flue gases, and as warm water that is released into Halifax Harbour.

In an effort to recapture some of the waste heat and convert it to usable energy, Nova Scotia Power Inc. has begun a feasibility study of applying district heating to metropolitan Halifax. In a district heating process, the steam cycle is modified so that the steam is extracted and used to produce hot water. The water then would be pumped through pipes to surrounding buildings (such as offices, schools, and shopping malls) to supply heat.

Producing two useful forms of energy from the same source is known as co-generation, or combined heat and power (CHP). To initiate a district heating project, a variety of infrastructure improvements must be made. A closed-loop delivery system of pre-insulated pipe must be constructed between the plant and selected buildings. This allows hot water to flow from the plant to customers and then to return for reheating. Receiving buildings also require a heat exchanger to transfer energy to their heating systems. Customers with steam heating also may have to retrofit their systems to use hot water. Because of the substantial costs associated with such improvements, district heating systems normally are phased in over time.

District heating can produce a number of economic and environmental benefits. Proponents of the Halifax study cite that the more efficient use of fuel will decrease dependence on burning fossil fuels, thereby reducing greenhouse gas emissions. Furthermore, oil-fired boilers in the buildings targeted for district heating would no longer be needed, reducing the risk of fires and soil cont- amination. Opportunities for co-generation are not limited to district heating. Epcor, Edmonton's utility company, uses co-generation from landfill methane as part of its power supply, and Box 11–5 considers research under- taken to produce electricity from sour natural gas. Employing the energy-saving benefits of co-generation would give North Americans a chance to catch up to Western Europe, where co-generation has been used extensively since the 1970s.

Home Efficiency

When we think of the ways human activities influence the release of greenhouse gases into the atmosphere, we commonly blame our cars and factories, not our homes and offices. However, the average Canadian home pro- duces four tonnes of carbon dioxide per year. Much of this can be avoided by improving the energy efficiency of our homes through insulation, air sealing, better-quality windows and doors, and more efficient heating of rooms and water. Natural Resources Canada (1996b) estimated that the average Canadian homeowner could reduce heating costs by up to 25 percent per year by performing a variety of reasonably priced home renovations.

Canada's foray into improved home energy efficiency began during the 1970s energy crisis. At that time, the fed- eral government began the Super Energy-Efficient Home program to promote improvements in residential heating, insulation, and ventilation. In 1982, the Department of Energy, Mines and Resources (now Natural Resources Canada) introduced the R-2000 program in partnership with the Canadian Home Builders Association (CHBA). The R-2000 "environmentally friendly" home can generate up to 40 percent savings in energy consumption over a tra- ditional house of the same size.

The central feature of an R-2000 home is the strict requirement for air sealing and the use of the heat recovery ventilator (HRV) or the air-to-air heat exchanger. Located near the furnace, an HRV unit draws fresh air into the house while expelling existing air. As the air streams pass each other, incoming air is either heated or cooled by outflowing air, resulting in improved heating and cooling efficiency. R-2000 standards require an HRV unit to exchange all the air in a house at least once every three hours. Although the placement of windows, doors,

BOX 11–5

CO-GENERATION: THE PRODUCTION OF ELECTRICITY FROM SOUR NATURAL GAS

Approximately 30 percent of Canada's natural gas contains more than 1 percent hydrogen sulphide (H_2S). Before these reserves can be sold as a useful energy source, the H_2S must be removed. The present technology used to remove H_2S is called the regenerative amine process. Byproducts of the process are elemental sulphur (S_8), sulphur dioxide (SO_2), and carbon dioxide (CO_2). For a long time, producers of natural gas could remove the sulphur and sell it for a profit. Now, the supply of sulphur far exceeds demand. This has lead to serious sulphur management and storage problems for natural gas producers. The process also contributes to climate change through the release of greenhouse gas emissions. In an attempt to reduce these problems, researchers at Alberta Sulphur Research Limited (ASRL) at the University of Calgary are working on a

process that eliminates the production of sulphur and CO_2 while co-generating electricity at the same time.

The research being tested by ASRL seeks to inject the H_2S and CO_2 back into the sour gas reservoir. A chemical reaction occurs to produce liquid sulphur, which sinks to the bottom of the reser- voir as a result of gravity. The recapture of sulphur is eliminated as it remains in the reservoir. Co-generation is possible because there is enough energy available in the sulphur and H_2S to produce elec- tricity to run the amine absorber and the injection system. Another advantage is that rather than increasing greenhouse gas emis- sions, reinjecting CO_2 turns the reservoir into a carbon sink (see Chapters 5 and 9). Although in the developmental stage, researchers are optimistic that this technology may lead to improved environmental efficiencies in the natural gas industry.

SOURCES: *SO_2 Injection: An Energy-Efficient Strategy for Dealing with H_2S in Sour Gas Production*, P. Clark & P. Davis, 1999, Alberta Sulphur Research Ltd., presented at Sulphur 2000, San Francisco; "Large Scale Electricity Production from Sour Natural Gas without Sulfur Production or Carbon Dioxide Emissions," P. Clark & P. Davis, 2000, *Alberta Sulphur Research Ltd. Quarterly Bulletin, 36*(3).

and roof overhangs is taken into consideration in energy-efficient homes, virtually any house design can accommodate the R-2000 approach. However, only houses built by R-2000 certified builders may be certified R-2000 homes. The features and benefits of an R-2000 home are identified in Box 11–6.

By 2000, about 8800 R-2000 homes had been certified in Canada, a small number considering that 150 000 to 200 000 new homes are built here each year. The 2 to 6 percent higher construction costs of R-2000 homes, as well as the administrative process imposed by R-2000 construction, may help explain the limited number of registered homes. Such homes must be constructed by the 800 licensed R-2000 homebuilders in Canada, and the dwelling must be tested during construction and inspected upon completion to be certified. Nevertheless, as a result of R-2000, manufacturers have developed many unique building products, such as HRVs (now a $50-million-a-year business), high-performance windows, and integrated mechanical heating and cooling systems. R-2000 technology also is exported abroad to assist other nations in improving residential energy efficiency.

Energy-efficient homes have captured the attention of homeowners in the Maritimes, where high-cost electric heating dominates. Heating costs in east coast R-2000 homes have been reduced by up to 50 percent, while resale values are estimated to be 5 percent higher than conventional houses (Lougheed, 1992).

Photo 11–15
The solar panels and SunPipe are two of the features of ASH House in Calgary.

"Leading Canadians to energy efficiency at home, at work and on the road" is the vision of Canada's Office of Energy Efficiency (OEE). Established in 1998, as part of Natural Resources Canada, the OEE manages 17 energy-efficiency and alternative fuels programs aimed at the residential, commercial, industrial, and transportation sectors. EnerGuide for Houses is one of the OEE programs aimed at improving energy performance in the residential

---

BOX 11–6
FEATURES AND BENEFITS OF AN R-2000 HOME

- A whole-house, continuous ventilation system resulting in better air quality and health advantages
- More environmentally friendly materials and equipment
- Advanced heating and cooling systems
- Energy-efficient appliances and lighting
- Energy-efficient windows and doors resulting in fewer drafts and cold spots
- A "tight" building envelope to reduce drafts and heat loss
- Less noise and dust
- High levels of insulation
- More comfortable, with lower energy bills

So why the name R-2000? The "R" is a value normally assigned to the insulation capacity of a building surface, such as R-35 walls, and the "2000" represents the futuristic building standard the federal government wanted to see in place by the year 2000.

Photo 11–16

SOURCE: *R-2000 Home Program,* Natural Resources Canada, Office of Energy Efficiency, http://www.oee.nrcan.gc.ca/english/newhousesr2000.cfm. Reprinted with permisiion. R-2000 is an official trademark of National Resources Canada.

sector. The program helps homeowners obtain individualized professional advice on how to improve the energy efficiency of their homes. Expanding on the EnerGuide labelling program for appliances, EnerGuide for Houses provides homeowners with facts to make informed decisions about energy efficiency, whether they are making improvements to their existing home or buying a new one.

Global efforts are underway to improve heating efficiency in homes and businesses. The World Wildlife Fund for Nature estimated that Britain's infamous drafty houses and apartments contributed one-third of the country's carbon dioxide emissions, while homes in the Netherlands were responsible for 10 percent of that nation's carbon dioxide emissions. On average, one-fifth of Europe's carbon dioxide emissions come from households.

In Britain, building regulations now require that the efficiency of water and space-heating devices in new homes be specified (on a scale of 1 to 100). This initiative places pressure on builders to meet purchasers' demands for lower heating costs and thereby encourages competition to make homes more energy-efficient (Russell, 1996).

In 1995, World Wildlife Fund for Nature–Netherlands signed an agreement with the country's five largest property developers to build 200 energy-efficient houses in a period of two years (Russell, 1996). Dutch developers must conform to a rigid set of building specifications, similar to construction standards for an R-2000 home, including the use of solar-powered boilers, solar cells, blown fibre wall cavity insulation, water tank jackets, double-glazed windows, and energy-efficient light bulbs. In return, WWF actively will promote the advantages of the homes to municipal authorities and prospective buyers and tenants.

Ground-source heating and an aquifer thermal energy storage project in Canada are considered briefly in Enviro-Focus 11.

ORGANIZED INITIATIVES

Canadians expect that heat and light for our homes, fuel for our vehicles, and power to operate our businesses and factories will be available on demand. In Western society, much of this energy comes from nonrenewable fossil fuels. Use of oil, coal, and natural gas is a major cause of air and water pollution, land disruption, and long-term changes in global climate. Similarly, use of other common energy sources, such as hydroelectricity, nuclear power, and biomass energy, also affect the natural environment and can have devastating effects on ecosystem sustainability.

In response to the detrimental environmental effects of energy use, Canada has developed a number of initiatives and has become a partner in international agreements aimed at conserving energy, reducing the output of greenhouse gases (discussed in Chapter 5), and developing renewable energy sources.

The United Nations Framework Convention on Climate Change, opened in 1992 at the Earth Summit in Rio de Janeiro, Brazil, acknowledged for the first time that human activities have an influence on global climate. The parties agreed to work toward preventing the predicted effects of climate change, such as sea-level rise and changes to agricultural production zones. In 1997, international commitments were strengthened when the Kyoto Protocol established targets for reducing greenhouse gas (GHG) emissions. The Kyoto Protocol, to which Canada was a signatory, committed nations to reduce GHG emissions to 6 percent below 1990 levels by 2008–2012. Many nations who signed the Kyoto Protocol were not on track to meet the reduction target. In 1999, Canada realized that a reduction of 26 percent in our GHG emissions from business-as-usual forecasts would be necessary if we were to reach the international objective.

The Conference of the Parties (CoP) is the primary decision-making body of the United Nations Framework Convention on Climate Change. Representatives from all signatories to the Kyoto Protocol have met annually to negotiate implementation measures, chart progress, review new technologies, and provide support in reaching Kyoto commitments. Recognizing that developed countries have contributed the most to the greenhouse gas problem, CoP requires industrialized countries to provide leadership through emission reduction as well as technological and financial support to developing countries. Implementation of the Kyoto Protocol is far from certain. The protocol needs the ratification of either the United States or the Russian Federation to be binding. The United States has declared it will not ratify Kyoto, and it appears the Russian Federation is unlikely to sign either, given that "adhering to the provisions of the Kyoto Treaty and achieving economic growth are incompatible" (Corcoran, 2003, p. C7; United Nations Framework of the Climate Change Convention, 2003).

Despite the difficulties in reaching an international agreement, Canada ratified the Kyoto Protocol in December 2002. Between 1998 and 2003, even without a formal plan, the federal government invested $3.7 billion to support the reduction of greenhouse gases by individual Canadians, industry and business, and governments and communities (Government of Canada, 2003). Some of the activities eligible for funding include new technology and alternative fuel development measures, energy-efficiency incentives, research, and public education programs (Government of Canada, 2003). All provinces and territories have announced action plans, as have all sectors of the economy.

Canada and the United States also are making efforts to reduce their contributions to climate change. In 1997, the two nations agreed to work together to encourage and adopt cleaner, more efficient transportation and to promote the use of innovative market strategies to reduce

Ground-Source Heating and Cooling

Both earth and water maintain a relatively constant temperature below the surface, making them an ideal heating source in the winter months or cooling source in the summer months. Tapping backyards, ponds, or lakes for geothermal energy is an environmentally friendly and efficient way to heat and cool many types of buildings.

Tapping geothermal energy is basically a way of moving (rather than generating) heat. A sealed or closed-loop system, in which a water-based solution is pumped through a polyethylene pipe extending vertically or horizontally below the Earth's surface, can operate effectively with ground temperatures ranging from –5°C to 38°C.

In the winter, for example, the solution absorbs heat and carries it to a geothermal unit, which compresses the heat to a high temperature (similar to the process used in conventional refrigerators) and delivers it to the home or building. A small electric pump operates the system, but the energy generated is often four or five times greater than the energy used. In comparison, if a conventional natural gas furnace achieves a one-to-one ratio, it is considered to be highly efficient.

The idea of geothermal energy use in Canada is not a new one. Open-loop heat pumps, which use the water and discard it, have been used in Canada since 1912, and closed-loop technology has been available since the 1940s. A 300-room hotel in Winnipeg has used geothermal power since the 1930s, and it is used extensively in British Columbia.

Although start-up costs of geothermal systems are about 40 percent higher than for conventional systems, savings in operating costs usually make up for the difference in three to five years. Because there is no combustion of fossil fuel, and no greenhouse gases, geothermal systems are safe. They produce less dust and provide better indoor air quality. Ground-source heating maintains a relatively constant heating in winter, greater control over humidity in summer, and is suitable for use in private residences and commercial and public buildings such as schools.

In Sussex, New Brunswick, an aquifer thermal energy storage (ATES) system at the Sussex Hospital complex began in 1992 and has been fully operational since the spring of 1996. The hospital uses the ATES system to reduce the cost of preheating air and saves approximately $86 000 in energy consumption annually.

The system works by using two well fields, one cool (7.5°C) and the other warm (10°C), located on either side of the hospital complex. During the cooling season, the groundwater is transferred from the cool well field through the exchange systems within the buildings to the warm well field; the process is reversed during the heating season. In the cooling season, heat is transferred from the air through, for example, a water-to-air coil to the groundwater, thus reducing the temperature of the air in the supply system.

Using a hospital to demonstrate the ATES concept is significant because hospitals depend on secure energy supplies for heating and cooling. Monitoring of this project is ongoing, but officials expect to be able to apply the ATES process to other community buildings and to demonstrate the potential for other community-based underground thermal energy sources.

SOURCES: "Ground-Source Heating: The Wave of the Future?" D. Burke, March 18, 1997, *Canmore Leader*, p. A14; "Underground Thermal Energy in Sussex, New Brunswick," F. Cruickshanks, 1997, *Water News*, 16(2), 3–6; "Underground Thermal Energy Rises to Future Challenges," F. Cruickshanks & J. L. Sponagle, June 1997, *Technical Bureau Supplement, Water News*, i–viii.

greenhouse gas emissions (Environment Canada, 1997). One strategy involves the use of tradable **emissions permits**, which allow companies to buy and sell from each other the right to release greenhouse gases. Tradable emissions permits also may be used between nations that export and import clean-burning fuels. For example, when the United States imports natural gas from Canada, the burning of oil and coal is reduced, subsequently lowering U.S. carbon emissions. Canada would receive a credit from the United States, which would offset the environmental cost of extracting and transporting the natural gas.

The federal government also has made strides in promoting energy efficiency and developing alternative energy sources. In 1993, the Energy Efficiency Act was introduced, giving the federal government the authority to make and enforce regulations related to energy efficiency and alternative energy. In addition to setting efficiency standards for energy-using equipment, the act requires labelling of energy-using products, and provides authority to the minister of Natural Resources to designate inspectors to ensure compliance with the act and regulations.

ENERGY FUTURES

What are the best options for meeting our energy requirements, and how do we meet these requirements without jeopardizing environmental sustainability? In the short term, we likely will continue to depend on fossil fuels. Oil is the lifeblood of our economy and, despite its environmental hazards, remains inexpensive (compared with its alternatives), has a relatively high net useful energy, and is easy to transport. However, the Earth's fossil fuels are diminishing. During the last century, we have consumed massive reserves of coal, oil, and natural gas—reserves that took eons to accumulate.

Combined with global deforestation, our burning of fossil fuels has upset the natural balance of the carbon cycle. We have been warned about the potential effects of human-induced climate change. Even so, most (if not all) of the world's richest economies will be unable to meet the Kyoto Protocol's target of reduced greenhouse gas emissions by 2008–2012. Our inability to meet this internationally negotiated objective emphasizes how attached we have become to fossil fuels, and how difficult it is to move away from their use.

Fortunately, we are trying to kick the fossil fuel habit—we have little choice. At current rates of use, fossil fuels are expected to become uneconomical at some point this century. Improvements in energy efficiency, increased use of coal, and co-generation efforts will help to extend the life of fossil fuels a little longer; however,

their use is finite. As the availability of fossil fuels decreases (and costs escalate), the use of alternatives will increase.

Alternatives may include sources we are using already, such as nuclear, biomass, and hydropower. As we have discovered, however, there are limits on these sources as well. The construction of hydroelectric dams results in widespread landscape and ecological changes, and few sites remain that would be accepted on both environmental and economic grounds. Nuclear power is an option that has suffered in popularity, especially in Western nations. The events at Chernobyl, potential for further accidents, and storage of radioactive wastes are issues detracting from the suitability of nuclear power to satisfy global energy requirements. Nuclear power also is not sustainable over the long term as uranium is a nonrenewable resource. Although renewable, the widespread use of biomass for energy would not be feasible unless harvest were equal to replacement. In addition, effective land use controls are required to maintain soil nutrients and prevent excessive soil erosion. Biomass also has a low net useful energy and is expensive to collect and transport.

In the longer term, it is likely that our communities will draw on a much wider variety of energy resources, depend much less on imported energy, and instead make optimal use of locally available renewable resources. Advances in technology and energy-efficiency measures are providing the bridge to shift us from nonrenewable fossil fuels to alternative, renewable energy sources.

Despite our appetite for oil and gas, we have begun making the transition to renewable energy. Although it is in its infancy, there are positive signs that this transition is taking place. International efforts to reduce the release of greenhouse gases are prompting countries to increase research budgets for renewable energy, levy fuel taxes, and offer incentives for energy conservation. Increased government funding for hydrogen research in the United States and the success of the Ballard hydrogen fuel cell battery are examples of the steps being taken toward wider use of alternative energy sources.

Chapter Questions

1. Identify the range of negative effects associated with fossil fuels. Which do you think is the most serious? Why?

2. An advantage of various forms of renewable energy (wind and solar energy, for instance) is that they cause no net increase in carbon dioxide. Is this true for biomass? Why or why not?

3. Excluding fossil fuels, what other forms of energy have the greatest potential where you live? What might be some of the barriers to developing these energy sources?

4. Discuss why or how energy conservation and improved energy efficiency might be considered major "sources" of energy.

5. Why is the permanent storage of high-level radioactive wastes such a problem in Canada and elsewhere?

6. What kinds of energy conservation measures could you adopt for each of the following aspects of your life: washing dishes, doing laundry, lighting, bathing, cooking, buying a car, driving a car?

references

Alberta Energy and Utilities Board. (2000). Guide 60: Upstream Petroleum Industry Flaring Guide.
http://www.eub.gov.ab.ca/bbs/products/newsletter/2000–01/atb_jan2000_fea02.htm

Ballard Power Systems. (2000a). *Ballard and DaimlerChrysler unveil direct methanol fuel cell demonstration vehicle.* http://www.ballard.com

Ballard Power Systems, Inc. (2000b). *Ballard, Xcellsis, Chicago Transit Authority conclude successful fuel cell bus demonstration program.* http://www.ballard.com/whatsnew_archive.asp

Ballard Power Systems, Inc. (2004). *Be informed.*
http//www.ballard.com/tB.asp?pgid=18&dbid=0

Bolger, L., & Isaacs, E. (2003). Shaping an integrated energy future. In A. Heintzman & E. Solomon (Eds.). *Fueling the future: How the battle over energy is changing everything.* Toronto: House of Anansi Press, 55–81.

Boras, A. (1995, November 18). Golden sands—amazing oilsands revival taking place. *Calgary Herald,* pp. C1, 2.

Boyd, D. R. (2001). *Canada vs the OECD: An environmental comparison.*
http://www.environmentalindicators.com/htdocs/about.htm

Butt, S. (2000a, November). Challenge victory is in the wind. Sustainable energy development and the environment. *Calgary Sun,* special supplement, p. 4.

Butt, S. (2000b, November). Going green at Enmax-imum speed. Sustainable energy development and the environment. *Calgary Sun,* special supplement, p. 26.

Canadian Association of Petroleum Producers. (n.d.). *Industry facts and information—Oil, oil sands and natural gas—Oil sands.* http://www.capp.ca

Canadian Association of Petroleum Producers. (2001). *Action on energy—Industry continues to surpass gas flaring targets.* http://www.capp.ca/default.asp?V_DOC_ID+816_

Canadian Energy Research Institute. (2003). *Economic impact of the nuclear industry in Canada.* Calgary: Author.

Canadian Environmental Assessment Agency. (1998). *Report of the Nuclear Fuel Waste Management and Disposal Concept.* Environmental Assessment Panel.

http://www.ceaa.gc.ca/panels/nuclear/reports/report_e.htm

Canadian Wind Energy Association. (2003). *Canada wind power: Installed capacity.* http://www.canwea.ca/pdfs/CanWEA_brochure.pdf

Chambers, A. (1996, June 5). Watchdog calls for strict control of emissions. *Calgary Herald,* p. C2.

Commissioner of the Environment and Sustainable Development. (2002). *The Sydney Tar Ponds.* www.oag-bvg.gc.ca/domino/reports.nsf/ html/c20021002se01.html

Corcoran, T. (2003, December 7). Eventually somebody is going to have to stand up and formally declare the Kyoto protocol to be dead. *Calgary Herald,* p. C7.

Dignard-Bailey, L. (2000). Photovoltaic technology status and prospects. *Canadian Annual Report 2000,* pp. 1, 4.

Energy Information Administration. (1999). *Brazil: Environmental issues.* United States, Department of Energy. http://www.eia.doe.gov/emeu/cabs/brazenv.html

Energy Probe Research Foundation. (1995). *1995 Annual Report.* Toronto: Energy Probe.

Enmax. (2001). *It's like planting 120 mature trees every year.* Brochure.

Environment Canada. (1996). Energy consumption. *State of the Environment Bulletin,* no. 96-3 (Spring).

Environment Canada. (1997). *Minister Marchi announces Canada–U.S. actions to protect the environment.* Press release.

Environment Canada. (2003). *Canada's national indicator series 2003.* http://www.ec.gc.ca/soer-ree/English/headlines/toc.cfmbut

Epp, J., Barnes, P., & Jeffrey, R. (2003). *The Report of the Pickering "A" Review Panel.* Ontario Ministry of Energy.

Fairley, P. (2003). The perfect energy: From earth, wind, or fire? In A. Heintzman & E. Solomon (Eds.), *Fueling the future: How the battle over energy is changing everything.* Toronto: House of Anansi Press, 153–177.

Feldman, D. L. (1995). Revisiting the energy crisis: How far have we come? *Environment, 37*(4), 16–20, 42–44.

Francis, W. (1997). Burning questions about gas flares. *Environment Views and Network News, 1*(1), 18–19.

Frenay, R. (1996). Water power. *Audubon, 98*(3), 24–26.

Gjertson, H. (1997). Still the worst. *Alternatives, 23*(3), 5.

Government of Canada. (2003). *Taking action on climate change: Government of Canada announced $1 billion toward implementation of the climate change plan for Canada.* http://www.climatechange.gc.ca/english/publications/announcements/news_release.html

Green Lane. (1997). *Canadian consumption of energy in national environmental indicator series,* No. 96-3 – updated, p. 2. http://www.ec.gc.ca/Ind/English/Energy/Tables/ectb01_e.cfm

Hamilton, G. (1997, February 16). Emblem of death. *Calgary Herald,* p. A13.

Hill, R., O'Keefe, P., & Snape, C. (1995). *The future of energy use.* London: Earthscan.

Honda Canada. (2001). *2001 Insight.* http://english.honda.ca/models/insight.asp

Honda Corporation. (2003). *Honda FCX hydrogen fuel-cell vehicle displayed at White House event.* http://hondacorporate.com

Hornung, R. (2003). *Canadian Wind Energy Association: Federal government's wind power production target in jeopardy.* http://www.canwea.ca/PressReleases.html

International Energy Agency. (2003a). *Key energy indicators in 2000: Canada.* http://www.iea.org/stats/files/selstats/ketindic/country/canada.htm

International Energy Agency. (2003b). *Renewables information: 2003 edition,* pp. 89–91. http://cetc-varennes.nrcan.gc.ca/eng/publication/2001-45e.pdf

International River Network. (1998). *The river dragon has come.* www.irn.org/programs/threeg/dragon.html

Joint Action Group for the Environmental Clean-up of the Muggah Creek Watershed. (2003). *The big picture and community recommendations.* http://www.muggah.org/

Laird, G. (2001). One last boom. *Canadian Geographic, 121*(3), 38–52.

Lau, M. (2000, July 24). Rancher's gas battle still flares. *Calgary Herald,* p. B1.

Lougheed, T. (1992, May/June). R-2000. *Canadian Consumer,* 24–28.

Macfarlane, A. (2003). Is nuclear energy the answer? In A. Heintzman & E. Solomon (Eds.), *Fueling the future: How the battle over energy is changing everything* (pp. 127–151). Toronto: House of Anansi Press.

Martin, D. (2003a). *Canada nuclear subsidies: Fifty years of futile funding.* Sierra Club. http://www.sierraclub.ca/national/nuclear/reactors/index.html

Martin, D. (2003b). *Nuclear restarts a major problem … August 19, 2003.* Sierra Club. http://www.sierraclub.ca/national/media/ontario-blackout-03-08-19.html

McLearn, M. (2003). *Canadian business. New nukes – or none?* Energy Probe. http://www.energyprobe.org/energyprobe/print.cfm?contentID=9048

Natural Resources Canada. (1996a). *Anderson announces funding for Ballard fuel cell technology.* http://www.nrcan.gc.ca/css/imb/hqlib/9619.htm

Natural Resources Canada. (1996b). *New initiative will encourage energy-efficient home renovations.* http://www.emr.ca/hqlib/9611.htm

Natural Resources Canada. (2000a). *Improving energy performance in Canada—Report to Parliament under the Energy Efficiency Act (1997–1999).* Ottawa: Author.

Natural Resources Canada. (2000b). *Energy in Canada 2000.* Ottawa: Author.

Natural Resources Canada. (2003a). *Canadian Renewable Energy Network: Programs.* http://www.canren.gc.ca/programs/index.asp?CaId=60&PgId=143

Natural Resources Canada. (2003b). *Energy use data handbook 1990 and 1995 to 2001: Canada's secondary energy use by sector.* http://oee.nrcan.gc.ca/neud/dpa/data_e/Datahandbook2003.pdf

Newman, L., and Brooks, D. B. (2004). The soft path holds up. *Alternatives Journal, 30*(1), 13.

Office of Energy Efficiency. (2004). *Buying a fuel-efficient vehicle.* http://oee.nrcan.gc.ca/englishp_transportation/index.cfm?Text=N&PrintView=N

Pape-Salmon, A., et al. (2003). *Low-impact renewable energy policy in Canada: Strengths, gaps and a path forward.* Drayton Valley, Alberta: Pembina Institute for Appropriate Development.

Pembina Institute. (2000, February 18). *Environmentalists disillusioned by PanCanadian decision.* News release. http://www.pembina.org/press/02182000.htm

Petro-Canada. (1999). *Reserves and production.* http://www.petro-canada.ca/eng/about/business/8946.htm

Public Safety and Sour Gas. (2000, December 18). *Provincial Advisory Committee on Public Safety and Sour Gas releases final report: Findings and recommendations.* http://www.publicsafetyandsourgas.org/FinalRec.htm. Accessed January 2001.

Russell, S. (1996, March). Making eco-friendly houses. *WWF Newsletter.* http://www.panda.org

Scanlan, L. (2001). Power switch. *Canadian Geographic, 121*(3), 54–62.

Sierra Club of Canada. (2003). *Fundamental solutions to prevent future blackouts.* http://www.sierraclub.ca/national/media/ontario-blackout-03-08-19.html

Sly, L. (1997, January 18). Upheaval on the Yangtze. *Calgary Herald,* p. C3.

Smith, D., & MacCrimmon, G. (1999). Environmental assessment review of Suncor Energy. In *Canadian Environmental Defence Fund. Project millennium.* http://www.cedf.net/ca/suncor.html

Statistics Canada. (2002). *Human activity and the environment: Annual Statistics 2002.* Ottawa: Minister of Industry.

Statistics Canada. (2003). Gross domestic product at basic prices by industry. CANSIM II Tables 379-0017 and 379-0020. http://www.statcan.ca/english/Pgdb/econ41.htm

Stothart, P. (1996). Nuclear electricity: The best option given the alternatives. *Policy Options, 17*(3), 14–16.

Suncor Energy Inc. (2000, October). *Canada's climate change and voluntary registry program, sixth annual progress report.*

Suncor Energy Inc. (2003). *Investor information.* http://www.suncor.com/bins/content_page.asp?cid=8

The Sydney Tar Ponds: A city's toxic legacy. (2003, June 21). *Toronto Star.* http://www.sierraclub.ca/national/sydney-tp-media.html

Sydney Tar Ponds Agency. (2004). Home page. http://gov.ns.ca/stpa/default.asp?T=1

Taylor, D. M. (1994). *Off course: Restoring balance between Canadian society and the environment.* Ottawa: International Development Research Centre.

U.S.–Canada Power System Outage Task Force. (2003). *Interim report: Causes of the August 14th blackout in the United States and Canada.* Governments of Canada and the U.S.

United Nations Framework of the Climate Change Convention. (2003). *Caring for climate.* Bonn, Germany: Climate Change Secretariat (UNFCCC).

Wallis, C., Klimek, J., & Adams, W. (1996). Nationally significant Sage Creek Grassland threatened by Express pipeline. *Action Alert, 7*(4), 4 pp.

Zich, A. (1997). China's Three Gorges: Before the flood. *National Geographic, 192*(3), 2–33.

"The endangered spaces that must be loved and protected are the irreplaceable landscapes and waterscapes whose mosaics contribute to the health, beauty, permanency, and productivity of the globe. To perceive native landscapes and waterscapes, parks and wildernesses, as beyond price, as sacrosanct, is the saving goal that humanity must pursue."

J. S. Rowe, in Hummel (1989)

Chapter Contents

Chapter Objectives

After studying this chapter you should be able to

- understand the biodiversity concerns relating to Canada's species and spaces

- identify a range of human uses of wild species and natural spaces

- describe the impacts of human activities on wild species and their habitats and environments, including protected areas

- appreciate the complexity and interrelatedness of wild species, habitat, protected areas, and biodiversity issues

- outline Canadian and international strategies to protect wild species and spaces

- consider our role in protecting biodiversity in urban areas

- discuss challenges to a sustainable future for wild species and protected areas in Canada

INTRODUCTION

Once the most widespread frog species in North America, the northern leopard frog experienced a population crash in the mid-1970s all across the continent. "Piles of dead and dying frogs were reported from many Lake Manitoba shorelines, [and] heaps nearly a metre high were recorded from the major frog holes" (Zolkewich, 1995, p. 3). By 1989, the realization that a variety of frog species had disappeared almost simultaneously from large, well-protected national parks and nature reserves around the world raised particular concern. Golden toads, for instance, disappeared from the Monteverde Cloud Forest Reserve in Costa Rica, and gastric breeding frogs vanished from a remote national park in Australia. The golden toad has not been seen since 1989, and the gastric frog since 1981 (Pounds et al., 1999; Campbell, 1999). Recent research has shown that red-legged frogs have disappeared from pristine habitat near Yosemite National Park, California (Drost & Fellers, 1996), and apparently are extirpated (National Parks Service, 2000).

Most previously observed amphibian (frog, toad, salamander, and newt) population declines and extirpations were attributed directly to habitat destruction by logging, urbanization, and drainage of wetlands, and indirectly to pollutants. Local amphibian populations also could have been affected greatly by factors such as weather (particularly drought), predation, and extensive commercial harvest such as occurred with bullfrogs in Algonquin Provincial Park, Ontario (Brooks & MacDonald, 1996; Orchard, n.d.).

The loss of frogs from geographically dispersed and apparently pristine protected areas suggested one or more global agents might be adversely affecting amphibians. Possible candidates for the causes of these global declines include an increase in UV-B radiation resulting from ozone layer depletion; chemical contamination including the effects of acid precipitation, pesticides, herbicides, and fertilizers; introduction of exotic competitors and predators; and disease (Declining Amphibian Populations Task Force, n.d.; Dunn, 1996).

Frogs and other amphibians are highly sensitive to a variety of environmental stressors and are good indicators of ecological problems (see Chapter 3). Their permeable skins, through which they breathe, make them extremely vulnerable to both airborne and waterborne pollutants. In addition, their low mobility and complex life cycles (involving both aquatic and terrestrial habitats) make them vulnerable to subtle habitat changes. Of all the vertebrate classes, they may be the best indicators of ecological health, able to pinpoint degradation of terrestrial and aquatic habitats better than even the more commonly monitored bird populations (Bishop et al., 1994).

Photo 12–1a

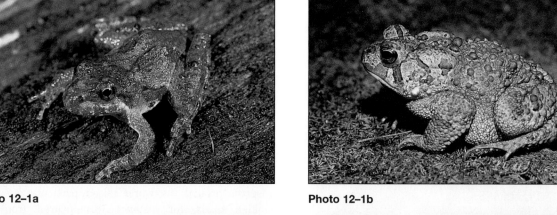

Photo 12–1b

Photo 12–1c

Photo 12–1d

Amphibians, such as frogs (12–1a), toads (12–1b), newts (12–1c), and salamanders (12–1d), are highly sensitive to a variety of environmental stressors. Because they do not migrate, because they reside in aquatic and terrestrial environments, and because their skins are permeable, they are excellent indicators of chemical changes in the environment.

Why does the loss of amphibians matter? Vanishing amphibians not only signal a loss of biodiversity, a cause of concern in itself, but also may indicate "profound environmental change affecting all life on earth" (Dunn, 1996, p. 4). There is a sense that if frogs are in trouble, humans are not far behind (Zolkewich, 1995). In addition to their significance as a measure of the health of the environment, amphibians are an important part of the ecological balance of many habitats. As predators, frogs and salamanders consume many times their weight in invertebrates, including many pest species, and, as prey, they control the abundance, distribution, and health of numerous aquatic and terrestrial predators (Bishop et al., 1994). Amphibians also have biomedicinal value: for example, we are just beginning to appreciate the potential and value that the skins of amphibians have in yielding drugs useful to medicine.

In spite of their ecological significance, scientists have only a rudimentary understanding of amphibian population dynamics, and baseline data on the populations of the 140 types of salamanders and 90 frogs and toads occurring in Canada and the United States are almost nonexistent (Bishop et al., 1994). Since the populations of most amphibian species exhibit large natural fluctuations, extensive research and long-term monitoring are necessary to determine whether the extirpations and large population declines are following natural patterns or are accelerating.

In response to international declines in frog populations, the IUCN (the World Conservation Union, formerly known as the International Union for the Conservation of Nature and Natural Resources) Species Survival Commission struck the Declining Amphibian Populations Task Force (DAPTF) in 1991. Over 3000 scientists and conservationists now belong to the network of 90 working groups around the world (including one in Canada) attempting to determine the nature, extent, and causes of global declines of amphibians and to promote means by which the declines can be halted or reversed and species diversity maintained (Declining Amphibian Populations Task Force, 2000).

The documented disappearance of amphibian species emphasizes the urgency of establishing reliable inventories and long-term monitoring programs in general, and of establishing them in national parks and other protected areas in particular. Inventories in Mount Revelstoke and Glacier national parks in British Columbia, for example, have revealed the presence of two amphibian species (western toads and spotted frogs) that have undergone severe population declines elsewhere. Monitoring these species within the parks should contribute data to a national database and help analysts identify and understand both the internal and external threats to biodiversity and ecological integrity (Dunn, 1996). (For further information about amphibian monitoring programs in Canada, see Box 12–1.)

Saving remaining populations and their habitat also is critical to the survival of some amphibian species, and Ducks Unlimited is one organization that has been active in this area. Their Prince's Spring project, a northern leopard frog breeding colony located about 190 kilometres southeast of Hanna, Alberta, is a protected site where this species has a good chance for survival. The frog population has remained stable since the early 1980s, when Ducks Unlimited constructed two dykes to create a spring-fed freshwater marsh, and built five nesting islands within the marsh (Zolkewich, 1995).

In addition to amphibians, there are many other wild animal and plant species in Canada whose continued healthy existence is threatened by loss and degradation of habitat through human economic development activities. Unrestricted clear-cut logging, agricultural expansion, wetland drainage, urban growth, pollution, global climate change, unsustainable use of some wildlife populations, and poaching and illegal trade are among the key threats to species biodiversity.

BOX 12-1
AMPHIBIAN MONITORING IN CANADA

In 1992, volunteer observers began monitoring the mating calls of male amphibians in Canada. By 1995, over 400 observers were involved in large-scale monitoring programs in Nova Scotia, Quebec, Ontario, Manitoba, and Saskatchewan. Each night for three minutes during the April to July mating season, observers report on the species calling and the intensity of the calling. Observers choose their own monitoring sites, which can be a favourite pond or marsh, a rural road, or even their own backyard. This program has become very popular with the general public, partly because volunteers are sent tapes of the frog calls, and people enjoy the opportunity to learn the calls of animals other than birds.

In Ontario, the Long Point Bird Observatory and Environment Canada's Marsh Monitoring Program combine sight and sound surveys of marsh birds and amphibians in Great Lakes wetlands to assess the need for rehabilitation of a marsh or to determine whether a rehabilitation project has been successful. In 1995 and 1996, 5502 amphibian choruses were recorded on 256 routes.

In Nova Scotia, the monitoring program focuses on the spring peeper; volunteers call the provincial museum on the first date in spring when they hear the spring peepers. These data collected by volunteers not only provide information on the presence or absence of the spring peeper, but also promote awareness because the locations of the first dates of calling are reported on television weather maps.

There is also an extensive cooperative amphibian monitoring program between Canada and the United States. The North American Amphibian Monitoring Program (NAAMP), established in 1994, is a collaborative effort of the amphibian research and conservation community in North America.

The broad goal of NAAMP is to develop a statistically defensible program to monitor the distribution and abundance of amphibians. So far, provinces and states have focused on implementing calling surveys and establishing herpetological (reptile and amphibian) atlases. Collaboration among researchers and volunteers is important to the success of projects such as Saskatchewan's Amphibian Monitoring Program and Ontario's Herpetofaunal Atlas. The NAAMP provides data reports to such atlas projects, and the atlas projects assist monitoring programs through analysis of data, development of additional survey routes, and "frog watch" programs.

Participants in NAAMP's monitoring programs come from federal governments, provincial and state natural resource groups, national parks, wildlife refuges, academia, ENGOs, and the public. Both the Canadian and North American amphibian monitoring programs rely to a considerable extent on partnerships between professional biologists and volunteer observers. These partnerships also serve as educational and training opportunities, as volunteers learn how to identify the calls of adult amphibians and develop other skills in order to help collect high-quality data (Orchard, personal communication, 1997).

The Canadian Amphibian and Reptile Conservation Network (CARCNET) is a proactive group, principally of biologists, working to reverse the trends in habitat loss and to better understand frogs, toads, salamanders, turtles, snakes, and lizards. This organization helps coordinate public involvement in frog and toad monitoring programs across Canada and recently has developed a system to identify Canada's most critical and valuable amphibian and reptile habitats. This system is used to forewarn Canadians of places that have special significance for the conservation of amphibians and reptiles so that these places can be protected. By December 2000, two sites had been identified: Pelee Island in Ontario and the south Okanagan Valley in British Columbia. For additional information about CARCNET's important amphibian and reptile areas (IMPARA) project, see their website: http://www.carcnet.ca

SOURCES: *Important Amphibian and Reptile Areas,* Canadian Amphibian and Reptile Conservation Network, 2003. Reprinted by permission of CARCNET, http://www.carcnet.ca

The value of monitoring biodiversity is a recurring theme in this text. See also Chapters 9 and 14 for the relationships of biodiversity to forestry and meeting environmental sustainability challenges.

In May 2000, the Committee on the Status of Endangered Wildlife in Canada (COSEWIC), whose primary mandate is to develop a national listing of Canadian species at risk, identified a total of 353 mammal, bird, reptile, amphibian, fish, mollusk, lepidoptera (butterfly and moth), and plant species (including mosses and lichens) that were at risk across Canada. In a May 2003 meeting, COSEWIC raised the number of plants and animals at risk of extinction to 431. This is an increase of more than 22 percent in two years. Species on the list are found in every part of Canada, but most are found in the Okanagan Valley of south-central British Columbia; on the prairies of Alberta, Saskatchewan, and Manitoba; in southwestern Ontario; in southern Quebec; and on Nova Scotia's Atlantic coastal plain (World Wildlife Fund Canada, 2000a). Figure 12–1 illustrates areas of risk to biodiversity in Canada.

Just as important as the threatened wild species themselves are the wild spaces (landscapes) they inhabit. From the coasts of Newfoundland to the rain forests of British Columbia, and from the prairie grasslands to the Arctic tundra, Canada is losing wilderness at the rate of more than one acre every 15 seconds (World Wildlife Fund Canada, 2000b). Since all species rely on a healthy natural world that includes intact wild places, the consequences of this loss for wild species and humans potentially are far reaching.

THE IMPORTANCE OF BIODIVERSITY

The diversity of life in Canada and on this planet supports vital ecological processes, including oxygen production, water purification, conversion of solar energy into carbohydrates and protein, and climate moderation. Even though society has not completely understood that human health depends on these ecological processes, and has not fully valued them, conserving biodiversity has

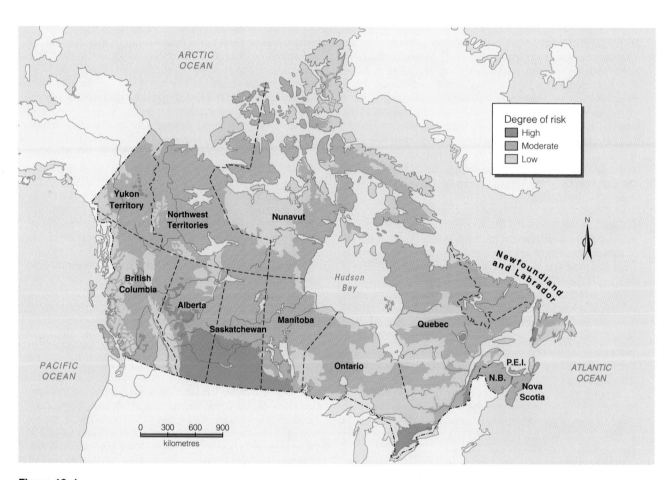

Figure 12–1
Risk to biodiversity for Canada
SOURCE: Adapted from *Protect Canada's Biodiversity*, Western Canada Wilderness Committee, Spring 1995, *14*(4), 4–5. Based on *Biodiversity in Canada: A Science Assessment for Environment Canada*, Minister of Supply and Services, 1995, p. 2.

become an essential part of Canada's efforts to achieve sustainability. Beyond its ecological processes, biodiversity is important for at least three additional reasons: employment, spiritual inspiration and identity, and insurance for the future.

The diversity of Earth's life forms enables people to satisfy many of their needs, including that for gainful employment. Millions of people who work in the fishing, forestry, agriculture, tourism and outdoor recreation, pharmaceutical, and biotechnological industries depend directly on high-quality biological resources to earn their living. Particularly in the north of Canada, many indigenous communities derive a large portion of their food and income from harvesting of biological resources.

As we have seen in the cases of east coast fishers and indigenous and First Nations peoples, loss of wildlife species or their habitats affects traditional lifestyles and reduces the quality or availability of country foods, as well as opportunities to undertake other hunting, gathering, and guiding activities. The attendant economic repercussions are important also.

In part, Canada's cultural identity has been shaped by the wild, elemental beauty of our natural landscapes. Indigenous cultures developed intimate relationships with nature, and many nonindigenous Canadians also have found that the country's diversity of species and spaces provides spiritual, emotional, and artistic inspiration. Painters, writers, and musicians have captured landscape and wildlife elements in their work and have helped define Canada domestically as well as internationally.

Many Canadians believe that each wild species that inhabits our landscape has **intrinsic value**—a value based on the inherent qualities of that species, independent of its value to humans. This suggests that human society should be built on respect for the life that surrounds us, and that biodiversity should be conserved for its own sake, regardless of its economic or other values. Loss of biodiversity in wild species and spaces means that certain intangible values also are lost, ranging from relationships with the natural world to aesthetics. Conversely, maintaining biodiversity increases the genetic variation among individuals within populations and provides an insurance policy for species survival when environmental conditions extend to threshold limits (see Chapter 3).

A distinct advantage of maintaining Canada's (and the Earth's) biodiversity and of using biological resources sustainably is that we can maximize our options in the event that we need to respond to unforeseen changes in environmental conditions. "Maintaining our potential as a country to be creative, productive, and competitive will also provide us with opportunities for discovering and developing new foods, drugs, and industrial products" (Government of Canada, 1996). One example of this is the potential to use the genetic material contained in many of our native plants (that enables them to endure both cold

winters and hot summers) to develop agricultural crops that can withstand even greater temperature ranges. This point is raised in Chapter 6 in the section on genetically modified foods; see Box 6–8.

If Canadians fail to conserve biodiversity, we foreclose future options, flexibility, and economic opportunities, and pass on the enormous costs of this failure to future generations. Given the current rate of environmental change, and the likelihood that species of unknown (but valuable) ecological, medical, and economic potential are vanishing before their existence has even been confirmed, striving to conserve biodiversity is an investment in the future and makes good business sense. Conserving biodiversity also is one way to help provide for intergenerational equity.

Loss of habitat or loss of access to habitat reduces potential employment, income, or trade opportunities. In the recreational field, increased use of wildlands for both consumptive (hunting, fishing, resort development) and nonconsumptive uses (wildlife viewing, photography, hiking) may have negative local impacts. When coupled with infrastructure requirements to support outdoor recreational activities (e.g., roads, shelters), such impacts can compound the potential for destruction of the resources on which these activities depend.

Given the desirability of preventing a reduction in the benefits of biodiversity, Canada has an important stewardship role for major portions of the world's tundra, temperate forest, and aquatic ecosystems, as well as in the protection of smaller areas of grassland and cold-winter desert ecosystems. The fact that relatively fewer species inhabit Canadian landscapes (compared with other nations) does not diminish the importance of the species

Photo 12–2
First Nations people on Canada's west coast derive distinctive art forms from their close relationship with wildlife species.

that are present, particularly if they are endemic species found only in particular geographic regions of Canada and nowhere else in the world.

A major global threat to species diversity is that many of the areas richest in endemic species—such as Vancouver Island and the Queen Charlotte Islands in British Columbia—also are prime targets for intensive economic development, including forestry. In other areas, such as the large sand dunes on the south shore of Lake Athabasca, Saskatchewan, the 10 endemic plant species that occur there could be put at risk by large-scale removal of sand, or unrestricted use of all-terrain vehicles. Recall from Chapter 3, protecting biodiversity involves protecting function, structure, and processes that support the life of plants and animals. Therefore, it is necessary to support the protection of habitats in order for flora and fauna to co-exist with humans.

Having briefly reviewed some reasons to protect the biodiversity and sustainability of Canada's species and spaces, the following sections note the nature of Canada's wildlife species and protected areas, and highlight relevant concerns regarding biodiversity and sustainability. As we shall see later in the chapter, national and international efforts are being made to address the loss of species diversity and wild spaces.

HUMAN ACTIVITIES AND IMPACTS ON CANADIAN SPECIES AND NATURAL ENVIRONMENTS

HUMAN ACTIVITY AND BIODIVERSITY

Globally, the numbers of species in most **taxonomic** groups tend to decrease from equatorial to polar latitudes; thus, in spite of its large size, and because it consists mainly of polar and temperate ecosystems, Canada has less species richness than many other nations. Brazil, for instance, has close to 55 000 known species of flowering plants, while Canada has only about 4039 known species of higher plants, including about 2980 species of native flowering plants and about 930 introduced flowering plants.

With regard to the total wildlife population in Canada (that is, the total of all plants, animals, and microorganisms, of both terrestrial and aquatic ecosystems), estimates are that there are more than 138 000 species in Canada (Government of Canada, 1996). This total includes the 4039 plants noted above, plus nearly 1800 vertebrate animals and more than 44 000 invertebrates (see Table 12–1).

Beginning with European settlement, Canada's ecosystems have been altered significantly through settle-

ment, cultivation, industrial and domestic pollution, and harvesting of commercially valuable life forms such as fish and trees. Habitats have been altered through physical changes, competition from non-native biota, and other cumulative agents of change. Figure 12–2 provides an overview of the level of human activity occurring on Canada's ecosystems; note the overlap with areas of risk to biodiversity shown in Figure 12–1 (refer also to Figure 3–6). Table 12–2 indicates in summary form some of the key areas, characteristics, and environmental stresses relating to biodiversity within Canada.

The state of wild species and natural spaces in Canada reflects natural changes, such as periodic fluctuations in populations due to disease, weather events, naturally occurring fires, competition among species, and predation, as well as human activities that alter habitats or introduce non-native species. Major types of habitat alteration, as well as their causes and effects, are outlined briefly in the following sections.

Habitat Alteration Due to Physical Changes

Forestry, Agriculture, and Other Human Activities
Physical changes to habitats have resulted from activities in forestry, agriculture, and other human pursuits. In forestry, logging and clear-cutting result in loss or fragmentation of habitat for certain species but may enhance habitat for other species. In addition, replanting and reseeding are based on commercial preferences of the forest industry and are not likely to reproduce the original species mix and balance. This results in simplified, evenly aged forests of the tree-farm variety. Fire suppression activities deter the return to early or pioneer stages of natural succession; this has implications for species reliant on young-growth forest. Fire suppression also may change the species composition of the climax forest in the area affected.

Agriculture has distorted the original habitat balance of forests, grasslands, and wetlands as they were converted to agricultural lands. Over time, however, previously farmed lands that proved to be marginal for agriculture have been allowed to return to their wild state, while other marginal agricultural lands have been restored through conservation programs such as the North American Waterfowl Management Plan.

Other human activities, including urbanization, large- and small-scale hydroelectric facilities, multi-lane highways, railways and transportation infrastructure, and various industrial activities, have impacts on wildlife habitat through physical removal of land. Hiking, use of off-road vehicles, and other recreational pursuits may seem environmentally benign as their effects may not be visible, but even they may result in widespread damage, especially if wildlife populations are stressed already.

TABLE 12-1
THE DIVERSITY OF WILD SPECIES IN CANADA

Kingdom	Major Subdivisions and Common Names	Estimated Number of Species: Reported to Date	Suspected but Not Reported
Procaryotae	bacteria, cyanobacteria (blue-green algae), chloroxybacteria	2 400	20 800
Protista	algae, diatoms, protozoa	6 303	2 980
Fungi (Eumycota)	zoosporic fungi, mushrooms, rusts, smuts, lichen, etc.	11 800	3 831
Plantae	mosses and liverworts	965	50
	club mosses, horsetails, and ferns	141	11
	conifers and kin	34	0
	dicots and monocots[a]	3 864	75
Animalia	molluscs (snails, bivalves, octopus, squid, etc.)	1 500	135
	crustaceans (crabs, lobsters, shrimp, etc.)	3 139	1 411
	arachnids (spiders, mites, ticks, and kin)	3 275	24 653
	insects	29 913	24 653
	other invertebrates	6 444	4 417
	sharks, bony fishes, and lampreys	1 091	513
	amphibians and reptiles	84	2
	birds	426	0
	mammals	194	0
Total		71 573	66 609

[a] Of the dicots and monocots, about 77 percent (2980) are native and the rest are exotic. Dicots are a group of flowering plants that have two seed leaves. They include many herbaceous plants and most families of trees and shrubs. Monocots are flowering plants with a single seed leaf, including grasses and lilies.

NOTE: Viruses also contribute to Canada's biological diversity. Almost 150 000 species are suspected to exist in the country, but only 200 species have been reported to date.

SOURCE: *The State of Canada's Environment—1996,* Environment Canada, 1996, Table 10.17. Reprinted with permission of the Minister of Public Works and Government Services Canada, 2004.

Fragmentation Human activities (such as those described above, and others) result in varying degrees of habitat fragmentation. Divided highways and logging and recreational access roads cut off usable portions of some animals' home ranges and create barriers to movement; other animals are reluctant to cross extensive open areas. In both cases, individual animals that attempt to cross highways or large open areas are exposed to greater threats of predation and mortality.

Chemical Changes Terrestrial and aquatic habitats can be degraded by hydrocarbon spills, industrial and municipal effluents, and application of pesticides, herbicides, and other chemical compounds. Acidic deposition, such as from long-range transport of atmospheric pollutants, can degrade life-support systems and reduce overall productive capacity of ecosystems. Some chemical

Photo 12–3
Wildlife viewing and nature photography may cause local environmental stress.

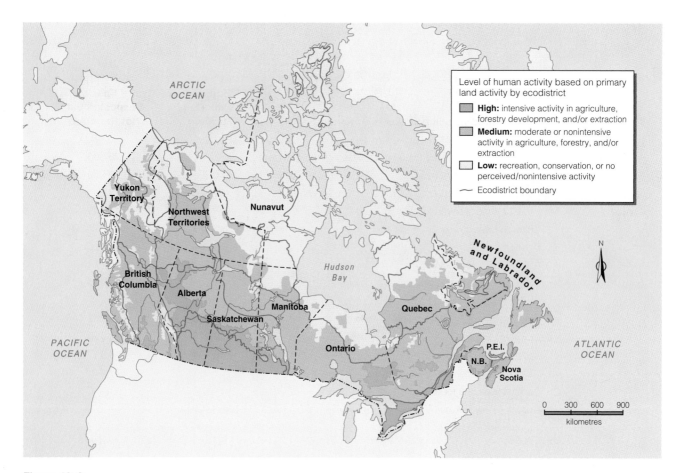

Figure 12–2

Degree to which human activity has changed Canada's ecosystems

SOURCE: *The State of Canada's Environment—1996,* Environment Canada, 1996, Figure 14.5. Reprinted with permission of the Minister of Public Works and Government Services of Canada, 2004.

TABLE 12–2
KEY AREAS, CHARACTERISTICS, AND ENVIRONMENTAL STRESSES RELATING TO CANADA'S BIODIVERSITY

Area	Key Characteristics and Environmental Stresses
Arctic diversity	• Arctic ecosystems occupy about 10 percent of the world's land area; Canada has jurisdiction over about 20 percent of these lands. • Arctic regions sustain a wealth of lichens and plant species. • Seasonally high productivity of the marine ecosystem attracts sea birds and marine animals to Arctic waters. • Atmospheric and oceanic currents carry pollutants to polar regions, causing toxic contamination of the food chain up to marine mammals, birds, and people; over time this places species at risk of decline and possible extinction.
Forest diversity	• Close to 10 percent of the world's forests, and a much greater percentage of all boreal forests, grow in Canada. • Forests provide a valuable economic resource and a vital habitat for temperate organisms. • On the global level, forest biodiversity is being compromised seriously by continuing loss of tropical rain forests.

TABLE 12–2

(CONTINUED)

Area	Key Characteristics and Environmental Stresses
	• Forest ecosystems filter and purify the atmosphere; provide a living substrate for complex communities of organisms (such as insects, birds); protect watersheds against soil erosion; and moderate impacts of extremes of temperature, wind, and precipitation. Living trees contribute oxygen to the atmosphere, and breakdown of their leaves and other organic debris sustains microorganisms and invertebrates that contribute to soil formation.
	• In Canada, forest biodiversity is threatened by reductions in the extent of forest types (such as southern Ontario's Carolinian species, northern boreal forests, and British Columbia's old-growth forests that provide habitat for specialized species such as the spotted owl).
	• Logging and clear-cutting result in fragmentation of forested habitat areas so that forests can no longer support species that rely on large, contiguous forest habitat.
Grassland diversity	• Grassland biomes represent about 20 percent of Earth's land area; most have been altered dramatically by human activities.
	• At the time of European settlement, prairie grassland ecosystems covered about one-half million square kilometres of western Canada, sustaining a rich, highly specialized floral and faunal community.
	• Since then, more than 80 percent of Canada's grasslands have been converted to agricultural use; their original diversity has been altered profoundly—only about 1 percent of Canada today remains in grassland ecosystems.
	• An important part of the conservation challenge is to protect and enhance native prairie biodiversity within modern agricultural landscapes.
Wetland diversity	• Canada has about 24 percent of the world's wetlands as well as a much greater proportion of its boreal fens and bogs.
	• Wetlands are among the world's most varied and biologically productive ecosystems, supporting a great diversity of plants and providing ideal breeding and feeding sites for many kinds of invertebrates, fish, amphibians, and waterfowl.
	• Wetlands also perform important ecological functions, including water retention and purification, and flood and erosion control.
	• Removal of wetlands for agricultural or urban development eliminates habitat for wetland-dependent wildlife species and can seriously disrupt the seasonal supply of freshwater over large areas.
Freshwater diversity	• Canadian runoff to the sea is about 9 percent of the world's freshwater runoff; freshwater ecosystems cover more than 7 percent of Canada's surface area, having significant climatic effects (cooling the summer heat, moderating the winter cold).
	• Canada's freshwater ecosystems sustain about 180 fish species, a rich variety of aquatic plants, and many invertebrates, including some groups of arthropods that appear to achieve greatest diversity in temperate latitudes.
	• Pollution by industrial activities and urban wastes is a significant threat to the quality of freshwater habitats; point-source release of industrial effluents, discharges from urban settlements, and airborne pollutants that fall as wet (acid rain) or dry deposition are of concern.
Marine diversity	• Globally, the range and diversity of marine ecosystems is only beginning to be understood and measured.
	• Canada has five major marine ecozones spread over five million square kilometres of ocean; these ecozones contain some of the most productive marine ecosystems in the northern hemisphere; the total marine food web contains thousands of different kinds of organisms.
	• Threats to Canada's marine ecosystems include overfishing, by-catch (nontargeted species in the catch), destruction of habitat, pollution, and global environmental changes such as climate, temperature, and ocean current changes. Thinning of the ozone layer globally places phytoplankton (the biological foundation of the marine ecosystem) at risk.

SOURCE: *The State of Canada's Environment—1996*, Environment Canada, 1996, Chapters 10, 14. Reprinted with permission of the Minister of Public Works and Government Services Canada, 2004.

alterations, including increases in nitrogen and phosphorus, can enhance life-support systems for certain species and result in an increase in overall productive capacity of ecosystems. In the case of eutrophication, however, oversupply of nutrients results in detrimental effects on ecosystem functioning.

Climate Change   Climate change has the potential to lead to dramatic alterations in ecosystem structure over the long term. Some species could find evolving climatic conditions favourable, while others—especially those at the limits of their ranges—could disappear. Climate change could result also in a shift in location of many wildlife habitats. For instance, coastal wetlands could be displaced or created as sea levels rise; aquatic and semi-aquatic habitats could recede as wetlands dry up; and boreal forest plants and animals could shift northward.

Habitat Alteration Due to Competition from Non-native Biota

Non-native (alien, exotic, nonindigenous, or introduced) species can become established in Canada intentionally or accidentally; in either case, native species must compete with non-native species for space, water, food, and other essentials of life. Virtually every region of the country has an introduced species that is actively displacing a native, rare, or endangered species. Released in Central Park, New York, the European starling has spread throughout North America, displacing native species that require tree holes or other cavities for nests.

Similarly, raccoons released on the Queen Charlotte Islands (to provide local trappers with a new species to harvest) have put sea bird colonies at risk, killing about 10 percent of the breeding colonies of burrow-nesting alcids (auklets, guillemots, and ancient murrelets—the latter is listed as a vulnerable species) in only five years. Plants such as Scotch broom, purple loosestrife, and crested wheatgrass also have competed aggressively with and replaced native species. Currently, about 23 percent of wild flora (mostly weeds) and about 1 percent of fish in Canada are exotic species.

Habitat Alteration Due to Harvesting

Harvesting has the potential to cause short- and long-term changes in populations and species composition, regardless of the type of wildlife harvested or the methods employed. Logging, for example, causes temporary (sometimes permanent) reduction in tree populations, which affects trees' delivery of ecological functions such as habitat provision, carbon fixation, and oxygen production. Logging also affects the structure of the forest community as older growth stands are replaced with younger trees. Harvesting of fish not only targets

large, older specimens, but also extracts enormous numbers from the population pools of aquatic species. In 1999, for example, commercial landings of fish were more than 1 million tonnes with a commercial value of almost two million dollars (Statistics Canada, 2000). Harvesting of species for collection purposes cannot be overlooked; some butterfly species and at least nine nationally endangered plants already at risk face some threat from collectors.

Habitat Alteration Due to Toxic Contaminants

Toxic substances may occur naturally (mercury, lead) or **anthropogenically** (industrial discharges, municipal waste, agricultural and forestry activities, etc.); these contaminants may spread over great distances through air and water currents. Mercury and lead do not appear to have any nutritional or biochemical function; at higher concentrations, however, they can adversely affect plant and animal growth and health.

Lead poisoning of waterfowl and other birds occurs when ducks and other bottom-feeding species ingest the lead pellets from lead shot used in hunting ammunition. Lead enters the gizzard and becomes available for absorption into the body. Ten to 15 percent of the mortality of golden and bald eagles has been attributed to secondary lead poisoning (from feeding on waterfowl carrying lead pellets embedded in their bodies). As of 1997 there is a national ban on the use of lead shot for all migratory game bird hunting.

Use of lead sinkers in recreational fishing also causes secondary lead poisoning; when fish escape from anglers with the "hook, line, and sinker" attached, loons or other fish-eating birds can swallow the fish whole, including the lead sinkers, and subsequently die of lead toxicosis. Starting in the fall of 1996, it was illegal to use lead fishing sinkers or jigs (weighted lead hooks) in Canada's national parks and national wildlife areas.

Persistent organochlorines and metals are two notable groups of toxic substances. Organochlorines include pesticides, industrial chemicals, and byproducts of certain industrial processes that take decades or centuries to break down naturally (thus their persistence). This persistence and their high solubility in fat leads to bioaccumulation in animal tissues, which are then passed on through food webs, reaching very high concentrations in the tissues of predators at the top of the food web (biomagnification). Bioaccumulation and biomagnification are discussed in more detail in Chapter 3.

Flooding for hydroelectric developments frequently results in mercury contamination; this is because naturally occurring inorganic mercury in submerged organic material is converted into methylmercury by anaerobic bacteria. Methylmercury easily bioaccumulates in many organisms and also biomagnifies through food webs.

Advisories have been issued regarding human consumption of fish above certain sizes from both reservoirs and natural lakes in Canada.

Habitat Alteration Due to Urbanization

While we do not frequently identify wildlife with urban areas, in fact, many wild animals do live in or near cities. Certainly, they live in and about small towns that are located in natural settings such as national parks (e.g., Banff, Alberta; Waskesiu, Saskatchewan; Clear Lake, Manitoba). Urban settlements and small "resort" municipalities are under great pressure to expand their populations and services. Processes of urbanization place enormous pressures on wildlife and habitat. Clearance of land to build more housing for primary residences or second homes may reduce habitat for indigenous flora and fauna, alter natural waterways, and destroy trails and/or migration routes used by animals. Pollution of the air, water, and land may affect the ability of species to thrive.

When urban areas alter local habitats, some species actually thrive. In these cases, however, the numbers of wild animals can become out of synch with the availability of food and other resources they need to survive. For example, in the townsite of Banff, conditions have become favourable for large ungulates like elk. A visitor to the townsite might be delighted that the elk are so numerous. However, in their search for food, elk destroy much vegetation and ultimately become "problem animals," vulnerable to programs established to cull animals and/or remove them from areas of high human population. In a related situation, in January 2002, the Ontario Ministry of Natural Resources introduced a 30-month moratorium on wolf killing in townships surrounding Algonquin Park. Environmental organizations suggest that two-thirds of wolf deaths are human-caused, as wolves are shot, snared, trapped, hit by vehicles, and sometimes even poisoned. The development of new subdivisions for primary housing or cottages can also have devastating effects as habitats are destroyed and domesticated pets and exotic plants are introduced and compete with and even disrupt conditions necessary for the maintenance of indigenous species.

Habitat Alteration Due to Cumulative Agents of Change

Wildlife and habitat changes frequently are the result of a combination of factors that act directly or indirectly to cause change. Insect populations, for example, can be reduced through habitat loss, pesticide applications, and other factors. Loss of insects also can have a domino effect on the total functioning of the ecosystem and can have ecological and economic repercussions. For example, use of the insecticide fenitrothion (used to control spruce budworm) appears to have caused native wild bee populations to plummet, affecting the success of pollination and cross-pollination.

SPECIES AT RISK

It is clear that without adequate and healthy habitat, wildlife (terrestrial and aquatic plants, animals, and microorganisms) cannot survive. This section uses several illustrations of human-related changes in the Arctic and elsewhere to highlight how species may be put at risk by habitat change.

Resource extraction, such as mineral and diamond mining in the Northwest Territories and oil and gas exploration in the Arctic, raises a variety of concerns related to wildlife. The Arctic National Wildlife Refuge (ANWR) in Alaska, often called the Serengeti of North America, sustains vast wildlife populations and also is a potential source of crude oil. Concerned about the fate of the Porcupine caribou herd that over-winters in Canada and migrates each spring to the coastal plain of the ANWR, Canadian conservationists have debated with U.S. government officials about whether oil exploration should be permitted there. Caribou show reduced calf production and lower calf survival rates around oil development areas; oil exploration could cause a 20 to 40 percent decline in the size of the herd, and cause economic hardship to the 7000 Aboriginal people who depend on the caribou for their survival ("Oil Exploration," 1997).

According to the Canadian Wildlife Federation, if the United States did open up the wildlife refuge to oil exploration and drilling, it would be violating four international accords: the Migratory Birds Convention (1916), the Agreement on the Conservation of Polar Bears (1976), the North American Waterfowl Management Plan (1986), and the Porcupine Caribou Agreement (1987) ("Oil Exploration," 1997). President Clinton respected his campaign pledge to protect the refuge from drilling. However, in September 2000, the U.S. House of Representatives Committee on Resources Issues stated that President Clinton's policy was responsible for the energy crisis in the United States. Oil consumption had increased by 14 percent and domestic production decreased by 17 percent since 1993, and reliance on oil imports had grown to 56 percent. President George W. Bush has made opening the refuge a cornerstone of his administration's energy policy, and hopes to create up to 700 000 new jobs to tap the estimated 31 billion barrels of oil (United States House of Representatives, 2000). While Canada has always supported protection of this ecosystem, there is concern that the negative effects of rising energy prices on Canada's economy may soften the federal government's stance on the issue. However, in February 2001, during his first official visit with the newly elected president, Canada's Prime Minister Jean Chrétien expressed

Canada's disagreement with President Bush's intention to open the ANWR to oil and gas exploration. Although President Bush included the refuge in his national energy plan and Arctic drilling revenues in his 2005 budget, both the U.S. House of Representatives and the Senate dropped this item from the budget. Nevertheless, instability in global supplies of oil and gas may reopen this issue in the future.

Canada's first diamond mine, the BHP Billiton diamonds project at Lac de Gras, NWT, received regulatory approval from the federal government in January 1997. In 2000, the government of the Northwest Territories was committed to facilitating a value-added diamond industry. By 2003, the Ekati and Diavik diamond mine projects were in full production (see chapter 10). De Beers' Snap Lake project is expected to begin production by 2006, with other diamond, base, and precious-metal mines under consideration in the north. By 2002, mining and minerals contributed $1.3 billion to the economy of the Northwest Territories (Natural Resources Canada, 2002).

The Canadian Arctic Resources Committee and the Canadian Wildlife Federation, among other organizations, are concerned about the cumulative effects that these future developments will have on wildlife and other resources as well as on Aboriginal people. The Minister of Indian Affairs and Northern Development indicated to the Canadian Wildlife Federation that his and other federal departments would cooperate to develop a protected areas strategy by 1998, and provide $750 000 per year over five years for collection of baseline data that could help determine the cumulative effects of other potential mineral developments in the area (Canadian Wildlife Federation, 1997). However, while a Protected Areas Strategy for the Northwest Territories was approved in September 1999, the strategy does not prescribe a target amount of protected area or a date for meetings goals. Indeed, by 2003, no increase in protected areas had been realized since the strategy had been established. In June 2003, the Canadian Arctic Resources Committee's executive director noted that there were no permanent protected areas in the region between Yellowknife and the Arctic coast (where most diamonds were being found), a cumulative impact monitoring program required by legislation and land claims agreements were several years overdue, and a cumulative effects assessment and management framework that formed part of the approval of the Diavik mine was behind in implementation and had no long-term funding (Wristen, 2003).

The Arctic already has experienced the environmental impacts of actions taken by other people thousands of kilometres away. These impacts include pollutants, borne by wind and water currents, that enter the food chain, biomagnify, and end up in Inuit and other Aboriginal people who eat what they hunt. Pollutants such as pesticides from southeast Asia, PCBs from

Photo 12–4
Inuit hunters during an annual whale hunt.

eastern Europe, mercury from the United States, along with other persistent organic pollutants, heavy metals, and radionuclides have been found in the snow, ice, soil, animals, and even people of the Arctic. For example, mirex, a pesticide never registered for use in Canada, has made its way into breast milk of Inuit women. Nevertheless, if Inuit people turn away from the use of country foods, they risk becoming less physically active, less connected to cultural practices associated with hunting and eating local food, and they are more likely to get "southern" diseases such as heart disease, diabetes, and cancer. As well, the predicted severe climatic consequences of global warming for the Arctic could result in a shift in location of many wildlife habitats.

Economically, the Arctic is connected to global events also. For many years the Inuit, Dene, and Métis people have relied on the sale of pelts and skins to world markets as a source of cash to purchase market goods and hunting equipment. In turn, the hunting equipment was used to provide food for their families. In 1976, seal skins contributed $1.5 million to the economies of Inuit communities in the Northwest Territories, Quebec, and Labrador. During the 1980s, as Western European and United States governments banned or restricted imports of fur and skins, the economic effect on small communities in the Canadian north was devastating. In 1980, for instance, one seal skin could bring in $23; in 1985, the value dropped to $7. All types of social pathologies increased—suicides, spousal assault, and alcoholism (Fenge, 1996).

As industrial development proceeds in the North, protecting habitat important for wildlife is likely to be an issue of increasing importance. A comprehensive network of parks, reserves, and protected areas will be needed to ensure environmental protection and wildlife sustainability, as well as cultural and social sustainability. Protecting habitat does not mean excessive hunting by Aboriginal people will occur. In fact, cooperative management of aquatic, marine, and terrestrial animals by government and Aboriginal peoples has been undertaken pursuant to land claim and self-government agreements. These efforts are attracting interest from many nations because they are aimed at ensuring that harvesting remains within sustainable limits. Co-management also encourages people to integrate traditional and scientific knowledge in a quest for better management of the environment, including protection of wild species and spaces.

Northern Aboriginal people consider sustainability as a means to safeguard their cultures and economies as well as the natural environment on which they depend so heavily. Even though involvement of the federal government provides some assurance that sustainability may be implemented domestically, there is a concern that action within Canada alone will not be sufficient to achieve the desired ends. In this case, the eight Arctic Inuit nations in Canada, Greenland, Alaska, and Chukotka are implementing an environmental protection strategy that they agreed on during their 1991 Circumpolar Conference. In effect, they are undertaking a global effort to ensure sustainability Arctic-wide, including the future of the Canadian Arctic and its wildlife (Fenge, 1996).

The Arctic is not the only area where wild species are at risk because of human activities. A variety of threats face wildlife in ecosystems across Canada; while some species have been impacted negatively by agricultural and industrial activities, acid precipitation, or contaminants, other species have been affected by hunting pressure and recreational and residential impacts. Some species, such as the peregrine falcon, have been brought back from the brink of extinction, while others, such as the whooping crane, remain at risk in their Canadian ranges.

Brief stories of four selected species at risk illustrate the threats from human actions facing wild species in ecosystems across Canada (the stories are based on information from Canadian Geographic Enterprises [1997]). The threatened marbled murrelet is a small sea bird, a member of the auk family, whose nesting behaviour had been a mystery to researchers for nearly 100 years. It was not until 1993 that the first occupied murrelet nest in Canada was located in the dense coastal rain forest of British Columbia. Marbled murrelets forage for small schooling fish in nearshore areas (where they are at risk from gill nets and oil spills), and nest only on broad, moss-covered branches of mature conifers within 30 kilometres of the ocean (many similar species of sea birds nest in colonies on coastal cliffs). Only one egg is laid each year; to elude predators, the nest is relocated every year. As logging in the old-growth forests in valley bottoms continues, murrelet numbers have dropped as their prime nesting habitat has disappeared. As forest cover shrinks, opportunistic ravens and Stellar's jays find the murrelet eggs and chicks easy prey; in Clayoquot Sound, the murrelet population has declined by about 20 percent since 1982.

First identified in Canada in 1920, the endangered pink coreopsis is a pink and yellow perennial herb that is scattered along the shorelines of Nova Scotia's southwestern lakes. In Canada, this coastal plain plant occurred exclusively in the Tusket River valley at six separate locations, all but one on private land. Three of those sites have been lost to residential development. The remaining sites are at risk because the hardy and stress-tolerant coreopsis thrives in gritty, sandy, and seasonally flooded soil on the shorelines of lakes. This locational preference makes the plant vulnerable to activities of cottagers and off-road drivers. Efforts are under way to educate landowners about this plant and to protect the pink coreopsis and other biologically associated plant species in remote areas where it is still found.

Living on the Queen Elizabeth Islands in the Northwest Territories, endangered Peary caribou are the only members of the deer family adapted to life in the Canadian High Arctic. During the summer, the Peary caribou graze river valleys and plains for grasses, sedges, herbs, and willows. During winter they move to higher areas where winds sweep snow away from vegetation; some travel across the frozen sea to search for food. Caribou numbers have dropped from 25 000 in 1961 to fewer than 3000 in 1997 because heavier snowfall and freezing rain during the past decades have increased the caribou's vulnerability to starvation. On some islands, hunting and wolf predation have accelerated the herd's decline. Local Inuit communities that have relied on the caribou for generations voluntarily have reduced their hunting. A national park on Bathurst Island has been created to protect calving areas, and a captive breeding program at the Calgary Zoo also was initiated (but was not implemented because inclement weather prevented capture of the caribou).

The aurora trout is a colourful brook trout known to live only in two remote lakes (Whirligig and Whitepine) in the Temagami region north of Sudbury. By 1961, the aurora trout had been extirpated from both lakes because their high elevation and their sensitive geological setting made them prone to acidification that killed trout fry within weeks of hatching. Fortunately, the aurora trout escaped becoming the first casualty of acid precipitation because, a few years prior to 1961, researchers had collected over 3600 eggs to form the basis of a brood stock. This stock has been maintained for about 40 years. Attempts to introduce the species into lakes other than the species' native ones failed; rehabilitation of the trout's native lakes began prior to 1990 when water in the 11-hectare Whirligig was neutralized with 21 tonnes of powdered lime. In 1990, 950 fish were reintroduced to Whirligig, and today both lakes are successfully stocked with hatchery-raised aurora trout. As we know, emissions causing acid precipitation have been reduced substantially, but nevertheless the water quality of the fish habitat is monitored constantly. Twice since 1990

Photo 12–5a

Photo 12–5b

Photo 12–5c
Among Canada's threatened and endangered species are the pink coreopsis (12–5a), aurora trout (12–5b), and Peary caribou (12–5c).

the recovery team has had to intervene and add more lime to Whirligig.

Toxic contaminants in the environment may occur naturally (mercury and lead in bedrock and soils) or through human activities; wild species may be affected by both sources of toxins. Concerns raised in the early 1960s about the bioaccumulation and biomagnification of persistent organochlorines resulted in a ban or severe restrictions on their release into North American ecosystems. Interestingly, certain wildlife species are used as environmental indicators to monitor levels of organochlorines and toxic contaminants in ecosystems. Researchers use the eggs of one of these species, the double-crested cormorant, to monitor concentrations of selected compounds in ecosystems (see Box 12–2 for further information on use of wildlife indicators to track toxic contaminants).

Perhaps one of the best illustrations of the cumulative impacts of many human activity pressures is found in the case of grizzly bears in the Banff–Bow Valley area of Alberta. This saga, outlined briefly in Enviro-Focus 12 and Box 12–3, emphasizes the competition bears face with humans for

critical montane habitat, and the habitat fragmentation and barriers that human development, land use, and transportation systems have on bear populations located within protected areas (in this case, a national park).

Initiated in 1977, the Committee on the Status of Endangered Wildlife in Canada (COSEWIC) is the agency responsible for identifying the status of all wild species in Canada. COSEWIC employs a seven-category set of status definitions (outlined in Table 12–3); these designations help to monitor the condition of wild species known to be at risk of eventual extinction from one or more agents of change. Table 12–4 provides a summary of the numbers of species at risk in Canada in 2003, as well as those known to be extinct.

SPACES AT RISK

The health, biodiversity, and sustainability of wild species is closely connected to the availability and quality of the natural spaces they occupy. Principally because of the extent and natural character of its grasslands, as well as the concentration of rare, threatened, and endangered species it supports, Sage Creek, Alberta, is a nationally significant environmental area.

Hundreds of species of prairie plants and animals occur in this contiguous, 5000-square-kilometre area in the southeastern part of Alberta, including such familiar threatened wildlife species as the burrowing owl, swift fox, and sage grouse, as well as rare plant species such as Pursh's milk vetch and the plains boisduvalia. Sage Creek also is recognized as one of the most diverse areas in North America for breeding grassland birds, and is home to the threatened mountain plover, fewer than 10 breeding pairs of which exist in Alberta (Wallis, Klimek, & Adams, 1996). By 2000, the mountain plover's status had been uplisted by COSEWIC to endangered (Canadian Wildlife Service, 2000).

BOX 12-2

THE USE OF WILDLIFE INDICATORS TO TRACK TOXIC CONTAMINANTS IN ECOSYSTEMS

Indicators are selected key statistics that represent or summarize some aspect of the state of ecosystems. By focusing on trends in environmental change, they convey how ecosystems are responding to both stresses and management responses.

The insecticide DDT was widely used in Canada between 1947 and 1969 to control agricultural and forest insects. The main breakdown product of DDT is dichlorodiphenyldichloroethylene (DDE), a compound that interferes with enzymes necessary for the production of calcium carbonate in female double-crested cormorants, bald eagles, peregrine falcons, and other birds. Eggshells with less calcium carbonate are thinner and more likely to crack or break during incubation. Because of these and other toxic properties, most uses of DDT were banned in Canada by the mid-1970s. Concentrations of DDE have declined substantially since the 1970s. In recent years, declines have levelled off, possibly as a result of the slow release of contaminant residues from bottom sediments or long-range atmospheric transport from countries still using DDT.

Certain toxic **organochlorines**, including DDE, PCBs, and some dioxins and furans, have been monitored in species of wildlife since the early 1970s. Tracking concentrations in wildlife simplifies the detection of some chemicals that are present in extremely low concentrations in air and water and are therefore difficult to measure directly. Levels of some organochlorine contaminants in the eggs of fish-eating birds, for example, may be as much as 25 million times the concentrations in the waters in which the fish live, because of the processes of bioaccumulation and biomagnification.

The double-crested cormorant has been selected as a national indicator species for organochlorine levels in wildlife because of its broad distribution across southern Canada, especially in areas of concentrated human activity, and because it is a top predator that eats live fish. A disadvantage of using the double-crested cormorant as an indicator, however, is that, like many Canadian birds, it migrates south in the winter. It is therefore not known what proportion of the contaminants measured in its eggs comes from non-Canadian sources.

Photo 12–6

The double-crested cormorant (*Phalacrocorax auritus*) is an excellent swimmer and diver. In pursuit of its fish prey it may remain underwater for 30 seconds or longer, sometimes using its wings for propulsion in addition to its webbed feet. It nests in colonies and builds its nest on the ground or in a tree.

SOURCE: *The State of Canada's Environment—1996,* Environment Canada, 1996, Reprinted with permission of the Minister of Public Works and Government Services Canada, 2004.

In March 1996, the Alberta Energy and Utilities Board (AEUB) recognized that Sage Creek's native prairie grassland ecosystems were important, vulnerable, and disappearing rapidly. In spite of awareness that the cumulative long-term impacts of all development in such grasslands, including that of oil and natural gas, can be very significant, the AEUB recommended to the National Energy Board (NEB) that Express Pipeline be granted approval to construct and operate a crude oil pipeline in the Sage Creek area. The NEB granted that approval. When the pipeline construction began in August 1996, one of the most extensive grasslands left in North America began to be fragmented.

Having opposed the approval of this pipeline proposal, the Alberta Wilderness Association (AWA) was "very concerned about further fragmentation and loss of threatened grassland ecosystems and is very disappointed in the apparent rubber stamping that seems to occur in regulatory hearing processes" (Wallis et al., 1996, p. 4). Calling the area one of the "crown jewels of prairie biodiversity," the AWA believes that the failure of the National Energy Board and the Canadian Environmental

TABLE 12–3

THE COMMITTEE ON THE STATUS OF ENDANGERED WILDLIFE IN CANADA (COSEWIC): STATUS DEFINITIONS (REVISED 2000)

Status Category		Definition
SC	special concern	• A species of special concern because of characteristics that make it particularly sensitive to human activities or natural events.
T	threatened	• A species likely to become endangered if limiting factors are not reversed.
E	endangered	• A species facing imminent extirpation or extinction.
XT	extirpated	• A species no longer existing in the wild in Canada, but occurring elsewhere.
X	extinct	• A species that no longer exists.
NAR	not at risk	• A species that has been evaluated and found to be not at risk.
DD	data deficient	• A species for which there is insufficient scientific information to support status designation.

TABLE 12–4

THE COMMITTEE ON THE STATUS OF ENDANGERED WILDLIFE IN CANADA (COSEWIC): SUMMARY OF SPECIES AT RISK IN CANADA, 2003

Group	Number of Species					
	Special Concern	Threatened	Endangered	Extirpated	Extinct	Totals
Mammals	23	13	22	4	2	64
Birds	22	8	21	2	3	56
Fish	32	22	19	2	5	80
Mollusks	2	2	11	2	1	18
Vascular plants, mosses and lichens	43	38	65	3	1	150
Reptiles and amphibians	19	16	10	5	0	50
Lepidoptera	2	3	5	3	0	13
Totals	143	102	153	21	12	431

SOURCE: *Canadian Species at Risk,* Committee on the Status of Endangered Wildlife in Canada (COSEWIC), May 2003, http://www.cosewic.gc.ca/htmlDocuments/CDN_SPECIES_AT_RISK_May2003_e.html#results. Reprinted by permission of The Committee on the Status of Endangered Wildlife in Canada.

Assessment Agency's hearing process to deny Express Pipeline's application and reroute the pipeline through less sensitive terrain "demonstrates clearly the need to have a legislated system of areas that are protected from industrial activity" (Wallis et al., 1996, p. 4).

In 1999, the World Wildlife Fund, through its Endangered Spaces Progress Report Program, assigned Alberta an F grade. Although the Alberta government introduced new protected areas legislation, industrial development was still included in all categories of provincial protected areas. This legislation places in doubt the future of Alberta's threatened ecoregions (World Wildlife Fund Canada, 2000c).

Most biologists agree that the best way to protect species is to protect native habitats; that saving species starts with saving spaces for them. Island biogeography is a sub-field of conservation biology. It was developed in the early 20th century to understand the rate of colonization by bird species in tropical islands in the South Pacific. This theory has been adapted and applied to many kinds of problems to predict the effects of fragmentation of habitat on the population success of different

Grizzly Bears and Humans in Banff National Park and Area

In spite of the fact that national parks were created in part to protect wildlife, "Banff National Park is a dangerous place to live, if you are a grizzly bear. Researchers have learned that of the 73 grizzlies known to have died in the park from 1971 to 1995, 52 were either destroyed or removed in the interests of public safety. Ninety percent of the grizzlies died close to developed areas, and 56 percent of those were females, since females with curious cubs are the most likely to run into trouble with people" (Marty, 1997, p. 37).

These few statistics highlight some important dimensions of the competition between wild species and humans for habitat. Grizzly bears, whose numbers are estimated to be between 60 and 80 in Banff National Park, are 80 percent vegetarian and range widely in search of food sources such as berries, roots, grasses, and other plants, as well as carrion. Male bears require up to 2000 square kilometres to survive and females need between 200 and 500 square kilometres.

The Banff–Bow Valley area contains **montane** habitat (valley bottom and open forest of trembling aspen and Douglas fir) that is critical for wildlife survival, including that of grizzlies. Only 3 percent of Banff National Park's 6641 square kilometres is montane habitat, and half of that portion lies in the Bow Valley, where wildlife compete for the most productive habitat with humans and their developments. For instance, in the Bow Valley, people have built towns such as Banff, Lake Louise, and Canmore, and associated ski resorts, golf courses, shopping malls, industrial parks, campgrounds, and an airstrip, leaving only a remnant of their former range for bears to roam.

In placing their infrastructure within the valleys, humans have obstructed the traditional north–south and east–west links or corridors that wild species have used for centuries to move north and south within their range between the Yukon and Yellowstone ecosystems. Genetic pathways for regional gene flow, these movement corridors also are fractured by the Trans-Canada Highway, which carries about 20 000 vehicles per day (one every six seconds), and the Canadian Pacific Railway. Bears do not like crossing highways or using the underpasses that Parks Canada provided for big

(continued)

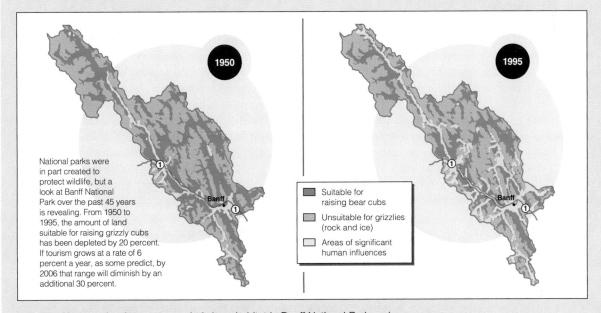

National parks were in part created to protect wildlife, but a look at Banff National Park over the past 45 years is revealing. From 1950 to 1995, the amount of land suitable for raising grizzly cubs has been depleted by 20 percent. If tourism grows at a rate of 6 percent a year, as some predict, by 2006 that range will diminish by an additional 30 percent.

Suitable for raising bear cubs

Unsuitable for grizzlies (rock and ice)

Areas of significant human influences

Impacts of human development on grizzly bear habitat in Banff National Park and area.

SOURCE: "Homeless on the Range: Grizzlies Struggle for Elbow Room and Survival in Banff National Park," S. Marty, 1997, *Canadian Geographic, 117*(1), 35, 36. Reprinted by permission of The Royal Canadian Geographic Society.

game animals (see Box 12–3 for further information on highway and railway impacts on wildlife).

Biologists working on the Eastern Slopes Grizzly Bear project are concerned about habitat fragmentation in Banff National Park and the Bow Valley area and have sensed a general decline in bear numbers. Sensitive to human incursions, grizzly bears are an indicator species (when they are in trouble, the entire ecosystem usually is out of balance). Essentially, the Banff–Bow Valley area now provides only a fraction of its former potential for large mammals. While bears, wolves, and cougars still travel through the Banff–Bow Valley movement corridors, biologists warn that more development will cause wary animals to avoid this area, potentially placing the continued survival of these populations at risk (Canadian Parks and Wilderness Society, 1997).

Human development and land use practices relating to "world class tourism" and the millions of visitors it draws annually to Banff National Park have had significant impacts on wild species, their behaviour and survival, and their habitat in the Banff–Bow Valley area. For instance, until improved garbage management systems were available in 1981, habituated bears (human-food addicts) frequently were relocated or destroyed. Although tourists feed bears less frequently than they did 20 years ago, finding finances and workers to deal effectively with habituated wildlife remains a problem.

Another example of human development impacts on wildlife is that, until recently, the policy of fire suppression to protect visitors and property in the park reduced the amount of feeding habitat for grizzly bears. The species remains slow to recover because mortality rates for female bears remain high (females do not breed until they are between four and eight years old, and average 0.5 cubs per year after that). Also, there is concern that the grizzly bear population on the east slopes of the Rocky Mountains could become genetically inbred and isolated from other subpopulations farther south as mate selection and other biological necessities are constrained by human barriers to bear movement.

In addition to the relatively recent developmental pressures being put on bears in the Banff–Bow Valley area, grizzlies were extirpated from the Prairies near the turn of the century. Current population estimates for

Photo 12–7

Alberta suggest there are between 500 and 800 grizzlies in the province; the provincial goal is to reach 1000. Even though grizzly bears are identified as a vulnerable species in the province, and are considered to be at risk of becoming endangered, the Alberta Fish and Wildlife Department allows 150 individuals to obtain hunting licences for the annual Alberta grizzly bear hunt (Lunman, 1997). In 1996, 18 grizzlies were harvested in this sport or trophy hunt (it is illegal to sell bear parts, such as gall bladders or paws). Even though some parties support this hunt, public concern has raised questions about the wisdom of continuing it. The Alberta Sustainable Resource Development department has an ongoing program to manage human–bear interactions. This program includes educating landowners and recreationists to minimize food sources that may attract bears, collaborating with nongovernmental organizations that place conservation easements in key habitat, using aversion techniques in chronic problem areas, and placing road-killed ungulates in favourable locations in spring.

Harmonizing human developments with wild species' habitat needs requires people to realize that some of their attitudes and actions are directly detrimental to the long-term survival of grizzly bears and other wild species. If we cannot make room for the grizzly bear in our national parks and surrounding lands, then where can the wilderness that sustains the grizzly find protection? If wild places cannot find protection, then biodiversity—and humans, too—clearly are threatened.

SOURCES: *Banff–Bow Valley: At the Crossroads. Technical Report of the Banff–Bow Valley Task Force,* Banff–Bow Valley Study (R. Page, S. Bayley, J. D. Cook, J. Green, & J. R. B. Ritchie), 1996, Ottawa; *The Bow Valley: A Very Special Place,* Canadian Parks and Wilderness Society (CPAWS), 1997, Calgary: Author; "Province Defends Grizzly Bear Hunt," L. Lunman, March 28, 1997, *Calgary Herald,* p. B3; "Homeless on the Range: Grizzlies Struggle for Elbow Room and Survival in Banff National Park," S. Marty, 1997, *Canadian Geographic, 117*(1), 28–39.

B O X   1 2 – 3

WILDLIFE, THE TRANS-CANADA HIGHWAY, AND THE CANADIAN PACIFIC RAILWAY: COLLISION COURSE?

As a place where one can view wild animals close up, Banff National Park has become a favourite vacation destination for Canadian and international visitors alike. However, the pressure on the transportation network that brings about five million visitors to Banff annually is increasing, as are collisions and congestion on the highway. To cope with the expected increase in visitors and the anticipated 3 to 4 percent annual increase in traffic on the Trans-Canada Highway through 2015, a decision was made to extend the twinned (expanded from two to four lanes) portion of the highway.

The Trans-Canada Highway has serious effects on wildlife populations in the park: not only does the highway fragment habitat and act as a barrier to natural movements in the Bow Valley, it also is a significant factor in wildlife mortality. About half of the reported wildlife deaths in the park can be attributed to highways—statistical records confirm that known wildlife losses are in the thousands for several species. This is ironic, given the park's role as a core refuge for wildlife protection.

Upgrading and twinning of the Trans-Canada Highway in Banff National Park began in 1980 and continued in phases throughout the 1990s. Wildlife exclusion fencing was erected to keep animals off the highway right of way, as records showed that fencing the Trans-Canada Highway reduced ungulate (principally elk) mortality by 96 percent. Bears, however, infrequently crossed the divided highway to reach other portions of their fragmented habitat, and the effectiveness of the highway crossing structures for bears and other species was questionable.

The latest section of the highway that was twinned in 1997 incorporated two experimental overpasses to enable wildlife to cross the highway. Costing over $2 million each, these 50-metre-wide overpasses were based on successful European overpasses. Within terrain limitations, they were constructed along known wildlife corridors and are being blended into the land-

Photo 12–9

An example of a wildlife underpass.

scape using ground cover, shrubs, and other vegetation selected for wildlife security. Whether bears will use these overpasses to cross the highway remains to be seen, but if the overpasses are successful, two more will be considered.

Another transportation corridor passing through Banff National Park is the Canadian Pacific Railway line and right of way. The railway, too, is a source of considerable wildlife mortality, although little research has been done on railway impacts on wildlife populations. At a 1997 "Roads, Rails and Environment Workshop" in Revelstoke, British Columbia, a locomotive engineer presented some observations about train-killed wildlife on the main line between Field and Revelstoke, British Columbia. The collected statistics indicated that trains killed two species of birds (bald eagle and great horned owl) and 12 species of mammals, including black bear, cougar, beaver, bighorn sheep, elk, wolf, and wolverine. Numerous other small mammals and birds were killed also.

It appears that birds are attracted to the railway track to feed on the mice that are eating the various grains that spill from passing grain trains. Other wildlife come to the tracks to feed on the carcasses of large animals killed by trains (carcasses are not removed as they are along the Trans-Canada Highway and in national parks). For example, the engineer observed that all bald eagles and almost all coyotes and wolves were killed near carcasses. Other factors such as snow depth, wildlife using the railway right of way as a travel corridor because it is plowed in winter, and the type of seed with which cleared slopes and right of ways are planted (such as clover and other grasses—nutrient-rich food sources that attract bears) contribute to wildlife kills by trains. Changing some of the current management strategies, such as removing train-killed animals from the railway right of way, would help reduce the number of wildlife killed.

Photo 12–8

A wildlife overpass under construction (before landscaping).

SOURCES: "Highway Effects on Wildlife in Banff National Park," A. Clevenger, 1997, *Research Links, 5*(1), 1, 6; *Wildlife Mortality on the Canadian Pacific Railway between Field and Revelstoke, British Columbia,* P. Wells, 1997, paper presented at the Roads, Rails and Environment Workshop, April 9–10, Revelstoke, B.C.

Photo 12–10
The black-tailed prairie dog—listed by COSEWIC as a species of special concern—is protected in Grasslands National Park, Saskatchewan.

Photo 12–11
Point Pelee National Park, on the north shore of Lake Erie, is known around the world for its bird and monarch butterfly migrations. The marsh boardwalk provides people of all ages with the opportunity to discover nature.

plant and animal species. Contemporary theories associated with island biogeography now attempt to predict the minimum viable area required by certain species and to determine whether and in what ways geographically separated groups should be interconnected to avoid population loss and species extinction. These predictions are very difficult to make and are subject to controversy. Biologists agree that fragmentation of habitat can have severe consequences, but the specific details of these consequences are not well understood. Furthermore, calculations of the habitat needs of one species may not reflect the needs of another species living within the same landscape. For species that move on a seasonal or other basis, protection may require a more extensive approach to conservation than simply setting individual areas aside. In these cases connectivity of habitat patches (e.g., corridors) may be beneficial. Yet in other cases, isolation of species may be required to maintain their populations,

and thus questions remain about how large these protected areas should be. Despite these uncertainties, it is clear that to protect biological diversity and promote sustainability, habitat conservation must be a priority nationally, as well as locally. In the following section of this chapter, some of the responses to the need to conserve biodiversity, both globally and in Canada, are noted.

STRATEGIES TO ADDRESS ENVIRONMENTAL IMPACTS AND CHANGE

There is broad international recognition that wild species, ecosystems, and natural spaces are under increasing pressure from a growing human population and a wide range of human activities. If, as predicted, as much as one-quarter of the world's species could be extinct within 40 years, our future options for food, fibre, and medicines could be compromised severely. This prospect lends additional credence to the need for protection and preservation of biological diversity, and of the quality and quantity of suitable habitat, as vital international goals. Accordingly, numerous international agencies, national governments, and other parties have instituted treaties, policies, strategies, legislation, and programs in response to the need to protect biodiversity.

INTERNATIONAL CALLS TO ACTION

Since 1980, a series of important documents have influenced international thinking about and responses to the

Photo 12–12
The common pincushion cactus reaches the north end of its range on the Canadian prairie. The ball-like stems occur singly or in small clusters, protruding partway through native prairie sod.

issues of conservation of biodiversity and sustainable use of biological and other resources. Six of these documents are the following: *World Conservation Strategy* (1980); *Our Common Future* (1987); *Caring for the Earth* (1991); *Global Biodiversity Strategy* (1992); **Agenda 21** (1992); and the Earth Charter (2000). These documents are covered in more detail in Box 1–1 and Table 1–2. The Canadian government has been influenced by these documents, too, and our national efforts to conserve biodiversity involve international treaties, in situ conservation, ex situ conservation, sustainable use policies, and improvement in understanding of biodiversity. Each of these approaches is considered briefly in the sections that follow.

International Treaties

The United Nations Environment Programme–World Conservation Monitoring Centre (UNEP–WCMC) is an international centre of excellence devoted to the location and management of information on conservation and sustainable use of the world's living resources. Located in Cambridge, England, the Centre is guided by a high-level Scientific Advisory Council (UNEP–WCMC, 2001). Of the more than 80 multilateral treaties the WCMC lists relating to conservation and management of biodiversity, Canada is involved in about 20, including 9 out of 10 global conventions. The objectives of these nine international treaties are noted briefly in Table 12–5.

The 1992 United Nations Convention on Biological Diversity is the basis for much of the current international action on biodiversity. The objectives of the convention are to conserve biological diversity; to attain sustainable use of ecosystems, species, and genetic material; and to ensure the fair and equitable sharing of benefits arising from genetic resources.

As a leading proponent of the Convention on Biological Diversity, Canada (like all the other signatories) was responsible for developing a national strategy for the conservation and sustainable use of biological resources. The outcome—the Canadian Biodiversity Strategy—was derived by a working group comprising federal, provincial, territorial, and nongovernmental representatives.

The working group identified five strategic goals for the Canadian Biodiversity Strategy: to conserve biodiversity and use biological resources in a sustainable manner; to improve Canada's understanding of ecosystems and increase our resource management capacity; to promote public understanding of the need to conserve biodiversity and use biological resources in a sustainable manner; to maintain or develop incentives and legislation that support these goals; and to work with other countries to achieve the objectives of the convention (Environment Canada, 2000). The Canadian Biodiversity Information Network (CBIN) provides support for the Convention on Biological Diversity and Agenda 21.

Photo 12–13
Bears killed illegally to satisfy a specific market are confiscated by authorities.

Another important international treaty that deals with protection of wild species is the Convention on International Trade in Endangered Species of Wild Fauna and Flora (CITES). As signatories to CITES, Canadian and international law enforcement agents are able to target poaching, smuggling, and illegal trade in wildlife species, their parts and derivatives. For instance, the Canadian Wildlife Service is responsible for implementing CITES in Canada against poachers who illegally kill bears to obtain gall bladders and paws for sale to international markets.

Enforcement of CITES internationally is necessary to eliminate trade in illegally obtained animals and plants. CITES enforcement is essential to reduce the pressure of poaching and smuggling of rhino horn that has reduced the world's rhinoceros population by close to 85 percent since 1970. The black rhino has been hardest hit; its numbers dropped from at least 65 000 to fewer than 4000 in only 20 years.

A Canadian federal law proclaimed in 1996, the Wild Animal and Plant Protection and Regulation of International and Interprovincial Trade Act (WAPPRIITA) protects Canadian and foreign species from illegal trade. It also protects Canadian ecosystems against the introduction of designated harmful species.

In Situ Conservation

In situ conservation is the conservation of ecosystems and the maintenance and recovery of viable populations of species in their typical surroundings. Three main methods are associated with in situ conservation: protection of spaces that are large enough to support ecosystem processes and species diversity; re-creation of those spaces through species recovery programs; and sustainable use of biological resources outside of protected areas.

TABLE 12-5

CANADA'S PARTICIPATION IN SELECTED GLOBAL TREATIES THAT HELP PROTECT BIODIVERSITY

Global Treaty	Treaty Objectives
International Plant Protection Convention (Rome), 1951	• To maintain and increase international cooperation in controlling pests and diseases of plants and plant products; to prevent introduction and spread of these pests and diseases across national boundaries.
Convention on Fishing and Conservation of the Living Resources of the High Seas (Geneva), 1958	• To improve conservation of the living resources of the high seas and prevent overexploitation.
Convention on the High Seas (Geneva), 1958	• To codify the rules of international law relating to the high seas.
Convention on Wetlands of International Importance Especially as Waterfowl Habitat (Ramsar), 1971	• To stem the progressive encroachment on and loss of wetlands; to recognize ecological functions of wetlands and their economic, cultural, scientific, and recreational value.
Convention Concerning the Protection of the World Cultural and Natural Heritage (Paris), 1972	• To establish an effective system of collective protection of cultural and natural heritage of outstanding universal value, organized on a permanent basis and in accordance with modern scientific methods.
Convention on International Trade in Endangered Species of Wild Fauna and Flora (Washington), 1973	• To protect, via import/export controls, certain endangered species from overexploitation.
United Nations Convention on the Law of the Sea (Montego Bay), 1983	• To set up a comprehensive, new legal regime for the sea and oceans as far as environmental provisions are concerned; to establish material rules concerning environmental standards as well as enforcement provisions dealing with pollution of the marine environment.
International Tropical Timber Agreement (Geneva), 1983	• To provide an effective framework for cooperation and consultation between countries producing and consuming tropical timber; to promote the expansion and diversification of international trade in tropical timber; to improve structural conditions in the tropical timber market; to promote research and development; to promote sustainable utilization and conservation of tropical forests and their genetic resources; to maintain the ecological balance in regions concerned.
United Nations Convention on Biological Diversity (Rio de Janeiro), 1992	• To conserve biological diversity, the sustainable use of its components, and the fair and equitable sharing of the benefits arising out of the utilization of genetic resources, including, by appropriate access to genetic resources and by appropriate transfer of relevant technologies, taking into account all rights over those resources and technologies, and by appropriate funding.

SOURCE: Canada's National Environmental Indicators Series, Government of Canada, 2003, http://www.ec.gc.ca/soer-ree/English/Indicators/default.cfm

Protected Areas Protected areas are known by a number of names, including ecological areas, wildlife management areas, parks, and conservation areas. Each type of protected area has a different set of management objectives, ranging from almost complete protection of biotic and abiotic components from human disturbance to protection that is offered only to selected ecosystem components (such as wildlife or soils). Conserving biodiversity through the use of protected areas involves careful management of human activities and prohibition of activities that could harm ecological processes or ecosystem integrity.

The World Conservation Union (IUCN) is the only global organization that unites nations, government agencies, and ENGOs in conserving the integrity and diversity

TABLE 12-6
IUCN CATEGORIES OF PROTECTED AREA

IUCN Category	Defining Characteristics	Management Goals or Practices	Canadian Example
I. Natural reserve or wilderness area			
a. Nature reserve	• Areas possessing some outstanding or representative ecosystems, geological or physiographic features, or species	• Scientific research or ecological monitoring primarily	Oak Mountain Ecological Reserve (New Brunswick)
b. Wilderness area	• Large areas, unmodified or slightly modified, retaining their natural character and influence, without permanent or significant habitation	• Preservation of natural conditions	Bay du Nord Wilderness Area (Newfoundland)
II. National park (or equivalent)	• Areas designated to sustain the integrity of one or more ecosystems, exclude exploitation or intensive occupation, and provide a foundation for scientific, educational, recreational, and visitor opportunities, all of which must be ecologically and culturally compatible	• Ecosystem protection and recreation	Banff National Park (Alberta)
III. Natural monument	• Areas containing one or more specific natural or cultural features of outstanding or unique value because of inherent rarity, representation of aesthetic qualities, or cultural significance	• Protection of specific outstanding natural features, provision of opportunities for research and education, and prevention of exploitation or occupation	Parrsboro Fossil Cliffs (Nova Scotia)
IV. Habitat/species management areas	• Areas important for ensuring the maintenance of habitats or for meeting the requirements of certain species	• Securement and maintenance of habitat conditions necessary to protect species and ecosystem features where these require human manipulation for optimum management	Watshishou Migratory Bird Sanctuary (Quebec)
V. Protected landscape or seascape	• Areas where interactions of people and nature have produced a distinct character with significant cultural or ecological value and often with high biodiversity	• Conservation, education, recreation, and provision of natural products aimed at safeguarding the integrity of harmonious interactions of nature and culture	Algonquin Provincial Park (Ontario)
VI. Managed resource and protected areas	• Predominantly natural areas that are large enough to absorb sustainable resource uses without harming long-term maintenance of biodiversity	• Long-term protection and maintenance of biodiversity and other natural values and the promotion of sound management practices for sustainable production purposes	Battle Creek Community Pasture (Saskatchewan)

SOURCE: *The State of Canada's Environment—1996,* Environment Canada, 1996, Box 14.4. Reprinted with permission of the Minister of Public Works and Government Services Canada, 2004.

of nature. The IUCN has categorized protected areas into six types (I to VI). Table 12–6 identifies these categories, their defining characteristics, and the management goals or practices associated with each type. Comparison of various types of protected areas found in Canada with those found in other countries shows slightly more protected areas in Canada are in IUCN classes IV to VI than in categories I to III, which involve the highest level of protection. This means the Canadian system affords a slightly lower level of biodiversity protection than might be suggested by simple designation of protected areas.

Most of Canada's protected areas traditionally have been held in public ownership. Provincial parks and our national park system (see Figure 12–3) constitute the largest portion of protected land in Canada. ENGOs, however, have made important contributions to conserving biodiversity. The success of the North American Waterfowl Management Plan in protecting thousands of hectares of productive wetland ecosystems has been generated partly through incentives to private landowners to conserve wetlands for wildlife, and partly by the stewardship actions of partner ENGOs such as Ducks Unlimited and the Nature Conservancy. By 2000, ENGOs owned or managed over eight million hectares of conservation lands in relation to Canadian biodiversity efforts (Nature Conservancy, 2000, personal communication).

Growth in protected areas is one measure of the commitment to protect ecosystems. From the mid-1970s until 1994, protected lands more than doubled in area and represented approximately 8 percent of our total land mass. According to an IUCN-WCMC document produced in 1997 and reanalyzed in 2000, more than 9.5 percent of Canada, or more than 950 000 square kilometres, was protected area. Only about half that land will not be subject to major resource extraction activities, however, as less than half the area is protected under designations that correspond roughly to IUCN categories I to III (Government of Canada, 1996; World Conservation Monitoring Centre, 2000). Proportions of different types of federally and provincially protected areas, and other conservation areas, are shown in Figure 12–4.

In November 1992, federal, provincial, and territorial ministers responsible for environment, parks, and wildlife issued a joint commitment that they would attempt to complete Canada's network of protected areas for land-based natural regions by the year 2000. This initiative was called the Endangered Spaces Campaign. At the end of the campaign, 132 of Canada's 486 natural regions (or 6.8 percent of Canada's total area) were considered adequately or moderately protected (World Wildlife Fund Canada, 2000b; 2000d). Yet information from different sources raises questions about discrepancies in the data and highlights some of the difficulties involved in measuring the success of protecting natural regions (see Table 12–7).

In part, the impetus to protect areas representing all of Canada's natural regions came from the 1989 Canadian Wilderness Charter. Part of the Endangered Spaces Campaign of World Wildlife Fund Canada, the Canadian Wilderness Charter has a broad base of support—by the end of the campaign in 1999, more than 600 000 individuals and 300 organizations had signed the charter (World Wildlife Fund Canada, 2000d). Following the recommendation in *Our Common Future,* the charter specified that at least 12 percent of Canadian lands and waters should be protected. (The Endangered Spaces Campaign is considered in more detail later in this chapter.)

Canada's National Parks Canada has relied on provincial and national parks as cornerstones of protecting biodiversity. Although national parks take up only 3 percent of Canada's land mass, "they contain over 70 percent of the native terrestrial and freshwater vascular plant species and over 80 percent of the vertebrate species" (Dearden & Dempsey, 2004, p. 233). The Canadian national parks system protects environments representative of Canada's natural and cultural heritage. Since Canada's first national park was established in Banff, Alberta, in 1885, the system has evolved to include parks, reserves, and marine conservation areas. These protected spaces are public lands administered by the federal government under the provisions of specific legislation. For instance, the new Canada National Parks Act (and regulations), proclaimed in February 2001, provides for the protection and management of these natural areas of Canadian significance for the long-term benefit, education, and enjoyment of present and future generations.

In the past, parks management principles promoted natural resource development and permitted multiple-use activities such as forestry, mining, and agriculture to take place within park boundaries. The new Parks Act strengthens and places first priority on the ecological integrity of parks. This means that protecting **ecological integrity**—those conditions that are characteristic of a park's natural regions and likely to persist, including abiotic components, native species and biological communities, rates of change, and supporting processes—is the prime focus of park management efforts (Parks Canada Agency, 2001). A parallel concept important in parks management is that of **ecological health** of an ecosystem in which native species are present at viable population levels (Searle, 2000). So that human activity does not impair ecological integrity, extractive industries are now prohibited within park boundaries (but, as noted in Box 10–5, such developments adjacent to park boundaries can prove problematic).

In early 2004 there were 41 national parks and park reserves in Canada, located in all parts of Canada and varying in area from 8.7 km^2 to more than 44 800 km^2 (see Figure 12–3). While over 224 000 km^2 (about 2 percent)

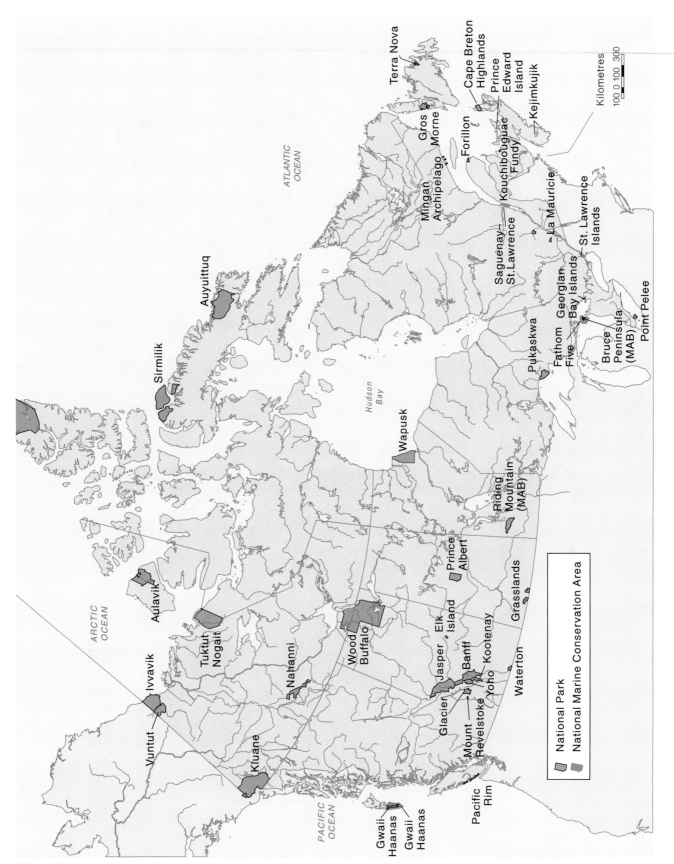

Figure 12–3a

National Parks and Marine Conservation Areas of Canada

CHAPTER 12: WILD SPECIES AND NATURAL SPACES

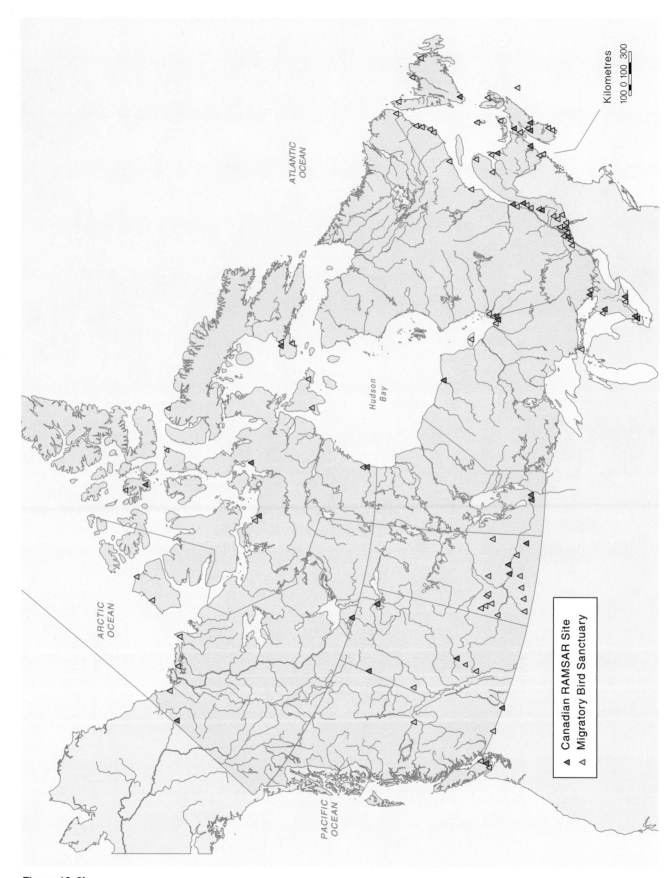

Figure 12–3b

Ramsar Sites and Migratory Bird Sanctuaries of Canada

Canadian RAMSAR Site
Migratory Bird Sanctuary

ATLANTIC
OCEAN

Hudson
Bay

ARCTIC
OCEAN

PACIFIC
OCEAN

Kilometres
100 0 100   300

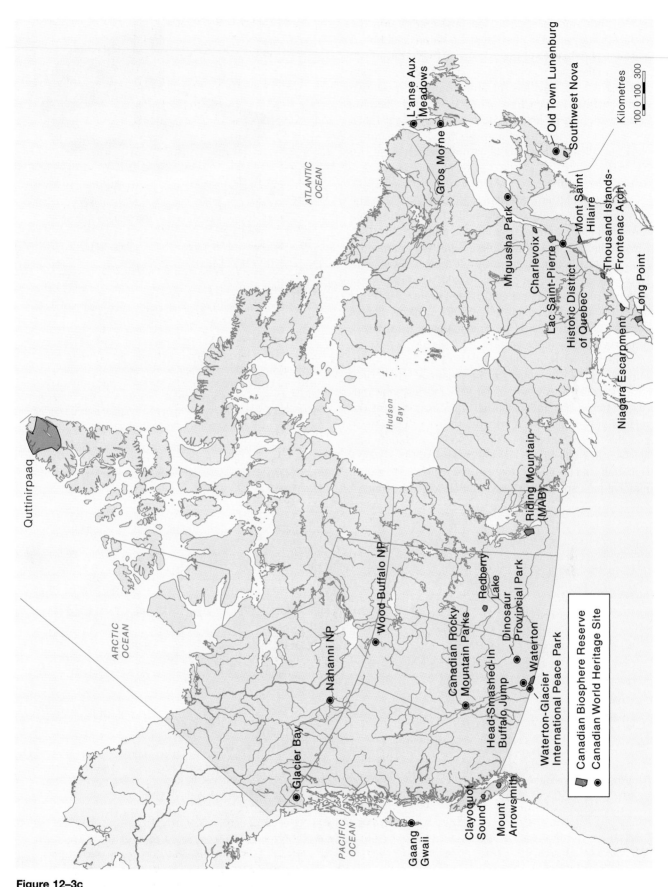

Figure 12–3c

Biosphere Reserves and World Heritage Sites of Canada

of Canada's land mass has been protected within the national park system, to be complete, there should be at least one park in each of the existing 39 terrestrial and 29 marine regions, or about 364 000 km² under protection (Eidsvik, 1989). This idea of protecting representative environments derives from a 1971 Systems Planning Framework that set out a process by which new national parks could be designated and established. This framework classified Canada into 39 natural regions, each with its own combination of climate, vegetation, landforms, and internal biodiversity. The idea was that if each of the 39 regions could be represented adequately within the parks system, Canada's diversity of life forms, natural features, and natural processes would be protected. In 1992, Environment, Parks, and Wildlife ministers endorsed a Statement of Commitment to Complete Canada's Networks of Protected Areas; in 2001, Parks Canada Agency (2001) recommended that implementation of that statement should be completed by 2003. While new areas were set aside after 2001, implementation of the network has not yet been completed.

The natural areas that are best represented within our national park system are the montane, boreal, and Arctic cordillera ecozones. Major gaps in ecosystem representation occur in the southern prairies, most of the regions of the southern Arctic, the three taiga ecozones, and the most densely populated regions of Ontario and Quebec. In Nunavut, six natural regions still lack representation, and in Quebec, four natural regions still require national park status. It is very difficult to protect under-

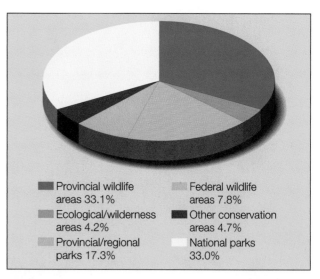

Figure 12–4
Proportion of federal, provincial, and other protected areas in Canada in the 1990s

NOTE: These numbers are now outdated; however, there is no longer central reporting and collating of these data. They illustrate the heavy reliance we have placed on provincial and national parks to be the repositories of wild species and natural spaces. There is now a growing emphasis on environmental protection through increased involvement of private organizations (for-profit and nonprofit) in stewardship programs and projects.

SOURCE: *The State of Canada's Environment—1996,* Environment Canada, 1996, Figure 14.8. Reprinted with permission of the Minister of Public Works and Government Services Canada, 2004.

TABLE 12–7
MEASURING SUCCESS?

Organization	Date	Area Protected (square km)	Percentage of Canada's Total Land Mass
Government of Canada	2003	820 000	8.2
Government of Canada	2001	660 003	6.6
Government of Canada	1996	800 000	8.0
WCMC	2000	950 000	9.6
WWF Canada	2000	683 000	6.8

NOTE: These statistics highlight the importance of being a critical thinker (see Chapter 2). The data above reflect inconsistencies in measuring Canada's protected areas. All the data are from respected sources. The numerical values indicated above could suggest possible differences in the mandates and stated values of the three agencies represented. Therefore, it is important to ask the following questions: What is the nature of the data? How were they collected? How were they interpreted? How were "protected areas" and "natural regions" defined?

SOURCES: *Canada's National Indicators Series, 2003,* Government of Canada, 2003, http://www.ec.gc.ca/soer-ree/English/ Indicators/default.cfm; *The State of Canada's Environment—1996,* Government of Canada, 1996, Ottawa: Supply and Services Canada; *Human Activity and the Environment, Annual Statistics 2003,* Statistics Canada, 2003, Ottawa: Ministry of Industry; *State of the World's Protected Areas,* Annex 1: Summary of protected areas, World Conservation Monitoring Center (WCMC), 2000, http://www.unep-wcmc.org/ protected_areas/albany_a1.pdf; *Endangered Spaces: The Wilderness Campaign That Changed the Canadian Landscape: 1989–2000,* World Wildlife Fund, 2000, pp. 22–23.

represented ecosystems in southern Canada because frequently they have been modified highly through human settlement and economic activities. Nevertheless, in 2002, the federal government announced the creation of the Georgia Basin National Park to incorporate the islands and waterways located between Vancouver and Vancouver Island, in the most densely populated region in British Columbia. In addition, many candidate lands are in private ownership; in these cases, the role of private stewardship is vital. For example, the Nature Conservancy of Canada has worked with landowners adjacent to Waterton Lakes National Park to establish conservation easements on land title to ensure that ranching and farming activities in the immediate area do not jeopardize important animal habitats or other environmental values. Conservation easements are purchased by the Conservancy and attached to the property titles. They specify the location and/or types of activities that may be restricted on land owned by private landowners. The Conservancy has worked throughout Canada and the United States using this approach to promoting private stewardship. The federal government also has a small program to purchase conservation easements from private landowners in an effort to encourage the protection of important ecosystems and species at risk. In northern Canada, in areas set aside as national parks pending settlement of any outstanding Aboriginal land claims, national park reserves permit traditional hunting, fishing, and trapping activities to continue.

Protection of marine ecosystems is as important as protection of terrestrial ecosystems, but Canada has lagged behind some other developed nations in achieving the desired level of protection. Parks Canada Agency has identified 29 marine regions (including the Atlantic, Pacific, and Arctic oceans as well as the Great Lakes) in Canada. The agency is seeking to ensure that each region will be represented in its network with at least one national marine conservation area. Marine conservation areas do not confer the same level of protection as national parks. The three existing national marine conservation areas—Gwaii Haanas (Queen Charlotte Islands in British Columbia), Fathom Five National Marine Park (Georgian Bay, Ontario), and an area at the confluence of the Saguenay and St. Lawrence rivers in Quebec—are intended to protect whales, marine mammals, and endemic plant specifies. Sustainable-use zones permit commercial fishing to continue. It is not clear whether future commercial activities will be allowed. In addition to Parks Canada Agency initiatives, the Canada Oceans Act provides for establishment of marine protected areas, and the provincial government of British Columbia also has established ecological reserves in marine areas. In 2002, the federal government announced the creation of five new national marine conservation areas in the next five years in the following regions: Gwaii Haanas, British

Columbia, western Lake Superior, Ontario, and the southern Strait of Georgia, British Columbia. Regions for the remaining two marine conservation areas had yet to be chosen.

The emphasis on **ecosystem management** within our national parks was reinforced by recommendations from two federal government task forces that examined the health of Canada's national parks. The Banff–Bow Valley Task Force (1996) and the Panel on the Ecological Integrity of Canada's National Parks (Parks Canada Agency, 2000) reported that our parks were under serious threat. Stressors (including habitat loss and fragmentation, losses of large carnivores, air pollution, pesticides, exotic species and over-use by people) clearly had diminished the ecological integrity of these special places. Many stressors resulted from visitor or tourism facilities. These were difficult to address as Parks Canada's budget was reduced by 25 percent between 1995 and 2000, and greater emphasis was placed on gaining revenue from visitors (Dearden & Dempsey, 2004).

In order to resolve such problems, Parks Canada's action plan for the future focuses on ways to (1) make ecological integrity central in legislation and policy that direct park management, (2) build partnerships with park neighbours and other stakeholders to cooperate in maintenance and protection of ecological integrity, (3) plan for ecological integrity by defining the direction for maintenance or restoration of ecological integrity and guiding appropriate public use, and (4) renew the Parks Canada organization to support more effectively the mandate of ecological integrity (Parks Canada Agency, 2001).

Other Protected Areas in Canada Some of Canada's protected areas are designated under the terms of international treaties or agreements. The Ramsar Convention (see Chapter 7) protects wetlands of international importance within Canada, and sites (including areas of particular scientific or aesthetic value as well as the habitats of endangered species) are designated under the Convention Concerning the Protection of World Cultural and Natural Heritage. Ramsar and World Heritage sites may be designated without being protected formally (this helps identify candidate sites for formal protection in the future). In 2003, there were 13 World Heritage Sites in Canada (two shared with Canadian national parks), 12 biosphere reserves, and 36 Ramsar sites (see Figure 12–3). In addition, in British Columbia, a system of ecological reserves offers the highest degree of protection offered to any protected area. Ecological reserves were established in the 1970s for the protection of biodiversity. Their primary purpose is research and education, not outdoor recreation, although many are open to the public for nonconsumptive, observational uses such as nature appreciation, wildlife viewing, and photography. While large in number, they are very small in size.

Efforts by governments and nongovernmental organizations to increase the number and geographic extent of protected areas continue. In 2002–2003, Quebec designated 27 new areas for protection, covering about four million hectares and increasing the proportion of protected areas from 2.9 percent to 5.3 percent. In April 2003, the Protected Natural Areas Act came into force in New Brunswick, providing more comprehensive legislation to manage and administer the province's entire network of protected natural areas. In February of the same year, Nova Scotia announced that four parcels of land will now be protected in an agreement that involved the Nature Conservancy of Canada (a nonprofit, nongovernmental organization) and Bowmater Mersey Paper Company Limited (a forestry company).

Other models for the protection of biodiversity include biosphere reserves. These places are intended to be "living landscapes" where people work within local landscapes and demonstrate how they can earn a livelihood by living sustainably within their ecosystems. Biosphere reserves are established under the United Nations Educational, Scientific and Cultural Organization's (UNESCO) Man and the Biosphere (MAB) Programme. These reserves have a wide range of objectives in addition to conservation, including scientific research, training, monitoring, and demonstration. Biosphere reserves require a core area that is strictly protected by national legislation (e.g., national park, protective treaty, or convention) surrounded by a buffer in which activities that are consistent with the objective of conservation are permitted. Surrounding the core and buffer is a large "zone of cooperation," where a wider range of human activities occurs and where ecological restoration and sustainable resource use can be demonstrated (see Figure 12–5). Biosphere reserves are important because they recognize and support cultural as well as biological diversity. They are created by local initiative, coupled with provincial and federal support, and finally recognized by the United Nations. Once designated, they become part of an international network, and members participate in national and international meetings to discuss their common issues and progress. In 2000, there were 8 biosphere reserves; by 2003, this number had increased to 12. Nevertheless, part of the challenge of the biosphere reserve concept is that outside the core area, the biosphere reserve is merely an area of recognition. This means that there are no laws or regulations that can be used to enforce or curb particular practices. Efforts to pursue conservation practices in support of sustainability are entirely voluntary and financial and logistical support of provincial and federal governments has been minimal. Instead, volunteer members of biosphere reserve committees assist researchers and

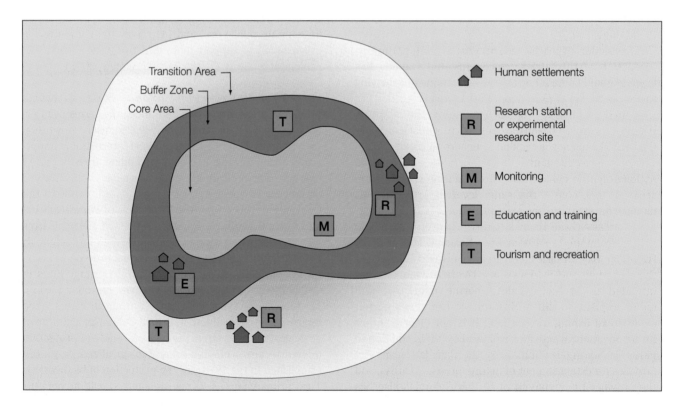

Figure 12–5

The three zones of a biosphere reserve

SOURCE: *Solving the Puzzle: The Ecosystem Approach and Biosphere Reserves,* UNESCO, 2000, Paris: Author, p. 6. Reprinted by permission of UNESCO.

engage local communities in outreach activities that will help to build understanding and commitment to the conservation objectives of the biosphere reserves.

Restoration and Rehabilitation   There is not much point in expending time, energy, and finances on species recovery or restoration if the ecosystem that is meant to support them is not rehabilitated also. This is why Canada reviews the status of certain classes of native flora and fauna under COSEWIC. A complementary program, Recovery of Nationally Endangered Wildlife (RENEW), develops recovery plans for more than 75 species that have been designated by COSEWIC as extirpated, endangered, or threatened. RENEW and similar programs recognize the need, but do not have the knowledge base, to respond effectively to threats to plants, invertebrates, fungi, and algae, even if many of these species perform more vital ecological functions than the vertebrates.

Encouraging results have been attained from several species protection or reintroduction programs. Sea otters have been reintroduced to Vancouver Island's west coast kelp forests; peregrine falcons have been restored to territories from which they had been extirpated; the ferruginous hawk has been downlisted from threatened to special concern, and the wood bison has been downlisted from endangered to threatened. The American white pelican and the prairie long-tailed weasel, two prairie species that had been listed as threatened, are among those that have been removed from the COSEWIC list.

In addition, private organizations such as Ducks Unlimited Canada (DUC) also undertake restoration of degraded habitats and encourage good stewardship practices of private landowners. While originally established to protect habitat for waterfowl species desirable for hunting, Ducks Unlimited has expanded its role to habitat protection of wetlands. In some cases DUC buys land outright to ensure that the property is not drained for agricultural or other purposes. In other cases, DUC works with current owners to identify important habitat and engage in agreements (some of which are cost-sharing agreements) to ensure that farming or ranching activities do not jeopardize important nesting, rearing, or overwintering areas.

Photo 12–14b

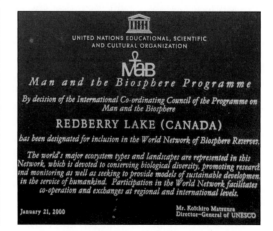

Photo 12–14a   **Photo 12–14c**   **Photo 12–14d**

The residents of the Redberry Lake Biosphere Reserve in Saskatchewan seek to protect their ecological and cultural heritage for present and future generations.

Ex Situ Conservation

Ex situ conservation of biodiversity is the conservation of species or genetic materials under artificial conditions, away from the ecosystems to which they belong. As the supply of natural habitats has declined, maintaining captive populations (or cultivated populations in the case of plants) has been accepted as one of the few ways that survival of increasing numbers of species can be ensured until the long-term goal of restoring them to the wild (in at least some of their original ranges) is attained.

Among the advantages of ex situ facilities such as zoos, aquariums, aviaries, and botanical gardens are opportunities for scientific study as well as public education opportunities about species on display. In turn, this can help build broader public support (and provide fundraising opportunities) for species conservation in situ. The majority of conservation biology opinion, however, indicates that captive breeding programs are not the answer to conservation needs.

There are many reasons why not, including the fact that ex situ protection is expensive, and, as natural habitats continue to disappear, too many species require housing. Within ex situ facilities, animals often become domesticated and lose their wild behaviours. Lack of natural selection of mates is an issue that also may contribute to inbreeding problems such as depression. Another loss associated with ex situ facilities is the inability to pass on to captive-born young their hunting strategies or migration routes.

Plants About 800 of the world's 1500 botanical gardens are committed to conserving rare, endemic, or other threatened plant species. Canada's 60 botanical gardens contain close to 40 000 plant species native to Canada and from around the world, including at least 11 endangered, rare, or vulnerable native plant species. In 1995, no plants were being propagated specifically for reintroduction.

Seed banks can be important ex situ facilities for conservation of plant species and genetic diversity. Many of the world's 528 seed bank facilities are coordinated through the International Board for Plant Genetic Resources. Both Agriculture and Agri-Food Canada and the Canadian Forest Service maintain seed stocks for use in plant breeding or in sustainable use programs. The Heritage Seed Program is a nongovernmental initiative that maintains a large variety of vegetable species, and "Seedy Saturday" (a concept that began in Vancouver and since has spread throughout British Columbia, Alberta, and recently Ontario) promotes springtime sharing of seeds and seed stories among backyard gardeners, farmers, and others. This is an informal but effective way to preserve old varieties of garden vegetables and family farm livestock (Penstone, 1997).

Animals In the 1990s, Canada had about 50 zoos, 15 aquariums, and 15 aviaries; these facilities hold hundreds of species of mammals, birds, reptiles, fish, and invertebrates. More than 870 zoos and aquariums worldwide, housing over one million vertebrate specimens, are involved in global efforts to preserve both wild and native animal diversity ex situ. The focus in many of these institutions is no longer on entertainment but on providing living conditions that simulate various species' natural habitats accurately enough to promote breeding. This is seen to be particularly important for species whose numbers in the wild are so low that survival depends on captive breeding programs, called Species Survival Programs, to build up population numbers to the point where species can be returned to their native ecosystems.

Canadian zoos hold at least 12 endangered, threatened, or vulnerable animal species, including the whooping crane, ferruginous hawk, eastern massasauga rattlesnake, spotted turtle, grizzly bear, and polar bear. Canadian captive breeding programs have provided swift foxes and peregrine falcons for reintroduction programs (Government of Canada, 1996; Henry, 1994). Some Canadian zoos participate in international breeding programs of non-Canadian species for reintroduction to native ecosystems in other countries. Not all species can be propagated successfully in captivity, however.

At the University of Alberta, a poultry conservation program to conserve the genetic diversity of six breeds of chickens that contributed to our past or present meat and egg stocks has been ongoing since 1992. This program is a bit like an insurance policy; it is a conscious effort to sustain the genetic variability of the old-fashioned birds against the danger that today's commercially bred chickens are becoming more genetically uniform and may develop traits that will limit their viability (Robinson, 1997).

Photo 12–15

Captive breeding programs in zoos help conserve genetic diversity but are not a substitute for natural, functioning ecosystems. The black-footed ferret (above) is being reintroduced to North American grasslands.

In spite of some benefits obtained through ex situ protection, it is not sufficient merely to save seeds, extract and freeze sperm, and protect some species in "stationary arks." Rather, we need to save wild species and natural ecosystems in intact, natural spaces if we hope to protect biodiversity and promote sustainability.

Sustainable Use of Biological Resources Growing human populations and increases in resource consumption continue to intensify the pressures on biodiversity, in situ and ex situ conservation efforts notwithstanding. Given the consensus that current and projected consumption levels are unsustainable (World Commission on Environment and Development, 1987), the best way to satisfy basic human economic needs without compromising biodiversity is the rigorous adoption of sustainability policies in all resource sectors.

In Canada, this means that agriculture, forestry, and fisheries are of particular significance in developing sustainable-use practices that will conserve genetic resources. In forestry, for instance, conserving the natural diversity of trees (and associated wildlife) requires not only better knowledge of forest ecosystems and species, but also less aggressive and more selective harvesting as well as protection of old-growth forests.

Improving Understanding of Biodiversity If we are to be successful in conserving biodiversity and using biological resources sustainably, adopting ecological management principles is necessary. Ecological management directs human activities so that the biotic and abiotic components of ecosystems, and the processes that sustain them, continue. To manage effectively, it is necessary to have adequate understanding of ecosystems, species, and human impacts on them; unfortunately, the current state of this knowledge is inadequate. This means there needs to be a strong commitment to research.

Research needs include reliable baseline data, indicators for conserving biodiversity, standardized protocols for conducting surveys and inventories, and indicators of biodiversity change (for trend prediction). Gap analysis is a research tool used to evaluate and complete representative protected areas networks (Rowe, Kavanagh, & Iacobelli, 1995). Using ground surveys, inventories, satellite imagery, and geographic information system (GIS) technology, gap analysis identifies missing dimensions in the representation of biodiversity. Hot spot analysis is a related tool that locates areas of species or endemic richness that should have priority for protection, sustainable use, or new biotechnologies. Biodiversity prospecting— the search for wild species that may yield new medicines or add vigour to food crops—provides economic reasons for conservation.

Efforts to monitor the status of biodiversity are ongoing. The federal Ecological Monitoring and Assessment Network (EMAN) program, for example, monitors a wide variety of ecosystem and species parameters to assess the interactions and sustainability of regional ecosystems. In addition to the research needs noted above, it is necessary to develop a dynamic program of public education that promotes awareness of biodiversity and develops strategies to reduce resource consumption levels.

CANADIAN LAW, POLICY, AND PRACTICE

With the growth of international concern and action relating to biodiversity issues (species and spaces), and Canada's active participation in global efforts to preserve biodiversity, the national response has involved strengthening and augmenting existing laws, policies, and practices as well as instituting new partnerships. Table 12–8 highlights selected examples of Canada's approach to protecting biodiversity. To illustrate Canadian efforts to protect Canadian species and spaces, the following sections comment briefly on the Endangered Species Protection Act and the Endangered Spaces Campaign of World Wildlife Fund Canada.

Protecting Canadian Species

Wildlife management in Canada is "a complex, almost precarious system of shared responsibilities between the federal government and the provinces and territories" ("Protecting Canadian Species," 1996). The legislative basis of shared responsibility places migratory birds, fish, and marine mammals within federal jurisdiction; the majority of wildlife protection responsibilities lie with the provincial and territorial governments. In spite of this, interjurisdictional cooperation (such as is seen in COSEWIC) and informal agreements have been a hallmark of wild species protection in Canada.

As useful as COSEWIC's designations of species at risk are in raising both public awareness and the political will to protect wildlife, the designations are less effective than they might be simply because they have no legal status. In 1996, a Canada Endangered Species Protection Act was tabled in Parliament. The original act was controversial, engendering opposition because some conservationists (including leading scientists) did not believe it was sufficient to protect endangered species and particularly their habitat, while landowners believed they risked losing their property and livelihood because of it. Not until December 2002 did a revised Species at Risk Act (SARA) receive royal assent. It came into full effect in June 2004. The act is part of a three-part strategy for protecting biodiversity. The other components include a habitat stewardship program, and an Accord for the Protection of Species at Risk—a Canada-wide agreement on

TABLE 12-8

PROTECTING BIODIVERSITY IN CANADA: EXAMPLES OF POLICIES, STRATEGIES, AND LEGISLATION

National/regional	Canadian Biodiversity Strategy
	Bill C-33, Species at Risk Act
	Federal Policy on Wetland Conservation
	Canadian Environmental Protection Act
	National Accord for the Protection of Species at Risk
	Committee on the Status of Endangered Wildlife in Canada
	North American Waterfowl Management Plan
Wildlife	RENEW (Committee on the Recovery of Nationally Endangered Wildlife)
	Wildlife Policy for Canada
	Canada Wildlife Act
	CITES (Convention on International Trade in Endangered Species of Wild Fauna and Flora)
	WAPPRIITA (Wild Animal and Plant Protection and Regulation of International and Interprovincial Trade Act)
	Endangered Species Recovery Fund
Birds	Migratory Birds Convention Act
Plants	Canada Forest Accord
	National Forest Strategy
	Agricultural Weed Biocontrol Program

federal–provincial–territorial cooperation. Environment Canada, Fisheries and Oceans Canada, and the Parks Canada Agency of Heritage Canada are the core departments responsible for implementing SARA. The act prohibits killing species listed under the act and destroying their critical habitat. The act also established COSEWIC as a legal entity at arm's length from the government to ensure that species were designated by a scientific panel independent of government.

Protecting Canadian Spaces

Protecting our "natural capital" is essential if future generations of Canadians are to experience a healthy and prosperous environment. This is why the World Wildlife Fund Canada's Endangered Spaces Campaign continued to push governments to complete the promised nationwide system of protected areas (see Hummel, 1989). In this 10-year campaign, begun in 1989, conserving representative areas of the country's diverse lands and waters was considered a critical first step toward the necessary protection of biodiversity in support of sustainability (Hackman, 1996).

Setting aside ecologically representative terrestrial and marine areas, whether they are in Ontario's hardwood forests or on Newfoundland's Grand Banks, means that these areas are to be free from activities likely to cause large-scale or long-term habitat disruption. Wilderness benefits, which in general are inadequately considered during land use decisions, support the case for protection of representative spaces. Aside from the previously discussed intrinsic values associated with wild places, Canada's wilderness areas, like tropical rain forests, contain genetic materials that may provide new opportunities for the pharmaceutical industry. In addition, wilderness-related activities, such as hiking, fishing, camping, and canoeing, inject millions of dollars into regional economies. In 1996 in British Columbia, for example, the parks system contributed about $400 million per year to the provincial gross domestic product, and sustained more than 9000 direct and indirect jobs over widely distributed communities and regions (Hackman, 1996). In Ontario, tourists spend $8 million annually in the region of the Bruce Peninsula National Park and Fathom Five Marine Park (Bayly & Ives, 1997).

Establishing parks and protected areas can help to diversify regional economies. In Alberta, for instance, a study for Alberta Environmental Protection found that parks and recreation created more than three times more jobs than the forestry sector (Engman, 1997). This is an important benefit given that jobs in forestry, mining, and other resource industries are being lost to mechanization and other factors (Hackman, 1996). However, jobs in tourism and recreation are typically seasonal in nature, they do not pay well, and frequently they do not offer the social benefits (e.g., leave provisions, pensions) Canadians have come to expect. In addition, new jobs may require different kinds of education and training than jobs in resource industries. Therefore, jobs available in tourism and recreation are unlikely to offer direct replacements to jobs based on resource extraction. Instead, changes will likely be realized over a generation or more rather than offer direct and immediate relief to workers displaced in resource sectors. Nevertheless, there is potential for creating regional economies that are built from learning about and demonstrating sustainable use of our environment through nature protection and appropriate recreation.

In the bigger picture, as global markets shift from traditional resource-extraction to knowledge-based industries, ecosystems offer new opportunities for research, tourism, and sustainable economies. Canada's large share of the world's diminishing wilderness potentially gives us a competitive advantage in developing these business opportunities. Most Canadians understand this connection, too: a 1996 national survey by the Environics Research Group revealed that 86 percent of the respondents believed that a network of protected natural areas would make a major or moderate contribution to Canada's long-term economic health (cited in Hackman, 1996).

Such positive statements about the ability of our protected areas to contribute economically, socially, and ecologically to sustainable development should proceed with caution. Many Aboriginal peoples have been displaced by the creation of national parks, and they contest the Euro-Canadian perspective that national parks should not include resource extraction activities such as hunting. In addition, there is growing concern that the ecological integrity of national parks located in high tourist areas (e.g., Banff, Prince Edward Island) is being compromised by the large number of visitors (national and international) and the pressure on Parks Canada to provide facilities to accommodate them. For example, in the late 1990s, it was estimated that Canadian national parks averaged over 14 million visits per year, with an anticipated annual growth rate of 4.5 percent. This means that the number of visitors to national parks could double in 15 years (Parks Canada Agency, 2000). These increasing visitor numbers, coupled with efforts by national parks to build and/or approve new facilities to accommodate them, have raised serious questions about the extent to which the national parks are able to meet their mandate to protect ecological integrity (Parks Canada Agency, 2000; Searle, 2000).

Several nongovernmental organizations attempt to keep track of Canadian efforts. The Endangered Spaces Campaign was a 10-year campaign of World Wildlife Fund Canada, ending in 2000. It had a strong scientific base, rooted in concepts such as ecological integrity and ecosystem management. The healthy state, or integrity, of an ecosystem is one in which the ecosystem is complete and functions properly. To have integrity, an ecosystem must have all its native species, complete food webs, and naturally functioning ecological processes, and it must be able to persist over time. Ecosystem management is a holistic approach, recognizing that the well-being of parks and human communities alike depends on the ecological state of larger landscapes.

Leading conservation biologists, major corporations, and organizations such as the Canadian Council on Ecological Areas assisted in developing the scientific approach of the Endangered Spaces Campaign. Each year, to measure progress in the Endangered Spaces Campaign, World Wildlife Fund Canada graded the provincial, territorial, and federal governments on their progress toward completing the goal. The 1998/1999 Report Card shows the final grades assigned to each jurisdiction for

BOX 12–4
THE 1998/1999 ENDANGERED SPACES REPORT CARD

The Endangered Spaces Campaign had a specific measurable goal: to help conserve Canada's biological diversity by protecting a representative sample of each of the country's terrestrial and one-third of its marine natural regions by the year 2000, and to complete the marine protected areas system by 2010. Following are the "grades" received by federal, provincial, and territorial governments from 1995/96 to 1998/99.

Federal Grades

Jurisdiction	1995/96	1996/97	1997/98	1998/99
Terrestrial	C	A–	D	C
Marine—Pacific	C	C–	D+	N/A
Marine—Arctic	D–	D–	D–	N/A
Marine—Atlantic	D+	D	D+	N/A
Marine—Great Lakes	D	D	D	N/A

Provincial and Territorial Grades

Jurisdiction	1995/96	1996/97	1997/98	1998/99
Alberta	B	D+	F	F
British Columbia	A	C	C+	C
Manitoba	D–	B+	C	B–
New Brunswick	F	F	D	D
Newfoundland	D	C–	D	D+
Nova Scotia	A	C–	C+	C+
Northwest Territories	D	C–	C	C–
Ontario	F	C–	D+	B+
Prince Edward Island	C+	B	B	C
Quebec	C–	D–	F	F
Saskatchewan	C	F	B–	C
Yukon	D	C–	C+	C–

NOTE: The campaign is now over, so a national grading update is not available.

SOURCES: *The Endangered Species Campaign: 1997/98 Marine Grades and 1997/98 Terrestrial Grades,* World Wildlife Fund Canada, 1998, www.wwfcanada.org/wwf_spaced.htm; *Grades: The 1998–99 Endangered Spaces Progress Report on Canada's Wild Lands,* World Wildlife Fund Canada, April 27,1999, www.wwfcanada.org/news-room/map.htm

their progress in establishing the protected areas system (see Box 12–4). Not only did this simple measure capture year-to-year differences in provincial, territorial, and federal progress in setting aside protected areas, it also was a means for Canadians to compare performance across and within political jurisdictions over time. From 1997/98 to 1998/99 World Wildlife Fund downgraded British Columbia from C+ to C primarily because the government commitment to preserve 12 percent of British Columbia's lands did not allow for adequate representation of all of the province's natural areas: overall almost 20 percent remains inadequately represented (World Wildlife Fund Canada, 1999).

Every year since 1993, the Sierra Club of Canada has used a similar report card approach to rate the federal, provincial, and territorial governments on their progress in keeping the commitments they made at the 1992 Earth Summit regarding climate change and protection of biodiversity. In 2000, they gave the federal government a C for its commitment to reduce greenhouse gases and an F in protecting biodiversity (Sierra Club of Canada, 2000). By 2003, this rating had improved to A for its commitment to reduce greenhouse gases and B+ for its commitments to protect biodiversity (Sierra Club of Canada, 2003). Individual provinces faired far worse, however, with British Columbia and Alberta receiving failing grades on both counts, while Quebec rated most highly, receiving a B for biodiversity protection and an A– for its commitment to climate change.

PARTNERSHIPS FOR THE FUTURE

Growing concern over wildlife and wild spaces issues has caused a broad range of groups—government agencies, private-sector interests, ENGOs, and individuals—to work together to preserve and protect wildlife and to sustain Canada's ecosystems. The involvement of this range of groups helps ensure that resources are used wisely and that efforts are directed to the best possible use.

The policies, strategies, and legislation identified in Table 12–8, as well as numerous provincial and territorial conservation and sustainability strategies, wetland and wildlife policies, forest management plans, and protected areas strategies, are evidence that cooperation and collaboration can provide positive outcomes for wildlife and wild spaces. The Canadian Biodiversity Strategy, Canada's first effort to implement the United Nations Convention on Biological Diversity, sets the stage for all jurisdictions in Canada to identify the actions required, individually as well as in partnership with others. The Canadian Biodiversity Strategy also establishes a comprehensive planning framework within which Canadians can participate in the effort to conserve biodiversity (Government of Canada, 1996).

While every province and territory in Canada has management policies, programs, and laws in place to protect wildlife, 431 species remained on COSEWIC's 2003 list of species at risk. In order to prevent any species from becoming extinct as a consequence of human activities, and to institute a harmonized national approach to conservation of species at risk, federal, provincial, and terri-

Photo 12–16

The whooping crane is one species that has benefited from wildlife protection policies and strategies. It nests in Wood Buffalo National Park, in the Northwest Territories, and winters on the gulf coast of Texas.

torial governments have signed a National Accord for the Protection of Species at Risk in Canada.

An underlying precept of this approach is that all Canadians must share responsibility for the conservation of wildlife at risk. A second precept is that if it is to be effective and complete, a national endangered species framework must be able to address all nondomesticated living organisms native to Canada. RENEW (see Table 12–8) coordinates recovery programs for COSEWIC. Under RENEW's umbrella, between 1999 and 2000, federal, provincial, and territorial government agencies and ENGOs spent nearly $14.4 million on recovery programs for 75 species at risk. Whooping cranes, swift foxes, and piping plovers are among the species to have benefited from activities of RENEW partners (Environment Canada, 2000).

Canada and the United States have signed several agreements to improve the management of birds that migrate between the two countries, and to deal with questions of fairness in regulating waterfowl harvests among Alaskan and Canadian Aboriginal peoples. A Framework for Cooperation to protect shared endangered species was signed by the Canadian Environment Minister and the U.S. Interior Secretary. Probably the best-recognized cooperative effort, the billion-dollar North American Waterfowl Management Plan (NAWMP) is the combined initiative of Canada, the United States, and Mexico. NAWMP is implemented through regionally based joint ventures that, in turn, involve federal, provincial, territorial, and state government agencies, ENGOs such as Ducks Unlimited and Wildlife Habitat Canada, the private sector, and landowners all cooperating together. Regional habitat management in Canada is undertaken through the Prairie Habitat, the Eastern Habitat, and the Pacific Coast joint ventures. Initially conceived as a waterfowl plan, the NAWMP has expanded to include multispecies and biodiversity objectives.

The Canadian Coalition for Biodiversity, the Natural Heritage League, the Canadian Parks Partnership, Wildlife Habitat Canada, and the Canadian Parks and Wilderness Society (CPAWS) are among many citizen-based ENGO groups across Canada whose collaborative and cooperative efforts to preserve and protect wild species and spaces have made a difference at the national level. The current Yellowstone to Yukon Biodiversity Strategy project involving CPAWS and the reintroduction of the swift fox to the southern Alberta and Saskatchewan prairie and the timber wolf to Montana are examples of projects that exemplify the range of actions possible through various forms of cooperation and collaboration.

On the provincial scale, organizations such as the Federation of Ontario Naturalists and the Nature Trust of British Columbia have helped coordinate energies of multiple groups to magnify the effectiveness of local biodiversity conservation efforts. For example, acting as a partner with the Nature Trust of British Columbia, the Pacific Estuary Conservation Program was able to purchase Englishman River Flats on the east coast of Vancouver Island in 1992. This partnership, and the cooperation of the provincial government, enabled protection of a key feeding and resting area along British Columbia's coast for up to 30 000 brant (geese) that fly from Baja to Alaska and Russia. The Englishman River Estuary also is a wintering area for the trumpeter swan and other waterfowl; more than 110 bird species have been recorded in the estuary, which also provides essential rearing habitat for steelhead trout and salmon (Nature Trust of British Columbia, 1993). Parks Canada Agency is now working with Aboriginal organizations to co-manage parks such as Haida Gwaii and Pacific Rim National Park Reserve.

Two trends offer hope and challenge. First, no longer is biodiversity protection the sole responsibility of government. The federal habitat stewardship program offers funding for private landowners to conserve biodiversity on private lands. Nongovernmental organizations (e.g., Nature Conservancy and Ducks Unlimited) and private landowners are now participating alone and in combination with one another and with governments to provide education, incentives, and programs that invite individuals to undertake stewardship and conservation practices. Many banks and credit unions have environment funds that also support stewardship activities. For example, Vancity Credit Union, the largest credit union in the country, provides no-interest loans to landowners who seek to protect riparian habitats that occur on their property.

Second, biodiversity protection is not only to be achieved in wilderness areas far distant from the urban places in which we live. The protection of wild species and natural spaces may also occur in our own neighbourhoods and cities. For example, community gardens in Vancouver and Montreal (among other cities) are places that cultivate heritage apples, thereby protecting biodiversity of a species important for human consumption. Similarly, across the country, community organizations have undertaken campaigns to "liberate" streams (uncover streams that had been organized into systems of structured waterways and culverts for water provision and sewage treatment) and/or demonstrate the effect of effluent disposal on fish populations of urban waterways. The movement to uproot grass lawns and promote indigenous plants also promotes biodiversity across the country. While initially rejected by municipal councils who enforced strict "weed" controls, governments now support such initiatives. For example, the Saskatchewan Ministry of Environment publishes a book on planting species to attract birds and maintain "backyard biodiversity." The Trans Canada Trail, an initiative of nongovernmental organization, winds its way through urban parks and wilderness, reminding Canadians of their common natural and cultural heritage. These initiatives suggest that the protection of species and spaces does not solely take place in distant wilderness areas. By contrast, urban areas are vital places for learning about and cultivating a love of nature and engaging in stewardship activities to protect biodiversity. Activities in urban places may help engender attitudes and commitments to halt the degradation of national parks and other protected areas, ensuring they are not "loved to death."

FUTURE CHALLENGES

How important are wildlife and wilderness to the future of Canada and Canadians? As the rich heritage and diversity of Canada's wild species and natural landscapes continue to be placed at risk through human demands for expansion of urban, industrial, agricultural, recreational, and

Photo 12–17

The sustainability of timber wolves depends partly on the cooperative and collaborative actions of groups and individuals.

other activities, the importance of wildlife and wild spaces to the future sustainability of Canadian society cannot be underestimated. In September 2000, Saskatchewan's Environment Minister, the Hon. Buckley Belanger, wrote that surveys continually indicate that environmental concern is a core Canadian value. Continued stewardship and actions on the part of individual Canadians, ENGOs, the private sector, and governments are vital if Canadians are to achieve completion of the national parks system and establishment of the network of protected areas across the country (Noss, 1995).

In part, the achievement of these goals depends on continued development, application, and enforcement of legislation and regulation. International and national legislation, including Canada's Species at Risk Act, RENEW, CITES, WAPPRIITA, and the Migratory Birds Convention Act, provide important directions for protecting the future of biodiversity. These directions require continued monitoring, not only of the effectiveness of the legislation, but also of the species and spaces that are the subject of this legislation. As well, continued development of creative conservation programs and conservation research efforts (such as endangered species and spaces campaigns, and gap analysis) that include but are not limited to species currently protected under legislation will be necessary. Local stewardship initiatives are important in this regard.

If Canadians are to achieve balanced sustainable use and protection of wild species and spaces, in national parks or elsewhere, there is a need to shift from individual species management to a more holistic landscape management approach. Here, the model of the "biosphere reserve" is an important contribution to such an approach. The philosophy of sustainable and living landscapes adopted by biosphere reserves is a useful approach to contemporary conservation. The knowledge-building requirements to meet that challenge are ongoing, from use of traditional and local ecological knowledge, to data collection by individual volunteers and scientists, to analysis and interpretation of data by academics in universities and research institutes, to the new information that GIS and other remote-sensing technologies generate. The contributions that both physical and social sciences have to make to wild spaces and species in Canada must be supported and incorporated in actions from local to international scales if the global mission of conservation of biodiversity is to be attained.

However, actions do not have to be big and expensive to be significant. As scores of Canadians have already found out, making a big difference to Canada's species and spaces can start in our own backyards. Even urban environments contain important habitats for genetic diversity, and for plant and animal species in need of protection. Learning more about these spaces and species and translating that knowledge into action is a vital component of biodiversity protection. Such learning may also translate into a shift in societal priorities that will be necessary to remain vigilant in protecting "wilderness" spaces beyond urban borders. What we believe and what we do influence others around us; each of us can be stewards of Canada's wild species and natural spaces.

Chapter Questions

1. Review the ways in which habitat alteration can occur. In your neighbourhood, identify examples of current habitat alteration. Which types or sources of habitat alteration do you think are most important in your region? Justify your choice.

2. Go to your backyard (or the backyard of a friend). What plants or animals can you identify? What plants or animals are indigenous to the bioregion in which you live? What ones are introduced? What do you need to know to bring more indigenous species back to the yard?

3. Discuss the reasons why saving larger tracts of habitat are necessary to protect (endangered) species. Does "saving" mean not allowing any human activity? If not, what types and levels of human activities are acceptable? What types and levels are not?

4. Using Environment Canada's lists of Canadian species at risk (check the URLs in the References section) and other sources that are available to you, try to identify examples of species that are at risk in the region where you live. Are there any activities underway to try to protect, preserve, or rehabilitate these species and their habitats? If so, who is undertaking these efforts, what specific actions are being taken, and what is the rationale for these actions? If not, describe what you think needs to be done to achieve protection for these species.

5. If you are "pro-nature," does that mean you are "anti-development"? Discuss your reasoning.

6. Discuss the pros and cons of in situ and ex situ conservation approaches. In what circumstances is ex situ conservation justified?

7. In what ways would modifying your consumption habits help protect wild species and natural spaces?

references

Banff–Bow Valley Study. (1996). *Banff–Bow Valley: At the crossroads*. Technical Report of the Banff Valley Task Force (R. Page, S. Bayley, J. D. Cook, J. Green, & J. R. B. Ritchie). Ottawa.

Bayly, S., & Ives, S. (1997). Seeing the forest for the jobs. *Encompass, 1*(3), 6–7.

Bishop, C., et al. (1994). *A proposed North American amphibian monitoring program*. http://www.im.nbs.gov/amphib/naampneeds.html

Brooks, R. J., & MacDonald, C. J. (1996). Ranid population monitoring in Algonquin Provincial Park. *Froglog, 16* (February). http://acs-info. open.ac.uk/info/newsletters/FROGLOG-16-5.html

Campbell, A. (1999). Declines and disappearances of Australian frogs. *Environment Australia,* 229. http://www.environment.gov.au/bg/threaten/information/frogs/frogs.pdf

Canadian Geographic Enterprises. (1997). *Wildlife at risk.* Vanier, ON: Author.

Canadian Wildlife Federation. (1997). *Diamond mining in the Northwest Territories.* http://www.cwf-fcf.org/current/curmain.htm

Canadian Wildlife Service. (2000). *Species at risk.* http://www.speciesatrisk.gc.ca/Species/English/SearchDetail.cfm?SpeciesID=27

Committee on the Status of Endangered Wildlife in Canada. (2000a). *Canadian species at risk, May 2000.* 23 pp. http://www.cosewic.gc.ca/COSEWIC/Default.cfm

Committee on the Status of Endangered Wildlife in Canada. (2000b). *COSEWIC species assessments—November 2000.* http://www.cosewic.gc.ca/COSEWIC/Default.cfm

Dearden, P., & Dempsey, J. (2004). Protected areas in Canada: A decade of change. *The Canadian Geographer, 48*(2), 225–239.

Declining Amphibian Populations Task Force. (n.d.). *What are amphibian declines and their causes?* http://www.open.ac.uk/OU/Academic/Biology/J_Baker/DAPTF.What_are_ADs.html

Declining Amphibian Populations Task Force. (2000). Home pages. http://www.open.ac.uk/daptf/index.htm

Drost, C. A., & Fellers, G. M. (1996). Collapse of a regional frog fauna in the Yosemite area of the California Sierra Nevada, USA. *Conservation Biology, 10,* 414–425.

Dunn, P. (1996). The need for amphibian monitoring in protected areas. *Research Links, 4*(3), 4, 10.

Eidsvik, H. (1989). Canada in a global context. In M. Hummel (Ed.), *Endangered spaces: The future for Canada's wilderness* (pp. 30–45). Toronto: Key Porter Books.

Engman, K. (1997, April 21). Fewer jobs in forestry than in parks—study. *Calgary Herald,* p. A4.

Environment Canada. Canadian Biodiversity Network. (2000). Home page. http://www.cbin.ec.gc.ca/cbin/HTML/en/default.cfm

Environment Canada. Canadian Wildlife Service. (1997). Endangered species in Canada. http://www.ec.gc.ca/cws-sct/es/default.htm

Fenge, T. (1996). The Arctic comes in from the cold. *World Conservation, 3,* 11–12.

Government of Canada. (1996). *The state of Canada's environment—1996.* Ottawa: Supply and Services Canada.

Hackman, A. (1996). Protecting wild places. *World Conservation, 3,* 13–14.

Henry, J. D. (1994, August). Home again on the range. *Equinox, 76,* 46–53.

Hummel, M. (Ed.). (1989). *Endangered spaces: The future for Canada's wilderness.* Toronto: Key Porter.

National Parks Service. (2000). *Yosemite Valley Plan: Final.* Vol. II. Existing Environments Chapter 4, Part 1. Table K 2. http://www.nps.gov/yose/planning/yvp/seis/vol_II/appendix_k_c4p1.html

Natural Resources Canada. (2002). *Northwest Territories mining facts: Minerals and mining statistics on-line.* http://www.mmsd1.mms.nrcan.gc.ca/mmsd/facts/canfact_e.asp?regionid=11

Nature Trust of British Columbia. (1993). PCEP purchase of Englishman River Flats "jewel" triggers creation of rare wildlife management area. *Natural Legacy, 6,* (Summer), 1–3.

Noss, R. (1995). *Maintaining ecological integrity in representative reserve networks.* Toronto: World Wildlife Fund Canada/World Wildlife Fund United States.

Oil exploration in the Arctic. (1997). *Canadian Wildlife Federation Bulletin, 6,* 7.

Orchard, S. (n.d.). *Why are most amphibians declining?* http://www.cuug.ab.ca:8001/~animal/AD.html

Parks Canada Agency. (2000). *Unimpaired for future generations? Protecting ecological integrity within Canada's national parks: Vol. 1, A call to action. Vol. 2, Setting a new direction for Canada's national parks.* Report of the Panel on the Ecological Integrity of Canada's National Parks. Ottawa.

Parks Canada Agency. (2001). *Parks Canada: First priority-progress report on implementation of the recommendations of the Panel on the Ecological Integrity of Canada's National Parks.* Ottawa: Public Works and Government Services Canada.

Penstone, S. (1997). Seedy Saturday. *Environment Views/Environment Network News, 1*(1), 14.

Pounds, A. J., et al. (1999). Biological response to climate change on a tropical mountain. *Nature, 398,* 611–615.

Protecting Canadian species. (1996). *World Conservation, 3,* 9.

Robinson, F. E. (1997). Where have all the chickens gone? *Environment Views/Environment Network News, 1*(1), 15.

Rowe, S., Kavanagh, K., & Iacobelli, T. (1995). *A protected areas gap analysis methodology: Planning for the conservation of biodiversity.* Toronto: World Wildlife Fund Canada.

Searle, R. (2000). *Phantom parks: The struggle to save Canada's national parks.* Toronto: Key Porter Books.

Sierra Club of Canada. (2000). *Eighth annual Rio report card.* http://www.sierraclub.ca/national/rio/rio00-summary.html

Sierra Club of Canada. (2003). *Rio + 11: The eleventh annual Rio (Report on International Obligations) report card, 2003 – summary of grades. Grading the Government of Canada and the provinces on their environmental commitments.* http://www.sierraclub.ca/national/rio

Statistics Canada. (2000). *Nominal catches and landed values of fish.* http://www.statcan.ca/english/Pgdb/Economy/Primary/prim45.htm

Statistics Canada. (2003). *Human activity and the environment, annual statistics 2003.* Ottawa: Ministry of Industry.

United Nations Environment Programme–World Conservation Monitoring Centre (UNEP-WCMC). (2001). *About UNEP-WCMC.* http://www.wcmc.org.uk /reception/whoare.htm

United States House of Representatives. (2000, September 29). *Committee on Resources: ANWR oil and natural gas vital to resolving US energy crisis.* http://www.house.gov/resources/press/2000/20000929anwrvital.htm

Wallis, C., Klimek, J., & Adams, W. (1996). Nationally significant Sage Creek grassland threatened by Express Pipeline. *Action Alert, 7*(4), 4 pp.

World Commission on Environment and Development. (1987). *Our common future.* Toronto: Oxford University Press.

World Conservation Monitoring Center (WCMC). (2000). *State of the world's protected areas.* Annex 1: Summary of Protected Areas. http://www.unep-wcmc.org/protected_areas/albany_a1.pdf

World Wildlife Fund Canada. (1999). *Endangered Spaces campaign: 1998–1999 Endangered Spaces progress report on Canada's wild lands, Number 9, British Columbia.* www.wwfcanada.org

World Wildlife Fund Canada. (2000a). *Species at Risk campaign.* http://www.wwfcanada.org/

World Wildlife Fund Canada. (2000b). *Endangered Spaces campaign.* http://www.wwfcanada.org/

World Wildlife Fund Canada. (2000c). *Endangered Spaces campaign: 1998/1999 Endangered Spaces progress report on Canada's wild lands, Alberta.* http://www.wwfcanada.org/

World Wildlife Fund Canada. (2000d). *Endangered spaces.* Toronto: World Wildlife Fund.

Wristen, K. (2003). *Diamonds and sustainable development?* Adapted presentation to the workshop for sustainable development in the diamond mining sector, hosted by Environment Canada, Ottawa, June 5–6, 2003. http://www.carc.org/2003/diamonds_and_sustainability.htm

Zolkewich, S. (1995). Leopard frogs abound at DU project. *Ducks Unlimited Canada Conservator, 16*(1), 3.

Getting to Tomorrow

CHAPTER 13

Sustainability and the City

Chapter Contents

Chapter Objectives

After studying this chapter you should be able
to

- understand the main issues and concerns
  relating to urbanization and ecosystems in
  Canada

- identify the effects of urbanization on the
  environment around us

- appreciate the complexity and interrelatedness of environmental, social, and economic
  issues relating to urbanization and their
  effects on our environment

- appreciate the effects our lifestyle choices
  have on our environment

- understand the types of efforts Canadians
  have made toward more sustainable communities

INTRODUCTION

The issue of urbanization and sustainable cities has emerged as one of the most critical environmental and developmental challenges of the new millennium. As we enter the 21st century, more than half the human population lives in urban areas. Although Canada's cities occupy only about 0.2 percent of the country's total land area, already almost 80 percent of Canadians live in communities of over 1000 persons. We often seem to forget that our cities are part of the ecosystem and that urbanization has had a high impact on productive land, aquatic systems, forested lands, and other valued components of regional and local ecosystems. In the 21st century we will continue to face challenges in balancing the demands of our urban lifestyles and the subsequent threat to ecosystem health thresholds.

In this chapter we identify some of the major kinds of effects our urban ways of life have on the environment, consider what efforts have been made to take the sustainability of the environment into account in urban planning, and comment on future directions for more sustainable communities.

URBAN ENVIRONMENTAL CONDITIONS AND TRENDS

In 1999, for the seventh consecutive year, Canada ranked number one among all countries in the world on the United Nations' Human Development Index (HDI). In 2002, Canada slipped to third place, and in 2003 ranked eighth. In 2004, Canada achieved a fourth-place ranking, this time achieving our highest-ever absolute score (0.943/1.000). First constructed in 1990, the HDI measures average achievements in basic human development in one simple composite index and produces a ranking of countries. This index measures people's well-being according to three indicators: life expectancy, knowledge (derived from adult literacy and mean years of schooling), and standard of living (the per capita gross national product adjusted for the local cost of living) (United Nations Development Program, 2000).

As useful as they are, neither the HDI nor other similar measures take into account the environmental stresses that are involved in maintaining this high quality of life. When we realize that Canadians use water and energy and generate household garbage at rates similar to those of residents in other developed countries, and when we realize that lifestyle choices are influential elements in achieving environmental sustainability, the importance of new technologies as well as new attitudes becomes clear. A high quality of life with reduced waste, energy use, and pollution is both a goal and a challenge that can be assisted by new technologies that improve the resource and energy efficiencies of our activities. At the same time, attitudes that espouse less wasteful choices and less stress on the environment are necessary. This is apparent when we acknowledge that environmental factors as well as economic (material) and health factors contribute to our high quality of life.

If we consider environmental, material, and health factors, we realize that Canadian urban areas create environmental stresses that have interrelated and cumulative effects on atmosphere, water, energy, materials, and land, on both regional and global scales. These and other stresses are discussed briefly below, as are illustrations of efforts being made to enhance sustainability of our communities.

ATMOSPHERE AND CLIMATE

Since our ancestors learned how to use fire, humans have been adding pollutants to the atmosphere and affecting its quality. Today, as the threat of global warming is realized, not even climate is immune from human influence. Cities, because of their structure and form, and because of the results of various activities taking place within them, modify the local climate and create their own microclimates.

Microclimate

Five main factors shape a city's microclimate. These are the storage and reradiation of heat by buildings and streets, reduction of wind speed (which reduces the wind's cooling effect in summers), human-made sources of heat, rapid runoff of precipitation (which reduces the cooling effect of evaporation), and the effects of atmospheric pollutants. Most of these effects are the result of the loss of vegetation and the creation of landscapes made of nonpermeable materials. The net result of the influence of these factors is the urban **heat island** effect, in which temperatures are one to two degrees Celsius higher in the city than in the surrounding rural area. Two implications of the heat island effect are that cities need less energy for heating but more for cooling (see Box 13–1), and that in using cooling devices that contain chlorofluorocarbons (CFCs) or CFC substitutes, we generate emissions with known (and unknown) impacts on the ozone layer.

Air Quality

Although many sources of air pollution are located outside cities (refineries, pulp mills, forest fires), there is a concentration of many different pollution sources in cities. In 2000, for example, air emissions in the Greater Vancouver Regional District (GVRD) totalled 525 000 tonnes, including 300 000 tonnes of carbon monoxide (CO) (21 percent less than 1990), 75 000 tonnes of volatile organic compounds (VOCs) (12 percent less than 1990), 54 000 tonnes of nitrous oxides (NO_x) (the same as in 1990), 8500 tonnes of sulphur oxides (SO_x) (an increase of 6 percent from 1990), 88 000 tonnes of particulate matter (26 percent increase over 1990), and many other hazardous air pollutants such as benzene and lead. Figures from the GVRD demonstrate that air pollution in the area is decreasing for some constituents and increasing for others. Figures for total pollutants are down 13 percent from 1990 levels (City of Vancouver, 2002). Almost 80 percent of this pollution was produced by motor vehicles. One of the challenges facing policymakers is that even though motor vehicles are less polluting than they were a decade or two ago, there are more vehicles on the road due to local population increases and the fact that there are more vehicles being used on a per capita basis.

The wind cannot always disperse the pollutants produced in the Greater Vancouver area; the result is **photochemical smog** that can stretch more than 50 kilometres out to Abbotsford and beyond. This smog can have serious health effects on humans, wildlife, and livestock, and damage natural vegetation and buildings (recall Box 5–5). Asthmatic attacks can worsen, risk of contracting respiratory diseases such as bronchitis can increase, and the danger of developing certain types of cancer can increase as well (Greater Vancouver Regional District, 1994). Although there is the immediate benefit of convenience when we hop into a car and drive wherever we want to go (rather than face the inconvenience of waiting for the bus), we are beginning to appreciate that the results of our personal choices have diffuse and sometimes hidden environmental costs (such as in diseases related to air pollution).

Canada's National Air Pollution Surveillance (NAPS) network measures atmospheric levels of five common pollutants in cities. These pollutants are sulphur dioxide (SO_2), suspended particles, ground-level ozone, carbon monoxide (CO), and nitrogen dioxide (NO_2). Atmospheric pollutant data from NAPS are compared with the National Ambient Air Quality Objectives (NAAQOs) set out in the Canadian Environmental Protection Act.

The NAAQOs define the maximum desirable, maximum acceptable, and maximum tolerable levels for each pollutant. Maximum desirable levels specify the long-term goal; maximum acceptable levels specify those at which there is adequate protection of human comfort and well-being, as well as adequate protection of soils, water, and vegetation; and maximum tolerable levels are those

Cities create heat, and Toronto is heating up. Since 1975, summer temperatures in the city have increased steadily. With the rising temperatures come health risks to residents. These include:

- heat-related illness and mortality during heat waves, especially for Toronto seniors
- exacerbation of pulmonary disease due to concentrations of ozone and particulate matter
- increased risk of some infectious diseases, such as encephalitis, which spread more readily in hotter weather

The City of Toronto Health Department and the Toronto Atmospheric Fund (TAF), with the financial assistance of the Government of Canada Climate Change Action Fund (CCAF), have embarked on a project to put in place municipal policies and practices that will help protect Torontonians from the adverse effects of summer heat. The project has three parts.

Part 1: A Monitoring and Alert System

The Heat–Health Watch/Warning System, tailored to Toronto's unique climate, is based on an analysis of climate and mortality data for Toronto. It provides 48–60 hours' notice of the arrival in the city of an oppressive air mass—an air mass associated with predicted or actual morbidity and mortality rates well above the mean value for a given period. Developed by Laurence Kalkstein of the Center for Climatic Research at the University of Delaware, the system has been in place for several years in Philadelphia and Washington, D.C. In those two cities the system is credited with saving hundreds of lives during heat waves. Toronto's Health Department uses the system to implement mitigation plans based on estimates of the predicted number of people at risk of illness and death, a number that varies according to the number of consecutive days of the oppressive air, the time of year, the minimum temperature, and other criteria.

As an additional part of the project, the City of Toronto became a participant in the United Nations Showcase Project, along with Rome and Shanghai and other cities, to develop this system for vulnerable cities around the world.

Part 2: Longer-Term Adaptation

The second part of the project focuses on mitigating the heat island effect in Toronto through the use of lighter-coloured sur-

faces on streets and buildings and the strategic placement of urban vegetation. Toronto is warmer than surrounding rural areas due to the urban heat island effect. The city's dark surfaces and infrastructure amplify the heating capacity of incoming solar radiation. Over the long term, municipal policies and measures that encourage urban reforestation and more reflective roofs and streets can cool ambient temperatures, creating more healthful microclimates for people, reducing the hot air that air conditioning pumps into the environment, and ameliorating heat-induced smog levels (see Enviro-Focus 13 for information on "green roof" infrastructure).

This part of the project will provide greater scientific understanding of Toronto's heat island and the benefits of mitigation strategies. To this end the project will:

- quantify the direct and indirect benefits of shade trees, vegetation, and reflective (high-albedo) surfaces for several residential and commercial building types, using computer simulations of typical Toronto days in various seasons and under various conditions
- conduct simulations that quantify the impact of heat island mitigation measures on building energy use and on Toronto's smog
- develop the scientific basis for new municipal practices and policies that will help cool the city and reduce smog in the long term

Part 3: North American Summit on Urban Adaptation

The third part of the project aims to increase understanding of climate change on urban health with a conference on the urban heat island. In May 2002, leading scientists from Canada and the United States were invited to a conference to present the scientific basis of intervention and mitigation strategies, and municipal representatives from North American cities were asked to describe effective implementation practices.

The summit provided a status report on the latest research related to urban heat islands. It explored how research can inform policy and practices designed to reduce and respond to extreme summer heat. It also identified challenges and barriers to change, as well as best practices. Finally, the summit examined methods of addressing urban heat island and health issues.

SOURCE: Toronto Atmospheric Fund, 2002, http://www.city.toronto.on.ca/cleanairpartnership/uhis_summit.htm. Adapted with permission from Eva Ligeti, Manager, Cool Toronto Project. Telephone 416-392-1220.

---

beyond which action is required to protect human health. The NAAQOs do not include standards for greenhouse gases such as carbon dioxide that contribute to global warming.

Environment Canada indicated that, from 1980 to 2000, average levels of most pollutants measured by the NAPS network (calculated as a percentage of the respective maximum acceptable levels) declined: CO by 19 percent, SO_2 by 18 percent, NO_2 by 18 percent, ground-level ozone concentrations by 22 percent, and airborne particles by 33 percent (Government of Canada, 2003).

The number of hours when pollution exceeded maximum acceptable levels declined for each of SO_2, CO, airborne particles, and, to a lesser extent, NO_2. Much of this reduction was achieved through the use of catalytic converters in cars, cleaner-burning engines, cleaner gasoline,

Photo 13–1

Motorcyclists wear masks in Pontianuk, on the Indonesian island of Kalimantan, in September 1997, to filter the polluted air caused by forest fires on neighbouring islands.

more and better scrubbers on industrial smokestacks, and greater overall energy efficiency. Generally, air quality is good in Canadian cities for most of the year, and from 1979 to 1996, air quality in the larger Canadian cities (where vehicles are a significant source of air pollution) improved noticeably. Nevertheless, in cities such as Vancouver and Toronto, summers continue to be characterized by smog alerts.

While very small airborne particles may be cause for concern about human health (because they are thought to contain unburned pieces of carbon originating from fossil fuel combustion in vehicles and heating of buildings), ground-level ozone is Canada's most serious urban air pollution problem (see Box 13–2). The seriousness of the problem varies regionally across the country, depending on each city's land use, industrial base, commuting pat-

terns, topography, weather conditions, prevailing winds, and location relative to other sources of ozone. Given these factors, settled rural areas downwind from large cities generally experience higher average annual ozone levels than do urban areas themselves.

As we know from previous chapters, not all air pollution in every Canadian city is derived entirely from activities taking place within the city. Air quality in some Canadian cities is affected by the long-range transport of pollutants from other areas, principally the United States. The Lower Fraser Valley (affected by Vancouver-area ozone), the Windsor–Quebec corridor (affected by local and U.S. Great Lakes and Midwest sources), and the Fundy region of southern New Brunswick and western Nova Scotia (affected by sources in the northeastern United States) are the three ozone problem areas in Canada (Canadian Council of Ministers of the Environment, 1990). Neither Victoria, British Columbia, nor Prairie cities have the density of vehicles or the prevailing winds from industrial areas needed for ozone problems to develop. Figure 13–1 illustrates, for selected Canadian cities, the number of hours air-quality objectives for ground-level ozone were exceeded from 1985 to 1994.

As a precautionary measure, other toxic chemicals in the air such as benzene (a VOC) are being monitored in large urban areas. Benzene, a known carcinogen that initiates tumours and has been linked to a specific form of leukemia, has been monitored since 1989. Transportation sources account for more than 85 percent of the benzene released into the atmosphere in urban areas. However, since 1989, average airborne benzene concentrations have fallen by one-third, due largely to better emission controls on vehicles and more efficient engines. Although Canada does not have a health standard for airborne ben-

---

B O X 1 3 – 2
GROUND-LEVEL OZONE

While stratospheric ozone serves an important protective function, tropospheric or ground-level ozone is harmful to plants and animals because of its extremely strong oxidant properties. Ground-level ozone is considered the criterion for measuring photochemical smog. It is the component most responsible for smog-related respiratory problems and eye irritations.

Ground-level ozone is produced by a series of chemical reactions involving hydrocarbons, nitrogen oxides, and sunlight. Almost all anthropogenic NO_x arrives in the troposphere as automobile exhaust, the result of incomplete combustion of fossil fuels in internal combustion engines. NO_x and sunlight react to produce ozone. Because of the reaction's dependence on sun-

light, ground-level ozone problems are most severe during Canadian summers.

In December 2000, the United States and Canada committed to vigorous reductions in transboundary NO_x emissions as part of the Ozone Annex of the Canada–United States Air Quality Agreement. Canada will implement new regulations for vehicle and fuel standards that are aligned with more exacting U.S. standards, and the new Canada-Wide Standard for Ozone. Projections are that Canada's NO_x emissions in the transboundary area (which includes the Windsor–Quebec corridor) will be reduced by 44 percent by 2010.

SOURCES: *Environmental Chemistry* (6th ed.), S. E. Manahan, 1994, Boca Raton, FL: Lewis Publishers; *SOE Bulletin* No. 99-1, 1999, Environment Canada. Reprinted with permission of the Minister of Public Works and Government Services Canada, 2004.

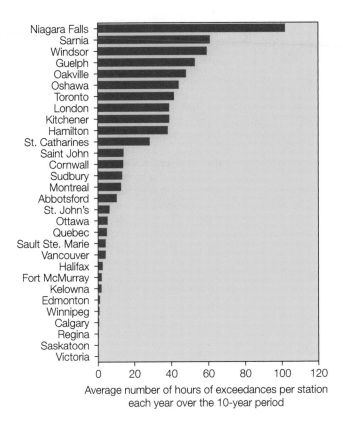

Average number of hours of exceedances per station
each year over the 10-year period

Figure 13–1a

Ground-level ozone in selected Canadian cities, 1985–1994:
Average number of hours in excess of the maximum acceptable
level

Note: Refers to the average number of hours in excess of the maximum acceptable level at stations within each urban area. Values in
centres with more than one station have been averaged for each
year first. Stations in rural areas or small towns, and with insufficient records over the 10-year period, have not been included.
Measurements are generally taken hourly, April through September.
Measurements have been normalized to 100% of readings per station to compensate for missing readings, for each month during
this period each year. Maximum acceptable level was 82 ppb (1h).

SOURCE: *The State of Canada's Environment—1996,*
Environment Canada, Figure 12.10. Reprinted with permission of
the Minister of Public Works and Government Services Canada,
2004.

zene, and very little is known about the human health
effects of prolonged exposure to trace amounts of benzene in city air, breathing city air is the most common
form of exposure. (This excludes cigarette smoking,
which is by far the principal source of adult exposure to
benzene.) New regulations designed to reduce the
amount of benzene in gasoline to less than 1 percent by
volume were announced in November 1997. These regulations, effective July 1, 1999, were predicted to reduce
the amount of benzene released to the air by 3000 tonnes
per year (Environment Canada, 2001b). Indeed, ambient
benzene concentrations in air have decreased nationally

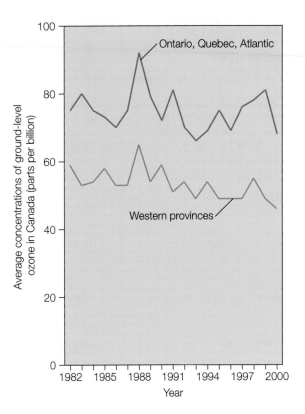

Figure 13–1b

Average concentrations (in parts per billion) of ground-level ozone
in Canada, 1982–2000

SOURCE: National Air Pollution Surveillance Network.

by almost 70 percent since 1992. They have declined by
45 percent since 1998, the year prior to the regulation
(Environment Canada, 2001c).

But as long as car ownership and use continue to rise
and traffic congestion continues to grow (at a pace that
outstrips per vehicle reductions in energy use and airborne emissions), it is unlikely that overall levels of air
pollution will continue to decline as rapidly as they have
during the past decade.

NOISE

Modern society is getting louder. From high-powered
stereo systems in cars to leaf blowers, vacuums, dish-
washers, highway traffic, personal watercraft, helicopters,
snowmobiles, "surround-sound" big-screen TVs, and
portable stereos and earphones, the Canadian population
is exposed to more noise than ever before. Noise-induced
hearing loss is the most common occupational health
hazard in industry today, affecting more men than women.
The British Columbia Workers' Compensation Board
stated that one-quarter of all B.C. workers are exposed to
occupational noise loud enough to damage their hearing
(Workers' Compensation Board of B.C., 2003).

One of the factors leading to increased hearing loss is the ability to make much more powerful sound equipment; the sound at rock concerts today is more powerful than it was 20 or 30 years ago. Rock concerts frequently are measured at 110 to 128 decibels—about the same level as a pneumatic drill or military jet, respectively. The higher the decibel level, the less time it takes before sound receptor cells start dying and permanent hearing damage occurs. At 130 decibels, after 75 seconds, you are at risk for permanent damage to your hearing; at 135 decibels, exposure time for permanent damage drops to 37.5 seconds (Shideler, 1997). Sound levels from car stereos have been measured at up to 138 decibels. In addition to loudness, both the length of exposure and the proximity to the source increase the damaging effects.

Most noise-induced hearing loss is preventable with proper use of protective ear devices. However, because hearing damage may not show up until years later, promoting safe hearing is a "tough sell," particularly among high school students, and especially when advertisers encourage young adults to play music and video games at loud levels. Even though the idea of earplugs at rock concerts has little appeal, Hearing Education and Awareness for Rockers (H.E.A.R.) worked with radio stations around the United States and Canada to give out over 60 000 earplugs during the 1996 Lollapalooza Tour, the first major music tour ever to give out earplugs. Some rock bands now sell ear protectors at their concerts (see Box 13–3).

Noise, or unwanted sound, "is the 'silent' environmental issue of the 1990s" (City of Vancouver, 1997). Second-hand noise (experienced by people who did not produce it), just like second-hand smoke, can have negative effects on people without their consent. Although people do not bleed, limp, or get sick as a result of noise exposure, there is evidence that noise causes increased stress and blood pressure levels, disrupted sleep patterns, altered heart function, and difficulties in concentrating (Fried, 1996; Patterson, 1995). Hearing loss caused by noise and the natural aging process is cumulative over a lifetime, so people over 50 years of age are impacted particularly severely.

Although noise pollution often is not thought of as an air quality problem, particularly in comparison with the ozone hole or the greenhouse effect, it is an increasingly serious problem for urban populations. Warning labels on noisy appliances are virtually nonexistent, and while we are admonished about wearing safety equipment while riding a bike we rarely think to wear earplugs while mowing the lawn. Other noise issues in rural areas, such as the concern of the Innu in Labrador and Quebec regarding noise impacts on wildlife and human health from low-level military training flights over their traditional lands, or the concerns of backcountry hikers about noise from helicopters ferrying tourists into prime wilderness areas, also are increasing (Canadian Environmental Assessment Agency, 1995).

WATER

Cities and the Hydrologic Cycle

The relationship of cities to the hydrologic cycle is a good illustration of how cities are linked to the larger ecosystem. Generally, a city withdraws water (which may be contaminated) from a lake or river and treats the water to make it potable. As it is used, the water receives pollutants, including human wastes, that require the water to be treated again prior to its return to the hydrologic system (where natural processes further clean the water). Downstream, other communities that depend on the river or lake for their water supply put the water they withdraw through a similar succession of treatment processes. Some, perhaps, discharge untreated wastewater. Eventually the river carries the water and any remaining pollutants into the ocean. There, the natural processes of evaporation, transportation (as clouds), and precipitation continue the hydrological cycle (recall Figure 3–16).

Water Supply and Water Quality

Surface water supplies most Canadian cities, although nearly 10 percent of the population is served by municipal systems that rely on groundwater. As we saw in Chapter 7, supplies of both surface water and groundwater are susceptible to problems of availability and quality. Residents of cities that rely on surface water, such as Victoria, sometimes face seasonal and temporary shortages of or restrictions on water use, whereas residents of cities that rely on groundwater, such as in Prince Edward Island and southern Ontario, may face the possibility of a long-term decline in supply. In Kitchener–Waterloo, Ontario, for example, the search for alternative water supply sources has led to consideration of a 120 kilometre water pipeline from Georgian Bay, part of Lake Huron. Cities such as Regina, where supply sources of surface water are of questionable or declining quality, also may have to find a new supply source or raise the treatment levels of the water they use.

Many sources of contamination may affect both surface water and groundwater supplies: drainage or seepage from industrial, commercial, residential, and recreational land uses, including waste disposal sites; runoff or seepage from chemical fertilizers and other agricultural chemicals; spills and discharges from shipping; and deposition of atmospheric pollutants. The Walkerton case is instructive here: to be safe for human consumption, water that is likely to be affected by such contamination must be filtered and treated chemically before it enters a city's distribution system.

Although water quality in Canadian cities is generally good, quality does differ across the country since it is regulated at the provincial rather than the national level. On August 26, 2000, the province of Ontario's new Drinking Water Protection Regulation came into effect (Government of Ontario, 2001). The regulation established a new Ontario Drinking Water Standard (ODWS) that is legally enforceable. As per the regulation, the City of Toronto has committed to issuing quarterly reports on water quality. (For additional information on new provincial and territorial water quality initiatives, see Chapter 14.)

While urban water quality may be good generally, there are numerous instances in which water treatment plants serving residents of smaller Canadian communities either do not meet provincial guidelines on treated water quality or on treating bacteria, or have not performed sufficient testing for bacteria and toxic chemicals to determine if the guidelines have been met. Even in some of the largest communities—Vancouver, for instance—tap water is safe to drink but water treatment improvements are needed to provide the best water quality. Construction of new water treatment facilities in Greater Vancouver began in the fall of 1997. New facilities include secondary disinfection stations, upgrading of chlorination and corrosions systems, and a new ozone disinfection facility. The improvements will address four specific areas of concern to health: inadequate disinfection of water sources; seasonal coliform bacteria; seasonal turbidity; and natural acidity or corrosiveness (Greater Vancouver Regional

Photo 13–2a

Photo 13–2b

Leaking water supply systems and overwatered lawns and gardens create economic and environmental costs for Canadians.

District, 2000). Despite the recent improvements, turbidity remains a problem. The issue of turbidity is of concern to public health officials because it is associated statistically with increased complaints of gastrointestinal illness.

Water Use and Wastewater Treatment

Over 10 percent of the water withdrawn from natural sources in Canada supplies municipal systems used by residents, businesses, and some industries. More than half the water in municipal systems supplies residential consumption; between 1989 and 1996, residential consumption increased by 103 percent! As noted in Chapter 7, most water within the home is used in the bathroom, but on peak days in the summer, lawn and garden watering and car washing can drive water use up by 50 percent. This is just one way in which Canadians have become the second highest per capita consumers of water in the world.

Our low water prices and flat-rate pricing contribute to our profligate use. Since water consumption declines as its cost increases, it may be possible to reduce consumption by having water charges reflect the amount of water used. In the past, for example, when most Calgary residents paid a flat rate for their water, they consumed up to 60 percent more water per capita than did residents of Edmonton, a similar-sized city that used water meters to charge for water according to the volume people used. Since 1989, however, Calgary waterworks officials have been encouraging reduced wastage through a water meter incentive program. Through the program, Calgary Waterworks personnel install a water meter with a remote readout in single-family homes. Residents are billed for

Photo 13–3
The effects of localized urban flooding may range from the inconvenience of being unable to use a pathway to paying for costly repairs to residences and public property.

water and sewer charges for 12 consecutive months based on meter readings. After 12 months, a comparison is sent to the resident outlining the difference between the metered charge and the flat rate charge for the period. If the flat rate is more economical, residents can revert back and the difference between the rates is credited to their account. If the metered service proves to be more economical, the resident continues to be billed on the amount of water used. Overall, the percentage of Canada's municipal population with water meters increased from 52 percent in 1991 to 54 percent in 1994 (Environment Canada, 2001a). In addition, many cities have imposed by-laws requiring that residents not water their lawns and only use water for gardens or car washing on alternate days.

The construction of cities affects the water cycle, which, in turn, affects soils, plants, and animals. Cities may receive 5 to 10 percent more rainfall than the surrounding areas because the particulates and dust above cities provide nuclei for condensation of raindrops. But the concrete, stone, and asphalt streets, and other impervious building materials, prevent water infiltration and induce rapid runoff directly into stormwater systems. Local flooding events may increase, and the frequency of downstream flooding may increase also. It is estimated that because of the surfaces such as pavement and rooftops, there is nine times more runoff from a city block than from a woodland area of the same size (Taus & McClure, 2002). In addition, hard surfaces prevent water in the soil from evaporating; in natural ecosystems evaporation is an important process that cools the surface. Changes in the built form of the city and the activities within a city increase the runoff and the pollutant loads that are introduced into urban waterways.

Most of the water used in urban areas is employed to remove domestic, industrial, and commercial wastes, including human sewage. Treatment to remove impurities is needed to safeguard human health, but the large and increasing volumes of treated wastewater may stress aquatic ecosystems. Some cities such as Vancouver do not separate their storm and sanitary (sewage) waters. Even if separate storm and sewer systems exist, heavy storm drainage flows may overload a treatment plant's capacity and cause either or both storm and sewage discharges to leave the plant untreated or insufficiently treated.

The percentage of Canadians served by municipal wastewater treatment plants has been increasing steadily, but the level of treatment varies widely across regions (see Figure 13–2a). For instance, many coastal cities as well as those on the lower St. Lawrence River discharge their wastewater directly into oceans or rivers. Most Ontario cities discharge into smaller rivers or the Great Lakes, while Prairie cities discharge solely into river systems. By 1999, about 71 percent of Ontario residents and about 75 percent of residents in Alberta and Saskatchewan living in communities with treatment sys-

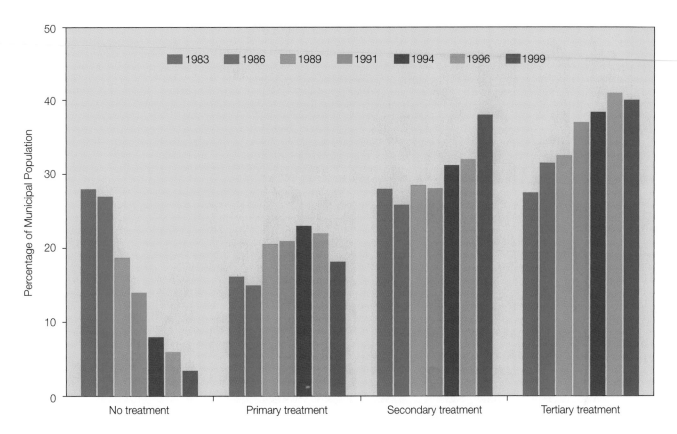

Figure 13–2a

Municipal population served by type of sewage system in Canada

SOURCE: *The State of Canada's Environment—1996,* Environment Canada, Figure 12.13; Municipal Water Use Database (MUD) Survey, 2000, Environment Canada. Reprinted with permission of the Minister of Public Works and Government Services Canada, 2004.

tems were served by tertiary treatment systems (see Figure 13–2b). The high level of tertiary treatment in Ontario is due to provincial regulations that require phosphate removal systems, while on the Prairies, the highest level of treatment is in Alberta. Manitoba reported no tertiary treatment in 1999. Nevertheless, by 1999, virtually all residents living in the Prairies, 94 percent of Ontario residents, and 78 percent of the Canadian municipal population were served by secondary and/or tertiary treatment facilities.

In contrast, a minimal percentage of the municipal population in the Atlantic region (less than 1 percent of Nova Scotia's population) was served by tertiary treatment in 1999. Large volumes of wastewater, including sewage from cities such as St. John's, Newfoundland, and Halifax, Nova Scotia, are discharged into the ocean untreated or after only primary treatment. As noted previously in Chapters 7 and 8, such practices have adverse effects on local shellfish grounds, coastal wetlands, and shorelines.

In Nova Scotia, 30 percent of all sewage generated is released untreated into the environment, while another 45 percent is treated by household sewage disposal sys-

tems of varying quality. Inadequacies have been identified with the treatment facilities that handle the remaining 25 percent. Halifax discharges sewage directly into its harbour, and has made many attempts during the past two decades to complete a sewage treatment system. Only recently are citizens of the Halifax Region seeing a regional strategy implemented. Halifax Regional Council has approved the Halifax Harbour Solutions project, a $315 million, 10-year undertaking to eliminate the flow of raw sewage and other contaminants into the harbour with a four-treatment-plant plan (City of Halifax, 2001).

Water and Recreation

Recreational pursuits—even those that seem environmentally friendly—can have effects on the environment. These impacts range from use of resources such as land for golf courses and ski areas, to chemicals used to purify water in swimming pools, to energy consumed and emissions produced by the vehicles that get us to the cottage or recreational event. Clearly, recreational pursuits are important in Canadians' health and quality of life. It is

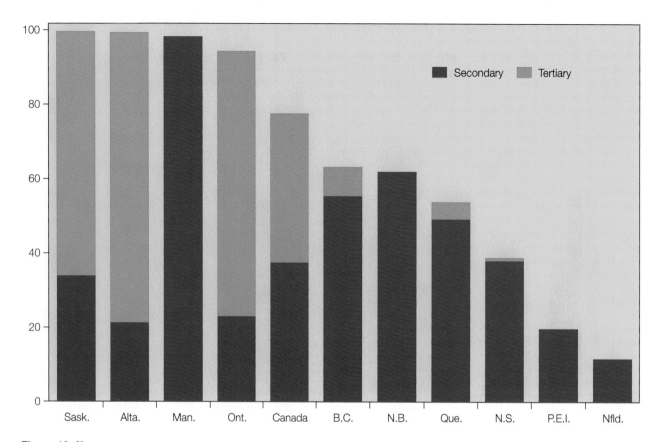

Figure 13–2b

Total population served by secondary and tertiary wastewater treatment in Canada, 1999

NOTE: Insufficient data to include Nunavut, the Northwest Territories, and Yukon.

SOURCE: *The State of Canada's Environment—1996,* Environment Canada, 1996, Figure 12.14; Municipal Water Use Database (MUD) Survey, 2000, Environment Canada. Reprinted with permission of the Minister of Public Works and Government Services Canada, 2004.

equally important, however, that more of us learn how we can minimize the impacts of our recreational activities on the environment.

One way to consider the sustainability of our recreational activities is to examine how much energy, use of materials, emissions, and waste are involved. In general, activities that are fuel-dependent (snowmobiling, motorcycling, and power boating) are more harmful to the environment than are human-powered activities (hiking, skiing). If hikers stay on designated trails and follow other environmentally friendly practices, they can enjoy the outdoors without causing much damage to it. However, the cumulative effects of human-powered activities on the environment are potentially significant.

In some recreational pursuits, fuel-dependent activities predominate. Pleasure boating in Canada, for example, involves more than five times as many motor boats and yachts (693 500) as sailboats (124 000). Power boats are estimated to consume 1373.90 litres of fuel per vessel; a typical 5.2-metre (17-foot) runabout consumes about 25 to 30 litres of fuel per hour and averages about one to two kilometres of travel per litre of fuel (automobiles attain an average of over eight kilometres per litre)

(Government of Canada, 1996). In addition, power boats cause noise and air/water pollution; as much as 8 percent of total hydrocarbon emissions in the Quebec–Windsor corridor are estimated to be released from power boats. When aging or improperly maintained engines are included, air pollution from power boat engines may be 25 to 30 percent higher.

Water pollution may be exacerbated by power boat use in some areas, and the wakes (waves) from these vessels can disturb aquatic wildlife and erode vulnerable shorelines. The practice of releasing raw sewage from boat heads (toilets), particularly into sheltered bays and coves where boats tend to anchor and people swim and fish, also is problematic. In the past, and prior to improved regulations, houseboats on Shuswap and Okanagan lakes were among the pleasure craft that contributed to serious sewage contamination problems.

Golf courses do not rely on fuels directly, but they are often heavily reliant on pesticides, herbicides, and water to maintain their greens. In addition, more recent golf courses have been located in ski destinations to help improve prospects for year-round tourism. New courses in these locations have displaced animal habitat and have

Photo 13–4
Regulations prohibit houseboat users from discharging raw
sewage wastes directly into Shuswap Lake, B.C.

increased traffic, thereby reducing conditions favourable
to biodiversity protection.

We can choose less environmentally damaging recreational activities, and we can choose to carry out leisure
activities in less damaging ways. Doing so increases the
likelihood that we can maintain the benefits of leisure
activities at the same time as we reduce their long-term
environmental impacts.

ENERGY

Canadians use considerable energy to cope with our cold
climate, to travel the long distances between population
centres, and to satisfy our lifestyle choices (such as our
preference for detached, single-family houses). Indeed,
our overall energy consumption is not declining. Data
indicate that between 1980 and 1997, Canadian consumption of energy grew by 20 percent, slightly higher
than the 18 percent average of the OECD countries. As
noted in Chapter 11, the production and consumption of
fuels and electricity in cities lead to local and global environmental stresses. Automobile use, home heating,
resource and manufacturing industries, and other commercial enterprises cause local air pollution from emissions of NO_x, VOCs, SO_2, and particulate matter; as well,
they contribute to global warming through release of CO_2.
Ecosystems also experience stress from production,
transport, and use of energy.

While it is difficult to compile a picture of municipal
energy use in Canada (because the data are not collected
by municipalities), it is known that per capita use of
energy in the inner city is lower than in suburban areas,
and lower still than in small towns and rural areas. This
seems to suggest that, given its higher population density,
the inner city is a more energy-efficient form of settlement, although this suggestion is tempered by the fact

that many city neighbourhoods contain the poorest members of urban populations. The extensive, enclosed environments in large cities mean that people can travel from
their homes and jobs to go shopping, dining, or to attend
to business without ever going outdoors. While these environments provide convenience and comfort, they require
heavy energy consumption for heating, cooling, and ven-
tilating the system. Typically, industrial energy users do
not locate in large cities, but in Toronto and Ottawa commercial and institutional sectors account for one-third of
all energy use. This is double the national average and
suggests that these businesses are good targets for energy
management efforts.

Sustainable Housing

Examples of sustainable housing in Canada are growing.
The concept of an ecologically friendly house has
expanded beyond single dwellings and is being taken up
by co-housing units as well as on some Canadian university campuses. Co-housing, a community planning model
developed in Denmark more than 25 years ago, offers an
alternative to current housing options. Co-housing projects are cooperative neighbourhoods designed, developed, and managed with a high degree of owner/resident
participation (Kerr, 1998). Co-housing projects come in
many forms. Some communities consist of single-family
houses, although most are townhouse developments; a
few are apartment buildings.

In many cases, sustainable housing also means
affordable housing. Particularly in large urban centres
where property values are high, affordable housing is a
key issue that points to the need to pay attention to the
social dimensions of sustainability. Housing cooperatives,
Habitat for Humanity, and other initiatives provide opportunities to integrate design features that are environmentally friendly with those that are affordable, making urban
living a sustainable option for a wider range of residents.

The Alberta Sustainable Home/Office in Calgary is a
three-bedroom, 170-square-metre (1820-square-foot)
house designed and built to demonstrate sustainability in
cold-climate housing. "An inventory of ideas," the house
reflects concern for environmental stewardship, occupant
health, resource conservation, the use of appropriate technologies and alternative energy sources, and self-
sufficiency (Checora, 1996). The project was undertaken
on the initiative of a small group of individuals who are
partners in the business ASH—Autonomous and
Sustainable Housing, Inc. To demonstrate its marketability
and financial feasibility, the project was funded by a conventional mortgage without government assistance.

The project has three distinct phases: the sustainable
stage, the autonomous stage (when the house no longer
needs any city water or sewage treatment), and the
energy-credit stage (when surplus electricity produced by

the photovoltaic panels on the roof will be sold to the power company). The house was built to achieve these goals in Calgary's cold climate without a conventional forced-air furnace or boiler—the house is not even connected to natural gas. Instead, the passive solar design of the home allows its occupants to take advantage of Alberta's year-round sunny climate.

Even in Calgary's cold winters, about 60 percent of the home's space heating requirements are met simply by allowing the sun to shine through the expanse of south-facing windows onto the dark ceramic and recycled-glass floor tiles. Placed on top of a 12.5-centimetre-thick concrete slab, these materials constitute a substantial thermal mass that helps store the heat from the sun. To keep the heat stored in the thermal mass in the house, the walls are insulated with cellulose to R-50 and the roof cavity to R-74 (well beyond the current building code). Any backup heat required comes from the highly efficient, wood-burning masonry heater that can be used for baking as well as future electrical generation. In 2002 the ASH house spent $2500 per year less on utility costs than a conventional house of equivalent size (Ostrowski, 2003, personal communication). Designers estimate that the value of the house is contained in its higher resale value compared to its neighbours and in its security against the vagaries of climate change, government policy shifts, and links to infrastructure (e.g., water, sewage, energy) that are not required.

High-performance windows throughout the house, including one with an R-16.6 value (developed in Winnipeg), complete the building's envelope. Some of the other technology in the project, much of which was donated by more than 220 leading-edge companies from around the world, includes nonadditive concrete, "Eco-stud" wall trusses, nontoxic drywall mud, waterless (composting) and ultra-low-flush toilets, a solar hot water collector, two types of greywater heat exchangers, an air-to-air heat exchanger, and radiation-shielded, full-spectrum, and electroluminescent lighting.

As of 2002, the ASH house is completely independent of public water, sewer, and gas utilities. The project is expected to be completely autonomous once a solar electric system is completed. This will include a 2.5 kW photovoltaic system that will be able to supply the electrical requirements of the ASH house and export enough green energy to supply three other homes designed like the ASH house and one conventional home (Ostrowski, 2003, personal communication).

There is no oven in the house; most of the cooking is done in a solar oven (made in Saskatoon) that sits on the front porch facing the sun. Since the house is not connected to the city's water supply, all water for the house comes from rainwater collected from the roof and stored in a 14 500-litre cistern buried in the backyard. Experiments on a number of filtration and purification processes (to produce potable water) are ongoing, as are

Photo 13–5
A straw bale house under construction near Millarville, Alberta.

strategies to treat and reuse the greywater produced in the household. The waterless composting toilet saves about 200 000 litres of purified drinking water from being polluted every year. Education being a large part of ASH Inc.'s mandate, the house is open to the public every third and fourth Saturday of the month.

A similar demonstration project—the Toronto Healthy House—is a three-bedroom infill home that harvests its own energy, collects rainfall and purifies it for drinking, and biologically treats its own waste. Part of the Canada Mortgage and Housing Corporation's (CMHC's) Healthy Housing initiative, the house has low operating costs and is affordable (see Figure 13–3). In Red Deer, Alberta, in 1994, Healthy Housing principles were applied to the renovation of a 1905 home, demonstrating environmental responsibility through such elements as material selection, energy efficiency, airtightness, and equipment selection. One of the challenges of bringing these kinds of housing initiatives from demonstration into practice is that they cost more in the short run to design and build. If sustainability also means making them affordable to a wide range of income earners, then elements of environmental design need to be coupled with social policies to ensure that the environmental benefits are equitably distributed.

EcoResidence is an ecological living experiment at McGill University's agricultural campus. The renovated residence reuses many of the materials in the existing structure. Greenhouses attached to the fronts of all units are an essential feature of EcoResidence. Not only do they

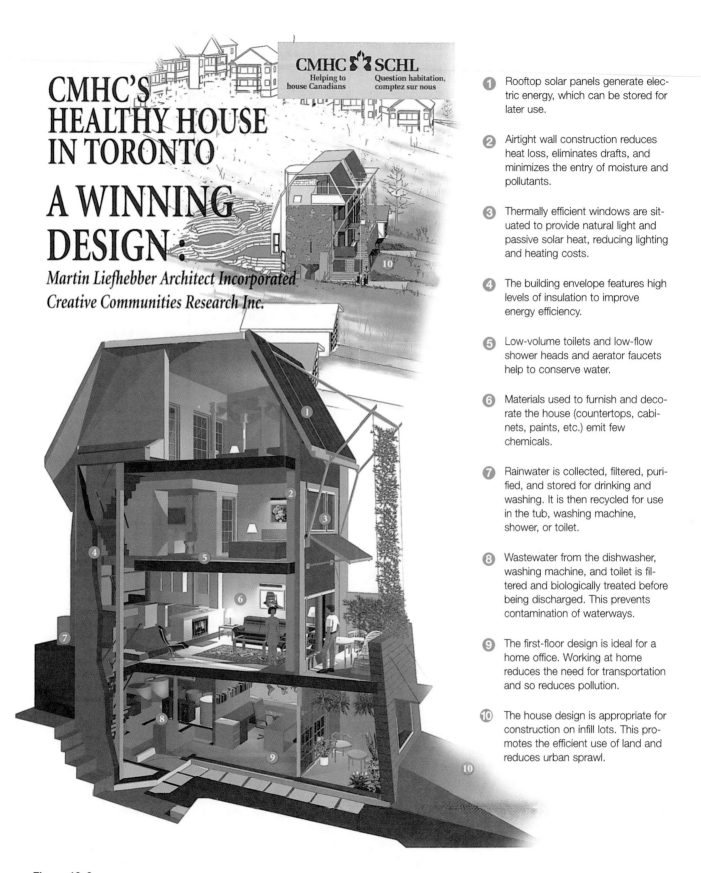

CMHC'S HEALTHY HOUSE IN TORONTO

A WINNING DESIGN:

Martin Liefhebber Architect Incorporated
Creative Communities Research Inc.

CMHC ✹ SCHL
Helping to house Canadians
Question habitation, comptez sur nous

1. Rooftop solar panels generate electric energy, which can be stored for later use.

2. Airtight wall construction reduces heat loss, eliminates drafts, and minimizes the entry of moisture and pollutants.

3. Thermally efficient windows are situated to provide natural light and passive solar heat, reducing lighting and heating costs.

4. The building envelope features high levels of insulation to improve energy efficiency.

5. Low-volume toilets and low-flow shower heads and aerator faucets help to conserve water.

6. Materials used to furnish and decorate the house (countertops, cabinets, paints, etc.) emit few chemicals.

7. Rainwater is collected, filtered, purified, and stored for drinking and washing. It is then recycled for use in the tub, washing machine, shower, or toilet.

8. Wastewater from the dishwasher, washing machine, and toilet is filtered and biologically treated before being discharged. This prevents contamination of waterways.

9. The first-floor design is ideal for a home office. Working at home reduces the need for transportation and so reduces pollution.

10. The house design is appropriate for construction on infill lots. This promotes the efficient use of land and reduces urban sprawl.

Figure 13–3

Toronto Healthy House

SOURCE: Canada Mortgage and Housing Corporation brochure, provided courtesy of CMHC and Martin Liefhebber Architects. Reproduced by permission. Actual house design may vary from design shown.

capture and store solar energy for redistribution by the passive solar heating system, but they also act as a natural air filtration system. The roofs are designed to collect rainwater that is filtered into a central area in the buildings and stored in tanks. The next big step for the EcoResidence is an ecological wastewater treatment facility that would treat sewage using plants, animals, and microorganisms to purify wastewater in much the same way that wetlands do (Dupuis, 2000; McGill University, 1999).

Among other trends in sustainable housing are the use of straw bales, recycled tires, packed dirt, and mud. Straw bale construction is an old technology, used in Europe 300 years ago and pioneered in North America in the sandy, treeless Nebraska prairie in the 1890s. At that time, building homes from bales of straw was a necessity; now, as the world is concerned about possibly running short of wood, straw bale construction of residences, workshops, and garages creates a new use for straw (normally an almost useless byproduct of such grains as wheat, oats, flax, and barley).

Straw bale construction has many advantages over conventional wood frame construction, including the fact that straw is a low-cost, renewable, easy-to-use material with high insulation value (see Table 13–1). For example, buying and transporting the bales for a 2200-square-foot home built in 1996 near Millarville, Alberta, cost $1000 (King, 1996). Stacked like huge bricks, straw bales are easy to build with and walls go up quickly. The frame of the Millarville-area house used corner posts and beams, and the walls were built by stacking straw bales horizontally and pushing them into floor-to-ceiling threaded metal rods, clamped at the top. The insulation value of the straw bale walls approaches R-50. A two-storey straw bale workshop in Nova Scotia made construction history when it became the first "code-approved" load-bearing straw building in Canada (Wood, 1997).

"Earthships," built from recycled tires, packed dirt, and mud, are another type of sustainable housing that finds a new use for "waste" material (tires, cans, and bottles). More common in the United States than in Canada (although a 4000-square-foot, $500 000 earthship exists outside Edmonton), these homes use passive solar heat to reduce utility bills by about 75 percent compared with conventional construction methods (Hope, 1995). Typically, water is supplied from catchment systems in the earth roof (stored in cisterns for future use); other environmental technology, such as solar toilets that turn sewage into dust, also are common. Some earthships also include domestic food production.

Sustainable houses leave a much smaller footprint on the ecosystem than do conventional single-family homes. Depending on their design and construction, sustainable houses may increase efficiency in land use and energy and water consumption, and provide healthy indoor environments. By overcoming most of the environmental prob-

TABLE 13–1
ADVANTAGES OF STRAW BALE CONSTRUCTION

- Straw bales are cheap to buy. Considered an agricultural waste product, straw is available annually. Rather than plow it under or burn it in the fields and thus create air pollution, straw can be baled and turned into an energy-efficient resource.

- Straw bales have a high insulation value (R-2.7 per inch; an 18-inch-wide bale has an R value of 48).

- Straw bale buildings have lower heating and cooling requirements, resulting in reduced fossil fuel use and reduced CO_2 emissions.

- Straw bale construction is low-tech, easy, and requires few power tools.

- Lumber use is reduced.

- Straw bales are nontoxic and, when finished with natural plaster, allow a gradual transfer of air through the walls, promoting good indoor air quality.

- Straw bale buildings are soundproof: one Nebraska pioneer family, playing cards in the kitchen, was unaware that a tornado had just roared through the town.

- Straw bales resist combustion: because of the thickness and lack of oxygen available, it takes two hours to burn through a plaster, straw, and stucco wall (double the resistance of most wood-frame homes).

- With the proper foundation, roof, and finish plaster, straw bale buildings can last indefinitely (some Nebraska historic homes are still standing).

- Anecdotal evidence indicates that there are no problems with bugs in straw bale buildings.

SOURCES: *Straw Bale Construction,* Black Range Films, (n.d.), Kingston, NM; "Hay, There's a New Way to Build a House!" by F. King, May 11, 1996, *Calgary Herald,* p. I10; "Piggy's Idea Recycled: Straw Replaces Scarce Wood," N. Oosterom, October 25, 1995, *Calgary Herald,* p. A2.

lems of single-family homes (such as high energy use and high costs of infrastructure services), as well as by being affordable, sustainable housing may become even more common in the future.

Depending Less on Our Cars

Opportunities exist to reduce the environmental effects of housing, including planning for better use of land, transit, and infrastructure systems. The spread-out nature of Canadian suburban development helps keep us dependent on the automobile. This means that use of energy for transportation in suburban and fringe areas is particularly high, which in turn highlights the need to ensure that vehicles are maintained (to reduce smog),

Photo 13–6
Canada's millions of cars and light trucks emit over 4 tonnes of pollutants into the atmosphere every year.

that public transit systems are promoted and are convenient alternatives to driving, and that bicycle paths are provided for commuting and recreational use.

Although the single most important contribution that individuals can make to reducing environmental impacts of cars is to use them less, all over Canada people have taken various actions to reduce energy use in urban transportation. Environment Canada's Action 21 "Down-to-Earth Choices" program identified a range of actions taken by community groups, environmental groups, individuals, municipalities, and employers. From special events to promote alternative transportation and "clean-air commuting," to a group of volunteers who devised a "walking school bus" (children are walked to school holding a loop on a rope instead of using a seat on the bus), many Canadians have shown they are willing to reduce their car use. Some have even worked together to halt road construction (see Verrall, 1995).

Every year, each of the 15 million cars and light trucks on Canada's roads emits over 4 tonnes of pollutants into the atmosphere. In 2004, for the nineteenth consecutive

Photo 13–7
Bicycle paths for commuters and recreational cyclists promote pollution-free travel and physical fitness.

year, Environment Canada held its voluntary emissions inspection clinics. The program is aimed at raising awareness about the importance of proper vehicle maintenance in reducing air pollution and protecting the environment and the health of Canadians. Motorists can have checkups of the antipollution equipment on their vehicles.

Sustainable transportation systems are being explored by municipalities across Canada. In the Greater Vancouver Regional District, for example, the member municipalities adopted a positive vision for reducing auto dependence in the region called Transport 2021. The road to achieving the vision, however, is not an easy one. Although the plan has been endorsed in principle by the province of British Columbia, the GVRD does not have the province's financial support. Continuing provincial funding cuts to transit may force a retreat from the plan (Raad & Kenworthy, 1998).

Photo 13–8
Although many Canadians are highly dependent on their vehicles, others are choosing to reduce their energy consumption by driving less and increasing their use of bicycles and public transit.

Canadian transit systems already have made considerable efforts to improve their infrastructure, including new fleets of buses, bus shelters, and computerized information systems. Future efforts to evaluate the costs and benefits of providing transit systems would help to demonstrate the true costs of automobile use and provide a comparison between the two systems (Pucher, 1998).

MATERIALS USE

Solid Waste

Throughout Canada, generating and managing solid wastes is an expensive environmental and social issue. From an ecosystem perspective, waste materials that enter landfills or are incinerated represent energy and resources that have not been used fully and that could have been recycled, reused, or reduced at their source, thereby reducing the need to extract and process new resources. From an economic perspective, solid wastes cost municipalities more than $3 billion annually for their collection, transportation, and disposal. In community and environmental terms, continuing generation of waste leads to the need to find new landfill sites when the old ones fill up. Finding and approving new sites is becoming harder, in part because of the "NIMBY" (not in my backyard) syndrome.

One of the most controversial aspects of solid waste disposal is the location of disposal sites. Vancouver, for example, trucks waste to Cache Creek, while Toronto's shipment of garbage to the United States has generated much debate, encouraging Toronto to look to new sites within Canada. In the United States, activists have protested against "environmental racism." They point out that solid and hazardous waste sites are more likely to be located in or near neighbourhoods of poor people of colour than in or near white, middle-class areas. This is not solely an American problem. In the 1990s, the Halifax Regional Municipality (composed of Halifax, Dartmouth, Bedford, and Halifax County) was mired in a debate over environmental racism when it sought to replace its landfill site. African-Canadian residents had been subjected to years of residential segregation, systematic racism, deplorable living conditions, and ultimately, relocation without consultation. They feared that a similar story would be repeated with the creation of a new landfill. Their activism to protest the site became part of the push for Halifax to create among the most comprehensive and successful waste resource management strategies in the country.

Landfilling and incinerating wastes almost always are controversial land use decisions because both incur environmental (and socioeconomic) effects. Even though technology has improved, landfills emit methane (a greenhouse gas) and other gases (some toxic), and there is as yet no method that is 100 percent efficient in capturing and containing these gases. Inadequate engineering, too, can result in leachates (leaking liquid) contaminating surface water or groundwater. Incineration produces both air emissions that contain toxic contaminants and particulates, and solid residues that are hazardous and require further disposal in specialized facilities. With a combination of high-temperature incineration methods and emission controls, problems of air emissions could be solved (Environment Canada, 1991). However, because high volumes of waste are required to make incineration economical, municipalities could have less incentive to "reduce, reuse, and recycle" their waste materials.

Although comparisons with other countries are difficult, Canadians have the dubious distinction of being among the world leaders in per capita waste production. In a study completed in 2001, Canada ranked 18 of 29 in waste production compared to the other member countries of the OECD, based on estimates of 490 kilograms of waste per person in 1997. In 1992, Canadians generated 18.1 million tonnes of municipal solid waste. This is 637 kilograms per person per year, or 1.7 kilograms daily; about one-half of that is from residential sources. If we add construction and demolition waste, the estimates of solid waste in Canada rise to 29.3 million tonnes, or 1030 kilograms per person (Government of Canada, 1996). Waste disposal rates remained high in 2000. Canadians disposed of 23.1 million tonnes of solid waste, or 750 kilograms per person per year. This number is actually an increase over the amount of solid waste disposed on a per capita basis in the 1990s. In comparison, waste generation in Sweden is estimated at only 0.8 kg per capita per day. Given these statistics, the Canadian Council of Ministers of the Environment established a Waste Resource Management Strategy, agreed on by all the provinces, that set a national target for a 50-percent reduction of solid

Photo 13–9
Of the material we send to landfills, how much could we reuse, recycle, or compost?

Photo 13–10

Easily accessible community facilities are one way Canadians may be encouraged to increase their recycling of solid wastes.

wastes from 1988 levels (on a per capita basis) by 2000. At the end of 2000, Nova Scotia was the only province that had reached that goal. The goal was achieved through implementation of a recycling program, landfill reduction, and green bin doorstep pickup composting. It is estimated that the new system in Halifax has reduced annual greenhouse gas emissions by about 1.4 tonnes per resident. In 1999–2000, residents and businesses diverted 43 percent of the waste that would have normally gone to landfill, including 36 000 tonnes of organics and 22 000 tonnes of other recyables (including appliances). The new system resulted in 125 new permanent jobs in a variety of occupations, and there are now 3000 jobs in Nova Scotia's recycling industry.

Nevertheless, the challenge remains for Canada as a whole. While Nova Scotia has the lowest rate of annual per capita waste disposal at 460 kilograms, British Columbia and Quebec "boast" disposal levels of 910 and 940 kilograms respectively (in 2000). The challenge remains to reduce the waste stream. In 1992, only about 17 percent of municipal waste was recycled or composted while the remaining 83 percent was landfilled (78 percent) and incinerated (5 percent) (Government of Canada, 1996). By weight, the main materials in municipal solid waste were paper and paperboard (25 percent), food wastes (19 percent), yard wastes (13 percent), and plastics (12 percent). Asphalt, concrete, rubble, and wood dominate the construction and demolition waste stream and suggest the importance of transportation infrastructure in cities. The Halifax example illustrates that with commitment, these levels can be dramatically reduced.

Waste management can also occur before materials get sent to the disposal facility. (See Box 13–4 for examples of waste management efforts in the workplace.) Availability

and accessibility of community recycling facilities, effectiveness of recycling technologies, markets for manufactured postconsumer waste products, and levels of awareness and willingness to recycle on the part of consumers, businesses, and institutions are the main reasons for variations in recycling or in composting of waste materials.

Land Contamination

An important urban waste management problem has arisen on old industrial and other sites where toxic materials were deliberately dumped or accidentally spilled or leaked from underground storage tanks. On these sites, contamination levels of certain persistent compounds and metals, such as lead, cadmium, chromium, and nickel, prevent redevelopment for residential, recreational, or even commercial use without a costly cleanup. Numerous examples exist across the country, including the Expo 86 site in False Creek in Vancouver, and the Ataratiri housing project in Toronto (dropped in the late 1980s when cleanup costs were estimated to be over $30 million).

This problem is not confined to large cities; soils in the town of Trail, British Columbia, may be contaminated by deposition of dust containing lead from the town's smelters. In northern Canada, contaminated sites are associated with abandoned U.S. military facilities such as the 21 Distant Early Warning (DEW) Line radar sites. High levels of PCBs, heavy metals, and POLs (petroleum, oil, and lubricants) constitute a toxic legacy, and are known to be entering the Arctic food chain. Cost estimates to clean up only four U.S. military installations range from $350 million to more than $1 billion, depending on the standards used for cleanup (Canadian Arctic Resources Committee, 1997).

As we saw in the case of creosote contamination of the Bow River (Chapter 7), toxic substances also may migrate from old industrial sites to contaminate other areas through groundwater movement. Chemical fertilizers and pesticides, often applied to lawns and gardens in quantities greater than are applied in agricultural operations, enter the ecosystem by leaching into the ground or entering the wastewater disposal system. In response to public concerns about use of pesticides and herbicides, many municipal governments have developed integrated pest-management plans and reduced or eliminated use of chemicals on public lands such as parks.

URBANIZATION OF LAND

Green Space in the City

Until relatively recently, green space in the city was not accorded much value other than for purposes such as space for buildings and roads. As a result of this lack of

Provided that a company workplace is supportive of employee efforts to implement environmental initiatives, employees can have important effects on company practices. Whether as individuals, members of committees, or managers, many people have helped their companies to become greener in such areas as purchasing, environmental codes of practice, nonsmoking buildings, and conservation strategies for water, energy, and waste.

Support for bicycle commuting (including provision of shower facilities in the workplace) and contributions to events such as community tree planting and river cleanups have become meaningful company activities. Many of these activities have resulted from people sharing information about the importance of individual attitudes, actions, and impacts on our environment.

Within the workplace, individuals and environmental committees can influence other employees to participate in initiatives such as recycling programs. With the advent of computers in the workplace, the paperless office was predicted as the way of the future. However, as a result of people using more paper by printing more drafts of their work, paper use actually increased in 50 percent of Canadian companies surveyed in 1994. Fortunately, many companies and employees now participate in efforts to reduce paper use and to recycle used paper. In the same survey, it was noted that 83 percent of respondents participated in efforts to recycle paper, 60 percent used two-sided photocopying, 56 percent used the backs of paper sheets, and 55 percent used the computer for revisions to their documents rather than printing a new hard copy (Government of Canada, 1996; Pitney Bowes, 1994).

Telework or telecommuting is a trend that may be positive for the environment. As some corporations offer their employees the opportunity to work at home and connect to the office via computer and telecommunication technologies, reductions of air emissions and other environmental stresses due to commuting are expected. In addition, as employees spend less time in the office, telework can allow more people to share office space. This reduces the need for land for buildings as well as the heating and cooling costs associated with such real estate.

Environmental stewardship also is reflected in Canadian Pacific Hotels and Resorts (CPH&R) Green Program. In the fall of 1990, Canadian Pacific Hotels embarked on the project of developing a set of environmental standards for all of its hotels in Canada. The company conducted a detailed audit of all its operating hotels, asking hotel departments to submit information on every aspect of their impact on the environment. The program's goals are ambitious. Canadian Pacific Hotels is reducing the amount of waste sent to landfill by 50 percent across the chain, running an extensive recycling program (including blue boxes for collecting recyclables in guest rooms), redesigning its purchasing policies to ensure that waste is reduced at source,

and ensuring that supplies used in the hotels are nature-friendly. While every one of Canadian Pacific Hotels' properties has an environmental program, many individual hotels have developed some creative and unusual ways to become more nature-friendly. For example, Le Château Montebello uses its own composted soil as fertilizer; Hotel Vancouver now uses baking soda and salt instead of chlorine to maintain water clarity in its pool; and the Royal York donates leftover food to Second Harvest—a Toronto organization that collects and redistributes food to relief agencies.

In order to assess the effects of the environmental program and how its properties were responding to the 16-point plan, Canadian Pacific Hotels hired an outside environmental consultant to audit each property on overall compliance to the program. The results were very encouraging. Here are some of the highlights:

- Canadian Pacific Hotels has placed blue recycling boxes in every one of its hotel rooms at every property throughout Canada and realized 100 percent compliance with its objective.
- 90 percent of all used soap is made available to local charities in Canada and the Third World. The Queen Elizabeth Hotel alone sent 4200 pounds of soap to humanitarian agencies in underdeveloped countries.
- 86 percent of all paper used in Canadian Pacific Hotels properties is recycled or kraft paper that meets or exceeds Canadian Environmental Choice Standards. Over 80 percent of properties have succeeded in reducing their paper consumption by the 20 percent objective.

In 1996, Anne Checkley, director of communications and environmental affairs for CPH&R, indicated that there was "absolutely no down side to having an environmental program" (Beale, 1996, 24). Even though hotel occupancy has increased during the past decade, waste management costs have decreased. In fact, CPH&R publishes "The Green Partnership Guide," a collection of tips from its environmental committees that is sold around the world. This is another indication that greening of business can be profitable.

Over the past decade, growing numbers of companies have made changes in their operations and management in order to reduce the environmental impact of their activities. At the same time, companies are recognizing that there are business benefits to be realized by greening their operations. Business decision making that reflects environmental and social concerns and values is being termed corporate social responsibility (CSR).

SOURCES: "Canadian Pacific Hotel's Enviro-Initiatives," J. Beale, 1996, Winter, *Ecolutions,* 24–25; Canadian Pacific Hotels, home page, http://www.cphotels.com; *The State of Canada's Environment—1996,* Government of Canada, 1996; *Pitney Bowes Fourth Annual Green Office Survey,* Pitney Bowes, 1994, Toronto: Author.

value accorded to green space, many natural areas and stream valleys were used as convenient locations for dumping garbage and for highway routes, many watercourses were used as storm sewers, and many waterfronts were cut off from public access for use by industries, railways, and roads (Turner, 1996).

Even the creation of parks in the city caused destruction of the natural environment as, for example, when productive wetlands were transformed into sports fields, playgrounds, manicured lawns, flower displays, and other planted areas using exotic species. Golf courses and other formal parks often are fertilized, treated with pesticides, and irrigated. Such treatments often eliminate remaining natural characteristics and add polluted runoff to lakes, streams, and rivers.

Apart from its important social benefits, most urban outdoor recreational space has very little conservation, ecological, or environmental value. Only when large wild areas are preserved (such as Stanley Park in Vancouver or Nose Hill Park in Calgary) is there a potential for protection of environmental values. Even then, heavy use of Stanley Park has resulted in severe erosion of some of the area's original natural characteristics. Certain wildlife found in Nose Hill Park may not survive over the long term as their corridors into the park now are virtually surrounded by urban development.

Other species of wildlife, wildflowers, and trees, however, are able to survive and even thrive in urban environments. Ravines, woods, and other vegetated areas within cities are important in providing habitat for a variety of wildlife and plants and in reducing or preventing soil erosion. In addition, natural areas help moderate urban microclimates, decrease air pollution (trap particles and absorb carbon), and reduce storm flows (thus reducing the overload on sewers and treatment plants). Urban forests also are important recreational resources. Almost 80 percent of Canadians live in cities and, perhaps unknowingly, derive considerable environmental, economic, and social benefits from the trees that grow in their parks, on their street boulevards, and in their own backyards. Some of the benefits trees bring to human health and well-being are identified in Box 13–5.

As the ecosystem approach to green space within cities has developed, many communities have established systems of interconnected natural areas that permit reproduction and migration of many species of plants and wildlife and continuation of some or most ecosystem functions. While natural green spaces provide residents and visitors with aesthetically pleasing surroundings for recreation, relaxation, or contact with nature, the longer-term view is that such natural heritage systems would become the focus around which cities develop, rather than providing just a backdrop for development.

In spite of growing awareness of the multiple values of urban green space, city land use controls do not always

Photo 13–11
While socially important, manicured green spaces in cities have fewer environmental benefits than natural areas.

afford green areas or natural spaces the protection they require to remain functional in an ecological or ecosystem sense. Many private developers and public agencies continue to view green space as unused or underused land and as prime sites for buildings and other facilities. Additional pressure on urban green space comes from efforts to increase the intensity of occupancy of urban land in order to reduce infrastructure costs and increase energy efficiency. From 1966 to 1986 (the last year for which Canada-wide data are available), the rate of conversion of rural and prime agricultural land to urban uses varied according to the population size of urban centres— the nine largest centres (over 500 000 population) accounted for 42.9 percent of the conversion (Warren, Kerr, & Turner, 1989).

Growth in urban populations during the coming decades is inevitable, as is continued urban expansion. The effects of continued expansion could be enormous; for example, a conservative estimate indicates that by the year 2021, even if the most compact growth is pursued, the Greater Toronto area is expected to grow an additional 23 percent or 350 square kilometres (from a 1988 area of 1520 square kilometres). If current, dispersed trends were to continue, the increase would be 900 square kilometres or 59 percent (IBI Group, 1990).

One of the major contributing factors to this rate of land occupation is the popularity of the detached, single-family house surrounded by its own lot. Curvilinear street patterns (rather than the more compact grid layout) also have reinforced the spread-out form of urban development. Single-family detached housing remains popular. Prairie cities tend to have the highest proportion of single-family detached housing starts, with west coast and most Quebec centres slightly lower. However, there has been an increase in the average density of new residential developments, in part because frontage of single-family

Trees greatly influence the health of the urban environment; they are a source of beauty and help to purify the air, abate noise, modify heat, stabilize soil, reduce flood risks, and provide wildlife habitat as well as recreational settings for residents. For its regulatory services, an urban forest is an integral part of the hydrologic cycle in urban environments. The urban forest includes not only large stands of native trees, such as are found in Stanley Park in Vancouver, B.C., and in C.A. Pippy Park in St. John's, Newfoundland, but also the millions of native and exotic trees lining our streets, in our yards, parks, and plant nurseries, and on the edges of some of our newest neighbourhoods.

One reality of life for urban trees is that many get a great deal less water than trees in rural settings do (streets, gutters, and sewer systems channel water away quickly). Urban trees also experience higher average temperatures, air and soil pollution, and the constant threat of root or stem damage from human traffic or heavy equipment. As a result, their life spans are shorter than trees in rural areas. A tree planted in the black-top jungle of downtown Vancouver will survive an average of 13 years, while those planted in large treed gardens may survive up to 60 years (Forest Alliance of British Columbia, n.d., p. 2). While an average life expectancy of 32 years for trees in the Vancouver area is significantly less than the average life span of trees in rural areas, trees in urban areas naturalize built landscapes and provide important psychological benefits also.

The benefits of trees (rural and urban) to human psychology are difficult to quantify, but they contribute noticeably to human health and well-being. Research has shown, for instance, that patients recover more quickly if their hospital room has a view of trees and natural landscapes (Ulrich, 1979). Urban trees shade and cool streets and buildings in summer, moderating the higher temperatures experienced in cities. If coniferous trees are placed strategically to buffer winter winds, they can help reduce heat loss from buildings in winter and contribute to fuel savings of 20 percent or more (Forest Alliance of British Columbia, n.d.).

Each tree in the urban forest removes pollutants and other particulates from city air. Carbon, chlorine, fluorine, ozone, sulphur dioxide, peroxyacetylnitrate (a component of photochemical smog), and other gases are absorbed by these trees. Trees function as carbon sinks. Every year, each city tree removes about 6 kilograms of carbon dioxide from the atmosphere. Because they are located in highly emitting areas, urban trees are 5 to 15 times more beneficial than wilderness trees with regard to the purification of the air we breathe. Similarly, trees filter airborne particulates, resulting in 27 to 42 percent less ground-level dust in treed areas than in open areas, an important health benefit for people sensitive to dust or allergic to pollen (Forest Alliance of British Columbia, n.d.). Ultimately, however, air pollution will damage forest health; the decline in the health of German forests (noted in Chapter 9) and the loss of ponderosa pine forests near Los Angeles due to smog attest to this fact.

For all of the above reasons, as well as the fact that they increase property values by 5 to 20 percent or more, "trees are not mere niceties, they're necessities" (Krakauer, 1990). We need to remember this as our cities grow—in Canada every day, forests are cleared to make way for new subdivisions, shopping centres, roadways, agriculture, and grazing lands. Population growth within cities means city trees also are lost to development changes designed to accommodate new residents. While not a total substitute for natural forests, urban tree planting, such as that encouraged by the Green Streets Canada Program (an initiative of Tree Plan Canada), is an important greening activity. Green Street Canada's objective is to create a partnership between the Tree Canada Foundation and municipalities across Canada to help improve their urban forests and provide citizens with a greater appreciation of how trees can contribute to the environment and the overall quality of life in their community (Tree Canada, 2001). By 1994, more than 40 million trees had been planted in rural and urban areas (Natural Resources Canada, 1995).

SOURCES: "The Urban Forest," Forest Alliance of British Columbia, (n.d.), *Choices, 4*(1), 2; "Trees Aren't Mere Niceties—They're Necessities," J. Krakauer, April 1990, *Smithsonian, 21,* 160–171; *The State of Canada's Forests 1994: A Balancing Act,* Natural Resources Canada, Canadian Forest Service, 1995; "Visual Landscapes and Psychological Well-Being," R. S. Ulrich, 1979, *Landscape Research, 4,* 17–23; Tree Canada Foundation, home page, 2001, http://www.treecanada.ca/index_e.htm

---

detached lots has decreased from the 15 to 18 metres (50 to 60 feet) common up until the 1970s to about 9 to 12 metres (30 to 40 feet) in the 1990s.

LOSS OF AGRICULTURAL LAND

Both historically and today in Canada, much of the expansion of settlement has occurred on fertile agricultural land. About 25 percent of Canada's high-capability (class 1 to 3) agricultural land is located within 80 kilometres of the 23 largest cities; this includes more than 50 percent

of our prime, class 1 agricultural land. While it may seem that the amount of agricultural land lost to urbanization is quite small, there are two important factors that we need to remember. One factor in this complex issue is that in some parts of Canada urbanization affects specialty crop areas. The Okanagan Valley in British Columbia, the Niagara Peninsula in Ontario, and the horticultural lands adjacent to Vancouver and Montreal are areas that account for only a tiny proportion of the total amount of productive land in Canada. In these areas, however, urbanization has far more significant impacts than in other parts of Canada.

Photo 13–12
Increasingly heavy use of even large protected areas, such as Stanley Park in Vancouver, presents sustainability challenges.

The second factor is that urban growth affects agriculture in many indirect ways. Agricultural regions experience significant economic and social impacts when extensive industrial sites, gravel pits, golf courses, recreational facilities, and residential estates are developed. Sometimes called urban shadow effects, these impacts extend over large areas and cause declines in agriculture in urban regions (Gertler, Crowley, & Bond, 1977). The problem seems to be that once agricultural production is discouraged, even land that is not needed for urban growth comes to be occupied by nonagricultural uses or is abandoned. In efforts to combat this problem, Quebec and British Columbia have had protective agricultural zoning in place for years, and other provinces have adopted policies that regulate urban expansion in agricultural areas.

In addition to agricultural land loss, removal of woodlands, disruption of wildlife corridors, destruction of habitat, and accelerated soil erosion, urban expansion has resulted in the loss of wetlands. By 1981, for instance, 98 percent of the original wetlands in the vicinity of Windsor, Winnipeg, and Regina had been converted to other uses. In the Toronto and Montreal areas, 88 percent of wetlands had been filled or drained by 1981, and a similar fate had claimed 78 percent of wetlands in the vicinity of St. Catharines–Niagara Falls, Calgary, and Vancouver (Environment Canada, 1988).

Other environmentally sensitive areas may be threatened (and sometimes are destroyed) by urban expansion, including aquatic habitats if groundwater and surface water bodies are polluted by storm runoff or septic tank and landfill seepage. Air quality, too, may be affected by the operation of gravel pits and landfills, including the frequent truck traffic they generate. These and previously noted effects of human activities associated with the city remind us that our ecological footprint (see Chapter 1) is impressed on the productive output of a land area many times larger than the geographic size of our cities. If we are to continue to support people's demands for food, water, forest products, and energy, and to assimilate the wastes resulting from our urban activities, then actions to ensure sustainability of our communities and to protect the environment that maintains them are vital. (See the Enviro-Focus box for information on urban agriculture.)

Toward Sustainable Communities

If cities did not grow, many of the ecological impacts noted above might not occur. However, we have come to appreciate that many consequences of urban growth could be avoided, or their impacts reduced, if we paid more careful attention to and incorporated stricter controls in land use planning. Similarly, if our attitudes toward our consumption patterns and lifestyles consciously reflected an awareness of our dependence on the natural environment for our own (and the planet's) well-being, our decisions, behaviours, and choices might be different and more environmentally friendly than they are now.

Before we consider examples of actual actions undertaken to move toward more sustainable communities, it is worthwhile outlining what is meant by sustainability in an urban context.

CITIES AND SUSTAINABILITY

The density, form, and structure of Canadian cities, as well as the activities of people within cities, are the principal sources of stress to the ecosystem in urban areas. The relationships between Canadians and other citizens around the world are very complex, particularly as global trade continues to accelerate. City living tends to provide little sense of this intimacy with the Earth. We probably buy most of our foods—imported from all over the world—from grocery stores, and flush our wastes down the toilet or place them in garbage cans for collection. Ironically, we might even be able to shop at a wilderness store in the mall!

Recall Box 1–3, which briefly describes the ecological footprint concept using the footprint of the Lower Fraser

Urban Agriculture

For most Canadians, food is plentiful, and we are among the best-fed people in the world measured by caloric intake. However, Canadians living in urban areas rarely think about where their food comes from or about the environmental issues associated with the abundance of food seen in the supermarket. For instance, food prices tend not to incorporate fully the long-term environmental costs of production and transportation—it has been estimated that it takes three times as much energy to truck a head of lettuce from California to Toronto as it does to grow it locally in season (cited in Government of Canada, 1996).

Concerns about the sustainability of Canada's resource-intensive food production system, health concerns surrounding the use of additives and preservatives, the availability of fresh food for lower income residents, as well as newly emerging issues concerning the use of genetically modified organisms (GMOs) as foodstuffs (see Chapter 6) have sparked increased interest in organic farming, and in gardening in the city (urban agriculture). Since 1978, City Farmer, a non-profit society, has promoted urban food production and environmental conservation from their small office in downtown Vancouver and from their demonstration food garden in a residential Kitsilano neighbourhood.

In 1996, Annex Organics of Toronto retrofitted an old warehouse building in order to grow tomatoes, peppers, eggplant, and herbs—on its roof! Rooftop gardening is an example of "green roof infrastructure," touted as a truly sustainable development technology (Kwik, 2000). Common in Europe, not only do rooftop gardens help improve urban air quality and moderate the urban heat island effect, they also insulate buildings and reduce the energy costs of heating and cooling. Urban food production is increased by the use of gardens on rooftops, where there is more space and sunlight than at ground level. City farmers find it advantageous to be so close to their customers; transport costs are minimized, and so are associated pollutants.

The Hilton Montreal Bonaventure hotel boasts a 2.5-hectare landscaped rooftop, with winding brooks, birch trees, and many species of plants and animals, where guests can relax.

Photo 13–13
Community gardens bring people together to help meet economic, social, and environmental sustainability objectives.

Vertical gardens (where vines and other vegetation are placed on or adjacent to interior or exterior walls) have many of the same benefits as horizontal rooftop gardens. These include reducing the cooling loads of buildings, moderating internal temperature variations, beautifying, assisting in food production, and providing additional green space (Bass & Hansell, 2000).

Community gardening is another strategy to help revive inner-city neighbourhoods. From an environmental standpoint, community gardening offers a way for urban residents to grow their own food organically, thereby reducing dependence on current methods of food production. Community gardens are often places where new immigrants to Canada can cultivate foods from their countries of origin and share in the cultural diversity of the country. The Cultivating Communities project in Calgary is an initiative that began in 1996. The idea behind the establishment of a community garden within the grounds of a community association was to create a project in which disabled people, seniors, schoolchildren, and other neighbourhood residents could take part. Participants celebrate their success with a harvest of fresh vegetables (Pezzi, 1998).

Montreal's Community Gardening project is recognized as the largest and best-organized city gardening program in the country. The program's work in the 1990s involved composting research, food donations to community kitchens, access for disabled gardeners, and horticultural therapy projects. Vancouver's Strathcona

Community Garden contains a heritage apple project, contributing to retaining the diversity of our food sources. Saskatoon combines a community garden program with its Child Hunger and Education Program, to encourage the production of healthy and inexpensive foods for the city's most vulnerable populations. Similar community garden initiatives have taken root in small and large centres across the country. Is there one in your community?

SOURCES: "Climbing the Walls: Vertical Gardens Can Cool Buildings and Clear the Air," B. Bass & R. Hansell, 2000, *Alternatives Journal, 26*(3), 17–18; *Montreal's Community Gardening Program,* City Farmer, 1997a, http://www.cityfarmer.org./Montreal13.html#ontreal; *Urban Agriculture Notes,* City Farmer, 1997b, http://www.cityfarmer.org/urbagnotes1.html#notes; *Connections: Canadian Lifestyle Choices and the Environment,* Environment Canada, 1995, State of the Environment Fact Sheet No. 95-1; *The State of Canada's Environment—1996,* Government of Canada, 1996, Ottawa: Supply and Services Canada; "Gardens Overhead: Rooftop Culture Sprouts in North American Cities," J. Kwik, 2000, *Alternatives Journal, 26*(3), 16–17; "Community Gardens: Growing Communities," B. Pezzi, April 1998, *Encompass,* 11.

Valley in British Columbia as an example. The ecological footprint model suggests that achieving more sustainable communities will require that stresses on natural resources be reduced. This does not mean that cities should strive to be self-sufficient within their own boundaries, but rather that cities should meet the needs of society while simultaneously using resources and generating wastes at levels that are compatible with ecological sustainability within their region. In addition, cities should strive for no overall loss of environmental capital within both the nation and the planet (Mitlin & Satterthwaite, 1994). Moreover, as not all residents consume resources in the same quantities, city planners must be sensitive to these differences by attempting to curb consumption for some while attempting to meet basic needs for others who live in cities.

The Principles of Sustainable Cities as outlined by the Sustainable Calgary project are to (1) maintain or enchance ecological integrity, (2) promote social equity, (3) provide the opportunity for meaningful work and livelihood for all citizens, and (4) encourage democratic participation of all citizens. This means that cities need to perform not only the economic functions that are the basis of their existence but also evolve to meet changing social and economic needs. In practice, citizens need to place a high priority on the condition of their environment and ensure that city administrators not only recognize the need for continuing economic and physical development and revitalization in the city but also provide water and sewage treatment plants, clean up contaminated land, and preserve open spaces. In short, environmental protection and resource conservation must be recognized as integral components of urban form and function.

Potentially, cities may be better for environmental protection and resource conservation than dispersed settlement patterns because cities may achieve economies and efficiencies in water, sewage, and waste disposal (including recycling and reuse), in energy use (through district heating), in use of land (through compact development), and in transportation (substituting walking, bicycling, and transit for car use). However, if urban sustainability is to be an objective of Canadian cities, we need to recognize that some obstacles exist.

Among the major obstacles is the fact that we still lack certain kinds of information on which to base long-term decisions about the future of our urban areas. For instance, we do not know the long-term consequences of climate change, or the implications of certain air and water pollutants for human and ecosystem health, nor the best physical form or appropriate density for residential occupation. Lack of full knowledge, however, is never an excuse for inaction on any of these issues.

We do know that Canadians continue to prefer single-family detached housing over higher-density housing, and that they prefer to use their cars rather than public transit. Changing ingrained social values, personal lifestyles, and economic expectations of individuals is always difficult. And, as the oldest and relatively prosperous segment of the Canadian population continues to grow in number, resistance to change may increase (Government of Canada, 1996). Such social and demographic factors suggest that, to date, the concept of environmental sustainability has not been a significant influence for change in the complex field of urban development.

Another reason we have been slow to move to a sustainable communities approach to urbanization is because of the great expense and long life of the buildings, expressways, sewage treatment plants, public transit systems, and other facilities that make up our urban fabric. It would be very costly, both socially and financially, if we decided to quickly and radically alter this urban fabric. Particularly given current fiscal constraints, slow evolution toward more sustainable communities is the most likely scenario for change.

Another obstacle is that political control and administration of cities in Canada is not well suited to achieving sustainable communities. Different federal, provincial, and municipal policies and programs, the lack of coordination among them, and the lack of cooperation among the three levels of government affect communities differentially. Sometimes, for instance, government policies make it difficult to balance social, economic, and environmental aspects of urban development, or to weigh the specific interests of a small part of the population against more diffuse interests of everyone. In addition, during the 1990s, governments began to withdraw from providing services, or higher orders of government provided less money to provincial levels, which, in turn, provided fewer resources to municipalities. In many cases, critical infrastructures—e.g., roads, sewage systems—were not maintained on a regular basis and now require considerable investment to meet new demands for sustainability. Fortunately, government departments and agencies are working to improve their cooperation and coordination so that progress toward sustainable communities may advance. Several federally supported sustainable community development initiatives attest to the changing scene. Interestingly, support for these initiatives has come from a variety of government departments, including Environment Canada, Health Canada, Natural Resources Canada, and Fisheries and Oceans Canada (New Economy Development Group, 2001). In addition, the federal government is now setting money aside to provide municipalities with resources needed to maintain and upgrade their transportation and other urban infrastructure. These investments are critical if we are to move toward more sustainable urban forms.

MAKING CANADIAN CITIES MORE SUSTAINABLE

In spite of the difficulties in striving for urban sustainability, there are signs of progress in urban form, conservation, reduction of environmental impacts, transportation, and planning. In the sections that follow,

we consider some examples of actions that have been taken to help make Canadian cities more sustainable.

Urban Form

Current thinking is that urban sustainability will be advanced through a more compact urban form. A more compact urban form is expected to lead to more economical use of land, water, energy, and materials, and to reduce our dependence on cars in the city. Several municipalities across Canada, including the city of Halifax, the regional municipality of Hamilton-Wentworth, Metropolitan Toronto, and the city of Regina, have revised their planning policies deliberately to support the shift to more compact urban forms. The kinds of changes that are envisioned for sustainable cities in Canada in the future are identified in Table 13–2.

Some of these changes are being put in place now; for example, in Greater Toronto and Greater Vancouver, suburban town centres on high-capacity transit networks are being developed to house commercial and institutional activities that once were concentrated downtown. Kitchener and Winnipeg have adopted urban growth boundaries. In communities such as Bamberton (near Victoria), McKenzie Towne in Calgary, and Cornell (near Toronto), new community developments have been planned explicitly to achieve conservation and environmental protection goals in addition to social goals. Each of these developments features relatively high densities and encourages movement by foot, bicycle, and transit rather than by car. Also, the Historic Properties in Halifax, Montreal's Vieux Port, Toronto's St. Lawrence neighbourhood, The Forks in Winnipeg, and False Creek in Vancouver are examples of the cleanup and transformation of obsolete industrial, rail yard, and warehouse areas for different purposes, including housing, office buildings, and parkland.

Photo 13–14
Urban redevelopment has transformed Vancouver's False Creek.

TABLE 13-2
CHARACTERISTICS OF CANADIAN "SUSTAINABLE CITIES" IN THE FUTURE

Changes to the form of the city (the pattern and density of its physical fabric) are expected to advance urban sustainability. In the future, sustainable cities in Canada would be expected to have the following characteristics:

- A substantially higher average density than today's city
    - Land is used more fully; abandoned or underused sites are redeveloped.
    - Obsolete industrial and commercial buildings are converted to residential use.
    - Single-family detached housing (low-density) is largely replaced by more dense, compact forms.
    - Compact, affordable, and adaptable housing, such as the narrow, two-storey rowhouse (Grow Home) designed at McGill University's School of Architecture, provides ground-level access and some private outdoor space.
- A network of viable, linked subcentres
    - These subcentres and the downtown area are linked by mixed-use corridors and high-capacity rapid transit.
    - Each centre provides a range of services plus employment opportunities and moderate- to high-density housing.
- A mixed land use pattern
    - Residential and other uses are mixed (more than we see now) with compatible, nonpolluting industries located in or adjacent to subcentres.
    - This mixed-use development, including a mix of different housing types and sizes, reduces the need for travel.
- A citywide transit system and a network of bicycle routes
    - A convenient, efficient, and reliable city public transit system results in restrictions on the use of private cars, particularly in the downtown area and core of the subcentres.
    - Safe bicycle routes also are constructed.
- A range of housing choices in every residential district
    - All houses are built to high standards of energy and water efficiency.
    - Residents walk safely and conveniently to transit and cycle routes and to local schools, shops, and parks.
- Corridors of open space
    - Open-space corridors are left in their natural condition and run through the entire city.
    - These provide recreational opportunities and ecological links to parks and the open countryside.
- A well-defined edge of the city
    - Carefully planned urban expansion is compact and has no scattered urban infiltration into the surrounding rural area.
    - Expansion is mainly in the form of physically separate satellite communities developed around their own subcentres and served by the city's rapid transit system.

Planners feel that such reshaping of the city is attainable and would go a long way toward conserving land, natural ecosystems, energy, and water, and would reduce air pollution. As well, it would provide a highly livable urban environment.

SOURCE: *The State of Canada's Environment—1996*, Environment Canada, 1996, Chapter 12. Reprinted with permission of the Minister of Public Works and Government Services Canada, 2004.

A redevelopment opportunity in Vancouver involving Southeast False Creek, a patch of former industrial land east of the present False Creek redevelopment, demonstrates the integration of sustainable development into urban design. In 1991, Vancouver's City Council gave direction to explore the lands of Southeast False Creek as a model of sustainable development. In 1996 the city contracted a study of possible forms of redevelopment. The plans that were originally presented suggested an approach that, to many, seemed to be "more of the same"—that is, unsustainable development with little emphasis on the principles of sustainable communities.

An environmental lobby organized around the proposed redevelopment plans and ultimately was able to pressure the city to conduct another study exploring how sustainability could be defined in an inner-city context.

The report that emerged treated ecological, social, and economic considerations as complementary aspects of a single package of sustainability. The report ultimately served as the basis for the creation of the city's policy statement governing development of the site. The policy direction was further strengthened in 1998 when sustainable design concepts were invited for the site. In October 1999 Vancouver council approved the sustainable community policy for Southeast False Creek that included five environmental plans for waste management, water management, energy, transportation, and urban agriculture. In May 2003, an official development plan proposal was prepared, followed by a public consultation process in the spring and summer 2003. After comments were received, a revised proposal was resubmitted in February 2004 for public discussion. As these plans are ongoing documents, a final deadline for completion has not been established. This is an example of sustainability principles being incorporated into specific design elements.

Intensification of land use (developing or redeveloping land at higher densities) is another change that is being worked out through building conversions, neighbourhood rehabilitation, and infill construction. In cities as diverse as Toronto, St. John's, Sudbury, Montreal, and Vancouver, places of employment and residence are being integrated in mixed-use neighbourhoods. Higher-intensity land use is being achieved also through the use of infill housing, smaller lots in new suburban areas, and by Main Street initiatives that encourage residential development above retail establishments. These initiatives are also aimed at making housing affordable.

Conservation

Water Conservation   Water and energy conservation programs have been put in place in many Canadian municipalities. Water conservation programs typically involve city programs such as leak detection and repair, metering, retrofitting, public education, and use restrictions. Major centres such as Toronto, Ottawa–Carleton, Laval, and Edmonton have broad water management plans in place, as do many smaller urban centres such as Cochrane, Ontario, and Rosemère, Quebec. In light of chronic water supply problems in Kitchener–Waterloo, a university student wrote a thesis that outlined a program for water conservation. His thesis was so convincing that he was awarded a new position with the municipal council—that of water conservation officer! Communities also contemplate alternatives: in considering how their area could move toward sustainability, the Greater Vancouver Regional District proposed that rainwater be collected for flushing toilets and for use in gardens (Balcom, 1997). Both municipal and community actions are important in water conservation.

Energy Conservation   Like water conservation programs, energy conservation efforts focus both on reduction of city costs for fuel and electricity and on community-wide initiatives that contribute to economic, social, and environmental objectives such as local economic development, improved air quality, and reduced CO_2 emissions. Common energy conservation initiatives include changing streetlighting, adopting energy standards

for buildings, retrofitting municipal buildings, and converting municipal vehicles to alternative fuels (Federation of Canadian Municipalities, 1995).

Energy audits and energy controls in municipal buildings, and energy reviews of newly designed municipal facilities, are among the ways both large and small communities can identify energy-saving possibilities. Quebec, for example, has had a provincial law on energy efficiency in new buildings since 1983. In other parts of the country, community partnerships with utility companies such as Ontario Hydro have enabled smaller communities to identify opportunities for managing energy demand and to implement energy and cost-saving measures.

Increasingly, municipalities have become involved in urban energy management, and have linked with broader community environmental efforts. Ontario's Green Communities program is one example in which the focus was on smaller centres such as Guelph, Sarnia, and Peterborough. Green Communities are nonprofit, community-based, multi-partner organizations (made up of businesses, institutions, and governments) that bring environmental solutions to homes. They are in the business of selling environmental action through behaviour change and uptake of green products and services (Green Communities Association, 2001). Domestic energy conservation measures also are being adopted both through retrofitting of older buildings and in new construction (including R-2000 standards; see Chapter 11).

Conservation of Materials   Many Canadian municipalities now offer recycling programs, compost collection, and other waste reduction services. Typically, emphasis is on the "3Rs"—reduction of the volume of waste at its source, reuse of materials, and recycling and composting. Some cities, including Sherbrooke, Montreal, Toronto, and Vancouver, have firm waste reduction targets; some, such as Toronto and Vancouver, are trying to establish controls on packaging; and still others have placed limits on the amount of garbage the city will collect from each dwelling.

Access to recycling programs of various types has increased throughout Canada. In 1994, 69.6 percent of Canadian households had access to a paper recycling program either through curbside collection or recycling depots. While paper recycling is the most widely available program, access to recycling programs for glass bottles (67.4 percent), metal cans (67.2 percent), and plastics (62.8 percent) has also increased. Also, 40.2 percent of communities offer special hazardous-waste disposal opportunities for paints, chemicals, and batteries. Since the early 1990s, when large cities generally had programs in place, access to recycling facilities has increased the most in larger urban areas in Quebec and Atlantic Canada as well as in smaller centres in Ontario and western Canada. At least through 1994, more than 81 percent of households with access to recycling programs continued to use them. In

Photo 13–15
Readily available and affordable, recycled-plastic compost bins are an increasingly common sight.

Photo 13–16
Collecting rainwater in barrels or other containers is an effective way for individuals and communities to augment municipal water supply systems.

spite of its strong potential to reduce the amount of organic material entering the municipal waste stream, composting has not been accepted as readily as other recycling programs; in 1994, only 23 percent of households used a compost heap (Statistics Canada, 1995).

Canada's experience with recycling is also being shared internationally. The International Centre for Sustainable Cities (ICSC), a nongovernmental organization established to promote sustainable urban development around the world, is leading a consortium of four Canadian partners to help Polish municipalities acquire Canadian technology and knowledge in the field of composting and sludge management. The project began in 1996 and has two main elements: the development of a municipal composting/sorting plant for urban and industrial waste in the city of Katowice, Poland; and the transfer of state-of-the-art concepts and practices in sludge management to Polish decision makers to improve their regulating capabilities in this area. The project is being supported by the Canadian International Development Agency (Recycling Council of Ontario, 2001).

Conservation of Ecosystems and Natural Features

Cities are moving away from the traditional view of urban parks as green areas for recreation and toward conserva-tion of natural areas, environmentally sensitive areas, and natural habitats. Although the biodiversity values of wetlands, shore zones, forests, and wildlife corridors have begun to be recognized, their protection is just beginning at the local level.

To encourage wildlife population growth and restoration of natural ecosystems, such cities as Ottawa, Edmonton, Toronto, and Montreal have begun to create or restore green corridors by linking small natural areas, cemeteries, waterfronts, transmission line right of ways, and other open spaces. Nationally, the Trans Canada Trail Foundation is overseeing the creation of a cross-country trail that conserves and preserves our natural heritage (see Box 14–4). The city of Calgary employs its Natural Area Management Plan to help protect existing natural environments and to identify potential areas for future conservation prior to their development. Many cities now practise urban forestry, encourage naturalization of vegetation, and substitute integrated pest management for chemical pesticides. From Gander, Newfoundland, to Fort Saskatchewan, Alberta, experiments have been undertaken with sheep and other herbivores to replace gasoline-powered lawnmowers (Federation of Canadian Municipalities, 1995).

Some recent municipal legislation has focused on achieving ecosystem-based planning. The ecosystem approach—land use planning that is based on natural boundaries and respect for ecosystem integrity (Table 13–3)—is more of a promise than a reality at the moment. However, outstanding examples of the application of the ecosystem approach are the Fraser River Estuary Management Program, Saskatoon's Meewasin Valley Authority, and the Hamilton Harbour Remedial Action Plan (Tomalty et al., 1994).

Reduction of Environmental Impacts

Ideally, cities should be free of activities that harm the health of other people or natural ecosystems. In reality, however, financial, jurisdictional, and behavioural barriers arise that affect progress toward air and water quality, waste management, and cleanups of harmful sites or substances.

Air Quality The main way to lower concentrations of ground-level ozone (the leading urban air quality problem in Canada) is to reduce the emissions of its precursors (nitrogen oxides and VOCs) by motor vehicles. Federal and provincial governments largely control air pollution action—such as banning leaded gasoline and regulating vehicle emissions—through the federal Motor Vehicle Safety Act and other provincial regulations. Federal and provincial governments continuously review quality standards and regulations (the NAAQOs), and some cities are taking action to improve air quality.

Recently the Federation of Canadian Municipalities' (FCM) "20 percent club" merged with the International Council for Local Environmental Initiatives' (ICLEI's) Canadian Cities for Climate Protection campaign. This united effort is now named Partners for Climate Protection: For a Better Quality of Life. The goal of the merged program remains to support Canadian municipal governments and to prepare and implement local climate action plans. FCM has responsibility for all formal relations with the Canadian federal government. ICLEI is the technical partner, assuming direction for activities such as software tools, technical manuals, and monitoring and verification activities.

The ultimate goal is to reduce greenhouse gas emissions from municipal operations 20 percent below 1990 levels within 10 years of joining the program, and to reduce community-wide greenhouse gas emissions at least 6 percent below 1990 levels within 10 years of joining the program. The priorities of the new program are to build capacity, support champions, provide up-to-date information, ensure access, create model plans, facilitate participation, generate feedback, and build partnerships (Federation of Canadian Municipalities, 2001).

Several cities have policies and programs relating to the use of alternative fuels as well as the reduction of emissions of sulphur dioxide, carbon dioxide, and nitrogen dioxide. Vancouver directs attention toward reduction of greenhouse gas and ozone-depleting emissions, Montreal's strategy focuses on reducing emissions and use of CFCs and halons, and Toronto has specific targets for reducing SO_2 emissions by 2006.

Water Quality Ideally, when treated water is released to the receiving body of water, the treated water should be at least as clean as when it entered the supply system or fell as natural precipitation. This idea that there should be no increase in the net level of impurities in the hydrologic cycle has been interpreted to mean that cities should (1) provide secondary or tertiary treatment of all wastewater, including storm drainage; (2) prevent overloading of treatment plants in heavy storms; and (3) make necessary improvements to their collection system to prevent leakages resulting from age and deterioration.

Although variable, the level of wastewater treatment has increased across Canada. Montreal recently completed a major sewage treatment scheme, while smaller centres such as Banff, Portage la Prairie, and others use innovative methods of sewage treatment, including biological processes and ultraviolet disinfection. However, on the argument that health risks and harm to the environment are negligible if the discharge pipe is located at sufficient depth and distance from shore, and if currents are suitable, some coastal cities discharge sewage into the sea after only low levels of treatment. Victoria's practice of discharging untreated sewage waste into the Strait of Juan de Fuca has, for years, generated continuing complaints from Washington state.

Waste Management and Cleaning Up Good urban waste management programs help protect the environment and conserve materials by reducing the total quantity of waste that must be assimilated into the environment. Because everything is connected to everything else, any disposal method will have some adverse environmental effects. While we cannot eliminate them, we can work to minimize these effects. On this basis, in addition to promoting the 3Rs, many cities locate their landfills in such a way that (1) truck traffic is minimized, (2) their location, design, and construction prevent leaching and gas emis-

TABLE 13-4
ACTIONS TO ACHIEVE MORE SUSTAINABLE URBAN TRANSPORTATION

- Shift budget and program priorities from road systems to transit systems.
- Provide diverse transit options such as rapid transit, commuter rail, and surface transit networks; special facilities for high-occupancy vehicles such as car and van pools; and cycle and pedestrian pathways.
- Improve efficiency, speed, reliability, and general attractiveness of transit through dedicated lanes for buses, transit priority at intersections, schedule reliability, provisions of up-to-the-minute travel information, and integration of fares and schedules between routes and systems.
- Encourage more efficient use of infrastructure and vehicles such as travelling at off-peak time, combining trips, substituting transit for car use, sharing vehicles, and using less-congested routes.
- Discourage unnecessary automobile use in local residential areas by such means as traffic-calming street design.
- Provide public outreach, awareness, and education programs.
- Improve vehicle technology by designing lighter, more aerodynamic vehicles and smaller, more fuel-efficient engines, and by developing alternative fuels.
- Reduce the need for vehicular movement by implementing urban designs in which land use and transportation are integrated to create pedestrian-friendly streets and compact centres of intensive mixed activity linked by mixed-use corridors.

SOURCE: *The State of Canada's Environment—1996,* Environment Canada, 1996, Chapter 12. Reprinted with permission of the Minister of Public Works and Government Services Canada, 2004.

sions and permit restoration as usable land, (3) they employ the best incineration technology (if used), and (4) the public is involved in planning of disposal programs.

Cleaning up past environmental legacies of activities within a city sometimes entails dealing with contaminated land (by removing, treating, or disposing of the soil appropriately), or finding ways to use the site that will not endanger human health. Outside the city, it might be necessary to deal with old garbage dumps and tire disposal sites that may affect groundwater and pose health risks or fire hazards. Vancouver, Calgary, Toronto, Ottawa, and Montreal have developed comprehensive inventories of contaminated sites and have review processes in place to control development on or near these sites (Federation of Canadian Municipalities, 1995).

Transportation

Apart from walking and human-powered vehicles such as bicycles, all forms of transport entail significant consumption of resources and impacts on the environment, particularly on air quality. Because transportation decisions affect urban environments in so many ways, transportation is a key area for action. Some of the many ways cities can and have achieved greater sustainability in their transportation systems are outlined in Table 13-4. In 1993 it was determined that if we could apply these kinds of measures, even without changes in urban structure or improved emission control and fuel efficiency, the average energy use could be reduced by 29 to 36 percent by the year 2000 (IBI Group, 1993). In addition, we have the lessons from northern European ecological community design to assist us in reducing dependency on our vehicles (Saunders, 1996).

In fact, in some cases, residents are showing initiative in this regard. In large cities, including Vancouver, car cooperatives are beginning to emerge. Car cooperatives typically operate on a nonprofit basis. Individuals buy a "share" at the outset and then book a vehicle and pay for it on an as-use basis. Users are charged based on the amount of time the vehicle is booked and the distance travelled. These charges then pay for the insurance, maintenance, and other costs of the vehicles. For urban residents who do not require a car every day for commuting purposes, car cooperatives provide an affordable and convenient way to go shopping or take a day trip that is economical as well as environmentally and socially responsible.

Throughout the country, in both larger and smaller urban centres, a variety of government-led initiatives to improve transportation have been undertaken and are under way. These initiatives may be categorized as urban structure and urban design policies, transportation infrastructure, demand management, traffic and transit management, and cleaner vehicle technology (Irwin, 1994). Specific municipal policies and programs to improve transit services have been implemented in St. John's, Dartmouth, Montreal, and Toronto; reserved bus lanes have been established in Vancouver, Toronto, Ottawa, and Montreal; parking-related measures have been proposed in Montreal, Toronto, and Vancouver; and bicycle and pedestrian networks exist in Calgary, Vancouver, and other cities.

Planning

Deliberate movement toward urban sustainability implies "a more holistic, ecosystem-based approach to urban policy and governance than has been usual in Canada" (Government of Canada, 1996). This kind of approach means that attention would be paid at the local level to developing and implementing urban sustainability goals, standards, and criteria, and that progress toward sustainability would be monitored and reported on in regular, municipal state-of-the-environment reports. In addition, sustainability goals would be supported by appropriate provincial (and sometimes federal) fiscal policies and economic development policies that are consistent with sustainability goals and principles. This approach also implies integrated, effective planning for the entire urban region.

Examples of municipal cooperation in planning for sustainability are found in the official plan of the Regional Municipality of Ottawa–Carleton, the land reclamation program of the Regional Municipality of Sudbury, and the Greater Vancouver Regional District's Livable Region Strategy. Other municipalities are moving toward more comprehensive sustainable development or environmental policies. The city of Halifax, and the Regional Municipality of Hamilton–Wentworth, among others, are implementing their broad visions of sustainability for the future (Pearce, 1995). Many cities now have environmental offices and interdepartmental environmental committees composed of municipal staff. Others, such as Calgary, have developed environmental advisory committees that value public representation and input. Severe budgetary cutbacks in some municipalities, however, portend reductions in environmental programs.

Several municipalities have produced (or are in the process of preparing) state-of-the-environment reports, while other cities have completed reports on the quality of life, state of the city, or environmental issues. One such report was developed by Sustainable Calgary as part of its Sustainability Indicators Project. Sustainable Calgary is a citizen-led, nonprofit society formed in 1996 as a result of a series of public forums. Its objectives are to encourage and support action to influence policy, planning, and community processes. All of Sustainable Calgary's projects are guided by the following sustainability principles: ecological integrity, social equity, sustainable economic development, and democratic participation in decision making.

Sustainable Calgary believes that quality of life in the city can be sustainable only if the processes that support it generate long-term health and vitality for all. Although Calgary residents enjoy a relatively high quality of life, from a global ecological perspective the city is relatively unsustainable. Canadians in general and Calgarians in particular consume a high percentage of the natural wealth of the planet to live as they do. Sustainable Calgary wants to develop the necessary tools and processes to help the city move toward greater sustainability.

In 2000, approximately 2000 Calgarians were involved in the Sustainability Indicators Project—attending workshops, presentations, and think tanks. Other successful endeavours include the creation of a Calgary Green Map, a program to assess Calgary's ecological footprint, and an educational program of Tools for Citizenship in the 21st century.

The project established a broadly accepted set of sustainability indicators, which provide information for understanding and enhancing the relationships between ecological, social, and economic factors. To date, two "State of Our City" reports have been published.

Regarding citizen action, the persistent, frequently well-informed, and skillful lobbying of environmental advocacy organizations often has helped bring about new or improved policies and programs within cities. In addition, some municipalities try to involve residents in environmental actions ranging from community cleanups, to tree-planting and park naturalization efforts, to bird counts. Similarly, the healthy communities movement in Quebec, Ontario, and British Columbia has the common (and provincially supported) goal of socially, economically, and environmentally healthy communities. Literally hundreds of communities across Canada have developed a variety of achievable and practical sustainability strategies through involvement of local citizens and municipal representatives. Even universities are getting on board. In the past few years, several universities across Canada are beginning or have developed sustainability plans to ensure that universities demonstrate leadership in sustainable land and resource use including pesticide and water use, energy generation and use, waste use and handling, emissions, and land use. Plans for upgrading of infrastructure and construction of new infrastructure often have to meet new sustainability standards as specified through their sustainability objectives.

Sustainability reporting initiatives such as the one being implemented by Sustainable Calgary demonstrate a growing movement to look beyond a narrow set of economic indicators currently employed to make decisions. By its very nature, sustainability reporting requires that a broad range of social, ecological, and economic indicators be brought into the decision-making process.

Alternatives to the conventional approach of using gross domestic product (GDP) as a measure of a nation's economic health have been proposed. Developers of the genuine progress indicator (GPI) argue that as a measure of economic health, GDP is badly flawed. By counting only monetary transactions as economic activity, the GDP omits much of what people value and activities that serve basic needs. The GDP also ignores the value of leisure time spent in recreation and relaxation, or with family and friends (Cobb, Goodman, & Wackernagel, 1999). The GPI adds up a nation's expenses (GDP), factors in sectors that are usually excluded from the market economy, such as housework and volunteering, and then subtracts social ills: crime, natural resource depletion, and loss of leisure time (Baker, 1999).

"We want people to rethink what progress is all about," says Mathis Wackernagel, director of Indicator Programs at Redefining Progress, the San Francisco–based policy organization that developed GPI and other social progress indicators. "What the GDP deems as growth is merely increased spending. It does not tell us if the spending is good or bad. GPI, on the other hand, differentiates between what most people perceive as positive and negative economic transactions, and between the costs of producing economic benefits and the benefits themselves. It adds up the value of products and services consumed in the economy—whether or not money changes hands" (Cobb et al., 1999). In Nova Scotia, GPI Atlantic, a nonprofit research agency, was formed in 1997 to construct an index of sustainable development for that province. The Nova Scotia GPI consists of 22 social, economic, and environmental components under categories of time use, natural capital, environmental quality, socioeconomic indicators, and social capital. These kinds of efforts illustrate alternative ways to track our performance and to consider policy options related to sustainability.

Progress toward Urban Sustainability?

Although it is not possible to provide a definitive answer to the question of whether Canadian cities are being planned and managed in ways that are moving them toward greater sustainability, there are signs that some resource demands and some stresses on both local ecosystems and the global ecosphere are being reduced. If we accept various reservations and exceptions, progress is evident in air and water quality and waste management areas. Although more can be done with regard to the conservation of water, energy, and materials, progress also has been made in these areas. Municipalities are continuing to make substantial efforts to conserve and protect environmentally sensitive areas, green space, and natural systems.

As intensification has become a major policy goal, there has been increasing restraint on low-density expansion in the urban fringe. Cities are trying numerous approaches to have citizens use transit, bicycles, and their feet instead of cars. And new techniques and approaches such as environmental assessment, sustainable development policies, and state-of-the-environment reporting are appearing more frequently in municipal operations.

In spite of these successes, however, there is concern that continuing improvement in energy conservation and in air quality depends on attaining a more compact urban form and a substantial shift in modes of transportation. Municipal and provincial action currently being taken may not be sufficient to achieve the required changes. For instance, in light of the cuts to transit expenditures being experienced by many cities, how likely is it that costly, substantial improvements to transit systems will be authorized? Given current and foreseeable future market forces and consumer preferences, how likely is it that large numbers of people will choose more compact types of housing, or switch from cars to buses or subway trains? Demographic trends do not suggest encouraging changes, and yet our cities, and the number of cars within them, continue to grow.

Potential impacts of global warming for Canadian cities include flooding, depletion or degradation of water supply, increased energy demand, deteriorating air quality, and even the potential for serious impacts on the economic base in some areas. Given that cities in Canada (and elsewhere) depend on the depleting and increasingly less diverse resources from the "global hinterland," they are made even more vulnerable to these long-term threats. In short, while there are signs of progress, there also are signs of concern that the recent municipal movement toward an ecosystem-based approach to land use planning may not come to full fruition. The words of Roberta Bondar (1994), Canada's first female astronaut, are worth remembering: "Good intentions and expressions of concern are not enough. Protecting the environment is the most serious challenge we face today. Every Canadian can and must get involved, not just today, but every day." Long-term change comes about when we all care about the future world we will inhabit and act on our environmental convictions.

Chapter Questions

1. How do cities change their own environment and affect the environment of surrounding areas? How can we plan cities to minimize some of these effects?

2. In what ways is a high-quality life, with reduced waste, reduced energy use, and reduced pollution, both a goal and a challenge?

3. Discuss the ways in which sustainable housing can leave a smaller footprint on the ecosystem than conventional single-family homes. Are there ways in which existing housing in your community could be made more sustainable, environmentally, socially, and economically?

4. Comment: Natural area habitats in city parks will become more important as wilderness decreases.

5. All over Canada, people are taking action to encourage and promote healthy neighbourhoods. Consider your neighbourhood. What are some of the actions that have been taken—or could be taken—to reduce the impact of "the car culture"? What could you do to take the lead in your neighbourhood to reduce the effects cars have on the environment and on our health?

6. One of the challenges facing Canadians in urban areas is to reduce the waste stream. How would you design a program for your community that would achieve a 50 percent reduction of the waste stream? What would be the most important factors to take into account?

7. Critically examine the city or town where you go to university. What are the top three priorities to meet economic, social, and environmental criteria for sustainability? Do priorities in one category make priorities in another category difficult to attain? Can you identify actions (e.g., tax systems, policies, programs) that can be used to help meet multiple priorities simultaneously?

8. Do an inventory of community gardens in the city or town where you currently live. How might this program be introduced or expanded? How might gardens help meet economic, social, and environmental objectives?

9. What elements might you include in a sustainability audit of your university?

references

Baker, L. (1999, May-June). The genuine progress indicator could provide an environmental measure of the planet's health. *emagazine*. http://www.emagazine.com/may-june_1999/

Balcom, S. (1997, April 12). West Coast experiments with sustainable housing. *Calgary Herald*, p. I13.

Bondar, R. (1994, April 18). Earthweek only first step in campaign to save environment, Dr. Roberta Bondar says. *Canada News Wire*.

Canadian Arctic Resources Committee. (1997, Spring). Letter from Chair.

Canadian Council of Ministers of the Environment. (1990). *Management plan for nitrogen oxides (NOₓ) and volatile organic compounds (VOCs). Phase 1*. Winnipeg: Author.

Canadian Environmental Assessment Agency. (1995). *Military flying activities in Labrador and Quebec*. Report of the Environmental Assessment Panel. Ottawa.

Checora, G. (1996). Calgary house goes beyond sustainability. *Alternatives, 22*(4), 5–6.

City of Calgary. (2001). www.gov.calgary.ab.ca/cww/watermeterincentiveprogram.html

City of Halifax. (2001). www.region.halifax.ns.ca//harboursol/index.html

City of Vancouver. (1997). *City noise: Report of the Urban Noise Task Force*. Vancouver: Author.

City of Vancouver. (2002). www.city.vancouver.bc.ca/commsvcs/licandinsp/inspections/environment/trends/air.htm

Cobb, C., Goodman, G. S., & Wackernagel, M. (1999, November). Why bigger isn't better: The genuine progress indicator—1999 update. *Redefining Progress*.

Dupuis, O. (2000). The green housing effect. *Alternatives Journal, 26* (1) (Winter), 7–8.

Environment Canada. (1988). *Wetlands of Canada*. Ecological Land Classification Series No. 24. Ottawa: Environment Canada, National Wetlands Working Group, Canada Committee on Ecological Land Classification.

Environment Canada. (1991). *The national incinerator testing and evaluation program.* Ottawa.

Environment Canada. (2001a). *Urban water: Municipal water use and wastewater treatment.* http://www.ec.gc.ca/Ind/English/Urb_H20/Bulletin/uwind2_e.cfm

Environment Canada. (2001b). *The Green Lane. Vehicle Emissions Inspection Clinic Program.* http://www.ec.gc.ca/special/emissions_e.htm

Environment Canada. (2001c). *The 2001 report on benzene in gasoline.* http://www.ec.gc.ca/energ/fuels/reports/benz_2001/ben_2001_e.pdf

Federation of Canadian Municipalities. (1995). *Canadian municipal environmental directory: Municipal government actions for a sustainable environment: A compendium of environmental initiatives.* Ottawa.

Federation of Canadian Municipalities. (2001). *Partners for climate protection program.* http://www.fcm.ca

Fried, J. J. (1996, January 6). Noise. *Calgary Herald,* p. B4.

Gertler, L. O., Crowley, R. W., & Bond, W. K. (1977). *Changing Canadian cities: The next 25 years.* Toronto: McClelland & Stewart.

Government of Canada. (1996). *The state of Canada's environment—1996.* Ottawa: Supply and Services Canada.

Government of Canada. (2003). *Environmental signals: National Indicator Series 2003,* http:www.ec.gc.ca/soer-ree/English/Indicator_series/

Government of Ontario. (2001). Environmental compliance reports. http://www.ene.gov.on.ca/

Greater Vancouver Regional District. (1994). *Let's clear the air: Draft air quality management plan—Summary document.* Vancouver.

Greater Vancouver Regional District. (2000). *Drinking Water Treatment Program. Program overview.* Issue No. 15. http://www.gvrd.bc.ca/services/water/projects/Program_overview_070100.pdf

Green Communities Association. (2001). http://www.gca.ca

Hope, M. (1995, June 3). Earthship Alberta: Construction method uses recycled tires, compacted soil. *Calgary Herald,* p. H3.

IBI Group. (1990). *Greater Toronto area urban structure concepts study—Summary report.* Toronto: Greater Toronto Coordinating Committee.

IBI Group. (1993). *Initiatives to limit transportation energy consumption and emissions in Canadian cities.* Ottawa: Natural Resources Canada.

Irwin, N. (1994). A new vision for urban transportation: Current Canadian initiatives. In Transportation Association of Canada, Federation of Canadian Municipalities, and Canadian Institute of Planners, Ottawa, *Proceedings: New visions in urban transportation, Part 3.* Ottawa: Transportation Association of Canada.

Kerr, S. (1998, April). Cohousing: Build your own community. *Encompass Magazine,* 6–7.

King, F. (1996, May 11). Hay, there's a new way to build a house! *Calgary Herald,* p. I10.

McGill University. (1999, Summer). McGill news. *Alumni Quarterly.* www.mcgill.ca/alumni/news/s99/ecoresidence.htm

Mitlin, D., & Satterthwaite, D. (1994). *Cities and sustainable development: Background paper prepared for Global Forum '94.* London: International Institute for Environment and Development, Human Settlements Program.

New Economy Development Group. (2001, January). *Progress report on PRI's Sustainability Project on Sustainable Communities.*

National Round Table on the Environment and the Economy (NRTEE). (2003). *Environmental quality in Canadian cities: The federal role.* Ottawa: NRTEE.

Patterson, B. (1995, November 14). Sound off. *Victoria Times Colonist,* p. C1.

Pearce, B. (1995). Hamilton–Wentworth region's sustainable community initiatives. *Plan Canada, 35*(5), 26–27.

Pucher, J. (1998). Back on track. *Alternatives Journal, 24*(1) (Winter), 26–34.

Raad, T., & Kenworthy, J. (1998, Winter). The US and us: Canadian cities are going the way of their US counterparts into car dependent sprawl. *Alternatives Journal, 24*(1).

Recycling Council of Ontario. (2001). http://www.rco.on.ca

Saunders, T. (1996). Ecology and community design. *Alternatives, 22*(2), 24–29.

Shideler, K. (1997, March 15). Common noise and "boom cars" cause hearing loss. *Calgary Herald,* p. C9.

Statistics Canada. (1995). *Households and the environment 1994.* Statistics Canada Catalogue No. 11-526. Ottawa: Statistics Canada, Household Surveys Division.

Statistics Canada. (2002). *Human activity and the environment, Annual Statistics 2002.* Ottawa: Ministry of Industry.

Sustainable Calgary. (2001). *Sustainable Calgary.* http://www.telusplanet.net/public/sustcalg/

Taus, M., & R. McClure. (2002). When it rains, it pours pollutants into the waters. *Seattle Post.* http://seattlepi.nwsource.com/local.95883_sound20.shtml

Tomalty, R., Alexander, D. H. M., Fisher, J., & Gibson, R. B. (1994). *Planning with an ecosystem approach in Canadian cities.* Toronto: Intergovernmental Committee on Urban and Regional Research Press.

Turner, J. (1996, August 28). Tide is turning against industrial waterfront. *The Globe and Mail,* p. A2.

United Nations Development Program (UNDP). (2000). *Human Development Report 2000: Human Rights and Human Development.* http://www.undp.org/hdr2000/English

Verrall, C. (1995). A tale of community activism: Residents work together to halt the Brantford Southern Access Road. *Alternatives, 21*(4), 47–49.

Warren, C. L., Kerr, A., & Turner, A. M. (1989). *Urbanization of rural land in Canada 1981–1986.* SOE Fact Sheet No. 89-1. Ottawa: Environment Canada, State of the Environment Directorate.

Wood, G. (1997). The first little pig was right. *Alternatives, 23*(3), 7–8.

Workers' Compensation Board of B.C. (2003). *Hear for good: Preventing exposure at work.* http://www.worksafebc.com/publications/Health_and_Safety_Information/by_topic/assets/pdf/hear_for_good.pdf

CHAPTER 14

Meeting Environmental Challenges

Chapter Contents

*"Our past experience has demon-
strated that successful environmental
management depends on our ability to
fully engage individuals and communities in defining the problems, finding
the solutions and taking action to
improve the quality of our environment. In the end—whether the source
of a problem is in our backyard or on
the other side of the globe—the
problem becomes a local one. As such,
individuals and communities must do
their part, and governments and
others must support their efforts to do
so."*

David Anderson, Minister of the Environment,
Canada (personal communication, 2003)

Chapter Objectives

After studying this chapter you should be able to

- appreciate the general trends in environmental issues and sectors
- outline the broad range of Canadian efforts to safeguard our environment
- identify those challenges that remain in our quest for sustainability
- identify current models and practices that can help to achieve sustainability
- appreciate some of the Earth-sustaining actions we can take

INTRODUCTION

Every day, each Canadian makes lifestyle choices that have substantial environmental effects, choices that relate to the kind of housing we live in, the foods we eat, the appliances we use, the household products we select, and the means of transportation we favour. Every one of these choices may seem insignificant, but cumulatively (especially if many people make choices with negative impacts) these choices may result in serious damage to, and may influence the sustainability of, our future environment. The same is true of the kinds of decisions that local organizations, private firms, governments at all levels, and public agencies make in their daily operations.

Protecting and sustaining the quality of our environment is a serious challenge for all Canadians. Balancing economic and social well-being and the integrity of the ecological systems that support our economy and society requires all of us to come to grips with the fundamental issues involved in achieving sustainability. Not only do we need to think globally and act locally, cognizant of the impacts our actions may have on others and on our environment, but also we need to work together in a cooperative and collaborative fashion. Achieving sustainability goals requires new, greener ways of thinking, making decisions, and acting on these choices. Will we meet these challenges?

In the sections that follow, we briefly review the trends in key environmental issues and sectors and indicate the kinds of actions taken and progress made in safeguarding Canada's environment. The chapter ends with a review of the challenges that remain, indicates models, and offers some suggestions about what we as individuals

Photo 14–1
Have we done everything we can to "reduce, reuse, and recycle"? From individual efforts come advances in our collective well-being.

and together in groups can do to live in a more Earth-sustaining manner.

PROGRESS IN SAFEGUARDING CANADA'S ENVIRONMENT

Since the 1970s, environmental awareness, conservation, and protection have become increasingly important elements in economic and social decision making. As scientific understanding of ecosystem dynamics and complexity has grown, the need to take an ecosystem approach to environmental issues has been impressed on many decision makers. As a result, in addition to legislative and technical solutions, Canadians began to promote cooperation among governments, industry, ENGOs, and communities. Public education and action also were fostered, various economic instruments developed, and voluntary codes of conduct established. These changes enabled progress on certain environmental issues, but in other areas we have fallen short of our goals.

In 1989–1990, in the wake of heightened concern about Canada's environmental health, the government of Canada began developing environmental indicators based on four themes: ecological life-support systems, human health and well-being, natural resources sustainability, and pervasive influencing factors. The indicators measured were developed to be relevant to the Canadian public, useful in daily decision making, and catalysts to promote behavioural and ethical changes. They were also meant to be flexible enough to respond to changes in sci-entific data and public opinion. After 10 years of tracking, there is now a usable database for assessing Canada's environmental trends. In 2003, Environment Canada published *Environmental Signals: Canada's National Environmental Indicator Series 2003* in collaboration with Agriculture and Agri-Food Canada, Health Canada, Natural Resources Canada, and Statistics Canada. The purpose of the document is to provide information on progress made in addressing key environmental issue areas. The document also describes indicators evaluated, actions taken to date, and challenges remaining.

This publication is significant because it is one of the first governmental initiatives that assesses the effectiveness of Canada's environmental policies over the longer term. Also, it highlights how long it takes to evaluate and address environmental concerns. After more than a decade of efforts aimed at mitigating environmental degradation, there has been improvement in some issue areas, but additional degradation has occurred in others. The environmental indicator series is significant because it exists—this document represents progress made in tackling Canada's environmental concerns. However, Environment Canada cautions that the assessment is by no means exhaustive and recognizes that there are indicator gaps, particularly in human health and ecological effects. Table 14–1 provides a summary of the key points in the publication regarding air and water quality issues.

AIR QUALITY ISSUES

Regulatory changes by the federal and provincial governments, supported by technological advances in the private sector as well as by partnership actions involving

TABLE 14–1
SUMMARY OF CANADA'S ENVIRONMENTAL SIGNALS, 1990–2000

Environmental Issue Area	Improvement Trend and Indicator[1]	Challenges
Air quality		
Acid rain	Trend: Improving (+15%) Indicator: Trend in total emissions	• Critical loadings lower than originally understood, requiring further reductions. • NO_x deposition rate threatens to override gains made by SO_x reduction. • Interaction of acid deposition, climate change, and stratospheric ozone depletion emphasizes need for multi-issue approach.
Stratospheric ozone	Trend: No change Indicator: Trend in Canadian values	• Global implications make problem resolution more difficult, despite Canada's proactive approach. • Climate change may drive ozone loss. • Lack of reporting. • Smuggling of ODSs by some countries.

TABLE 14-1
(CONTINUED)

Environmental Issue Area	Improvement Trend and Indicator[1]	Challenges
Urban air quality	Trend: No change Indicator: Regional trends of ground-level ozone, total suspended particulate, NO_x and SO_x emissions	• Better understanding of chemistry of pollutants and consequences of synergistic effects is needed.
Passenger transportation	Trend: Deteriorating (−10%) Indicator: Trend in automobile use	• Government initiatives to increase use of public transportation have been virtually ineffective. • Automobile use has increased by 9% since 1990. • Decrease in this sector by 1.1% attributed to popularity of SUVs and minivans.
Water quality		
Municipal water use	Trend: Improving (+4%) Indicator: Percentage change in per capita water use	• Cost to municipal consumers does not reflect supply costs—Canadian water costs are currently among the lowest in the world. • Increase in use efficiency is needed. • Climate change effects on water quality and quantity are unknown. • Most coastal municipalities have no water treatment.
Municipal wastewater treatment	Trend: Improving (+20%) Indicator: Percentage change in proportion of population with secondary or tertiary sewage treatment	• Effects of treatment chemicals released into the environment are unknown. • Existing water treatment infrastructure is "faltering."
Biological diversity	Trend: Improving (+70%) Indicator: Increase in strictly protected areas	• 64% of strictly protected areas are under 10 km^2 and are inadequate for large mammal protection. • More than half of Canada's identified ecoregions have little or no protection. • Trend assessment does not include other important indicators including COSEWIC data. • No reliable baseline data exist against which to measure habitat loss, species status, range, or population sizes, indicating the need for more research. • Existing databases are difficult to compare.
Climate change	Trend: Deteriorating (−20%) Indicator: Percentage change in greenhouse gas emissions	• Changing the energy and other resource consumption habits of individuals. • Program tracking for successful initiatives. • Better understanding of regional impacts.
Sector industries		
Forestry	Trend: Improving (+30%) Indicator: Percent of strictly protected area in all four forest ecozones	• Forest contribution to overall ecosystem health not well quantified. • Forest health indicators not encompassed in trend evaluation include change in forest species (no change) and harvesting rates (still increasing).
Agricultural soils	Trend: Improving (+20%) Indicator: Number of days soil left unprotected by vegetation	• Data collection excludes identification of smaller-scale degradation problems. • Many soil health indicators are not included in trend assessment (e.g., residual nitrogen, soil organic matter, salinization, nutrient levels).

(continued)

TABLE 14–1
(CONTINUED)

Environmental Issue Area	Improvement Trend and Indicator[1]	Challenges
Fisheries	Not included	
Mines and minerals	Not included	
Energy	Trend: Deteriorating (–10%) Indicator: Canadian energy consumption	• Energy efficiency is improving, but energy use is still increasing. • Fundamental shift in thinking is necessary to decrease fossil fuel dependency and expand use of alternative energies.

[1] Numerical values are a description of one indicator only and are not necessarily a reflection of total issue trends. They are used by Environment Canada to highlight the rate of progress occurring on a specific issue.

SOURCE: *Environmental Signals: Canada's National Environmental Indicator Series 2003,* Environment Canada, 2003, http://www.ec.gc.ca/soer-ree/English/Indicator_series/default.cfm

industries, ENGOs, and communities, are among the efforts that have brought about reductions in the production and emission of many air contaminants. On an international level, for instance, Canada has exceeded its commitments to reduce ozone-depleting substances, and is meeting or exceeding both domestic and international targets for emissions that contribute to acidic precipitation.

Canada participates in other international efforts that are expected to improve air quality, including control of the long-range transport of pollutants (heavy metals, pesticides such as DDT, and persistent organic contaminants such as PCBs). Domestically there have been some important improvements in air quality such as the virtual disappearance of lead from Canadian air following the 1990 phase-out of lead as a gasoline additive for road vehicles.

Nevertheless, smog levels and particulate matter remain important air quality issues. The Bay of Fundy, Lower Fraser Valley, and Windsor–Quebec corridor are subjected to elevated smog levels partly because of their geographic locations and partly because of pollutants generated in industrial, transportation, and energy production activities. The expectation that use of the more than 15 million cars in Canada will increase means that ground-level ozone problems could worsen. Another important issue is the particulate matter in our air that constitutes a known public health issue. In order to have Canadian objectives for particulate levels in our air reflect current understanding of health effects, a federal–provincial working group recommended new objectives. In May 2000, additional national efforts to reduce particulate matter were announced. The federal ministers of both Environment and Health jointly announced their intention

Photo 14–2
Education is central to the making of appropriate and sustainable decisions.

to declare toxic all particulate matter less than 10 microns in size. As well, federal, provincial, and territorial governments ratified the Canada-wide Standard for Particulate Matter less than or equal to 2.5 microns in diameter, and agreed to meet the standard by the year 2010 (Green Lane, 2001).

WATER QUALITY ISSUES

Both freshwater bodies and oceans have benefited from efforts to clean up and prevent pollution from cities, industries, and agriculture. As a result of stronger laws, increasing demands for greener products, and behavioural changes, there have been important reductions in the levels of many emissions. The forest industry's reduction in discharges of dioxins and furans is an example of the improvements made. Some impacts of the minerals and metals industry on watersheds have been reduced also. Public participation in watershed management and rehabilitation efforts has been an important factor in reduction of pollutants entering fresh and ocean waters.

However, significant stresses on our aquatic ecosystems continue to challenge us. Our per capita levels of water use remain among the highest in the world, even though commercial and industrial users have improved their efficiency and reduced consumption. We still have untreated municipal and industrial wastewater entering water bodies, and both surface water and groundwater supplies continue to be subject to contamination. Fish and wildlife are known to be experiencing reproductive problems as a result of endocrine disruptors entering water bodies; this could endanger the survival of some species (Colborn, vom Saal, & Soto, 1993).

The Walkerton experience (see Chapter 7) provided an important impetus for Canadian jurisdictions to (re)assess their drinking water policies and standards (see Table 14–2 later in the chapter). Some provinces strengthened drinking water laws; however, no binding national standards exist (Boyd, 2003).

BIOLOGICAL DIVERSITY

Loss of biodiversity is a complex issue and is a consequence of most of the impacts discussed in this text. The decline in biodiversity in Canada is a continuing concern: in 2003, COSEWIC listed 431 species at risk, up from 380 in 2000 and from 402 in 2002. Thirty-three of these species are extirpated or extinct (see also Chapters 9 and 12) (COSEWIC, 2003). The World Wildlife Fund conducted its first Nature Audit in 2003. The goal of the project was to assess Canada's success in meeting its national and international commitments to conserve biological diversity. The audit concluded that Canada had made significant commitments to conserve nature but that we

Photo 14–3
Each action has an environmental effect, including the cutting down of trees by beavers.

have struggled to achieve "on-the-ground success at the scale of intervention required to adequately respond to the conservation need of the nation" (World Wildlife Fund, 2003, p. 2).

Although Environment Canada's (2003a) Environmental Signals document indicates that there has been a 70 percent increase in strictly protected area (the indicator for biological diversity), 64 percent of that protected area is under 10 square kilometres and is inadequate for large mammal protection. The Canada Species at Risk Act (SARA), ratified in 2002, in conjunction with the National Accord for the Protection of Species at Risk, may help to provide the requisite protection for wildlife and their habitat. To achieve its aims, SARA emphasizes cooperation among all parties, from the territorial and provincial governments to Aboriginal peoples, farmers, scientists, environmental groups, and industry. Presently, about 278 endangered, threatened, and extirpated species are in recovery programs under the Recovery of Nationally Endangered Wildlife (RENEW) Program that is part of Canada's species-at-risk plan (Environment Canada, 2003a).

SARA is intended to fulfill part of Canada's obligations under the United Nations Convention on Biological Diversity. (Note that the Permanent Secretariat of the United Nations Convention on Biological Diversity is located in Montreal.) Building on a range of existing initiatives, the Canadian Biodiversity Strategy draws on the commitment of a broad range of interests. Quebec and British Columbia were the first provinces to report formally on how they were implementing the strategy and the convention. However, there are inadequate baseline habitat data against which to measure changes in biological diversity, insufficient information on how humans impact ecosystem processes, and little knowledge of what the critical thresholds are. Indeed, information is lacking on most of Canada's species, their status, behaviours, ranges, population sizes, and trends (Environment Canada, 2003a).

Another trend in Canadian efforts to protect biodiversity has been the increase in protected space. In 1970, 23 parks made up the Canadian national parks system; by 2000, there were 39, including the most recent addition, Sirmilik, in Nunavut (Canadian Parks Ministers Council, 2000). The federal government announced the intention to create several new parks and marine conservation areas (MCAs) in 2002. In May 2003, the Gulf Islands National Park Reserve of Canada was formally established (Parks Canada, 2003). The Canada National Marine Conservation Areas Act was passed in 2002 and there are now two marine conservation areas in Canada, Fathom Five in Ontario and Saguenay–St. Lawrence in Quebec. Through the National Park and National Marine Conservation Areas Action Plan, Southern Okanagan National Park (terrestrial), Southern Strait of Georgia, and Gwaii Haanas National Marine Conservation Areas (marine) were proposed as future national parks and MCAs. Canada's national parks are discussed in more detail in Chapter 12.

Growth in establishment of protected areas in Canada is illustrated in Figure 14–1. Individual provinces have added or expanded over 285 park areas to their protected spaces systems; between 1995 and 2000, for instance, British Columbia added about 11.7 million hectares as part of its effort to preserve species at risk and conserve representative ecosystems (B.C. Parks, 2001). Although federal tax law changes now encourage donations of ecologically sensitive land, there are many ecosystems (particularly in the more heavily populated parts of the country) that continue to be unprotected. *Environmental Signals* (Environment Canada, 2003a) indicates that more than half of Canada's identified ecoregions have little or no protection. We continue to lose lands due to urban expansion, and the protection of wetland habitats remains uncertain.

In response to increasing global awareness and international agreements such as the United Nations Convention on Biological Diversity, and initiatives from ENGOs such as the World Wildlife Fund's Endangered Species Campaign, three Canadian ministerial councils jointly signed a Statement of Commitment to Complete Canada's Networks of Protected Areas in 1992. The councils involved are the Canadian Parks' Ministers Council, the Canadian Council of Ministers of the Environment, and the Wildlife Ministers' Council of Canada. The goal of the commitment was to make a sincere effort to complete a network of protected areas that represented each of Canada's terrestrial and marine natural areas. Despite these gains, the goal has not been met. Private land ownership of some of the remaining terrestrial ecosystems, jurisdictional conflicts, and lack of knowledge about

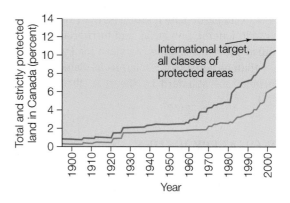

Figure 14–1
Growth in establishment of protected areas in Canada

NOTE: Strictly protected areas include nature reserves, wilderness areas, national parks, many provincial parks, and natural monuments. Other protected areas include habitat/species management areas, protected landscapes and seascapes, and managed resource areas.

SOURCES: *Building Momentum: Sustainable Development in Canada,* Government of Canada, Department of Foreign Affairs and International Trade, 1997, Ottawa, p. 5; Environmental Signals: Canada's National Environmental Indicator Series 2003, Environment Canada, 2003, p. 2, http://www.ec.gc.ca/soer-ree. Reprinted with permission of the Minister of Public Works and Government Services Canada, 2004.

marine systems have hampered progress. According to Canada's Auditor General, other problems include inadequate monitoring of progress by two of the three councils and lack of accountability in the federal government (Auditor General of Canada, 2000; Canadian Parks Ministers Council, 2000).

CLIMATE CHANGE

Canada's concerns regarding climate change have been addressed frequently throughout this text, particularly in the chapters on atmosphere (Chapter 5) and energy (Chapter 11). Climate change and sustainability are the focus here.

The Intergovernmental Panel on Climate Change has reported the international consensus that human activities clearly influence global climate. If projections are correct, and Canada experiences greater temperature changes than most regions of the world, the implications could be numerous. Among the potentially most severe consequences are more heat waves; increased storms, floods, and droughts; major shifts in fisheries, forestry, and agricultural resource bases; and damage to northern ecosystems.

The challenges we face in reducing anthropogenic sources of greenhouse gases are significant. With approximately 89 percent of total greenhouse gas emissions in

Photo 14–4
Healthy communities and healthy environments are interdependent.

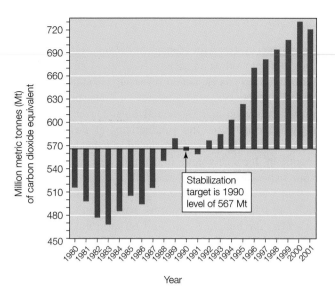

Figure 14–2
Greenhouse gas emissions 1980–2001: comparison to stabilization target

SOURCES: *Building Momentum: Sustainable Development in Canada*, Government of Canada, Department of Foreign Affairs and International Trade, 1997, p. 5; *Canada's Greenhouse Gas Inventory, 1990–1998: Final Submission to the UNFCC Secretariat*, Pollution Data Branch, Environment Canada, Greenhouse Gas Division, October 2000, pp. vii, ix, http://www.ec.gc.ca/pdb/ghg/english/ Docs/CGHGI_Vol1_Web_Eng.pdf; *Canada's Greenhouse Gas Inventory, 1990–2001*, Greenhouse Gas Division, Environment Canada, August 2003, p. vii, http://www.ec.gc.ca/pdb/ghg/ 1990_0:_report/ 1990_01_report_e.pdf; *1990–2001 GHG Emission Estimates for Canada*, Greenhouse Gas Division, Environment Canada, July 2003, Summary Tables, http://www.ec.gc.ca/pdb/ghg/ canada_2001_e.cfm; *Canada's Greenhouse Gas Inventory: 1997 Emissions and Removals with Trends*, F. Neitzert, K. Olsen, & P. Collas, April 1999, Pollution Data Branch, Greenhouse Gas Division, Environment Canada, p. 73, http://www.ec.gc.ca/pdb/ ghg/english/Docs/gh_eng.pdf

Canada attributed to transportation and fossil fuel production and consumption, and with our large land mass, cold climate, and increasing population, reducing use of fossil fuels is difficult. While progress has been made in fuel efficiency, Canada's greenhouse gas emissions have risen steadily since the 1980s (see Figure 14–2). By the end of 2001, Canada's greenhouse gas emissions were up 18.6 percent over 1990 levels and were 26.9 percent higher than target levels under the Kyoto Protocol, precluding Canada from meeting its commitment under the UN Convention on Climate Change. However, there are indications that the rate of growth of emissions is slowing. The 2001 greenhouse gas level of 720 megatonnes (Mt) marks the first decline in annual greenhouse gas emissions in more than a decade. Furthermore, the 1.3 percent decline occurred during a time of economic growth, unlike the previous decline, which occurred during an economic recession (Olsen et al., 2003).

In an effort to improve Canada's performance in greenhouse gas reduction, several new initiatives have been announced to promote research and education, provide emission reduction incentives, and introduce new regulatory measures. Action Plan 2000 provides funding and education for individuals and businesses to help reduce their greenhouse gas emissions (Green Lane, 2003). Federal–provincial–territorial cooperation remains an important component in strengthening and expanding the Climate Change Plan for Canada. Changing the energy consumption habits of individuals, however, remains a considerable challenge. There is a real need to start tracking the effects of Canada's national climate change programs in order to understand what actions have been successful and why (Environment Canada, 2003a). We also need to have a better understanding of the regional effects of climate change (see Chapter 5).

SECTOR INDUSTRIES

All of Canada's primary sector industries depend on the existence and continuation of natural capital stocks. All of these industries have been subject to efforts to improve sustainability and reduce negative environmental impacts. Here, too, there have been varying levels of success in attaining sustainability goals. Sector industry issues relevant to the following information have been covered in detail in Chapters 5 (atmosphere), 6 (agro-ecosystems), 8 (oceans and fisheries), 9 (forests), 10 (mining), and 11 (energy).

Agriculture

Healthy ecosystems, and the quality of soil, water, and air, are fundamental to agriculture. While Canada's total farmland area has remained relatively constant over the past 25 years, prime agricultural land has been lost to urban expansion and other activities. Through increasingly intensive use, the productivity of farmland has increased during the past two decades, aided by improvements in fertilizers and integrated pest management approaches.

As awareness of the linkages between farm operations and the larger ecosystem have been recognized, concerns for human health and off-farm environmental effects have begun to alter Canada's approach to agriculture. In particular, water quality has become an important concern. Efforts to maintain and enhance water quality have entailed improvements in management of soils, nutrients, manure, and pesticides. In addition, sustainable land management practices that reduce soil erosion are being adopted, such as conservation tillage methods and reduced livestock access to riparian habitat (see Chapter 6), and through new legislation. However, as the *Environmental Signals* document notes, many soil health indicators are not included in trend assessment (for example, residual nitrogen, soil organic matter, salinization, nutrient levels) and there is a lack of attention paid to small-scale degradation problems (Environment Canada, 2003a).

Forests

A highly significant component of Canada's environment, economy, and culture, forests also have benefited from efforts to improve sustainability. Although disagreements remain about the appropriateness of harvesting practices such as clear-cutting, efforts have been made to reduce logging impacts on watersheds and to protect some sensitive areas. While pollution from pulp and paper mills has decreased in general, concern continues regarding specific emissions such as organochlorines.

The 1992 National Forest Strategy, protection legislation and policies, the Canadian Council of Forest Ministers' criteria and indicators of sustainable forest management, and various industry codes of practice and standards are among the actions taken to improve forest management and move toward sustainability. The model forests are important testing grounds for sustainable practices. When initiated, the model forests focused primarily on biological research, but an increasing emphasis is on social dimensions of forest use. However, more research is needed regarding forest ecosystems and ecosystem management, assessment of harvesting practices, and community adaptation to fire hazards and climate change. Among the challenges facing the forest

Photo 14–5
Do we know where our water comes from and where our wastes go?

sector are that the forest contribution to Canada's overall ecosystem health is not well quantified, and that forest health indicators, such as forest species (no change) and harvesting rates (still increasing), are not included in Environment Canada's assessment of environmental signals (Environment Canada, 2003a).

Minerals and Metals

Canada is a world-leading mineral exporter, and the minerals and metals industry is an important contributor to the Canadian economy. Frequently, however, Canada's mining, smelting, and refining processes cause environmental impacts such as surface disruption, toxic and nontoxic air emissions, and discharge of liquid effluents. The multistakeholder Aquamin process recommended that a national environmental protection framework be established and that effluent quality regulations be revised. As a result, the new Metal Mining Effluent Regulations came into force in 2002, which impose strict limits on the releases of contaminants and metals from Canada's 100 metal mines (Environment Canada, 2003b).

The minerals and metals industry has made technological advances, changed regulatory regimes, and developed voluntary measures in efforts to improve their mining practices. The environmental policy that commits members of the Mining Association of Canada to sustainable development is one example. In addition, the Whitehorse Mining Initiative illustrates the industry's effort to involve stakeholders in defining what is meant by a socially, economically, and environmentally sustainable mining industry. Nevertheless, continuing concerns regarding environmental and social impacts of mining, such as those affecting Aboriginal people, remain to be resolved.

Thousands of tonnes of highly toxic chemicals have accumulated in abandoned northern mines. The hazard represented by these toxic deposits was addressed by the Commissioner of the Environment and Sustainable Development in her 2002 report. The cost estimate for site remediation is at least $555 million. Development of remediation plans has been challenging, but the Department of Indian and Northern Affairs Canada has now released an action framework that includes timelines for completion of the remediation efforts. In addition, the Department of Indian and Northern Affairs Canada has increased security deposits required from mining companies to cover cleanup and closure costs (Gélinas, 2002).

Energy

Canada's energy sector contributes significantly to the national economy and economic well-being through jobs. The energy sector is also the largest contributor to carbon dioxide emissions in Canada. For this reason, strategies to manage the energy sector in a sustainable manner have been developed, including environmental management systems, federal tax law changes to promote investment in energy efficiency, and green procurement practices. Response to these initiatives has been mixed. In all sectors but transportation, energy efficiency has increased by 4 percent since 1990. However, despite an improvement in efficiency, total energy use increased by 13.8 percent in the same time period (Office of Energy Efficiency, 2003). The latter trend points to the fundamental shift in thinking that is necessary to decrease fossil fuel dependency and to expand use of alternative energy sources (Environment Canada, 2003a).

If sustainability is to be achieved, additional efforts and research are required in such areas as development of alternative energy sources and technologies, identification of successful means of achieving reductions in energy demand and consumption levels, and consumer support for energy efficiency (as in R-2000 homes) (National Energy Board, 1994). Toward these goals, in 2003, the federal government announced the Energuide for Houses Retrofit Incentive, which provides some funding to Canadian homeowners who improve the energy efficiency of their houses (Natural Resources Canada, 2003).

Fisheries

Coastal and northern communities in particular receive important economic and social benefits from access to fisheries resources. However, as experience has shown, it is critical to practise sustainable use of these resources. Unsustainable harvesting practices and changing oceanic conditions are among the reasons for declines in the fish resource that subsequently led to the closure of some key fisheries, and resulted in severe socioeconomic effects

Photo 14–6

Consulting residents with knowledge of local ecosystems is a critical component of the decision-making process.

within coastal communities. The seriousness of unsustainable harvesting practices became apparent in early 2003 when, after 10 years of closed commercial cod fisheries, two populations of Atlantic cod had failed to recover. In May 2003, COSEWIC placed both populations on the endangered species list (COSEWIC, 2003).

Governments, the fishing industry, and communities have undertaken a variety of initiatives to promote more sustainable fisheries in the future. Canada's Oceans Act and the Code of Conduct for Responsible Fishing are among the efforts undertaken to help ensure ocean resources are used sustainably. Notably, the Oceans Act is built on the precautionary principle.

GOVERNMENTAL EFFORTS TO SAFEGUARD OUR ENVIRONMENT

Moving toward sustainability is a challenging endeavour. Every level of government in Canada is involved in sustainability actions, as are a multitude of nongovernmental organizations, other agencies, and individuals. To illustrate governmental action and concern for our environment, we asked each provincial and territorial Minister of Environment (or equivalent), the federal Minister of the Environment, and opposition members of each government to comment on the most serious environmental issues facing their respective jurisdiction and how they were responding to these issues. The Environment ministers also were asked about the success of their responses and how they proposed to achieve sustainable environments in the longer term. All but two governments

(Quebec and Alberta) responded to the brief survey that was conducted in the summer of 2003. One opposition party accepted the opportunity to comment on its province's environmental policy (British Columbia). Summarized results of the survey are presented in the following text and tables.

The Environment ministers' responses contribute significantly to the contents of the following sections of this chapter. Not unexpectedly, given Canada's vast space and political diversity, the environmental issues identified as "major" differed from province to province. Even so, a number of common environmental problems affected most regions of Canada.

Canada's ratification of the Kyoto Protocol has brought climate change to the forefront of environmental issues in 9 of the 12 jurisdictions that replied to the survey. In the wake of contaminated water supplies in Walkerton, Ontario, and North Battleford, Saskatchewan, water quality initiatives continue to be a priority (see Table 14–2). The survey also revealed widespread concern regarding protection of biological diversity, waste management, and reduction of toxic contaminant releases.

Survey results suggest governments are realizing that there are few "quick fixes" to environmental degradation. Respondents indicated that their jurisdictions are beginning to move toward the prevention of environmental

TABLE 14-2
EXAMPLES OF PROVINCIAL AND FEDERAL WATER QUALITY REGULATORY ACTIVITY SINCE 2001

Province	Drinking Water Quality Action
Newfoundland	A Water Resource Management Act has been introduced.
P.E.I.	Alteration of Environmental Protection Act; sewage disposal regulations have been amended to further protect drinking water quality.
New Brunswick	Following New Brunswick's Clean Water Act, regulations promulgated in 2001 now enable the Minister of Environment to protect land adjacent to water courses in community watersheds that serve as groundwater sources. As of 2003, the law ensured that 30 municipalities in the province could receive some level of protection for their drinking water.
Nova Scotia	Strategy to manage and protect drinking water was released in 2002.
Quebec	The Quebec Water Policy was released in 2002; this policy regulates aquatic ecosystems, dams and hydrology, drinking water, ground water, surface water, and waste water.
Ontario	The Safe Drinking Water Act and the Sustainable Water and Sewage Systems Act were both adopted in 2002 following the recommendations of the Walkerton Inquiry. Ontario now requires the publication of quarterly "right-to-know" reports, summarizing the results of water quality analyses.
Manitoba	The Drinking Water Safety Act, 2002, encompasses associated policy in six areas, including water quality, conservation, use and allocation, water supply, flooding, and drainage.
Saskatchewan	The province established the Operation Certification Board to ensure safety of drinking water. Saskatchewan has new regulations that require annual reports on water quality testing. Ninety-three percent of municipal drinking water now receives at least primary treatment.
Alberta	*Water for Life: Alberta's Strategy for Sustainability* was adopted in 2003. The province is looking into curbing water use in the oil and gas sector that currently uses about 25 percent of Alberta's ground water for its enhanced recovery process. $1 billion in funding is available to upgrade deteriorating infrastructure and to mount a campaign to reduce water consumption.
B.C.	Amendments to the Drinking Water Protection Act and groundwater legislation are expected to improve drinking water source quality. Implementation of a $16 million drinking water action plan is underway.
Canada	In the 2003 federal budget, $600 million was allocated to improve drinking water quality in Aboriginal communities.

SOURCES: *Environmental Trends in British Columbia 2002*, British Columbia Ministry of Water, Land and Air Protection, 2002, Victoria: British Columbia Ministry of Water, Land and Air Protection; *Quebec's Water: A Resource to Be Protected*, Environnement Quebec, 2003, http://www.menv.gouv.qc.ca/eau/inter_en.htm; *The Manitoba Water Strategy*, Manitoba Minister of Conservation, 2003, Winnipeg: Manitoba Conservation; *Environment Policy and Law*, 14(4) and 14(7), P. Menyasz, K. Pole, & R. Ray, eds., 2003; "Alberta Unveils $1B Water Strategy," T. Olsen, 2003, Nov. 27, *Calgary Herald*, p. A1; Ontario Ministry of the Environment, 2003, http://www.ene.gov.on.ca/; *State of the Environment Report 2003*, Saskatchewan Environment, 2003, http://www.se.gov.sk.ca/pdf/SOE_Report_2003.pdf

degradation, rather than focusing solely on cleanup of existing problems. This is a positive step because it acknowledges the need for long-term solutions that coincide with sustainability concepts. Public involvement, education, more cooperative approaches, and accountability are identified as actions being adopted to promote sustainable environments. Summaries of survey responses from government jurisdictions regarding specific plans and mechanisms for incorporating sustainability are discussed below and presented in Tables 14–3 to 14–6.

CANADA

In his response to the questions regarding Canada's major environmental issues and the initiatives being taken to resolve them, David Anderson, the federal Minister of the Environment, noted that three areas of concern—reducing health impacts of environmental threats, climate change, and sustaining the natural environment—received commitments of long-term funding in the 2003 federal budget. Ratification of the Kyoto Protocol, proclamation of the Species at Risk Act (SARA), additional investment in the Clean Air Agenda, funding to implement Steps II and III of the *Climate Change Plan for Canada*, and the modernization of Canada's weather forecasting capabilities are some of the actions the federal government has taken recently to address these three critical areas. To help tackle some of Canada's international environmental priorities, the federal government passed the International Boundary Waters Treaty Act, which prohibits large-scale export or diversion of waters along our international borders.

Canada was the first country to ratify the *Stockholm Convention on Persistent Organic Pollutants*.

Minister Anderson indicated the need for new thinking in Canada's approach to environmental issues. In his response to our 2003 survey, he stated, "Key measures have been put in place to deal with the impacts of environmental neglect but there is still too much emphasis on cleanup. We need to continue to build towards an environmental management system for Canada that would develop lasting solutions to address the root causes of problems" (Anderson, 2003).

Table 14–3 identifies some of the "lessons learned" by the federal government regarding resolution of environmental issues as well as some points regarding the nature of federal leadership required to contribute to sustainable environments and futures for Canadians.

ATLANTIC CANADA

Atlantic Canada's Environment ministers identified a wide variety of environmental concerns in our survey. The Newfoundland and Labrador Minister of the Environment identified waste management and public concerns associated with pesticide use as major environmental issues. A Waste Management Strategy is now in place and a working group to investigate concerns surrounding pesticide use has been established. Development pressures have resulted in increased water consumption, creating an urgency to protect representative landscapes in Nova Scotia. Other environmental issues include sewage management, climate change, and degraded air quality via long-range transport of airborne

TABLE 14-3
SELECTED EXAMPLES OF FEDERAL SUSTAINABILITY ACTIONS, 2001–2003

Successes and Lessons Learned	Leadership Roles
• In order to protect and sustain Canada's environment, the federal government is shifting from a "cleanup approach" to one that is "preventative, integrated, innovative and long-term." Preventative, long-term solutions will require not only financial resources but will also require "changing how Canadians think about and respond to environmental challenges" • The spring 2004 federal budget committed significant funding to develop better environmental indicators that will support integrated, long-term decision making and environmental technical innovation	• Achieving environmental goals in an increasingly integrated fashion • Addressing environmental challenges through voluntary actions, consumer education, and incentives so that environmental objectives parallel Canada's social and economic objectives • Engaging individuals and communities to find solutions and improve environmental quality

SOURCE: D. Anderson, personal communication, 2003, 2004.

TABLE 14–4

SELECTED EXAMPLES OF PROVINCIAL SUSTAINABILITY ACTIONS, ATLANTIC CANADA, 2001–2003

Province	Successes and Lessons Learned	Leadership Roles
Newfoundland and Labrador	• Too early to assess success of waste management initiatives, but public awareness is high. • Information is being distributed on appropriate pesticide use. • Need more trained water system operators.	• Creation of public–private partnerships to implement the provincial waste management strategy will be challenging, but will hopefully enhance opportunities for success. • Public consultation to develop an appropriate response to climate change.
Prince Edward Island	• Successes are attributed to cooperation among provincial government departments and different levels of government. • Use of renewable energy has increased by 8.5 percent. • Water and waste-water treatment has increased by at least 7 percent. • Almost half of citizens have been involved in some form of activity aimed at environmental improvement. • Concerted effort to provide information to citizens regarding new legislation to generate understanding, compliance, and eventual support.	• 66 percent of farmland is under an environmental farm plan. • Wind research is being conducted at North Cape wind test site. • Province-wide source separation waste management diverts approximately 80 percent of waste. • Tax relief offered to landowners who guarantee to protect their land under a restrictive covenant. • Leaders in surface water buffer zones and legislated crop rotation. • Only province that requires beer and pop be sold only in refillable bottles; this policy has resulted in a 97 percent return rate on empties.
Nova Scotia	• Cooperative approaches and improved regulatory management are proving to be successful ways for Nova Scotia to provide greater environmental protection while involving stakeholders. • Specific examples include Tobeatic Wilderness Area, supporting the environmental industries and innovations sector, involvement in the New England Governors/Eastern Canada Premiers initiatives to address transboundary air pollution, and risk-based regulatory management.	• World leader in waste resource management; successfully diverted more than 50 percent of solid waste from landfill disposal and, in doing so, created more than 3000 jobs. • Support of environmental industries and innovations to promote "made in Nova Scotia solutions." • Emphasis on the health aspect of environmental protection. • Continued focus on cooperative approaches to issues and regulatory management that shifts responsibility for environmental management to those who generate the pollution through voluntary and cooperative tools that are proven effective.
New Brunswick	• Initiatives are not yet in the implementation stages—"… for a new program to be effective and achieve its objectives in a sustainable manner, years of planning and consultation are required."	• "… continuing to develop some of the best water source protection programs in the country." • Programs still under construction, the Coastal Areas Protection Program and the Wetland Areas Protection Program, will complement what is already in place.

SOURCES: Newfoundland and Labrador Minister of Environment and Conservation, personal communication, 2003; New Brunswick Environment and Local Government, personal communication, 2003; Nova Scotia Environment and Labour, personal communication, 2003; Prince Edward Island Fisheries, Aquaculture and Environment, personal communication, 2003.

Photo 14–7
How well are we monitoring our impact on the environment?

particles (LRTAP). The Sydney Tar Ponds remains a major problem (see Chapter 11). On a positive note, construction has begun on a long-awaited sewage treatment facility for Halifax; in addition, drinking water, energy, and protected area strategies are under development and the province has implemented the ClimAdapt program. Development pressures also are a concern in New Brunswick, particularly development impacts on coastal zones and wetlands. Two policies are under development to resolve both matters.

Prince Edward Island's Environment minister identified soil erosion as a primary concern. Consequences of soil erosion provide a good example of the interconnectedness of degradation in ecosystems. Soil is eroding from agrochemical-laden field crops, and mismanaged woodlots and riparian habitats, into river systems, resulting in high fish mortality. P.E.I.'s Environment minister also is concerned that nitrates from heavily fertilized potato crops are leaching into groundwater in unsafe quantities. Forest conservation and protection of biodiversity are increasingly difficult because 90 percent of P.E.I.'s land base is privately owned. The government of Prince Edward Island has employed a multifaceted strategy to address these problems that includes legislative action, heavy fines for noncompliance, funding support for environmental farming practices, education, and public involvement (see Chapter 6). The provincial government also has implemented a Climate Change Action Plan and a Renewable Energy Strategy to address climate change. Further details on sustainability actions being undertaken in Atlantic Canada are found in Table 14–4.

CENTRAL CANADA

The Minister of the Environment for Ontario responded to our survey by referring us to their website. Quebec was in the midst of an election when we sent the survey and did not respond. The following information is taken from their respective government websites (but these websites do not provide sufficient information to develop tables parallel to those provided for other regions of Canada).

Ontario's major environmental issues fall into three categories: water, air, and land. Water quality remains a chief concern of the Ontario Ministry of the Environment. Actions taken to alleviate those concerns are the Safe Drinking Water Act and the Sustainable Water and Sewage Systems Act, both implemented during 2002. Ontario has continued to implement the Walkerton Inquiry recommendations to protect its drinking water. Other protective actions include the adoption of the Waste Diversion Act as well as implementation of mandatory air emissions monitoring (Ontario Ministry of the Environment, 2003).

Quebec's Ministère de l'Environnement operates in several fields of activity, including ecosystem protection, preservation of biological diversity, contaminant reduction, management of the public water domain and dams, research, and interjurisdictional cooperation to promote environmental protection. For the past 10 years, Quebec has been actively encouraging all government departments to incorporate sustainability into their activities. In 2003, Quebec mandated the formulation of its Green Plan to give added impetus to the government's directions in sustainable development (Quebec Ministère de l'Environnement, 2003).

After the 1995 election, one of the first environmental initiatives undertaken by Ontario's Conservative government was to call for a review of all environmental regulations in an effort to reduce the barriers that government actions posed to doing business in Ontario. After less than one year in office, the Harris government dismantled the Interim Waste Authority and returned power to the municipalities; ended the previous government's

Photo 14–8
Is our technology environmentally friendly?

policy prohibiting incineration by municipalities; released a new policy establishing limits on government access to self-initiated environmental evaluations; and released a new regulation that was designed to stimulate growth in the mining sector, protecting prospectors from cleaning up historic environmental contamination. Since then, concern has been expressed that the government eliminated environmental safeguards too quickly and that cost-cutting compromised the ministry's ability to protect the environment (Ontario Ministry of the Environment, 2000). These concerns resurfaced in the wake of Walkerton's problems (see Chapters 7 and 13) and perhaps will increase pressure on the government to take environmental issues seriously.

WESTERN CANADA

Based on the results of our survey, Manitoba is concerned primarily with climate change issues, water quality, and the implementation of sustainable development. To that end the Manitoba Water Strategy and the Lake Winnipeg Action Plan have been developed. Eutrification is an increasing problem in Lake Winnipeg (see Chapter 7). *Kyoto and Beyond* commits Manitoba to exceed the national Kyoto commitment by 12 percent. Following identification of provincial environmental indicators, sustainability reporting is now in place.

Saskatchewan identified sustainability as its key environmental concern and intends to address the issue through its quality monitoring and reporting system. To achieve sustainability, Saskatchewan's Environmental Protection Branch has identified three core goals: a clean, healthy environment; protection of ecosystem health and natural abundance of renewable resources; and provision of fair opportunity for sustainable use, development, and enjoyment of natural resources. Saskatchewan Environment must also address water quality issues, particularly in rural areas, where old infrastructure is "faltering." Saskatchewan faced a water quality crisis in 2001 when several residents of North Battleford fell ill following contamination of the water supply. Saskatchewan, like other provinces, has now strengthened its drinking water regulations.

Alberta Environment identified five strategic priorities: air, water, climate change, resource planning, and regulatory systems. *Albertans and Climate Change: Taking Action* describes Alberta's climate change strategy, and the Climate Change and Emissions Act is the legislative instrument adopted to address climate change. Drought mitigation and the establishment of an online air quality system are other highlights identified in its annual report (Alberta Ministry of Environment, 2003).

Like Nova Scotia and New Brunswick, British Columbia is faced with a growing population and development pressures that are taxing the province's environ-

mental quality. Declining water and air quality problems, increasing greenhouse gas emissions, and species vulnerability are attributed to an expanding population. Some of the measures implemented to mitigate these pressures include the Forest and Range Practices Act to protect critical habitat, a $2 million Living Rivers Trust Fund, and a $16 million drinking water action plan. British Columbia's official opposition responded to our survey. They voiced concerns over wildlife protection in general and, more specifically, the negative effects of aquaculture on wild salmon stocks (see Chapter 8). They have introduced legislation requiring future fish farms to use closed containment technology in order to reduce negative impacts of aquaculture on marine environments. See Table 14–5 for specific information about sustainability actions in Manitoba, Saskatchewan, and British Columbia.

NORTHERN CANADA

Canada's territories have been actively working toward transferring the authority for land and resource management, water rights, and other provincial-type programs and services from the federal Department of Indian and Northern Affairs to territorial governments. Devolution is complete in the Yukon. Nunavut has authority through the Nunavut Land Claims Agreement. The Northwest Territories is actively negotiating these matters. The Yukon government prepared mirror legislation to continue environmental protection, identifying climate change, maintaining biodiversity, and associated protection of representative landscapes as key environmental concerns. The Yukon's Climate Change Action Plan is being finalized and the Northern Climate Exchange Centre has been established as a clearinghouse for climate change studies in the north. Species-at-risk legislation has been drafted and identification of representative natural regions for protection is currently under way.

The Northwest Territories is experiencing unprecedented economic growth, largely through new mining projects and expansion of the Mackenzie Pipeline. The challenge is to balance that economic growth with traditional lifestyles and environmental protection. Long-range transport of airborne pollutants is negatively impacting the NWT. Species protection has been promoted through revisions to the Wildlife Act and the development of species-at-risk legislation. A Waste Reduction and Recovery Act has been developed and a Northern Greenhouse Gas Strategy has been drafted.

Nunavut expressed concerns about climate change and land use planning. They have developed a climate change strategy that respects *Inuit Qaujimajatuqangit* (Inuit knowledge) while incorporating scientific research (see Box 14–1 for additional information on incorporating traditional knowledge to achieve sustainability). Implementation of the Nunavut Land Claims Agreement involves the devel-

TABLE 14-5

SELECTED EXAMPLES OF PROVINCIAL SUSTAINABILITY ACTIONS, WESTERN CANADA, 2001–2003

Province	Successes and Lessons Learned	Leadership Roles
Manitoba	• "While we may be able to quickly address certain aspects of problems associated with climate change, water management, and sustainable development implementation, we must take a long-term and broad view of the issues in order to meet the challenges as they arise...." • "Too early to determine if these strategies are completely successful, but we are confident we are taking the right approach by consulting with Manitobans and taking direction from the public, experts and other stakeholders involved in the issues."	• Emphasizes partnerships with businesses, individuals, and other stakeholders. • Environmental protection industries contributed $1.8B to 1998 economy.
Saskatchewan	• 93% of municipal drinking water systems now meet minimum treatment requirements. • Waste management initiatives: have a provincial beverage container recycling program; oil, oil filter, and oil container collection centres; scrap tire collection. • Improved best management practices have resulted in measurable reductions of benzene.	• Leadership in monitoring and reporting systems: integrated format of data collection to facilitate ecosystem approach to resource management. • Actively developing and implementing monitoring programs in surface and groundwater and drinking water; air quality and soil quality. • Monitoring is the first stage in assessing trends and continuing policies devoted to the protection and sustainability of natural renewable resources.
Alberta	*Did not respond.*	
British Columbia	• Significant solid waste reduction through recycling. • Water contamination was reduced by an increase in secondary and tertiary water treatment facilities. • Pulp mills have reduced release of chlorine and other toxic byproducts.	• Continue to work through national organizations, such as the Canadian Council of Ministers of the Environment (CCME), toward positive environmental and conservation measures. • Recognize that watersheds, airsheds, wildlife, forests, and fish do not respect borders ... we need to manage the environment collectively.

SOURCES: Personal communications from British Columbia Ministry of Water, Land and Air Protection, 2003; Manitoba Conservation, 2003; and Saskatchewan Environment, 2003.

opment of planning regions that account for cumulative impacts and are consistent with sustainability ideals. Table 14–6 conveys additional results from our survey.

ENGO ACTIONS TO SAFEGUARD OUR ENVIRONMENT

The federal government established a Commissioner of the Environment and Sustainable Development position in 1996. The commissioner's first annual report noted that in many areas the federal government's performance fell short of its stated environmental and sustainable development objectives. Unfortunately, the commissioner's reports echoed similar problems in subsequent years. In his 2000 report, the commissioner commented that in spite of repeated statements about commitments to sustainable development, the federal government seemed to have difficulty turning commitments into action (Commissioner of the Environment and Sustainable Development, 2000). Such an *implementation gap*, or failure to translate policy direction into effective action, is common to many levels of government.

Sustainable solutions invoke knowledge, innovation, action, interjurisdictional cooperation, and the participation and involvement of key stakeholders, affected communities, and individuals. Involvement of local communities and the incorporation of traditional knowledge in plans for sustainability are acknowledged to be critical to successful strategies for sustainability. Two agencies in Northern Canada, one a nongovernmental organization (NGO), the other a territorial government, are successfully involving local communities and incorporating traditional knowledge to achieve sustainability objectives.

Devolution was completed in the Yukon in 2003. That means the authority for land and resource management, water rights, and other provincial-type programs and services was transferred from the federal Department of Indian and Northern Affairs to the Yukon Territorial government. The NGO Council of Yukon First Nations (CYFN) was instrumental in the negotiations that achieved the necessary transfer of power.

The Council of Yukon First Nations is the "central political organization for the First Nations people of the Yukon" (CYFN, 2003). Originally convened to negotiate land claims and self-government agreements, the CYFN has facilitated First Nations' control over significant proportions of traditional lands. In essence, control of First Nations' traditional lands is now more firmly in the hands of the people who inhabit them, First Nations people.

The CYFN has expanded its mandate to work in partnership with other organizations that aim to protect and restore the natural environment and promote responsible resource development, particularly in northern Canada. Currently the CFYN is working with other agencies to ensure traditional foods are safe from contaminants introduced into the food chain by long-range transport of airborne pollutants. They are cooperating with other stakeholders to address climate change. Internationally, working with other indigenous groups, the CFYN successfully lobbied to have the rights of local communities as holders of traditional knowledge recognized at the World Summit on Sustainable Development in South Africa in 2002.

The impacts of climate change are predicted to have profound impacts on northern communities. The CFYN and many other organizations and governments are addressing the climate change issue at many levels. The territorial government of Nunavut has identified climate change as one of the two most important environmental issues it faces. In response, the government of Nunavut has developed the Nunavut Climate Change Strategy that will address climate change through scientific research and the use of *Inuit Qaujimajatuqangit* (Inuit knowledge). The strategy embraces the dimensions of sustainability and includes the following principles:

- *Pijitsirniq*—action to control emissions
- *Aajiqatigiingniq*—opportunities for affected parties to share ideas and be involved in decision making
- *Piliriqatigiingniq*—wise resource use through balance of Inuit knowledge and science
- *Pilimmaksarniq*—enhanced capacity, self-reliance, and empowerment through community involvement and plan implementation

Other principles of the strategy include an approach that is comprehensive, phased and balanced, engages effective mitigation, and honours the precautionary principle.

Nunavut's other major environmental issue is land use planning. The primary purpose of the Nunavut Land Claim Agreement is "to protect and promote the future well-being of the residents and communities of the Nunavut Settlement Area … and where necessary, to restore the environmental integrity of the Nunavut Settlement Area" (Nunavut Sustainable Development, 2003). Nunavut is using the opportunity to learn from mistakes made in southern Canada in developing its land use strategy. "A major flaw throughout much of Canada (with the exception of the north) has been the attempt to address cumulative effects through project specific review in the absence of regional land use plans that *a priori* would have established land use goals and thresholds by which project assessments could be judged as to their acceptability at all, and if acceptable, then under what conditions" (Nunavut Sustainable Development, 2003).

Land use planning to offset the negative implications of cumulative effects will include effective land use zoning; delineation of thresholds against which proposed project impacts (including small, less regulated projects) will be measured; identification of environmentally sensitive areas; stakeholder consultation to determine which environmental and cultural indicators will be subject to assessment monitoring; and broadened jurisdictional responsibility to reduce interference of wider regional mitigation measures.

SOURCES: Council of Yukon First Nations, personal communication, 2003; Nunavut Minister of Sustainable Development, personal communication, 2003.

Although governments play an important role in environmental protection, government actions (or lack thereof) are often criticized by environmental groups, important watchdogs who confront the criticisms voiced by the Commissioner of the Environment and Sustainable Development. At the same time as Ministers of the Environment were asked for information, selected ENGOs also were sent a survey form seeking information about their objectives and activities, how well they thought government initiatives promoted environmental sustainability, and how they and other ENGOs could show leadership in contributing to environmental sustainability in Canada. Of the 32 groups that received surveys, 12 replied, while many other ENGOs sent apologies citing funding shortages

TABLE 14-6

SELECTED EXAMPLES OF TERRITORIAL SUSTAINABILITY ACTIONS, YUKON, NORTHWEST TERRITORIES, AND NUNAVUT, 2001–2003

Territory	Successes and Lessons	Leadership Roles
Yukon	• Devolution has occurred; mirror legislation is providing for development applications and permitting. • Northern Climate Exchange Centre is providing public education, outreach, impact studies coordination, links to outlying communities, and discussions about adaptation. • Energy Solutions Centre has generated award-winning solutions for alternative energy (cold-climate wind energy is an example) and energy conservation. • Representative natural-area protection will be implemented when land claims process is complete.	• Experience and success with devolution will benefit both the Northwest Territories and Nunavut. • The *Yukon Environmental and Socio-Economic Assessment Act* has designated local offices that will provide the first step in the screening process in the development application process. This innovation "represents a quantum leap in local empowerment. Sustainability is one of the cornerstones of this legislation.... This law will apply equally to all lands and all developments." • "Yukon First Nations Final Agreement and Self-Government Agreements are the most comprehensive in Canada and provide a framework for Canada's other aboriginal groups to address aboriginal rights including their inherent right to self-government."
Northwest Territories	• Lessons learned: Control of resources and decision making must be in hands of those most affected. • The will to find consensus among apparently divergent positions must be present among interest groups for success to be a possibility.	• NWT leadership will demonstrate that development can take place in a sustainable manner. • Devolution will demonstrate that natural resources are best managed by those who are potentially the most affected.
Nunavut	• The climate change strategy and associated business plan is still in its early stages and will be judged successful if it meets its targets in the prescribed time frame and can see, in both the near and long term, positive mitigating effects and effective adaptation strategies to counter the effects of climate change. • Climate change initiatives must be evaluated in national and global contexts.	• Awareness of impacts of climate change in Arctic environments and implementation of adaptive programs designed to offset climate change impacts in northern environments. • "Acting as a global litmus for evaluating and reporting on the effectiveness of climate change actions by reporting accurately and quickly on the effects of climate change in the northern and Arctic regions." • "By ensuring that land use planning addresses cumulative effects, Nunavut could become one of the only jurisdictions in North America that would have a holistic and integrated land use planning system ... that could become a model for all other jurisdictions in the world in terms of co-mingling land use planning with effective environmental stewardship and management."

SOURCES: Northwest Territories Resources, Wildlife and Economic Development, personal communication, 2003; Nunavut Sustainable Development, personal communication, 2003; Yukon Environment, personal communication, 2003.

TABLE 14-7

SELECTED ENGO RESPONSES TO THE QUESTION "HOW WELL DO GOVERNMENT INITIATIVES PROMOTE ENVIRONMENTAL SUSTAINABILITY FOR THE FUTURE?"

ENGO	How Well Do You Think Government Initiatives Promote Environmentally Sustainable Futures?
Alberta Wilderness Association	• The Alberta government has not demonstrated interest in moving toward an environmentally sustainable future. • Alberta's economy is dependent on resource extraction, often at the expense of the environment. • The federal government is beginning to show signs of a new commitment as demonstrated by the ratification of the Kyoto Protocol and the new Species at Risk Act.
Canadian Institute for Environmental Law and Policy	• Not well.
Federation of Ontario Naturalists	• Government policy is often short term, which makes long-term planning for environmentally sustainable futures difficult. • Through partnerships, the NGO perspective strengthens the final outcomes of government initiatives.
Mountain Parks Watershed Association	• Some provincial and federal initiatives, especially education and restoration initiatives, have been an important part of moving toward environmentally sustainable futures. • Government initiatives lend credibility to ENGOs. • Government funding contributes to ENGO ability to act effectively. • Canada lacks strong environmental laws and regulations to control pollution and protect habitats allowing industry to act with very little environmental responsibility.
Nature Saskatchewan	• "Some government initiatives are well conceived, but budget constraints and socioeconomic considerations tend to impede implementation." • "Ambitious conservation, education, and biodiversity action plans drafted by the provincial government are ignored when revenue-generating development is given top priority, or qualified employees in the Department of the Environment are laid off."
Sustainable Calgary	• Both levels of government are tied to a philosophy of economic growth at all costs. • Measures taken so far are remedial for ecological damage done through our current economic system. • A fundamental shift has not yet taken place within government. One promising initiative at the federal level is the work of the National Round Table on the Environment and the Economy (NRTEE).
Tree Canada Foundation	• Governments inform. However, they are "weak in the delivery and sustained funding of their initiatives, which tend to come with limited funding and short time frames ... and are detrimental to sustainability."
Western Canada Wilderness Committee	• "Lack of resources allocated to environmental initiatives, a lack of progressive policy direction, and an unwillingness to take a stand which may be unpopular among certain key sectors of the voting public contribute to forming governments whose initiatives do not adequately protect wilderness and biodiversity or address the significant environmental problems Canada is facing."
Wildcanada.net	• Because governments are "gripped with the mentality that economic growth has to come at a cost to the environment," there is no motivation for provincial governments to embrace long-term environmental protection. • Alberta and Ontario are poor at enforcing legislation and regulatory mechanisms that protect the environment because of their reliance on natural resources to support their economies. • Federal government has been very slow to fulfill election promises.

TABLE 14-7
(CONTINUED)

ENGO	How Well Do You Think Government Initiatives Promote Environmentally Sustainable Futures?
Wildlife Habitat Canada	• Governments are oriented toward reaction rather than prevention. • Efforts tend to be concentrated where humans live, rather than the habitats of nonhuman species. • Government departments focus on portions of ecosystems but are not set up to provide information at the ecosystem level. • Mechanisms for integration of governments' information limit integrated land use planning, and holistic ecosystem management is lacking. • The jurisdictional boundaries that governments operate within do not necessarily align with ecosystem/habitat boundaries. • Rules of authority tend to override practicality in habitat conservation.

SOURCES: Personal communications (2003) from Alberta Wilderness Association; Canadian Institute for Environmental Law and Policy; Federation of Ontario Naturalists; Mountain Parks Watershed Association; Nature Saskatchewan; Sustainable Calgary; Tree Canada Foundation; Western Canada Wilderness Committee; Wildcanada.net; Wildlife Habitat Canada.

Photo 14–9
Have we introduced our children to the wonders of nature?

and time constraints that prevented them from responding.

In response to the question about how well ENGOs felt government initiatives promoted environmental sustainability for the future, many respondents concurred with the Commissioner of Environment and Sustainable Development's comments above. A common criticism among ENGOs was governments' inclination to defer to short-term economic growth over long-term sustainability. While the selected responses highlighted in Table 14–7 do not constitute a representative sample of ENGO opinion, the nature and consistency of their comments point to the need for governments to enhance their sustainability actions. However, as noted in previous chapters, reduced fiscal resources have resulted in lower levels of spending on environmental services. This trend provides clear challenges for both governments and ENGOs.

The ENGOs identified in Table 14–7 employ a variety of approaches in reaching their goals. Increasingly, the expertise within ENGOs is being used in participatory planning mechanisms in which industry, government, community, and ENGOs work together to resolve environmental and land use issues. Many ENGOs view their major contributions to the environmental movement in terms of advocacy for environmental responsibility and leadership, education and building awareness of the possibility of various environmental initiatives, and research and monitoring. Others serve as watchdogs, catalysts, information providers, and nonpartisan observers. One strength identified by many respondents was the ability of ENGOs to work at various scales. Small-scale and grassroots initiatives often serve as models for other agencies. ENGOs are not as constrained by political frameworks and can sometimes be effective where politically restrained governments cannot. One organization is looking for new ways to acquire conservation areas; another provides research into environmental law and policy that promotes the public interest and sustainability. Clearly, ENGO actions incorporate public values in political decision making, sometimes through roundtables dealing with environmental, social, and economic issues, sometimes through formal hearings, and sometimes through local "cleanup" or other "action" days.

Many of the ENGOs in Canada have been in existence for decades and are likely to remain active in a variety of ways to raise public awareness of environmental issues, to challenge governments and industries to deal with the root causes of ecosystem problems, and to mount partnerships and promote stewardship activities that will lead to sustainable practices. These grassroots activities are important given that Canadians continue to leave a large ecological footprint on the planet (Wackernagel & Rees, 1996). In fact, if all humans on Earth used the same levels of energy and other resources as Canadians do, it would take 4.6 planet Earths to supply the resources for them (see Figure 1–5).

The following sections briefly review the challenges that remain in Canada's progress toward sustainability.

CHALLENGES FOR THE FUTURE

Several major types of challenges face Canadians as we try to respond effectively to current and future environmental sustainability issues. These challenges are found in such concerns as our approach to resources management; the need for conservation of resources, species, and ecosystems; waste reduction requirements; the need to design urban areas and transportation systems for greater sustainability; pollution control requirements; the cleanup of past environmental problems; and changes in decision-making processes. Each of these challenges is considered briefly below.

Integral to these challenges is the link between international, national, regional, community, and individual contributions to sustainability. Recognition of challenges and implementation of solutions are equally important and are elaborated further in discussions of ecological footprints (Chapter 1) and genuine progress indicators (Chapter 13).

RESOURCE MANAGEMENT

We know that resource management must involve affected stakeholders in the planning and implementation of resource development projects. Not only does public involvement reduce the risk of community opposition, it also reduces the likelihood of raising legal challenges to projects. Though not perfect, environmental review procedures in many provinces and territories now provide considerable opportunity for public involvement. Citizen involvement permits compromise and development of a consensus on appropriate solutions to environmental issues (and may be particularly important in contentious issues such as climate change). Public participation also may enhance explicit consideration of the needs of future generations.

Public involvement also may lead to initiatives that promote sustainable resource management. Examples include the decision to use co-management approaches and practices in northern communities, the Model Forest Program, and the Aquamin and MEND programs in mining. These and other resource management initiatives not only reflect public input but also provide insight into the effects of adaptive resource management processes (helpful in improvement of future decisions).

While public participation promotes integration of environmental, social, and economic values, decisions often take a long time to be reached when the public is involved. Some developers have complained about these delays; one of the challenges we face is to develop a process that minimizes delays and yet ensures adequate time to make appropriate and effective decisions. Other challenges include the need for further research into the effects of resource management practices, the need to ensure that public involvement exercises are meaningful ones for all participants, and the need to create partnerships that can address key issues.

CONSERVATION

Collapse of the northern cod stocks provided concrete evidence of how important it is to "live within our means"—to have resource management strategies and regulations that act to conserve resources. This is a challenge that, to some degree, has been reflected in the Atlantic Groundfish Strategy and the Oceans Management Strategy, as both have given important emphasis to conservation. Living within our means—finding economically and environmentally sound ways to use resources—results in challenges for the future such as generating adequate and scientifically sound knowledge on which to base decisions about sustainable use, undertaking appropriate monitoring, and strictly enforcing regulations regarding specific resources. If solutions to these sustainability challenges can be met, then it should not be necessary to take such drastic measures as closure of the fishery.

Another challenge is to ensure that more people understand the need for effective conservation across the spectrum of natural resources. To some extent awareness is developing: we can see the effectiveness of such changes as urban water pricing based on volume consumed, the Permanent Cover program returning marginal cropland to pasture, and the stewardship of landowners supporting wildlife habitat conservation (through such organizations as the Alberta Wilderness Association, Western Canada Wilderness Committee, and the Nature Conservancy of Canada, as well as through programs such as the North America Wetlands Conservation Council and North American Waterfowl Management Plan). Part of this chal-

Photo 14–10
Do we strive for consistency in our daily actions?

lenge to improve public understanding of the importance of conservation involves extending awareness of the need for conservation to nonrenewable resource sectors also.

In the past, a variety of governmental and other programs resulted in unintended consequences because the quest for increased production was encouraged without full consideration or understanding of environmental side effects. Examples include subsidizing the removal of natural areas, encouraging cultivation on slopes, and promoting wetland drainage. The challenge of restoring the biological diversity of areas affected by these kinds of decisions and removing inappropriate subsidies to prevent these impacts from occurring is beginning to be met.

WASTE REDUCTION

Canadians are among the world's leading producers of domestic waste. National, per capita nonhazardous solid waste generation has increased by 10 percent since 1998 (Environment Canada, 2003a). However, since the early 1990s there have been gains locally in recycling through blue box and other community programs, and small improvements in the amount of home composting. One challenge is to ensure there are markets for recycled materials so that the rates of recycling of paper, glass, metals, and plastics can increase. Another challenge is to ensure recycling opportunities are convenient and readily available to all residents.

At the national level, a variety of targets have been established to stimulate waste reduction by households, industries, and government. For instance, the Canadian Council of Ministers of the Environment has developed the National Packaging Protocol and the National Solid Waste Management Program. By 1996 the National Packaging Protocol goal of a 50 percent reduction in packaging had been achieved by all signatories (Canadian Council of Ministers of the Environment, 2000). The National Solid Waste Management Program had as its objective the reduction in Canadian per capita output of solid waste to 50 percent of the 1988 level by the year 2000. Unfortunately, success has not come as quickly in the Solid Waste Management Program. At its conclusion in 2000, only Nova Scotia had achieved the target (Menyasz, Pole, & Ray, 2000). Nevertheless, the case of Halifax illustrates how a single community can make a difference. Faced with the need to create a new landfill, residents and city officials opted to develop a comprehensive waste management strategy that involved a recycling program, landfill reduction, and green bin doorstep pickup composting. Halifax's efforts have earned it the distinction as the most innovative in the country and possibly one of the best in North America. And by 2003, Prince Edward Island had diverted approximately 80 percent of its solid waste from landfills through a source separation program.

One of the challenges in the waste reduction field is to have more firms reduce risks and potential liabilities by adopting environmentally sound waste management practices (which may be economically advantageous as well). In response to this challenge, the private sector has developed some highly innovative waste management programs. The 3M Corporation, for example, saved itself $1 million by reducing waste disposal volumes (from 2800 to 115 tonnes per year) at an Ontario manufacturing plant. A U.S. company, Malden Mills, developed a Recycled Series of fleece fabrics in which at least 89 percent of the material comes from postconsumer recycled pop bottles. Not only does it take less energy to make the fibres that go into this fabric, it creates 17 times fewer air pollutants. An average-sized jacket keeps 25 two-litre pop bottles out of landfills; every year a pile of pop bottles "the size of a few dozen Boeing 747s gets recycled instead of going to waste" (Malden Mills Industries, Inc., 1994).

Ecosystem health is threatened directly by hazardous wastes; in 2000, approximately 90 percent of the 292 participants in the federal government's Accelerated

Reduction/Elimination of Toxins (ARET) program were involved in manufacturing and mining industries. The success of the ARET program provides some indication that these industries are taking their responsibilities regarding hazardous waste more seriously than they have in the past. This voluntary program helps operations that generate hazardous waste improve their environmental performance. Hazardous wastes require special treatment and disposal to make them less dangerous.

URBAN CENTRES AND TRANSPORTATION SYSTEMS

Urban centres supply their residents with food, clothing, housing, recreation, and other amenities; in providing these goods and services, Canadian cities often are the source of many environmental stresses. The water, food, fibre, energy, and minerals that flow into our cities come from surrounding ecosystems; flowing back out into those ecosystems are wastewater, garbage, and air contaminants. Also flowing out of urban areas are residents seeking recreation and natural areas.

Cities can be catalysts in finding solutions to environmental stresses. For instance, providing clean water, effective sewage treatment systems, and recycling programs is easier, more efficient, and less costly when people are concentrated in smaller areas rather than dispersed. Recycling programs, for example, can run more efficiently in densely populated centres because large amounts of used glass, paper, plastic, and aluminum cans can be collected from a small area. Large urban popula-

Photo 14–11
Do we know which elements of our lifestyles are harmful or beneficial to the Earth?

tion bases also provide a larger group of people who can provide environmental leadership; this is one reason why municipalities often have led the way in developing and implementing new initiatives and programs to promote environmental sustainability.

As we saw in previous chapters, there has been mixed success in efforts to improve the sustainability of urban environments. Some of the encouraging results include the trends toward smaller, more energy-efficient homes, more home-based businesses (reduced transportation demands), tighter land use controls, green infrastructure and community planning processes, and growing support of mixed land uses and intensification policies (such as infill and redevelopment of land to support higher population densities). Also, the increasing importance attached to protecting ecologically sensitive areas in urban landscapes has grown even though government resources have declined.

Some of the less encouraging trends include the continued Canadian preference for single-family, detached housing located in low-density neighbourhoods; the establishment of many new businesses in low-density business parks; and the resultant reliance on our vehicles and a transportation infrastructure focused less on public transit and more on the automobile. In fact, numerous efforts have been made by municipalities to increase use of public transportation services. Many of these actions, including higher parking fees and use of preferential lanes for buses and car pools, have not been very effective in changing people's behaviour. As essential as transportation is to Canada's socioeconomic well-being, Canadians' transportation practices seem particularly environmentally unsustainable with regard to pollution and fuel supply. Despite urban initiatives to reduce our reliance on cars, automobile use has increased 9 percent since 1990 (Environment Canada, 2003a).

The importance of environmentally sustainable transportation has been reflected in international agreements (such as the Nitrogen Oxide Protocol, the Volatile Organic Compounds Protocol, and the Canada–United States Air Quality Agreement), national efforts (such as the Canadian Environmental Protection Act [CEPA] regulations regarding benzene content of gasoline), as well as regional and local-level initiatives (such as the Twenty Percent Club and the Air Care Program in British Columbia's Lower Mainland). Commuter Connections (2001) is a website that provides advice to people wanting to carpool as an alternative to other commuting methods. They also have a registry that matches would-be carpoolers at various postsecondary institutions, including one each in Alberta, Saskatchewan, Nova Scotia, and New Brunswick, nine in British Columbia, four in Ontario, and two in Quebec. The emergence of car cooperatives in larger urban centres (see Chapter 13) is also a positive step toward reducing the number of vehicles on the road.

These actions, however, may be insufficient to reverse long-standing trends toward increased transportation activity. In the short term, it appears that even technological changes (such as lighter materials in vehicles, improved fuel efficiency, use of alternative fuels, hydrogen cells, and electric vehicles) will have only limited effects on fuel use and emissions. Fortunately, research is continuing in these areas.

POLLUTION CONTROL

The Canadian public continues to place high priority on a clean and healthy environment, and expects that public and private institutions will take the necessary actions to achieve environmental quality. Since the early 1990s, higher fines for pollution offences, concerns over personal and corporate environmental liability, considerations of corporate image, and demands from insurance and lending institutions have been important elements in pollution reduction. Perhaps additional education is required to encourage corporate officials to act less out of fear of reprisal and more out of understanding of the need for and importance of pollution control measures in attaining environmental sustainability.

Canada's participation in the United Nations Convention on Climate Change, our Climate Change Plan for Canada, and federal initiatives such as the Efficiency and Alternative Energy Program have helped to reduce energy-related greenhouse gas emissions, but have proved insufficient to reach established goals. Clearly, further action is necessary.

In relation to acidic deposition, Canada has surpassed its emission reduction target for sulphur dioxide, but this target has not proved adequate to achieve the desired reduction in acid precipitation. Additionally, a National Strategy on Acidifying Emissions for beyond the year 2000 will help protect air clarity and human health as well as acid-sensitive areas. Local air pollution control efforts, too, are being assessed with an eye to required changes for the future.

The Ocean Dumping Control Action Plan was initiated in 1991 to strengthen regulations, improve surveillance, and set up a national program to reduce persistent plastics in the marine environment. This plan enhances national efforts to regulate disposal at sea (by permit under CEPA and the Ocean Disposal Regulations) and to comply with the 1972 London Convention. Continued diligence in application and enforcement of this and other regulatory initiatives is an important challenge in achieving a sustainable future.

The toxic pollutants list in the National Pollutant Release Inventory (NPRI) is part of the large task of identifying, monitoring, and controlling toxic pollutants in Canada. One of the benefits of collecting data on pollutants being released into Canada's environment is that, over time, these data will provide a solid basis for comparison and enable trends to be identified and analyzed. For instance, the 1994 data (released in 1996) indicated that there was a 16 percent decrease in the amount of pollutants discharged to Canada's water and air compared with 1993. Through 1996 and 1997, there was an overall increase in the pollutants released, but in 1998 that trend reversed slightly from 1997 levels. In 1998, Ontario had the highest on-site releases (63 960 tonnes) but recorded a 5 percent decrease from 1997. Alberta ranked second in releases (46 644 tonnes); this represented a 6 percent increase over 1997. Of the 10 provinces, Prince Edward Island released the least (221 tonnes), and of the territories, Yukon had the lowest on-site release (National Pollutant Release Inventory, 1998).

Given the difficulty in determining the safety of the large number of chemical products already in existence—more than 110 000—and because of the 1000 or so new substances added each year, regulations under the New Substances Notification Regulation of CEPA now require manufacturers and importers to provide toxicological information on new commercial substances introduced into Canada. Knowledge of these substances is important as it forms the basis for action and regulation directed at their control. More industrial responses such as the Responsible Care program of the Canadian Chemical Producers' Association—which manages chemicals from "cradle to grave"—will be necessary to help reduce total pollution from toxic substances (Canadian Chemical Producers' Association, 1999).

New approaches to managing toxic substances are being developed; the life-cycle concept (see Box 14–2) and "industrial ecology" (which manages industrial impacts from a more holistic perspective) are among these new developments. Site assessments and environmental auditing also are becoming more common. In fact, the federal government has adopted an overall pollution prevention strategy that incorporates these ideas.

Policies such as the Toxic Substances Management Policy (a precautionary, proactive, and science-based management framework to be applied to all areas of federal responsibility) and the National Pollutant Release Inventory are making it easier to track pollutant releases, to identify pollution sources, and to more effectively promote management of polluting processes and their conversion to more sustainable processes. The 2000 Auditor General's audit criticized the lack of action in implementing the Toxic Substances Management Policy. The audit team cited a lack of implementation plans and management strategies in federal jurisdictions as reasons for inaction (Government of Canada, 1995a; Office of the Auditor General, 1999).

BOX 14 – 2
THE LIFE-CYCLE CONCEPT

The life-cycle concept is a "cradle to grave" approach to thinking about products, processes, and services. The concept recognizes that all life-cycle stages (from extracting and processing raw materials to manufacturing, transportation and distribution, use and reuse, and recycling and waste management) have environmental and economic impacts.

Public policymakers as well as industrial and private organizations can use the life-cycle concept to help them make decisions about environmental design and to make improvements in resource efficiency and pollution prevention. In addition, the life-cycle approach can be used as a scientific tool for gathering quantitative data to inventory, weigh, and rank the environmental burdens of products, processes, and services.

In contrast to the specific approaches to environmental management that occur at the "end of the pipe" or "within the plant gate," decision makers can apply the life-cycle approach to all of the upstream and downstream implications of site-specific actions. For instance, decision makers could examine the changes in emission levels that would result from changing a raw material in the production process.

A variety of specific life-cycle tools have been developed to help decision makers make a difference, including life-cycle assessment (LCA), design for environment, life-cycle cost accounting, total energy cycle assessment, and total fuel cycle assessment. Industries increasingly are using LCA to improve their environmental performance. A life-cycle assessment quantifies energy and resource inputs and outputs at all stages of a life cycle, then determines and weighs the associated impacts to set the stage for improvements.

The accompanying figure shows the breakdown of a product life-cycle inventory (LCI) into inputs and outputs for material and energy, as well as environmental releases. An environmental engineer might use an LCI to baseline the operation's performance against generic data and help guide pollution prevention and process improvements. Similarly, a manufacturer might provide consumers with environmental profiles of finished products based on input/output accounts of its own activities, its materials and energy use, and external data. This could influence product users and also help a manufacturer to meet changing customer requirements.

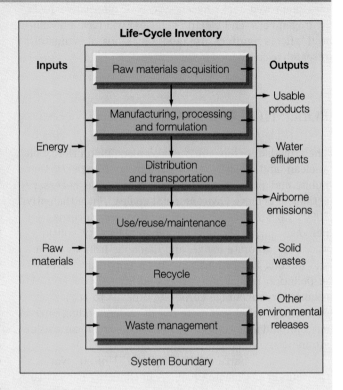

Box Figure 14–1

The life-cycle concept has been developed into a framework called EPR, or Extended Producer Responsibility. EPR extends the responsibility for discarded byproducts and end products to the producer rather than to the individual and ultimately to governments. EPR has been used to create successful legislation in 15 Western European countries; in Poland, Hungary, and the Czech Republic; and in Korea, Japan, and Taiwan. The legislation is usually applied first to packaging—similar to the National Packaging Protocol in Canada—and then to electronic equipment and finally to vehicles (Fishbein, n.d.).

SOURCES: *What Is Life-Cycle Management?* Environment Canada, 1996, http://www.ec.gc.ca/eco cycle/whatis/index.html; *Preventing Pollution by Extending Producer Responsibility,* B. N. Fishbein, (n.d.), http://www.inforinc.org/eprarticle.htm; "LCM across the Life Cycle— Considering Your Role in the Total Life Cycle of Products and Services," S. B. Young, 1996, *Ecocycle, 4,* http://www.ec.gc.ca/ecocycle/ issues/issue4/index.html

CLEANUP OF PAST ENVIRONMENTAL PROBLEMS

Environment Canada reported that in 1989 an estimated 10 000 sites in Canada potentially were contaminated with environmentally harmful substances (Government of Canada, 1991). However, this estimate apparently has remained an estimate. Contaminated sites generally fall under provincial jurisdiction, and, in 1998, there was still no consolidated national assessment of contaminated sites. In 1995–96, the office of the Auditor General estimated that potential cleanup costs were in the range of $2 billion (Office of the Auditor General, 1996; Pilgrim, 1998). Harbour bottoms, fuel storage areas, former gasoline stations, closed metal mines, former industrial facilities, railyards, former military bases, Distant Early Warning Line sites, and waste disposal areas are among these contaminated sites. Many are orphan sites; no responsible party can be found that is capable of paying for remediation. The remaining sites require rehabilitation by governments and responsible local landowners.

As the soil under many of these sites must be cleaned or removed if the site is to be sold or used for other purposes, rehabilitation can be very costly. Costs of cleaning up leaking underground storage tanks, for example, have averaged about $150 000; some sites cost millions of dollars, particularly where aquifers are contaminated by hydrocarbons or other chemicals.

Owners of properties with past and current environmental liabilities, who are facing stronger laws regarding contamination and rising civil settlements for damages, are turning to environmental auditing to identify these liabilities. Environmental auditing allows remedial or preventive actions to be taken before civil damages or noncompliance with regulations result in high costs. Governments, too, are using remedial action plans to clean up areas where the risk is greatest (such as the remaining Areas of Concern on the Canadian side of the Great Lakes).

CHANGES IN DECISION-MAKING PROCESSES

Decision-making processes that integrate environmental, economic, and social considerations in the management of natural resources and ecosystems are the basis of achieving sustainability in Canada. (See Enviro-Focus 14 for some Canadian-made models of promoting sustainability.) Examples of the kinds of environmentally sensitive decision making required range from the Crombie Commission in Toronto; to Vision 2020 statements from such diverse communities as Hamilton, Ontario, and Canmore, Alberta; to the University of British Columbia's Task Force on Healthy and Sustainable Communities. These planning efforts, as well as concepts championed in such documents as the World Conservation Strategy, *Our Common Future*, and Agenda 21, have resulted in improved understanding of the need to make changes in consumption patterns, wealth distribution, and resource planning. Some of this understanding is reflected in Canada's annual Report(s) to the United Nations Commission on Sustainable Development.

National and provincial roundtables (see Box 14–3), community-initiated programs, and voluntary initiatives by private industry are important signs that at least some environmental ethics have been internalized among Canadians and in the business world. In addition, there is more of an effort now to anticipate and prevent environmental problems (rather than simply react to them). Part

BOX 14–3

THE NATIONAL ROUND TABLE ON THE ENVIRONMENT AND THE ECONOMY

Created in 1988 as one of Canada's main institutional responses to the challenges of sustainable development, the National Round Table on the Environment and the Economy (NRTEE) consists of a chairperson and 24 distinguished Canadians appointed by the prime minister. The NRTEE was established to generate awareness of environment–economy issues by identifying, explaining, and promoting the principles of sustainability to government, business, and the broader community.

Proclamation of the Round Table Act in 1994 established the NRTEE as a key forum for discussion of sustainability issues. The NRTEE also helped governments to address public policy questions relating to sustainable development, provide a neutral meeting ground where stakeholders together could tackle natural resource and environmental issues, and produce a broad range of information and publications to encourage grassroots initiatives.

In 2001, Canada lost its United Nations ranking as the best place in the world to live, although the difference in the Human Development Index values between first-place Norway and eighth-place Canada is only .007. NRTEE cautioned that Canada's ranking as the best place in the world to live is threatened further unless environmental concerns are addressed immediately. In its report, *The Millennium Statement,* NRTEE cited the following concerns: threats to biodiversity, decline of urban environments, global economic changes, and accumulation of contaminants. NRTEE is encouraging the federal government to change how it measures progress by supplementing the GDP and other macroeconomic indicators with measures of human and natural capital. Suggested indicators are air quality, freshwater quality, greenhouse gas emissions, forest cover, wetlands, and educational attainment.

SOURCES: "Quality of Life Threatened," January 30, 2001, *Calgary Herald,* p. A4; *Changing How Canada Measures Progress,* National Round Table on the Environment and the Economy, 2003, http://www.nrtee-trnee.ca

of this new way of acting is the result of much more consistent public consultation on the design and implementation of decision-making strategies. Also, with better ways of disseminating information (such as Environment Canada's Green Lane website), both the public and local decision makers are better informed and able to contribute significantly to decisions.

Biosphere reserves are models in this regard. To create a biosphere reserve, local communities must come together and demonstrate the environmental, cultural, and economic significance of their area. Biosphere reserves must incorporate areas of environmental protection with areas in which sustainable development of local

Promoting Sustainability via "Canadian-Made" Models

Here are some "made in Canada" models that promote sustainability. Consider their contributions and effects. What other models might you identify within your community? How can citizens work together to enhance the efficacy of individual choices?

Conceptually, the ecological footprint is a powerful metaphor to demonstrate the impact of our choices on the ecological carrying capacity of the Earth. When the ecological footprint is calculated according to different socioeconomic status, it also illustrates how increasing wealth leads to increasing consumption and pressures on ecological systems. This metaphor is now taught from elementary school to university to encourage students to think in new ways about how consumption and sustainability are incompatible objectives.

This conceptual model has spawned new ways to measure our impact on the Earth's systems. The efforts of GPI (Genuine Progress Index) Atlantic in Nova Scotia to establish a set of "genuine progress indicators" for the province has gained provincial and national attention. These measures, which span a range of environmental, social, and economic characteristics, have demonstrated new ways to move toward sustainability. Similarly, the Endangered Spaces Campaign of the World Wildlife Fund Canada (1989–2000) placed pressure on the Canadian government to increase the number and size of protected areas across the country. The amount of

land granted protected status over the course of the campaign increased from 3.0 percent to 6.8 percent of the total land mass in Canada. The percentages seem small, but the total represents a 127 percent increase (more than double), or an increase from 29.4 million to 68.3 million hectares.

The community garden movement, begun in the 1970s, has grown larger in the past decade. Community gardens promote ecological and social sustainability as citizens work together to provide healthy, safe, and abundant food for themselves and others. They also promote cultural understanding as groups from different cultural backgrounds share seeds, gardening tips, recipes, and a love of the outdoors in the city.

Many universities are now examining ways that they can become sustainable. Universities as institutions have been slow to realize the impact they have as large-scale consumers of goods and as models for the built environment. New initiatives across university campuses are now directed toward reducing wastes (including hazardous wastes), improving commuter practices, demonstrating appropriate designs, and even modelling sustainable agricultural practices.

At local to regional scales, biosphere reserves are places where people seek to live out and demonstrate their commitment to sustainability. The increased number of biosphere reserves since the outset of the 21st century indicates a growing interest in putting sustainability into practice. Similarly, when faced with the need to establish a new landfill, the Halifax Regional Municipality opted for a comprehensive waste management strategy that encouraged residents to reduce waste and reuse waste products rather than simply throw it all away.

resources is promoted. Reserves are then recognized by provincial and federal levels of government before they gain their status conferred by the United Nations Environmental, Scientific, and Cultural Organization (UNESCO). During the 1970s and 1980s, Canada had six biosphere reserves; however, since 2000, six more have been created. This model, then, is having increasing appeal as a way to promote sustainable development and to implement its ideals within living and working landscapes. However, as biosphere reserves are areas of "recognition" not "regulation," they still require financial

and logistical support from other levels of government in order to become functioning models of sustainability.

Education is also a major influence on how Canadians understand and react to environmental issues and is a key to ensuring sustainability will become a reality. State-of-the-environment reports (from provinces, municipalities, corporations, and other bodies), environmental modules in school curricula, and other educational materials help ensure that future decision makers will be better informed about potential consequences of actions than were their predecessors.

Environmental indicators (or indicators of sustainability) are also tools for improving decision making. They provide concise, understandable, scientifically credible information that profiles the state of the environment and helps measure progress toward the goal of sustainability. Additionally, they give us solid quantitative information, rather than impressions, allowing us to assess strategies and ask questions: Does this work? Can it be adapted successfully in other arenas? If not, what can we change to get better results? Economic instruments such as pollutant emission charges ("green taxes") have been implemented much more slowly, and have been used mostly to pay for the environmentally safe disposal of a product (Macdonald, 1996).

Government cutbacks and reductions or terminations to programs have resulted in reduced availability of environmental data, reduced monitoring capability, and the loss of national baseline data and ecosystem-specific information required for decision making. This makes it difficult to know whether Canada is on an environmentally, economically, and socially sustainable path. Clearly, challenges remain, including increasing population, a continuing dependence on natural resource use to promote economic growth, and our generally unchanged consumption patterns. Each type of problem requires continuing effort, commitment, dialogue, and innovation to be resolved. Some comments about how individual Canadians and groups of Canadians can address sustainability concerns are provided in the following section.

The Importance of Individuals and Collectives

From the outset of this book, we have identified the ecological realities of our individual actions and the challenges we face in shifting our thinking and actions toward sustaining our environment. Because ultimately we are inextricably connected to the Earth's ecosystems, we noted that individual and combined actions do make a difference to the environmental, social, and economic sustainability of our environment. Even though many people look to governments to show leadership and commitment in these areas, the responsibility for sustainability is shared among all members of Canadian society.

Canada's national environmental indicators tell us that some of our initiatives are working: we have seen improvements in the indicators for acid rain, water quality, protected spaces, energy efficiency, and soil management. However, Canada has more species at risk than ever, we emit more greenhouse gases, use more energy, and generate more solid waste than previously. Given these realities, each one of us faces a number of funda-

mental choices as we look toward the future of the Canadian environment: What kind of future do we want? How can we live a life that helps to sustain the Earth? How can we get where we want to be in the future? What kind of legacy will we leave for our children, grandchildren, and succeeding generations? Knowing that each small action we take is important—because, fundamentally, healthy human communities and healthy environments are interdependent—we need to understand our own attitudes and values. If need be, we must change our view of the world and adopt new ways of thinking and acting. In our survey, Canada's Minister of Environment, David Anderson, stated: "Our past experience has demonstrated that successful environmental management depends on our ability to fully engage individuals and communities in defining the problems, finding the solutions and taking action to improve the quality of our environment. In the end—whether the source of a problem is in our backyard or on the other side of the globe—the problem becomes a local one. As such, individuals and communities must do their part, and governments and others must support their efforts to do so."

As stewards of our planet, we must ensure not only that we are (re-)connected with the natural world, but also that we learn to understand, respect, and work with one another toward sustainability objectives. Out of the exercise of our individual and composite intelligence, insight, and innovation, we should be able to create many successful approaches to sustaining our Earth. The Trans Canada Trail, spanning the country from coast to coast, illustrates how individuals and groups can work together over a period of time to build a positive legacy that improves environmental and social well-being for both present and future generations (see Box 14–4).

What other models might we advance? In this text, we have offered several avenues that suggest there are many creative ways to envision and to implement sustainability.

Photo 14–12
What legacy do we want to leave for succeeding generations?

BOX 14–4
THE TRANS CANADA TRAIL: A LEGACY FOR FUTURE GENERATIONS

When the main trunk of the 18 000 kilometre Trans Canada Trail is completed, it will be the longest trail in the world. A shared-use recreational trail that will link all provinces and territories and touch all three of our ocean shores, the Trans Canada Trail is a community-based project that will preserve and protect the environment, promote physical fitness and well-being, provide a safe and secure place for recreational activity, act as a stimulus for local economies (such as bed-and-breakfast operations), educate people by bringing them closer to nature and their historical roots, and foster eco-tourism opportunities.

Since 1994, when the Trans Canada Trail Foundation was launched publicly as an independent registered charity, more than $12 million (of the estimated $42 million required) has been raised toward the building of the trail. Donations mainly come from private individuals. The goal of the foundation is to receive 83 percent of the donations required from individuals, 12 percent from corporations, 2 percent from merchandising and governments, and 1 percent from foundations. Through local trail and community groups, about 1.5 million people across Canada are volunteers with the trail councils in their region and their organizations have united to fulfill a shared vision of making the Trans Canada Trail a reality.

Built on provincial and federal park and Crown lands, on abandoned railway lines, alongside railway lines, and on private land, the trail will accommodate five core activities: walking, cycling, horseback riding, cross-country skiing, and (where possible or desired) snowmobiling. In 1996 about 800 kilometres were dedicated to the Trans Canada Trail, including the Galloping Goose Trail in Victoria, British Columbia (a 60 kilometre former rail line, considered the first "rails to trails" conversion in Canada); the Caledon Trailway, Jackson Creek Kiwanis Trail, Elora Cataract Trailway, and Grand River Trails in Ontario; several sections of trail in the National Capital Commission area (Ottawa–Hull); Le Petit Temis (between Cabano, Quebec, and Edmundston, New

Photo 14–13

Brunswick); Guysborough Trail in Nova Scotia; and Confederation Trail in Prince Edward Island. By 2001, about 62 percent of trail length had been dedicated. Each year, about one dozen new projects across the country are added to the trail network.

On February 19, 2000, the 2000 Relay began in Tuktoyaktuk, Northwest Territories, to commemorate the official opening of the trail in the Ottawa area on September 9, 2000. Relay participants walked, cycled, cross-country skied, rode horseback, and sometimes snowmobiled the trail as an acknowledgment of the multi-use nature of the trail.

Although it is not yet complete, metre by metre Canadians are making the Trans Canada Trail happen. If you would like to be a part of this important undertaking, you can find more information by e-mail (info@tctrail.ca) or by phone (1-800-465-3636), as well as at the website noted below.

SOURCE: *The Trans Canada Trail,* Trans Canada Trail Foundation, 2001, 2003, http://www.tctrail.ca. Reprinted by permission of the Trans Canada Trail.

Because there is no one right approach to achieving sustainability of our environment—indeed, in diversity we find the greatest potential to adapt to Earth's ever-changing conditions—our choices will be predicated on individuals learning about the place where they live, caring about the air, water, soil, wild plants, wild animals, wild places, and people of the place where they live (as well as beyond the immediate area), and acting on that caring. Since acting on one's own is not always sufficient or conducive to long-term change, many people find it important to get involved with locally based organizations or even local workplaces that are dedicated to helping and empowering local people bring about change in their own lives and communities.

Taken together, Figure 14–3 and Table 14–8 provide a summary overview of major components of Earth-sustaining actions. The elements of Figure 14–3 are expanded on in Table 14–8; while not comprehensive, the table is meant to demonstrate the ways in which sustainability tools support sustainability approaches and, in turn, how sustainability approaches help achieve sustainability objectives (that reflect at least some of the characteristics of sustainability that stewards attempt to practise and achieve).

Individual and cooperative actions matter when it comes to environmental sustainability. For instance, our individual and collective health may rely on simple choices, such as commuting by foot or bicycle. Such a

TABLE 14-8

MAJOR COMPONENTS OF EARTH-SUSTAINING ACTIONS

Sustainability Characteristics	Sustainability Objectives	Sustainability Approaches	Sustainability Tools

• an ethical principle
• commitment to equity
• quality of life and well-being
• integrated approach to planning and decision making
• legacy left to future generations
• an international concept
• respect for others (both humans and other elements of nature)
	• sustain our natural resources
	• protect the health of Canadians and ecosystems
	• meet our international obligations
	• promote equity
	• improve quality of life and well-being
		• integrated approach
		  – involvement of all affected parties in decision making
		  – full-cost accounting (social, economic, and environmental)
		  – environmental assessment (social, economic, and environmental)
		• ecosystem management
		• monitoring
		• sound science and analysis (regarding key issues, goals, etc.), including traditional ecological knowledge
		• collaboration and cooperation among
		  – citizens and organizations
		  – private sector
		  – governments and Aboriginal people
			• policy tools
			  – voluntary actions of groups, individuals, industry, or communities
			  – economic instruments
			  – government expenditure
			  – legal tools
			  – environmental auditing
			• information and awareness tools
			  – labelling programs
			  – technology sharing
			  – sustainability indicators
			• quality standards
			• demonstrating sustainable practices

SOURCE: Adapted from *A Guide to Green Government*, Government of Canada, 1995, pp. 4–17.

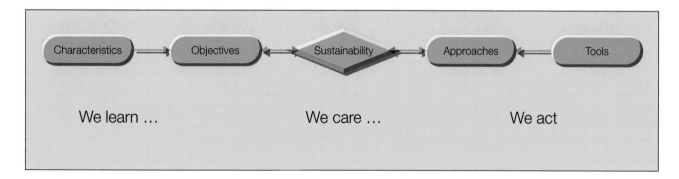

Figure 14–3

Toward environmental sustainability

choice will reduce traffic congestion and greenhouse gases while improving the individual's level of physical fitness and mental well-being. Recent evidence about human longevity places a high value on lifetime physical activity (and a low-fat diet). The interrelationship of environmental and social sustainability may induce people to make choices to improve their social well-being while improving conditions for the environment as well.

Whether we lead by example (for instance, in picking up litter from our streets, showing a neighbour how our backyard compost works, or volunteering for a position in an ENGO), or whether we lead by working within existing economic and political systems, the power of one-on-one communication to influence environmental change within political elites, among other elected officials, and in our circle of friends, family, and neighbours should not be underestimated.

To ensure that we live as sustainably as we can, leave the lightest possible footprints on the planet, and make good decisions with respect to environmental sustainability, a number of practical and philosophical considerations may guide our thinking and actions. These considerations include the following:

- *Critical assessment to determine how development proposals can be made sustainable:* Do they deplete Earth's capital? Do they preserve biodiversity? Do they enhance cultural diversity? Do they promote self-reliance on the part of individuals and communities? Do they address equity issues that may arise by their proposals? Is the precautionary principle useful to help ensure we use resources efficiently and live off Earth's income?
- *Use of appropriate technology:* To what degree are design-with-nature concepts employed? Are they simple, resource-efficient, and culturally adaptable technologies? Do they rely mostly on local sources of resources (recycled) and labour?
- *Information and education:* Are we teaching people

(and learning, ourselves) how to think holistically, in a systematic, integrated, interdisciplinary fashion about planet Earth? Are we listening to the variety of sources of knowledge about Earth? Do we know where our water comes from, where our wastes go, what kinds of soils support our food production, and how long our growing season is?
- *Demonstrating sustainability:* What models can we identify that are useful to demonstrate sustainability? Do they incorporate the dimensions of social equity, environmental sustainability, economic viability, and cultural diversity?
- *Simplicity:* Have we reduced, reused, recycled, and refused (unnecessary products) wherever possible? Have we eliminated unnecessary consumption and waste of energy and other resources? Do we know what elements of our lifestyles are harmful or beneficial to the Earth?

Each small action we take in support of environmental sustainability is important in helping to sustain Earth's life-support systems for ourselves and all life. Individual actions, however, must be coupled with strategic actions at higher levels to establish the conditions for effective societal change. Sustainability also means taking actions that maintain important elements of our social and cultural well-being. Protecting the environment is not a goal that can be achieved without recognizing the social context and cultural identity of people who live and work and are part of ecological systems. Actions that are taken without considering the effects on people who are affected by those decisions will simply be unsustainable. Perhaps, in working with others in our neighbourhoods and communities, our actions will help promote stewardship and encourage an acceptance by all groups of their role in promoting and implementing sustainability. It is not too late to learn how to work with Earth's ecosystems; if we really care, together our thoughts and actions will help achieve a sustainable society.

Chapter Questions

1. Looking at the area where you live, identify examples of where progress has been achieved in reaching environmental goals, as well as examples where less progress has been achieved. What might some of the reasons be for the differential success? How have social and cultural issues been considered when deciding on sustainability strategies?

2. Identify any ENGOs or other community organizations that are active in your region. What are their objectives and what are some of the environmental issues they are concerned about? What kinds of actions do they take? Have they been successful in achieving their goals?

3. In October 1996, at the World Conservation Union's first World Conservation Congress in Montreal, then Deputy Prime Minister Sheila Copps said: "We believe in good, tough environmental regulations. We need good laws to protect the gains we have made." Do you agree with these statements? Why or why not?

4. Comment on this statement: "A strong correlation is emerging between environmental and economic success." If you were to develop a complete response to this question, what additional information would you need to determine whether or not this statement is correct?

5. Using the information contained in Tables 14–1 through 14–8, identify which issues and which sectors received the greatest attention from regulators. Which issues or sectors received the least attention? Are there issues or sectors of concern that you feel have not been recognized by the regulators? What actions could you take to bring these issues to their attention?

6. Draw up a two-column list of the economic goods you use; in one column identify those economic goods that meet your basic needs and in the other column, those that satisfy your wants. Which of these economic wants would you be willing to give up? Are there some you would like to give up but are unwilling to? Are there any of these wants that you hope to satisfy in the future? Will any of these improve the quality of your life? If not, what would? Relate the results of your analysis to your personal impact on the environment. How do your results compare with others in your class?

references

Alberta Ministry of Environment. (2003). *Annual Report.* http://www3.gov.ab.ca/env/dept/reports/annual/2002-03/2002-2003_Annual_Report.pdf

Auditor General of Canada. (2000). *2000 Report of the Commissioner of the Environment and Sustainable Development.* http://www.oag-bvg.gc.ca/domino/reports.nsf/html/c007ce.html

B.C. Parks. (2001). *Doubling the legacy.* http://www.env.gov.bc.ca/pac/foreverbc/home.html

British Columbia Minister of Environment, Lands and Parks. (1999). *Business plan, 2000–2001.*

Boyd, D. (2003). *Unnatural law: Rethinking Canadian environmental law and policy.* Vancouver: UBC P.

Canadian Chemical Producers' Association. (1999). *Responsible care: The picture is getting brighter.* http://www.ccpa.ca/english/library/RepDocsEN/RC99EN.pdf

Canadian Council of Ministers of the Environment (CCME). (2000). *National Packaging Protocol 2000. Final report.* http://www.mbnet.mb/ccme/pdfs/NaPPFinalJun22_e.pdf

Canadian Parks Ministers Council. (2000, August). Working together: Parks and protected areas in Canada. *FPPC Report,* p. 7. http://www.gov.ab.ca/env/parks/fppc/workingtogether.html

Colborn, T., vom Saal, F. S., & Soto, A. M. (1993). Developmental effects of endocrine-disrupting chemicals in wildlife and humans. *Environmental Health Perspectives, 101*(5), 378–384.

Commissioner of the Environment and Sustainable Development. (2000). *2000 Report of the Commissioner of the Environment and Sustainable Development.* The Commissioner's Observations. http://www.oag-bvg.gc.ca/domino/reports.nsf/html/c0menu_e.html

Committee on the Status of Endangered Wildlife in Canada (COSEWIC). (2000). *COSEWIC Species Assessment—November 2000.*

Committee on the Status of Endangered Wildlife in Canada (COSEWIC). (2003). *Committee on the Status of Endangered Wildlife in Canada: Results of the May 2003 COSEWIC Species Assessment Meeting.* http://www.cosewic.gc.ca/eng/sct0/index_e.cfm#sct0_2

Commuter Connections. (2001). *Find a carpool.* http://www.carpool.ca/carpool_find.asp

Environment Canada. (1995). *Canadian Biodiversity Strategy.* Ottawa.

Environment Canada. (2003a). *Environmental Signals: Canada's National Environmental Indicator Series 2003.* http://www.ec.gc.ca/soer-ree/English/Indicator_series/default.cfm

Environment Canada (2003b, March/April). Protecting water from mine waste. *The Science and the Environment Bulletin.* http://www.ec.gc.ca/science/sadefeb03/printversion/p4_e.html

Gélinas, J. (2002). *Report of the Commissioner of the Environment and Sustainable Development 2002.* Ottawa: Office of the Auditor General.

Government of Canada. (1991). *The state of Canada's environment—1991.* Ottawa: Supply and Services Canada.

Government of Canada. (1995a). *Toxic substances management policy.* Ottawa.

Government of Canada. (1995b). *A guide to green government.* Ottawa.

Green Lane. (2001). *Particulate matter (PM<10).* http://www.ec.gc.ca/air/p-matter_e.shtml

Green Lane. (2003). Fact Sheet: *Government of Canada Initiatives.* http://www.ec.gc.ca/press/2002/020403_f_e.htm

Greenhouse Gas Division. (2000, October). *Canada's greenhouse gas inventory, 1990–1998: Final submission to the UNFCC Secretariat.* Pollution Data Branch, Environment Canada, pp. vii, ix. http://www.ec.gc.ca/pdb/ghg/english/Docs/CGHGI_00Vol1_Web_eng.pdf

Macdonald, D. (1996). Beer cans, gas guzzlers and green taxes. *Alternatives, 22*(3), 12–19.

Malden Mills Industries, Inc. (1994). *Cool stuff to know about Polartec® fabrics.* Lawrence, MA: Author.

Menyasz, P., Pole, K., & Ray, R. (Eds.). (2000). Nova Scotia: Province alone in meeting goal of 50 percent garbage recycling by 2000. *Environment Policy and Law, 11*(8), 108.

National Energy Board. (1994). *Canadian energy supply and demand 1993–2010, trends and issues.* Calgary: Author.

National Pollutant Release Inventory. (1998). *The national pollutant release inventory, 1998.* http://www.ec.gc.ca/pdb/npri/1998.report.html

Natural Resources Canada. (2003). *Energuide for houses retrofit incentive launched.* http://www.nrcan-rncan.gc.ca/media/newsreleases/2003/200388_e.htm

Nickerson, M. (1990). *Planning for seven generations—guideposts for a sustainable future.* Merrickville, ON: Bakavi School of Permaculture.

Office of the Auditor General. (1996, November). *1996 Report of the Auditor General.* Chapter 22. http://www.oag-bvg.gc.ca/domino/reports.nsf/html/9622ce.html#0.2.Q3O5J2.O25UY6.E9TLQE.DP

Office of the Auditor General. (1999). *1999 Report of the Commissioner of the Environment and Sustainable Development.* http://www.oag-bvg.gc.ca/domino/reports.nsf/html/c904ce.html#0.2.2Z141Z1.NBS3AG.T8WQBF.52

Office of Energy Efficiency. (2003). *Energy Use Data Handbook 1990 and 1995 to 2001.* Total End Use Sectors June 2003. M92-245-2001e. http://oee.nrcan.gc.ca/publications/infosource/home/index.cfm?act=category&category=03&PrintView=N&Text=N

Olsen, K., Wellisch, M., Boileau, P., Blain, D., Ha, C., Henderson, L., Linag, C., McCarthy, J., & McKibbin, S. (2003). *Canada's greenhouse gas inventory 1990–2001.* http://www.ec.gc.ca/pdb/ghg/1990_01_report/1990_01_report_e.pdf)

Ontario Ministry of the Environment. (2000). *MOE business plan, 2000–2001.*

Ontario Ministry of the Environment. (2003). http://www.ene.gov.on.ca/

Parks Canada. (2003). *Gulf Islands National Park Reserve of Canada.* http://parkscanada.pch.gc.ca/pn-np/bc/gulf/index_E.asp

Pilgrim, W. (1998). *The northeastern states and eastern Canadian provinces mercury study.* http://www.ceiw.ca/eman-temp/reports/publications/mercury/page78.html

Saskatchewan Environment. (2003). *State of the environment report 2003.* http://www.se.gov.sk.ca/pdf/SOE_Report_2003.pdf

Taylor, D. M. (1994). *Off course: Restoring balance between Canadian society and the environment.* Ottawa: International Development Research Centre.

Wackernagel, M., & Rees, W. (1996). *Our ecological footprint: Reducing human impact on the earth.* Gabriola Island, BC: New Society.

World Wildlife Fund Canada. (2003). *The nature audit: Setting Canada's conservation agenda for the 21st century.* Report No.1–2003. Toronto: Author.

Table of the Elements

Element Name	Symbol	Atomic Number	Atomic Mass	Element Name	Symbol	Atomic Number	Atomic Mass
Actinium	Ac	89	(227)	Neon	Ne	10	20.1797
Aluminum	Al	13	26.981539	Neptunium	Np	93	(237)
Americium	Am	95	(243)	Nickel	Ni	28	58.6934
Antimony	Sb	51	121.757	Niobium	Nb	41	92.90638
Argon	Ar	18	39.948	Nitrogen	N	7	14.00674
Arsenic	As	33	74.92159	Nobelium	No	102	(259)
Astatine	At	85	(210)	Osmium	Os	76	190.2
Barium	Ba	56	137.327	Oxygen	O	8	15.9994
Berkelium	Bk	97	(247)	Palladium	Pd	46	106.42
Beryllium	Be	4	9.012182	Phosphorus	P	15	30.973762
Bismuth	Bi	83	208.98037	Platinum	Pt	78	195.08
Boron	B	5	10.811	Plutonium	Pu	94	(244)
Bromine	Br	35	79.904	Polonium	Po	84	(209)
Cadmium	Cd	48	112.411	Potassium	K	19	39.0983
Calcium	Ca	20	40.078	Praseodymium	Pr	59	140.90765
Californium	Cf	98	(251)	Promethium	Pm	61	(145)
Carbon	C	6	12.011	Protactinium	Pa	91	(231)
Cerium	Ce	58	140.115	Radium	Ra	88	(226)
Cesium	Cs	55	132.90543	Radon	Rn	86	(222)
Chlorine	Cl	17	35.4527	Rhenium	Re	75	186.207
Chromium	Cr	24	51.9961	Rhodium	Rh	45	102.90550
Cobalt	Co	27	58.93320	Rubidium	Rb	37	85.4678
Copper	Cu	29	63.546	Ruthenium	Ru	44	101.07
Curium	Cm	96	(247)	Samarium	Sm	62	150.36
Dysprosium	Dy	66	162.50	Scandium	Sc	21	44.955910
Einsteinium	Es	99	(252)	Selenium	Se	34	78.96
Erbium	Er	68	167.26	Silicon	Si	14	28.0855
Europium	Eu	63	151.965	Silver	Ag	47	107.8682
Fermium	Fm	100	(257)	Sodium	Na	11	22.989768
Fluorine	F	9	18.9984032	Strontium	Sr	38	87.62
Francium	Fr	87	(223)	Sulfur	S	16	32.066
Gadolinium	Gd	64	157.25	Tantalum	Ta	73	180.9479
Gallium	Ga	31	69.723	Technetium	Tc	43	(98)
Germanium	Ge	32	72.61	Tellerium	Te	52	127.60
Gold	Au	79	196.96654	Terbium	Tb	65	158.92534
Hafnium	Hf	72	178.49	Thallium	Tl	81	204.3833
Helium	He	2	4.002602	Thorium	Th	90	232.0381
Holmium	Ho	67	164.93032	Thulium	Tm	69	168.93421
Hydrogen	H	1	1.00794	Tin	Sn	50	118.710
Indium	In	49	114.82	Titanium	Ti	22	47.88
Iodine	I	53	126.90447	Tungsten	W	74	183.85
Iridium	Ir	77	192.22	Unnilennium	Une	109	(267)
Iron	Fe	26	55.847	Unnilhexium	Unh	106	(263)
Krypton	Kr	36	83.80	Unniloctium	Uno	108	(265)
Lanthanum	La	57	138.9055	Unnilpentium	Unp	105	(262)
Lawrencium	Lr	103	(262)	Unnilquadium	Unq	104	(261)
Lead	Pb	82	207.2	Unnilseptium	Uns	107	(262)
Lithium	Li	3	6.941	Uranium	U	92	238.0289
Lutetium	Lu	71	174.967	Vanadium	V	23	50.9415
Magnesium	Mg	12	24.3050	Xenon	Xe	54	131.29
Manganese	Mn	25	54.93805	Ytterbium	Yb	70	173.04
Mendelevium	Md	101	(258)	Yttrium	Y	39	88.90585
Mercury	Hg	80	200.59	Zinc	Zn	30	65.39
Molybdenum	Mo	42	95.94	Zirconium	Zr	40	91.224
Neodymium	Nd	60	144.24				

Atomic masses in parentheses are the mass number of the longest-lived isotope of the element.

Selected Canadian Population Statistics

TABLE 1
REGIONAL POPULATION GROWTH IN CANADA, 1950–2003

Region	Population (thousands)							Increase (%)
	1950	1960	1970	1980	1990	2000	2003	1950 to 2003
Canada	13 712	17 870	21 297	24 593	27 791	30 750	31 630	131
Newfoundland & Labrador	351	448	517	574	579	539	520	48
Prince Edward Island	96	103	110	124	131	139	138	44
Nova Scotia	638	727	782	855	913	941	936	47
New Brunswick	512	589	627	708	743	757	751	47
Quebec	3 969	5 142	6 013	6 528	7 021	7 372	7 487	89
Ontario	4 471	6 111	7 551	8 770	10 342	11 669	12 238	174
Manitoba	768	906	983	1 037	1 108	1 148	1 163	51
Saskatchewan	833	915	941	970	1 011	1 024	995	19
Alberta	913	1 291	1 595	2 201	2 556	2 997	3 154	245
British Columbia	1 137	1 602	2 128	2 755	3 300	4 064	4 147	265
Yukon	8	14	17	24	28	31	31	289
Northwest Territories	16	22	33	47	59	42	42	162
Nunavut	n/a	n/a	n/a	n/a	n/a	28	29	n/a

SOURCE: *Population, Provinces and Territories,* Statistics Canada, 2003, http://www.statcan.ca/english/Pgdb/demo02.htm. Reprinted by permission of Statistics Canada.

TABLE 2
POPULATION DYNAMICS IN CANADA, YEAR 2003

Total Population	Rate of Natural Growth[1] (% annual)	Doubling Time (in years at the current growth rate)	Crude Birth Rate[2]	Crude Death Rate[2]	Infant Mortality Rate[3]	Total Fertility Rate[4]	Average Life Expectancy (year)	Dependency Ratio (%)
31 629 700	0.3	231	10.5	7.2	5.3	1.5	79	45.2

[1] Rate of natural increase is expressed as birth rate minus death rate.
[2] Birth and death rates per 1000 population.
[3] Infant mortality rate is calculated as the number of deaths of children less than one year of age per 1000 live births.
[4] Average number of children born to a woman during her lifetime.

SOURCES: *Births and Birth Rate, 2002–2003,* Statistics Canada, 2004, http://www.statcan.ca/english/Pgdb/demo04b.htm; *Deaths and Death Rate, 2002–2003,* Statistics Canada, 2004, http://www.statcan.ca/english/Pgdb/demo04b.htm; *Infant Mortality Rate, 1999,* Statistics Canada, 2004, http://www.statcan.ca/english/Pgdb/health21a.htm; *World Population Data Sheet,* Population Reference Bureau, 2003, http://www.prb.org

TABLE 3

REGIONAL CANADIAN POPULATION DISTRIBUTIONS BY AGE GROUP AND GENDER, YEAR 2003 (THOUSANDS)

Region	0–14 Years Males	0–14 Years Females	15–65 Years Males	15–65 Years Females	65 Years and Over Males	65 Years and Over Females	All Ages Males	All Ages Females
Canada	2 960.60	2 821.00	10 947.40	10 840.50	1 753.70	2 306.40	15 661.70	15 967.90
Newfoundland	43.5	41.2	183.1	186.5	29.0	36.2	255.6	264.0
Prince Edward Island	13.1	12.6	45.7	47.2	8.2	10.9	67.0	70.8
Nova Scotia	81.5	77.9	321.4	324.9	55.6	74.6	458.5	477.5
New Brunswick	65.3	61.5	262.1	260.0	43.3	58.4	370.7	379.9
Quebec	658.1	626.0	2 619.10	2 582.90	416.1	585.0	3 693.30	3 793.80
Ontario	1 177.70	1 129.40	4 195.10	4 190.20	671.4	874.5	6 044.30	6 194.00
Manitoba	120.2	114.9	389.1	380.5	67.1	90.9	576.4	586.4
Saskatchewan	104.3	98.7	326.1	318.1	64.2	83.4	494.5	500.3
Alberta	320.3	302.6	1 129.10	1 076.30	144.7	180.7	1 594.20	1 559.60
British Columbia	362.8	342.9	1 440.20	1 439.40	251.7	309.6	2 054.70	2 091.90
Yukon	3.0	3.0	11.6	11.5	1.0	0.9	15.7	15.4
Northwest Territories	5.4	5.3	15.3	14.2	0.9	0.8	21.6	20.3
Nunavut	5.3	5.0	9.5	8.8	0.4	0.3	15.2	14.2

SOURCE: *Population by Sex and Age Group,* Statistics Canada, 2003, http://www.statcan.ca/english/Pgdb/popula.htm. Reprinted by permission of Statistics Canada.

TABLE 4

REGIONAL CANADIAN ABORIGINAL POPULATIONS, 2001

Region	Total Population	Aboriginal Population[1] Total	North American Indian	Métis	Inuit	Non-Aboriginal Population	Percentage of Total Population
Canada	29 639 030	976 305	608 850	292 305	45 070	28 662 725	3.3
Newfoundland	508 080	18 775	7 040	5 480	4 560	489 300	3.7
Prince Edward Island	133 385	1 345	1 035	220	20	132 040	1.0
Nova Scotia	897 565	17 010	12 920	3 135	350	880 560	1.9
New Brunswick	719 710	16 990	11 495	4 290	155	702 725	2.4
Quebec	7 125 580	79 400	51 125	15 855	9 530	7 046 180	1.1
Ontario	11 285 545	188 315	131 560	48 340	1 375	11 097 235	1.7
Manitoba	1 103 700	150 045	90 340	56 800	340	953 655	13.6
Saskatchewan	963 155	130 185	83 745	43 695	235	832 960	13.5
Alberta	2 941 150	156 225	84 995	66 060	1 090	2 784 925	5.3
British Columbia	3 868 875	170 025	118 295	44 265	800	3 698 850	4.4
Yukon	28 520	6 540	5 600	535	140	21 975	22.9
Northwest Territories	37 100	18 730	10 615	3 580	3 910	18 370	50.5
Nunavut	26 665	22 720	95	55	22 560	3 945	85.2

[1] Includes the Aboriginal groups (North American Indian, Métis, and Inuit), multiple Aboriginal responses, and Aboriginal responses not included elsewhere.

NOTE: The Aboriginal identity population comprises those persons who reported identifying with at least one Aboriginal group, that is, North American Indian, Métis, or Inuit, and/or who reported being a Treaty Indian or a Registered Indian, as defined by the Indian Act of Canada, and/or who reported being a member of an Indian Band or First Nation.

SOURCE: "Aboriginal Peoples of Canada: Highlight Tables, 2001 Census," Statistics Canada, Cat. no. 97F0024XIE2001007. Reprinted by permission of Statistics Canada. http://www12.statcan.ca/english/census01/products/highlight/Aboriginal/

GLOSSARY

abiotic. The nonliving components of an ecosystem such as water, air, solar energy, and nutrients necessary to support life in a given area. Compare *biotic*.

acclimation. The adjustment of a species to slowly changing conditions in an ecosystem, such as temperature. See also *threshold effect*.

acid mine drainage. Acidic water that drains from mine sites and sometimes enters streams and lakes.

adaptation. Any genetically controlled characteristic—structural, physiological, or behavioural—that enhances the chance for members of a population to survive and reproduce in its environment. See also *mutation*.

aerobic respiration. A complex chemical process that drives the life processes of living things, by using oxygen to convert nutrients such as glucose back into carbon dioxide and water. The opposite of *photosynthesis*.

aesthetic arguments. A rationale for the conservation of nature based on its beauty and aesthetic qualities. Compare *ecological justification, moral justification, utilitarian justification*.

age-specific fertility rate. The number of live births per 1000 women of a specific age group per year.

agroecosystems. Communities of living organisms, together with the physical resources that sustain them (such as biotic and abiotic elements of the underlying soils and drainage networks), that are managed for the purposes of producing food, fibre, and other agricultural products.

agroforestry. The raising of trees or shrubs together with crops and/or animals on the same parcel of land.

alternative energy. Renewable energy sources, such as wind, flowing water, solar energy, and biomass, which create less environmental damage and pollution than fossil fuels, and offer an alternative to nonrenewable resources.

alternative livestock. The raising together of non-native animal species and domesticated native species.

anadromous. Fish that are born and develop in rivers and streams, migrate out to sea for as long as seven years or as short as a few months (depending on the species), and then return to their birthplace to spawn and die.

anthropogenic. Human-induced changes to the environment.

aquaculture. The breeding and raising of fish under controlled conditions, with the goal of high-level production for food or recreational purposes.

aquifer. Underground zone or layers of porous rock saturated with water from which an economically significant amount of groundwater can be obtained from a well.

assimilative capacity. The ability of a water body to accept sewage and other substances without significant harm to plants, organisms, and animals, human health, or other water uses.

atmosphere. A thin layer of gases consisting mostly of nitrogen and oxygen that completely surrounds the solid and liquid earth. See also *troposphere, stratosphere*.

atomic number. The number of protons in an atom's nucleus, which distinguishes it from the atoms of other elements.

atoms. The smallest particles that exhibit the unique characteristics of that particular element.

autotrophs. See *producers*.

background extinction. The continuous, low-level extinction of species that has occurred throughout much of history. Compare *mass extinction*.

barrier islands. Long, low, offshore islands of sediment that run parallel to much of North America's Atlantic and Gulf coasts and help protect coastal wetlands and habitats from storm damage.

bellwether species. See *indicator species*.

benthic environment. The ocean floor, one of the two main divisions of the open sea environment. See also *pelagic environment*.

bioaccumulation. The uptake and retention of substances in organisms.

biodiversity. The diversity of life on earth, consisting of genetic diversity, species diversity, and ecosystem diversity.

biofuels. See *ethanol*.

biogeochemical cycles. See *nutrient cycles*.

biological evolution. The change in inherited characteristics of a population from generation to successive generation.

biological oxygen demand (BOD). The amount of oxygen needed during the time it takes for waste material to be oxidized. Water quality is directly affected by this; some organisms thrive on a higher BOD and some suffocate for lack of oxygen.

biomagnification. The accumulation and concentration of certain substances in organisms, such as chlorinated organic compounds (DDT and PCBs) in the fatty tissues of predators in the Arctic marine system.

biomass. The dry weight of all organic matter contained in plants and animals in an ecosystem.

biomass burning. Using plant materials and animal wastes as fuel.

biome. A broad, regional type of ecosystem characterized by distinctive climate and soil conditions and a distinctive biological community adapted to those conditions.

bioremediation. A process that involves using naturally occurring or genetically modified microorganisms to break down or degrade hazardous substances into less hazardous or nontoxic substances.

biosphere. That part of the Earth inhabited by plants and animals, and their interactions with the atmosphere, hydrosphere, and lithosphere.

biotechnology. The use of a living organism (or a part thereof) to create some different product, whether cheese to eat, a vaccine to combat disease, or a plant or animal with novel attributes. Genetic engineering is a more recent aspect of biotechnology.

biotic. The living components of an ecosystem, including plants, animals, and their products (secretions, wastes, and remains) and effects in a given area. Compare *abiotic*.

biotic potential. The maximum rate a population can increase under ideal conditions.

bitumen. A black oil rich in sulphur that is found in oil sand. It can be treated and chemically upgraded into synthetic crude oil, though the net useful energy yield is lower than for conventional oil because more energy is required to extract and process it. See also *oil sand*.

carcinogen. A cancer-causing agent.

carnivores. Organisms that feed indirectly on plants by eating the meat of herbivores. Most carnivores are animals, but a few examples are in the plant kingdom, such as the Venus flytrap. See also *herbivores, omnivores*.

carrying capacity. The number of organisms that an ecosystem can support indefinitely while maintaining its productivity, adaptability, and capability for renewal.

cash crop. A crop grown to be traded in a marketplace.

catadromous. Species that spend most of their life cycle in fresh water, but enter the ocean to spawn.

chemical change. A change in which a chemical reaction is produced and a new substance created, as when gasoline is burned to produce carbon dioxide. Compare *physical change*.

chemical contamination. The presence of a chemical that makes something unfit for its intended use. Pesticide contamination, for example, makes soil unfit for food production.

chemical formula. A shorthand way to show the number of atoms (or ions) in the basic structural unit of a compound. Examples include $NaCl$, H_2O, and $C_6H_{12}O_6$.

chemosynthesis. The process in which some organisms (usually certain types of bacteria) convert, without sunlight, inorganic chemical compounds into organic nutrient compounds—food energy for their own use. Contrasts with *photosynthesis*.

chemotrophs. Producers, including algae and bacteria, that convert the energy found in inorganic chemical compounds into more complex energy without the use of sunlight. See also *consumers, producers*.

clear-cutting. A system of tree harvesting that removes all the trees in a given area, as opposed to selective cutting that leaves some trees standing. Replanting after clear-cutting can be difficult.

climax community. The mature stage of succession in a particular area, in which all organisms and nonliving factors are in balance.

closed system. See *systems*.

coal. The most abundant fossil fuel in the world, with reserves four to five times that of oil and gas combined. It has a relatively high net useful energy yield and is highly effective for providing industrial heat.

coastal wetlands. Coastal area that provides breeding grounds and habitats for many marine organisms as well as for waterfowl, shorebirds, and other wildlife.

coastal zone. The area where the ocean meets the land, which constitutes 10 percent of the ocean's area but contains 90 percent of all marine species.

co-generation. The production of two useful forms of energy from the same source, such as heat and power. See also *district heating*.

commensalism. An interaction between species in which one benefits and the other is neither helped nor harmed. See also *mutualism, symbiosis*.

community. An area where different species interact, such as an alpine community or a prairie community. See also *habitat*.

competitive exclusion principle. When two species are competing for the same resources, one must migrate to another area if possible, shift its feeding habits or behaviour, suffer a sharp decline in population numbers, or become extinct.

compounds. One of the basic forms of chemical composition, which involves two or more different elements held together in fixed proportions by the attraction in the chemical bonds between their constituent atoms. See also *elements, mixtures*.

conservation tillage. A soil conservation practice in which most of the crop residue is left on the soil surface to protect against erosion, reduce soil crusting, and increase the organic matter content of soil.

consumers. Those organisms that eat the cells, tissues, or waste products of other organisms. Animals are common consumers. Also called *heterotrophs*. See also *chemotrophs, producers*.

continuous clear-cutting. In timber harvesting, cut blocks are located adjacent to each other in successive years, which rapidly lays bare much larger areas.

controlled experiment. An experiment designed to test the effects of independent variables on a dependent variable by changing one independent variable at a time.

coral reefs. Found in warm tropical and subtropical oceans, these formations are rich in life and may contain more than 3000 species of corals, fish, and shellfish.

country food. Food grown by people in small communities living in harmony with their local environment.

crude birth rate. The annual number of live births per 1000 population, without regard to age or sex composition.

crude death rate. The annual number of deaths per 1000 population.

crude growth rate. The net change, or difference, between the crude birth rate and the crude death rate.

decomposers. See *microconsumers*.

deductive reasoning. Drawing conclusions from observations of the natural world by means of logical reasoning. Compare *inductive reasoning*.

deforestation. To clear an area of forests or trees, usually for lumber or agricultural uses.

demographic transition. A four-stage model of population change that links industrial development with zero population growth, and suggests a postindustrial phase that would focus more on sustainable forms of economic development.

demographic trap. A state in which a nation or population is stuck in the second stage of demographic transition, with a low death rate, a high birth rate, and increasing demand on available resources.

demography. The study of the characteristics and changes in the size and structure of human populations.

dependency ratio. A measure of the number of dependants, young and old, that each 100 people in their economically productive years must support. It is used to forecast the condition of the future human population.

dependent variable. See *responding variable*.

desertification. A combination of human-induced environmental degradation (such as overgrazing of livestock) superimposed on a natural drought situation, causing expansion of desert conditions into areas that previously were more humid.

detritus feeders. See *detrivores*.

detrivores. Consumers that ingest fragments of dead organic material. Examples are earthworms and maggots.

differential reproduction. The ability to produce more offspring with the same favourable adaptations as the parents, which will allow them to survive under changed environmental conditions.

discharge. Refers to the amount of water returned to the original source. See also *withdrawal uses*.

dissolved oxygen content. The amount of oxygen dissolved in a given volume of water at a particular temperature and pressure. This can be a limiting factor on the growth of many aquatic populations.

district heating. An effort to maximize energy efficiency in power generating stations, which involves a steam cycle that is modified so that the steam is extracted and used to produce hot water. The water is then pumped through pipes to surrounding buildings to supply heat. See also *co-generation*.

Dobson unit. One Dobson unit is equivalent to a layer of pure ozone 0.01 mm thick at standard temperature (0°C) and pressure (101.3 kPa) spread over Earth's surface.

doubling time. The length of time required for a population to double in size.

drainage basin. The area of land that contributes water and sediment to a river.

driftnetting. The placing of very long gillnets (2.5 km and longer) that drift with currents and wind for the purpose of entangling fish in webbed panel(s).

ecological diversity. The variety of biological communities, such as forests, deserts, grasslands, and streams, that interact with one another and with their physical and chemical (nonliving) environments. See also *species diversity*.

ecological footprint. A link between human lifestyles and ecosystems that allows people to visualize the impact of their consumption patterns and activities on ecosystems.

ecological health. An ecosystem where native species are present at viable population levels.

ecological integrity. A condition in which the structure and function of an ecosystem are unimpaired by human activity and are likely to persist into the future.

ecological justification. A rationale for the conservation of nature based on the idea that the environment provides specific functions necessary to the persistence of our life. Compare *aesthetic arguments, moral justification, utilitarian justification*.

ecological niche. The role an organism plays within the structure and functions of an ecosystem, and the way it interacts with other living things and with its physical environment.

ecology. The study of the interactions of living organisms with one another and with their nonliving environment of matter and energy.

ecosphere. See *biosphere*.

ecosystem. A community and its members interacting with each other and their nonliving environment.

ecosystem management. Concentrates on managing entire ecosystems rather than isolating only parts of the systems.

electrons. Negatively charged ions that continually orbit the nucleus of an atom and are held in orbit by attraction to the positive charge of the nucleus. See also *neutrons, protons*.

elements. One of the basic forms of chemical composition. All matter is built from the 109 known chemical elements; these are the simplest building blocks of all matter. See also *compounds, mixtures*.

emissions permits. A strategy developed to reduce greenhouse gas emissions in which companies buy and sell from each other the right to release greenhouse gases.

endemic species. A species that is native to a particular geographic region.

energy. The ability or capacity to do work. Energy enables us to move matter and change it from one form to another.

energy quality. The measure of an energy source's ability to perform useful work, such as running electrical devices or motors. See also *high-quality energy, low-quality energy*.

entropy. A measure of randomness or disorder. The higher the entropy, the greater its disorder. See *high-quality energy, high-quality matter, low-quality energy, low-quality matter*.

environment. The surroundings in which plants and animals live, affected by various physical factors such as temperature, water, light, and food resources.

environmental ethics. A new discipline that analyzes the issues regarding our moral obligations to future generations with respect to the environment.

environmental impact assessment (EIA). A process that aims to provide decision makers with scientifically researched and documented evidence to identify the likely consequences of undertaking new developments and changing natural systems. See also *environmental impact statement*.

environmental impact statement (EIS). A key component of an environmental impact assessment, an EIS provides a nontechnical summary of the study, including the main project characteristics, aspects of the environment likely to be affected, possible alternatives, and suggested measures and systems to monitor or reduce any harmful effects. See also *environmental impact assessment*.

environmental resistance. The limits set by the environment that prevent organisms from reproducing indefinitely at an exponential rate.

epiphytes. Plants that use their roots to capture nutrients and moisture from the air and to attach themselves to other plants, particularly in tropical forests (and some in temperate rain forests).

estuaries. A body of coastal water partly surrounded by land, with access to the open sea and a large supply of fresh water from rivers. These conditions provide excellent conditions for many important shellfish and fin fish species.

ethanol. A fuel converted from biomass materials and used to power motor vehicles, either directly as fuel or as an octane-enhancing gasoline additive. Ethanol can reduce carbon monoxide emissions from regular gasoline blends.

eukaryotic. Cells with a high degree of internal organization, including a nucleus (genetic material surrounded by a membrane) and several other internal parts surrounded by membranes. See also *prokaryotic*.

eutrophic. A lake enriched with nutrients in excess of what is required by producers. See also *mesotrophic, oligotrophic*.

eutrophication. An increase in the concentration of plant nutrients in water. Natural eutrophication is a slow process, but human-induced eutrophication (as from fertilizers used in agriculture) may accelerate the process and make water unfit for human consumption and for aquatic organisms.

exajoule. 10^{18} joules. See also *joule*.

exclusive economic zone. An area of exclusive fishing rights granted to Canada in the 1982 United Nations Convention on the Law of the Sea. It came into force in 1994.

exotic. A species that enters an ecosystem from a different part of the world through introduction (deliberately or accidentally) by humans.

exponential growth. Growth in a species that takes place at a constant rate per time period. When plotted on a graph, the exponential growth curve is J-shaped.

ex situ conservation. Conservation of species or genetic materials under artificial conditions, away from the ecosystems to which they belong.

extinction. The process whereby a species is eliminated from existence when it cannot adapt genetically and reproduce successfully under new environmental conditions. See also *background extinction, mass extinction*.

fact. An observation that all (or almost all) scientists agree is correct.

first law of energy. See *first law of thermodynamics*.

first law of thermodynamics. During a physical or chemical change, energy is neither created nor destroyed. See also *second law of thermodynamics*.

flaring. A method of disposing of unwanted, unprocessed natural gas. Gas is burned to release hydrogen sulphide (sour gas) and to avoid the buildup of potentially explosive levels of gas at work sites.

food chains. The sequence of who feeds on or decomposes whom in an ecosystem.

food web. A complex network of feeding relationships in which the flow of energy and materials through an ecosystem takes place. That flow occurs on the basis of a range of food choices on the part of each organism involved.

fossil fuels. The remains of prehistoric animals, forests, and sea floor life that have become buried in layers of sediment and decomposed very slowly, eventually being converted into crude oil. See also *hydrocarbons*.

fundamental niche. The full range of physical, chemical, and biological factors each species could use if there were no competition from other species. See also *interspecific competition*.

gene pool. The sum of all genes possessed by the individuals of a population.

general fertility rate. The number of live births per 1000 women of childbearing age per year.

generalist species. Species with the ability to live in many different places while tolerating a wide range of environmental conditions. Humans are considered a generalist species. See also *specialist species*.

genes. Segments of various deoxyribonucleic acid (DNA) molecules found in chromosomes. Genes impart certain inheritable traits to organisms.

genetic diversity. The diversity within a given population that shares common structural, functional, and behavioural traits but varies slightly in genetic makeup and so exhibits slightly different behaviours and appearances.

ghostfishing. The consequence of fish becoming entangled and drowning in lost and/or unmanned fishnets.

global warming potential. A concept developed to take into account the differing times that gases remain in the atmosphere and their individual radiative forcings (see Box 5–1), in order to evaluate the potential climate effects of equal emissions of each of the greenhouse gases.

gross primary productivity. The rate at which producers in an ecosystem capture and store chemical energy as biomass. Compare *net primary productivity*.

gross water use. The total amount of water used (intake + recirculation).

groundwater. Water that has accumulated beneath the Earth's surface in underground aquifers (reservoirs) and in the saturation zone below the water table. Water percolates down through soils, gravel, and rock and rises up from below to slowly replenish aquifers and saturation zones. Compare *surface water*.

habitat. The place where an organism or population lives, such as an ocean, a forest, or a stream. See also *community*.

half-life. The time required for one-half of a substance to disappear—for example, the time required for one-half of a toxic substance to be converted to some other form. Also, the average time it takes for one-half of a radioisotope to be transformed to some other isotope.

halocarbons. Any compound of carbon and a halogen (one of the chemical elements fluorine, chlorine, bromine, iodine, astatine) used especially as a refrigerant and propellant. CFC-11, for instance, was used widely in plastic foam blowing and CFC-12 was used in vehicle air conditioners and refrigerator coolants. Now being phased out because of potential to cause harm to stratospheric ozone layer.

halons. Compounds related to chlorofluorocarbons that contain bromine and are used in fire extinguisher systems. Implicated as ozone-destroying gases in the stratosphere.

heat island. A microclimate in which the air temperature is slightly higher than in the surrounding area. In an urban heat island, for example, the temperature in the city is 1–2°C higher than in the rural area around it.

herbivores. Organisms that eat green plants directly as a source of nutrients. Deer are common herbivores. See also *carnivores, omnivores.*

heterotrophs. See *consumers.*

highgrading. An unsound practice associated with selective cutting techniques that involves logging the highest-quality and most accessible timber first.

high-quality energy. Concentrated energy sources such as electricity, gasoline, and some food types that enable people and machines to perform useful tasks. See also *energy quality, low-quality energy.*

high-quality matter. Material such as coal or salt deposits commonly found near the Earth's surface in an organized or concentrated form, so that its potential for use as a resource is great. See *low-quality matter.*

human cultural diversity. The variety of human cultures that represent our adaptability and survival options in the face of changing conditions.

hydrocarbons. Any of a class of compounds containing only hydrogen and carbon, which include fossil fuels. See also *fossil fuels.*

hydroelectric power. Electrical power generated from the energy of falling water or any other hydraulic source.

hydrogen power. A source of energy that converts hydrogen to electricity to provide heat, light, and power. Though hydrogen is readily available, the production of hydrogen power is expensive and not yet commercially viable.

hydrologic cycle. The movement of water between the atmosphere, terrestrial systems, and the oceans, through evaporation, runoff from streams and rivers, and precipitation.

hydrosphere. The Earth's supply of moisture in all its forms: liquid, frozen, and gaseous. This includes surface water, underground water, frozen water, water vapour in the atmosphere, and moisture in the tissues and organs of living things.

hypothesis. An explanation that is based on testable observations and experiments, and that can be accepted until it is disproved.

igneous rock. Rock formed from molten materials crystallizing at the Earth's surface (such as lava from volcanoes), or beneath the surface (such as granite). See also *metamorphic rock, sedimentary rock.*

immigrant species. Those species that migrate into or are introduced into an ecosystem, deliberately or accidentally, by humans.

impact-benefit agreements. Agreements that are undertaken in large-scale resource developments among industry, government, and affected communities. Normally they are undertaken where Aboriginal communities may be affected by resource extraction and production. Agreements may cover a range of social and economic concerns, including employment and training, economic development and business opportunities, community and social support, as well as implementation, coordination, and funding.

independent variable. A condition that is deliberately manipulated by scientists to test the response in an experiment. See also *operational definitions, responding variable.*

indicator species. Those species that provide early warnings of environmental damage to communities or ecosystems.

inductive reasoning. Drawing a general conclusion based on a limited set of observations. Compare *deductive reasoning.*

infant mortality rate. The ratio of deaths of infants under 12 months per 1000 live births.

inferences. Conclusions derived either by logical reasoning from premises and/or evidence, or by insight or analogy based on evidence.

inorganic compounds. Any compound not classified as an organic compound. Compare *organic compound.*

in situ conservation. Conservation of ecosystems and the maintenance and recovery of viable populations of species in their typical surroundings.

instream uses. Water used in its natural setting for hydroelectric power, transportation, fisheries, and other applications. See also *withdrawal uses.*

intake. The quantity of water withdrawn or used. See also *withdrawal uses.*

interspecific competition. Competition from other species for one or more of the same limited resources of food, sunlight, water, soil, nutrients, or space. See also *fundamental niche.*

intrinsic value. A value placed on the inherent qualities of a species and/or an ecosystem, independent of its value to humans.

invertebrates. Animals without backbones, such as jellyfish, worms, insects, and spiders. Compare *vertebrates.*

ions. Subatomic, electrically charged particles in an atom. See also *protons, neutrons.*

joule. The unit for measuring the amount of heat required to raise 1 g of water from 14.5 to 15.5 degrees C in the Système Internationale (SI). There are 4.184 joules in 1 calorie. The calorie is the unit used to measure heat energy in the English system.

keystone species. Those species that play a crucial role in helping to maintain the ecosystems of which they are a part, by pollination, regulation of populations, or other activities.

kinetic energy. Energy associated with the movement of matter and mass. A moving air mass such as wind has kinetic energy, as do flowing streams, moving cars, heat, and electricity. See also *potential energy.*

law of conservation of matter. Matter is neither created nor destroyed, but is combined and rearranged in different ways.

law of tolerance. The presence, number, and distribution of a species in an ecosystem are determined by whether the levels of one or more physical or chemical factors fall within the range tolerated by the species. See also *limiting factor principle.*

limiting factor principle. Too much or too little of any abiotic factor can limit or prevent growth of a population even if all other factors are at or near the optimum range of tolerance. See also *law of tolerance.*

limnetic zone. The open water area away from the shore of a lake or pond, with less light penetration and fewer producers. See also *littoral zone, profundal zone.*

lithosphere. The upper zone of the Earth's mantle and the inorganic mixture of rocks and mineral matter in the Earth's crust.

littoral zone. The shallow water and vegetated area along the shore of a lake or pond, and the most productive zone of the lake. See also *limnetic zone, profundal zone.*

logistic growth curve. The idea that the population increases exponentially at the outset and then levels out as the carrying capacity of the environment is reached.

long-distance commuting. The practice of flying miners into a mine to work for a designated period and then flying them back to their homes in larger communities for another period.

low-quality energy. Dispersed energy, such as the heat stored in the oceans, with little capacity to perform useful tasks. See also *energy quality, high-quality energy.*

low-quality matter. Hard-to-reach matter, such as that dispersed or diluted in the atmosphere or oceans. See also *high-quality matter.*

macroconsumers. Organisms that feed by ingesting or engulfing particles, parts, or entire bodies of other organisms, living or dead, including herbivores, carnivores, omnivores, scavengers, and detrivores.

macronutrients. The main constituents of the complex organic compounds required by all living organisms. The six major macronutrients are carbon, oxygen, hydrogen, nitrogen, phosphorus, and sulphur. See also *micronutrients.*

mangrove swamp. A collection of tropical evergreen trees with stiltlike aerial roots that cause thick undergrowth and provide habitat for marine organisms, waterfowl, and other coastal species.

manipulated variable. See *independent variable.*

mass. The amount of material in an object.

mass extinction. The disappearance of numerous species over a relatively short period of geological time. See also *background extinction.*

mass number. Sum of the number of protons and the number of neutrons in the nucleus of an atom. This sum gives the approximate mass of that atom.

matter. Anything that has mass and takes up space, including everything that is solid, liquid, or gaseous.

mesotrophic. A lake that falls in the mid-range between the two extremes of nutrient enrichment required by producers. See also *eutrophic, oligotrophic.*

metamorphic rock. Rock formed when existing rocks lying deep below the Earth's surface are subjected to high temperatures, high pressures, chemically active fluids, or a combination of these agents, causing the rocks' crystal structure to change. See also *igneous rock, sedimentary rock.*

microconsumers. Organisms that live on or within their food source, completing the breakdown of complex molecules into simpler compounds (which we call rot or decay).

micronutrients. The trace elements of complex organic compounds required by all living organisms. These include boron, copper, zinc, and others. See also *macronutrients.*

milling. In the processing of minerals, the crushing and grinding of ores to separate the useful materials from the nonuseful ones. See also *tailings.*

mineral exploration. Finding geological, geophysical, or geochemical conditions that differ from those of their surroundings.

mineral fuels. Crude oil and equivalents, including natural gas, coal, and natural gas byproducts. In 2002, they accounted for approximately 77 percent of the total value of Canadian mineral production.

mixtures. One of the basic forms of chemical composition. A combination of elements, compounds, or both.

molecules. Particle formed when two or more atoms of the same or different elements combine.

monoculture. Planting and cultivation of a single crop (or even a single strain or subspecies), usually on a large area of land.

montane. An ecozone in the Rocky Mountains dominated by coniferous trees. The region is important for its biodiversity and its comparatively mild microclimate that provides winter habitat for many mountain-dwelling species.

moral justification. A rationale for the conservation of nature based on the idea that elements of the environment have a right to exist, independent of human desires. Compare *aesthetic arguments, ecological justification, utilitarian justification.*

mutation. The random and unpredictable changes in DNA molecules that can be transmitted to offspring and produce variability. See also *adaptation.*

mutualism. A symbiotic relationship in which both interacting species benefit, as when honeybees pollinate flowers as they feed on the flower's nectar. See also *commensalism, symbiosis.*

native species. See *endemic species.*

natural capital. Earth's natural resources and ecological systems that provide vital life-support services, such as maintenance of soil fertility, flood control, and stabilization of climate. Natural capital can be depleted through human actions, such as poor agricultural practices, overharvesting, water pollution, toxic contamination, development, and other activities.

natural gas. A gaseous hydrocarbon mixture of methane combined with smaller amounts of propane and butane. The conventional or "associated" type is located underground above most reserves of crude oil, while the nonassociated type is found on its own in dry wells.

natural selection. The tendency for only the best adapted organisms to survive and reproduce in a given environment.

net primary productivity. The rate at which organic matter is incorporated into plant bodies so as to produce growth. See also *gross primary productivity*.

net useful energy. The usable amount of energy available from an energy source over its lifetime.

neutrons. Uncharged or electrically neutral ions, which cluster with protons in the centre of an atom and comprise its nucleus. See also *electrons, ions, protons*.

nitrogen fixation. A part of the nitrogen cycle in which atmospheric nitrogen is converted into other chemical forms available to plants.

nonfuel minerals. Metallic minerals such as copper, gold, iron ore, nickel, and zinc, as well as nonmetallic minerals such as potash, sand, and gravel.

nonpoint sources. Pollutants discharged in an unconfined manner.

nonrenewable resources. Resources such as coal, oil, and other fossil fuels that are finite in supply or replaced so slowly that they are soon depleted. Compare *renewable resources*.

nuclear energy. The energy released by reactions within atomic nuclei, such as nuclear fission or nuclear fusion. See also *radioactive wastes*.

nutrient cycles. The means by which the nutrient elements and their compounds cycle continually through Earth's atmosphere, hydrosphere, lithosphere, and biosphere.

nutrients. The materials that an organism must take in to enable it to live, grow, and reproduce.

observations. Information gathered through any of our five senses or instruments that extend these senses.

oil sand. A combination of clay, sand, water, and bitumen. Canada is home to the largest known oil sand deposits in the world. See also *bitumen*.

oil shale. Rock that contains a solid mixture of hydrocarbon compounds called kerogen. Once crushed and heated, kerogen vapour is condensed to form heavy, slow-flowing shale oil.

oligotrophic. A lake with minimal levels of nutrients required for producers. See also *eutrophic, mesotrophic*.

omnivores. Consumers that eat both plants and animals, such as black bears, pigs, and humans. See also *carnivores, herbivores*.

one-industry town. A community whose existence depends on the exploitation of a single resource.

open-pit mining. A type of mining in which minerals are extracted from the earth by digging that leaves a large pit in the surface. Compare *strip mining*.

open system. See *systems*.

operational definitions. Set of criteria that tell scientists what to look for or what to do in order to carry out the measurement, construction, or manipulation of variables. See also *independent variable, responding variable*.

organic compound. Molecule that contains atoms of the element carbon, usually combined with each other and with atoms of one or more other elements such as chlorine, fluorine, hydrogen, nitrogen, oxygen, phosphorus, and sulphur. Compare *inorganic compound*.

organic farming. Producing crops and livestock naturally by using natural soil-forming processes including organic fertilizer (compost, manure, legumes) and natural pest control (plants that repel bugs, bugs that eat harmful bugs, and environmental controls such as crop rotation) instead of using commercial inorganic fertilizers and synthetic pesticides and herbicides.

organism. A complex organization of cells, tissues, organs, and body systems that work together to create a multicellular individual such as a bear, whale, human, or orchid.

organochlorines. Carbon–hydrogen compounds in which one or more hydrogen atoms have been replaced by a chlorine atom.

organohalides. Carbon–hydrogen compounds that are bonded to a halogen (fluorine, chlorine, bromine, or iodine).

overburden. The layers of rock and soil that overlay mineral deposits. These layers are removed during surface mining.

ozone. An oxygen gas (O_3) that is an air pollutant in the lower atmosphere but beneficial in the upper atmosphere. See also *ozone layer*.

ozone layer. The layer of ozone in the stratosphere that filters out harmful ultraviolet radiation from the sun.

parasitism. A symbiotic relationship in which the parasite benefits by obtaining nourishment from the host and the host is weakened or killed by the parasite.

pelagic environment. The ocean water, one of the two main divisions of the open sea environment. The marine environment from the low-tide mark to the open ocean within the vertical division from the surface to floor. See also *benthic environment*.

permafrost. A permanently frozen layer of subsoil, characteristic of the tundra biome.

persistence. In reference to chemical compounds, those substances that do not break down easily in ecosystems and remain in the environment for long periods of time. Many organochlorine compounds are persistent.

petajoule. A unit for measuring heat energy in the Système Internationale. One petajoule is equal to 10^{15} joules. See also *joule*.

phenology. The study of periodic occurrences in nature (such as timing of first and full bloom of plants, ripening of fruit, and migration of birds) and their relation to climate.

photochemical smog. The product of chemical reactions involving hydrocarbons and nitrous oxides in the presence of sunlight.

photovoltaics. The direct conversion of sunlight into electricity.

physical change. A change from one state to another, as when water changes from ice to its liquid state. See also *chemical change*.

point sources. Pollution sources that discharge substances from a clearly identifiable or discrete pathway such as a pipe, ditch, channel, tunnel, or conduit.

polar stratospheric clouds. Formed in extremely cold temperatures within the polar vortex as it matures, cools, and descends, these clouds have been linked to depletion of the ozone layer.

polar vortex. An atmospheric condition that occurs during the polar winter night when the Antarctic air mass is partially isolated from the rest of the atmosphere and circulates around the pole.

polynyas. An area of unfrozen sea water, created by local water currents in northern oceans. They act as biological hotspots and serve as vital winter refuges for marine mammals.

population. A group of individuals of the same species living and interacting in the same geographic area at the same time.

population age structure. The distribution of the population by age, used in analysis of demographic trends.

population lag effect. See *population momentum*.

population momentum. When a population achieves replacement fertility, that population continues to grow for several generations before stabilizing.

potential energy. Energy stored and potentially available for use, such as the chemical energy stored in gasoline or food molecules. See also *kinetic energy*.

precious metals. Metals, such as gold and silver, that are valuable to humans because of their rarity or appearance.

precursor chemicals. An early stage or substance that precedes or gives rise to a more important or definitive stage or substance.

predation. When members of a predator species feed on parts or all of an organism of a prey species.

predator. An organism, usually an animal, that feeds on other organisms, as when a turtle eats a fish in a freshwater pond ecosystem.

predator–prey relationships. The most obvious form of species interaction, which occurs when one organism (the predator) feeds on another (the prey).

prey. The organism consumed by a predator.

primary consumer. See *herbivores*.

primary energy use. The total requirements for all uses of energy, including energy used by the final consumer, energy in transforming one energy form to another, and energy used by suppliers in providing energy to the market.

primary succession. The development of biotic communities in a previously uninhabited and barren habitat with little or no soil. Compare *secondary succession*.

primary treatment. The lowest level of treatment in the management of municipal wastes that involves the mechanical removal of large solids, sediment, and some organic matter. See also *secondary treatment, tertiary treatment*.

principle of connectedness. Everything in the natural world is connected to and intermingled with everything else, and a change in environmental conditions will have multiple effects.

producers. Those self-nourishing organisms that perform photosynthesis by converting relatively simple inorganic substances such as water, carbon dioxide, and nutrients into complex chemicals such as carbohydrates, lipids, and proteins. Green plants and phytoplankton are common producers. See also *chemotrophs, consumers*.

profundal zone. The deepest zone of a lake, where lack of light means that no producers can survive. See also *limnetic zone, littoral zone*.

prokaryotic. Cells that lack a nuclear envelope and other internal cell membranes, including bacteria. Compare *eukaryotic*.

protons. Positively charged ions that cluster with neutrons in the centre of an atom and comprise its nucleus. See also *electrons, ions, neutrons*.

qualitative data. Non-numerical records of independent and dependent variables kept during experiments.

quantitative data. Numerical records of independent and dependent variables kept during experiments.

radioactive wastes. Radioactive byproducts from the operation of a nuclear reactor or from the reprocessing of depleted nuclear waste. See also *nuclear energy*.

realized niche. That portion of a fundamental niche actually occupied by a species, which results from the sharing of resources in a given ecosystem.

recirculation. Water that is reused in a particular distribution system. It may be used more than once in a specific process or used once and then recycled to another process.

reclamation. The rehabilitation of a site (a disused mine, for example) in order to make it a viable and, if possible, self-sustaining ecosystem that is compatible with a healthy environment.

regional sustainability. An alternative to the globalization of the food production system in which developing countries would be encouraged to grow food first for themselves and then for export.

renewable resources. Resources such as forests, solar energy, and fisheries that can be replaced by environmental processes in a time frame meaningful to humans. Also known as renewable natural capital. Compare *nonrenewable resources*.

replacement fertility. The fertility rate needed to ensure that the population remains constant as each set of parents is replaced by their offspring.

resource partitioning. The division of scarce resources in order that species with similar requirements can use the resources in different ways, in different places, and at different times.

responding variable. A condition that responds to changes in the independent variable in an experiment. Also referred to as a dependent variable. See also *independent variable, operational definitions*.

riparian area. The "thin green line" along streams, rivers, and wetlands, formed as the result of water, soil, and vegetation interacting with one another. Part of extensive drainage basins, these productive green areas provide forage, shelter, fish, wildlife, and water.

salinity. The amounts of various salts dissolved in a given volume of water. This can be a limiting factor on the growth of aquatic populations.

salinization. The accumulation of salts in soil—a process that may result in soil too salty to support plant growth.

scavengers. Consumers such as vultures and hyenas that eat dead organic material.

scientific method. Systematic methods used in scientific investigations of the natural world, which include designing controlled experiments, gathering data, developing and testing hypotheses.

secondary consumers. See *carnivores*.

secondary energy use. Energy used by final consumers for residential, agricultural, commercial, industrial, and transportation purposes.

secondary succession. The development of biotic communities in an area where the natural vegetation has been removed or destroyed but where soil is present. See also *primary succession*.

secondary treatment. The second level of treatment in the management of municipal wastes that employs biological processes by which bacteria degrade most of the dissolved organics, about 30 percent of the phosphates, and about 50 percent of the nitrates. See also *primary treatment, tertiary treatment*.

second law of energy. See *second law of thermodynamics*.

second law of thermodynamics. With each change in form, some energy is degraded to a less useful form and given off into the surroundings, usually as low-quality heat. See also *first law of thermodynamics*.

sedimentary rock. Rock formed when small bits and pieces of matter and sediments are carried by wind or rain and then deposited, compacted, and cemented to form rock. See also *igneous rock, metamorphic rock*.

silviculture. The theory and practice of controlling the establishment, composition, growth, and quality of forest stands.

soil compaction. A form of structural degradation in soil in which soil is packed so tightly that its air spaces are closed, reducing aeration and infiltration and thus reducing the ability of the soil to support plant growth. Caused mainly by the repeated passing of heavy machinery over wet soil.

solar energy. Energy derived from the sun in the form of solar radiation.

specialist species. Species with the ability to live in only one type of habitat, eat only a few types of food, or tolerate a narrow range of climatic or environmental conditions. See also *generalist species*.

speciation. The formation of two or more species from one as the result of divergent natural selection and response to changes in environmental conditions.

species. A group of organisms that resemble one another in appearance, behaviour, chemical makeup and processes, and genetic structure, and that produce fertile offspring under natural conditions.

species diversity. The number of different species and the relative abundance of each in different habitats on Earth. See also *ecological diversity*.

stewardship. The concept that mankind has an ethical responsibility to care for plants, animals, and the environment as a whole, due to our superior intellect and power to change the natural world.

stratosphere. The layer above the troposphere that contains the ozone layer and protects life on Earth's surface by absorbing most incoming solar ultraviolet radiation. See also *atmosphere, troposphere*.

strip mining. Surface mining in which heavy machinery strips away the overlying layer of rock and soil to create a trench that exposes the mineral resource below. Compare *open-pit mining*.

succession. The process of community development over time, in which the composition and function of communities and ecosystems change.

summerfallow. Land left unsown and unharvested, usually for one season, to conserve moisture in the soil and to allow accumulation of nitrogen and other nutrients.

surface runoff. Precipitation that flows on the land (instead of soaking into it) and into bodies of surface water. May carry contaminants.

surface water. All bodies of water, such as lakes, rivers, streams, and oceans, that lie on the surface of the Earth. Compare *groundwater*.

sustainability (environmental). The ability of an ecosystem to maintain ecological processes, functions, biodiversity, and productivity over time. See also *sustainable development*.

sustainable development. Maintaining environmental resources so that they continue to provide benefits to living things and the larger environment of which they are a part. See also *sustainability*.

sustained yield. The practice of harvesting renewable resources so that an even flow of resources in perpetuity may be obtained.

symbiosis. Any intimate relationship between two or more different species. The fur of the three-toed sloth is often occupied by algae and insects that feed on the algae. See also *commensalism, mutualism*.

synergistic effects. Outcomes in which the effects of two or more substances or organisms acting together are greater than the sum of their individual effects (they are multiplicative, not additive).

systems. Systems may be open or closed. A system that is open in regard to some factor exchanges that factor with other systems. An example of an open system is the ocean that exchanges water with the atmosphere. A system that is closed in regard to some factor does not exchange that factor with other systems. Earth is an open system in relation to energy and a closed system in regard to material.

tailings. The nonuseful materials removed from the mill after the recoverable minerals have been extracted in the processing of minerals. See also *milling*.

taxonomic. The classification of organisms according to evolutionary relationships.

tertiary consumers. Carnivores that eat other carnivorous (or secondary) consumers.

tertiary treatment. The third level of treatment in the management of municipal wastes, which involves a chemical process that removes phosphates, nitrates, and other contaminants not removed during secondary treatment. See also *primary treatment, secondary treatment.*

theories. Models based on currently accepted hypotheses that offer broadly conceived, logically coherent, and well-supported concepts.

threshold effect. The harmful or even fatal reaction to exceeding the tolerance limit of a species in a given ecosystem. See also *acclimation, law of tolerance.*

tillage erosion. The movement of soil downhill during plowing operations that contributes to soil erosion and degradation on rolling or hummocky land.

total allowable catch (TAC). A limit set by the Northwest Atlantic Fisheries Organization (NAFO), an agency of the United Nations Food and Agricultural Organization, to ensure that groundfish stocks were not depleted.

total fertility rate (TFR). The average number of children expected to be born to a woman during her lifetime.

trophic level. The feeding level to which each organism belongs depending on whether it is a producer or a consumer and on what it eats or decomposes.

troposphere. The lowest layer of the atmosphere and the zone in which most weather events occur. See also *atmosphere, stratosphere.*

urban shadow effects. Urban impacts that extend over large areas and cause declines in agriculture in urban regions.

utilitarian justification. A rationale for the conservation of nature based on the idea that the environment provides individuals with direct economic benefits. Compare *aesthetic arguments, ecological justification, moral justification.*

vertebrates. Animals with backbones, including fish, amphibians, reptiles, birds, and mammals. Compare *invertebrates.*

volatile organic compound (VOC). Compounds that result primarily from the combustion of fossil fuels in motor vehicles. Most VOCs are hydrocarbons, such as methane, propane, chlorofluorocarbons, and benzene. They are also found in the vapours of substances such as gasoline, solvents, and oil-based paints.

water resources. The network of rivers, lakes, and other surface waters that supply water for food production and other essential human systems.

watershed. An area of land in which all of the water under it or draining from its surface ends in the same place. Watersheds are separated from each other by the highest points of elevation in a region, for example, the Continental Divide, formed by the Rocky Mountains.

wetlands. Transitional areas between aquatic and terrestrial ecosystems, usually covered with fresh water for part of the year, with characteristic soils and vegetation.

withdrawal uses. Water removed from its natural setting by pipes or channels for a particular human use (human consumption, mineral extraction, irrigation, and other applications). Compare *instream uses.*

worldview. A set of commonly held values, ideas, and images concerning the nature of reality and the role of humanity within it.

PHOTO CREDITS

Cover and title page: Daryl Benson/Masterfile
Preliminary pages (except title page): © Corel

Part 1 opener: © Corel. **Chapter 1:** opener: © Corel; Photos 1–1a, b, c: © Dianne Draper; 1–1d: PhotoDisc/Getty Images; 1–2a, b: © Dianne Draper; 1–3: F. Lanting/firstlight.ca; 1–4: CIDA Photo/Roger Lemoyne; 1–5: CIDA Photo/Peter Bennett; 1–6a: CIDA Photo/S. Maslowski; 1–6b: AP/CP Picture Archive; 1–7: © Al Harvey/The Slide Farm; 1–8a: UN/DPI photo; 1–8b: Ricardo Mazalan/CP Picture Archive; 1–9, 1–10: © Al Harvey/The Slide Farm; 1–11: Blaise Edwards/CP Picture Archive.

Chapter 2: opener: © Corel; Photo 2–1: William Armstrong/National Archives of Canada/C-040293; 2–2: © Mary Evans Picture Library; 2–3: Ryan Remiorz/CP Picture Archive; 2–4: © Bill Banaszewski/Visuals Unlimited; 2–5: Associated Press/CP Picture Archive; 2–6: Clement Allard/CP Picture Archive; 2–7: © Mary Evans Picture Library; 2–8: © National Archives of Canada/C-061557; 2–9: © Todd A. Gipstein/CORBIS/MAGMA; Figure 2–2: Hulton Archive/Getty Images (J. Muir, G. Pinchot), Turofsky/CP Picture Archive (C. Sifton), University of Wisconsin–Madison Archives (A. Leopold), and Getty Images (R. Carson); Photo 2–10: Kevin Lamarque/Reuters; 2–11: Al Harvey/The Slide Farm; 2–12: © Dianne Draper; 2–13: © Charles E. Rotkin/CORBIS/MAGMA.

Part 2 opener: © Corel. **Chapter 3:** opener: © Corel; Photos 3–1a, b: © Dianne Draper; 3–2: © PhotoDisc/Getty Images; 3–3: © PhotoDisc/Getty Images; 3–4: © Raymond Gehman/CORBIS/MAGMA; 3–5a: © Tom J. Ulrich/Visuals Unlimited; 3–5b, c: © PhotoDisc/Getty Images; 3–5d: Kevin Schafer/firstlight.ca; 3–6: © Dianne Draper; 3–7: © D. Cavagnaro/Visuals Unlimited; 3–8: © Michigan Sea Grant; 3–9: CP Picture Archive; 3–10: Peter K. Ziminski/Visuals Unlimited; Figure 3–11 (from top to bottom): Arthur Morris/Visuals Unlimited, S. Maslowski/Visuals Unlimited, S. Maslowski/Visuals Unlimited, S. Maslowski/Visuals Unlimited, Rudolf G. Arndt/Visuals Unlimited; Photo 3–11: Joe McDonald/Visuals Unlimited; 3–12: National Wildlife Federation; 3–13: © Dianne Draper; 3–14: Al Harvey/The Slide Farm; 3–15: F. Lanting/firstlight.ca; 3–16: Dianne Draper; 3–17: Al Harvey/The Slide Farm.

Chapter 4: opener: © Corel; Photo 4–1a: Paul Henri/National Archives of Canada/PA–028944, 4–1b: Paul Henri/National Archives of Canada/PA–143463; 4–1c: Martin Chamberland/CP Picture Archive; 4–2: CIDA Photo/Pierre St-Jacques; 4–3: © Dianne Draper; 4–4: © Reuters NewMedia Inc./CORBIS/MAGMA; 4–5: Time Life Pictures/Getty Images; 4–6: Saurabh Das/AP/CP Picture Archive; 4–7: CIDA Photo/Nancy Durrell McKenna; 4-8: UN/DPI photo; 4–9: CIDA Photo/David Barbour; 4–10: UN/DPI photo.

Part 3 opener: © Corel. **Chapter 5:** opener: © Corel; Photo 5–1a, b: Victor Last/Geographical Visual Aids; 5–2: © Ted Streshinsky/CORBIS/MAGMA; 5–3: CIDA Photo/Roger Lemoye; 5–4a: Sylvan H. Wittwer/Visuals Unlimited; 5–4b: John Meuser/Visuals Unlimited; 5–5a, b: Jasper National Park/Parks Canada; 5–6: Science VU/Visuals Unlimited; 5–7: © Patrick Ward/CORBIS/MAGMA; Enviro-Focus 5: Canadian Dermatology Association; 5–8: Victor Last/Geographical Visual Aids; 5–9: © Al Harvey/The Slide Farm; 5–10: Dick Hemingway; 5–11: Courtesy of the City of Toronto.

Chapter 6: opener: © Corel; Photo 6–1a: © Al Harvey/The Slide Farm; 6–1b: © Dianne Draper; 6–2: Agriculture and Agri-Food Canada; 6–3: © Dianne Draper; 6–4: R. Arndt/Visuals Unlimited; 6–5: Agriculture and Agri-Food Canada; 6–6a: Agriculture and Agri-Food Canada; 6–6b: PhotoDisc/Getty Images; 6–6c: PhotoDisc/Getty Images; 6–7: Ken Mantyla/CP Picture Archive; 6–8: Charlie Heidecker/Visuals Unlimited; 6–9: © Ducks Unlimited Canada; 6–10: © Ducks Unlimited Canada; 6–11: CIDA Photo/Virginia Boyd; Photo 6–12: © IRDC; 6–13: © Dianne Draper; 6–14: Len Rue, Jr./Visuals Unlimited; 6–15: © Dianne Draper; 6–16: Ducks Unlimited Canada.

Chapter 7: opener: © Corel; Photo 7–1: Al Grillo/AP/CP Picture Archive; 7–2: © Tourism Saskatchewan; 7–3: © Maureen G. Reed; 7–4: Tom Hanson/CP Picture Archive; 7–5: John Lehmann/CP Picture Archive; 7–6: Joe Traver/CP Picture Archive; 7–7: © Neil Rabinowitz/CORBIS/MAGMA; 7–8: © Hydro-Québec; 7–9: Frank M. Hanna/Visuals Unlimited; 7–10a: Courtesy of Department of Fisheries and Oceans Canada; 7–10b: Bob Semple; 7–11: © Vicki Gould; 7–12: © S.H. Draper.

Chapter 8: opener: © Corel; Photo 8–1: CP Picture Archive; 8–2: CP Picture Archive; 8–3: © Al Harvey/The Slide Farm; 8–4: Vance Rodewalt/The Calgary Herald, 15 March 1995, p. A4. Reprinted with permission of The Calgary Herald; 8–5: Richard Drew/AP/CP Picture Archive; 8–6, 8–7, 8–8: © Dianne Draper; 8–9: Courtesy of Department of Fisheries and Oceans Canada; 8–10: © Natalie Forbes/CORBIS/MAGMA; 8–11: Peter Ziminski/Visuals Unlimited; 8–12: Courtesy of Department of Fisheries and Oceans Canada; 8–13: Ray Smith/Victoria Times Colonist, 17 September 1995, p. A1; 8–14: Andrew Vaughan/CP Picture Archive; 8–15: © Al Harvey/The Slide Farm.

Chapter 9: opener: © Corel; Photo 9–1: Associated Press/CP Picture Archive; 9–2: John Oohlden/Visuals Unlimited; 9–3: Ivy Images; 9–4: Natural Resources Canada; 9–5a, c: Steve McCutcheon/Visuals Unlimited; 9–5b: Brooking Tatum/Visuals Unlimited; 9–5d: Kirtley-Perkins/Visuals Unlimited; 9–5e: Berndt Wittich/Visuals Unlimited; 9–6: Courtesy of the Museum of Anthropology, UBC; 9–7: © Laurie Wierzbicki; 9–8, 9–9, 9–10: Photo sequence compiled by Richard G. Thomas, Alberta Environmental Protection – Air Photo Services, Alberta Sustainable Resource Development, Edmonton; 9–11: © IRDC; 9–12: Arthur R. Hill/Visuals Unlimited; 9–13: © Dianne Draper; 9–14: Allen H. Benton/Visuals Unlimited; 9–15: Mary Cummins/Visuals Unlimited; 9–16: Vancouver Sun; 9–17: P. Marck/The Calgary Herald, 23 June 1997, p. C4. Reprinted with permission of The Calgary Herald.

Chapter 10: opener: © Corel; Photo 10–1: © Image A-00355, BC Archives; 10–2: Victor Last/Geographical Visual Aids; 10–3: Natural Resources Canada – Photolibrary; 10–4: © Al Harvey/The Slide Farm; 10–5: © Erik Schaffer; Ecoscene/CORBIS/

ACAP, Atlantic Coastal Action Program

ACRE, Association for a Clean Rural Environment

AECL, Atomic Energy of Canada Limited

AEUB, Alberta Energy and Utilities Board

AIDS, Acquired Immune Deficiency Syndrome

ANWR, Arctic National Wildlife Refuge

AOC, Areas of Concern

AOGCM, atmospheric-ocean general circulation models

ARET, Accelerated Reduction/Elimination of Toxins

ARNEWS, Acid Rain National Early Warning System

ASH, Autonomous and Sustainable Housing

ASRL, Alberta Sulphur Research Limited

ATES, aquifer thermal energy storage

AWA, Alberta Wilderness Association

BHP, Broken Hill Proprietary

BOD, biological oxygen demand

BSE, Bovine spongiform encephalopathy

CANMET, Canada Centre for Mineral and Energy Technology

CARE, Conservation of Agriculture, Resources and Environment

CBIN, The Canadian Biodiversity Information Network

CBR, crude birth rate

CCAF, Climate Change Action Fund

CCFM, Canadian Council of Forest Ministers

CCIW, Canada Centre for Inland Waters

CCME, Canadian Council of Ministers of the Environment

CCREM, Canadian Council of Resource and Environment Ministers

CDM, Clean Development Mechanism

CDR, crude death rate

CEAA, Canadian Environmental Assessment Act

CEPA, Canadian Environmental Protection Act

CFC, chlorofluorocarbon

CFIA, Canadian Food Inspection Agency

CGR, crude growth rate

CHBA, Canadian Home Builders Association

CHP, combined heat and power

CHRS, Canadian Heritage Rivers System

CIDA, Canadian International Development Agency

CITES, Convention on International Trade in Endangered Species of Wild Fauna and Flora

CLI, Canada Land Inventory

CMHC, Mortgage and Housing Corporation

COG, Canadian Organic Growers

CoP 6, Sixth Conference of the Parties

CORE, Commission on Resources and Environment

COSEWIC, Committee on the Status of Endangered Wildlife

CPAWS, Canadian Parks and Wilderness Society

CPH&R, Canadian Pacific Hotels and Resorts

CRC, Cardinal River Coals

CSA, Canadian Standards Association

CWD, chronic wasting disease

CZM, national coastal zone management

DAPTF, Declining Amphibian Populations Task Force

DDE, dichlorophenylethylene

DDT, dichlorodiphenyltrichloroethane

DEW, Distant Early Warning

DFO, Department of Fisheries and Oceans

DGD, Deep Geological Disposal

DNA, deoxyribonucleic acid

DNAPL, dense nonaqueous phase liquids

DSD, Duales System Deutschland

DU, Dobson units

dwt, deadweight tons

EA, Environmental Assessment

EARP, Canadian Environmental Assessment Review Panel

EC, European Community

EDC, Endocrine-disrupting chemicals

EEZ, Exclusive Economic Zone

EIA, Environmental Impact Assessment

EIS, Environmental Impact Statement

ELC, Environment Liason Center

ELV, end-of-life vehicles

EMAN, Ecological Monitoring and Assessment Network

EMCBC, Environmental Mining Council of British Columbia

EMS, environmental management system

ENGO, Environmental nongovernmental organizations

ENSO, El Niño and the Southern Oscillation

EPL, Environment Policy and Law

EPR, Extended Producer Responsibility

EPRF, Energy Probe Research Foundation

ESC, Ecological Science Cooperative

EU, European Union

EV, electric vehicles

FAO, Food and Agriculture Organization

FBMB, Fraser Basin Management Board

FCCC, United Nations Framework Convention on Climate Change

FCM, Federation of Canadian Municipalities

FPB, Forestry Practices Board

FRAP, Fraser River Action Plan

FRCC, Fisheries Resources Conservation Council

FREMP, Fraser River Estuary Management

GATT, General Agreement on Tariffs and Trade

GBEI, Georgia Basin Ecosystem Initiative

GBS, gravity base structure

GCM, general circulation model

GDP, gross domestic product

GE, genetic engineering

GHG, greenhouse gases

GIS, Geographic Information Systems

GLWQA, Great Lakes Water Quality Agreements

GMF, Genetically Modified Foods

GMO, genetically modified organisms

GNWT, Government of the North West Territories

GPI, Genuine Progress Indicator

Gt, Giga tonnes (billion tonnes)

GVRD, Greater Vancouver Regional District

GWP, Global Warming Potential

H.E.A.R., Hearing Education and Awareness for Rockers

HBFC, hydrobromofluorocarbons

HCB, hexachlorobenzene

HCFC, hydrochlorofluorocarbons

HCHC, hexachlorocyclohexane

HDC, highly developed country

HDI, Human Development Index

HIV/AIDS, Human Immunodeficiency Virus/ Acquired Immune Deficiency Syndrome

HRV, heat recovery ventilator

ICLEI, International Council for Local Environmental Initiatives

ICME, International Council on Metals and the Environment

ICPD, International Conference on Population and Development

ICSC, International Centre for Sustainable Cities

IDRC, International Development Research Centre

IGBP, International Geosphere Biosphere Program

IGU, International Geographical Union's Commission on Climatology

IHDP, International Human Dimensions Program

IJC, International Joint Commission

ILO, Intensive livestock operations

IPCC, The Intergovernmental Panel on Climate Change

IPM, Integrated pest management

ISO, International Organization for Standardization

ITTA, International Tropical Timber Agreement

ITTO, International Tropical Trade Organization

IUCN, International Union for the Conservation of Nature and Natural Resources

KEY, Knowledge of the Environment for Youth Foundation

KPMG, Klynveld Peat Marwick Goerdeler

KWh, kilowatt hours

LCA, life-cycle assessment

LCI, life-cycle inventory

LDC, less developed country

LOS, Law of the Sea

LRTAP, long-range transport of airborne pollutants

LRTP, long-range transport of pollutants

MAB, Man and the Biosphere

MDC, more developed country

MEND, Mine Environmental Neutral Drainage

MNR, Natural Resources

MOE, Minister of the Environment

MOH, Minister of Health

MREAC, Miramichi River Environmental Assessment Committee

Mt, mega tonnes, (million tonnes)

MW, megawatts

NAAMP, North American Amphibian Monitoring Program

NAAQO, National Ambient Air Quality Objective

NAEBA, North American Elk Breeders Association

NAFO, Northwest Atlantic Fisheries Organization

NAFTA, North American Free Trade Agreement

NAPCC, National Action Program on Climate Change

NAPS, Canada's National Air Pollution Surveillance

NASA, National Aeronautics and Space Administration

NAWMP, North American Waterfowl Management Plan

NEB, National Energy Board

NGO, Nongovernmental Organizations

NMCA, National Marine Conservation Areas

NPRI, National Pollutant Release Inventory

NPS, non-point source

NRBS, The Northern River Basins Study

NRC, Natural Resources Canada

NRTEE, National Round Table on the Environment and the Economy

NWRI, National Water Research Institute

ODS, ozone-depleting substance

ODWS, Ontario Drinking Water Standard

OECD, Organization for Economic Cooperation and Development

OEE, Office of Energy Efficiency

OIG, Office of the Inspector General

OPEC, Organization of Petroleum Exporting Countries

OSEC, Oil Sands Environmental Coalition

PAH, polyaromatic hydrocarbon

PAS, Protected Areas Strategy

PCB, polychlorinated biphenyl

PCP, pentachlorophenol

PDO, Pacific Decadal Oscillation

PECOS, Prairie Ecosystem Study

PEM, proton exchange membrane

PFC, perfluorocarbons

PFRA, Prairie Farm Rehabilitation Administration

POL, petroleum, oil, and lubricant

POP, Persistent Organic Pollutant

ppbv, parts per billion by volume

ppmv, parts per million by volume

PSC, polar stratospheric clouds

PUC, Public Utilities Commission

PV, photovoltaics

RAN, Representative Area Network

RAP, Remedial Action Plan

RCMP, Royal Canadian Mounted Police

RENEW, Recovery of Nationally Endangered Wildlife

SARA, Species at Risk Act

SCOPE, Scientific Committee on Problems of the Environment

SLDF, Sierra Legal Defense Fund

SUV, Sport Utility Vehicle

TAC, Total Allowable Catch

TAF, Toronto Atmospheric Fund

TCDD, tetrachlorodibenzo-p-dioxin

TCDF, tetrachlorodibenzo-furan

TCH, tetra hydro canabinol

TEK, traditional ecological knowledge

TFL, tree farm licenses

TFR, total fertility rate

THM, trihalomethanes

TPC, Technology Partnerships Canada

TSS, Total Suspended Solids

UN, United Nations

UNAIDS, the UN agency that tracks the AIDS epidemic

UNCED, United Nations Conference on Environment and Development

UNCLOS, United Nations Convention on the Law of the Sea

UNDP, United Nations Development Program

UNEP, United Nations Environment Programme

UNEP–WCMC, The United Nations Environment Programme–World Conservation Monitoring Centre

UNESCO, United Nations Educational, Scientific and Cultural Organization

UNICEF, United Nations Children's Fund

UV, ultraviolet

UV-A, ultraviolet- alpha

UV-B, ultraviolet-beta

VCR, Voluntary Challenge and Registry

VOC, volatile organic compound

WAPPRIITA, Wild Animal and Plant Protection and Regulation of International and Interprovincial Trade Act

WCMC, World Conservation Monitoring Center

WCS, World Conservation Strategy

WHO, World Health Organization

WMI, Whitehorse Mining Initiative

WMO, World Meteorological Organization

WLED, white light emitting diode

WWF, World Wildlife Fund

scavengers, 66
science, 29–38
  assumptions in, 31
  complexity, values, worldviews, and, 36–37
  critical thinking and, 36
  deductive reasoning and, 31
  defined, 30
  environmental decision making and, 37–38, 541–43
  inductive reasoning and, 31
  language use and, 34
  measurement in, 31–32
  methods of, 33–34
  misunderstandings about, 34–37
  probability and, 31, 32
  research models and, 34
  theory and, 34
  value-free, 34–35
scientific method, 36, 37
sea levels, 140, 279
sedimentary rock, 78
self-realization, principle of, 45
sewage treatment. See under water
Sierra Legal Defence Fund
  drinking water and, 240, 244
  Temagami Forest and, 339
Sifton, Clifford, 40, 41, 42
silviculture, 330–31
smog alerts, 484
social norms, 23
soil
  agriculture and, 519, 524
  capability, 180–81
  carbon content of, 203–4, 205
  contamination of, 190–91, 209, 211
  desertification and, 191, 192
  erosion of, 187–89
  forests and, 320
  levels of organic matter in, 187
  productivity of, 7
  salinization of, 182, 189–90
  structure of, 189
solar energy, 421–23
  autotrophs and, 64–65
  radiation and, 58, 60, 126
solar intensity, 125, 126
songbirds and coffee growers, 200
Southern Oscillation, 151–52
specialist species, 69
speciation, 89
species, 437–74. See also individual species
  amphibians, environmental sensitivity of, 437–40
  Arctic, 273–75
  Atlantic Ocean, 277–78
  biodiversity of, 14, 63, 198–203, 342–44, 443
  birds (see birds)
  Canadian, at risk, 342–44, 447–50
  captive breeding programs for endangered, 468
  Carolinian forest, 353
  ecological succession and, 90
  endangered, 440, 452, 467
  endocrine systems of, human health

and, 245
  exotic, introduction of, 234, 235, 446
  in forests, 334–35, 339, 342–44, 359–60
  generalist, 69
  humans vs. other, 9
  interaction between, 69
  intrinsic value of, 441
  keystone, 68–69
  loss of, 7–8, 437–40
  natural selection, adaptation, and, 88–89
  number and organization of, 60, 61
  Pacific Ocean, 276–77, 288–93
  plants (see plants)
  population of, 60, 86–90
  protecting Canadian, 469–70
  reproduction strategies of, 87
  speciation and extinction of, 89–90, 91
  tolerance ranges of, 67–68
  in tropical forests, 344
  types and roles of, in ecosystems, 68–72
  types of, in biomes, 80–86
  ultraviolet radiation and, 147–48
Species at Risk Act (SARA), 304, 469, 470, 472, 474, 521, 527
St. Lawrence Action Plan/Vision, 296–97
standard error, 32
starvation, 11
statistical significance, 32
stewardship
  environmental, 24–25, 41
  mining and, 394–95
stratosphere, 57
succession, 90
sun sensitivity test, 150
supply and demand, 24
sustainability, 14–25
  actions and objectives for, 18, 527–28, 531, 533–36
  agriculture and, 204, 206–17
  Arctic ecosystems and, 449
  of biological resources, 469
  Cheviot Mine and, 395, 396–98
  cities, urban form, and, 504–6
  of cities, 481, 501–11
  components of, 545
  costs, benefits, and, 24
  defined, 14
  ecological (see ecological sustainability)
  economic (see economic sustainability)
  ecosystem approach to, 22–23
  energy, 410, 420–21, 525
  environmental stewardship and, 24–25
  First Nations and, 532
  forests and, 329–34, 339, 341, 352, 354–61
  fresh water and, 265, 267
  future challenges for, 536–43
  housing and, 491–94
  life-cycle concept and, 540

made-in-Canada models for, 542
  milestones in, 17
  monitoring for, 25
  of northern cod, 284–88
  of ocean environment, 279–82, 311–12
  precautionary principle of, 24
  principles of, 16–25
  of resources, 6
  social, 19, 23
  transportation and, 494–96
sustainable development, 14–15, 17
  deep ecology, green alternatives, and, 45–46
  mining and, 377–78, 391–94
  objectives, government actions, and, 531–46
  worldviews and, 38–42
Sustainable Development Technology Fund, 422
Swan Lake, B.C., 265, 266, 347
Sydney Tar Ponds, 413–14, 528
symbiosis, 70

taiga. See boreal forest
taxol, 71
Technology Partnerships Program, 38
Temagami Forest, 338–39
Tembec Inc., 359
temperate deciduous forests, 81
temperate rain forests, 81, 338
  collapse of, 5
temperate shrub lands, 81
temperate woodlands, 81
temperature data, 136, 138, 140
terrestrial biomes/ecosystems, 79–86
theory, 34
thermodynamics, laws of, 56–57
Thompson, William Boyce, 35
Three Gorges dam, 253, 416
Three Mile Island, 4, 419
threshold effect, 67–68
tillage. See agriculture, conservation tillage
Torrie, Ralph, 171
tourism and recreation
  forests and, 349–51
  water and, 253, 255–56, 489–91
toxic substances, 537–41
  cities and, 484–85
  endocrine disrupters and, 245
  Great Lakes and, 75, 247–49
  habitat alteration and, 446–47
  natural, 108–9
  reduction/elimination of, 392
  scientific research and, 38
  Sydney Tar Ponds and, 413–14
  U.S. military and, 97
trade
  agriculture and, 204, 206–8
  coffee growers, songbirds and, 200
  forest products and, 323–25, 340
  illegal, in wildlife species, 457
  sustainable agriculture in Canada and, 208–9, 211–14, 217
Trans Canada Trail, 473, 507, 544